Plants and Habitats of European Cities

John G. Kelcey · Norbert Müller
Editors

Plants and Habitats of European Cities

Springer

Editors
John G. Kelcey
Čečkovice 14
Bor u Tachova 348 02
Czech Republic
johnkelcey@hotmail.com

Norbert Müller
Department of Landscape Management
and Restoration Ecology
University of Applied Sciences Erfurt
Leipziger Str. 77, 99085 Erfurt, Germany
n.mueller@fh-erfurt.de

ISBN 978-1-4939-5098-0 ISBN 978-0-387-89684-7 (eBook)
DOI 10.1007/978-0-387-89684-7
Springer New York Dordrecht Heidelberg London

Printed on acid-free paper

Springer is part of Springer Science+Business Media (www.springer.com)

Contributors

Michael J. Crawley, Prof.
Imperial College London, Silwood Park, Ascot, Berkshire SL5 7PY, England
m.crawley@imperial.ac.uk

Elias D. Dana
G.I. Transferencia I+D+i Recursos Naturales, Universidad Almería, Spain
edana@ual.es

Dimitar Dimitrov Assoc. Prof. Dr.
National Natural History Museum, 1 Tsar Osvoboditel Blvd., Sofia, 1000, Bulgaria
dimitroff@mail.bg

Viera Feráková, & Assoc. Prof., RNDr., CSc.
Institute of Botany, Slovak Academy of Sciences, Dúbravská cesta 9,
SK-84523, Bratislava, Slovak Republic
viera.ferakova@savba.sk

Halina Galera, Dr.
Department of Plant Ecology and Environmental Conservation, University
of Warsaw, Al. Ujazdowskie 4, 00-478, Warsaw, Poland
h.galera@uw.edu.pl

Sandrine Godefroid, Dr.
National Botanic Garden of Belgium, Domein van Bouchout,
1860 Meise, Belgium
sandrine.godefroid@br.fgov.be

Manuel A. Guerrero
Bda. Virgen del Espino 10. El Pedroso (Sevilla)
maguered@yahoo.es

Maria Ignatieva, Prof.
Swedish University of Agricultural Sciences, Division of Landscape Architecture,
Department of Urban and Rural Development,
P.O. Box 7012, SE-750 07 Uppsala, Sweden
maria.ignatieva@slu.se

Dobri Ivanov, Dr.
Medical University Varna, 55 Marin Drinov Str, Varna, 9002, Bulgaria
dobrii@hotmail.com

Ivan Jarolímek, RNDr., CSc.
Institute of Botany, SAS, Dúbravská cesta 9, SK-84523,
Bratislava, Slovak Republic
ivan.jarolimek@savba.sk

Bogdan Jackowiak, Prof. Dr.
Department of Plant Taxonomy, Faculty of Biology, Adam Mickiewicz University,
Umultowska 89, 61-614 Poznań, Poland
bogjack@amu.edu.pl

John G. Kelcey
Čečkovice 14, Bor u Tachova, 348 02, Czech Republic
johnkelcey@hotmail.com

Galina Konechnaya
Komarov Botanical Institute of the Russian Academy of Sciences,
Popova St. 2, St. Petersburg, Russia 197376
galina_konechna@mail.ru

Elias Landolt, Prof. em. Dr.
Institut für Integrative Biologie (Swiss Federal Institute of Technology),
ETH Zürich, Universitätsstr. 16, CH-8092 Zürich, Switzerland
landolt@env.ethz.ch

Juan García-de-Lomas
Depto. de Biologia, Facultad de Ciencias del Mar y Ambientales,
Universidad de Cádiz. Pol. San Pedro s/n. 11510-Puerto Real (Cádiz)
juan.garciadelomas@uca.es

Alexander Mrkvicka
Gruppe Stadtwald, Erholungsgebiete und Nationalpark, MA 49 – Municipial
Forestry Office Vienna, A – 1082 Wien, Volksgartenstraße 3, Austria
alexander.mrkvicka@wien.gv.at

Norbert Müller, Prof. Dr.
Department of Landscape Management and Restoration Ecology,
University of Applied Sciences Erfurt, Leipziger Str. 77, 99085 Erfurt, Germany
n.mueller@fh-erfurt.de

Marilena Onete, Dr.
Ecology, Taxonomy and Nature Conservation Centre, Institute of Biology,
Romanian Academy, Spl. Independentei 296, Sector 6, 060031
Bucharest, Romania
m_onete@yahoo.com; marilena.onete@ibiol.ro

Mihaela Paucă-Comănescu, Dr.
Ecology, Taxonomy and Nature Conservation Centre, Institute of Biology, Romanian Academy, Spl. Independentei 296, Sector 6, Bucharest, 060031, Romania
mcpauca@gmail.com

Alexander Shvetsov, Ass. Prof. Dr.
Main Botanical Garden Russian Academy of Sciences, Botanicheskaya St., 4, 127276 Moscow, Russia
floramoscow@mail.ru

Glenn Stewart
Department of Environmental Management, Faculty of Environment, Society & Design, P.O. Box 84, Lincoln University, Lincoln 7647, Christchurch, New Zealand
Glenn.Stewart@lincoln.ac.nz

Maya Stoyneva, Ass. Prof. Dr.
Sofia University 'St Kliment Ohridski', 8 Dragan Tsankov Blvd., BG-1164, Sofia, Bulgaria
mstoyneva@abv.bg

Barbara Sudnik-Wójcikowska, Prof. Dr.
Department of Plant Ecology and Environmental Conservation, University of Warsaw, Al. Ujazdowskie 4, 00-478, Warsaw, Poland
barbara.sudnik@uw.edu.pl

Herbert Sukopp, Prof. Dr.
Technical University Berlin, Institute of Ecology, Rüdesheimer Platz 10, 14197 Berlin, Germany
herbert.sukopp@tu-berlin.de

Eduard J. Weeda, Dr.
Alterra Wageningen UR, P.O. Box 47, 6700AA Wageningen, The Netherlands
ejweeda@hotmail.com

Contents

Preface

John G. Kelcey

I wish to open this preface with an acknowledgement that would normally appear at the end, but the circumstances are exceptional, which is why it has to be done here. As newspaper editors know only too well most readers only read the first 1–2 column inches. And so it is that I wish to thank my partner Liz Colville for her sacrifices, tolerance and the immense effort she has applied in ensuring the completion of this book. Without Herbert Sukopp, Norbert Müller and Clive Stace, the book would not have been started, without Liz it would never have been finished – in reality a third editor.

This book and its companion volume on birds are two of many European seeds that were probably sown in my sub-conscious childish mind 50–60 years ago – they are the first two to bear fruit having been carefully nurtured and cultivated for the last 30 years or so by Herbert Sukopp. Whether the rest of the fruit matures or falls to the earth rotten remains to be seen. The journey from seed to fruit has been long and convoluted and subjected to many influences, a few are outlined in the following paragraphs – in no particular order.

The European peninsula is a fascinating natural unit to which, with the benefit of experience, the Middle East (the Fertile Triangle of the European homeland) should be added. The peninsula is divided into two opposing factions, human beings constrained by raging and desperate nationalism (despite the European Union) and plants and other animals that are not constrained by administrative/political boundaries. The former has had and continues to have a very serious effect on the understanding of the latter exacerbated by the failures in cultural understanding.

The issue is not one of federalism but of considering Europe as a single but diverse unit, the problem is the lack of understanding of the European condition, the solution is education and the provision of information. My early childhood interests in the cultures (a terrible word) of Europe in general but especially central and eastern Europe, in particular, were frustrated by the British attitude to Europe, the inability to travel and the lack of books. The former, which continues, raises the issue of why 50–100 million people died between 1914 and 1945; the freedom of movement has substantially improved since the early 1990s but the lack of information remains.

Back to natural history, the component nations appear to have considered and continue to consider that interest in it should stop at the national boundaries and yet those boundaries have been changing continually since the "establishment" of the nation state 1,000 years ago or so. Parts of present France were once part of Britain. Then, there were countries such as Persia and Prussia and the "units" of the Habsburg Empire. Czechoslovakia did not exist until the early 1920s, the Czech and Slovak Republics not until 1993. Transylvania is now part of Romania, much to the annoyance of Hungary. Yugoslavia has been divided into seven or so States although Bulgaria lays claims to Montenegro. And so it goes on. In short, it is more appropriate to study the flora of Europe than that of its component countries. Whether a species is endemic or an archaeophyte is dependent upon the political boundaries of the time.

It is a sad reflection on European botanists that there is no current, reliable flora of Europe and no standardization of the nomenclature or, it appears, the spelling of some names. It should be possible for an academic botanist or even someone with a general interest in plants to read or buy a book in Lisbon and find that the same names are used in Moscow, Athens and Helsinki. It is also a sad fact that for the most part national floras are confined to angiosperms or vascular plants – algae and bryophytes are rarely considered. The same applies to fungi and lichenised fungi (lichens), which although no longer within the Plant Kingdom, for practical purposes, should be considered as "plants".

A botanist wishing to enjoy the plants of Europe (for serious study or general interest) will find it impossible to do so, not only for the reasons considered in the previous paragraph but because there are no national floras for some countries, for example France. Observations of the physical form of the European landscape and its vegetation suggests that probably only a small proportion of it has ever been explored by botanists, who may well have missed much of botanical interest. For example, consideration of the logistics of undertaking a botanical survey of the Carpathians between Sibiu and Cozia in Romania indicate the magnitude and impossibility of the task – and that is a small area compared with Romania, even smaller if Bulgaria is included and smaller still when the whole of Europe is considered. However, these issues have much more serious implications in relation to the assessment of the status of species and the need or otherwise for statutory protection and the inclusion of a species within European protection measures.

My appointment in 1972 as the ecologist with Milton Keynes Development Corporation (a Government Agency) was the first of it kind in Britain occurred well before I was able to travel throughout Europe as I wished to do. The appointment was a baptism of fire – continual cross examination by cynical engineers, architects, estate managers, lawyers and finance officers was only just less traumatic than the refusal of the Government's nature conservation agency and academics to show any interest on the grounds of the presence of people and the need for a 5-year research programme, respectively.

In the early 1970s, the Corporation developed a particular interest in the approach of the Dutch to the provision, design and management of existing and new green space in Amsterdam, Almere, Delft and Lelystadt. In addition, the Corporation bought virtually all (if not all) of its trees from The Netherlands. Consequently, attending ecological conferences and undertaking a few days of study were easily approved – in addition, the Corporation saw considerable opportunities for publicity, "Look, we have an ecologist." The Garden Festival projects in Germany and the

British Government's interest in them provided the opportunity to go east, even as far as Berlin – but no further.

And so it was that in the early 1970s, I was thrilled and relieved to meet Hebert Sukopp, Maciej Luniak, Tj Deelstra and many others who were previously unknown to me but understood the natural history of urban areas. It was a revelation that impressed my employers – here were ecologists who actually worked in cities and who could be turned to for advice and to whom people were just part and parcel of life. They were immensely helpful and influential – in those days moving 1% forward was a major step.

Working in an urban development project with the objective of transforming more or less 9,000 ha of rural land into a "city" for 250,000 in 20 years sharpens the mind. Ecological arguments were subject to stringent testing while the arguments of other disciplines were tested although less vigorously. In order to understand and argue a case, it was necessary to understand the other disciplines involved and to be ready, almost immediately, with a credible answer. A week is a long time in urban development – stopping a job will cost thousands of Euros per day, changing a contract or instructions may cost hundreds of thousands if not millions of Euros and then there is the small matter of who pays and professional liability. In short, it was a matter of "using best endeavours."

As a consequence, the mind was broadened to include an appreciation of architecture, landscape design and other applied arts and from there to the other arts. The debates raged from the intellectual to the elementary. Sadly, what emerged was disappointing – as a rule, democracy only results in the mediocre – it requires authoritarianism to produce the outstanding but often at a social cost.

With some exceptions, there was little published information (in English) about urban natural history, there were some notable exceptions such as Berlin and Warsaw, and therefore "being prepared" was difficult, often impossible. This stimulates the mind to identify what is needed to solve the problem. The answer, "technical information that is easily understood and available quickly." However, there was another equally important approach and one that I had frowned on for a decade or so, namely, 'curiosity' the pursuit of knowledge out of sheer pleasure and interest. For decades, botanists had been describing and analyzing the flora of countries and units of countries woodlands, grasslands, coasts, road verges and many other habitats but not cities. For some reason, cities were exempt as was former industrial land, which botanists were intent on re-grading and turning bright green using *Lolium* or multi-coloured using *Lupinus.*

We needed some books about the natural history of cities, especially European cities. "Why in English?" asked Herbert Sukopp – a good point. He also asked me why most ecological/botanical papers published in Britain did not contain references to German papers. The same principle applies to papers in Italian, French and Spanish. Of course, the answer was and is easy – the British do not understand other languages, which is different from France where the French do but prefer not to. This diversion turned out to be interesting because the comment ultimately led to my reaching the view that European-wide books should be published in German and/or French. The suggestion was unanimously rejected by the contributors to the book "Birds in European Cities," (edited by Goetz Rheinwald and myself and published by

Ginster-Verlag in 2005). I was told, very firmly, that it is inappropriate to publish a scientific book these days unless it is written in some form of English, which is why that and this book are in English.

My initial aspiration (which remains to be achieved) was to write a series of volumes about the natural history of specific cities, including the geology, geomorphology, soils, origin, evolution, architecture, landscape, engineering, human values and behaviour, public health and such design, planning and management as there may have been, climate, air and water quality, habitats and the organisms from viruses to vertebrates of some of the major cities of Europe. This assumed there were sufficient data and that it was in English or I could get it translated – both rather naïve, with the benefit of hindsight.

Even trying to write a book about the natural history of a city proved impossible, and so it was at 2 a.m. one morning the obvious dawned on me. If the aspiration was to make any progress, I would need to consider cities in terms of the major groups of organisms starting with plants (or more precisely vascular plants), followed by birds. Time to discuss these matters with Herbert and Maciej, for reasons that are not at all clear now, I decided to reverse the order and start with birds. May be it was because birds are more popular, there are less of them (in terms of species), there are more ornithologists and bird watchers, and therefore likely to be more data. Encouraged by my two mentors, a delightful joint editor and enormous support from those who agreed to contribute the book, which was published at the end of 2005.

The preparation of, and the enthusiastic response to the bird book encouraged me to discuss again with Herbert Sukopp the possibility of a companion volume on the plants and habitats of European cities. He suggested a collaborative venture with Norbert Müller. Herbert and Clive Stace kindly provided us with lists of first class contributors most of whom graciously and generously accepted our invitation to write a chapter about "their city." Sadly, some were too busy, some did not reply and one dropped out after two years stating that he knew nothing about the subject, and therefore he could not write the chapter.

Norbert and I have tried to ensure that the preparation of the book has been a democratic exercise in which we have acted as enzymes – simply trying to make something worthwhile happen. I shall be eternally grateful to all those who have given their time without any financial reward (so far) to assist me in my endeavours to spread the understanding of the natural history throughout Europe in general and European cities in particular.

Many other companion volumes about the plants, habitats and birds of one or more other European cities remain to be written, as they do about cities elsewhere in the world. It is evident that most of the chapters in this book could be expanded into whole books. There are volumes to be written about the vertebrates and the invertebrates – as a whole or in "groups" such as the insects as a whole or individually Lepidoptera, Odonata and Orthoptera and then the molluscs. And of course, the original aspiration still remains to be achieved.

John G. Kelcey
A restless itinerant of Europe
November 2009

Introduction

Herbert Sukopp, Norbert Müller, and John G. Kelcey

The flora and vegetation of some European cities (mainly in central Europe) have been described and mapped over several decades. As a result, it is now possible to draw conclusions about environmental changes that have occurred by comparing the historical data with present conditions. The investigations demonstrate that the recent spatial distribution of spontaneous plants in cities has different causes:

1. Land use, care and substrate play an important role as well as the climate. Thus, the distribution patterns mirror building, economic and social structures.
2. The distribution and dispersal of archaeophytes and neophytes in cities have changed during the last centuries resulting in the present higher percentage of species that are sensitive to frost and cold.
3. The meso- and macro climatic characteristics overlie the distribution patterns by forming a gradient from the centre to the periphery.
4. The spatial differences caused by the special temperature conditions are reflected in the systematic phenological investigations, which enable heat islands and cooler areas to be mapped easily.

In many cases, the phenological phases start several days earlier in the centre of the city than in the periphery or in large parks. In the urban core, the first flowers can be observed 8 days earlier than in the outskirts, where one day equates to 1°C, which correlates well with the distribution of phenological phases in Europe in general. In cool valleys and wetlands, the flowering may even start two days later than on the city margins. The steepest gradient is found at the boundary between forests and built up areas.

As the result of the long-term studies of the urban heat island and the related effects on the flora and vegetation in European cities since the middle of the nineteenth century, it is possible to use cities as models for the effects of climate change on flora and vegetation.

"Urban vegetation" in its narrowest sense is the vegetation of ruderal places, for example rubble, railway and port areas, ruins, walls and waste areas. The occurrence of new non-native species in areas that are subject to human influence has stimulated a large number of investigations of the adventive flora for many decades. The newcomers were recorded and categorized according to the time of introduction, the way of introduction and the degree of naturalization. The highlights of such studies were the rapid colonization of the bombed areas of cities in the 1940s (for example, London and Warsaw) and the climatic changes of the last decade. It is important to recognise that, as well as the typical urban habitats, many cities contain several pristine natural and semi-natural habitats.

A question frequently asked is, "whether or not the warmer city climate influences the flora and vegetation of cities?" However, as mentioned above there are not only climatic changes, but also other ecological and socio-economic factors that influence urban vegetation. Transport, trade and horticultural activities reasons are the main "vectors" for the introduction and subsequent dispersal of non-native thermophilous species. In many cases, this has resulted in the breakdown of biogeographical barriers. In cities, native species become associated with species that would never have reached them without human activities.

The presence of a large number of non-native species is a characteristic feature of urban areas. Urban areas not only show a decline in the number of native species and archaeophytes, they are and will continue to be the starting point and centre of dispersal of non-native species, especially from the warmer regions of Europe, Asia and America.

Many native species (called apophytes) are able to colonize new urban sites, which raises interesting questions about dispersal (especially over long distances), colonization and succession. In Berlin, 63% of the species established on urban sites are native; some botanists consider that all native plants could exist as apophytes in one form or another.

Non-native plant species in urban areas have a relatively slow rate of dispersal into the surroundings of cities; the colonization process can continue for several decades, centuries, or millennia. The dispersal and naturalization varies between species. Of all the introduced taxa, on average only 10% are able to colonise areas temporarily (casuals), 2–3% can exist in man-made habitats and become a permanent member of the flora whilst only 1% is able to survive in natural ecosystems (for example, forests).

This is the first book to describe, in one volume, the flora and habitats of European cities and the changes that have occurred in them as a consequence of urban development. The 16 cities were not selected but "chose" themselves based on two criteria; first, the availability of sufficient information and second, the willingness of an author with sufficient expertise to write the chapter. As will be discussed in the conclusions, there are some obvious omissions. The book is another step forward in the wider recognition of the long tradition of research on urban flora and vegetation of Europe that has been undertaken in central Europe since the 1960s. It is hoped that it will provide the basis for more extensive and intensive botanical research throughout Europe.

The chapters, which comprise a series of individual essays, are idiosyncratic in reflecting the similarities and differences in the approaches of the authors. Nevertheless, the chapters follow a general pattern starting with a description of the natural features of the city, including the geology, topography, soil and climate. Then follows a general account of the history of the physical, economic and political development of the city from the earliest human settlement to present and how urbanization has affected the environment (for example, pollution and the heat island effect) and how the environment has affected the city (for example, climate change).

The recorded botanical history of the city is followed by an account of the flora which is mainly concerned with flowering plants and ferns. The chapters go on to describe the most frequent spontaneous plants and the frequently planted trees along roads and in parks and gardens.

Each of the chapters considers the evolution of the urban flora from the earliest agricultural period to the introduction of a very large number of ornamental species from the sixteenth century onwards. The past, present and future implications of the mixing of non-native with native plants, which would not otherwise occur naturally, are discussed.

Where sufficient information is available, the chapters include summary descriptions of the algae and bryophytes and because of their general visual affinity with plants the lichens (lichenized fungi) and fungi are also included. Because of the imbalance in knowledge and information, the accounts of the non-vascular plants and fungi are not as comprehensive as they should be; this deficiency has been compensated for by the provision of references to other literature.

The plant communities and species composition of the major natural and semi-natural habitats within the municipal area are described followed by accounts of the species found in more typical urban habitats, including housing areas of different types and densities, industrial zones, unused and previously developed land, parks, cemeteries, allotments and similar habitats, transport routes (for example, railway land and road verges) and various aquatic habitats, including rivers and flooded mineral workings.

The chapters end with an account of the environmental planning, protection and education aspects of the particular city, including Red Data List species, statutory habitat and species protection with special emphasize on the European Union Habitats Directive. Educational programmes such as nature trails are also described.

Finally, it is necessary to explain some inconsistencies in the number of references quoted and listed. Originally, the contributors were asked to provide only about eight publications for further reading. Some contributors did their best to comply whilst others were unhappy because they considered that the absence of references and/or a short reference list would indicate to their academic colleagues/potential readers that they were not familiar with the literature and that this would adversely affect their academic credibility. The difficulty was resolved by the editors succumbing to contributor pressure, and the writing this "health warning."

Almería

Elias D. Dana, Juan García-de-Lomas, and Manuel A. Guerrero

Fig. 1 Fishing port of Almería, with La Alcazaba in the background (Photograph by Maria del Mar Bayo)

Abstract The Province and city of Almería are important areas of the Mediterranean basin with respect to their cultural heritage and environmental values. However, unfortunately, the recent and dramatic landscape transformation has resulted in the destruction of natural habitats and their component species – in the last 50 years,

Elias D. Dana (✉)
G.I. Transferencia I+D+i Recursos Naturales, Universidad Almería, Spain
e-mail: edana@ual.es

J.G. Kelcey and N. Müller (eds.), *Plants and Habitats of European Cities*,
DOI 10.1007/978-0-387-89684-7_1,

the area of the city has increased from about 2.5 km^2 to about 10 km^2. The present population is 190,000. The predominant traditional agricultural system of family farms has been replaced by intensive agriculture, greenhouse and built development. The urban area of the city supports about 200 vascular plant species, most of which are associated with ruderal or weed communities. The overall proportion of non-native taxa is 5% but in densely urbanised areas it increases to 12%. The non-native taxa include several genotypes and undescribed varieties of *Phoenix dactylifera* from North Africa and the highly invasive ornamental species *Pennisetum setaceum*. Most of the non-native taxa develop during the late spring and autumn, whereas the native taxa develop between the winter and early spring. In addition, the typical urban habitats in the city contain ten important natural or semi-natural coastal, scrub and grassland habitats of plant communities.

Natural Environment of the City

The continental areas of Spain are divided into Regional Governments (Comunidades Autónomas) and Provinces, which contain municipalities (101 in the Almería Province). Each province has an administrative city of the same name; hence, "Almería" is the name of the Province and the capital city, which are within the Regional Government of Andalusia (Comunidad Autónoma of Andalucía). More information on the political, administrative and social aspects can be found at http://en.wikipedia.org/wiki/Almeria and http://en.wikipedia.org/wiki/Provinces_of_Spain.

Unless otherwise stated, "Almería" refers to the city, which is located in south-eastern Spain (Latitude 36°50′ north, Longitude 2°28′ west). The locations of the Province and the city are shown in Fig. 2.

Most of the city is only a few meters above sea level. The geology of Almería comprises mainly calcareous and sedimentary deposits of marine and freshwater (river) origin (Aguirre 1998). These materials have determined the features of the main soil classes, which are characterised by small particle sizes, with a relatively high content of sodium (Na) and potassium (K) and in the inner parts of the city, calcium (Ca) derived from clay and limestone. The basal mountain areas on the western side of the city have poorly developed soils with calcareous outcrops. Sandy soils are generally restricted to the sea fringe (Fig. 2).

The general climate of the area is Mediterranean, semi-arid xeric, characterised by strong aridity and mild winters with high relative humidity. The annual mean rainfall is about 145 mm and the annual mean air temperature is 18°C. Meteorological data for the period 1929–2009 show mean maximum and minimum daytime temperatures of 23°C and 15°C, respectively, with a maximum of 44°C and a minimum of −1°C. These conditions, together with a high level of evapo-transpiration, impede the development of coastal forests, which are replaced by tall shrub communities, including *Maytenus senegalensis* ssp. *europaeus, Ziziphus lotus, Pistacia lentiscus* and *Olea europaea* ssp. *sylvestris*. Figure 2 shows that the landscape is dominated by perennial grasslands and medium to low scrub.

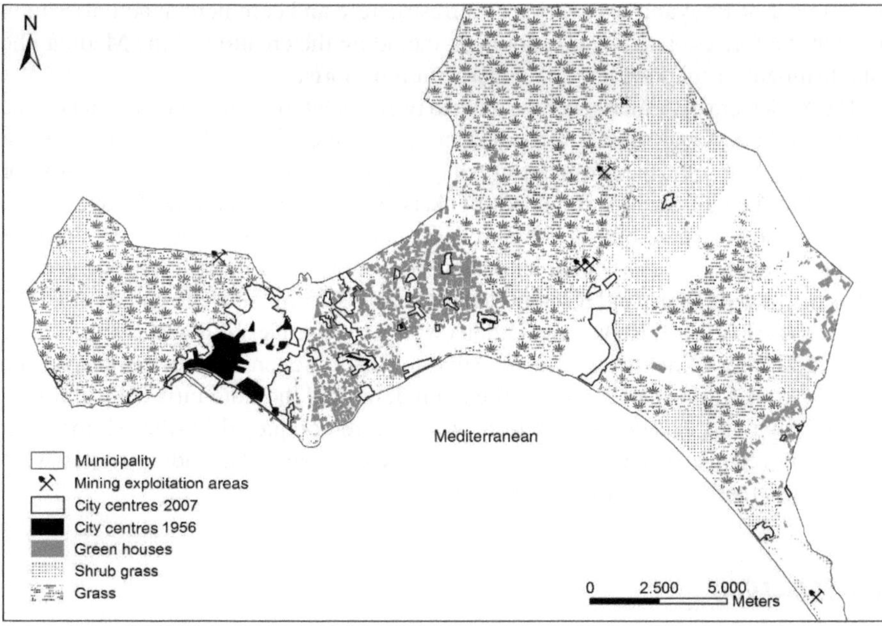

Fig. 2 Main landscape features and vegetation types of Almería city and administrative areas

The Andarax river has the hydrological characteristics of North African wadies and has superficial freatic layers. In former times, this river was associated with the development of riverine tree plant communities dominated by *Tamarix* spp. and occasionally by *Populus alba*. Currently, most of the non-riparian and riparian communities have disappeared or their extent has been drastically reduced. As described below, the latter are of considerable conservation importance.

Historical Development of the City

Until AD 1200

Almería and its surroundings have been inhabited since the Neolithic period, when the area was colonised from the east. During these times, the main settlements were located close to the current city boundaries and the Andarax river. Typical riparian forest communities dominated by *Salix* spp. and *Populus* spp. and the associated riparian fauna have been found in nearby archaeological sites. Although Roman culture left some heritage, the most important legacy came from the Moors civilisation, which made Almería the administrative capital and political centre of the area in AD 955. A fortress with watch towers was built to defend the previous administrative capital of Pechina (formerly called Bayyana), which is about 10 km from Almería. At that time, Almería was called Al Mariyyat Bayyana (meaning “the watchtower of Bayyana"). A series of infrastructure and commercial activities took place at the fortress and its surroundings, including the creation of the Medina, the establishment of markets and the construction of ports.

The golden era of Almería was the tenth to twelfth centuries. In that period, it became one of the largest and most important commercial cities for Mediterranean maritime traffic. The Arab influence resulted in many important philosophers being born in or attracted to the city, such as important thinkers of Al-Andalus and the Mediterranean areas. The area became a centre of excellence for contributors to the development of geography, agronomy (for example, Al Idrisi), theology and poetry (for example, Al Mutasim, the poet Emir) and for important leaders of Sufism (a mystic branch of Islam) such as Ibn Al Arif and Ruayni of Córdoba who were disciples of Ibn Masarra.

In 1147, the by now rich city of Almería was conquered by the troops of the Catholic king Alfonso VII and his allied kingdoms and the state-cities of Catalonia, Pisa, Geneva and France. Although it was soon re-conquered by the Almoravids, Almería never recovered its previous importance because the trading routes were diverted to other coastal cities of the Mediterranean.

AD 120–1900

Soon after the occupation (in 1489) by the Catholic monarchs Isabel and Fernando, the city and its inhabitants were plunged into a period of considerable poverty (both material and cultural) until the mid-nineteenth century, when iron and silver mines were discovered by British and French companies. This resulted in a rapid population increase and a temporary reduction in poverty in the city and the province. However, it was accompanied by extreme environmental degradation, which is a typical consequence of sudden population growth. The degradation included deforestation, soil erosion and lack of food (mainly proteins), which ultimately resulted in a general poverty vortex that reached its maximum during and just after the civil war of 1936–1939. Figure 3 shows the demographic trend since the existence of reliable census data.

The architectural development of Almería is as complex as its history. After the conquest by the Catholic monarchs, the city was almost completely destroyed at different times by earthquakes. It was not until the nineteenth century that the city really emerged from the influence of the ancient Medina and expanded to the size it is today (see Dana et al. (2002)).

1950s to the Present

Between 1950 and 1960, the landscape surrounding the city was typical of the Mediterranean irrigated agricultural system, comprising small pieces of land, each (or a few of them) maintained by families. Figures 4 and 5 show the intense and rapid transformation of the urban nucleus into the surrounding areas during the last 60 years or so.

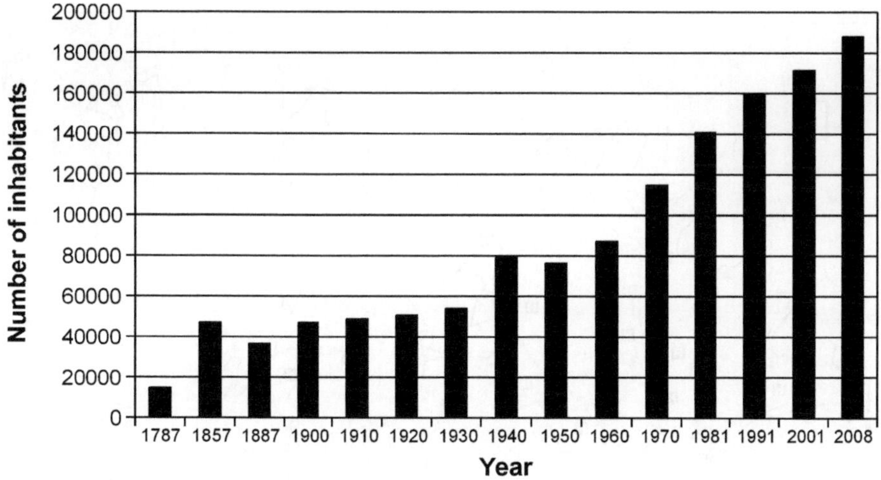

Fig. 3 Census data 1787–2008 for Almería city

Fig. 4 Aerial image of the area in 1956 showing the evolution of urbanised areas. USA National Cartographic Service

In the 1950s, there were only minor extensions of the urban area into the adjacent agricultural land, which contained only two relatively important villages. Since that time, the cultivated lands have been gradually urbanised and the traditional farm systems converted into agricultural structures based on intensive food production,

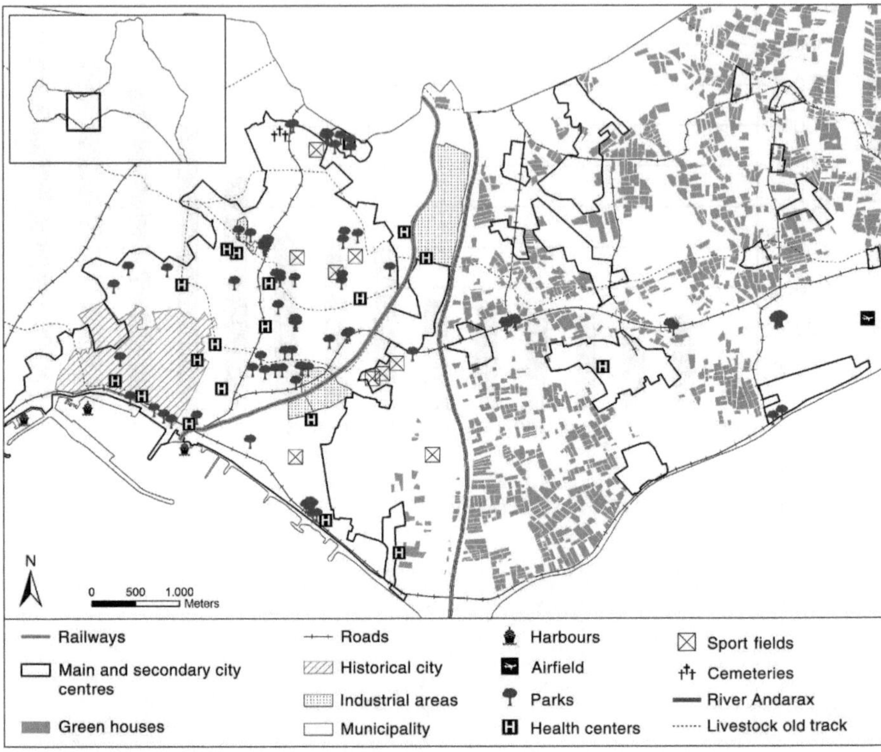

Fig. 5 Urban habitat and services. Note the scarce number of services in the secondary urbanised areas

mainly greenhouses. The analysis of aerial photographs indicates that the city occupied about 9.95 km^2 in 2007, almost a 300–400% increase since 1956, when the main urban area occupied 2.26 km^2 (the total urbanised land occupied 3.5 km^2). The images show the gradual appearance of new densely populated areas that were absent in 1956. It is important to note the gradual urbanisation of coastal habitats and the growth of existing settlements, especially since 2000.

Changes of the Environment Due to City Growth

Unfortunately, there are no published studies relating specifically to the environmental changes caused by the growth of the city. The most complete and interesting study investigated the socio-economic changes of the province and their effects on environmental and natural values (see García-Latorre and García-Latorre 2007). Among other sources, the authors used historical documents, archaeological evidence, place names, interviews with older farmers and field studies to gather information about how the landscape has probably been modified by people and their values. In the administrative area of Almería, most of the land has been occupied by irrigated land and pasture. Some of the irrigation systems were inherited from previous Arab cultures, while others were created during the nineteenth century as an answer to the increasing demand for food due to the expanding population. The greatest environmental change took place during the twentieth century when most of the cultivated land around the city became occupied by dense urban development, particularly since 1970. The most recent changes and their environmental implications can be assessed by comparing aerial images of 1950 with those of the present day (see Figs. 4 and 5). They can be summarised as the typical destruction of agricultural and grazing habitats and the associated species (in the former case, many of the taxa were linked to water availability), changes in traditional agricultural practices and the increase in urbanisation.

Agriculture has changed from "family farms" with *Lycopersicon esculentum, Capsicum* spp., *Solanum tuberosum* varieties, *Medicago sativa* spp. *sativa,* cereals, and fruits such as citric orchards, *Olea europaea, Armeniaca vulgaris, Prunus dulcis, Punica granatum, Eriobotrya japonica, Mespilus germanica* and *Ficus carica* were cultivated to intensive agricultural production (greenhouses with few modern varieties of *Lycopersicon esculentum* and *Capsicum* spp.). This change has involved large inputs of agrochemicals and has resulted in the homogenisation and impoverishment of landscape features and the associated biocoenosis. The traditional agricultural habitats that supported many typical animal species, especially birds (for example, *Upupa epops, Athene noctua, Tyto alba, Hirundo rustica* and *Erinaceus europaeus*), have declined dramatically, causing a local decrease in their populations (Manrique and De Juana 1991).

The city has extended eastwards; so far, both sides of the Andarax river have been totally developed. During the last 2 decades, several adjacent rural villages and towns have been incorporated into the city as it has expanded. Since 2000, a large area of semi-natural dunes on the outer edge of the Natural Park "Cabo de Gata" has been replaced by greenhouses and the associated infrastructure, including roads, car parks and recreational areas. The dunes were characterised by psammophilous communities (dominated by species such as *Ammophila arenaria* or *Pancratium maritimum*) and by tall thorny shrub communities dominated by *Ziziphus lotus*. The latter communities act as a refuge for several mammal and steppe bird species of avian taxa (Heath and Evans 2000).

In addition to increasing urbanisation, erection of greenhouses is the other cause of environmental change. The impacts of these intensive agricultural systems are mainly habitat loss, pollution (generation of large quantities of plastic sheeting and agricultural waste, and contamination of aquifers), groundwater imbalances (including marine intrusion) and severe landscape modification. Also, large pits have been created by the extraction of gravel and clay for use as a substrate for cultivation (Pulido-Leboeuf et al. 2003; Downward and Taylor 2007). It has been recently proposed that because of their greater albedo, these areas may be an important factor contributing to climate change, especially in territories where most of the available land is covered by greenhouses (Campra et al. 2008). However, their effect on the municipality of Almería has not been investigated.

It can be concluded that substantial environmental changes have occurred since the 1950s as a response to changes in socio-economic values; that is to say, the transition of a rural local society based on the exploitation of land by family members to an economic model based on the provision of homes, services (banks, commerce and trade) and recreational facilities. These changes have led to the increase in (i) population (the current population of Almería city is ca.190,000, see Fig. 3), (ii) urbanised area and consequently (iii) landscape fragmentation and edge effects.

Flora

Vascular Plants

The city and its immediate fringes support about 200 angiosperm species, and most of them are ruderal and weed species (more information on the species found and the ruderal/weed communities is provided in Dana and Rodríguez-Tamayo (1999) and Dana et al. (2002). Most of the native taxa develop in the late winter–early spring, while the development of most of the alien species occurs mainly from late spring to autumn. The 50 most common species in the city are given in Table 1.

The percentage of alien taxa found in the whole of the administrative territory is low (5%), probably due to climatic and historic reasons. However, the percentage

Table 1 The 50 most frequent species found in Almería (no neophytes)

Bromus diandrus	*Plantago coronopus*
Bromus rubens	*Plantago lagopus*
Atriplex halimus	*Parietaria judaica*
Avena barbata	*Phalaris minor*
Beta vulgaris	*Oryzopsis miliacea*
Calendula arvensis	*Plantago ovata*
Carrichtera annua	*Poa annua*
Chenopodium album	*Polycarpon tetraphyllum*
Chenopodium murale	*Polygonum aviculare* agg.
Chrysanthemum coronarium	*Polypogon monspeliensis*
Convolvulus arvensis	*Portulaca oleracea*
Catapodium rigidum	*Reichardia tingitana*
Dittrichia viscosa	*Salsola vermiculata*
Echium creticum	*Salsola oppositifolia*
Erodium chium	*Schismus barbatus*
Eruca vesicaria	*Setaria verticillata*
Fagonia cretica	*Sisymbrium irio*
Hedypnois rhagadioloides	*Solanum nigrum*
Herniaria hirsuta	*Sonchus oleraceus*
Hordeum murinum	*Sonchus tenerrimus*
Hymenolobus procumbens	*Spergularia bocconnii*
Lamarckia aurea	*Spergularia. diandra*
Malva parviflora	*Urospermum picroides*
Mercurialis annua	*Urtica urens*
Papaver hybridum	*Volutaria lippii*

is higher in the more densely urbanised parts of the city, where the average value is about 12%. Some of the most frequent alien species found in the territory are mainly neophytes such as *Amaranthus muricatus, Aster squamatus, Mesembryanthemum crystallinum, Atriplex semibaccata, Nicotiana glauca, Conyza sumatrensis, Conyza bonariensis, Coronopus didymus, Chamaesyce serpens, Heliotropium curassavicum* and *Zygophyllum fabago*. The archaeophytes present include *Arundo donax* (which is abundant in the administrative area and a casual in the main city) and *Oxalis pes-caprae*.

Some taxa that were frequently found on the fringes of the main city up to the early 1990s have gradually become infrequent or have suffered a reduction either in frequency or patch size as a consequence of urbanisation. This has resulted in a decrease in the availability of areas that provided a source of propagules for the species that are more associated with agricultural land, either cultivated or abandoned. Some examples are *Bassia hyssopifolia, Coyncia tournefourtii, Hyoscyamus albus, Anagallis arvensis, Asphodelus* spp., *Atriplex glauca, A. semibaccata, Chondrilla juncea* and *Hammada articulata*. The analysis conducted by Dana et al. (2002) showed some interesting species that can be considered as "relicts" of the traditional agricultural landscape such as *Linum ussitatissimum,* and are almost absent now.

Planted Trees and Shrubs

Taxa planted in urban public spaces are mostly low-maintenance species adapted to the aridity and salinity influences of the coast. The most common tree species used in roadside plantings are *Ficus retusa, Acacia saligna* and *Washingtonia* spp. More recently and because of the lax sanitary procedures, several genotypes and unknown varieties of *Phoenix dactylifera* have been imported from North Africa to satisfy, as quickly as possible, the aesthetic demands of people living in the recently urbanised areas. This has led to an increase in the occurrence of insect *Rhynchophorus ferrugineus* and the consequential loss of a large number of *Phoenix canariensis* and *P. dactylifera,* some of which were more than 150 years old. This organism is also destroying large numbers of ancient palm trees outside the city in typical agricultural areas and so adversely affecting the historic landscape of the Province's lowlands. It is also essential to stop the recent use of highly invasive ornamental species such as *Pennisetum setaceum*. This species, which has recently colonised open habitats, has the potential to colonise wadies and a number of habitats of conservation interest such as coastal habitats and permanent grassland to the detriment of the native flora.

There is no information about the non-vascular plants and the fungi of Almería.

Habitats

The administrative area still supports important plant communities as defined by the "Habitats Directive" (Directive 92/43/CEE as amended by Council Directive 97/62/EC of 27 October 1997, Regulation No 1882/2003 of the European

Parliament and of the Council of 29 September 2003 and Council Directive 2006/105/EC of 20 November 2006 modifications). Nine communities that occur in Almería municipality are listed in Annex 1 (of the Directive), of which three are "Priority Communities". These communities and their EC and Corine codes (* indicates a priority habitat type) are described below. However, not all the species and other aspects of the community as described in the Directive necessarily apply to Almería city or its municipality.

1. Southern riparian galleries and thickets belonging to *Nerio-Tamaricetea* and *Securinegion tinctoriae* (EUH code 92D0; Corine code 44.81).

   Tamarisk, Oleander and *Chaste* tree galleries, which are thickets and similar low-growing woody habitats associated with permanent and temporary streams and wetlands of the thermo-Mediterranean zone and south-western Iberia. The plant species commonly found are *Nerium oleander, Vitex agnus-castus, Tamarix* spp., *Securinega tinctoria, Prunus lusitanica* and *Viburnum tinus*.

2. *Crucianellion maritima* fixed beach dunes (EUH code 2210; Corine code 16.223).

   Coastal dune communities of the eastern Mediterranean of the alliances *Crucianellion maritima, Medicagini marinae-Triplachnion nitensis* and *Ammophilion arenaria*. The plant communities are rich in species of the genus *Silene,* together with *Euphorbia terracina* and *Pancratium maritimum* among others.

3. *Malcolmietalia* dune grasslands (EUH code 2230; Corine code 16.228).

   Therophyte communities of the coasts of the Mediterranean basin and the subtropical Atlantic that colonise deep sands in clearings of permanent communities of fixed or semi-fixed dune systems, and sometimes depressions of white dunes with several *Malcolmia* spp.

4. *Arborescent scrub with *Ziziphus lotus* (EUH code 5220; Corine code 32.251).

   Pre-desert deciduous brush of *Periploca laevigata, Lycium intricatum, Asparagus stipularis, Asparagus albus* and *Withania frutescens* with tall *Ziziphus lotus,* confined to the arid Iberian south-west under a xerophytic thermo-Mediterranean bio-climate; corresponds to the mature phase or climax of climatophile and edapho-xero-psammophile vegetation series: *Periplocion angustifoliae, Ziziphetum loti, Zizipho-Maytenetum europaei, Mayteno-Periplocetum*. Other species present include: *Lycium intricatum, Asparagus stipularis, A. albus, Calicotome intermedia, Chamaerops humilis, Maytenus senegalensis* ssp. *europaeus, Periploca laevigata* ssp. *angustifolia, Phlomis purpurea* ssp. *almeriensis* and *Rhamnus oleoides* ssp. *angustifolia.*

5. Mediterranean and thermo-Atlantic halophilous scrub (*Sarcocornetea fruticosi*) (EUH code 1420; Corine code 15.61).

   This scrubby, halophilous vegetation develops in the uppermost levels of salt marshes, often where there is a transition from saltmarsh to dunes or in some cases where dunes overlie shingle. The permanent vegetation is mainly composed of scrub species that have an essentially Mediterranean–Atlantic distribution, for example, *Salicornia, Limonium vulgare, Suaeda* and *Atriplex* communities belonging to the *Sarcocornetea fruticosi Class.* The associated species are *Atriplex portulacoides, Inula crithmoides* and *Suaeda vera.* The lower topographical level supports species such as *Sarcocornia perennis, S. perennis* ssp. *alpini, S. fruticosa* and *Arthrocnemum macrostachyum,* whereas the higher topographical level is dominated by *Limoniastrum monopetalum, Aster tripolium* and *Limonium* spp.

6. Halo-nitrophilous scrub (*Pegano-Salsoletea*) (EUH code 1430; Corine code 15.17).

   Halo-nitrophilous scrub belonging to the *Pegano-Salsoletea* class, typical of dry soils in arid climates, sometimes including tall, dense shrubs. The species found *are Peganum harmala, Artemisia herba-alba, Lycium intricatum, Capparis ovata, Salsola vermiculata, S. genistoides, S. oppositifolia, Suaeda pruinosa, Atriplex halimus, A. glauca, Camphorosma monspeliaca* and *Haloxylum articulatum.*

7. Annual vegetation of drift lines (EUH code 1210; Corine code 17.2).

   Formations of annuals or representatives of annuals and perennials, occupying accumulations of drift material and gravel that are rich in nitrogenous organic matter lying at or above mean high-water spring tides. These shingle deposits occur as fringing beaches that are subject to periodic displacement or overtopping by high tides and storms. The distinctive vegetation, which may form only sparse cover, is therefore ephemeral and composed of annual or short-lived perennial species such as *Cakile maritima, Salsola kali, Atriplex spp., Polygonum spp., Euphorbia peplis, E. paralias, Glaucium flavum, Matthiola sinuata, M. tricuspidata,* and *Eryngium maritimum.*

8. *Iberian gypsum vegetation (EUH code 1520; Corine code 15.912).

   Low, open *Thymus, Teucrium* and *Helianthemum* garrigues (= Mediterranean scrub), which colonise the poorly developed gypsiferous soils of the arid south-east of the Iberian Peninsula. The characteristic herbaceous elements comprise *Teucrium libanitis, T. polium, T. pumilum, T. carthaginense, Thymus longiflorus, T. antoninae, Helianthemum lavandulifolium, H. squamatum, Gypsophila hispanica, G. struthium* and *Astragalus alopecuroides*. The grass species present include *Lygeum spartum, Stipa tenacissima,* and *Brachypodium retusum. Artemisia barrelieri* and taxa in the Chenopodiaceae may be locally abundant.

9. *Pseudo-steppe with grasses and annuals of the *Thero-Brachypodietea* (EUH code 6220; Corine code 34.5).

   Includes a variety of xeric, thermophilic and mostly open permanent and annual Mediterranean grasslands usually growing on eutrophic soils but also found on oligotrophic soils. There are three major sub-types: (i) permanent basophile, rather hard short-grass communities, included in *Lygeo-Stipetalia;* (ii) very dense and short but highly productive, permanent swards that resist summer drought. These communities are created by intense and continuous livestock grazing and included in the *Poetalia bulbosae;* (iii) pioneer and ephemeral basophilous annual grasslands, included in the *Brachypodietalia (Trachynietalia) distachyae*. The diversity of plant, invertebrate and vertebrate communities is usually high. The most characteristic species are *Brachypodium distachyon, Bombycilaena erecta, Echinaria capitata, Polygala mospelliaca, Scabiosa stellata* and *Stipa capensis*.

The conservation of these communities and their characteristic species outside the protected areas is difficult to achieve, given the intense use of the territory and the expansion of the city towards the eastern and north-eastern natural and semi-natural areas.

Central and Residential Areas

City centre and high density housing areas (block- and ribbon-development, multi-storey buildings; historical city, business and shopping districts)

High density housing areas are characterised by the extensive presence of unnatural materials (for example, concrete, paving and tarmac), atmospheric pollution, frequent trampling by people and domestic animals and the compaction of the soil vehicles (Benvenuti 2004). Despite the absence of natural environments, high density housing areas support a mosaic of "extreme" micro-environments such as walls, cracks, slits and tiles; however, even these habitats are very scarce in Almería because of modern construction methods. The joints in the pavements and the bare soil depressions around trees contain short-lived species that are highly tolerant of trampling and have a wide distribution; they include *Aster squamatus, Conyza bonariensis, Amaranthus muricatus, A. viridis, Coronopus didymus, Malva parviflora, Cyperus rotundus, Poa annua,* and *Sisymbrium irio. Sonchus tenerrimus* appears as an epiphyte on palm tress (*Phoenix canariensis*) as do *Sonchus oleraceus* and *Parietaria judaica,* but more occasionally. *Nicotiana glauca* grows on open ground, for example, demolition sites. *Antirrhinum mollisimum* and *Sarcocapnos eneaphylla* often occur on walls as does *Umbilicus rupestris,* but only rarely. The three latter species have their origins in the rupicolous vegetation of the Gádor Mountains.

Low Density Housing Areas (Terraced and Detached Housing Areas, Villas)

There are virtually no low density housing areas in the city.

Industrial Areas

The industrial areas, which are generally highly developed, contain many exotic species, the most frequent being *Amaranthus muricatus, A. viridis, Aster squamatus, Conyza* spp, *Zygophyllum fabago,* and *Sisymbrium irio*. Some annual native species such as *Avena* spp. and *Malva parviflora* also occur.

Transport Routes and Areas

Railways

Railway land contains mainly ruderal and nitrophilous plants with life cycles adapted to arid conditions and lower disturbance regimes than urban areas. The plant species represented include native biennial and perennial species such as *Oryzopsis miliacea* and *Atriplex halimus* and the exotic species *Nicotiana glauca, Agave americana, Zygophyllum fabago* (which was introduced via cereal imports from eastern Europe) and *Amaranthus muricatus.*

Roads

The floristic composition of the roads, ditches and verges depends on their location, not least in relation to the source of propagules. In general, the roadside vegetation closest to the urban area generally contains ruderal species such as *Oryzopsis miliacea, Nicotiana glauca, Zygophyllum fabago, Hordeum murinum, H. leporinum* and *Dittrichia viscosa.* Roads that pass through or within the vicinity of the Cabo de Gata Natural Park generally contain native species such as *Launaea arborescens, Hyparrhenia hirta, Dittrichia viscosa, Oryzopsis miliacea* and *Carlyna corymbosa,* although some alien species such as *Nicotiana glauca, Agave americana, Arundo donax* and *Opuntia ficus-indica* are also present.

Airfield

Because of its small size (1 airstrip of 3,200 m long), it is not considered a habitat of particular floristic identity. There have been no detailed studies of the flora in the vicinity of the airstrip, although occasional observations revealed the presence of *Anacyclus clavatus, Atriplex halimus, Mesembryanthemum nodiflorum, Salsola oppositifolia, Halogeton sativus,* and the alien species *Mesembryanthemum crystallinum* and *Zygophyllum fabago.* The woody species are restricted to the unmanaged areas.

Recreation Areas

Parks

A good review of the ornamental plants of gardens and parks can be found in Leal and Guardia (1987). Exotic species such as the palm (*Phoenix dactylifera*) have been used frequently in Almería gardens. Among the larger trees, those in the genus *Ficus* are probably the most attractive, for example, *Ficus elastica, F. retusa* var. *nitida, F. macrophylla* and *F. rubiginosa*. Other tall trees include *Ailanthus altissima, Platanus* x *hispanica, Melia azedarach* and *Schinus molle* and *S. terebinthifolius*. The legumes are also widely represented with numerous specimens of the genus *Acacia* (including *Acacia saligna, A. retinoides, A. longifolia* and *A. dealbata*) together with other genera and species such as *Robinia pseudoacacia, Tipuana tipu, Albizia julibrissin* and *Cercis siliquastrum*. Climbers are also present, for example, the native *Hedera helix* and the exotic species *Jasminum officinale, Lonicera japonica, Bougainvillea* spp. and *Tecomaria capensis*. Considering the dry climate of Almería, some desert species such as *Agave americana, A. fourcroydes, A. sisalana* and *Yucca aloifolia* are also common. The ornamental flora also includes some species imported from the Canary Islands, such as *Phoenix canariensis, Pinus canariensis* and *Dracaena drago*. The native species that occur in this habitat include *Chamaerops humilis, Ceratonia siliqua* and some aromatic plants such as *Rosmarinus officinalis* or *Lavandula dentata*. Many spontaneous species have been recorded; they include *Coronopus didymus, Lolium* spp., *Conyza bonariensis, Aster squamatus, Poa annua, Chamaesyce serpens, Hordeum murinum, H. leporinum, Heliotropium curassavicum, Setaria viridis* and *Veronica persica* (less frequently).

Cemeteries

The cemeteries are very poor in species; except for *Cupressus sempervirens* (which is widely planted in Spanish cemeteries), they contain only a few ruderal species.

Open Land

Crops/Arable

The species that are commonly represented in the flora of the arable fields are *Lavatera cretica, Malva parviflora, Anacyclus clavatus, Heliotropium curassavicum, H. europaeum, Avena* spp., *Urospermum picroides, Sonchus asper, S. oleraceus,*

Lactuca serriola, Amaranthus viridis, A. blitoides, Oryzopsis miliacea, Hordeum murinum, H. leporinum, Setaria viridis, S. pumila, Conyza bonariensis, Aster squamatus, Arundo donax (along the margins of the boundaries and ditches), *Mesembryantheumum nodiflorum, M. crystallinum* and *Halogeton sativus*. The margins of the arable fields and fallow land contain species such as *Dittrichia viscosa, Atriplex halimus, Salsola vermiculata, Nicotiana glauca,* and *Zygophyllum fabago.*

Waste Ground

The waste ground habitat is less disturbed than the habitats within the high density housing areas; consequently, the vegetation shows characteristics similar to those of meadows. The habitat supports predominantly tall, annual species with some perennial herbs, such as *Chrysanthemum coronarium, Anacyclus clavatus, Plantago lanceolata, P. coronopus, Phalaris paradoxa, Oryzopsis miliacea, Avena* spp, *Malva parviflora, Lavatera cretica, Amaranthus blitoides, A. albus, Oxalis pes-caprae, Atriplex semibaccata, Moricandia arvensis, Eruca vesicaria, Sinapis alba, Brassica oleracea, Nicotiana glauca, Zygophyllum fabago, Mesembryanthemum crystallinum, M. nodiflorum,* and *Cyperus rotundus.*

Aquatic Habitats

Moving water habitats are very rare in Almería and confined to "ramblas" or "wadis" – dry river beds, which are typical of the driest areas of the Mediterranean and contain water following periods of heavy rain only. Many of these "watercourses have been culverted or are used for the disposal of rubbish (fly tipping).

Up to 20–30 years ago there were many long-established irrigation canals that were an essential part of the traditional agricultural system. In recent years, most of these canals and the land they irrigated have been destroyed by urbanisation and the construction of extensive greenhouse complexes. The vegetation of some of the few sections of canal that still exist comprises small pockets dominated by native species including *Tamarix* spp, *Phragmites australis,* and *Typha* spp. However, the vegetation of most of what is left of the canals has been replaced by exotic tall canes (*Arundo donax*) and more occasionally *Ricinus communis.*

Nature Conservation, Environmental Planning and Education

In Spain, nature conservation measures are mainly on the basis of (i) species protected by law, (ii) particular landscapes and (iii) regional endemic species. The environmental policy is based on the Treaty of European Union (Art. 2) to achieve a balanced and sustainable development. The applicable legal framework includes the Directive 92/43/EEC ("Habitat Directive"), the Spanish Law 42/2007 on Natural Heritage and Biodiversity and Regional Law 8/2003 on wild flora and fauna. In addition, economic projects that may adversely affect habitats and/or species of importance must be subject to environmental impact assessment according to European Directives 85/337 and 2001/42 and Spanish Law 9/2006 concerning the evaluation of the effects of certain plans and programmes on the environment. This European Community-born legal framework provides key advances in conservation.

In practice, the lists of protected species and habitats may be modified according to regular updates and improvement of knowledge of habitats. These updates involve primarily the IUCN criteria and the legal framework. However, political pressures and especially the resources available rather than the preservation of functionality of ecosystem processes are common constraints in the effective conservation of habitats and species.

Some examples may illustrate the limitations that are often found in conservation practice. Unfortunately, nature protection and management is not included in the priorities of the municipality or Provincial authorities. Regional and National Governments are the only public administrations that have developed specific albeit limited efforts to protect nature. The conservation approaches of the regional administrations are far from being effective or adequate. Much of the Municipality's income is obtained from the licensing of development proposals and other new economic activities. Therefore, politicians and Council officials often claim that nature conservation is opposed to economic development and employment. Consequently, legal restrictions are frequently ignored so that tourism and development can receive greater priority than nature conservation.

Another example comes from species conservation. Essentially global extinction is the sum of local extinctions. However, preservation of local populations is difficult when a species is quite common in several Provinces but is disappearing or there are historical records from a neighbouring Province(s). This situation is simply ignored by the Regional Government and politicians resulting in the occurrence of local extinctions.

Resource limitation to obtain essential scientific information about the status and distribution of a species is also a major problem. The conservation of high environmental values using the argument that "not everything can be protected" is of course a reflection of the difficulties that politicians and their officials must address when formulating environmental protection policies and trying to "do what they have to do." But how to achieve a more attractive and natural city when Councils are not only ineffective in protecting nature but are also the main perpetrators in the destruction of natural habitats through approving speculative developments? And how to achieve the protection of remnant patches of "what nature was" when conservation in practice is mainly designed by choosing priorities only in a regional context (eight Provinces with a total of nine million inhabitants)? Finally, how to protect remnants of private land to create a network of rural and semi-natural areas of bio-sociological interest? Only these types of approaches will lead to an improvement of the presence of nature (let us say more natural ecological processes and the habitats and species involved) in a city. Conservation of public land is difficult at best; conservation in private property is idealistic at least.

The EC Habitat Directive should help the design and maintenance of landscapes and biocoenosis of "Community Interest" that otherwise lack effective protection. Of course, this is only a partial solution that will allow the relationship between the effective protection of the many habitats, plant communities and their associated species mentioned in this chapter to be established. However, co-operation, co-ordination and involvement, and particularly the involvement of the regional and local administrations, should be a priority if nature conservation and sustainable development are to be achieved. Figure 6 shows the relationship between the natural areas (protected or not) and those areas of Almería in which the species cited in the Habitat Directive occur. It should be noted that most of the last fragments of coastal and riparian communities (see the Delta river in the figure) that are cited in the Habitat Directive, have no legal protection and are close to secondary urban centres – the implications are obvious.

An assessment of Figs. 1–6 indicates that the protected areas present within the city or its municipality are not preserved in the context of a more sustainable or at least an attractive environment (see the conceptual discussion by Rees and

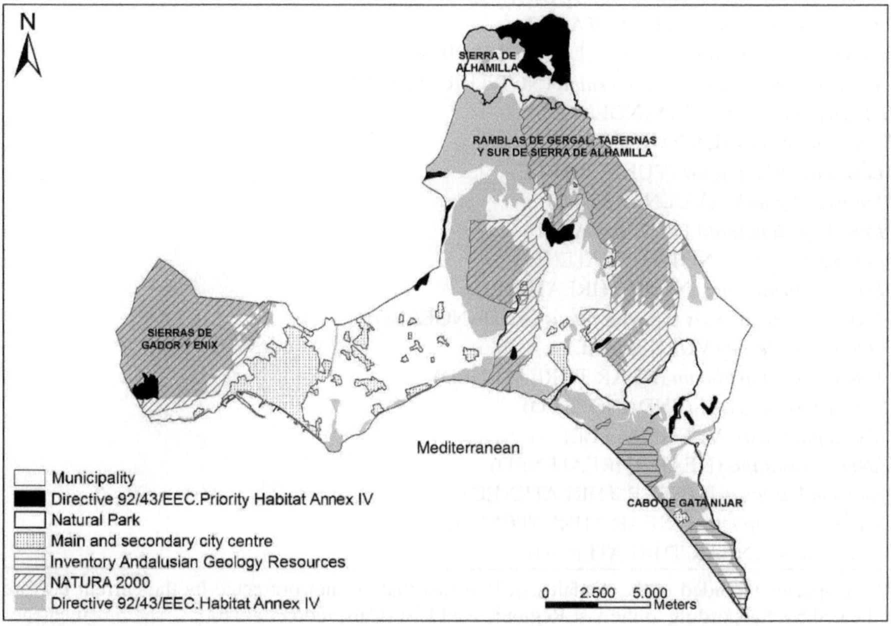

Fig. 6 Distribution of natural protected areas and their connection through the Natura 2000 Network

Wackernagel (1996) in relation to urban development). It is essential to stop the destruction of important areas of biodiversity in southern Spain and to ensure the conservation of threatened biodiversity at all ecological levels (including rare plant communities, habitats and landscape) in the regional network of protected areas. It is desirable to incorporate some natural areas in the urban design of the city. Also, the acquisition of a state of well being and environmental quality should consider nature and its importance to people. To this end, a city should not be designed solely on economic criteria but also on biological criteria. For example, the number of trees shown at each location in Fig. 4 is roughly the number of trees really present in the "park" – and they are mainly *Ficus retusa*!!

Several taxa of natural/semi-natural habitats that are protected by the Regional Government (Law 8/2003) and listed in the Andalusian Red List prepared and updated by Cabezudo et al. (2005) occur in Almería, see Table 2. In addition, an

Table 2 Some of the protected species and species listed in the last Regional Red List (Cabezudo et al. 2005) that occur in Almería

**Ammochloa palestina* (VULNERABLE)
Androcymbium gramineum (also listed in Annex IV of the "Habitats Directive")
**Caralluma europaea ssp. europaea* (ENDANGERED)
**Cistanche phelypaea* (DATA DEFICIENT)
Cynomorium coccineum (VULNERABLE)
Cosentinia vellea ssp. *bivalens* (NEAR THREATENED)
Euzomodendron bourgeanum (ENDANGERED)
* *Galium ephedroides* (VULNERABLE)
**Haplophyllum rosmarinifolium* (DATA DEFICIENT)
Herniaria fontanessii ssp. *almeriana* (NOT THREATENED)
Hypericum robertii (ENDANGERED)
Linaria nigricans (ENDANGERED)
**Lavatera oblongifolia* (VULNERABLE)
**Linaria oligantha* (VULNERABLE)
**Linaria pedunculata* (VULNERABLE)
Loeflingia baetica (NOT THREATENED)
**Lycium intrincatum* (NEAR THREATENED)
Maytenus senegalensis ssp. *europaeus* (ENDANGERED)
**Ononis talaverae* (VULNERABLE)
**Pancratium maritimum* (NEAR THREATENED)
Rosmarinus eriocalix (ENDANGERED)
Salsola papillosa (VULNERABLE)
Sideritis lasiantha (NEAR THREATENED)
Teucrium balthazaris (NEAR THREATENED)
**Teucrium compactum* (NEAR THREATENED)
T. intrincatum (NEAR THREATENED)

Those species included in the Andalusian Red List that are not protected by the current law are indicated by * according to the last Regional Red List (Cabezudo et al. 2005). The threat categories are those published by the IUCN Regional Red List (Cabezudo et al. 2005) and given in Appendix VIII of this book important number of species that are not protected by the current law but which are listed in the Andalusian Red List also occur in the city or its municipality (Table 2).

Tree and shrub species of semi-natural habitats with high conservation value for the maintenance of vertebrate fauna still occur in the city, sometimes as isolated small patches in areas that are subjected to development pressures. The species include *Quercus rotundifolia, Q. coccifera, Olea europaea ssp. sylvestris* and *Ziziphus lotus*. The survival and conservation of these species is absolutely incompatible with urbanisation and other intensive land uses.

Closing Comments

The Province and city of Almería are important areas of the Mediterranean basin with respect to their cultural heritage and environmental values. However, unfortunately, the recent and dramatic landscape transformation has resulted in the destruction of natural habitats and their component species. The fact that these processes take place in an area with an annual average rainfall of less than 250 mm slows down or even impedes the recovery of ecological processes, habitats and plant and animal populations. However, there are still important elements that should be totally preserved. Their long-term conservation is essential but there are no definitive measures to ensure that they will not be "colonised" by the expansion of urban development or by the intensive agriculture.

If it is true that the recognition of the origins of a society improves the quality of life of its members, then the destruction of its agricultural legacy (for example, domesticated varieties of plants and animals, irrigation channels and well-managed soils) and associated biodiversity by urban development in less than 4 decades should not be allowed by a modern society. The current configuration of urban areas may have also implications for personal developments patterns (Louv 2005). In addition, the conformation of current cities may be reflecting a global social change in respect of the relationship between the individual and natural open areas in modern societies (Pergams and Zadaric 2006). However, the case of Almería highlights some difficulties for integrating man and nature in urban areas. No sensitive changes of concepts in and approaches to urban design have been detected since the first declaration of the Cabo de Gata Natural Park in 1987 – 22 years ago. Therefore, no change in the *status quo* is expected. The only tool that is currently available is the "policing role" of regional administrations through the creation and maintenance of legal frameworks, especially the influence of European Directives, that give the national and regional government's additional support to develop protection measures. In short, human beings will only make responsible decisions if forced to do so by legislation. Maybe, the low social demand is a consequence of lack of awareness of the benefits and advantages of nature preservation and design of more sustainable cities. Experience has shown that 3 decades of continual work by scientists and managers in the design of urban areas using ecological and sociological principles have not been enough to reverse the general trend.

Therefore, active campaigns on sustainable development of urban areas are recommended to increase awareness and demand.

Finally, we invite readers to reflect on the real advances (if any) in the preservation and integration of biological interest in the design of urban networks. Let us direct this question to managers and environmentalists. Then let us ask city council officials and politicians about how their current and past projects have been designed to achieve any of the goals we have described in this chapter. It will be also interesting to know how urban citizens prioritise and rank living in a more "natural" city and whether they recognise the work of their city council officials and politicians.

Literature Cited

Aguirre J (1998) El Plioceno del SE de la Península Ibérica (provincia de Almería). Síntesis estratigráfica, sedimentaria, bioestratigráfica y paleogeográfica. Rev Soc Geol España 11(3–4):297–315

Benvenuti S (2004) Weed dynamics in the Mediterranean urban ecosystem: ecology, biodiversity and management. Weed Res 44:341–354

Cabezudo B, Talavera S, Blanca G, Salazar C, Cueto M, Valdés B, Hernández-Bermejo JE, Herrera CM, Rodríguez-Hiraldo C, Navas D (2005) Lista Roja de la Flora Vascular de Andalucía. Consejería de Medio Ambiente. Junta de Andalucía

Campra P, Garcia M, Canton Y, Palacios-Orueta A (2008) Surface temperature cooling trends and negative radiative forcing due to land use change toward greenhouse farming in southeastern Spain, J Geophys Res 113

Dana ED, Vivas S, Mota JF (2002) Urban vegetation of Almería City – a contribution to urban ecology in Spain. Landsc Urban Plan 59:203–216

Dana ED, Rodríguez-Tamayo L, Mota Poveda JF (1999) Los pastizales anuales semiáridos del sector Almeriense: *Spergulo fallacis-Plantaginetum ovatae,* una nueva comunidad endémica. Lazaroa 20:49–53

Downward SR, Taylor R (2007) An assessment of Spain's Programa AGUA and its implications for sustainable water management in the province of Almería, southeast Spain. J Env Manag 82:277–289

García-Latorre J, García-Latorre J (2007) Almería hecha a mano: una historia ecológica. Fundación Cajamar

Heath MF, Evans MI (2000) Important bird areas in Europe: priority sites for conservation. Vol 2: Southern Europe. BirdLife International (BirdLife Conservation Series No: 8), Cambridge

Leal F, Guardia ML (1987) Árboles y arbustos ornamentales de la ciudad de Almería. Boletín del Instituto de Estudios Almerienses. Ciencias 7:223–235

Louv R (2005) Last child in the woods: saving our children from nature-deficit disorder. Algonquin Books, Chapel Hill, NC

Manrique J, De Juana E (1991). Land-use changes and the conservation of dry grassland birds in Spain: a case study of Almeria province. In: Goriup PD, Batten L, Norton JA (eds). The conservation of lowland dry grassland birds in Europe. Joint Nature Conservation Committee, Peterborough

Pergams ORW, Zaradic PA (2006) Is love of nature in the U.S. becoming love of electronic media? 16-year downtrend in national park visits explained by watching movies, playing video games, internet use, and oil prices. J Environ Manag 80:387–393

Pulido-Leboeuf P, Pulido-Bosch A, Calvache ML, Vallejos A, Andreu JM (2003) Strontium, SO4 2–/Cl– and Mg2+/Ca2+ ratios as tracers for the evolution of seawater into coastal aquifers: the example of Castell de Ferro aquifer (SE Spain). C R Geoscience 335:1039–1048
Rees W, Wackernagel M (1996). Urban ecological footprints: why cities cannot be sustainable and why are a key for sustainability. Environ Impact Assess Rev 16:223–248
http://en.wikipedia.org/wiki/Almeria visited 15 September 2009
http://en.wikipedia.org/wiki/Provinces_of_Spain visited 15 September 2009

Further Reading

Dana ED (2002) Flora y vegetación urbanícolas de la ciudad de Almería: características taxómicas, biogeográficas, fitocenológicas y ecológicas de las especies vegetales no ornamentales de la ciudad. Instituto de Estudios Almerienses, Almería. In Spanish

Augsburg

Norbert Müller

Fig. 1 The City Hall – an important monument of the Renaissance period and an expression of the former prosperity of the city

Abstract Augsburg, which is considered to be the oldest city in Germany, occupies 149 km² and has 250,000 inhabitants. The vascular flora comprises 1,092 species which can be divided into 984 idiochorophytes (native) and archaeophytes and 108 neophytes. A total of 67 species are known to have become extinct since 1900. Of the municipal area, 25% is under nature protection with international responsibility (according the EU Habitat Directive) for several plants (for example *Gladiolus palustris*) and habitats (floodplain forests, semi-dry grasslands with orchids). In terms of biodiversity, Augsburg is one of the best researched cities in Germany with detailed studies of the urban area and the protected areas.

Norbert Müller (✉)
Department of Landscape Management and Restoration Ecology, University of Applied Sciences Erfurt, Leipziger Str. 77, 99085 Erfurt, Germany
e-mail: n.mueller@fh-erfurt.de

J.G. Kelcey and N. Müller (eds.), *Plants and Habitats of European Cities*,
DOI 10.1007/978-0-387-89684-7_2, © Springer Science+Business Media, LLC 2011

Natural Environment of the City

Augsburg is situated in the northern foothills of the Alps, which extend from Genf to Vienna. The Alps, which lie 100 km south of Augsburg, are a significant influence on the city's geology, morphology and climate. The city is situated mainly in the valley of the former Lech glacier, which was filled with gravel between and after the ice ages and subsequently eroded by the Lech and Wertach rivers. The historic city centre is built on a terrace formed in the Wurm Glaciation (Fig. 2). Shallow brown earths, loams and rendzina soils are dominant. The city is bounded in the west by mountains and valleys and in the east by a hilly topography that has been eroded by numerous rivers since the beginning of the Tertiary resulting in the deposition of gravel, sand and clay.

The average altitude of the city is 489 m a.s.l., and the highest point at 561 m a.s.l. is in the western part. The climate of the area in which the city lies is moderate Atlantic with a continental influence. The average long-term temperature is 8.1°C and the average annual precipitation is 831 mm.

Historical Development of the City

The fertile loess on the higher glacial terrace was the main reason for the first settlements in the Augsburg area. The Neolithic people were the first to settle along the glacial terrace where they practised cultivation and animal husbandry. The city of Augsburg was founded by the Romans in 15 BC as a fort "*Augusta vindelicorum*", situated at the strategic important "north–south" conjunction of the Via Claudia. By the Middle Ages, Augsburg had become one of the most important European centres of culture and trading. In the sixteenth century, the population of the city exceeded 20,000 and it was the home of some of the richest merchants in the world, including the "Fugger"- and "Welser" families.

At the end of the nineteenth century, the city expanded rapidly as the result of industrialisation, especially the textile industry. Urban development extended around the outside of the city walls, which were built in the Middle Ages. The presence of many watercourses to the east of the city centre provided water power for the "textile quarter".

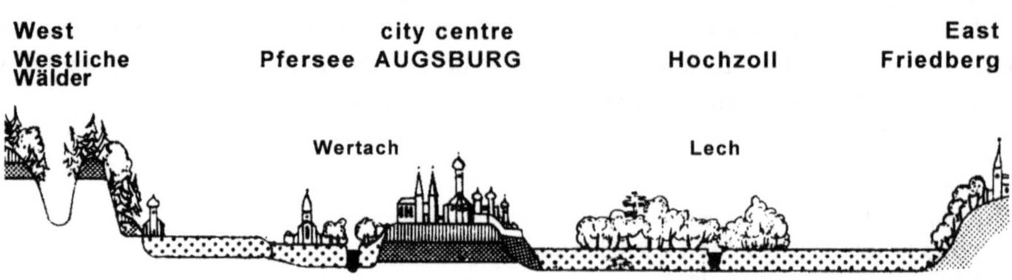

Fig. 2 Cross section of the municipal area from west to east

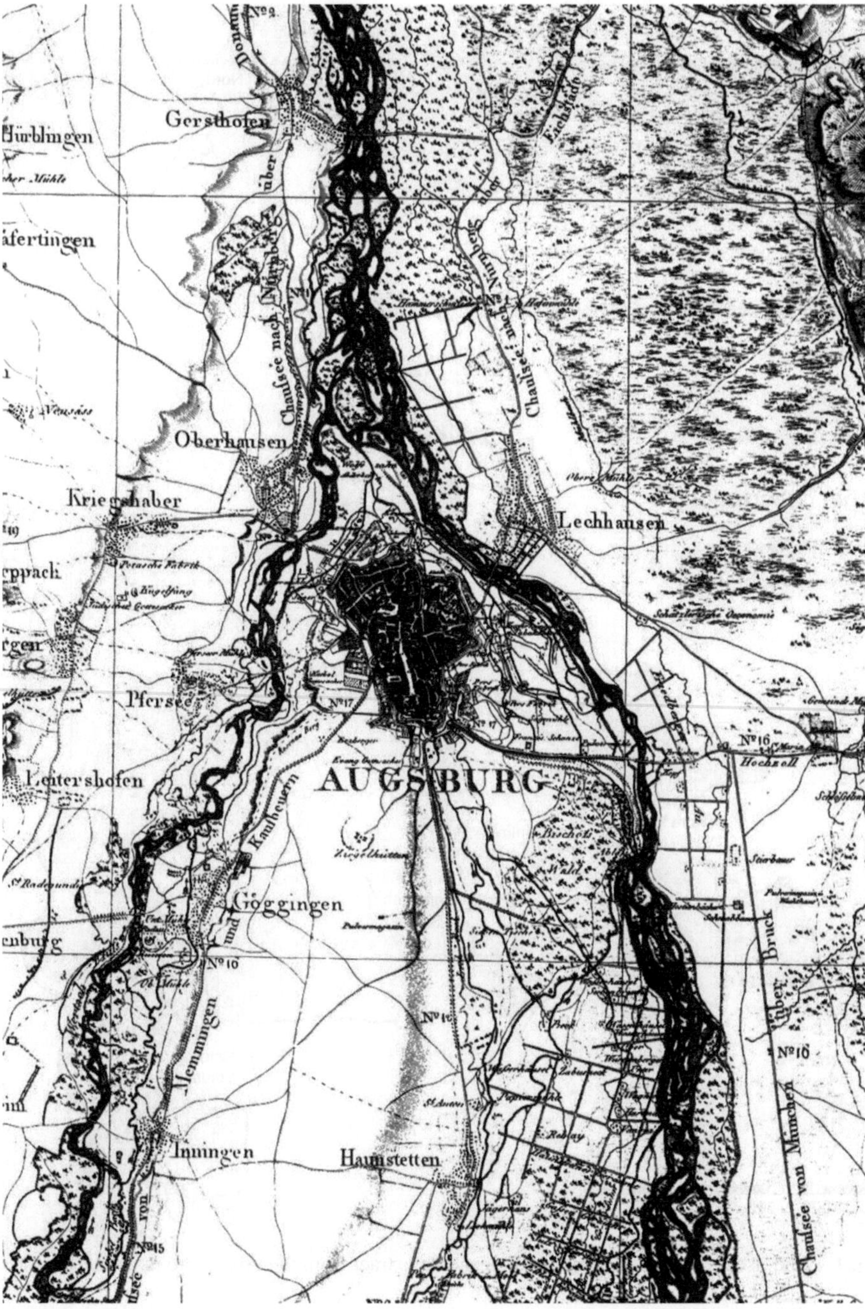

Fig. 3 Historical map from Augsburg in the year 1819 (from Müller and Schmidt 1988)

Between 1800 and 1970, the population grew from 25,000 to 250,000 (compare Figs. 3 and 4). After 1970, it was found that the number of people living in the urban core was declining, while the number living in the fringe was expanding.

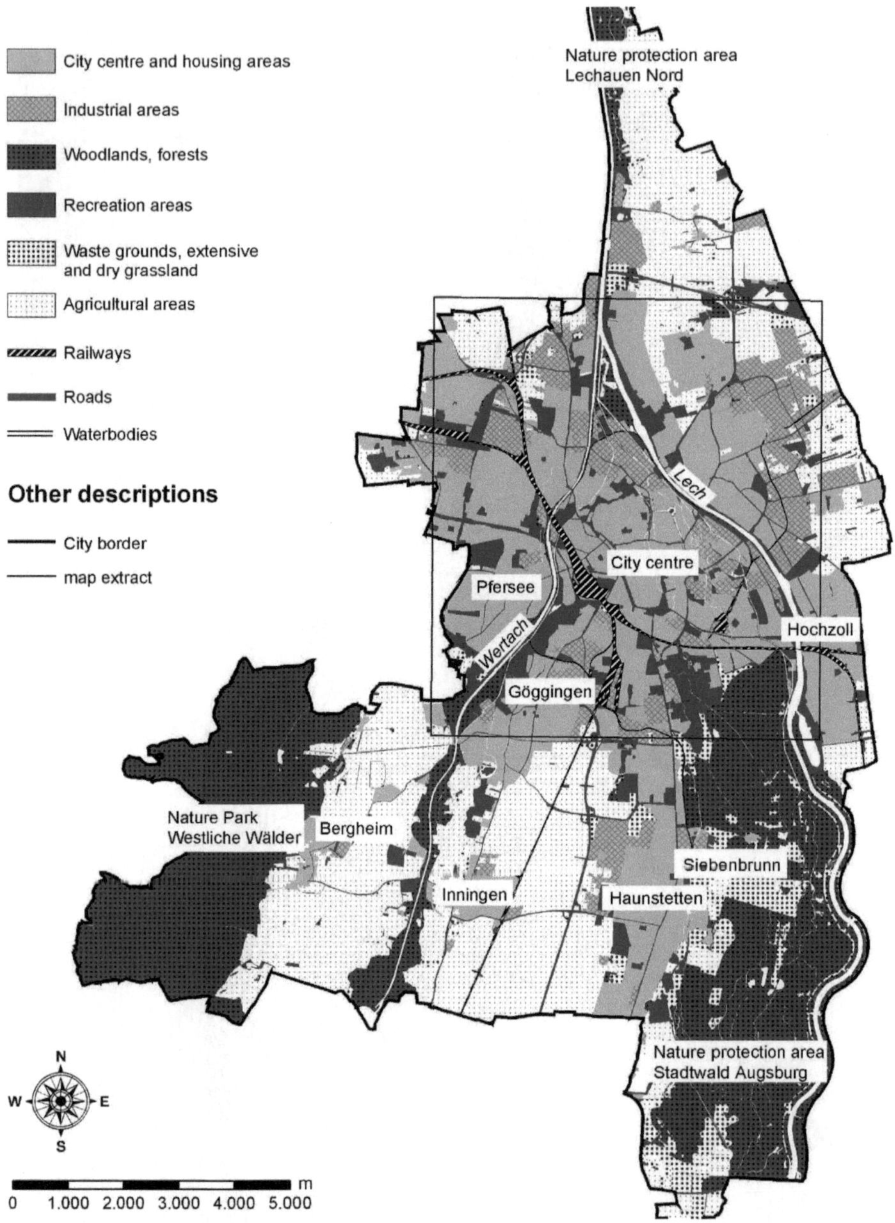

Fig. 4 Land-use types of the municipal area and an extract from the urban core mapping exercise

Today the city area comprises 147 km^2, of which 25% is occupied by woodlands and forests, 31% by agricultural land, 35% by housing and industrial areas and the associated infrastructure and 9% by recreation areas (for example parks and sport fields). The longest distance north to south is 22.2 km and 14.5 km east to west.

Extensive agricultural land and forests occur to the north, west and south of the urban area. To the south is the large nature protection area "Stadtwald Augsburg" in the west the nature park "Westliche Wälder" and in the north the nature protection area "Lechauen Nord."

Changes of the Environment Due to City Growth

As the result of urbanisation, the former natural soils have been changed to hortisols or even sealed, the latter dominating the city centre. The macro-climate has also been influenced by urban development, the heat island effect of the city centre being characterised by a longer period of growth and a lower frost frequency. For these reasons, thermophilous plants from Asia such as *Ailanthus altissima* and *Buddleja davidii* have become naturalised. In contrast to other German cities, air pollution in the urban core is low due to the flat topography and the good conditions for the movement of air.

Flora

Higher Plants

The first flora of the entire city area was published in 1822 (Alten 1822); other floras followed in 1848, 1898 and 1978 (Hiemeyer 1978). The first flora of the urban core and a study of all the planted trees and shrubs was completed in 1990. In 2004, the flora of the entire city area comprised 1,092 species, which can be divided in 984 idiochorophytes (native) and archaeophytes and 108 neophytes; 67 species are known to have become extinct since 1900.

The urban core comprises 637 plants, 573 of which are native, archaeophytes and apophytes, while 64 (10%) are neophytes (classification after Jäger and Werner 2005). Subsequently 47 ephemerophytes were recorded, including ornamental species such as *Crocus napolitanus* and *Eranthis hyemalis* in parks and *Forsythia x intermedia* and *Cotoneaster* spp. in domestic gardens.

The Most Frequent Species in the Urban Core

An important question in relation to the evolution and composition of an urban flora is, "which species are the most frequent and therefore best adapted to the specific urban-industrial conditions?" The most frequent species, which are listed in Table 1, are dominated by common grasses and herbs with a high capability of regeneration. Therefore, these plants are well adapted to frequent disturbances in their life cycle.

Table 1 The 50 most frequent species in the urban area and their origin

Achillea millefolium agg.	*Lolium perenne*
Aegopodium podagraria	*Matricaria discoidea*
Acer platanoides	*Plantago lanceolata*
Acer pseudoplatanus	*Plantago major*
Artemisia vulgaris	*Plantago media*
Bellis perennis	*Poa annua*
Betula pendula	*Poa pratensis*
Capsella bursa-pastoris	*Poa trivialis*
Cerastium fontanum	*Polygonum aviculare agg.*
Chenopodium album	*Prunella vulgaris*
Cirsium arvense	***Robinia pseudoacacia***
Cirsium vulgare	*Rumex crispus*
Convolvolus arvensis	*Rumex obtusifolius*
Conyza canadensis	*Sagina procumbens*
Dactylis glomerata	*Sambucus nigra*
Elytrigia repens	*Salix caprea*
Festuca rubra	*Senecio vulgaris*
Fraxinus excelsior	*Sinapis arvensis*
Galinsoga ciliata	***Solidago canadensis***
Galium aparine	*Sonchus oleraceus*
Galium mollugo	*Stellaria media*
Geranium robertianum	*Taraxacum officinale agg.*
Geum urbanum	*Trifolium pratense*
Glechoma hederacea	*Trifolium repens*
Lamium album	*Urtica dioica*

Normal = native or archaeophyte, **bold** = neophyte, underlined = apophytes

In contrast, woody species are rare; only *Acer pseudoplatanus, Fraxinus excelsior* and *Sambucus nigra* are abundant. In general, neophytes are frequent in cities; however, only four are included in the 50 most frequent species that occur in Augsburg – *Conyza canadensis, Galinsoga ciliata, Robinia pseudoacacia* and *Solidago canadensis*.

The high percentage of apophytes is interesting because it indicates that species have evolved during the long period of human activities in and influence on the European landscape. In general terms, these species are very successful invaders in cities of the Old and New worlds, for example, *Bellis perennis, Capsella bursa-pastoris, Chenopodium album, Plantago major, Poa annua, Polygonum aviculare* agg., *Senecio vulgaris, Stellaria media* and *Taraxacum officinale* agg.

Invasive Species

Until recently only a few of the 108 naturalised species in Augsburg have become invasive and therefore a threat to local biodiversity. Adjacent to the urban area

Solidago canadensis is colonising the dry grasslands, especially those that are little used or not used at all. *Solidago gigantea* has become established and is re-generating on the few remaining gravel bars of the Lech River, where it is replacing native species. Over the last 15 years, *Impatiens glandulifera* and *Heracleum mantegazzianum* have been spreading along the banks of the watercourses and ditches. As a result of its use (until the 1980s) in afforestations in the "Stadtwald Augsburg", *Robinia pseudoacacia* has become highly invasive throughout Augsburg.

Planted Trees and Shrubs

Sixteen thousand individual trees were studied between 1980 and 1985 comprising a total of 41 species, although 97% belonged to only 21 species (see Table 2). 55% of the trees that were along the streets were healthy with 36% being in slightly poor condition and 9% in very poor condition. As a result of urban stress and Dutch Elm Disease, 75% of the *Ulmus glabra* were in very poor condition – the highest of any species. Over 50% of *Corylus colurna, Acer pseudoplatanus, Aesculus hippocastanum, Tilia* x *euchlora, Sorbus aria,* and *Populus nigra* were in poor condition. In contrast, *Platanus* x *hispanica* (see Fig. 5), *Aesculus carnea, Robinia pseudoacacia, Sorbus aria, Tilia cordata* and *T. intermedia* were especially robust.

Table 2 Planted trees along streets and squares in the urban area and their percentage of the total number of trees investigated (last updated 1985)

Species	Percent (%)
Acer platanoides (including 'Fassens Black' and 'Globosa')	20
Tilia cordata	18
Fraxinus excelsior (including 'Globosa' and 'Pendula')	14
Aeculus hippocastanum	06
Acer pseudoplatanus	06
Tilia x *euchlora*	05
Betula pubescens	04.5
Tilia platyphyllos	04
Tilia intermedia	04
Populus x *canadensis robusta*	03.5
Robinia pseudoacacia	02
Platanus x *hispanica*	02
Ulmus glabra, Aesculus carnea, Populus nigra 'Italica', *Acer campestre, Sorbus aucuparia, S. aria* (including 'Magnifica'),	01
S. intermedia, Prunus serrulata, Corylus colurna.	00.5
Acer saccharinum, Ailanthus altissima, Carpinus betulus, Catalpa bignonoides, Crataegus monogyna, C. laevigata, Fagus sylvatica, Juglans regia, Parrotia persica, Pinus nigra, Pyrus malus, Prunus cerasifera 'Nigra', *P. mahaleb, Pterocarya fraxinifolia, Quercus robur, Salix alba, Sambucus nigra, Sophora japonica, Ulmus minor*	<00.5

Fig. 5 *Platanus x hispanica* is a frequent tree along roads in the perimeter block development of the "Wilhelminian" Quarter

For several years, the most frequently introduced species found in beer gardens, *Aesculus hippocastanum,* has been attacked by the alien leaf miner *Cameraria ohridella.* Since 2003, the fungi *Splanchnonema platani* has caused early die back of the branches of *Platanus* trees.

In addition to the trees along the streets, a study was carried out of the trees and shrubs in the different land-use types. The most frequent deciduous species were *Betula pendula, Fraxinus excelsior, Tilia cordata, Acer pseudoplatanus, A. platanoides, A. campestre, Carpinus betulus, Aesculus hippocastanum, Robinia pseudoacacia* and *Fagus sylvatica.* Twenty-four coniferous species were recorded, the most frequent were (in the order given) *Picea abies, Taxus baccata, Pinus nigra, Picea omorika* and *Pinus sylvestris.*

Bryophytes

By the end of the nineteenth century, 235 bryophyte species were described from the Augsburg area. In 1978, Klucniok recorded 202 bryophyte species including a list of probably lost and extinct species. The more frequent Musci that occur in the city include *Ceratodon purpureus* (open sandy soils), *Dicranum scoparium* (in forests), *Barbula unguiculata* (bare soils), *Syntrichia ruralis* (open soils and stones), *Tortula muralis* (walls and stones), *Funaria hygrometrica* (open soils), *Bryum caespiticium* (walls and stones), *B. capillare* (soil, stones and timber), *Mnium undulatum* (on timber in forests), *Polytrichum formosum* (forest soils), *Leucodon sciuroides* (timber), *Fontinalis antipyretica* (channels), *Thuidium tamariscinum* (forest soils), *Acrocladium cuspidatum* (wetlands), *Amblystegium serpens* (soils, stones and timber), *Eurhynchium swartzii* (open soils), *Scleropodium purum* (*Picea* and Pine forests), *Hypnum cupressiforme* (timber, soils, lawns and meadows), *Rhytidiadelphus triquetrus* (forests), *R. squarrosus* (forests, meadows and lawns), *Hylocomium splendens* (forests) and *Lophocolea bidentata* (forests).

Fungi (including Lichenised Fungi)

The investigation of the fungi has a long tradition and in the latest flora 1,500 species were described for the wider city area (Stangl 1985). Nothing is known about the lichenised fungi of the city.

Habitats

The plant communities are described following the phytosociological nomenclature of Jäger and Werner 2005; the European Union codes (EUR 2007) are referred to later in the chapter.

In so far as it is known, the original natural vegetation was dominated by deciduous forests comprising the *Galio-Carpinetum* on the loamy soils of the glacial terraces and the *Luzulo-Fagetum* and *Galio-Fagetum* on the western mountain ranges. Several types of floodplain forests once occurred along the Lech and Wertach rivers, including the *Alnetum incanae, Querco-Ulmetum minoris* and *Erico-Pinetum.*

Natural and Semi-Natural Habitats

Fagus Forests and *Picea* Plantations

The Nature Park "Westliche Wälder" is a large forest area in which *Picea abies* plantations are dominant. The remaining *Fagus* forest belongs to the *Luzulo-Fagetum.* In the species-poor herb layer, the characteristic species are *Luzula luzuloides* and

Oxalis acetosella; the rare species include *Blechnum spicant* and *Lycopodium annotinum.* On limestone soils, the herb layer of the *Galio-Fagetum* includes *Corydalis cava, Lamiastrium galeobdolon, Lilium martagon* and *Ranunculus ficaria.*

Vegetation on the River Gravel Bars

Major civil engineering works began along the Lech and Wertach rivers in 1910; up to that time, all the typical habitats of alpine braided rivers were present along them (see Fig. 6).

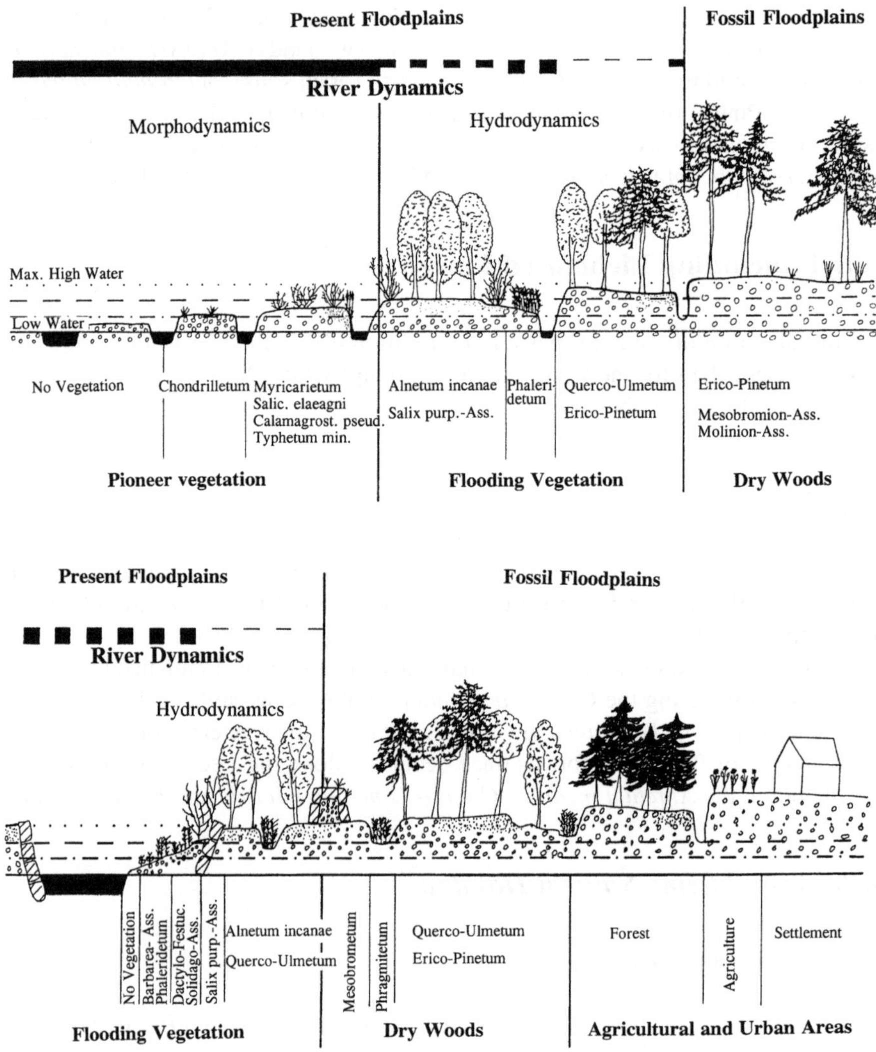

Fig. 6 Structure of the floodplain vegetation of the Lech river in the south of Augsburg before river regulations in 1910 (*above*) and today (*below*)

Fig. 7 The regulated Lech river with relicts of gravel bars in the inner city

The wide Lech floodplain, especially to the south of the city, contained many typical plant communities and species, such as pioneer herb communities (*Chondrilletum chondrilloides*) on newly created gravel bars, shrub communities (*Salicetum eleaegni*) on gravel and *Salici-Myricarietum* on sand. The 1 km wide floodplain was the widest outside the Alps. Now the remaining small gravel bars in the regulated river channel support herb communities such as *Barbarea vulgaris* and *Salix* scrub comprising *Salix purpurea* and *S. elaeagnus;* see Figs. 6 and 7. Some relicts of the *Calamagrostietum pseudophragmitis* are the only representatives of the original vegetation.

Floodplain Forests Including Pinus Forests

Because of river regulation, the river bed has started to erode, and as a consequence, the former floodplain forests *Salicetum albae* and *Alnetum incanae* are deteriorating because of the lowering of the water table and less inundation. They are replaced by *Querco-Ulmetum minoris* forests.

When the gravel bars are no longer flooded, *Pinus sylvestris* is able to colonise the more or less bare gravel. These *Pinus* forests (*Erico-Pinetum*) with a low vegetation cover are typical of the margins of alpine rivers and were maintained in former times by sheep and cattle grazing. In the early successional stages, they are characterised by *Erica carnea* and some orchids (*Epipactis atrorubens* and *Gymnadenia conopsea*) and later by *Calamagrostis varia,* which becomes dominant. Since the 1950s, the forest authorities tried to change these less productive and species-rich habitats into *Fagus-Acer* forests.

Still Water (Ponds and Reservoirs)

The still water habitat, which comprises a total of 73 ha, includes old river channels, disused, flooded mineral workings and ponds. The bed of the flooded mineral workings in the Lech and Wertach valleys is dominated by several species of the Characeae. *Phragmites australis* beds occupy the margins of those workings that are not used for recreation. Only a few of the once frequent old river channels with clear groundwater and supporting the pioneer species *Typha minima* now remain. As a result of the engineering works, which have changed the dynamics of the river over the last 100 years, the species associations have changed to those found in the more common wetlands, for example, *Phragmites australis* and *Alnus incana*. The ponds created north of Bergheim in the Middle Ages contain the very rare species *Elatine hydropiper, E. hexandra* and *E. triandra.*

Dry Grasslands

Species-rich grasslands have developed on the shallow gravel soils in the Lech and Wertach valleys by a combination of grazing and mowing since the Roman period. They are frequently found adjacent to the *Pinus* and *Alnus* forests; see Fig. 8. The dry grasslands (*Pulsatillo-Caricetum humilis*) are rare, while the semi-dry grasslands are frequent. Because of their unique species association, which includes *Brachypodium rupestre, Biscutella laevigata* ssp. *kerneri, Hieracium macranthum,*

Fig. 8 Species rich grasslands (*Mesobromion*) adjacent to Pine forests (*Erico Pinetum*) are typical on the Lech river terraces south and north of the city centre. The largest population of *Gladiolus palustris* within the European Union occurs in the nature protection area "Stadtwald Augsburg"

Ophrys fuciflora, O. insectifera and *O. sphegodes,* the semi-dry grasslands were described as the *Brachypodium rupestre* variation of the *Mesobrometum.*

Spring Swamps and Litter Meadows

In the "Stadtwald Augsburg", groundwater is seeping into the former channels of the Lech River. In addition to the "spring swamps" with *Carex davalliana* and *Juncus alpino articulatus,* there are several wet areas, which were once managed as "litter meadows" to provide bedding for cattle. The typical species found in this habitat are *Cirsium tuberosum, Tetragonolobus maritimus* and *Molinia arrundinacea.*

Fertilised Meadows and Arable Land

In the western part of the city and in the wider Lech floodplain, the agricultural land is managed as grassland, which belongs to the *Arrhenatherion elatioris*. These periodically fertilised meadows, which are species-poor, are mown three times a year. Before the first cut, *Anthrisus sylvestris* is flowering while *Heracleum sphondylium* is in flower before the second and third cuts. The arable land to the south and north of the urban area is used for crop production, including *Zea mays*. As a result of the intensive use of fertilisers and herbicides, the arable fields are species-poor. Typical species of calcareous arable land, for example, *Legousia speculum-veneris, Centaurea cyanus* and *Consolida regalis* are restricted to the field margins.

Urban Habitats

Detailed studies of the different land uses within the urban core provide an overview of the diversity of urban habitats (see Fig. 9).

City Centre and Perimeter Block Development

The city centre within the Medieval walls (175 ha) is the most densely developed part of the city (Fig. 10). Unsealed surfaces and places for weeds to grow are rare and mainly restricted to back yards, old brick walls and old pavements. The following are the most frequent communities found:

- On walls with *Cymbalaria muralis* (see Fig. 11) and *Asplenium ruta-muraria*
- Trampled communities with *Poa annua, Polygonum aviculare* agg., *Sagina procumbens* and *Eragrostis minor*
- Annual ruderal communities (*Sisymbrion*) and perennial ruderal communities (*Arction*) on the edges of courtyards and squares

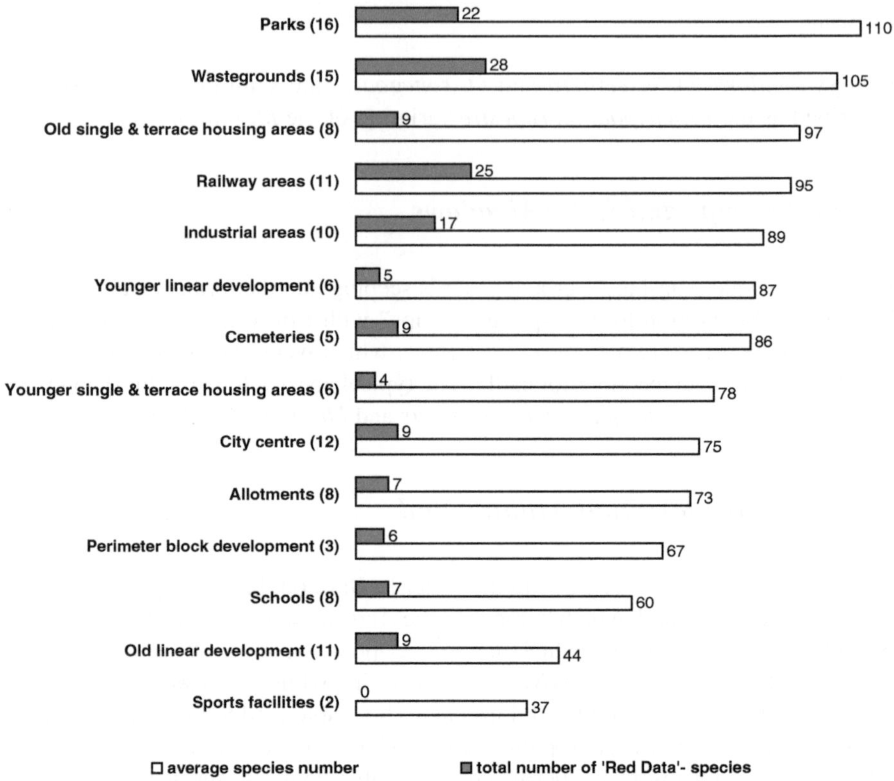

Fig. 9 Land-use types in the urban area with average number of vascular plants and number of "Red Data" species according to the Augsburg Red Data List (Müller 1985). The numbers in brackets behind the land-use types indicate the total number of researched areas

Fig. 10 In the densely developed city centre, spontaneous vegetation is rare and restricted to courtyards – the historic Roman road "Via Claudia" in the city centre with "St. Ulrich" Cathedral in the background

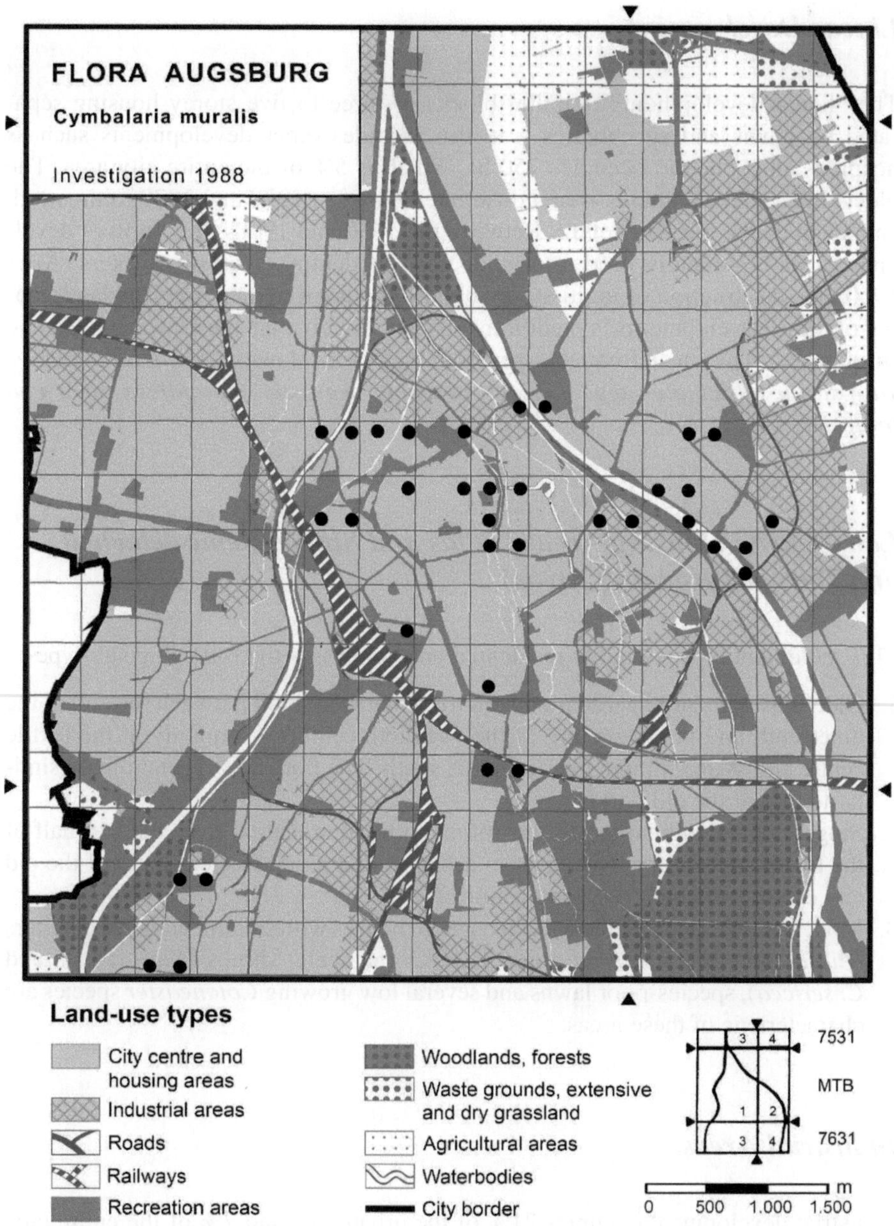

Fig. 11 Distribution map of *Cymbalaria muralis* – a typical species of old walls and buildings - is restricted to the urban core

The city centre is surrounded by 190 ha of block development that was constructed in the Wilhelminian period at the end of the nineteenth century. In addition to special designs and hard landscaping (ornamental fences and pavements), the gardens contain some typical ornamental non-native plants such as *Forsythia x intermedia, Symphoricarpus albus* and *Syringa vulgaris.*

Linear Development

The "linear development" (parallel blocks of three to five storey housing separated by lawns and shrubberies – it also includes other developments such as hospitals and schools) occupies 750 ha, which is 5% of the entire city area. The oldest linear developments were constructed between 1920 and 1930 to the north and south of the old city centre. Between the 1930 and 1960s, this form of development was frequently used to provide new residential areas for workers. After 1960, the construction of multi-storey buildings started to increase. The landscaping of these developments is predominantly intensively managed lawns with some shrubberies. The most frequent species found in the lawns are *Bellis perennis, Taraxacum officinale* agg, *Lolium perenne, Agrostis stolonifera* and *Poa pratensis.*

Low Density Housing Areas (Villas and Single, Semi-detached and Terraced Housing Areas)

This settlement type is mainly residential and comprises the following sub-types:

1. Old villas with large gardens and old trees that were built between the end of the nineteenth and the beginning of the twentieth centuries – mainly in the textile centre and Wilhelminian quarter in the south. The flora and vegetation are similar to that of the old parks.
2. Single houses with large gardens (often with old orchards) from the first half of the twentieth century. The quarters in the southern and western parts of the old city centre were built as "garden cities."
3. Modern single, semi-detached and terraced houses with smaller gardens. Conifers (*Picea omorika* and *Thuja occidentalis*), ornamental shrubs (*Cornus alba* and *C. sericea*), species-poor lawns and several low growing *Cotoneaster* species are characteristic of these areas.

Industrial Areas

Industrial development occupies 21% of the urban area and 7% of the entire city, mainly to the west of the old city centre (the old textile industry) and on the urban fringe. Until recently, many of these areas have not been used for varying periods; consequently, they contain a large number of plant species. *Echium vulgare* (Fig. 12) and *Reseda lutea* occur in the less frequently used areas, while *Diplotaxis muralis* only occurs in the textile industry quarter.

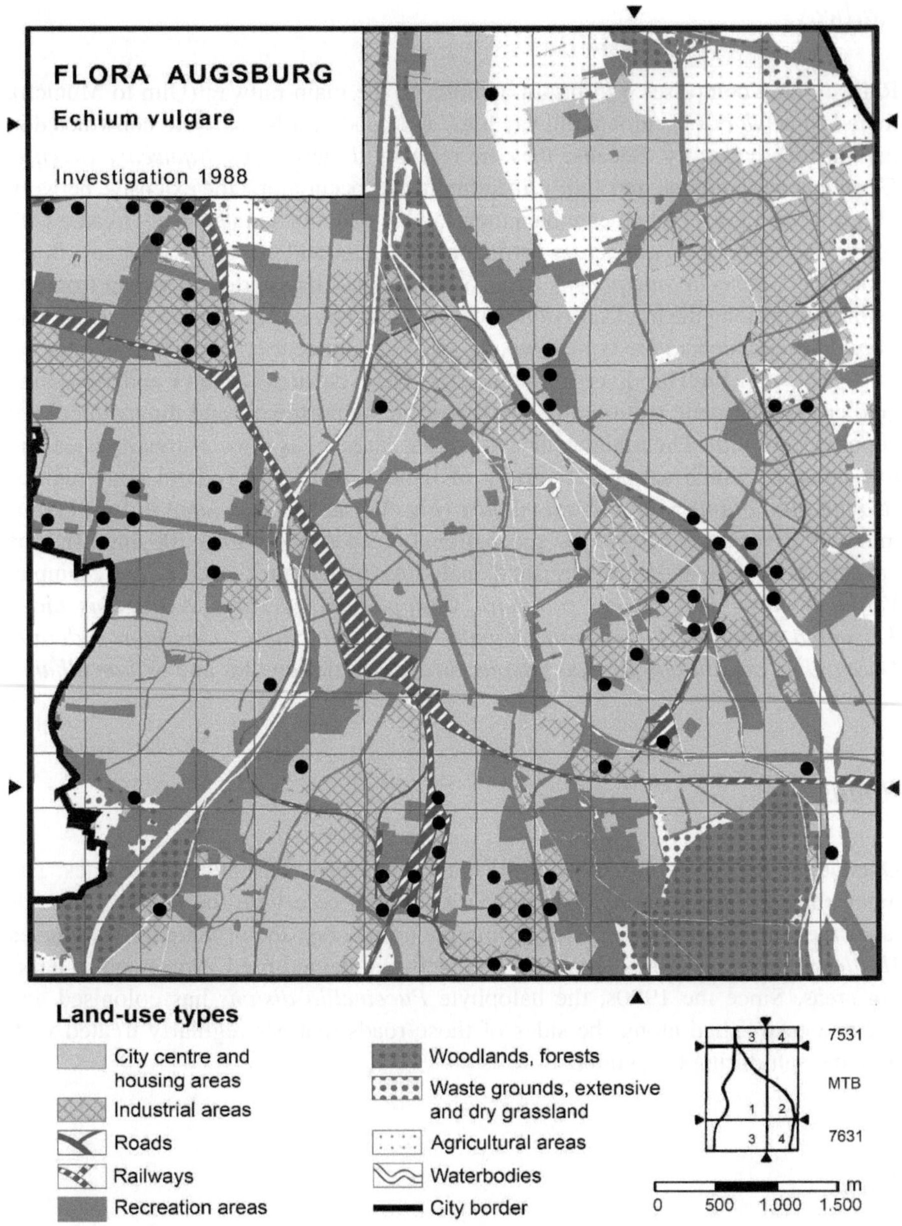

Fig. 12 Distribution map of *Echium vulgare* (naturally occurring on gravel bars) in the urban core; it is concentrated on dry sites in industrial areas and along railways

Railways

Railway land comprises 175 ha. In addition to the main railway (Ulm to Munich), there is a network of regional railway lines and goods yards that were constructed in the nineteenth century. Because they are resistant to herbicides, *Equisetum arvense, Geranium robertianum* and *Carex hirta* frequently occur along the extensive network of operational tracks. On the embankments, which are not so intensively treated with herbicides, *Solidago canadensis* forms dense stands and covers large areas. Since about 2000, *Senecio inaequidens* (from South Africa) has started to spread from the main station along the tracks. *Betula pendula, Buddleija davidii, Prunus mahaleb* and *Robinia pseudoacacia* are typical woody species that occur on the less used sites.

Areas with a high biodiversity include the extensive areas of track and associated land around the main station, the goods yard in the northwest and the maintenance depot in the south. The maintenance depot has large areas of occasionally used and unused land, which support a variety of habitats from dry ruderal communities (*Dauco-Melilotion*) with *Echium vulgare* (Fig. 12) and *Reseda lutea* to forest communities at various stages in the successional process. More than 200 angiosperms have been recorded on railway land including many rare species, for example, *Vulpia myuros, Petrorhagia prolifera, Clinopodium arvensis, Astragalus cicer, Anisantha tectorum, Descurainia sophia, Herniaria glabra, Lepidium ruderale, Onopordum acanthium, Reseda luteola, Saxifraga tridactylites* and *Sedum album.*

Roads

Outside the Medieval city, most of the main roads have trees along both sides. The nineteenth century avenues that connect the urban core with the surrounding villages are particularly impressive. Within the urban core, the urbanophilous species *Hordeum murinum* grows along the less well-maintained road verges and in parking areas. Since the 1980s, the halophyte *Puccinellia distans* has colonised and become established along the sides of those roads that are regularly treated with de-icing salt during the winter.

Parks

The parks comprise a total of 357 ha (2.4%) of the city. The larger parks, which were designed and constructed in the landscape garden style of the nineteenth century, are situated between the Medieval city centre and the nature protection area "Stadtwald," west of the railway main station and along the Medieval walls (Fig. 13). After 1945, several parks were developed along the Lech and Wertach rivers. With an average of 110 species per park, they are the most species-rich of all the land-use types.

Fig. 13 The larger parks were constructed in the landscape garden style in the nineteenth century and are the most species rich land-use type in the city. "Wittelsbacher" Park north of the main railway station

The number of species found in an individual park depends on its size, age, structural diversity and the intensity of management. The forest tree species that frequently occur in the old parks are *Acer platanoides* and *A. pseudoplatanus*. The more abundant herb layer species found in the old parks are *Brachypodium sylvaticum* (Fig. 14), *Carex sylvatica, Fragaria vesca, Allium ursinum, Arum maculatum, Circaea lutetiana, Corydallis cava, Galanthus nivalis, Hedera helix, Leucojum vernum, Ornithogalum umbellatum, Poa nemoralis, Primula elatior, Ranunculus ficaria, Scilla bifolia, Silene dioica* and seedlings of *Taxus baccata.* Forest edge species include *Alliaria petiolata, Anthriscus sylvestris, Chelidonium majus, Viola odorata* and *Veronica chamaedrys*. A rare species of areas frequently disturbed by the parks' gardeners is *Poa bulbosa,* which finishes its lifecycle by May.

The predominant plant community in parks is the frequently mown lawn – the species composition of the different areas of lawn depends on the location, age, management and soil type. The young lawns are dominated by sown grass species such as *Agrostis stolonifera, Festuca rubra, Poa pratensis* and *Lolium perenne*. After several years the lawns become colonized by other taxa. The spring flowering species include *Bellis perennis, Taraxacum officinale* and *Veronica filiformis* (Fig. 15). The non-native species *Veronica filiformis* is virtually restricted to lawns in places where it forms a characteristic plant community (*Triofolio repentis-Veronicetum filiformis*) which was described here for the first time in Germany (Müller 1990). Originally, *Veronica filiformis* was introduced from the Caucasus to England as an ornamental plant in 1780 and was first noticed as a naturalised species in Germany at a site near Munich in 1915; now it is widespread throughout the city, especially in lawns in public open spaces. Because the species does not

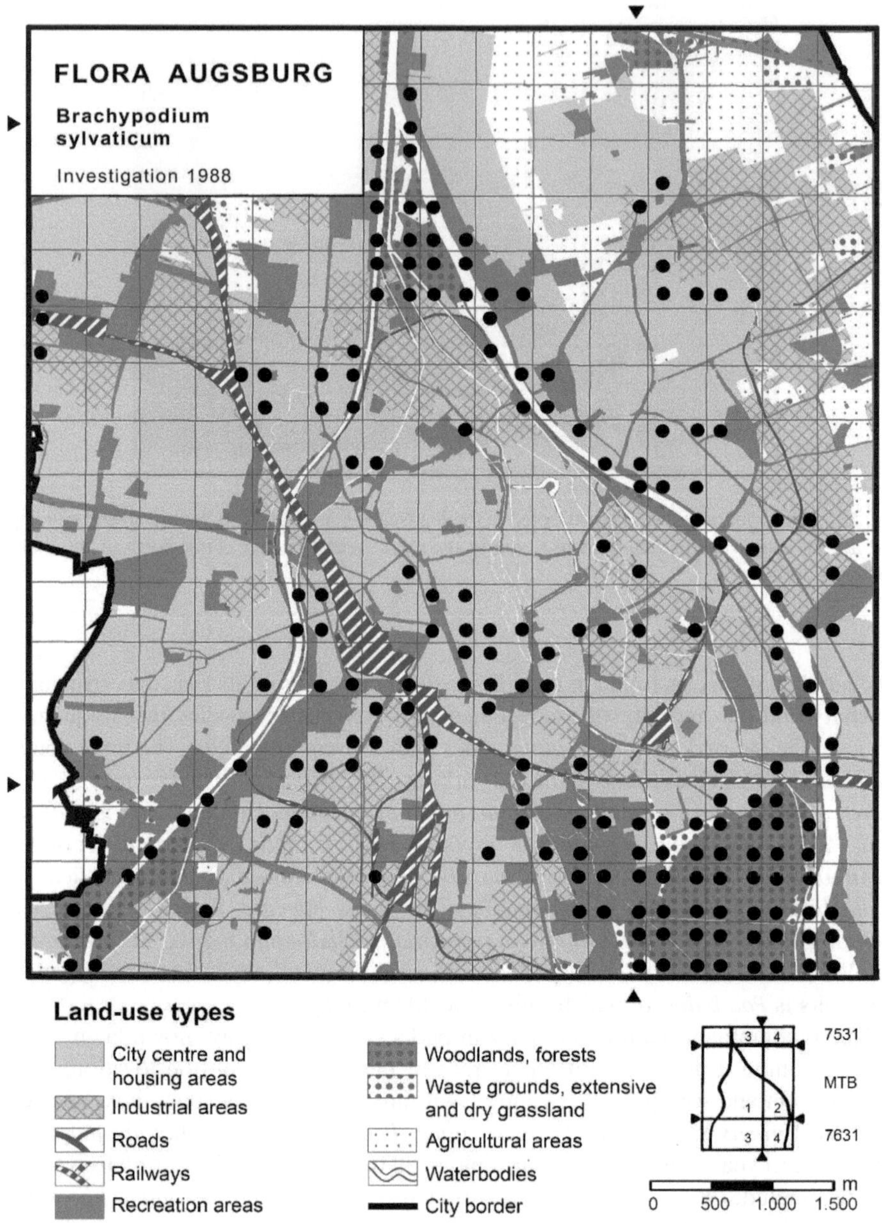

Fig. 14 Distribution map of *Brachypodium sylvaticum*, a typical forest species that grows in park forests and old gardens

produce seed in Germany, the main "vectors" are mowing machines. Until recently, *Leontodon saxatilis* was a rare species in Augsburg, but now it is typical of the newer lawns, having been introduced with grass seed mixes.

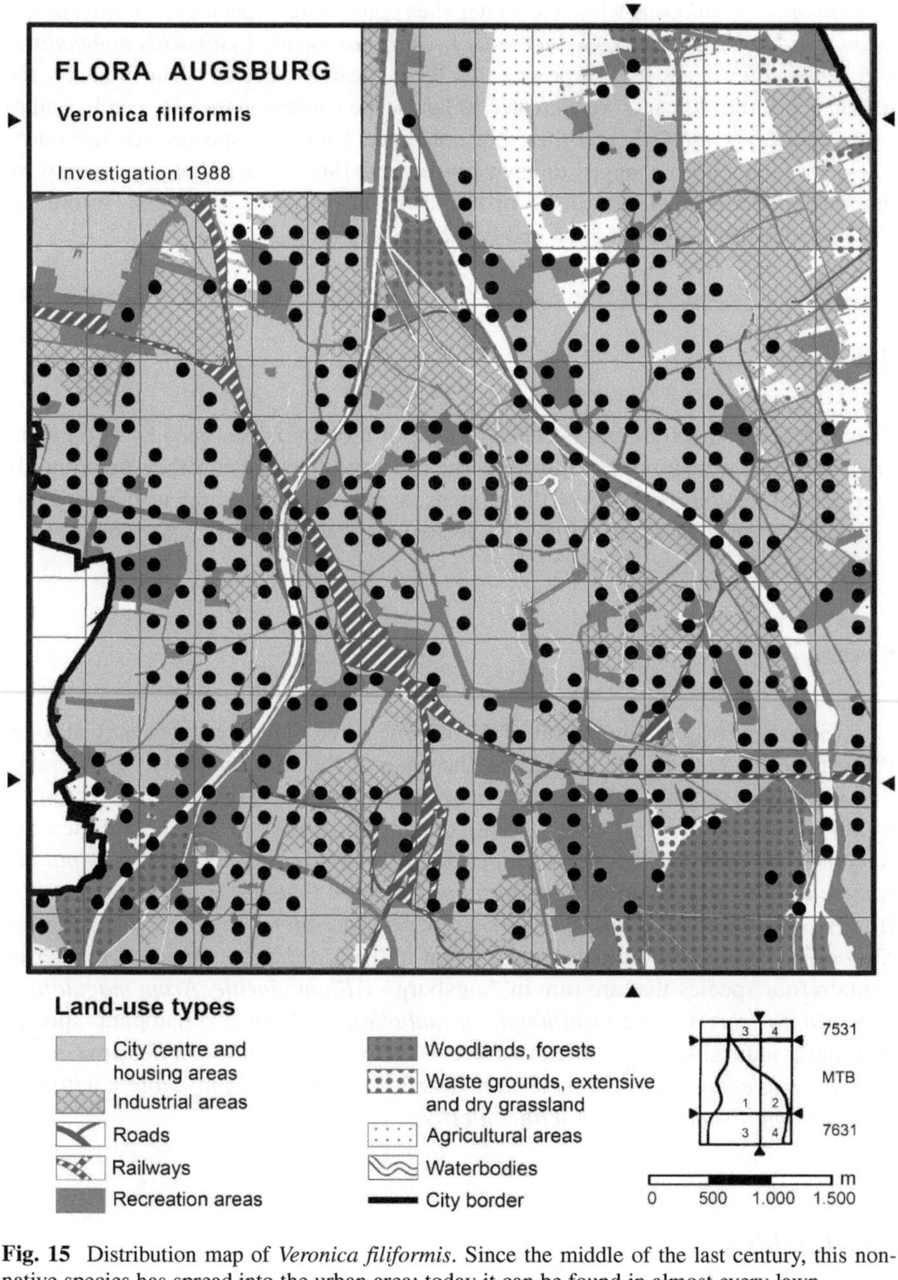

Fig. 15 Distribution map of *Veronica filiformis*. Since the middle of the last century, this non-native species has spread into the urban area; today it can be found in almost every lawn

The more frequent species of the older lawns include *Leucanthemum vulgare* and *Plantago media* and on the dry sites additionally *Helictotrichon pubescens, Brachypodium rupestre, Leontodon hispidus, Potentilla neumanniana, Salvia pratensis* and *Ranunculus bulbosus* while the wetter sites contain such species as *Cirsium oleraceum, Cardamine pratensis, Geum rivale, Lychnis flos-cuculi, Lysimachia nummularia* and *Ranunculus acris.* It is surprising that these meadow species occur in lawns; the reason is that the old parks were created as landscape gardens in the nineteenth century from semi-dry grasslands and therefore contained numerous species-rich meadows. These meadows were changed to lawns in the mid-1900s, when motor mowers were introduced. Since the 1980s, some of the lawns in the parks have been restored to meadows for nature conservation and economic reasons.

Allotments

Two hundred and sixty-eight hectares (1.8%) of the city is occupied by allotments. Originally they were used for cultivating fruit and vegetables but now they are primarily used for recreation. Because of their intensive use and management, they are less diverse (in terms of habitats and plant species) compared to other land-use types.

Cemeteries

The 13 cemeteries occupy a total of 91 ha. The three oldest cemeteries (Catholic, Protestant and Jewish) are located on the edge of the Medieval city. On average, the cemeteries contain 86 angiosperm species, which is lower than the number found in the parks. The species that occur frequently on the boundaries are *Aegopodium podagraria, Alliaria petiolata, Anthriscus sylvestris, Campanula rapunculoides* and *Epilobium montanum.* Species that occur frequently in the flower-beds include *Capsella bursa-pastoris, Cirsium arvense, C. vulgare, Euphorbia peplus, Urtica urens* and the non-native *Cardamine hirsuta.* Cemeteries contain four species that are rare in Augsburg - *Allium vineale, Arum maculatum, Moehringia trinerva* and *Ornithogalum umbellatum.* Typical ornamental species that have naturalised in cemeteries are *Crocus napolitanus, Helleborus niger, Muscari neglectum, Matteuccia struthiopteris, Tulipa gesnerana, Viola* x *wittrockiana, Sedum hispanicum* and *Scilla siberica.*

Sport fields

The city contains 81 sports fields; with a total size of 288 ha (2% of the city). As a result of their intensive use and management, they contain the lowest number of species compared with all the other land-use types.

Waste ground

The waste ground sites are concentrated in the urban fringe and in the former textile quarter. On open fertile soils, the succession from herb communities to scrub to forest is rapid. Sometimes different herb communities, especially the native *Calamagrostis epigejos* and the alien *Solidago canadensis* can slow down the succession by establishing extensive, stable populations. Casual disturbances of the substrate provide suitable conditions for the survival of some of the city's endangered species, for example, *Berteroa incana, Carduus acanthoides, Chenopodium glaucum, C. hybridum, Cynoglossum officinale, Descurainia sophia, Lappula squarrosa, Leonurus cardiaca, Onopordum acanthium* and *Reseda luteola*. The scrub habitat mainly comprises *Sambucus nigra,* which is eventually replaced by forests dominated by *Acer pseudoplatanus, A. platanoides* and *Fraxinus excelsior*. Initially, the drier sites are colonised by *Salix caprea* and *Betula pendula* scrub and later by *Robinia pseudoacacia*.

City Canals

Numerous canals emerge from the nature protection area "Stadtwald" and the Lech. In the Middle Ages and later in the industrial period, they were used to provide energy, via watermills. Today, these canals are important components of the urban core in providing recreation facilities and a healthy environment. In the city centre, the canals are totally culverted or contained within concrete channels while outside the Medieval city they are open and flow through natural banks lined with trees species such as *Fraxinus excelsior, Acer pseudoplatanus, Salix alba* and *Alnus glutinosa* with a herb layer comprising species typical of nutrient-rich soils, including *Filipendula ulmaria, Iris pseudacorus, Phalaris arundinacea, Phragmites australis* and *Carex acutiformis, Ranucunulus fluitans, Potamogeton pectinatus* and the moss *Fontinalis antipyretica* occurring in the clean and fast-flowing channels. As garden escapes from the nearby allotments, *Fallopia japonica* and *F. sacchalinensis* have colonised the edges of the watercourses. *Heracleum manteganzzianum,* which was first recorded in 1985, is growing along the Medieval channels despite considerable efforts to control it.

Nature Conservation, Environmental Planning and Education

Endangered and Extinct Species

Thirty-nine percent of the Augsburg flora has been identified as being extinct or endangered in the Augsburg Red Data List (Müller 1985). Most of the extinct and endangered species are those that occur in wetland habitats, for example, fens, spring swamps and "litter" meadows; 28 species have become extinct since 1900. Characteristic species of the former large fens in the Lech valley such as *Cladium mariscus, Utricularia intermedia* and *Viola persicifolia* were last seen at the end of the nineteenth century before the cultivation of the mires started. During the same period, the spring swamps in the nature park "Westliche Wälder" were also cultivated intensively; consequently, formerly frequent species such as *Eriophorum latifolium* are now extinct. The European-wide rare glacial relicts *Minuartia stricta* and *Betula humilis* grew in this area until the nineteenth century.

There has also been a major loss of habitats and species associated river gravel bars. As a consequence of river engineering works that have been carried out since the early 1900s, 11 species that are typical of the braided alpine rivers are now extinct including, *Chondrilla chondrilloides, Myricaria germanica* and *Typha minima* as well as the naturally rare alpine species *Pritzelago alpina, Kernera saxatilis, Linaria alpina* and *Poa alpina,* which used the river as a "migration corridor" from the Alps.

A high percentage of endangered species occur in the dry grasslands, 95% of which have been destroyed in the Augsburg area since 1900. The overall low number of extinct dry grassland species is notable because of the *refugium* provided by the "Stadtwald Augsburg" nature reserve, which is subject to continual monitoring and management by scientific societies concerned with natural history in co-operation with the city government.

There has been a rapid decline in the biodiversity of arable land due to the intensive use of fertiliser and herbicides since the 1900s; 14 species have become extinct, for example, *Agrostemma githago,* which is also extinct in Bavaria. On the other hand the number of endangered species in urban areas is low, although the abundance of *Chenopodium vulvaria,* a nitrophilous and characteristic urban species which was frequent in former times, has been declining throughout the whole of Germany and has become extinct in Augsburg.

The International Importance of the "Lech River" Corridor for Nature Conservation

From the biogeographical point of view the calcareous dry grasslands along the Lech River are of special value. They are important stepping stones of the internationally important Lech valley "corridor", which connects the calcareous habitats (dry grasslands and forests) of the Northern Alps in the south with those of the Swabian Jura Mountains in the north. These habitats are important for 80 species, including the following:

1. Internationally important populations of *Gladiolus palustris* and *Linum viscosum* in the Nature protection area "Stadtwald Augsburg." The populations, the largest within the European Union, have increased during the last 25 years as the result of effective management.

2. Nationally important alpine species, which have their northernmost distribution in the "Stadtwald Augsburg", for example, *Gentiana clusii* and *Gypsophila repens* or did so in former times until they became extinct – for example, *Saxifraga aizoides*.
3. Atlantic species – the German populations of *Aster amellus* grow near to the Alps.
4. Some continental species, including *Scabiosa canescens,* have their most eastern German distribution in Augsburg.
5. The distribution of some Mediterranean species in Germany is concentrated in the calcareous dry grasslands, for example, the magnificant *Ophrys fuciflora, O. insectifera* and *O. apifera* and the small shrub *Fumana procumbens.*

Nature Reserves

The nature protection area "Stadtwald Augsburg" has been protected since 1920 (Fig. 16). As the largest nature protection area in Bavaria outside the Alps, it is also important for supplying the drinking water for the whole city. For this reason the surrounding agricultural areas were changed to forests or low maintenance meadows in 1990. It was designated as a NATURA 2000 site "Lechauen between Augsburg and Königsbrunn" in 1998; see Table 3. The area contains populations of *Cypripedium calceolus* and the largest populations of *Gladiolus palustris* in the European Union, both species are listed in Annex II of the EC Habitats Directive.

In the 1980s, four sites along the Lech to the north of the city centre were designated as Nature Protection Areas. The area "Lechauen Nord" is protected as two NATURA 2000 sites, "Lechauen nördlich Augsburg" (401 ha) and "Höhgraben" (66 ha); see Tables 4 and 5, respectively.

The city contains several "Landscape Conservation Areas", the largest are the Nature Park "Westliche Wälder" and the floodplain forests along the Wertach River. In the urban core, the Medieval walls and the landscape parks "Wittelsbacher Park" and "Siebentischpark" are also designated as Landscape Conservation Areas. Numerous habitats recorded in the biotope mapping exercise are protected as Landscape Monuments and important old trees as Natural Monuments.

Ecological Research and Environmental Planning

A systematic investigation of the biodiversity of Augsburg was carried out in the late 1900s – one of the first studies of this type in Germany. In 1979, a selective biotope mapping exercise was undertaken, which was followed by a 15 year research programme of the vegetation, flora and fauna of the city. The programme included the comprehensive mapping of the vegetation and flora of all the land-use types in the urban core and in all the nature protection areas. The animal groups investigated included mammals, birds, reptiles, amphibians and insects (butterflies,

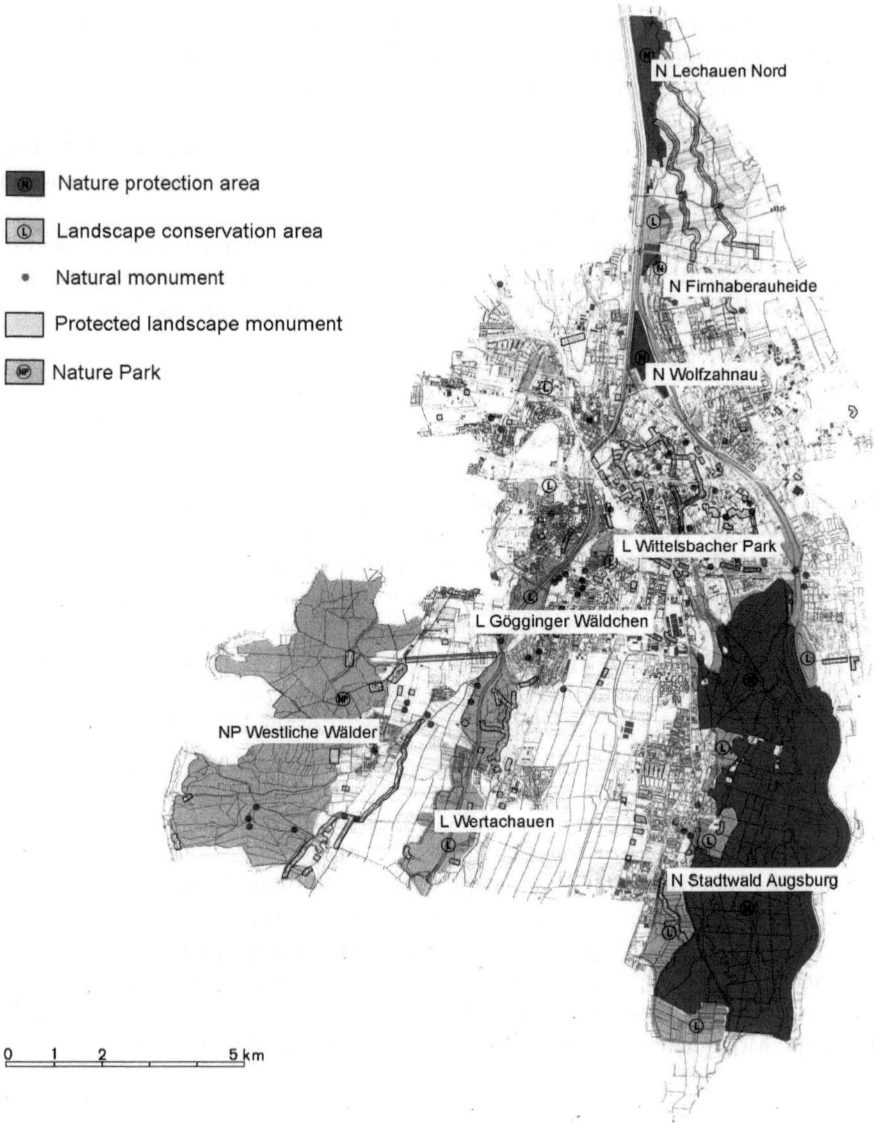

Fig. 16 Distribution of protected areas

bees, grasshoppers and dragonflies). The results of these investigations provided the basis for the preparation of the city's landscape and infrastructure planning policies, which were completed in 1995. Tree Preservation bye-laws for the urban core were made in 1989, and the system for the designation of Conservation Areas followed in 1995.

Table 3 European Union habitats (according EUR 2007) in the NATURA 2000 site "Lechauen between Augsburg and Königsbrunn" (2,304 ha)

Habitat no	Habitat name	Area %
91E0[a]	Alluvial forests with *Alnus glutinosa* and *Fraxinus excelsior* *(Alno-Pandion, Alnion incanae, Salicion albae).*	17
6510	Lowland hay meadows (*Alopercurus pratensis, Sanguisorba officinalis*)	4
3240	Alpine rivers and their ligneous vegetation with *Salix elaeagnos*	2
5130	*Juniperus communis* formations on heaths or calc. grasslands	2
6210	Semi-natural dry grasslands	2
3140	Hard oligo-mesotrophic waters with vegetation of *Chara* spp.	<1
3260	Water courses with the *Ranunculion fluitantis* and *Callitricho-Batrachion* vegetation	<1
6410	*Molinia* meadows	<1
7220[a]	Petriying springs with tufa formation (*Cratoneurion*)	<1
7230	Alkaline fens	<1

[a]Priority habitat

Table 4 European Union habitats in the NATURA 2000 site "Lechauen nördl. Augsburg" (401 ha)

Habitat no	Habitat name	Area %
91 F0	Riparian mixed forests of *Quercus robur, Ulmus laevis* and *Ulmus minor, Fraxinus excelsior* or *Fraxinus angustifolia*, along the great rivers (*Ulmenion minoris*)	8
[a]91 E0	Alluvial forests with *Alnus glutinosa* and *Fraxinus excelsior (Alno-Padion, Alnion incanae, Salicion albae*)	7
6410	*Molinia* meadows on calcareous, peaty or clayey-silt laden soils (*Molinion caeruleae*)	2
3150	Natural eutrophic lakes with *Magnopotamiom* or *Hydrocharition*-type vegetation	1
6210	Semi-natural dry grasslands and scrubland facies on calcareous substrates (*Festuco-Brometalia*)	1

[a]Priority habitat

Table 5 European Union habitats in the NATURA 2000 site "Höhgraben" (66 ha)

Habitat no	Habitat name	Area %
3260	Water courses of plain to montane levels with the *Ranunculion fluitantis* and *Callitricho-Batrachion* vegetation	11
6210	Semi-natural dry grasslands	2
[a]91E0	Alluvial forests	2

[a]Priority habitat

The more important results of the ecological studies were published by the city council in three books in the series "Augsburger Ökologische Schriften":

1. Biotope Mapping (Müller and Schmidt 1988)
2. The Lech river – alterations of a braided river (Müller and Schmidt 1991)
3. Trees in the city (Dobner et al. 1993)

Four applied projects, which became well known throughout Germany, emerged from the investigations

(a) "Wildflower meadows" – in all public parks the less frequently used lawns were changed to "wildflower meadows" in order to increase the biodiversity of the parks and that of the city.
(b) "Restoration of species-rich grasslands along the Lech valley" – over 20 years, the changes in several types of grassland were monitored after being restored.
(c) "Plant corridor Lech valley" – the objective was to re-connect those habitats in the Lech Valley that were important for biodiversity. In addition, sheep grazing was re-established in order to conserve the internationally important dry habitats in the valley.
(d) "Wertach vital" – the regulated Wertach river in the city was restored to a more natural flowing river with gravel bars.

The ecological research was accompanied by educational activities (for example, public information and excursions) to allow the city's inhabitants to become more sensitive to and understanding nature in the city.

Closing Comments

In terms of biodiversity, Augsburg is one of the best researched cities in Germany. Because the investigations were carried out by the city government, there is a long continuous history of application of the results in daily conservation practices. For this reason, numerous German-wide endangered species have stable populations in the city's nature protection areas, despite continual urban expansion. The examples described in the last section demonstrate that nature conservation in cities can have major positive effects on global biodiversity. Because cities are places of economic and scientific power, they offer good possibilities for ecological research and integration of the results into planning and nature conservation practice and management.

Literature Cited

Alten J W (1822) Augsburgische Blumenlese. J Wolffische Buchhandlung, Augsburg

Dobner M, Schmidt KR, Schneider U (eds) (1993) Bäume im Lebensraum Stadt. Augsburger Ökologische Schriften 3

EUR (2007) Interpretation Manual of European Union Habitats. http://ec.europa.eu/environment/nature/legislation/habitatsdirective/docs/2007_07_im.pdf access 10.10.2009

Hiemeyer F (ed) (1978) Flora von Augsburg. Ber Naturwiss Ver Schwaben Special edition

Klucniok B (1978) Laub- Torf- und Lebermoose aus Augsburg und Umgebung. In: Hiemeyer F (ed) Flora von Augsburg. Ber. Naturwiss. Ver. Schwaben

Müller N (1985) Rote Liste gefährdeter Farn- u. Blütenpflanzen in Augsburg und ihre Auswertung für den Arten- u. Biotopschutz - Ber. Naturwiss Vereins f Schwaben 89: 2–24

Müller N (1990) Lawns in German Cities. In: Sukopp H, Hejny S (eds) Urban ecology. SPB Academic Publishing, The Netherlands
Müller N, Schmidt KR (eds) (1988) Biotopkartierung Augsburg. Augsburger Ökologische Schriften 1
Müller N, Schmidt KR (eds) (1991) Der Lech. Augsburger Ökologische Schriften 2
Jäger E, Werner K (eds) (2005) Exkursionsflora Deutschland. Elsevier, München
Stangl J (ed) (1985) Pilzflora von Augsburg und Umgebung. Pilzverein Augsburg

Berlin

Herbert Sukopp

Fig. 1 The Brandenburger Tor (photo Huda Meiria)

Abstract Berlin, which occupies 900 km² and contains 3.4 million people, is unique in two respects, first it was politically and physically divided for many decades and second the flora of the western part is probably the most intensively and extensively studied of any city in the world. It is the city in which modern urban ecology was born. As with most cities Berlin can be divided into four abiotic and floristic concentric zones from the dense central core to the outskirts. Since the late eighteenth century, a total of 2,178 species have been recorded in the city, of these 1,392 species were present in 2000 – virtually 20% being non-native. Urban botany

---

Herbert Sukopp (✉)
Technical University Berlin, Institute of Ecology, Rüdesheimer Platz 10, 14197 Berlin, Germany
e-mail: herbert.sukopp@tu-berlin.de

J.G. Kelcey and N. Müller (eds.), *Plants and Habitats of European Cities*,
DOI 10.1007/978-0-387-89684-7_3, © Springer Science+Business Media, LLC 2011

allows us to watch and study evolution in action for example (a) the formation of new species, and plant communities, (b) hybridisation of native x non-native and non-native x non-native taxa that would never meet naturally and are concentrated in cities and (c) the gene/environment interaction – evolution of taxa in response to the unique environmental conditions that exist in urban areas.

Natural Environment of the City

Location

The geographical position of Berlin is 52° 30′ North and 13° East. The city, which now occupies ca. 900 km and has a human population of 3.4 million, is situated in eastern Germany. A diagrammatic map of the city is given in Fig. 2.

Geology, Geomorphology, Topography and Soils

Berlin is situated in the Weichselian glaciated area of the north Germany lowlands and within the older part of the young moraine landscape. The Warsaw-Berlin Urstromtal with the river Spree and the Havel lake-river system subdivides the Berlin region into ground moraine plateaux. The lowest point of the city is 30 m a.s.l., and the highest point is 115 m.

Geomorphology

Berlin is the first city in which urban soils have been systematically investigated. Two soil types are characteristic of the city. First, deeply cultivated garden soils (hortisols) with high nutrient content and high water-holding capacity. Second, anthropogenic rendzinas formed on rubble (for example, from demolished buildings). The latter are alkaline, dry, and well-aerated in the beginning followed by the accumulation of humus. They are usually low in nitrogen but have moderate to high contents of phosphorus, calcium, potassium, and other nutrients. These latter soils support characteristic ruderal vegetation with a succession leading to forest dominated by *Robinia pseudoacacia* and *Acer platanoides*.

Climate

The macroclimate of the city lies in the transition zone between oceanic and continental with mean annual temperatures of 8.9°C and mean annual precipitation of 590 mm. The wettest month is July and the driest is February; the hottest month is July, while the coldest is January.

Fig. 2 Diagrammatic map of Present-day Berlin with forests (*dark green*), parks (*green*) and arable land (*brown*)

Cities have particular climatic characteristics. Ecologically, the most important aspect of the urban climate is its elevated air temperatures compared with the surrounding rural area – cities form "heat islands" or "hot spots" on the surface of the earth. The annual mean temperature of the city of Berlin varies from 0.5 to 1.5°C higher than that of its surroundings but on clear days it may reach 9°C higher than that of the surrounding countryside. Climatic conditions within a city may vary considerably depending on its location, the distance to large vegetated areas, the type of construction and the extent of hard surface (for example, paving and buildings). According to these changes, different and more or less concentric climatic zones can be distinguished. The city heat island usually covers the same area as the built-up area, but changes in wind direction may raise the air temperature in adjacent areas.

Since the nineteenth century, trees have been appreciated as elements of urban planning and design; it was during this time that trees were first planted along the streets of most large European cities. Plants are able to exercise a positive influence on both the climate and the quality of the air of the urban and adjacent rural areas by ameliorating pollution caused by traffic, heating of buildings, power stations and industrial processes and trace gases (except low-level ozone), which can occur at concentrations 5–50 times higher than the normal ambient level. The influences of vegetation on the urban climate have been investigated by Stülpnagel; his results, which were published in 1987, showed a reduction in temperature not only in the green area but also up to 1.5 km away. This climatic influence increases with the size of a green area but it decreases where the green area is divided, for example, by a road.

Historical Development of the City

Prehistoric and Early Historic Times

The present flora and fauna and their biotic communities are the result of geomorphological and ecological processes and human activities over many millennia (Sukopp 2003). In central Europe, the natural forest development after the last Ice Age was not completed before the human influence began to cause local disturbances. The primeval vegetation, as reconstructed by pollen analysis, soil conditions, historical records and recent vegetation patterns comprise three main woodland types, namely: *Quercus-Carpinus* on arenosols, influenced by groundwater in the Urstromtal; *Pinus-Quercus* on dry arenosols in the sandy parts of the ground moraine plateaux and *Quercus* with *Pinus, Carpinus* and *Fagus* on luvisols of those plateaux.

Nitrophilous plants like *Urtica dioica, Artemisia vulgaris* and some of the Chenopodiaceae are found in association with the temporary camps of Palaeo- and Mesolithic hunters/gatherers. Cultivated plants and weeds of arable fields and gardens, for example, *Plantago lanceolata* (a typical indicator of grassland and fallow) occurred for the first time during the Neolithic period. During the Bronze and Iron Ages up to the Roman Migration and the Germanic Migration period secondary successions took place, with climax forest trees, for example, *Fagus* and *Carpinus* expanding to their maximum extent in the first millennium AD. The use of wild fruit-trees has been proven since early Neolithic times by archaeological and palaeo-ethnobotanical studies.

Historical changes (and their causes) to the flora and fauna are recorded in the substrate of the Tegel Lake. The beginning of woodland clearance within the vicinity of the lake during the middle to late Neolithic period resulted in an increase in primary production in the littoral and pelagic zones. The anthropogenic influences, which had been effective since that time in connection with relatively intensive land use in the early Bronze and Iron Ages, led to further inputs of allochthonous mineral substances and nutrients into the lake. More extensive changes in late Medieval times were an increase in the water level as a consequence of the Spandau mill barrage (constructed about AD 1232) and the extensive woodland clearances activities at that time, related to the establishment of Tegel village about AD 1230. on the north-east bank of the lake near the inflow of the Tegel brook.

Middle Ages

Large-scale disturbance, as shown by the increasing numbers of arable weeds and ruderal species, began with the extensive forest clearance (to provide more land for agriculture) and with the development of the cities of Spandau, Köpenick and Berlin about 800 years ago. At Spandau, a castle and a small settlement (occupied by craftsmen) were built in the eighth century by Slavic people. Plant remains (fruits and seeds) contained species of woodlands, grassland, arable and ruderal habitats (Fig. 3).

Fig. 3 The medieval fortified settlement of the Spandau Burgwall: archeological reconstruction of settlement phase 6a (AD 1000–1030), from Müller et al. 1993, with botanical sample sites from Brande 1999, modified. (1) plant macrofossil sample series; (2) pollen analytical sample series

The reconstruction of vegetation changes using pollen analyses of the cultural layers of the settlement has revealed intensive agricultural land use occupying an area of several square kilometers around Spandau since the eleventh century. The Berlin area included arable land (which was managed on the three-field system), pastures and forests. The main crop was *Secale cereale,* which was exported as far as Hamburg. Over-exploitation of the forests was because of their use for grazing, charcoal burning and bee-keeping. Vegetable cultivation was improved during the Middle Ages and also spice plants are documented from that period. The Plague of 1348, a depression in the crop market, exhaustion of the sandy soils and the steady influx of people into large villages and towns led to the partial abandonment of fields and villages.

Early Modern Times

Outside the city walls, especially along the river Spree, different industries such as slaughterhouses, tanneries and timber yards were built. Parts of the common land were transformed into gardens. In AD 1565, there were 70 vineyards, 26 wine gardens, one hop garden and 236 orchards and vegetable gardens in Berlin-Cölln and five sheep farms and 17 dairies outside the city. The number of human inhabitants was estimated at 7,000–8,000.

Since 1850 the cessation of agriculture has resulted in a strong decline in the characteristic nitrophilous flora of the former villages with livestock, including the complete loss of *Xanthium strumarium, Chenopodium urbicum, C. vulvaria, Coronopus squamatus, Marrubium vulgare, Malva pusilla, Anthemis cotula* and *Pulicaria vulgaris.* Some villages are now completely urbanised (Schöneberg, Lietzow, Schmargendorf and Steglitz); in the twentieth century only a few (for example, Gatow, Lübars) still contained typical village plants. Medicinal plants used in prehistoric and Medieval times have run wild and are now part of the ruderal flora and vegetation, including *Hyoscyamus niger, Datura stramonium* and *Leonurus cardiaca.*

Cultivated Flora in Gardens, Open Spaces and Parks

The first survey of the plant stock of German gardens was carried out in many cities in 1561 by Conrad Gesner, who discovered that introduced plants were more numerous than native ones. Whereas Medieval gardens were used for the growing of herbal and medicinal plants, the gardens of the seventeenth and eighteenth centuries were mainly created as ornamental features. Some of the species introduced by the horticultural trade now occur frequently throughout the city, for example, *Solidago canadensis, Erigeron annuus, Aster* spp., and *Oenothera* spp. The history of the introduction of ornamental trees and shrubs into Berlin shows that plants introduced intentionally for horticultural use were planted originally in monastery and peasant gardens and have spread spontaneously to the surrounding landscapes. Trees (mainly *Tilia* spp.) in central European cities generally occurred in front of the church or the town hall.

Wild Flora

As a result of the combination of favourable microclimatic conditions and disturbed soils, plants and animals from warmer regions are spreading into and within urban areas. In Berlin, 60% of all non-native taxa are from warmer regions; this is true for archaeophytes as well as neophytes. Ruderal species grow in places that have been substantially disturbed by human activities but not on cultivated areas – classical ruderata (from Lat. *rudus,* rubble, ruins). In 1751, Linnaeus considered "ruderal" to include ruins, wasteland, walls and pavements. In his "*Florae Berolinensis Prodromus*" of 1787, Willdenow describes the flora of Berlin *intra muros* at the end of the eighteenth century and the start of the industrial revolution. About this time Berlin had 110,000 inhabitants plus 30,000 soldiers, 5,000 refugees and 3,372 Jews and was bilingual.

The city wall enclosed 1,343 ha, of which 410 ha was in agricultural or horticultural use. The urban flora was not restricted to the area *intra muros,* but included areas far from the centre; of the 822 species found, Willdenow mentions only eight species "*in urbe ipsa*".

Early publications used the term "migration of plants" for the processes later described as the introduction and naturalisation of non-native plants and biotic invasions. In his book "Grundriß der Kräuterkunde", which was published in 1792, Willdenow, the founder of plant geography, makes no reference to urban conditions. In 1823, Schouw, introduced the term "*plantae urbanae*" for plants living near cities and villages, for example, *Onopordum acanthium* and *Xanthium strumarium.* He added: "*In most cases foreign origin is the reason why these plants are located only near cities and villages.*" In particular, he listed plants of walls, ruins, roofs and rubble and garden "weeds". Chamisso (1827) also described the conditions and effects of anthropogenic changes to the flora and fauna in settlements. The following quotation is taken from an instructional work "Botany for the non botanists" that Chamisso wrote for the Prussian Ministry of Culture. The book, which was written in the tradition of natural history, is subtitled "A survey of the most useful and harmful plants, whether wild or cultivated, which occur in Northern Germany. Including views of botany and the plant kingdom." Chamisso considers the effect of human activity as follows:

> *Wherever man settles, the face of nature is changed. His domesticated animals and plants follow him; the woods become sparse; and animals shy away; his plants and seeds spread themselves around his habitation; rats, mice and insects move in under his roof; many kinds of swallow, finch, lark and partridge seek his care and enjoy, as guests, the fruits of his labour. In his gardens and fields a number of plants grow as weeds among the crops he has planted. They mix freely with the crops and share their fate. And where he no longer claims the entire area, his tenants estrange themselves from him and even the wild, where he has not set foot, changes its form.*

The paragraph can be interpreted in modern terminology as "the introduction of non-native species, changes of biotopes, synanthropy, hemerochory, apophytes and agriophyty."

Since the studies of Ascherson published in 1864, the introduction and naturalisation of non-native plants (the so-called adventive plants) have been subject to frequent botanical investigations.

Modern Times: A New Type of Environment

Modern cities comprise a mixture of semi-natural areas including forests, parks, nature reserves, remnants of agro-ecosystems ("encapsulated countryside") and densely developed housing and employment areas including historic centres (Sukopp 1990). Urban ecosystems differ from non-urban ones in many ways, although most of the factors that affect urban ecosystems also operate in non-urban areas; it is the combination of these factors that have resulted in the development of the unique plant communities and species associations that occur in urban areas.

The landscape of the Middle Ages and its rural character have been totally changed since the second half of the nineteenth century by the development of a large city. Today, about 3.4 million people live in Berlin, which occupies 889 km^2; of this, 57% is covered by hard surfaces (for example, buildings and transport networks), while 43% is not built on.

To understand the plant–environment relationship, it is necessary to assess the present biotopes in terms of their historic development. From the beginning, towns have been places of shelter against nature and its dangers; nature was only allowed in them as an "ornament". Historically, cities have fought against nature, creating an artificial environment as opposed to the more natural environments prevailing outside the city.

In the course of the historic development of cities, natural conditions have been intentionally and unintentionally altered by human activity. Therefore, much of the present-day urban open spaces comprise or are based on older parks and similar areas. The similarities between former and present site conditions decrease with time and along a gradient from the periphery of the urban area to its centre. Within the built-up area, the original ecosystems are destroyed and many species become extinct but simultaneously new organisms and new biotic communities become established.

On the basis of their ecological characters (Fig. 4), Berlin has been divided into four concentric zones: (1) densely built up; (2) partly built up central areas; (3) inner and (4) the outer margins. These zones are, of course, not strictly concentric; their distribution and share of the total area vary according to the historical development of the city, see Fig. 4.

Small gardens, railway land and hills made of rubble and waste materials are typical of the inner zone (3) while forests and large parks occur in the outer margin zone (4). Air pollution, elevated temperatures, changes in the groundwater level and backfilling of large areas are the consequences of urban development. The volume of imported materials for construction, finished products and foodstuffs is greater than that of the waste materials removed; consequently, during the course of time the ground level of the city has risen several meters in some areas. In many locations eutrophication and the compaction or sealing of the soil are linked to the depth of the

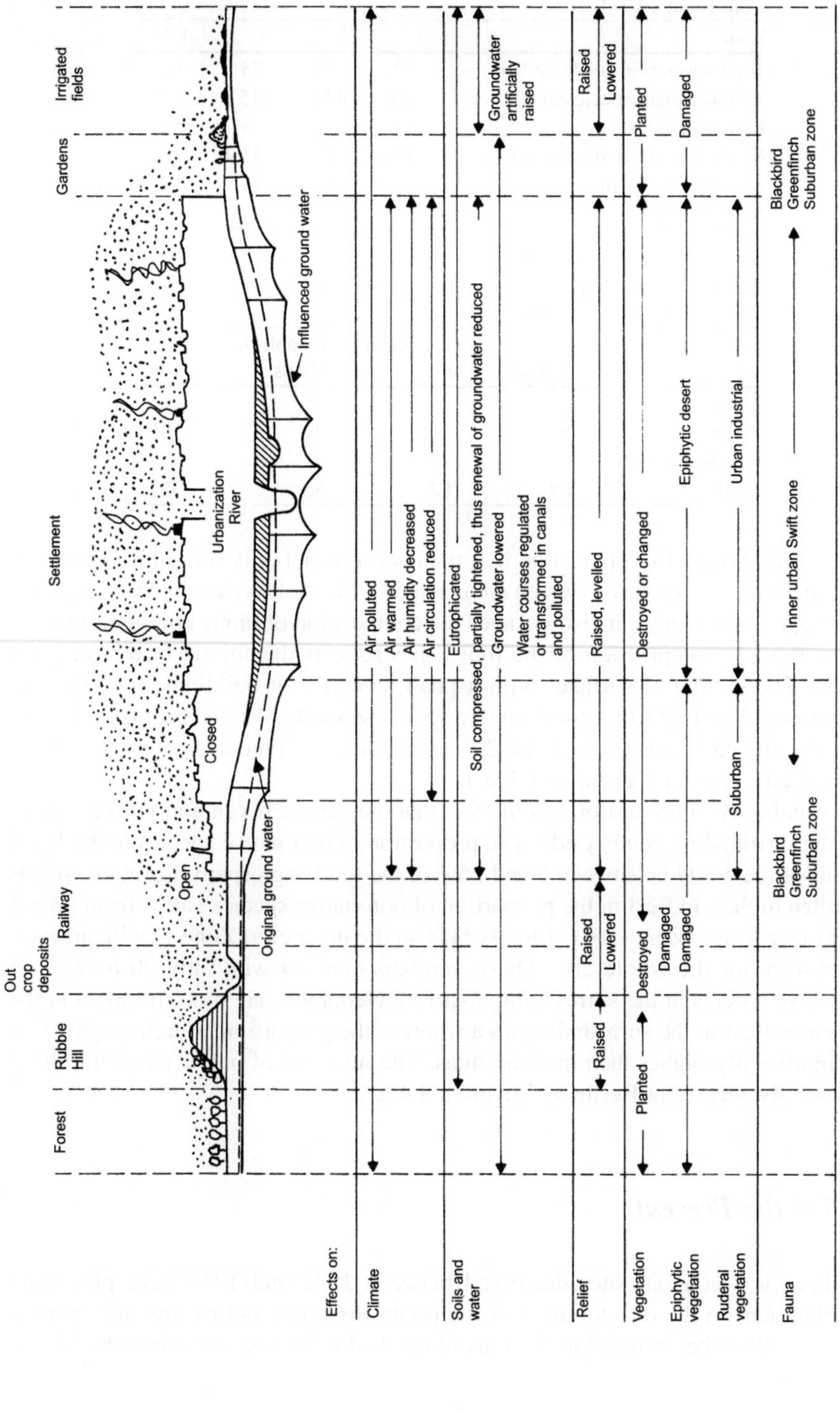

Fig. 4 Changes in the biosphere of Berlin

Table 1 Characteristics of floristic city zones in Berlin (Information from Kunick 1974, 1982)

Zone	1	2	3	4
Total vegetation cover (in %)	32	55	75	95
Vascular plant species/km^2	380	424	415	357
Rare species/km^2	17	23	35	58
Non-native plant species (in %)	49.8	46.9	43.4	28.5
Archaeophytes (in %)	15.2	14.1	14.5	10.2
Neophytes (in %)	23.7	23.0	21.5	15.6

Table 2 Native and naturalised alien woody species in Berlin (Information from Kowarik 1992)

	Natives	Introduced aliens	Established aliens	Total
Trees	30	68	25	123
Shrubs	57	109	29	195
Woody climbers	2	5	2	9
Total	89	182	56	327

accumulated layers. Eutrophication by urban waste is not only related to waste disposal sites and sewage fields (which are irrigated with waste waters in order to purify the sewage biologically), it also changes the composition of biotic communities.

The floristic composition of the four zones reflects the abiotic factors of plant growth. The sealing of surfaces with asphalt, concrete or buildings, for example, varies from 85 to 100% in Zone 1, to 0 to 15% in Zone 4. Also the heat island effect is more pronounced in Zones 1 and 2 than in 3 and 4. Some characteristics of the different city zones are given in Table 1.

A peculiarity of urban floras is the fact that they are richer in non-native species than the surrounding countryside. The proportion of non-native species in the floral regions of the world is between 5 and 25% of the total angiosperm flora; on islands it is often higher. In Berlin, the proportion of non-native species varies from 28.5% in the outer fringe (Zone 4) to virtually 50% in the inner city (Zone 1) with an average of 41% for the whole city. These numbers contrast with only 20 to 25% of non-native species in the surrounding districts. Neophytes are more frequent in the urban area than in the surrounding area whereas the proportion of archaeophytes is not significantly higher than in rural areas. The numbers of native and naturalised alien woody species in Berlin are given in Table 2.

1945 to the Present

In a historical context, sites destroyed between 1942 and 1945 have played an important part in the development of the urban flora and its investigation. Only a few years after the destruction, various plants had colonised the ruined buildings.

In many cities, the war damage and its effects gave rise to special studies of the rubble flora and fauna. Rubble, which provides warmer and drier conditions than natural habitats, provided ideal conditions for the colonisation of species from warmer regions of the world. This resulted in many plants that had been previously rather rare in Berlin becoming permanent members of the flora of the city. After 1945, an extensive and spontaneous development of vegetation began on the rubble, proceeding in more or less rapid succession from short-lived ruderal species to perennial herbaceous vegetation to shrub and forest-like stands.

The number of species found on such inner city wasteland was surprisingly large; for example, a site at the Lützowplatz (in Berlin-Tiergarten) supported 140 species of angiosperms and at least 200 insect taxa, whereas the frequently mown lawns and the shrubberies of the nearby Tiergarten Park support no more than 50 insect taxa on the same area (only 25% of the number found at Lützowplatz).

Most of this natural succession has been destroyed by site clearance and reconstruction. The areas cleared of rubble or destined for redevelopment and which have been colonised by woody plants are gradually disappearing as a result of construction works.

The dispersal of organisms and the development of vegetation can be studied on rubble sites on a large scale and in an environment that differs considerably from previously known ruderal habitats. The colonisation of rubble, created in many cities as a result of the destruction of buildings between 1939 and 1945, has unintentionally become a tremendous natural experiment, which with respect to its scale is comparable to the colonisation of new habitats created by volcanic activity.

The first peak in the ecological studies of Berlin was the investigation of the dynamics of vegetation development on rubble sites rather than wider studies of the city as a whole. Differing from adventive floristics, the investigations concentrated on dispersal strategies of a single species, succession related to site characteristics and the formation of new plant communities.

A major reason for the relative unpredictability of succession in urban ecosystems is the high degree to which these systems are subject to invasions by "aliens" – the biogeographical spectrum of species composition of cities is very different from that of the surrounding countryside. The main causes are first, the conditions for naturalisation (that is the high level of invasion of existing biocoenosis) and second, in the methods of dispersal (that is introduction and transportation). Disturbances generally increase invasibility and urban ecosystems are disturbed ones. Towns are open to invasions by alien species but the number is unpredictable.

Thellung, writing in 1917, gave the name "stratiobotany" (=polemobotany) to the destructive effects of war on plants and the subsequent formation of new plant communities with a characteristic flora and changes in the associations, comprising native and cultivated plants. Polemochores in Berlin that resulted from the war of 1939–1945 comprise two plants of eastern origin, namely, *Salsola collina* and *Artemisia scoparia* and one grass species of North American origin *Panicum lindheimeri*.

It is now generally accepted that disturbance encourages the establishment and spread of alien species. In urban environments, it is mainly the man-made components of disturbance that affect species composition and promote colonisation by non-native species; with naturalised non-native species being more commonly found on sites that are subject to a high level of disturbance. The highest percentage of alien species is found in ruderal vegetation where annuals are the dominant life-form.

Changes in the flora have been shown to reflect changes in economic and cultural history. According to Naegeli and Thellung (writing in 1905), the development of the ruderal flora "*essentially runs parallel to the size and the intensity of trade and industry; it is a direct standard of the technical culture*".

In addition to numerous non-native species, native organisms may also profit from urban habitats, especially if the conditions are similar to those of their original habitats. In the natural landscape, these plants and animals originally occurred on open sites such as gravel or mud banks along rivers and streams, in open places within forests and on ground disturbed by fire.

As most large cities are located adjacent to large rivers, erosion and sedimentation act as important initial habitats for urban apophytes. The native plants that have greatly expanded in Berlin during the last 50 years are *Cardamine hirsuta, Veronica hederifolia ssp. lucorum, Elytrigia repens, Calamagrostis epigeios, Calystegia sepium, Humulus lupulus, Rorippa sylvestris* and *Acer platanoides*.

Based on 6,000 vegetation "releves" made in West Berlin, the "hemeroby concept" was used to construct spectra of hemeroby for each species of the urban flora. This approach is particularly valuable in relating the response of a species to complex human influences. Human impact consists of many direct and indirect environmental factors, some of which (for example, stress and disturbance) cannot be easily measured.

In 1985, Wittig and his colleagues categorised species into three types; those that "liked" the urban environment, those that were neutral and those that "disliked" urban areas. They called such species urbanophilous, urbano-neutral and urbanophobous, respectively. Historically, ecological studies of the urban flora have investigated the distribution of plants in different districts of different ages. The Berlin Wall, started in 1961 and demolished in 1989, was the cause of differences in the distribution of plants. Surprisingly, *Iva xanthiifolia* (a species of North American origin) was mainly found in East Berlin. During the nineteenth and twentieth centuries, the species was introduced to eastern Europe and western Asia from North America and spread into arable land. Seed was imported to East Berlin in wheat from the Ukraine and Kazakhstan. Some of the seed spilled onto and germinated in the railway tracks from where it colonised and became frequent along the railway network of Eastern Germany.

Plant communities of the areas along the boundary between East and West Berlin were significantly influenced by the adjacent habitats. A typical example is the increased occurrence of ornamental species that have naturalised on sites adjacent to allotments. With the ending of the division of East and West Berlin in 1989, the ecological development of the city faced a new and interesting situation.

Species introduced as ornamentals will remain by far the largest group from which neophytes will emerge. In this regard, changes in human values and biological processes are important; predicting which species will become naturalised depends on the preference and fashions of people and the dispersal mechanisms of the species. Plants will continue to be imported from many places throughout the world. Understanding human behaviour in relation to plant dispersal will prove as important as understanding the biology of dispersed species.

A well-known example of the introduction and spread of plants by trade is the nineteenth century wool industry when diaspores of plants were transported in wool from other continents into Europe and were found growing around docks and wool mills. Over the last 30 years, *Senecio inaequidens* has become one of the most successful alien plants on ruderal sites in central Europe. The species (a native of South Africa) was repeatedly introduced by the importation of wool. After the initial colonisation in 1896, it became established in seven locations from where it successfully spread throughout Europe. In 1993, it was found for the first time in Berlin. Most of the locations in which it occurs are along motorways with a large number of sites being found on railway land. *Senecio inaequidens* continues to spread towards the north and the east and within the central area of Berlin.

Urban biocoenoses are an extreme example of communities produced by successive invasions and not by co-evolutionary development. In principle, the historic uniqueness of urban situations, that is, the combinations of environmental factors and organisms, differentiates urban ecosystems from most natural ones, even those subject to strong disturbance.

Flora

Vascular Plants

The altered climate, soil and water conditions in cities have a corresponding effect on the composition of the biotic communities occurring in urban areas (Scholz 1960). These changes are greater the larger the urban area is and the closer one gets to the city centre. Comparative studies of different cities, urban with rural areas and the inner city with periphery have become established methods for analysing these differences.

The number of fern and flowering plant species per unit area in cities with more than 50,000 inhabitants is higher than the number in the surrounding area. In central Europe, the number of fern and flowering plant species correlates closely with population size and/or density; see Table 3. The 50 most frequent species found in Berlin are listed in Table 4.

Table 3 Number of species found in cities with different population sizes

City size	No. of species
Small and medium-sized towns	530–560
Cities with 100,000 to 200,000 inhabitants	650–730
Older cities with 250,000 to 400,000 inhabitants	900–1,000
Cities with over a million inhabitants	>1,300

Table 4 The 50 most frequent vascular plants in Berlin (after data from RFKB 2001)

Acer negundo	*Glechoma hederacea*
Acer platanoides	*Humulus lupulus*
Acer pseudoplatanus	*Hypericum perforatum*
Achillea millefolium	*Hypochaeris radicata*
Aegopodium podagraria	*Linaria vulgaris*
Agrostis capillaris	*Lolium perenne*
Alliaria petiolata	*Plantago lanceolata*
Arrhenatherum elatius	*Plantago major*
Artemisia vulgaris	*Poa annua*
Ballota nigra	*Poa pratensis*
Bellis perennis	*Polygonum aviculare* agg.
Berteroa incana	*Quercus robur*
Betula pendula	*Ranunculus repens*
Calamagrostis epigejos	***Robinia pseudoacacia***
Capsella bursa-pastoris	*Rubus fruticosus* agg.
Chelidonium majus	*Rumex acetosella*
Chenopodium album	*Rumex thyrsiflorus*
Cirsium arvense	*Sambucus nigra*
Convolvulus arvensis	***Saponaria officinalis***
Conyza canadensis	*Sisymbrium loeselii*
Dactylis glomerata	***Solidago canadensis***
Festuca ovina	*Stellaria media*
Festuca rubra	*Tanacetum vulgare*
Geranium pusillum	*Trifolium repens*
Geum urbanum	*Urtica dioica*

Bold = neophytes

There are two main reasons for this; first, urban areas are heterogeneous, comprising a wide variety of structures, materials, land uses and small-scale habitats. This creates many specific and even unusual ecological conditions. Second, the introduction of non-native organisms.

Historically, species that have been introduced into an area by human activity have usually begun their dispersal in urban areas and therefore occur most frequently in them. With an increasing human population, the size, trade and traffic of the city expands with a consequential rise in the proportion of non-native species in the flora. The large number of non-native species is the result of direct planting or indirect introductions via imported materials (known as "hemerochores"). Today the number of naturalised plants far exceeds that of species that have become extinct in areas that have been urbanised, which is one of the reasons why introduced species comprise a higher proportion of the urban flora.

This is explained not only by considerable habitat heterogeneity (discussed above) and the role of big cities as centres of species immigration but also by the better adaptation of alien naturalised species to man-made disturbances.

Because of industrialisation, most introduced species reached central Europe in the second half of the nineteenth century. In addition to changes in species number, there have also been changes in species frequency. A comparison of species frequency in Berlin with data from the nineteenth century shows differences in the proportion of native and non-native species over time. Two-thirds of the species, which have become more frequent during the last 120 years, first established themselves after AD 1500.

A fundamental question is, "whether the process of species migration resulting from human activity has reached its zenith yet?" If not, a further enormous increase of species-richness in some areas, in particular urban ones, seems to be possible, but we do not know enough about the mechanisms of the invasion processes to make even a rough prediction.

The changes in the vegetation, species composition and site conditions in Berlin from the last Ice Age to the present have been thoroughly investigated. The number of fern and flowering plant species found in Berlin was 822 in the mid-eighteenth century, 1,130 in the mid-nineteenth century and 1,392 by AD 2000, of which 19.5% are non-native. A total of 2,178 spontaneous species have been recorded since AD 1787 – when the city expanded to its present size of 889 km^2. One thousand and fifteen of the 2,178 taxa are neophytes while 786 never became established; during the same period, 202 species became extinct.

In addition, there is a large number of cultivated plants in parks, gardens and churchyards, on small patches of ornamental green spaces, as street trees, and even on balconies and in flower pots. The number of taxa and their population size found in these areas far exceeds that of the spontaneous species.

Urban green spaces were enhanced or created to symbolise the superiority of man over nature (as in Baroque gardens) or to provide the dream of natural rural life (landscape gardens of the Romantics). In both cases, "green" stands for the contrast between nature and the city. On the other hand spontaneous urban vegetation, as a type of nature adapted to the urban conditions and capable of existing within them, symbolises the city. This provides the opportunity for a new approach to landscape design that does not accentuate the contrast between nature and the city as it used to do.

The green roofs of Berlin, which had become a well-known feature of the city during the twentieth century, have inspired many architects. Plants growing on buildings influence the climate in a positive way, for example, extensive investigations confirm that roof gardens and climbing plants on walls are of considerable benefit in ameliorating the "heat island" effect and air quality. A cover of plants on roofs can reduce the surface temperature, filter the air, fix harmful substances and reduce heat loss during the winter.

Although cultivated plants serve important functions in cities and provide habitats for animals, ecological investigations are generally concerned with spontaneous plants.

An unexpected result of early studies of plants in cities was the fact that they do not occur randomly but form distinct patterns of co-occurrence and plant communities in the same way as plants in more natural environments. The typical urban vegetation of Berlin is represented by the *Conyzo-Lactucetum serriolae* and *Echio-Melilotetum* associations. The characteristic plant communities and their component species found in the centre of Berlin include *Ailanthus altissima, Amaranthus albus, A. blitoides, Chaenorhinum minus, Chenopodium botrys, C. strictum, Commelina communis, Potentilla supina and Parietaria pensylvanica.*

One of the characteristic pioneer colonisers is the summer annual *Chenopodium botrys*. As the result of human influence, this southern Eurasian-Mediterranean plant has expanded its range into large parts of central and western Europe, North America and Australia. Introduced to Berlin in 1889, it became (after 1945) a characteristic and specific ruderal plant of the heat island of inner Berlin. The natural habitats of the plant are sandy and stony soils near riverbanks, that is, specialised habitats with little competition. Accordingly, roadsides, cultivated areas and fallow land are colonised as secondary habitats. Under natural conditions, the area covered by such open, low-competition habitats in central Europe is quite small. The open calcium-rich sandy to gravelly habitats, which have been created by man, have enabled the species to expand. However, after 1960, the colonies in Berlin have been decreasing, as the result of re-development and the consequential loss of suitable substrates.

Both the number of species of woody neophytes and their lush growth are surprising. Of the woody plants, stands of *Robinia pseudoacacia* cover a large area of Berlin. On the rubble-mortar substrate a calcium-rich loose syrozem develops into a para-rendzina under *Robinia. Ailanthus altissima* is another species that has spread and is spreading on inner city vacant land, especially when old trees (and therefore fruits) occur in the vicinity. This vigorous drought-tolerant species colonises "stressed" habitats such as railway land and areas adjacent to buildings, where it is able to establish large polycormons through suckering. It is also frequently found in green open spaces.

In many cases, it is evident that molecular/genetic changes occur following the introduction of some weed species. There are two theories about the origin of non-native weeds:

(a) Introduction and naturalisation of plants: archaeophytes and neophytes.
(b) Evolution of new taxa in secondary habitats of the cultural landscape, which Zohary (1962) has called "anecophytes". Processes frequently associated with genetic changes take place in the plants, permitting them to spread over large areas. Recent weed evolution after species introduction is evident in many cases, for example, *Oenothera, Corispennum, Xanthium* and *Fallopia.*

The creation of cultivated ecosystems and changes within cultivated landscapes result in faster evolutionary changes; consequently, the most naturalised aliens will persist and become part of the more permanent flora and create new ecological communities. More importantly, we would be wise to recall that it is the widely dispersed and fairly abundant species that are most likely to survive the stresses that cause extinctions and become the founder of new taxa and therefore diversity.

The high proportion of aliens in cities is partly due to the fact that cities are "centres of spread" as the result of the introduction of new species via railway stations and ports, landscape schemes and botanical gardens. On the other hand, anthropogenic changes resulting in more suitable conditions for growth facilitate their spread. Many non-native species originate from warmer regions and depend on the higher temperatures in cities, for example, *Chenopodium botrys* (Fig. 5) or *Ailanthus altissima.* On the other hand, as a result of the high level of human influence in urban ecosystems, many species have become or are in danger of becoming extinct.

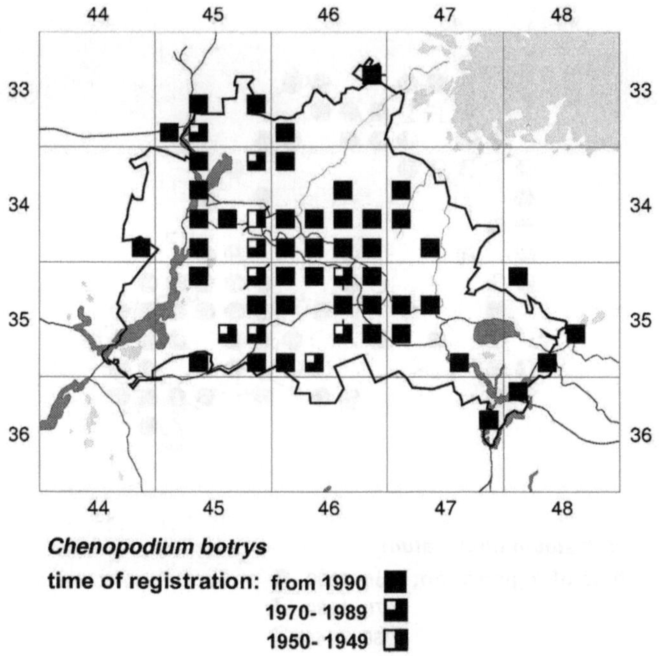

Fig. 5 Distribution map of *Chenopodium botrys* an example of the distribution of a neophyte spreading in inner-city warm settlement areas; source: RFKB 2009)

Habitats

Urban habitats differ from natural or rural ones in many obvious ways. Structurally and functionally, cities are a highly complex association of inter-connected habitats with the high consumption of energy and materials and the export of manufactured goods and waste. However, contrary to common expectations, cities may be quite rich in plants and animals, whose presence and survival are greatly dependent on human values and activities, which vary in time and space. The generally accepted method of studying urban biota and their habitats is to compare them along the rural-urban gradient. *Holosteum umbellatum* is a native pioneer species of sandy soils on the rural fringe; see Fig. 6. The organisms and communities along this gradient react to the different degrees of human influence in different ways enabling the effect of urbanisation to be understood. This in turn assists in the planning, design and management of habitats and their component species.

Urban Forests

There are about 16,500 ha of forest in Berlin plus 28,500 ha outside the city boundaries. The main types are *Quercus-Carpinus* on arenosols, influenced by groundwater in

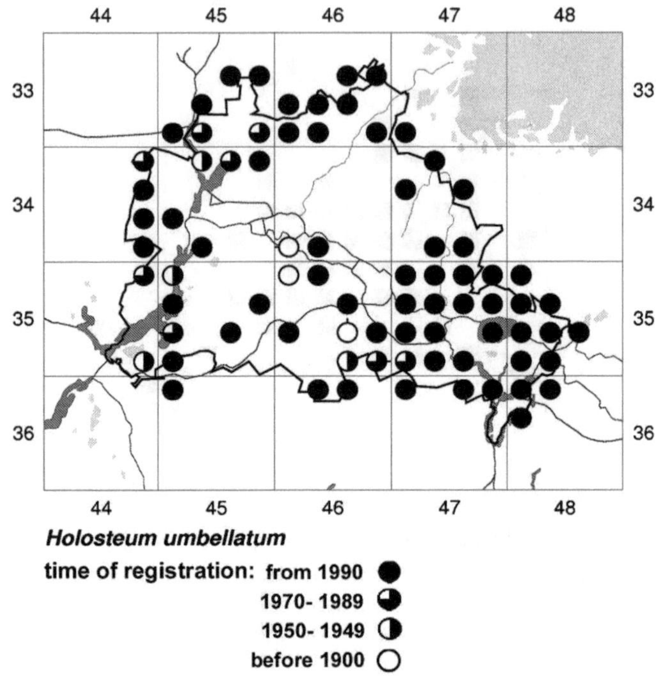

Fig. 6 Distribution map of *Holosteum umbellatum*, a native pioneer plant on sandy soils outside the built-up areas: source: RFKB 2009)

Fig. 7 Damp woodland vegetation in the Berlin glacial valley: *Quercus* forest Wuhlheide, Photo Bernd Machatzi

the Urstromtal; *Pinus-Quercus* on dry arenosols on the sandy parts of the ground moraine plateaux and *Quercus* with *Pinus, Carpinus betulus* and *Fagus* on luvisols of those plateaux. Damp *Quercus* woodland occurs in the glacial valley (Fig. 7).

The composition of the field layer varies according to the type of forest and its use; the more abundant species are *Avenella flexuosa, Dryopteris carthusiana, Viola riviniana,* and *Pleurozium shreberi.* Many rare and endangered native plants survive in the urban forests. Nitrophilous species and neophytes such as *Prunus serotina* have spread while *Acer platanoides* and *A. pseudoplatanus* (both frequent street trees) invade those forest edges that are enriched with nitrogen. Studies in Berlin have shown that some ornamental species such as *Mahonia aquifolium* and *Taxus baccata* are colonizing the forest from adjacent gardens.

The soils and vegetation of the forests in the former East Berlin are quite different from those in the former West Berlin. In the eastern part, where the production of timber was the main objective of forestry, *Pinus* spp. is dominant, accompanied by *Calamagrostis epigeios.* In the western part where recreation was most important, the forests comprise a mixture of deciduous and coniferous species.

Residential Areas

The vegetation in housing areas is virtually destroyed every time buildings are demolished and new ones built. Consequently, the soil and plant species found are related to the age of the housing area.

The soils of the oldest areas were richer in humus and in nutrients (nitrogen, phosphorus) while the soils of the landscape/open space areas in the more recent developments are mainly sandy. The plants found in the oldest parts of the city are typically native forest species, for example, *Hedera helix, Mycelis muralis* and *Sorbus aucuparia,* which have high nutrient and humidity requirements. On the more recently disturbed soils of the youngest ca. 25 years old housing areas, the flora contains a high proportion of non-native annuals such as *Geranium pusillum* and *Sisymbrium officinale.*

Transport Corridors

The main factor that limits vegetation in streets is the sealing of the surface with impermeable materials such as asphalt or paving stones. In other areas the vegetation is disturbed by vehicles, which also cause soil compaction. Nevertheless, the flora of roadsides can be quite rich. In a study of 61 streets in Berlin, Langer (1994) recorded a total of 375 spontaneous plant species (25% of the total flora of the city). Although the majority only occurred at a low frequency, a group of 13 species (mainly annuals with resistance to trampling) was present in all areas. The most widespread plant communities belonged to the phytosociological class *Plantaginetea,* a group which contains plant communities of heavily trampled swards. Plants can be useful as indicators of environmental conditions; for example, the presence of halophytes such as *Puccinellia distans* adjacent to some of the roads in Berlin is indicative of the use of de-icing salt. Oil derivatives are detected by the decrease of "petroleophobe" species, such as *Artemisia vulgaris* and *Arrhenatherum elatius* and the increase in "petroleotolerant" including *Urtica dioica* and *Cirsium arvense.*

Fig. 8 Abandoned goods depot with ruderal vegetation, Nordgüterbahnhof, photo Bernd Machatzi

Ecotones adjacent to roads can be important areas for plant evolution, for example, the frost-tolerant form of *Cynodon dactylon* from a Berlin street habitat was crossed with horticultural types resulting in a new cultivar – Tifton 44. Roads also act as migration corridors for plants. The annual seed rain caused by vehicles on the roadsides of an urban motorway in Berlin ranged from 635 to 1,579 seeds/m^2/year.

In the mid-twentieth century, large areas of railway land, especially goods depots became disused. This resulted in the formation of complex associations of vegetation types, including the establishment of trees and shrubs; see Fig. 8.

In many cities, such areas were commercially valuable and therefore subject to considerable re-development pressures. In Berlin, some of these areas were protected as landscape reserves; see Fig. 9.

Extensive areas of mestrophic grassland separate the runways and taxi-ways of airfields, as at Tempelhof Airport; see Fig. 10.

Parks and Cemeteries

Large parks (60–140 ha) contain between 250 and 450 species of vascular plants, city parks (10–25 ha) 120–280 species, small greens (1 ha) 50–140 species. The largest park in Berlin is the Tiergarten (Fig. 11). Only 40–120 woodland and aquatic species occur in the large parks of the inner city.

Parks and botanical gardens are centres for distribution of plants introduced unintentionally with grass seeds in former times (before 1950), for example, *Bromus erecta* from south-eastern France and *Poa chaixii* and *Luzula luzuloides* from southern Germany.

The first cemeteries were built outside the city walls during the Medieval plague epidemics. An ecological study of cemeteries of Berlin and an analysis of their

Fig. 9 The abandoned railway area "Schöneberger Bahngelände" was protected as a landscape reserve after the unification of the city, photo Norbert Müller

Fig. 10 Tempelhof, airfield with *Arrhenatherion* grassland, photo Bernd Machatzi

Fig. 11 "Tiergarten" Park – the largest public park in the city centre, photo Sascha Abendroth

flora and vegetation are described by Graf (1986). Cemeteries are important habitats for non-native ornamental plants with distribution and naturalisation in old gardens, parks and cemeteries, including *Galanthus elwesii, Scilla (species diversae)* and *Ornithogalum nutans*.

Aquatic Habitats

Most cities are located adjacent to major rivers. The form, chemistry and biology of rivers and lakes in urban areas have changed as a result of development, for example, canalisation, an increase in the impervious surface, reduction in water quality caused by surface water run-off (which contains pollutants) and a reduction in the richness of biotic communities. Drinking water supply and wastewater treatment became necessary relatively early in the development of the modern city. In 1902, Kolkwitz and Marsson devised a water quality standard by biological examination. The revision of this method resulted in the "saprobic system", which used plants and animals to indicate the quality of water according to their ability to survive under different saprobic conditions.

Berlin is a large city lying near many waterways and lakes – more than 100 lakes and approximately 150 small rivers, ponds and ditches. The Havel and Spree are the largest and most important rivers in the city. The area covered by water is 54 km^2 or 6% of the total area. The location close to rivers and on major trade routes was one of the main reasons for the founding of Berlin. Drinking water is taken from groundwater and from rivers.

The landscape around the lakes is dominated by reedbeds (*Phragmites australis*) varying from 10 to 100 m wide and associated with *Nuphar luteum* on the waterside. On the landward side, the lakes are bordered by *Alnus glutinosa* followed (on mineral soils) by alluvial woodland with *Ulmus laevis* and *Prunus padus* and neophytes such as *Acer negundo*. In 1969, a special law was adopted for the protection of the reedbeds, including the planting of *Phragmites australis* and the provision of mechanical protection against waves produced by motor boats.

The lakeside residential developments have had a significant adverse effect on the *Phragmites* beds. For example, from 1962 to 1987, the *Phragmites* of the Berlin Havel River decreased from 40% to 12% of the shoreline. The main causes were the mechanical erosion of the organic soil, eutrophication of the water and the artificial recharge of groundwater along the shore. The protection and restoration of the *Phragmites* beds required, among other measures, the lake margins to be protected and the level of eutrophication to be reduced.

The extinction of organisms as well as the introduction and dispersal of neophytes and neozoans in the Tegel Lake has been well documented over the last 200 years. *Elodea canadensis* (a native species of temperate North America) has been known in central Europe since 1859, when it was released from the Berlin Botanical Gardens into the ditches near Charlottenhof at Potsdam-Sanssouci. At this time, the species already existed outside the cultivated areas, for example, in Lake Tegel. The extensive growth of *Elodea* (known as the "Green Ghost") led to all kinds of difficulties for shipping, water management and fishing; however, after several years the mass expansion declined.

Nature Conservation, Environmental Planning and Education

Particularly valuable tools for urban nature conservation are the Red Data Lists of endangered flora and fauna. The first Red Lists for Berlin (then Berlin West) appeared in 1982, they were revised in 1991 and again in 2001 (Landesbeauftragter für Naturschutz und Landschaftspflege 2001). The Lists identify four categories:

Already Extinct

The species in this category include *Caldesia parnassifolia, Gentianella campestris, Linnaea borealis* and *Luronium natans*.

Threatened with Extinction

The category includes *Andromeda polifolia, Anthericum liliago, Botrychium matricariifolium, Botrychium multifidum* and *Campanula glomerata*.

Highly Endangered

Included in this category are *Betonica officinalis, Equisetum variegatum, Geranium sanguineum, Hepatica nobilis* and *Koeleria glauca.*

Endangered

The species include *Malva moschata, Melica nutans, Potamogeton crispus, Pulsatilla pratensis* and *Sedum rupestre.*

The proportion of threatened species in various groups of organisms in Berlin is often considerably higher than in areas with lower anthropogenic influence.

The Berlin species protection programme prescribes measures for the conservation of the flora and fauna for 54 types of biotopes, 36 biotope development areas and 18 groups of organisms.

School gardens have been created to provide plants for natural history lessons. The education system also involves the establishment of environmental centres and programmes for developing emotional links to nature through personal experience and considering the inter-relationships between individuals and society and between nature conservation economics.

In 1986, Henke and Sukopp devised a means of incorporating a natural approach in the planning system of Berlin. The Berlin Nature Conservation Act, which became law in 1979 divides subject planning into four programmes:

1. Nature conservation (areas, habitats, species)
2. Ecological processes (water, soil, air)
3. Landscape scenery (elements, landscape types, design units) and
4. Recreation (open spaces, neighbourhood recreation areas)

There are 37 nature reserves covering a total of 1957 ha (2.2% of the total Berlin area) and 12% covering a total of 11,944 ha.

Closing Comments

With the existing inadequate adaptation of society to its environment the problems can only be solved if due consideration is paid to ecological principles and the needs of society. Ecological criteria should serve primarily to create and maintain a high quality environment and make plain the consequences of actions for the environment, society and the individual. Nature should develop in close relationship with the local inhabitants and their customs. Organisms and biological communities should be conserved to allow people direct contact with the natural elements of their environment. Only such open spaces can lead to the experience of natural beauty which permits coexistence between nature existing in its own right and people who are free to determine their own actions in this space. Urban development needs reliable knowledge of the environment and the conditions for its protection.

Investigations indicate the need for a strategic plan based on sound scientific information and arguments for protecting and re-introducing wild plants and animals into the city. Close co-operation between natural scientists and many other disciplines (for example, planners, architects, engineers, economists, landscape planners and landscape managers) is required for this task. The conservation as well as re-integration of nature into the city – that part of the environment that man has transformed to the greatest extent – is an important endeavour, not only for the improvement of environmental quality, but also for human welfare.

Plants have evolved over hundreds of millions of years, and each species has a unique set of genes, which once lost via extinction can never be brought back. This thought along with ethical reasons should be sufficient reason to preserve all species in the world. On a smaller scale, the argument is less valid: because most species are not restricted to a single country or even a city, their vanishing from one of them would not mean their extinction (although taxonomic entities on a lower level such as subspecies or ecotypes may be more locally distributed and consequently more threatened).

In politics and public planning, the aim of conserving nature is often countered by economic arguments. Therefore, nature protection requires convincing arguments. From the above, it is clear that the aim of urban nature conservation is not so much the prevention of extinction of species but rather the preservation of diversity. Urban botany allows us to watch and study evolution in action, for example, as follows:

1. The formation of new species, and plant communities
2. Hybridisation of taxa (native × non-native and non-native × non-native) that would never meet in nature. In rural areas, the plants are probably too dispersed while in urban areas they are concentrated, hence causing a big increase in probability of interactions
3. Gene/environment interaction – evolution of taxa in response to the unique environmental conditions that exist in urban areas

Literature Cited

Ascherson P (1864) (Reprint 1990) Flora der Provinz Brandenburg, der Altmark und des Herzogthums Magdeburg, 2. Abt. Specialflora von Berlin, Berlin

Brande A (1999) Botanische Untersuchungen auf dem Burgwall Spandau – eine Übersicht. In Müller Av, Müller-Muci Kv (eds) Berliner Beiträge zur Vor- und Frühgeschichte. Neue Folge 9:130–140

Chamisso A (1827) Übersicht der nutzbarsten und der schädlichsten Gewächse, welche wild oder angebaut in Norddeutschland vorkommen. Nebst Ansichten von der Pflanzenkunde und dem Pflanzenreiche. Berlin

Gesner C (1561) Conradi Gesneri de hortis Germaniae liber recens. in: Valerius Cordus Annotationes in Pedacii Dioscoridis Anazarbei de medica materia libros V. Straßburg

Graf A (1986) Flora und Vegetation der Friedhöfe in Berlin (West). Verhandlungen Berliner Botanischer Verein 5:1–210

Henke H, Sukopp H (1986) A natural approach in cities. In Bradshaw AD, Goode DA, Thorpe E (eds) Ecology and design in landscape. 24th Symp. Br Ecol Soc Manch. 1983:307–324

Kolkwitz R, Marsson M (1902) Grundsätze für die Beurtheilung des Wassers nach seiner Flora und Fauna. Mitteilungen Königliche Prüfungsanstalt Wasserversorgung und Abwässerbeseitigung 1:33–72

Kunick, W (1974) Veränderungen von Flora und Vegetation einer Großstadt, dargestellt am Beispiel von Berlin (West). Diss. Technische Universität Berlin

Kunick, W (1982) Zonierung des Stadtgebietes von Berlin West. – Ergebnisse floristischer Untersuchungen. – Landschaftsentwicklung und Umweltforschung 14 Technische Universität Berlin. Berlin. 164 S

Langer A (1994) Flora und Vegetation städtischer Straßen am Beispiel Berlins. Landschaftsentwicklung und Umweltforschung. Sonderheft S 10:1–199

Liste der wildwachsenden Gefäßpflanzen des Landes Berlin mit Roter Liste, Senatsverwaltung für Stadtentwicklung/Der Landesbeauftragte für Naturschutz und Landschadftspflege (Hrsg.). Berlin 2001. 85 S

Müller Av, Müller-Muci Kv, Nekuda V (1993) Die Keramik vom Burgwall in Berlin-Spandau. Berliner Beiträge zur Vor- und Frühgeschichte. Neue Folge 8 (Archäologisch-historische Untersuchungen in Spandau 4):1–268

Naegeli O, Thellung A (1905) Die Flora des Kantons Zürich. I. Teil: Die Ruderal- und Adventivflora des Kantons Zürich. Vierteljahrsschrift Naturforschende Gesellschaft Zürich 50:225–305

RFKB (Regionalstelle für die floristische Kartierung Berlins, ed.) (2009) Verbreitungsatlas des wildwachsenden Farn- und Blütenpflanzen Berlins n. p

Schouw J F (1823) Grundzüge einer allgemeinen Pflanzengeographie. Berlin

Scholz H (1960) Die Veränderungen in der Ruderalflora Berlins. Ein Beitrag zur jüngsten Florengeschichte. Willdenowia 2:379–397

Sukopp H (1990) (ed) Stadtökologie – Das Beispiel Berlin. Berlin

Sukopp H (2003) Flora and vegetation reflecting the urban history of Berlin. Die Erde 134(3):295–316

Thellung A (1917) Stratiobotanik. Vierteljahrsschrift Naturforschende Gesellschaft Zürich 62:327–335

Willdenow CL (1787) (Reprint 1987) Florae Berolinensis Prodromus, Berlin

Wittig R, Diesing D, Gödde M (1985): Urbanophob – Urbanoneutral – Urbanophil. Das Verhalten der Arten gegenüber dem Lebensraum Stadt. Flora 177:265–282

Zohary M (1962) Plant life of Palestine, Israel and Jordan. Chronica Botanica 33:262 pp

Bratislava

Viera Feráková and Ivan Jarolímek

Fig. 1 Bratislava Castle above the Danube River

Abstract Bratislava was formed in 907 AD on the boundary of the Carpathian and Pannonian regions; in 2008, the city, which occupied 367.9 km^2 and had a population of 425,540. The first published records of the flora of Bratislava can be traced to that of Clusius in 1583, followed in 1791 by Lumnitzer and in 1830 by Endlicher, who recorded 1,192 vascular plants. In recent years almost 1,700 vascular species have been recorded. About 27% of the species and sub-species of the 1,270 included in the national Red List are or have been recorded in the city. Bratislava contains 2 Ramsar Sites, 10 Natura 2000 Sites, 1 National Nature Reserve, 1 National Nature Monument,

Viera Feráková (✉)
Institute of Botany, Slovak Academy of Sciences, Dúbravská cesta 9,
SK-84523, Bratislava, Slovak Republic
e-mail: viera.ferakova@savba.sk

J.G. Kelcey and N. Müller (eds.), *Plants and Habitats of European Cities*,
DOI 10.1007/978-0-387-89684-7_4, © Springer Science+Business Media, LLC 2011

8 Nature reserves, 20 Protected Areas and 3 Nature Monuments. In addition five Important Plant Areas have been identified in the surroundings of the city. So far as the non-vascular plants are concerned the algal flora has been thoroughly studied, many genera and species of Cyanobacteria and algae new to science have been described. 291 taxa of lichenised fungi have been recorded in the city; 139 are epiphytic, 98 epilithic and 54 terrestrial. Bratislava is rich in fungi, recent studies have recorded the occurrence of 113 taxa of parasitic micro-fungi on the leaves of vascular plants and 64 taxa of powdery mildew. It is expected that most, or even all, of the 479 taxa of macromycetes reported from the Podunajská nížina Lowland occur in Bratislava.

Natural Environment of the City

Location

As a result of changes in national boundaries, Bratislava is one of many European cities that is also known by several names (Prešporok, Pressburg and Pozsony). It is situated close to the centre of Europe, at the crossroads of the ancient Amber and Pannonian trading routes and on the boundary between the Malé Karpaty Mountains (the western most part of the Carpathian mountain range in Slovakia) and the Pannonian Lowlands; see Fig. 2.

The geographical co-ordinates of the centre of the city are 48°08′30" North and 17°06′30" East; the elevation ranges from 126 to 514 m a.s.l. The city lies within two phytogeographical regions, the Pannonian and the Carpathian and four phytogeographical districts (from the north Záhorská nížina Lowland, Malé Karpaty Mountains, Devínska Kobyla Hill and Podunajská nížina Lowland). The mighty River Danube connects Bratislava with the capitals of Austria (upstream) and Hungary (downstream). Both the Austrian and Hungarian borders are situated just a "stone's throw" from the city centre.

Geology

The geological bedrock of Bratislava varies. The oldest is the crystalline complex of the Malé Karpaty, which comprises mainly muscovite-biotite granite and granodiorite, with high mica content. Croan pegmatite and veins of aplites are abundant. Amphibolite and mica slate gneiss occur in the valley of the Vydrica brook while remnants of Lower Jurassic rocks occur in the municipal part of Devín and Devínska Kobyla Hill; these deposits include limestone, dolomite, sandstone and sand. In the Pleistocene, the watercourses that flowed south-east towards the Podunajská nížina Lowland formed huge sand-gravel alluvial fans on which part of Bratislava (Rača) is built. Loamy and loamy-rock deposits of the Holocene period outcrop in the valleys of the watercourses that drain the Carpathians.

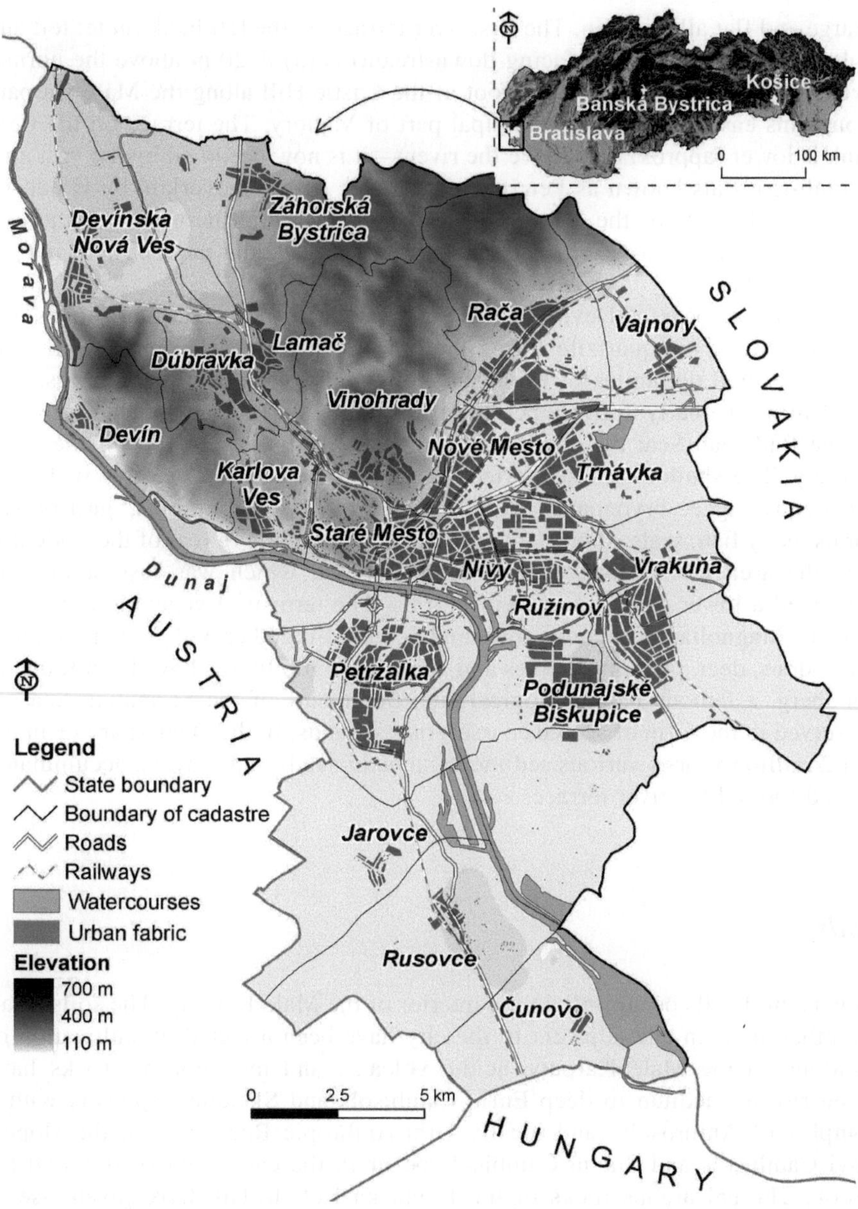

Fig. 2 Map of Bratislava

The eastern and southern parts of the city (which lie on the western part of the Podunajská nížina Lowland) are built on horizontally bedded calcareous clay, sand and gravel that overlie a crystalline complex. The Mesozoic deposits are overlain by a thick layer of Holocene sediments deposited by the River Danube, which after breaking through the Devínska brána (brána = gate) formed a large and flat alluvial fan. The first river terrace on the left bank (note: left and right banks are defined by facing downstream) is (5) 8–20 m above the normal river level and extends from the foot of the Castle Hill along the Malé Karpaty mountains eastwards to the municipal part of Vajnory. The terrace on the right bank is lower (approx. 5 m above the river) – it is now occupied by the vast area of high-rise flats known as Petržalka. Mesozoic deposits overlain by Holocene alluvial sediments of the Morava River occur in the northern part of the city between Devínska Nová Ves and Záhorská Bystrica – the most eastern part of the Vienna basin.

The oldest rocks of Devínska Kobyla are Palaeozoic crystalline schists and granites. Mesozoic quartzite forms the top of the Devínska Kobyla Hill. The limestones and dolomites of Devínska Kobyla also originated in the Mesozoic sea. During the early Tertiary period, a substantial part of the area was covered by the Badenian Sea; the shore was along the south-western foot of the Malé Karpaty. The shallow sea supported many plant and animal species including calcareous algae, bryozoans, worms, foraminifera, sea urchins, gastropods, corals, bony fish, seals, sharks and whales. Following the retreat of the Badenian Sea, the area was influenced by Sarmatian Sea, which was less saline and contained a lower diversity of living forms. The terrestrial vegetation included Cycads, Magnolias, Cinnamon trees and Sequoias together with animals such as Mastodons, deer, primitive horses and apes. Over 300 fossils have been found at Sandberg, a famous paleontological site. Sediments of the Sarmatian Sea are preserved in the form of scarce coarse-grained sands. In the Quaternary (± in the last 2 million years), various sediments, mostly sands and gravels, accumulated on and formed the river terraces.

Soils

Now natural soils occur only in the interior of the Malé Karpaty. The soils of all the other areas in and adjacent to the city have been affected by cultivation or building. In the Malé Karpaty, acidic volcanic and metamorphic rocks have given rise to medium to deep Euric Cambisols and Skeletic Leptosols with a complex of Anthrosols (and locally Urbi-Anthropic Regosols) on the slopes. Luvi Cambisols and Stagni-Cambisols occur in the catchment of the Vydrica Brook. The calcareous rocks of the Devínska Kobyla Hill have given rise to Rendzic Leptosols or locally Lithic Leptosols. Various Anthrosols are scattered throughout the whole city. In the Podunajská nížina Lowland, Haplic Chernozems and Gleyi-Haplic Chernozems prevail in the west and east with Eutric Fluvisols in the central area. Gleyic Fluvisols (locally Gleyisols) occur close to the Danube River. In the Borská nížina Lowland, Eutric Cambisols and Mollic Fluvisols are found. Near the Morava river, Gleyic Fluvisols have developed with Regosols occurring in small areas.

Climate

Bratislava lies in the North Temperate Zone and has a continental climate with four distinct seasons. It is often windy with a marked variation between hot summers and cold, humid winters. The city is situated in one of the warmest and driest parts of Slovakia. Recently, the transitions from winter to summer and summer to winter have been rapid, with short autumn and spring periods. The north-east/south-west orientation of the Malé Karpaty mountain range (400–500 m a.s.l.) has a major influence on the climate of the city. The ridge is at right angles to the prevailing north-west wind; consequently, the most extensive part of the city, which lies in the Podunajská nížina Lowland, is the warmest and driest with a mean annual temperature of 10.3ºC and a mean annual precipitation of 573 mm. Slightly less warm and dry is the Borská nížina Lowland, which is situated on the windward north-west of the city. The most humid and cold area is the Malé Karpaty with a mean annual temperature of 9.5ºC and mean annual precipitation of 661 mm. The coldest month is January (–1.1 to 2.0°C), and the warmest is July (19.3–21.0°C). The urban heat-island effect (which is more noticeable in the winter) results in a temperature increase of 1.0–1.5°C in the city compared with the surrounding area. The average duration of the sunshine is 50% = 2,100 h/ year (one of the longest in Slovakia); in the vegetation period, it is 60% = 1,600 h.

Drainage

The city is located within the catchment of the River Danube and its major tributary, the Morava (which forms the boundary between Slovakia and Austria), into which the smaller watercourses of the Malé Karpaty and the lowlands drain. The run-off from the city is discharged through a wastewater treatment plant into the Danube.

Historical Development of the City

First Settlement to AD 1200

People first settled in the general area in the Early Stone Age (Palaeolithic); however, the territory of the present city was not occupied until the early Iron Age. In the first century AD, Romans entered the area, using the Danube as their northern border. Their camp Gerulata, which is in today's Rusovce part of the city and on the right bank of the Danube remained there until the fourth century. Devín castle is mentioned for the first time in 864 in the Chronicle of Fulda as the impregnable fortress Dowina, which like Gerulata formed a part of the fortified frontier system called *Limes Romanus*. It served as a strategic outpost of the 14th and 15th Roman legion in Caraunt (Carnuntum) in today's Austria. Around AD 400, the area was settled by groups of indigenous people probably of Quadus origin. From the period of migration of nations (375–568), there is a precious archaeological find of European importance – a carbonised loaf of bread discovered in Devín in an outdoor oven. During the fifth and sixth centuries, Slavs and the nomadic Avars moved in. At the beginning of the ninth century, the Slavs established the Great Moravian Empire, which lasted up to 908. In the second half of the ninth century (AD 863), the brothers Kirillos and Methodius arrived from Thessaloniki and played an important role in the religious and cultural development of the area, including an attempt to introduce the Cyrillic alphabet that they had devised. The frontier stronghold at Devin survived throughout the tenth century and was still in existence in the thirteenth century; however, during the following centuries, it lost its strategic importance, although it remained quite well preserved until it was destroyed by Napoleonic forces in 1809.

According to historical sources, the first record of today's Bratislava (the official name since 1919) is most probably 907, when in the Annals of Salzburg (*Annales luvavenses*) *"Bellum pessiumum apud Preslavaspurce (Brezalauspurce) fuit 4, nonas Julii"* is mentioned. The name Preslava, from which evolved the Slovak "Prešporok" and the German "Pressburg", originates from the castle of Preslav (Braslav). Preslav was the third son of the King Svätopluk (871–894). The Hungarian name Pozsony was probably derived from the Slavonic duke Božaň, who owned the castle from 1052 to 1099. The latinised form *Posonium* (used up to 1526) and the Greek form Istropolis often occur in the literature.

AD 1200 to 1900

In 1291, Bratislava was recognised as a free royal city of 33,000 inhabitants. In 1465, Slovakia's first university, Academia Istropolitana, was founded by King Matthias Corvinus; unfortunately, it ceased to exist after his death in 1490. Following the Turkish victory over the Hungarians and the Turkish invasion of Buda in 1536, the capital of Hungary, the Hungarian Chamber and the Palatine's Council, the seat of the Diet (Parliament) and the place of coronations of Hungarian kings was relocated to Pressburg (11 kings and 8 consorts were crowned in the city between 1563 and 1830). The reign of Maria Theresa of Austria (1740–1780), who was also crowned in St. Martin's Cathedral in Pressburg in 1741, is regarded as the golden era of the city. In 1850/51, the population of the city was 42,238.

AD 1900 to 1945

In 1918, following the break up of the Austro-Hungarian Empire, the First Czechoslovak Republic was established, followed in 1938 to 1939 by the Second Czechoslovak Republic. During the period 1939 to 1945, Bratislava was the capital of the first Slovak state under the supervision of Hitler's Germany.

AD 1945 to 2000

After the Second World War and the re-establishment of the Czechoslovak Republic in 1945, especially as the result of the political change in 1948, the city started to expand. The historical city centre was declared an Urban Conservation Area in 1954. However, the greatest expansion occurred in the following decades with large housing estates being constructed in eastern and western parts of the city and later on the right bank of the Danube (an extensive forest of high-rise flats known as Petržalka). From 1960 to 1990 during the existence of the Czechoslovak Socialistic Republic (ČSSR) and in 1990 to 1992 of the Czechoslovak Federative Republic (ČSFR), Bratislava was the economic and political centre of Slovakia. On January 1, 1993, when Czechoslovakia was divided into two states (Slovakia and the Czech Republic), Bratislava became "an old capital of a new sovereign state". The building boom (partly with foreign capital) as well as the reconstruction of the historical centre that followed (and continues) have resulted and is resulting in the loss of open spaces. The current administrative system was introduced in 1996. The present Bratislava covers an area of 367.6 km^2 comprising 141.2 km^2 agricultural lands, 67.1 km^2 developed land and 159.4 km^2 mainly occupied by water and forests. The city is divided into 5 boroughs, 17 municipal parts, 20 cadastral territories and 262 urban units. In 2006, the population was 426,091 but by 2008 it had declined slightly to 425,540. The proportion of Slovaks in the population is about 90% with Hungarians and Czech comprising 4.6% and 1.6%, respectively. Other nationalities, including Croatians, make up the remaining 3.8%. The expansion of the city is shown in Fig. 3.

Changes of the Environment Due to the City Growth

A comparison of the map of natural potential vegetation (Fig. 4) with the map of the main groups of biotopes (Fig. 5) provides a clear indication of the effect of long-term human activities on the vegetation and flora of the city.

The comparison shows that only about 30% of the original forest cover has remained, namely, most of the *Fagus* forests in the highest parts and some *Carpinus betulus* forests on the upper slopes of the Malé Karpaty and elsewhere some small remnants of *Salix-Populus* and *Quercus-Ulmus-Fraxinus* forests. Two-thirds of the original forests have been cleared, about one-third for agricultural and about one-third for housing and associated developments. Large areas of the *Carpinus betulus* forests on the slopes of the Malé Karpaty have been converted to vineyards while the lowland forests were clear-felled and the land used for crop production. However, the total diversity of plant communities and species has increased. But it is "poisonous biodiversity" caused by the penetration of adventive species that now occur in anthropogenous habitats. In the worst cases, some neoindigenophytes invade remnants of the forest communities resulting in a significant detrimental change in their floristic composition. The most aggressive species in Bratislava forests are the herbs, *Impatiens glandulifera, I. parviflora, Aster lanceolatus, Solidago gigantea, S. canadensis* and three tree species *Acer negundo, Ailanthus altissima* and *Robinia pseudoacacia.*

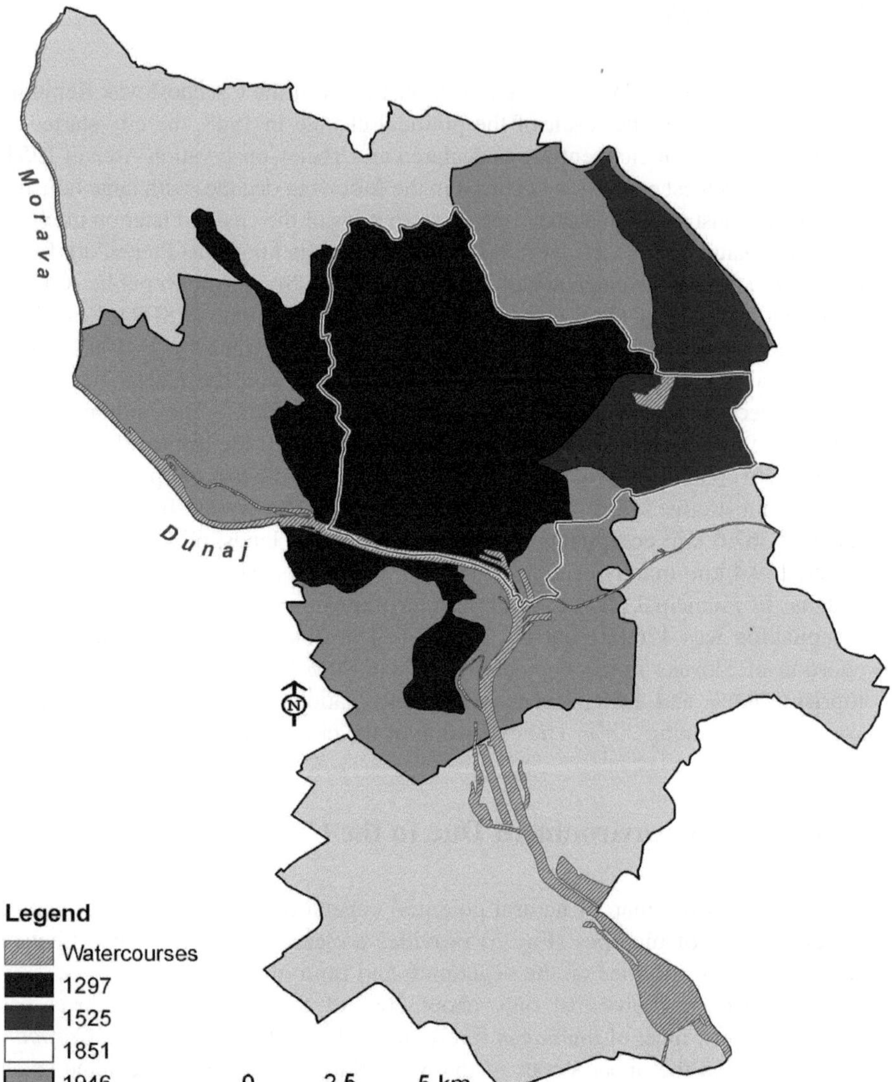

Fig. 3 Expansion of the city

Flora

Where taxa are not included in the 3rd edition of Stace's "New Flora of the British Isles", the nomenclature of the vascular taxa follows www.hear.org/gcw with some exceptions from Marhold and Hindák (1998). The classification of synanthropophytes corresponds with Holub (1971).

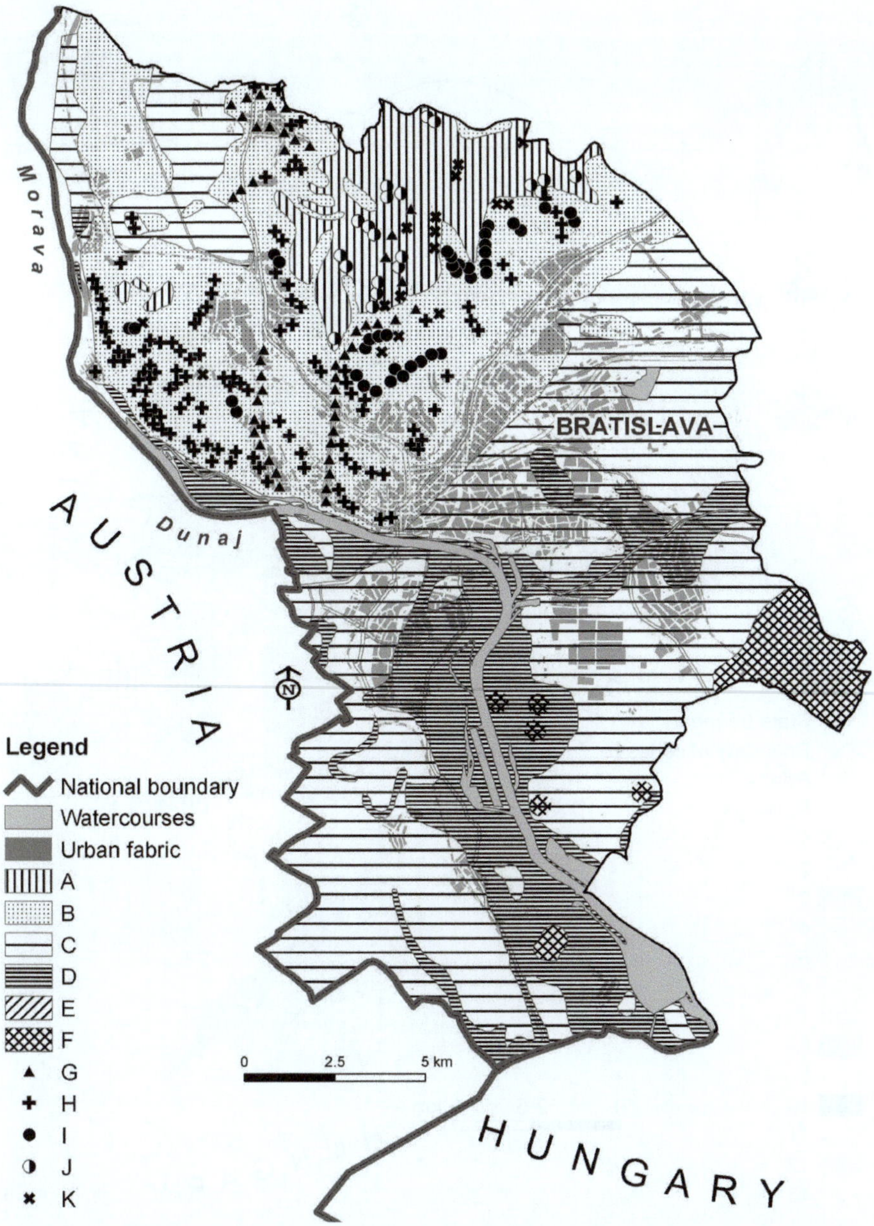

Fig. 4 Map of the natural potential vegetation. Key: A –*Fagus* forests (submontane, montane and calciphilous); B – Carpathian *Quercus-Carpinus* forests; C – *Fraxinus-Ulmus-Quercus* floodplain forests (hardwood alluvial forests); D – *Salix-Populus* floodplain forests (softwood alluvial forests); E – Pontic-Pannonian *Quercus* forests; F – *Alnus-Fraxinus* submontane floodplain forests; G – *Quercus* forests with *Quercus cerris;* H – Acidophilous *Quercus* forests; I – Acidophilous *Fagus* forests and J – *Tilia-Acer* forests and montane *Acer* forests

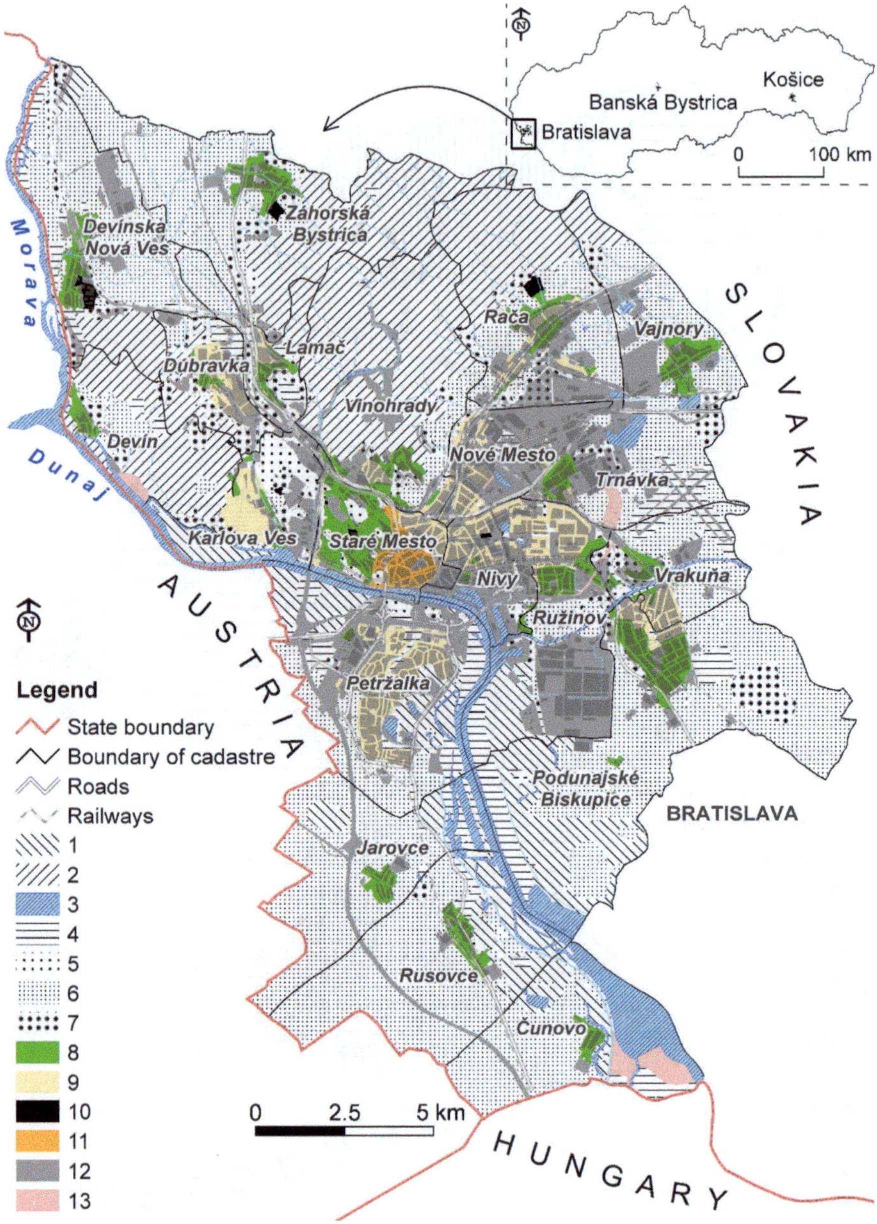

Fig. 5 Map of the main groups of biotopes in Bratislava (Modified from Hrnčiarová et al. 2006). Key: 1. Floodplain forest biotope, 2. Deciduous Carpathian forests biotope (mainly *Fagus* and *Carpinus-Quercus* forests), 3. Aquatic biotope, 4. Grassland biotope, 5. Park biotope, 6. Field, vineyard, balks and terrace biotopes, 7. Orchard, garden and "weekend house" biotopes, 8. Garden suburb and village biotopes, 9. High-rise residential biotope 10. Low-rise residential biotope, 11. Biotope of the historical centre, 12. Biotope of administrative-business and 13. Quarry and exposed ground biotopes

Vascular Plants

History of Botanical Studies

The oldest records of plant species are those of Clusius, who visited Bratislava in 1574. There are some unpublished eighteenth century data obtained by Windisch and Csapó; however, the main sources of early botanical information are the *Flora Posoniensis* (Lumnitzer, 1791 – a physician in Bratislava) and *Flora Posoniensis* (Endlicher, 1830 – Professor of Botany and Director of the Botanical Garden, University of Vienna). The city archive contains the manuscript of a third flora of Bratislava, which was written by Mergl in 1903. Between 1830 and 1899, Heuffel, Dietl, Bolla, Holuby, Kornhuber, Brancsik, Wiesbaur, Sabransky and others published many important papers. Twentieth century botanists including F. A. Novák, Domin, Scheffer, Szép, Gáyer, Ptačovský, Valenta, J. Novák, Eliáš, Feráková, Gojdičová, Jarolímek, Hajdúk, Hodálová, Kocianová, Kochjarová, Kothajová, E. Králik, T. Králik, Krippel, Krippelová, Letz, Lux, Májovský, Záborský, Bertová, Michálková, Schwarzová, Bartková, Kaleta, Maglocký, M. Michalko, Somogyi, Sadloňová and a large number of others added to the body of botanical information. The thorough botanical research of Hodálová and her co-authors (Hodálová et al. 1996, 1999) and Letz (2000) is of particular importance in contributing to the understanding of the Bratislava flora. Letz commented on the changes of the flora of Bratislava over the course of 200 years since the publication of the *Flora Posoniensis* by Lumnitzer using the present taxonomic treatment, converting the names of the sites from the original German or Hungarian to the current Slovak names and resolving some problematic and dubious records. As a result of the field studies by Ondrášek and the precise locations of Valenta's old records, several species considered to be extinct or on the verge of extinction were re-discovered, although unfortunately most were only small populations or scattered individuals (Ondrášek 2002).

For further information, see the bibliographies in Futák and Domin (1960), Feráková and Kocianová (1997) and Feráková (2002).

In *Flora Posoniensis,* which covers Bratislava and its wider surroundings, Endlicher recorded 1,574 species of which 1,192 were vascular plants. The approximate number of taxa at specific and sub-specific levels recorded in Bratislava in 1994 was 1,300; by 2004 a further 400 had been added. The 1,700 species and sub-species include also frequent ephemerophytes and 140 regionally extinct and probably extinct taxa. Twenty-two of the 140 were later re-discovered while the occurrence of a further 10 threatened and declining species was not confirmed; these include, *Aphanes arvensis, Artemisia pontica, Bassia laniflora* and *Xanthium strumarium.*

The exact number of plant taxa present in Bratislava is not known, many partial inventories and records have yet to be collated and indexed. Examples are as follows:

1. One thousand five hundred and seventy species and sub-species of vascular plants (including 110 casuals) have been recorded in the phytogeographical district of Devínska Kobyla (49 km^2).

2. Five hundred and thirty-five taxa were recorded in 1.5 km^2 of natural and semi-natural habitats in the lower part of the Morava floodplain.
3. Four hundred and fifty-five taxa have been reported from the 12.7 ha Nature Reserve of Štokeravská vápenka, a former limestone quarry and lime kiln.
4. Five hundred and fifty taxa have been recorded in the Nature Reserve Kopáčsky Ostrov.
5. Four hundred and ninety species were found in the Nature Reserve Ostrovné lúčky.

The number of taxa reported from different parts of the selected urban areas include the following:

1. Historic Centre – The Castle Hill of Bratislava – 292 taxa at specific and sub-specific levels including 32 bryophytes; the Castle Hill of Devín 453 taxa, mainly anthropophytes.
2. Zones of high-rise buildings and new housing estates – The housing estate of Ostredky 687 taxa, which include hemerophytes cultivated in the local parks. Petržalka, mainly ruderal sites and many abandoned plots in the housing areas – 370 taxa; subsequent surveys increased the number to 580.
3. Zone of family houses and the peripheral zone – 1,025 species (555 synanthropic plants: 305 apophytes, 146 archaeophytes, 97 neophytes (*sensu lato*) and 7 taxa of problematic classification) were recorded in the municipal part of Devín and the adjacent area. In the municipal part of Lamač, 551 taxa have been found.
4. Industrial zone – 30 years ago 442 taxa were recorded from an area of 3.5 km^2 occupied mainly by the chemical industry on the east side of Bratislava. Some of the species were facultative halophytes from wet, slightly saline meadows, which no longer exist.
5. Zones of peri-urban fields, vineyards and urban forests – these areas have not been thoroughly studied yet, but judging from preliminary investigations the number of vascular species in these biotopes varies around 400.

Since the political events in 1989, the parts of Bratislava with rural character have been subjected and are subject to continual change as a result of progressive urban development. Many gardens used for the cultivation of crops have been converted to the growing of plants for decorative purposes, abandoned areas partly disappeared and a rapid decline of urban and rural weeds has been observed. According to the investigations carried out by botanists in Austria between 1990 and 1995, some species of the rural flora have become extinct or their populations have substantially declined as a result of urbanisation. Holzner (in Holzner et al. 1994) estimated "*...dass die Tage der Dorfpflanzen gezählt sind*", for example, *Chenopodium murale* has become a rare species. Changes in the flora of Bratislava are some years behind those described in Austria, but that is likely to change as a result of rapid urbanisation, including the development of natural areas, the creation of uniform lawns and the planting of exotic species, all of which will impoverish the diversity of the native flora. The data given above indicate that no complete evaluation of the species-richness of the individual zones exists. The study areas were of different sizes and the depth of the taxonomic treatment differed as well. An approximate number of species based on the data from the urban area (peripheral zone and the zone of family houses) is up to 100/km^2, while in the protected areas with the highest landscape value it is > 600 species/km^2.

Most probably the following taxa have their only occurrence in Slovakia in Bratislava, namely, in the phytogeographical district of Devínska Kobyla; they are *Artemisia austriaca, Conringia austriaca, Ononis pusilla, Orobanche artemisiae-campestris, O. coerulescens, Peucedanum arenarium* ssp. *arenarium* and *Rhamnus saxatilis* ssp. *saxatilis.* Some species are important from the phytogeographical point of view, because they reach the limit of their natural distribution in Bratislava, for example, *Peucedanum arenarium* ssp. *arenarium* and *Rhamnus saxatilis* ssp. *saxatilis* are close to the margin of their natural distribution, for example, *Vinca herbacea* (with a site close to the absolute north-western limit of its broad distribution area). In terms of the altitude limits, *Fagus sylvatica* occurs in Bratislava at its lowest elevation in the west Carpathian mountains. Re-discovered taxa include *Typha shuttleworthii* already considered extinct in Slovakia (see the Black List in Čeřovský et al. 1999) and *Adonis flammea, Ophrys apifera, Tragus racemosus* and *Vitis sylvestris* which were categorised as extinct or missing at the regional level in 1994.

Taxa with *locus classicus* in Bratislava

The following taxa have their *locus classicus* in Bratislava, *Dianthus praecox* ssp. *lumnitzeri,* a west Carpathian-Pannonian sub-endemic (Fig. 6); *Hieracium echioides, Sempervivum hirtum* f. *glabrescens* and *Taraxacum danubium.*

Fig. 6 Devín Castle Rock (National Nature Monument), *locus classicus* of *Dianthus praecox* ssp. *lumnitzeri*, a species of the NATURA 2000 network, which was described as new to science by Wiesbaur in 1886

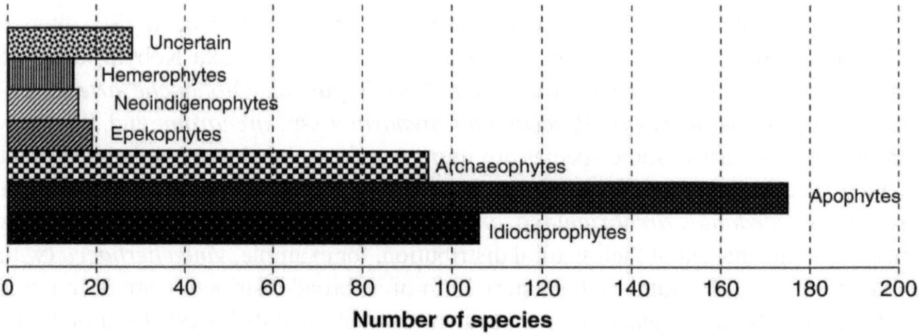

Fig. 7 Spectrum of vascular plants on Devín Castle Hill. Source, Feráková (1995). The total number of recorded species 453 (580 excluding extinct taxa, ephemerophytes, micro-species and hybrids)

Synanthropic Species

Hajnalová in 1994 analysed 69 archaeological sites in the old centre of the city and found the seeds and carbons of 82 species, of which 43 were apophytes and 49 were anthropophytes. An analysis of the present-day composition of the urban flora of the oldest parts of Bratislava has shown that it comprises an almost equal number of apophytes and anthropophytes, while in the newer quarters of the city anthropophytes, mainly neophytes, dominate. A high proportion of archaeophytes was recorded at the site Devín Castle Hill (including National Nature Monument Devín Castle rock); see Fig. 7.

Xenophytes

Archaeophytes

The fairly common archaeophytes in Bratislava include *Asperugo procumbens, Berteroa incana, Lepidium draba, Descurainia sophia, Lepidium ruderale, Mercurialis annua, Onopordum acanthium* and *Stachys annua.* The many archaeophytes that occur only rarely include species such as *Adonis flammea, Bifora radians, Bupleurum rotundifolium, Caucalis platycarpos, Chenopodium vulvaria, Glaucium corniculatum, Kickxia spuria, Kickxia elatine, Anchusa arvensis, Marrubium peregrinum, M. vulgare, Misopates orontium, Nigella arvensis, Papaver argemone, Ranunculus arvensis* and *Reseda phyteuma.*

Neophytes

Some of the neophytes that are the most widespread today are mentioned in the floras of Lumnitzer and Endlicher, for example, *Acer negundo* and *Erigeron annuus* (one of the first records for Europe). In 1856, Bolla recorded *Solidago canadensis*

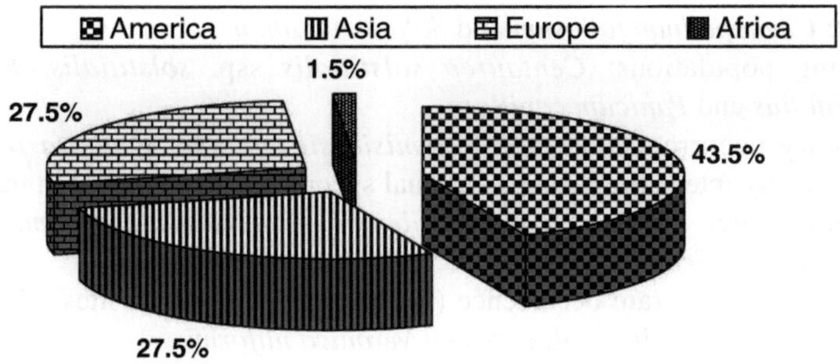

Fig. 8 Spectrum of neophytes of Bratislava, according to the geographical origin (Source, Feráková and Skrovná 1998)

only as a garden escape; today, its large populations together with those of *S. gigantea* and *Impatiens glandulifera* represent one of the main threats to the biodiversity of the Danube floodplain forests. On the basis of published records, the flora of the city contains 160 neophytes (mostly of North American origin) of which 79 are classified as epekophytes or neoindigenophytes, see Fig. 8. Data from the other parts of the city show that the number of taxa in the advanced stages of naturalisation increased to 125, comprising 76 epekophytes and 49 neoindigenophytes. When assessed in terms of Raunkiaer's life forms, therophytes are the most frequent (47.7%). There are now about 130 ephemerophytic taxa of which 75 occur frequently, mainly ergasiophygophytes.

One of the ephemerophytes, the Mediterranean-Central Asian species *Centaurea solstitialis* ssp. *solstitialis,* which disappeared from all historical sites of its secondary distribution in southern Moravia and Slovakia, has persisted as a naturalised species for over 200 years on the southern slope of the Bratislava Castle Hill. In recent years, only solitary individuals have been found; however, seed for restocking is being stored in the Botanical garden.

The ergasiolipophytes found in Bratislava include those used for dyeing: for example *Rubia tinctorum,* which was recorded in Bratislava by Clusius in 1583 but never confirmed again. On the other hand, *Isatis tinctoria* ssp. *tinctoria* (an archaeophyte from the Celtic period or from the Middle Ages) still persists and has recently been found spreading extensively in the ruderal semi-natural thermophilous communities at the Devín Castle Hill and elsewhere, especially on the banks of the Danube. *Chenopodium foliosum* and *Tetragonia tetragonoides* were both cultivated as vegetables, although rarely. The aromatic plant *Smyrnium perfoliatum,* which was known as a spice, is probably a naturalised relict from Roman times. The majority of its contemporary sites in Slovakia and the Czech Republic are situated in parks or in the vicinity of castle ruins although around Bratislava viable populations occur in the forests.

On the basis of an assessment of their contemporary occurrence, the alien species of Bratislava can be divided into six groups for which some examples are given:

1. Extinct: *Chenopodium foliosum* and *Solanum alatum.*
2. Declining populations: *Centaurea solstitialis* ssp. *solstitialis, Erechtites hieraciifolius* and *Panicum capillare.*
3. Temporary occurrence/casuals: *Artemisia sieversiana, Dinebra retroflexa.* A floristically interesting group of casual species are those found in the food of cage birds, such as *Fagopyrum esculentum, Panicum miliaceum, Phalaris canariensis* and *Setaria italica.*
4. Stable but of rather rare occurrence (present in less than ten sites): *Oxybaphus nyctagineus, Sisymbrium volgense* and *Veronica filiformis.*
5. Relatively stable and of frequent occurrence: *Cuscuta campestris* and *Juncus tenuis.*
6. Increasing in occurrence: *Ambrosia artemisiifolia, Asclepias syriaca, Aster lanceolatus, Echinocystis lobata, Erigeron annuus* ssp. *annuus* and ssp. *septentrionalis, Oxalis corniculata, Senecio vernalis* and *Veronica peregrina* ssp. *peregrina.*

Invasive, Potentially Invasive and Expansive Species

The classification of this group is in accordance with the "List of Alien, Invasive Alien and Expansive Native Vascular Plant Species of Slovakia" (Gojdičová et al. 2002). During the middle of the twentieth century, the main vectors of plant migration into Czechoslovakia were ships, trains and road traffic along three migration routes – west from the Ukraine through the railway station Čierna nad Tisou; the Elbe, from the German river ports to Bohemia and the Pannonian or the Danube route, by railway to the town of Štúrovo and by ship to the river ports of Komárno and Bratislava. Examples of species recorded at railway stations include *Iva xanthiifolia, Bassia scoparia* ssp. *densiflora, Lepidium virginicum, Panicum miliaceum* ssp. *ruderale, P. dichotomiflorum, Setaria faberi* and *Tribulus terrestris.* Alien species associated with the port of Bratislava include *Artemisia repens, Lactuca tatarica, Salsola collina* and *Sporobolus cryptandrus.* Probably the majority of well-established aliens that occur in large populations in the Danube floodplain reached Bratislava via barges, for example, *Bidens frondosa, Impatiens glandulifera, Echinocystis lobata, Xanthium albinum, Solidago canadensis* and *S. gigantea.* Other species such as *Archangelica officinalis, Brassica nigra* and *Mimulus guttatus* occur sporadically, while various cultivars of *Lycopersicon esculentum* and *Citrullus lanatus* can be found frequently, especially on the riverbanks. A new record of interest is the occurrence of *Datura inoxia* on the islands of the Danube. Some aliens, such as *Amaranthus viridis, Dinebra retroflexa, Rapistrum rugosum* and *Sisymbrium volgense,* were recorded around factories, mills and similar places, having been introduced with foreign goods.

In the late 1940s and early 1950s, Opluštilová began a floristical investigation of the synanthropic flora of Bratislava. In the following years, she devoted her studies to the phytosociology of segetal and ruderal communities.

Table 1 The most frequent epekophytes and neoindigenophytes found in Bratislava

Ailanthus altissima	*Fallopia × bohemica**
Amaranthus powellii	*Fallopia japonica**
Amaranthus retroflexus	*Geranium pyrenaicum*
Artemisia annua	*Helianthus tuberosus*
Asclepias syriaca	*Impatiens glandulifera**
Aster lanceolatus	*Impatiens parviflora*
Aster novi-belgii	*Iva xanthiifolia*
Ambrosia artemisiifolia	*Lycium barbarum*
Bidens frondosa	*Oxalis stricta*
Bunias orientalis #	*Robinia pseudoacacia*
Conyza canadensis	*Solidago canadensis**
Echinocystis lobata	*Solidago gigantea**
Erigeron annuus ssp. *annuus*	*Xanthium albinum*
Erigeron annuus ssp. *septentrionalis*	

*– Taxa included in the Order No. 24/2003 of the Ministry of Environment of the Slovak Republic Annex 2 "List of invasive species of plants and ways of their elimination as dangerous invasive plants", # –According to some authors an archaeophyte, which arrived in the Middle Ages

Table 2 Species frequently escaping from cultivation (ergasiophygophytes)

Amaranthus caudatus	*Nigella damascena*
Anethum graveolens	*Papaver somniferum*
Brassica napus ssp. *oleifera*	*Persicaria orientalis*
Calendula officinalis	*Phacelia tanacetifolia*
Citrullus lanatus	*Phalaris arundinacea*
Convolvulus tricolor	*Physalis* spp.
Cosmos bipinnatus	*Ricinus communis*
Duchesnea indica	*Rudbeckia hirta*
Fagopyrum esculentum	*Satureja hortensis*
Foeniculum vulgare	*Scilla luciliae*
Helianthus annuus	*Secale cereale*
Hemerocallis fulva	*Sedum hispanicum*
Lactuca sativa	*Sedum sarmentosum*
Lobularia maritima	*Silybum marianum*
Lunaria annua	*Tagetes patula*
Lycopersicon esculentum	*Tanacetum parthenium*
Melissa officinalis	*Triticum aestivum*
Narcissus spp.	*Viola* x *wittrockiana*

Of the neoindigenophytes that are widespread in the city, 11% represent an acute danger to the biodiversity of protected and sozologically important territories. The most frequent epekophytes and neoindigenophytes that occur in Bratislava are given in Table 1. Examples of the most frequent garden escapes, which are listed in Table 2, mainly occur in waste places such as rubbish tips, compost heaps and similar habitats. The alien taxa recorded in Bratislava since c.1970 include *Amaranthus deflexus, Archangelica officinalis* sensu lato, *Chenopodium pumilio, Chorispora tenella, Commelina communis, Datura inoxia, Dinebra retroflexa, Elodea nuttallii, Epilobium komarovianum, Juncus dudleyi, Lemna minuta, Mimulus guttatus, Panicum dichotomiflorum, Phytolacca esculenta, Senecio inaequidens, Solanum scabrum, Sporobolus cryptandrus, Typha laxmannii* and *Viola sororia.*

Plants and Public Health

Some neophytes are pernicious weeds of arable land and must be controlled by the Phytoquarantine Service; such species include *Amaranthus* spp., *Ambrosia artemisiifolia, Panicum miliaceum* ssp. *agricola* and *Cuscuta campestris;* some of these species also stimulate allergic responses, for example, *Ambrosia artemisiifolia, Artemisia* spp. and *Solidago* spp. Species with allergenic properties occur elsewhere in Bratislava, for example, Drábová-Kochjarová (1990) recorded 68 species of vascular plants in Petržalka that caused pollinosis. One hundred taxa (from a total of 535) that cause allergic reactions have been found in the floodplain area of the River Morava in the municipal part of Devín and Devínska Nová Ves by Feráková in the following years. This has resulted in good collaboration between botanists and medics (Hrubiško, Zlinská, Bartková) in relation to identifying allergenic species. For example, so far 64 taxa (26 trees, 24 grasses and 14 weeds) that are important sources of allergens have been subject to clinical tests.

Most allergenic species occur on derelict land that has been colonised by ruderal communities: *Lolio-Plantaginetum, Erigeronto-Lactucetum serriolae, Hordeetum murini, Tanaceto-Artemisietum, Elymo repentis-Sisymbrietum loeselii, Sisymbrio-Atriplicetum nitentis, Cynodonto-Atriplicetum tataricae, Echio-Melilotetum, Chenopodietum stricti* and *Odontito-Ambrosietum artemisiifoliae.* In the peripheral areas of the city, *Bromopsis erecta* (a species of secondary thermophilous grassland) is the most significant allergen. *Phalaris arundinacea* and *Phragmites australis* are the most frequent producers of allergenic pollen in wetland communities and along the Rivers Danube and Morava and as apophytes in new housing estates.

The sources of the most important pollen allergens (both native and introduced plants) in Bratislava can be divided into three seasonal main groups:

1. Vernal catkin-flowering woody species, for example, *Alnus glutinosa, Corylus avellana, Betula alba* and *Populus* spp.
2. Grasses and herbs in the city lawns and in semi-natural plant communities, including *Lolium perenne, L. multiflorum, Phleum pratense, Cynodon dactylon, Bromopsis erecta, Parietaria officinalis, Rumex* spp. and *Urtica dioica.*
3. Late summer flowering and autumnal species (dominantly neophytes) such as *Ambrosia artemisiifolia, Artemisia vulgaris, A. annua, Solidago canadensis* and *S. gigantea.*

Spontaneous Vegetation in the Urban Area

Spontaneous vegetation, which is vital and robust without demanding any special care in urban conditions, still occurs on the southern slope of Castle Hill and in the peripheral parts of the city. The most frequently occurring trees and shrubs in abandoned cemeteries, neglected parks, wasteland and similar areas are to a large extent the same as in other central European cities, *Ailanthus altissima, Clematis vitalba, Rubus fruticosus* agg., *Sambucus nigra, Salix caprea, Lycium barbarum* and *Robinia pseudoacacia*. The 50 most frequent taxa (all herbs) recorded in the city are given in Table 3.

Green Spaces and Protected Trees

According to Reháčková and Pauditšová (2006), Bratislava has the lowest human population and the highest area of green space of any comparable European city, for example, Dublin, Helsinki, Antwerp, Leipzig, Birmingham and Zürich. However, the city is the second lowest in terms of the provision of parks. The reason for this apparent paradox is that most of Bratislava's green space is provided by the peri-urban

Table 3 The 50 most frequent taxa recorded in Bratislava

Aegopodium podagraria	*Erodium cicutarium*
Alliaria petiolata	***Galinsoga parviflora***
Amaranthus retroflexus	*Galium aparine*
Amaranthus powellii	*Geranium pyrenaicum*
Anthriscus cerefolium ssp. *trichosperma*	*Glechoma hederacea*
Artemisia vulgaris	*Hordeum murinum*
Aster lanceolatus	*Lactuca serriola*
Ballota nigra	*Lamium maculatum*
Bellis perennis	*Lolium perenne*
Anisantha sterilis	*Melilotus officinalis*
Capsella bursa-pastoris	*Papaver rhoeas*
Carduus acanthoides	*Pastinaca sativa*
Chelidonium majus	*Picris hieracioides*
Chenopodium album agg.	*Poa annua*
Cichorium intybus	*Portulaca oleracea*
Cirsium arvense	*Plantago major*
Convolvulus arvensis	*Polygonum aviculare* agg.
Conyza canadensis	*Stellaria media* agg.
Crepis biennis	*Tanacetum vulgare*
Dactylis glomerata	*Taraxacum* sect. *Ruderalia*
Daucus carota	*Trifolium repens*
Echium vulgare	*Tripleurospermum inodorum*
Elytrigia repens	*Urtica dioica*
Eragrostis minor	*Veronica hederifolia* ssp. *lucorum*
Erigeron annuus agg.	***Veronica persica***

Bold = neophytes

forests with only about 1.5% provided by parks. The authors studied the species composition of the urban green spaces in the north-western part of Bratislava. They found that two-thirds of all of the trees comprise ten species (from a total of 112 taxa, excluding cultivars). The cultivated shrubs are strongly dominated by *Cotoneaster dammeri, Pyracantha coccinea* and *Lonicera pileata.* In the past *Mespilus germanica, Cydonia oblonga, Sorbus domestica* and *Morus nigra* were planted as solitary trees in vineyards, hedgerows and occasionally in sub-urban gardens. By the second half of the last century, the planting of these species became less popular and their distribution and frequency declined as a result. However, as often occurs in horticulture, they are becoming popular again and are being offered for sale in the catalogues of plant suppliers. The trees, shrubs and climbers that have been planted alongside roads and in the parks of Bratislava can be divided into three groups – native species, naturalised woody aliens and introduced and cultivated species. The list of relatively frequently occurring species (excluding cultivars) is given in Table 4. A relatively high proportion of introduced and naturalised species occurs in the peri-urban forests to the detriment of the natural vegetation. *Tilia cordata* (the national tree of the Czech and Slovak Republics) has been frequently planted as solitary trees, as rows and avenues and near to sacred and historic buildings.

Twenty six trees of native (various cultivars) and foreign origin are legally protected, with two exceptions; they all occur in the old core of the city. They include three trees of *Quercus robur;* one *Q. dalechampii;* one *Q. virgiliana;* one *Fraxinus excelsior;* one *Taxus baccata;* two *Sophora japonica;* one *Magnolia* × *soulangeana;* one *Sorbus domestica;* one *Platanus occidentalis* and one *Liriodendron tulipifera.*

Table 4 Trees, shrubs and climbers planted along the roadsides and in the parks and other green spaces of Bratislava

A. Native in Slovakia	
Abies alba	*Ligustrum vulgare*
Acer campestre	***Picea abies*** #
Acer platanoides*	*Pinus sylvestris*
Acer pseudoplatanus*	***Populus*** spp.
Betula pendula	*Prunus padus*
Cerasus avium	*Quercus* spp.
Carpinus betulus	*Sorbus aria* agg.
Colutea arborescens #	*Sorbus aucuparia**
Cornus mas	*Cornus sanguinea*
Euonymus europaeus	*Taxus baccata*
Fagus sylvatica	***Tilia cordata****
Fraxinus ornus #	*Ulmus glabra*
Fraxinus excelsior	*Viburnum lantana*
Hedera helix	*Viburnum opulus*
Larix decidua	

The seven most frequent species are given in bold type
* – Planted along roadsides and in avenues
# – Not native in the territory concerned

(continued)

Table 4 (continued)

B. Woody alien species naturalised

Acer negundo*	*Lonicera caprifolium* #
Ailanthus altissima !	*Lycium barbarum* !
*Aesculus hippocastanum**	*Mahonia aquifolium*
Amorpha fruticosa	*Morus alba*
*Celtis occidentalis**	*Prunus serotina*
Fallopia baldschuanica !	*Parthenocissus inserta*
Juglans nigra	*Prunus cerasifera**
Juglans regia*	*Robinia pseudoacacia* !
Laburnum alpinum	*Rubus armeniacus*
Laburnum anagyroides	

* – Planted along roadsides and in avenues

! – Invasively behaving mainly in the zone of low buildings and peripheral zone, where it penetrates into semi-natural communities

– Native occurrence not excluded

C. Introduced cultivated woody species

*Acer saccharinum**	*Koelreuteria paniculata**
Abies concolor	*Kolkwitzia amabilis*
Acer palmatum	*Laurocerasus officinalis*
Aesculus × *carnea*	*Liquidambar styraciflua*
Albizia julibrissin	*Liriodendron tulipifera*
Aucuba japonica	*Lonicera* spp.
Berberis spp.	*Maclura pomifera*
Biota orientalis	*Magnolia* spp.
Brousonettia papyrifera	*Mespilus germanica*
Buddleja davidii!	*Morus nigra*
Buxus sempervirens	*Paulownia tomentosa* <
Campsis radicans	*Philadelphus coronarius*
Caragana arborescens	*Physocarpus opulifolius*
Caryopteris × *cladonensis*	***Picea pungens***
Castanea sativa	***Pinus nigra* ssp. *nigra***
*Catalpa bignonioides**	*Platanus* × *hispanica**
Cerasus serrulata	*Populus* × *canadensis*
Cercis siliquastrum	*Prunus serrulata*
Chaenomeles japonica	*Pseudotsuga menziesii*
Chamaecyparis spp.	*Ptelea trifoliata*
Cornus alba	*Pyracantha coccinea*
Corylus maxima	*Quercus* spp.
Cotoneaster dammeri #	*Rhodotypos scandens*
Cotoneaster horizontalis	*Rhus typhina*
Crataegus spp.	*Rosa rugosa*
Deutzia scabra	*Sophora japonica*
Diospyros lotus	*Sorbaria sorbifolia*
Elaeagnus angustifolius	*Spiraea* spp.
Euonymus fortunei	*Symphoricarpos albus*
Ficus carica $	*Syringa vulgaris*
Forsythia spp.	*Tamarix* spp.
Fraxinus americana	*Tetradium danielli*

(continued)

Table 4 (continued)

Fraxinus pennsylvanica	*Thuja occidentalis*
Gleditsia triacanthos	*Tilia platyphyllos*
Hibiscus syriacus	*Viburnum farreri*
Hippophaë rhamnoides	*Viburnum rhytidophyllum*
Hortensia petiolaris	*Weigela* hybrids
Ilex aquifolium	*Wisteria sinensis* (invasive)
Jasminum nudiflorum	
Juniperus spp.	
Kerria japonica	

* – Planted along roadsides and in avenues

! – Seedlings often appear in gaps of walls and pavements

– Most often from all evergreen shrubs planted as ground cover plant

$ – Sparsely cultivated in gardens, some fruiting trees are found especially in front of south and west facing walls

< – Spreading especially in the old town by the lightweight seeds. Many young plants of various ages can be found on abandoned or neglected places on wall bases, margins of asphalt and concrete pavements, in the soil deposits. A pioneer plant tolerant for pollution, not enough hardy for cold winters. Reported as invasive tree from elsewhere

Non-vascular Plants

Cyanophytes and Algae

The first records of the algal flora of Bratislava were published by Endlicher in 1830, who found 15 species of algae (including two species of *Chara*). The diatoms of the Bratislava area were investigated by Pantocsek in the very early years of the twentieth century. However, comprehensive studies of the algal flora, mainly in connection with investigations of the River Danube and the flooded gravel pits, did not start until the 1950s/1960s. The investigations included floristic, taxonomic and hydrobiological studies and the effect of algae on water management. The first checklist of Cyanophytes and algae (the current taxonomic approach is to separate the algae (eukaryotes) from the Cyanophytes (prokaryotes)) found in the Slovak stretch of the Danube comprised 590 infrageneric taxa belonging to 191 genera. During the period 1982 to 1994, the presence of most taxa was confirmed again – 123 genera with 273 infraspecific taxa. Twenty-seven genera with 145 species, 17 varieties and 2 formas were recorded for the first time from this section of the river. Altogether 218 genera with 693 species, 53 varieties and 9 formas have been recorded in the Bratislava section of the Danube, which is about a fifth of all taxa found in Slovakia (525 genera with 3,441 species and infraspecific taxa).

The River Morava represents an important source of algal inoculum for the Danube. Marvan et al. (2004) recorded 25 genera of Cyanophytes with 58 species and 181 genera of several classes of algae with 634 species and infraspecific taxa in the Morava close to its confluence with the Danube. Algal diversity is relatively very rich; for example, in the small innundation lakes close to the Morava in Devín 578 species and 58 non-type varieties and formas of Cyanophytes and algae have been recorded recently (Hindák and Hindáková in Feráková and Kocianová 1997).

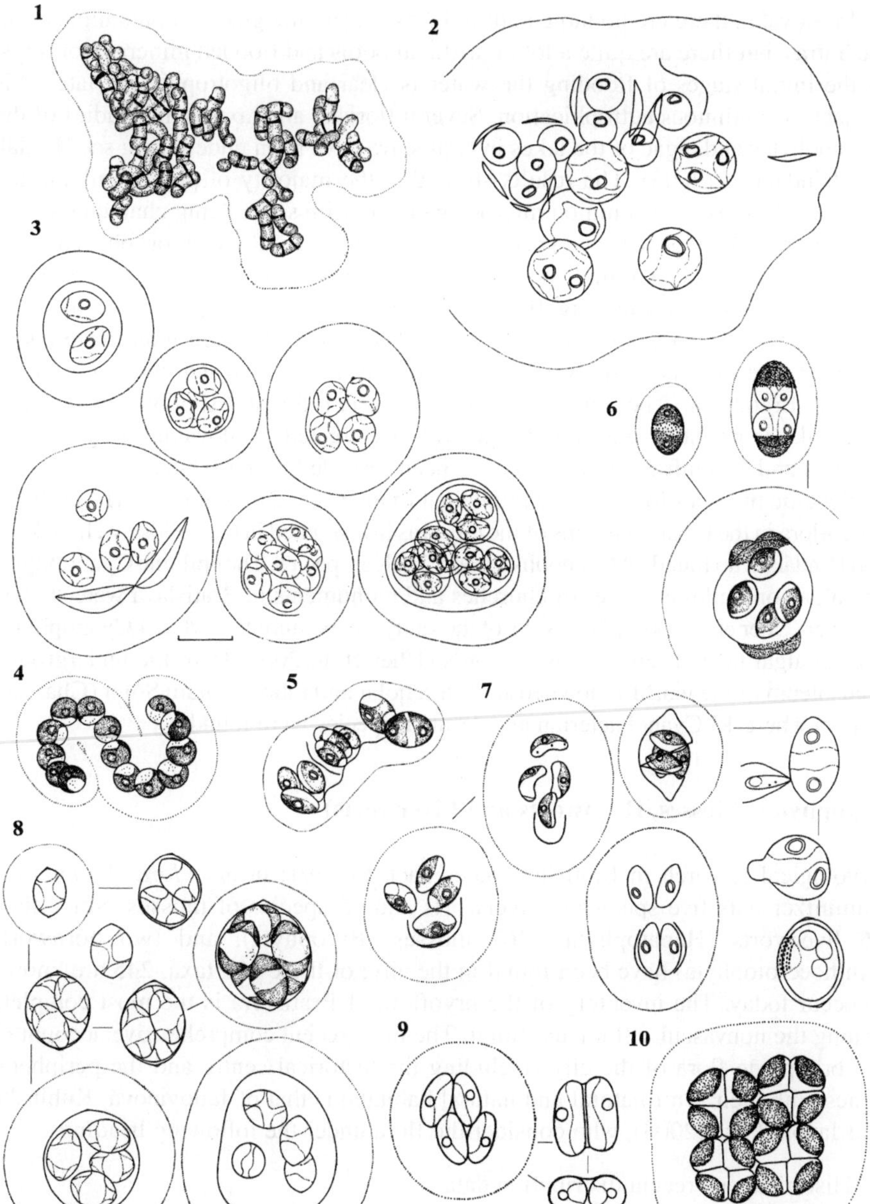

Fig. 9 Some taxa of Cyanophytes and algae described originally from gravel pit lakes in Bratislava. Key: 1. *Cyanocatena planctonica* Hindák, 2. *Coenochloris astroidea* Hindák, 3. *Coenocystis stellata* Hindák, 4. *Catenococcus tortuosus* Hindák, 5. *Coenocystis gerulata* Hindák, 6. *Granulocystis exuviata* Hindák, 7. *Oocystella oogama* Hindák, 8. *Oocystis biplacata* Hindák, 9. *Tetrachlorella incerta* Hindák, 10. *Tetrastrum komarekii* Hindák (From Hindák and Hindáková 2003)

In Slovakia there are no large natural lakes, except the glacial mountain lakes in the Tatras, but there are quite a lot of artificial pools and flooded mineral workings. In the initial stages of flooding the water is clear and oligotrophic but later it is subject to continuous eutrophication. Several floristic and taxonomic studies of the Cyanophytes and algae of the lakes in Bratislava have been undertaken; see Hindák and Hindáková (2003). The results show that the majority of phytoplankton and phytobenthos taxa are common micro-organisms with some being characteristic of the biotope. Many genera and species of Cyanophytes and the chlorophyceae new to science have been found in Bratislava while several taxa were recorded in Slovakia for the first time (Fig. 9). Altogether more than 600 species belonging to 187 genera of Cyanophytes and algae have been found in these man-made lakes. Another type of man-made water body is the fishpond, of which there are only a few in Bratislava. The microflora of four small forest fishponds at Železná Studienka is probably the most frequently investigated in Slovakia; 38 genera of diatoms and 308 infraspecific taxa have been recorded in these ponds.

Specific biotopes for algae are municipal fountains and monuments. The epilithic microflora in the urban fountains of the city was investigated in the 1990s by Hindáková and Hindák, who found 34 Cyanophytes and 23 algal species. The epilithic Cyanophytes and algae on building stones of churches and monuments in Bratislava were studied by Uher (Uher et al. 2006); the results of the study can be summarised as 4 Cyanophytes and 23 algal taxa. There are also records (Uher et al. 2005) from the underground mausoleum of the world-renowned Jewish scholar and rabbi Chatam Sófer (Chassam Sojfer) where 13 Cyanobacterian and 48 algal species were found.

Bryophytes (Mosses, Hornworts and Liverworts)

Bryological research in Bratislava dates back to 1791; in his *Flora Posoniensis,* Lumnitzer lists five species of liverworts and 65 species of mosses. Since then 66 liverworts (Hepatophyta), 267 mosses (Bryophyta), and two hornworts (Anthocerotophyta) have been found in the city; of these 337 taxa, 245 are known to occur today. The inventory of the bryoflora of Bratislava is the most complete among the nonvascular flora and fungi. The most recent comprehensive account of the bryophyte flora of the city (including the historical centre and the peripheral zones with their semi-natural and natural habitats) is that of Janovicová, Kubinská and Javorčíková (2003), who consider the flora under the following headings:

1. Historical and recent distribution data
2. Occurrence of threatened and rare bryophytes
3. Important habitats for bryophytes
4. Bryophyte communities
5. Results of karyological research

The threatened and rare species include *Asterella saccata, Athalamia hyalina, Acaulon muticum, Phascum curvicollum, P. floerkeanum, Rhynchostegium rotundifolium, Plagiothecium latebricola, Aphanorhegma patens, Pseudephemerum nitidum, Cinclidotus fontinaloides, C. riparius* and *Fissidens crassipes*. The presence of the rare liverwort *Asterella saccata* (which is listed in Annex II of the EC Habitat Directive) was taken into account when designating the Important Plant Area (IPA) Devínska Kobyla. The most important habitats for bryophytes in Bratislava are the xeric calcareous grasslands, floodplain forests in flooded and non-flooded areas, acidophilous oak forests, historical cemeteries and parks.

The characteristic species found in some of the habitats include the following:

1. Forests: *Atrichum undulatum, Polytrichum formosum, Plagiothecium* spp., *Plagiomnium* spp., *Dicranella heteromalla, Jungermannia gracillima, Lophocolea heterophylla, Buxbaumia aphylla, Dicranum polysetum, Diphyscium foliosum.*
2. Parks: *Plagiomnium* spp., *Eurhynchium hians, E. schleicheri, Brachythecium rutabulum, Calliergonella cuspidata, Climacium dendroides, Cirriphyllum piliferum.*
3. High density residential areas: *Tortula muralis, Bryum argenteum, Barbula convoluta, Pseudocrossidium hornschuchianum, Amblystegium serpens.*
4. Low density residential areas: *Tortula muralis, Bryum argenteum, Orthotrichum anomalum, Schistidium* spp., *Rhynchostegium murale, Cirriphyllum piliferum, Funaria hygrometrica, Bryum capillare.*
5. Historic centre: *Tortula muralis, Bryum argenteum, Funaria hygrometrica, Bryum capillare, Amblystegium serpens, Rhynchostegium murale, Ceratodon purpureus, Bryum ruderale, Eurhynchium hians, Phascum cuspidatum.*

Fungi (including Lichenised Fungi)

Fungi

Until relatively recently, the fungi were placed in the Plant Kingdom but they are now in a Kingdom of their own. There are about 100,000 scientifically recognised species. While some species are beneficial to human health (for example, species of *Penicillium* as producers of antibiotics, and yeasts for various human actitivities), some cause symptoms in plants and animals (including humans) called "disease", for example, "rusts" in cereal crops, Dutch Elm Disease and potato blight. For practical reasons, the fungi are frequently classified according to size into microscopic fungi – "micromycetes" (visible under a magnifying glass or microscope) and macroscopic fungi called "macromycetes" (visible by the naked eye; they are usually fungi that produce fruiting bodies and/or stromata).

Micromycetes

The first records of the microscopic fungi found in Bratislava and its environs were published in the floras of Lumnitzer and Endlicher. Endlicher listed 374 species of cryptogams including some species of the genus *Erysiphe.* Twenty-seven years later,

Bolla reported the presence of 258 taxa of fungi, including *Erysiphe aquilegiae, Microsphaera alni, Phyllactinia suffulta, Sphaerotheca fuliginea, Trichocladia euonymi, Uncinula salicis*. Most records were contributed by J. A. Bäumler, who published between the years 1887 and 1927 in the "Verhandlungen des Vereins für Natur- und Heilkunde zu Pressburg" five parts of "Beiträge zur Cryptogamen Flora des Pressburger Comitates". His collections consisted of 4,700 sheets including 77 species, which had not been previously found in Bratislava and its vicinity.

Parasitic microfungi in the environs of Bratislava have been subject to many studies; for example, 113 taxa of parasitic fungi were found on the leaves of vascular plants of Bratislava. Sixty-four taxa of powdery mildew were discovered in the plant communities of the Devínska Kobyla region. Another study in the same area reported 447 taxa and 11 Myxomycetes; a further 14 species were added as the result of subsequent research. A comprehensive account of the Erysiphales of Bratislava and its surroundings can be found in Flora Slovakia by Paulech in 1995.

As a result of recent global and regional ecological changes and an increase in the quality of the air, some species of microfungi, which were considered to be extinct or declined under anthropogenic pressure, have been re-discovered. Their re-appearance at sites known at the beginning of the twentieth century has been confirmed after 50–100 years; for example, *Taphrina rhizophora,* which was reported by Bäumler in 1899, was found in 2006 by Bacigálová, who also discovered *Protomyces buerenianus,* a species of the mycobiota that is new to Slovakia.

Macromycetes

Watching crowds of people with baskets getting on the train from Bratislava to visit the nearby pine forests in the Záhorská nížina lowland, it would seem that no macroscopic fungi can grow in the city. However, the contrary is true. Besides *Boletes,* which mainly lure mushroom-pickers to the Záhorie region, many other fungi grow there making Bratislava very rich in species of macromycetes. As described earlier, Bratislava is the meeting point of four phytogeographical districts, the different vegetation types of which represent suitable habitats for many common as well as rare and threatened fungi. While there are only scattered data on fungi of the Záhorská nížina Lowland in Bratislava, the evidence of mycological observations in the Devínska Kobyla region is more or less complete, resulting in 428 taxa of macroscopic fungi being recorded (Ripková and Ďuriška, 2009), most (379) reported by Záhorovská in Feráková and Kocianová (1997) using data from various previous collectors. *Helvella phlebophora* was recorded here for the first time in Slovakia and for *Flammulina fennae* it was only the second record. Considering the small area of Devínska Kobyla, the fungal diversity is rather high, 17.8% of the 2,609 taxa of macromycetes reported from all Slovakia (Adamčík et al. 2003). The Slovak "Red-listed species" that occur in the Devínska Kobyla region include *Coltricia montagnei* (Vulnerable), *Gyromitra fastigiata* (Lower Risk/Least Concern), *Leucopaxillus lepistoides* (Endangered) and *Urnula craterium* (Lower Risk/Near Threatened (categories according to Lizoň 2001).

The interest in fungi of the Podunajská nížina Lowland in Bratislava has increased since the fall of the "iron curtain" in 1989, when the guarded national borders along the Danube and Morava were opened to the public. A mycological inventory of Slovanský Ostrov and Sihoť Islands by Jančovičová in 2000 in her Ph.D. thesis, resulted in 211 taxa of macromycetes being found, some of which are included in the Slovak Red List of Fungi (Lizoň 2001), for example, *Abortiporus fractipes* (Vulnerable), *Crepidotus crocophyllus* (Vulnerable), *Entoloma byssisedum* (Lower Risk/Least Concern), *Hyphodontia latitans* (Endangered), *Hypoxylon ticinense* (Data Deficient), *Lentinus degener* (Lower Risk/Least Concern), *Phlebia ryvardenii* (Endangered and found for the first time in Slovakia on Sihoť Island), *Pluteus aurantiorugosus* (Lower Risk/Near Threatened), *Rhodotus palmatus* (Endangered) and *Stereum subtomentosum* (Data Deficient). Species-richness can be found also at other localities; for example, *Omphalina discorosea* (Vulnerable) was recorded on the Pečňa Island and the rare *Coprinus strossmayeri* and *Botryobasidium aureum* were recorded in the municipal parts of Rusovce and Petržalka, respectively. Because of the many other undisturbed or minimally disturbed areas in and adjacent to the city, for example, the Nature Reserve Kopáčsky Ostrov and in the Protected Landscape Area of Dunajské luhy, it can be expected that most or all of the 479 macromycetes reported from Podunajská nížina Lowland occur in Bratislava.

The most comprehensive data on the macromycetes of Bratislava (including those occurring in the Malé Karpaty) were published by Janitor in 1996 and 1997. He mentions ca. 350 taxa including some Slovak Red List species, for example, *Gyromitra fastigiata* (Lower Risk/Least Concern), *Boletus impolitus* (Vulnerable), *B. satanas* (Endangered) and *Xerocomus parasiticus* (Endangered).

Lichenised Fungi

The area of Bratislava is, from a lichenological viewpoint, quite important. The location of the city at the junction of the Pannonian lowland and Carpathian mountains as well as a great variation of substrates provides basic pre-conditions for a relatively large number of lichens – so far 284 taxa have been recorded. The first records of the presence of lichens in Bratislava are those of Lumnitzer (1791). Other researchers of the nineteenth century, for example, S. Endlicher, J. von Bolla and A. Zahlbruckner (1884, 1889) mainly concentrated on the adjacent forests of the Malé Karpaty mountains. In the twentieth century J. Suza, I. Pišút, E. Lisická, A. Lackovičová, A. Guttová, V. Slezáková contributed to the knowledge about the local lichens. A detailed treatment of the history and results of lichenological research of this area is given by Lackovičová (1978). At the end of the twentieth century, the studies were expanded to determine the diversity of lichen biota in noteworthy localities (for example, Lisická in Feráková and Kocianová 1997) and the use of lichens as indicators of pollution (see Pišút and Lackovičová 1991). Recent research has been related to the importance of epiphytic lichens in assessing the impact of changes in air quality on living organisms (Lackovičová et al. 2008).

Epiphytic Lichens

Suitable conditions for the 139 taxa (including Extinct: 2, Critically Endangered: 15, Endangered: 11 and Vulnerable: 12) occur in the uninhabited city outskirts, mainly in *Quercus* forests at lower altitudes and *Fagus* forests at higher altitudes of the Malé Karpaty mountains. Historical data and voucher specimens witness the presence of once rich communities of foliose and fruticose lichens, which are susceptible to microclimatic conditions and air quality, for example, *Anaptychia ciliaris, Lobaria pulmonaria, Ramalina fastigiata, R. fraxinea* and *Usnea florida*. As a result of poor air quality in the second half of the twentieth century (the concentration of pollutants peaked in the 1970s), these species became extinct while the frequency and distribution of many others, including *Evernia prunastri, Flavoparmelia caperata, Graphis scripta, Melanelia glabra, Pleurosticta acetabulum* and *Punctelia subrudecta,* declined significantly.

At that time the situation in the central parts of the city was even worse. The research in the municipal parks revealed the paucity of epiphytic lichens: Horský park (close to the city centre) – 4 species; Sad Janka Kráľa park (by the Danube) – 24 species and the park in Rusovce (on the outskirts of the town) – 34 species (Hajdúk et al.1975).

At the turn of the twentieth/twenty-first centuries, the first positive changes in epiphytic lichen flora were noted. The improvement is continuing with the recolonisation of habitats and a total increase in species diversity – 81 species have now been recorded. However, most of the species are nitrophilous such as *Physcia adscendens, P. dubia, Physconia grisea, Phaeophyscia orbicularis, P. nigricans, Amandinea punctata* and *Xanthoria parietina*. Medium-sensitive species, for example, *Hypogymnia physodes, Parmelia sulcata* and *Melanelia fuliginosa* are less frequent. Fruticose species (which are more sensitive to air quality), for example, *Evernia prunastri, Usnea* spp. continue to be rare (Lackovičová et al. 2008). It is the epiphytic lichens that provide the contemporary cover of the trees in the parks and alleys.

Terrestrial Lichens

More than 50 species of terrestrial lichens have been recorded in Bratislava. The following elements are important from a phytogeographical viewpoint – *Cladonia magyarica* (a Pannonian endemic), sub-Mediterranean elements *Fulgensia fulgens* and *Squamarina lentigera* or ephemeral species *Biatorella fossarum* and *Solorinella asteriscus*.

Epilithic Lichens

From the end of the nineteenth century to the present time, 98 species have been noted in Bratislava. However many former localities have and continue to be developed, trampled or overgrown. Such events adversely affect the less frequently recorded species, for example, *Lecania erysibe, Rhizocarpon geographicum, Candelariella coralizza,* as well as the common ones, for example, *Xanthoparmelia conspersa, Neofuscelia pulla* and *Physcia caesia.* On the other hand, many species found alternative possibilities for colonisation, especially man-made substrates including concrete and brick walls, columns, grave stones, which support, for example, *Protoparmeliopsis muralis, Caloplaca decipiens, Candelariella aurella* and *Lecanora dispersa.*

Habitats

The location of Bratislava at the interface of two phytogeographical regions and four phytogeographical districts, on the banks of a major European river (and a significant tributary of it), in an area of considerable geological and geomorphological diversity, a variable topography and subject to major human influences have combined to produce an immense variety of biotopes. The names of the habitats/biotopes are according to EC Habitats Directive 92/43 and the Corine Biotopes Manual published at web site:

(http://biodiversity-chm.eea.europa.eu/information/document/F1088156525/F1125582140)

Forests

The woodlands within the city occupy 8,100 ha (22% of the area). They include the remnants of the natural mountain (Carpathian) and floodplain forests and the commercial forests.

Carpathian Forests

The Carpathian forests comprise two main hypsometrical zones (*Quercus-Carpinus* and *Fagus* forest) and several types of forest depending on the geology and water regime, see Fig. 5, biotope 2).

Central European Acidophilous *Fagus* Forests with *Luzula* (*Luzulo-Fagion;* EUH code: 9110; CORINE code: 41.11)

Fagus forests on siliceous acidic rocks (gneiss and Bratislava granite) with shallow skeletal Leptosols on slopes above 25°. The floristic composition is relatively poor because of the dense canopy; mainly *Fagus sylvatica* and exceptionally *Betula pendula*, *Quercus petraea* and *Q. dalechampii* occur. The shrub layer is frequently absent or comprises scattered young trees. The patchy herb layer comprises oligotrophic acidophilous species such as *Avenella flexuosa, Hieracium murorum* agg., *Luzula luzuloides, L. pilosa, Poa nemoralis* and *Polygonatum verticillatum.*

Neutrophilous *Fagus* Forests (*Fagion, Eu-Fagenion;* EUH code: 9130; CORINE code: 41.13)

The most widespread of the Carpathian *Fagus* forests on deeper fresh moist soils (Eutric Cambisols) with a well-developed humus horizon (mul-moder) on flat or slightly sloping ridge plateaux of the Malé Karpaty mountains in the northern part of the city. *Fagus* dominates the canopy; the shrub layer is usually absent. The herb layer is rich in typical sciophilous forest species such as *Asarum europaeum, Carex pilosa, Cardamine bulbifera, Lamiastrum galeobdolon, Galium odoratum, Lilium martagon, Melica nutans, M. uniflora, Mercuralis perennis, Pulmonaria obscura, Salvia glutinosa, Tithymalus amygdaloides* and *Viola reichenbachiana*. The herb layer is limited in some places by the accumulation of leaf litter.

Limestone *Fagus* Forests (*Fagion, Cephalanthero-Fagenion;* EUH code: 9150; CORINE code: 41.16)

This is a less frequent forest type; it occurs in the western part of the city (Devínska Kobyla Hill) on calcareous rocks with Rendzic Leptosols. The dominant canopy species is *Fagus sylvatica;* the other canopy species include *Acer platanoides, A. pseudoplatanus, Quercus petraea* and *Tilia cordata,* with *Cornus sanguinea* and young trees abundant in the shrub layer. Together with *Campanula persicifolia, C. rapunculoides* and *Carex alba,* some orchids such as *Cephalanthera rubra* and *Epipactis* spp. can also be found.

Ravine and Slope *Tilia* Forests (*Tilio-Acerion;* EUH code: 9180 (Priority Biotope); CORINE code: 41.45 *pro parte* (Only Partially Corresponds to 41.45)

Scattered azonal forests on steep slopes or screes in ravines; typical tree species include *Tilia cordata, T. platyphyllos, Acer platanoides* and *Fraxinus excelsior* with the occasional *Carpinus betulus, Fagus sylvatica* and *Ulmus glabra.* The shrub layer is usually well developed. The herb layer is rich in nitrophytes including *Alliaria petiolata, Aruncus vulgaris, Chelidonium majus, Geranium robertianum, Lunaria rediviva, Mercuralis perennis, Parietaria officinalis* and *Urtica dioica.*

Quercus-Carpinus Forests (*Carpinion betuli;* EUH code: 91G0 (Priority Biotope); CORINE code: 41.26)

In the past, *Quercus-Carpinus* forests were widespread below the *Fagus* zone on all the slopes (especially on the Cambisols to the south-west and south-east) of the Malé Karpaty – see Fig. 4). The main canopy species are *Quercus dalechampii,*

Q. polycarpa, Q. cerris, Q. pubescens (probably more frequent in the past) and *Carpinus betulus,* which has become more frequent in recent decades. Other typical but less frequent species include *Acer campestre* and *Cerasus avium.* Meso- and thermophilous species occupy the shrub layer, including *Cornus mas, Crataegus monogyna, Ligustrum vulgare, Euonymus verrucosa* and *Cornus sanguinea.* Grass and sedge species such as *Carex pilosa, Melica uniflora* and *Poa nemoralis* are dominant in some stands. The species-rich herb layer comprises *Campanula rapunculoides, Convallaria majalis, Galium odoratum, G. schultesii, Lathyrus niger, L. vernus, Melittis melisophyllum, Polygonatum multiflorum* and *Stellaria holostea.* Some species such as *Anemone ranunculoides, Corydalis cava, Cardamine bulbifera* and *Ficaria bulbifera* form conspicuously colourful stands in spring. In the Devínska Kobyla Hill and a few other sites, *Smyrnium perfoliatum* also occurs. Most of these forests have been converted to vineyards or are gradually being developed for housing (see map Fig. 5, biotope 6).

Pannonian Woods with *Quercus pubescens* (*Quercion pubescenti-petraeae;* EUH code: 91H0 (Priority Biotope), CORINE code: not described)

This rare type of relatively xerothermophilous forest occurs in the north-west of the city on the south oriented calcareous slopes and ridges of the Devínska Kobyla Hill. The dominant canopy species is *Quercus* (*Quercus pubescens* and *Q. petraea*) with some *Acer campestre, Carpinus betulus* and *Tilia cordata.* Where the canopy is open, *Cornus mas* and *Viburnum lantana* are frequent in the shrub layer. The herb layer comprises grasses and sedges such as *Brachypodium pinnatum, Carex humilis, C. michelii, Dactylis polygama* and *Melica uniflora,* with herbs including *Clinopodium vulgare, Lithospermum purpureocaeruleum, Melittis melisophyllum, Stachys recta, Teucrium chamaedrys* and *Viola hirta.* Thermophilous fringe communities including *Geranium sanguineum* and *Dictamnus albus* occur along the woodland margins (see Fig. 5 biotope 2).

Medio European Stream *Fraxinus-Alnus* Woods (*Alnion incanae, Alnenion glutinoso-incanae;* EUH code: 91E0 (Priority Biotope); CORINE code: 44.3)

Azonal hygrophilous *Fraxinus-Alnus* woodland occurs in narrow strips adjacent to mountain brooks (Vydrica and its tributaries and other watercourses) and in damp-wet depressions in the Malé Karpaty (distribution is similar to that in Fig. 4 (F)). The limiting ecological factors are a high watertable and/or periodical inundation which induce the development of alluvial gley soils. *Alnus glutinosa* and *Fraxinus excelsior* are the most frequent trees, mixed with *Prunus avium, Populus tremula, Salix purpurea* and *Carpinus betulus.* At the higher altitude *Alnus incana* becomes dominant. *Viburnum opulus, Euonymus europaea* and *Rhamnus cathartica* occur in the sparse shrub layer. The herb layer comprises hygrophilous and nitrophilous species such as *Aegopodium podagraria, Caltha palustris, Carex brizoides, C. remota* (dominant in some stands), *Chrysosplenium alternifolium, Equisetum sylvaticum, Lysimachia nemorum, Stellaria nemorum* and *Urtica dioica.*

Lowland Forests

In the 1990s, 1,336 ha of floodplain forests were felled as part of the construction of the Gabčíkovo Dam on the Danube, downstream of Bratislava, on the Slovakian and Hungarian borders (see Fig. 5, biotope 1).

Riparian *Salix-Populus* Gallery Forests (*Salicion albae;* EUH code: 91E0 (Priority Biotope); CORINE code: 44.13 + *Populus* Plantations CORINE code: 83.3211, 83.3212)

The last remnants of the once extensive alluvial *Salix-Populus* forests (see Figs. 4 and 5) occur adjacent to the Danube on habitats with the highest watertable and subject to periodic inundation. Before the construction of the Gabčíkovo Dam, this type of old riparian forest occupied a much larger area of the Danube floodplain. In addition to the native *Salix alba, S. fragilis, Populus nigra* and *P. alba,* some of the stands are dominated by the non-native *Populus* × *canadensis*. The typical shrub layer species include *Salix purpurea, S. triandra, Cornus sanguinea* and *Sambucus nigra.* Eutrophic and nitrophilous species tolerant of inundation are abundant in the herb layer, for example, *Calystegia sepium, Carex riparia, C. acutiformis, Galium palustre, Lysimachia vulgaris, Lythrum salicaria, Persicaria hydropiper, Rubus caesius* and *Urtica dioica*. In some places the spring-flowering geophytes *Ficaria bulbifera, Galanthus nivalis* or *Scilla vindobonensis* form conspicuous carpets. Unfortunately, several neoindigenophytes, such as *Aster lanceolatus, Impatiens glandulifera, Solidago canadensis* and *S. gigantea* have colonised and are rapidly colonising the floodplain forests where they form dense populations that are detrimental to the native species.

Riparian Mixed *Quercus-Ulmus-Fraxinus* Forests along rivers (*Alnion incanae, Ulmenion;* EUH code: 91F0; CORINE code: 44.4)

These forests lie in close proximity to the previous forest type. They occur on the slightly elevated river terraces, which are subject to less frequent and shorter duration floods or no flooding at all. The soils are eutrophic alluvial or brown earths. The dominant trees are *Fraxinus angustifolia, Quercus robur* and in some places *Carpinus betulus* and *Tilia cordata*. Before the European Dutch Elm Disease epidemic in the 1970s, *Ulmus laevis* and *U. minor* occurred more frequently. Nitrophilous, mesophilous and hygrophilous species are the predominant herb layers species, which include *Aegopodium podagraria, Alliaria petiolata, Allium ursinum* (in some stands it forms dense carpets), *Anemone ranunculoides, Campanula trachelium, Convallaria majalis, Corydalis cava, Gagea lutea, Glechoma hederacea, Ficaria bulbifera* and *Polygonatum latifolium.*

Grasslands

Alluvial Cnidium Meadows (*Cnidion venosi;* EUH code: 6440; CORINE code: 37.23)

Small areas of periodically flooded meadows occur on the alluvial floodplain of the Morava River in the northwest part of the city near the municipal part of Devínska Nová Ves. The grasses *Alopercurus pratensis, Agrostis stolonifera, Elytrigia repens* and *Poa pratensis* and the less represented sedges *Carex acutiformis, C. melanostachya, C. praecox* and *C. vulpina,* determine physiognomy of these meadows. *Allium angulosum, Clematis integrifolia, Cnidium dubium, Leucojum aestivum, Oenanthe silaifolia* ssp. *silaifolia* and *Plantago altissima* occur rarely although they are more abundant in similar, but less intensively exploited meadows further north from the city (see Fig. 5, biotope 4).

Mesophilous Lowland and Submontane Meadows (*Arrhenatherum elatius, Alopercurus pratensis;* EUH code: 6510; CORINE code: 38.22)

These floristically heterogeneous meadows are scattered in both the lowlands and on the slopes and ridge plateaux of the Malé Karpaty mountains. In the lowlands, organic manure is applied to some of the meadows. Typical heliophilous and mesophilous common species, such as *Achillea millefolium, Alopercurus pratensis, Arrhenatherum elatius, Crepis biennis, Dactylis glomerata, Festuca rubra, Galium mollugo, Centaurea jacea, Knautia arvensis, Lotus corniculatus, Phleum pratense, Plantago lanceolata, Poa pratensis, Ranunculus acris, Trifolium pratense* and *Veronica chamaedrys* characterise the floristical composition of these meadows.

Subxero-thermophilous Grasslands, Inclusive Rupicolous Calcareous Swards, Central European Steppic Grasslands – Important Orchid Sites, *Alysso-Sedion albi*-(Priority Biotope), *Festucion valesiacae* (Priority Biotope), *Cirsio-Brachypodion, Geranion sanguinei;* EUH codes: 6110, 6210, 6240; CORINE codes: 34.1, 34.3121)

These rare biotopes occur on the north-west side of the city on the south and west slopes of the Devínska Kobyla Hill and cliffs of Devín Castle. They also occur sporadically on rocky outcrops between vineyards on south-east slopes of the Malé Karpaty mountains above the municipal parts of Krasňany and Rača. Shallow soils, usually on calcareous rocks exposed to the sun for most of the day favour xero-thermophilous and heliophilous stress tolerant species. They cannot grow in other biotopes and therefore they substantially enrich the biodiversity of the city. The calcareous rocks are occupied by pioneer communities comprising *Allium senescens, Alyssum montanum* ssp. *montanum, Arenaria serpyllifolia,*

Cerastium pumilum, Acinos arvensis, Saxifraga tridactylites, Sedum acre, S. album, S. sexangulare, Veronica arvensis and *V. triphyllos*. Habitats with shallow proto-rendzinas support *Carex humilis, Festuca pallens, F. valesiaca, Fumana procumbens, Iris pumila, Koeleria macrantha, Scorzonera austriaca, Stipa capillata, S. joannis* and *Taraxacum* sect. *Erythrosperma*. Concave relief with deeper loess soils provides suitable conditions for *Adonis vernalis, Astragalus onobrychis, Brachypodium pinnatum, Bromopsis erecta, Chamaecytisus austriacus, Dorycnium germanicum, Onobrychis arenaria, Seseli annuum* and *Veronica austriaca*. The Sandberg part of the National Nature Reserve Devínska Kobyla is an exceptional locality – it is covered by rare psammophilous sub-xerophilous vegetation comprising species such as *Crepis foetida, Erysimum diffusum, Gypsophila paniculata, Minuartia glaucina, Peucedanum arenarium* and *Silene conica*.

Herbaceous Clearings (CORINE code: 31.871)

Large areas of the *Fagus, Carpinus-Quercus* and *Salix-Populus* stands are managed for commercial purposes; consequently, forest clearings are a common component of the landscape. The vegetation that develops is determined by the nature of the underlying rock, the disturbed soil, soil-moisture and soil-chemistry (especially nutrients). On calcareous and nutrient rich soils, communities of the alliance *Atropion* occur with typical eutrophic and nitrophilous species such as *Atropa belladonna, Cirsium oleraceum, Eupatorium cannabinum, Fragaria vesca, Salvia glutinosa, Sambucus ebulus, Senecio ovatus, Urtica dioica* and *Verbascum thapsus*. On granite with nutrient poor soils, plant communities of the alliance *Carici piluliferae-Epilobion angustifolii* develop, including such species as *Avenella flexuosa, Calamagrostis epigejos, Chamerion angustifolium, Galeopsis pubescens, Gnaphalium sylvaticum, Rubus hirtus* and *R. idaeus*. Spontaneous forest trees and tall shrubs such as *Salix caprea, Sambucus nigra* and, in higher elevation, *S. racemosa* occupy the older clearings that have not been subject to intensive maintenance for about 10 years or more.

Aquatic and Watersides (see Fig. 5, Biotope 3)

Moving Water – Rivers and Brooks (CORINE code: 24)

The most important running water habitats in the Bratislava region are the rivers Danube and Morava with their oxbows and tributaries, especially the Malý Dunaj River and the Vydrica Brook in the Malé Karpaty. The northern part of the Hrušov Reservoir (part of the Gabčíkovo Dam) lies within the southern part of the city. Recent studies (Jursa and Oťaheľová 2005; Oťaheľová and Valachovič 2006) have found that the main channel of the Danube supports only a few macrophytes, namely,

Potamogeton nodosus, P. pectinatus, P. trichoides, Spirodela polyrhiza and *Zannichellia palustris*. The open oxbows are much richer; the riparian species include *Butomus umbellatus, Cyperus fuscus, Iris pseudacorus, Myosotis scorpioides, Persicaria amphibia, Phalaris arundinacea, Phragmites australis* and *Typha latifolia*. The floating-leaf zone contains *Callitriche cophocarpa* and *Lemna minor* while the submerged zone includes *Elodea nutallii, Myriophyllum spicatum, Najas minor* and *Potamogeton* spp. (*crispus, nodosus, pectinatus, perfoliatus, pusillus, trichoides*).

Separated Oxbows (CORINE code: 22.13)

Isolated oxbows, which have been cut off naturally or artificially from the main channel are an important source of Bratislava's biodiversity. The most separated of the oxbows (Rybárske rameno, Chorvátske rameno, Rusovské rameno, Čunovské rameno) are situated on the right side of the Danube with only Biskupické rameno on the left. In addition to the typical aquatic macrophytes, numerous riparian hydrophytes and algae grow in these oxbows. The most frequent species are *Ranunculus circinatus, Butomus umbellatus, Ceratophyllum demersum, Cyperus fuscus, Iris pseudacorus, Lemna minor, Myosotis scorpioides* and *Myriophyllum spicatum*. The oxbows also support several rare species including *Hippuris vulgaris, Myriophyllum verticillatum, Nymphaea alba* and *Nuphar lutea*. The aquatic moss *Fontinalis antipyretica* and the macro-algae *Chara vulgaris* and *Nitella syncarpa* and the green filamentous alga *Spirogyra* sp. also occur, together with many other non-vascular species.

Fishponds (CORINE code: 22.42)

The four fishponds, which are along the Vydrica Brook in the Malé Karpaty mountains and near the Šprinclov majer, are used for fish husbandry and periodically discharge eutrophic water into the watercourse. As a result of eutrophication, the ponds support a high diversity of algae; see the algal section above. The macrophytic vegetation is poor and mainly comprises the floating-leafed species *Lemna minor* and the submerged species *Lemna trisulca*.

Gravel and Sand Pits (CORINE code: 89.23)

Flooded gravel and sand pits occur on the Danube terraces in various parts of the central and southern part of the city. Most of them have been abandoned and are used for recreation and sporting purposes, for example, Kuchajda, Štrkovec, Rohlík, Zlaté piesky, Vajnorské jazero, Malý Draždiak and Veľký Draždiak. The marginal and floating-leafed vegetation is very poorly developed. The most abundant macrophyte is the submerged species *Myriophyllum spicatum* with *Potamogeton pectinatus, P. crispus* and *Zannichelia palustris* occurring less frequently.

Anthropogenic Habitats (CORINE code: 87.2)

Anthropogenic habitats are extensive and important habitats occupy two-thirds of the city; about a third is occupied by urban development with the other third being in agricultural use.

Residential Areas (CORINE code: 86.1 – Towns) (see Fig. 5, Biotopes 9, 10, 11)

Residential areas (excluding industrial zones) occupy about 6,170 ha. The main type of plant cover is ruderal vegetation, of which Jarolímek (1985) distinguished 45 ruderal plant communities. He considered these communities to be undesirable although on previously developed land they spontaneously formed the initial "green" carpet - at no cost until they are eventually replaced by other "free" spontaneous biotopes.

City Centre – Historical City, Business and Shopping Districts in the Centre (CORINE codes: 86.1; + 87.2) (see Fig. 5, Biotope 11)

The historic centre of the city is dominated by Bratislava Castle, which is situated on a high, steep south-facing escarpment covered by spontaneous vegetation dominated by the shrub *Lycium barbarum.* The more abundant trees are *Ailanthus altissima* and *Robinia pseudoacacia.* Thermophilous and nitrophilous species such as *Anthemis tinctoria, Asperugo procumbens, Anthriscus cerefolium* ssp. *trichosperma, Anisantha sterilis, Carduus acanthoides, Chaerophyllum temulum, Onopordon acanthium* and *Torilis japonica* are frequently occurring herbs. *Cymbalaria muralis* occurs on the walls in the city. In the old city, solitary trees and small parks with lawns represent the only "green islands" among the buildings. The spontaneous vegetation is restricted to the "trampled communities", especially in the gaps between paving stones. The most frequent species that are found in this situation are *Bryum argenteum* (a moss), *Poa annua, Polygonum arenastrum, Sagina procumbens* and in sunny places *Eragrostis minor* and recently *Portulaca oleracea.*

High Density Housing Areas – Old Buildings and New Block of Flats (CORINE codes: 86.1; + 85.4 – City Block Inner Spaces; + 87.2) (see Fig. 5, Biotope 9)

From the point of view of cultivated and spontaneous plants, these areas are the worst places. In the older parts of the city, close to the centre, houses are very close to one another and only have a small, generally enclosed, space. In the newer blocks of high-rise flats, situated away from the centre (for example, in Devínska Nová Ves, Dúbravka, Karlova Ves, Krasňany-Rača, Ružinov and Petržalka) the buildings are too tall and in some settlements also too close together; consequently, suitable places for lawns and parks are very limited. During the construction process, thermophilous therophytes occur including *Atriplex tatarica, Hordeum murinum* and *Sisymbrium loeselii* together with hemicryptophytes such as *Cynodon dactylon*.

Low Density Housing Areas – Villas, Village Areas (CORINE codes: 86.1; + 87.2) (see Fig. 5, Biotopes 10 and 8)

Areas of old villas occur mainly on the back slope of Bratislava Castle Hill, on the slopes of Koliba and on the south-east slopes of the Malé Karpaty. In the last century, several neighbourhood villages were incorporated into the city. In recent times, small gardens comprising mainly flowers and intensively managed lawns have become the most frequent component of the village areas. In spite of the rapid loss of the traditional rural lifestyle, some rural plant elements successfully survive, for example, *Arctium lappa, A. tomentosum, Chelidonium majus, Conium maculatum, Malva neglecta, M. pusilla, Marrubium peregrinum, Parietaria officinalis* and *Urtica urens*.

Industrial Areas (CORINE codes: 86.3 – Active Industrial Sites; 86.4 – Old Industrial Sites; 87.2 – Ruderal Vegetation)

The industrial zones, which occupy a total of about 1,500 ha, are distributed around the periphery of the city. The individual zones vary in size, age and structure. Managed lawns and parks are rare while various types of abandoned and neglected land are very frequent. Some of these areas have their own railways and distribution centres, which are important sources of adventive propagules. The intensively disturbed areas contain an abundance of therophytes including *Chenopodium strictum, Conyza canadensis, Lactuca serriola, Hordeum murinum, Sisymbrium loeselii* and *Panicum capillare* while *Lolium perenne* and *Plantago major*, among other species, occur in trampled habitats. Areas of slag support *Chenopodium botrys, Cynodon dactylon* and *Eragrostis minor* with hemicryptophyte ruderal species, such as *Artemisia vulgaris, Calamagrostis epigejos, Cichorium intybus, Daucus carota, Elytrigia repens* and *Picris hieracioides* with *Tanacetum vulgare* being found in areas that have been abandoned for a longer period.

Agricultural Land (CORINE codes: 82, 83) (see Fig. 5, Biotope 6)

The main plant communities of the agricultural land are cultivated crops, vineyards and spontaneous segetal plant communities.

Arable Land in the Lowlands (CORINE code: 82 – Crops)

A third (11,200 ha) of Bratislava is occupied by agricultural land, most or all of which occurs in the Podunajská nížina and Borská nížina Lowlands. Cereals, *Zea mays* and *Helianthus annuus* are the most frequent cultivated crops with the recent addition of *Brassica napus* ssp. *oleifera*. In the northern part (Záhorie), the traditional crop *Brassica oleracea* var. *capitata* is grown, cropped and then fermented for use by people during the winter. The segetal weeds include many archaeophytes and neophytes, *Viola arvensis, Anagallis arvensis, Myosotis arvensis, Papaver rhoeas, Consolida regalis, Apera spica-venti, Centaurea cyanus, Lathyrus tuberosus, Veronica persica, Amaranthus retroflexus, A. chlorostachys, Sinapis arvensis, Raphanus raphanistrum* and *Fallopia convolvulus*. *Panicum miliaceum* ssp. *agricola* has been recorded recently and is colonising fields of *Zea mays.*

Vineyards (CORINE codes: 83.212 – Intensive Vineyards + less 83.212 – Traditional Vineyards + 82.2 – Field Margin Cropland)

The city contains 800 ha of large-scale vineyards, most of which occur on the south-eastern and southern slopes of the Malé Karpaty mountains, usually between the bottom of the slope and the forest boundary. However, a study of old maps and photographs indicates that formerly the vineyards extended to a much higher altitude than they do today. Unfortunately (from the point of view of biodiversity), during the last few decades the majority of vineyards have been transformed from the traditional to large-scale commercial cultivation. Besides typical arable weeds, like *Fumaria officinalis, F. vaillantii, Lamium amplexicaule, L. purpureum, Stellaria media, Veronica hederifolia* agg. and *V. persica,* many spontaneous species grow on the steep slopes between vineyards and in other typical habitats, for example, heaps of rocks collected in the vineyards (known as "rúny"). The typical shrubs found in these areas are *Prunus spinosa, Crataegus* spp., *Sambucus nigra* and small trees, for example, *Cerasus avium, Pyrus pyraster* and *Prunus persica. Lepidium draba* - an aggressive adventive is spreading rapidly on the slopes.

Orchards, Gardens and Allotment Gardens (CORINE codes: 83.151 – Orchards; + 85.3 – Gardens; + 82.3 – Extensive Cultivation) (see Fig. 5, Biotope 7)

There are 480 ha of orchards and 1,800 ha of allotment gardens in Bratislava; the latter are usually located in the "protected zones" of railways, the airport, industrial areas, below high voltage transmission lines and at restored and re-vegetated waste dumps but less on the slopes of the Malé Karpaty mountains. The largest orchards are situated in the north part of the city between the municipal parts of Lamač and Záhorská Ves and in the south on the Danube terraces near Jarovce and Čunovo. *Malus domestica* cultivars are the most frequently planted trees, with some *Prunus armeniaca, P. persica* and *Ribes* spp. being planted, especially near Devín. Intensively managed orchards are subject to heavy chemical control, only few weeds survive and in some places neophyte *Conyza canadensis* is dominant. Some of the extensive orchards on the Devínska Kobyla Hill are suitable habitats for orchids, for example, *Orchis ustulata* and *Himantoglossum adriaticum.*

Green Space

The total amount of public green space in Bratislava is 46.8 km^2 (110 m^2 per inhabitant).

Parks (CORINE codes: 85.1, 85.2) (see Fig. 5, Biotope 5)

Ninety-one percent of the urban green areas (3,000 ha) is represented by forests with only 9% comprising the more 'traditional' urban parks. The forest parks, which are the most important of the public open spaces in the city, are remnants of the peri- or intraurban forests. "Horský Park" founded 1868 and now a Protected Area of 22.9 ha was created from a remnant of the *Quercus-Carpinus* forest by the introduction of some non-native species such as *Abies cephalonica, A. nordmanniana, Juglans nigra, Metasequoia glyptostroboides, Taxodium distichum, Pinus pungens, P. strobus* and *P. nigra.* "Vrakúnsky lesík", which is a remnant of the *Quercus-Ulmus-Fraxinus* forest, has been planted with a variety of ornamental trees and shrubs. Several "artificial" parks of different ages and sizes are situated in or close to the centre of the city. For example, "Sad Janka Kráľa" park (22 ha) founded between 1774 and 1776 as the first public urban park in central Europe has been incorporated within the densely developed high-rise flats of Petržalka – it is now protected as a historic green area. Most interesting are the 200-year-old *Platanus* trees and the "Sternallee"– a star-shaped alley comprising various species including *Alnus, Betula and Salix.* In the undergrowth can be found several forest herbs, for example, *Galanthus nivalis* and *Corydalis cava* and escaped ornamental plants such as *Epimedium alpinum, Eranthis hyemalis, Muscari armeniacum* and *Scilla luciliae.* Other parks include the "Medická záhrada", near the Medical Faculty of the Comenius University and "Grassalkovičova záhrada", which is behind the President's Palace. Smaller parks can be found in the squares – Americké námestie, Šafárikovo námestie, Račianske mýto, Námestie Slobody and Hviezdoslavovo námestie. A large and nice park joins the forest park behind the Rusovce castle where several escaped garden plants like *Anemone blanda, Brunnera macrophylla, Cyclamen purpurascens, Scilla luciliae* and *Veronica filiformis* (an invasive species) have been recorded. In the municipal forest on the Koliba Hill, there are remnants of the historical botanical garden. A special place among all the green open spaces of the city must be reserved for the Botanical Garden of the Comenius University, on the left bank of the Danube. The garden, which is open to the public, contains about 10,000 plant species and cultivars. A new city park is being planned for the municipal part of Petržalka between the Malý Draždiak and Veľký Draždiak lakes.

Cemeteries (CORINE – not described)

The oldest large cemeteries in Bratislava are Martinský cintorín and Ondrejský cintorín; they are now less used and have a park-like appearance. There are also two relatively modern cemeteries; in addition there are smaller cemeteries in villages that have been incorporated into the city. Cemeteries contain many native and non-native (archaeophytes and neophytes) tree and shrub species (and cultivars) such as *Aesculus hippocastanum, Buxus sempervirens, Catalpa bignonioides, Taxus baccata, Chamaecyparis* spp., *Platanus* spp., *Pinus* spp., *Salix* spp., *Thuja* spp. and *Tilia* spp. The lawns are now regularly mown, but even under these conditions a lot of "wild" species of the former village cemeteries can be found; they include in spring *Fumaria vaillantii, Gagea lutea, G. pratensis, Galanthus nivalis, Ornithogalum kochii, Ranunculus bulbosus* and *Thlaspi perfoliatum.* In summer are seen, *Anchusa officinalis, Arrhenatherum elatius, Anisantha tectorum, Campanula rapunculoides, Crepis tectorum, Cynodon dactylon, Dactylis glomerata, Falcaria vulgaris, Festuca* spp., *Lolium multiflorum, L. perenne, Poa annua, Tragopogon dubius, T. orientalis, Trifolium pratense* and *T. repens*. Some cultivated plants, such as *Campanula* cf. *poscharskyana, Cerastium tomentosum, Sedum hispanicum, S. sarmentosum, Tagetes patula* and *Viola x wittrockiana* occur between the graves.

Lawns and "Green" among blocks of flats (CORINE codes: 85.12 – Park Lawns; 85.4 – City Block Inner Spaces)

The lawns in the parks and other public open spaces in the city are mown several times but at irregular intervals during the growing season; consequently, the height of the vegetation varies from (7) 10–30 (50) cm. Between spring and autumn, the older lawns have conspicuously different appearances. In spring, there is a mixture of green (grasses and sedges), white (*Stellaria media* and *Erophila verna*), blue (*Veronica* spp.), red (*Lamium purpureum* and *L. amplexicaule*) and pink (*Erodium cicutarium,* which is locally abundant in places below the district heating pipes). In May, the abundance of *Taraxacum* sect. *Ruderalia* results in some lawns becoming predominantly yellow. During the summer and autumn, the presence of *Centaurea stoebe, Cichorium intybus, Daucus carota, Leontodon autumnalis, Medicago* spp., *Picris hieracioides* and *Trifolium* spp. causes some lawns to be a mixture of different hues of blues, whites and yellows.

Transport Routes and Areas

Railways and Roads (CORINE codes: 86.43 – Railroads; + 87.2)

The whole city contains a dense network of roads, tramways and railways. The vegetation of the railway tracks is controlled by the application of strong total herbicides; consequently, only a few resistant weed species are able to survive, for example, *Amaranthus retroflexus, Artemisia vulgaris, Bromopsis inermis, Conyza canadensis* and *Convolvulus arvensis*. About 10 years ago, the neophyte *Panicum capillare* was found along most of the railways but now it only survives in a much smaller number of localities. Dense stands of *Sambucus ebulus* occur along many sections of railways embankments and cuttings.

The use of de-icing salts along the roads in winter has provided favourable roadside conditions for the colonisation and spread of salt tolerant species such as *Atriplex tatarica, Lotus glaber, Puccinellia distans* and *Senecio vulgaris*. Warming of the asphalt during the summer enables xero-thermophilous species to survive along roadsides, for example, *Bothriochloa ischaemum, Crepis setosa, Cynodon dactylon, Portulaca oleracea, Sedum acre, S. album* and *S. sexangulare*.

Danube Harbour (CORINE code: 87.2)

The river harbour is an important point for the introduction of new adventive plant species to the city and potentially to the country. The Danube is a major commercial transport route with many barges carrying a variety of goods and material from the international ports on the North Sea coast (via the river Rhine) to the Black Sea ports. Therefore, each Danube harbour is a potential "entrance gate" for propagules of adventive plants. Jehlík (2008) distinguished 22 ruderal plant communities from two parts of the harbour. The intensively used parts of the harbour contain "trampled" plant communities of small prostrate species and tuft-forming grasses such as *Herniaria glabra, Lolium perenne, Plantago major, Poa annua* and *Polygonum arenastrum* with *Eragrostis minor* and *Portulaca oleracea* in the more xerophilous habitats. Small abandoned places support ruderal vegetation with biennial and perennial herbs including *Artemisia vulgaris, Berteroa incana, Calamagrostis epigejos, Carduus acanthoides, Cirsium arvense, Daucus carota, Elytrigia repens, Medicago lupulina, Pastinaca sativa* and *Poa angustifolia*. Adventive and apophytic therophytes grow in frequently disturbed habitats, for example, *Amaranthus retroflexus, Atriplex patula, A. tatarica, Anisantha tectorum, Hordeum murinum* and *Kochia scoparia* ssp. *scoparia*. The most frequent neophytes are represented by *Ambrosia artemisiifolia, Conyza canadensis* and *Panicum capillare*. The harbour is the only site in Bratislava for the neophytes *Lactuca tatarica, Senecio inaequidens* and *Sporobolus cryptandrus*.

Airport (EUH – not described CORINE code: 87.2)

The two runways and the various taxiways of the airport are contained in a matrix of short grass, which is mown periodically. The airport is mainly used for passenger aircraft with only a small amount of cargo arriving by air; consequently, it is not an important entry point for new adventive species.

Nature Conservation, Environmental Planning and Education

Planning and Related Matters

The public green spaces are being managed and maintained by several agencies including the local administrative authority for the district and public utilities. Extensive management measures are implemented in the majority of protected areas; this includes mowing, clearance of shrubs, such as *Crataegus* and the selective removal of some planted and invasive trees, for example, *Aesculus hippocastanum, Prunus serotina, Pinus nigra* and *Robinia pseudoacacia.* Management programmes prepared by the State Nature Conservation, Regional Centre Bratislava, the Administrations of the Protected Landscape Areas (Malé Karpaty and Záhorie) are in progress. It is essential to undertake appropriate and regular management works in the National Nature Reserve Devínska Kobyla and other sites (for example, Devínska lesostep, Nature Reserves Ostrovné lúčky and Kopáčsky Ostrov) to maintain optimal conditions for the survival of the xerothermic habitats and the thermophilous species they contain. From 2003 to 2007, various organisations collaborated in a project to manage the Protected Landscape Area of Dunajské luhy. Two of the seven areas selected were within Bratislava – the Nature Reserves Pečenský les and Starý háj. The operations included the selective clearance of invasive woody species such as *Acer negundo, Ailanthus altissima* and *Fraxinus americana* from an area of 430 ha. The cleared areas are being allowed to colonise by native species. Other improvement and management projects that have been carried out during the last 15 years include the restoration of the Biskupické rameno and Chorvátske rameno arms of the Danube, the Štrkovec disused gravel pit (used for recreation purposes since the 1960s) of the historical Calvary and dendrologically valuable sites Horský Park and Kochova záhrada. These projects were financed by the Association of Industry and Nature Conservation and its major contributors such as Slovnaft (a large petro-chemical company), Volkswagen (Slovakia) and Quelle (Slovakia). Proposals for 90 ha of new residential areas in Petržalka Southern City and up to 250 ha Lamač Gate (in the north-west part of the town) will include also the restoration and provision of public green space.

Protected Sites

International

There are two Ramsar Sites in Bratislava – the Protected Landscape Area Dunajské luhy and Morava River floodplain (Niva Moravy) a part of the Protected Landscape Area Záhorie and ten Natura 2000 Sites (Special Areas of Conservation = ÚEV – Územie európskeho významu in Slovak): Bratislavské luhy (including Protected Areas Chorvátske rameno, Pečniansky les, Sihoť, Soví les, Nature Reserves Slovanský Ostrov and Starý háj), Biskupické luhy, Devínska Kobyla, Devínske alúvium Moravy, Devínske lúky, Homolské Karpaty, Hrušovská zdrž, Ostrovné lúčky, Rieka Morava and Vydrica.

Following the international Plantlife project "Important Plant Areas in central and eastern Europe" in Slovakia, five IPAs (VBÚ – Významné botanické územie in Slovak) were identified in the surroundings of the city: Kopáčsky Ostrov (91 ha), Topoľové hony (61 ha), Ostrovné lúčky (676 ha), Devínska Kobyla (127 ha) and 2.140 ha of the Protected Landscape Area Malé Karpaty mountains within Bratislava.

National

The city includes 33 small-scale protected areas: one National Nature Reserve Devínska Kobyla, one National Nature Monument Devínska hradná skala, eight Nature Reserves, 20 Protected Areas and three Nature Monuments. It is proposed to enlarge some of the protected sites and create nine new ones. The protected areas of Bratislava are under the competence of four Administrations of Nature and Landscape Protection: PLA Malé Karpaty, PLA Záhorie, PLA Dunajské luhy and the majority of them under the Regional Administration of Nature and Landscape Protection of Bratislava.

Rare and Threatened Taxa

The number of taxa included in the List of extinct, probably extinct, endemic, threatened and rare taxa of ferns and flowering plants of the flora of Bratislava (Feráková et al. 1994) was 569; this was increased by additional data of several authors to 675 in 2004. About 140 autochthonous species and archaeophytes including some hybrids were classified as regionally extinct (see examples below). Sixty-two taxa of specific and sub-specific rank that are included in the Red Data Book of the Slovak and Czech Republics (Čeřovský et al. 1999) occur in Bratislava – of the 400 listed for the two countries, 335 of them recorded only in Slovakia.

Taxa listed in Appendix I of the Bern Convention – Convention on the Conservation of European Widlife and Natural Habitats 1979 (as amended): *Apium repens, Lindernia procumbens, Pulsatilla grandis, Typha minima* (extinct) and *Typha shuttleworthii.*

Taxa listed in the Annexes of the EC Habitats Directive 92/43 (as amended): *Apium repens, Dianthus praecox* ssp. *lumnitzeri, Echium russicum* (regionally extinct), *Gladiolus palustris* (regionally extinct), *Himantoglossum adriaticum,* (all II, IV), *Lindernia procumbens* (IV), *Lycopodium* spp. (V) and *Galanthus nivalis* (V).

Taxa in the Natura 2000 network of Slovakia: out of the total number 43 of vascular plants in the list, the following seven (and at the same time six without *Lindernia procumbens* out of 36 for which Special Areas of Conservation are designated) are or were recorded in Bratislava: *Apium repens, Dianthus praecox* ssp. *lumnitzeri, Echium russicum, Gladiolus palustris, Himantoglossum adriaticum, Lindernia procumbens* and *Pulsatilla grandis.* From the broader surroundings of Bratislava, as defined in the historical floras, *Angelica palustris* (*Ostericum palustre*) (extinct?), *Spiranthes aestivalis* (extinct) and *Cirsium brachycephalum* were given.

Taxa of pteridophytes and flowering plants of the flora of Slovakia occurring in Bratislava, included in the most recent version of the National Red List (Feráková et al. 2001) in the categories Extinct, Probably Extinct, Critically Endangered and Endangered. The threat categories are the version 3.1 of the IUCN Red List Categories of February 9, 2000, which are defined in Appendix VIII of this book. From the total number of 1,270 taxa included in the List, about 27% are or have been recorded in Bratislava; see paragraphs below.

Key to the Symbols in next two Paragraphs

* Taxa with implemented or actual rescue programme for critically endangered and protected species according to the Directive of the Ministry of Environment of Slovak Republic No. 8/1998-4.1.

x Taxa not protected in Slovakia (Order No. 24/2003 of the Ministry of Environment of the Slovak Republic, of January 9, 2003, which implements the Act of the National Council of the Slovak Republic No. 543/2002 code of laws On Nature and Landscape Protection, as amended 579/2008 of December 18, 2008). All species not marked 'x' are listed in the Annexes.

+ Taxa listed in Annex 5 of the Order only. All the other species are included in both Tables 4 and 5 of the Order. Table 4 lists species of European interest, species of national interest, bird species and priority species, for protection of which protected areas are designated. Annex 5 List of protected plants and priority plant species. Out of the taxa included in Annex 5, almost 100 were recorded in Bratislava.

BERN I = Listed in Appendix I of the Convention on the Conservation of European Widlife and Natural Habitats 1979 (as amended).

HD = Listed in EC Habitats Directive Annexes II and IV.

Taxa Regionally Extinct or Probably Extinct: (140 Taxa of which 22 have been re-discovered)

Some examples: *Androsace maxima, Armeria vulgaris, Bassia laniflora, Bromus secalinus, Echinops ritro* ssp. *ruthenicus, Eriophorum angustifolium, E. latifolium, Heleochloa alopecuroides, H. schoenoides, Lathyrus aphaca, L. palustris, Phlomis tuberosa, Scandix pecten-veneris, Scorzonera humilis, S. parviflora, Spergula pentandra* and *Teucrium scordium.* Nationally extinct or probably extinct species that once occurred in Bratislava are *Ceratocephala falcata, C. orthoceras, Lolium remotum, Moenchia mantica* and *Typha minima.*

Critically Endangered: (40 Taxa, which is 16.82% of the Slovak total of 243)

Adonis flammea x, *Alisma gramineum* ssp. *gramineum, Allium atroviolaceum* x, *Apera interrupta, Apium repens* BERN (I), HD (II, IV), *Artemisia austriaca* *, *Astragalus asper* *, *Blackstonia acuminata, Centaurea solstitialis* ssp. *solstitialis, Centaurium littorale, Gagea minima, Galium parisiense* ssp. *anglicum, Himantoglossum adriaticum* (HD II, IV), *Inula salicina* ssp. *sabuletorum, Lindernia procumbens* BERN (I), HD (IV), *Minuartia glaucina, Ononis pusilla* *, *Ophrys apifera, O. fuciflora, O. sphegodes, Orchis coriophora* ssp. *coriophora* *, *Orobanche artemisiae-campestris* +, *O. coerulescens* +, *O. teucrii* +, *Peucedanum arenarium* ssp. *arenarium* *, *Pholiurus pannonicus* (only one site at the limit of the area concerned), *Potentilla pedata, Ranunculus lateriflorus, Rosa arvensis* +, *Ruscus hypoglossum* +, *Schoenoplectus triqueter* +, *Senecio doria, S. paludosus, Spiranthes spiralis, Succisella inflexa, Trifolium retusum* (not confirmed at present), *T. striatum, Typha shuttleworthii* BERN (I), *Viola ambigua* and *Vitis sylvestris.*

Endangered (45 Taxa, which is 15.95% of the Slovak total of 282)

Agropyron pectinatum, Aira caryophyllea x, *Allium angulosum, Alyssum montanum* ssp. *gmelinii, Anacamptis pyramidalis, Asplenium adiantum-nigrum* +, *Bupleurum affine* +, *Bupleurum rotundifolium* +, *Campanula xylorrhiza, Ceratophyllum submersum, Chrysopogon gryllus* +, *Conringia austriaca, Consolida regalis* ssp. *paniculata, Dactylorhiza incarnata, Danthonia alpina, Dianthus collinus* ssp. *collinus, Dichostylis micheliana* +, *Gentiana pneumonanthe, Glaucium corniculatum, Gratiola officinalis, Groenlandia densa, Gypsophila paniculata* +, *Hippuris vulgaris, Limodorum abortivum, Limosella aquatica, Lycopodioides helveticum, Medicago monspeliaca, Oenanthe silaifolia* ssp. *silaifolia, Orchis tridentata* ssp. *tridentata, O. ustulata* ssp. *ustulata, Orobanche gracilis* +, *Phelipanche arenaria* +, *Potentilla rupestris, Reseda phyteuma* +, *Rhamnus saxatilis* ssp. *saxatilis, Silene conica, Stipa pulcherrima* +, *Stratiotes aloides, Taraxacum serotinum* +, *Ventenata dubia, Verbascum speciosum, Veronica anagalloides, V. catenata, Vinca herbacea* and *Viola pumila.*

The regional assessment made by Feráková et al. 1994, has not yet been revised in accordance with the IUCN recent criteria. However, a general assessment indicates that most of the taxa would remain in the same threat categories although down-grading may apply to some species on the regional or even the national Red Lists, for example, *Orobanche teucrii, Rhamnus saxatilis* ssp. *saxatilis* and *Smyrnium perfoliatum.*

Appreciating the importance of public education and awareness of nature, scientists from the Faculty of Nature Sciences of the Comenius University, Slovak Academy of Sciences, State Nature Conservation organisations and others have prepared and published many posters, guides and books (for example, two books about Devínska Kobyla) and prepared the text for information panels. Daphne – the Institute of Applied Ecology in Bratislava, in cooperation with Distelverein in Austria, has published the excursion guide "The Morava River Region" (Bizubová et al. 2000).

Education

There are many nature pedestrian trails in Bratislava varying in length from 3.5 km to 79 km, some combined with bike routes and therefore extend for many kilometers into the surrounding countryside. The nature trails can be found in many different landscapes and protected areas, including the following:

1. Protected Landscape Area – Malé Karpaty.
2. Forest park Lamač – Železná Studienka – Koliba, 6 km long through mainly *Quercus-Carpinus* and *Fagus* woodland.
3. Floodplain woods in the Podunajská nížina Lowland; for example, the nature trail Biskupické lužné lesy has three alternative routes up to 16 km long; the educational path at Rusovský channel is connected with a cyclist path, linking the area with the International Danubian cycle track.
4. Protected Landscape Area Dunajské luhy; it has 12 information panels starting at Pečenský les and ending at Veľkolélsky Ostrov (Danube Island) far from Bratislava.
5. Záhorská nížina Lowland 79 km nature trail from Devín to Moravský Sv. Ján and Sekule in the Protected Landscape Area Záhorie. There are 39 information panels in Slovak, German and English.
6. Devínska Kobyla Hill, from Devínska Nová Ves starting at the palaeontological site Sandberg through the National Nature Reserve Devínska Kobyla to Devín (3.5 km). There are seven information panels. The nature trail is probably the most visited, especially in spring.

A new nature trail is planned for the Castle Hill in Bratislava. The disused quarry in Devín contains a small open-air geological museum with six information panels. In Devínska Nová Ves there is a small dendrological park containing the native woody species that occur in the area of Devínska Kobyla. Lectures on and excursions to the biotopes of major interest are organised. They have been co-financed by the EC programmes INTERREG IIIA AT-SR and DAPHNE. According to its website, the Institute of Applied Ecology, Bratislava was established in 1993 to bridge the gap between science and conservation in order to conserve and enhance the richness and diversity of Slovakia´s natural heritage. The citizens of the capital of Slovakia are proud of its metaphoric name "the belle on the Danube", but for the phrase to be really meaningful a lot should be done to protect and enhance the existing green space and to create new green areas for people and nature.

Closing Comments

Bratislava is the name of the capital of Slovakia that has been used for 90 years; it was formerly known also as Preslava, Prešporok, Pressburg, Pozsony, Posonium and Istropolis. It was formed in AD 907 on the boundary of Carpathian and Pannonian regions. The altitude ranges from 126 m to 514 m a.s.l; the city is built on a variety of rock types, which give rise to different soil types. The climate is continental; the mean annual temperature and precipitation are 10.3°C and 573 mm, respectively. In 2008, the city, which occupies 367.9 km^2 had a population of 425,540.

The first published records of the flora of Bratislava can be traced to that of Clusius in 1583. In 1791 Lumnitzer published his *Flora Posoniensis* which was followed in 1830 by Endlicher's book (with the same title), which listed 1,192 vascular plants. At present, the number of recorded taxa of ferns and flowering plants is close to 1,700. No complete inventory exists so far, but several lists of species were compiled for the selected municipal parts and protected areas. The regional list of extinct, probably extinct, threatened, endemic and rare taxa of vascular plants contains 675 taxa (comprising 569 taxa published in 1994 by Feráková et al. augmented by additional data from others). According to the contemporary data 140 taxa are considered extinct or probably extinct, while 22 taxa originally listed in these two categories have been re-discovered. Bratislava contains 40 (16.82%) of the 243 critically endangered category of Slovakia and 45 (15.95%) of the 282 endangered taxa. About 27% of the species and subspecies of the 1,270 included in the national Red List are or have been recorded in the city, as have seven of the 43 Slovak species referred in the Natura 2000 network. Over 100 taxa are included in Annex 5 of the "List of Protected Plants and Priority Plant Species" of the Order No. 24/2003 (as amended) of the Ministry of Environment of the Slovak Republic. Some examples of phytogeograhically important species and taxa with *locus classicus* in the area concerned are also mentioned.

Among the protected areas of Bratislava there are two Ramsar Sites, ten Natura 2000 Sites, one National Nature Reserve, one National Nature Monument, eight Nature Reserves, 20 Protected Areas and three Nature Monuments. In addition five Important Plant Areas have been identified in the surroundings of the city.

Information is provided about the species spectrum of selected sites from the view point of synanthropisation: examples of archaeophytes, neophytes, invasive, potentially invasive and expansive plants and a list of newer additions to the alien flora of Bratislava. The majority of 160 naturalised neophytes originate from North America.

Bratislava and its surroundings represent an area of considerable geological and geomorphological diversity with immense variety of biotopes. The characteristic species of the individual biotopes (18 natural or semi-natural and 16 anthropogenous) in accordance with the EC Habitats Directive 92/43 and the Corine Biotope manual are described. The ruderal vegetation found in the settlement areas is considered, including the 45 ruderal communities that were distinguished by Jarolímek in the 1990s.

Bratislava belongs to the European cities with the largest proportion of periurban forests (91% of the city green areas) and the smallest area of urban parks. The total amount of public green space is 46.8 km^2 (110 m^2 per inhabitant). The diversity of the city green space in the inhabited parts of Bratislava is illustrated by results of a study by Reháčková and Pauditšová 2006 who found 112 taxa (excluding cultivars) in the northwestern area of the city. Twenty-six specimen trees in the city both native and of foreign origin are legally protected.

So far as the non-vascular plants are concerned, the algal flora has been thoroughly studied mainly by Hindák and Hindáková. In 1830, Endlicher listed 15 species of algae, while a recent study of two small inundation lakes in Devín recorded 578 species and 58 non-type varieties and formas (including Cyanophytes). About a fifth of all the algal taxa found in Slovakia have been recorded in the Bratislava section of the Danube. Many genera and species of Cyanophytes and algae new to science have been described, for example, from the gravel pit lakes. Several taxa were recorded in the city and its environs for the first time in Slovakia. In 1791, Lumnitzer listed five species of liverworts and 65 of mosses. Since then, 66 liverworts, 267 mosses and two hornworts have been found in the city; 245 of them occur today (Kubinská, Mišíková and Javorčíková (2003). Examples of characteristic species found in selected biotopes are described.

The first records on lichens date back to the *Flora Posoniensis*. Up to now, as reported by Lackovičová, Guttová and Pišút, a total of 291 taxa have been recorded in the city; 139 are epiphytic, 98 epilithic and 54 terrestrial. The deterioration in air quality in the twentieth century resulted in many sensitive and moderately tolerant species becoming extinct. At the turn of the twentieth/twenty-first centuries and as a result of improvements in air quality, it was discovered that epiphytic lichens were beginning to re-colonise; 45 species have been recorded so far, of which 36 are new to the area. However, 58 taxa that were previously recorded have not yet been found.

As the authors of the paragraph on fungi (Bacigálová and Jančovičová) state, Bratislava is very rich in species of both the micromycetes and macromycetes. The oldest records can be found also in Lumnitzer´s flora of 1791 augmented by the rich collections of Bolla and Bäumler in the nineteenth century. Recent studies have recorded the occurrence of 113 taxa of parasitic microfungi on the leaves of vascular plants and 64 taxa of powdery mildew. Twenty-five taxa of myxomycetes were found only in the phytogeographical district of Devínska Kobyla in the western part of Bratislava; this area of 49 km² contains 17.8% of the total fungal diversity of Slovakia. A mycological inventory of the Danube islands, Slovanský Ostrov and Sihoť belonging to the PLA Dunajské luhy resulted in the recording of 211 taxa. It is expected that most, or even all, of the 479 taxa of macromycetes reported from the Podunajská nížina Lowland occur in Bratislava. Some species of microfungi, which were considered to be extinct or declined under anthropogenic pressure, have been re-discovered after 50–100 years.

The section on Nature Conservation, Environmental Planning and Education considers, among other things the "Plant Red Lists", site protection and the management measures undertaken in some of those sites. It also describes examples of activities aimed at education and raising environmental awareness, including an overview of Bratislava's most important nature trails.

Acknowledgments The sections on non-vascular plants and the fungi have been written by the following authors:

1. Algae – Professor RNDr. František Hindák, DrSc.[1] and Mgr. Alica Hindáková, PhD.[1]
2. Bryophytes – RNDr. Anna Kubinská, CSc.[1], Assoc. Prof. Mgr. Katarína Mišíková, PhD.[2] and RNDr. Duňaša Javorčíková, CSc., Macharova 9, 85101 Bratislava
3. Lichens – RNDr. Anna Lackovičová, CSc.[1], Mgr. Anna Guttová, PhD.[1] and RNDr. Ivan Pišút, DrSc.[1]
4. Fungi – RNDr. Kamila Bacigálová, CSc.[1] and Mgr. Soňa Jančovičová, PhD.[2]

We are grateful to them for their enthusiastic collaboration.

Very special thanks are due to RNDr. Dušan Senko, PhD.[1], who kindly prepared maps and provided photographs and all other illustrations (except the figure on Algae). For the photograph No. 1, we thank Mr. Branislav Molnár. The help of colleagues from the Regional Centre of Nature Conservation, Bratislava, especially of RNDr. Helga Kothajová and Mgr. Radovan Michalka, of RNDr. Dobromil Galvánek, PhD. from DAPHNE, Mgr. Jana Jecková from the Council Department of Environment, municipal part Bratislava – Petržalka is cordially acknowledged. For providing valuable information on wetlands, we wish to express our gratitude to RNDr. Helena Oťaheľová, CSc.[1]

[1] Institute of Botany, Slovak Academy of Sciences, Dúbravská cesta 9, 84523 Bratislava, Slovak Republic

[2] Department of Botany, Faculty of Natural Sciences, Comenius University Bratislava, Révová 39, 81102 Bratislava, Slovak Republic

Literature Cited

Adamčík S, Kučera V, Lizoň P et al (2003) State of diversity research on Macrofungi in Slovakia. Czech Mycol 55: 201–213

Bizubová M, Bocek B, Čiampor F et al (2000) Niva Moravy. Manuál pre sprievodcov. Bratislava. DAPHNE – Inštitút aplikovanej ekológie. Bratislava

Čeřovský J, Feráková V, Holub J, Maglocký Š, Procházka F (1999) Červená kniha ohrozených a vzácnych druhov rastlín a živočíchov SR a ČR. 5, Vyššie rastliny. Príroda, Bratislava

Drábová-Kochjarová J (1990) Synantropná flóra sídlisk v Bratislave- Petržalke a niektoré jej prvky ako súčasné a potenciálne zdroje peľových alergénov. Acta Fac Rer Nat Univ Comen – Bot 37: 53–63

Endlicher S (1830) Flora Posoniensis, exhibes plantas circa Posonium sponte crescentes aut frequentius cultas, methodo naturali dispositas Bratislava (Posonii)

Feráková V (1995) Chránená prírodná pamiatka Devínska hradná skala a národná kultúrna pamiatka Devín- Slovanské hradisko –lokality významné aj z botanického hľadiska. In: Topercer J ml (ed) Diverzita rastlinstva Slovenska. Zborník referátov zo VI. zjazdu Slov Bot Spoločn pri SAV, Blatnica 6.–10. jún 1994, (ed.) Slov Bot Spoločn, Bratislava, p.121–124

Feráková V (2002) Floristic-taxonomic and plant conservation problematics of vascular plants in Bratislava and its surroundings. Acta Fac Rer Nat Univ Comen - Bot 41: 97–105

Feráková V, Kocianová E (eds) (1997) Flóra, geológia a paleontológia Devínskej Kobyly. Litera s.r.o. pre APOP Bratislava

Feráková V, Maglocký Š, Marhold K (2001) Červený zoznam papraďorastov a semenných rastlín Slovenska (december 2001). In: Baláž D, Marhold K, Urban P (eds), Červený zoznam rastlín a živočíchov Slovenska. Ochr Prír 20 (Suppl.): 44–78

Feráková V, Michálková A, Ondrášek I, et al (1994) Ohrozená flóra Bratislavy. Zoznam vyhynutých, nezvestných, endemických, ohrozených a vzácnych taxónov flóry Bratislavy. APOP Bratislava

Feráková V, Skrovná Ľ (1998) Spektrum neofytov vo flóre Bratislavy. In: Benčaťová B, Hrivnák R (eds), Rastliny a človek. Zborník z jubilejného seminára konaného k životnému jubileu Prof. Ing. M. Križu DrSc. Technická univerzita vo Zvolene, p. 103–108

Futák J, Domin K(1960) Bibliografia k flóre ČSR. Vydavateľstvo Slov akad vied Bratislava

Gojdičová E, Cvachová A, Karasová E (2002) Zoznam nepôvodných, inváznych a expanzívnych cievnatých rastlín Slovenska. List of Alien, Invasive Alien and Expansive Native Vascular Plant Species of Slovakia. (Second Draft). Ochr Prír, Banská Bystrica 21: 59–79

Hajdúk J, Lisická E, Pišút I (1975) Rozšírenie epifytických lišajníkov v niektorých parkoch v oblasti Bratislavy. Häufigkeit epiphytischer Flechten einiger Parkanlagen im Gebiet von Bratislava. Zborn Slov Nár Múz Prír Vedy 21: 75–117

Hindák F, Hindáková A (2003) Cyanophytes and algae of gravel pit lakes in Bratislava, Slovakia. Hydrobiologia 506: 155–162

Hindáková A, Hindák F (1998) Green algae of five city fountains in Bratislava (Slovakia). Biologia, Bratislava 53: 481–493

Hodálová I, Letz D, Janovicová K (1999) Výskyt niektorých zaujímavejších taxónov v mestskej časti Bratislava-Lamač. Bull Slov Bot Spoločn, Bratislava 21: 89–97

Hodálová I, Šipošová H, Matisová V, Kothajová H (1996) Príspevok k flóre prírodnej rezervácie Štokeravská vápenka (Devínska Kobyla). Ochr Prír 14: 17–27

Holub J (1971) Notes on the terminology and classification of synanthropic plants with examples from the Czechoslovak flora. Saussurea 2: 5–18

Holzner W, Ries Ch, Geisselbrecht-Taferner L et al (1994) Unkraüter. Begleiter und Freunde des Menschen. Grüne Reihe des Bundesministeriums für Umwelt, Jugend und Familie. Band 4. Styria Medienservice Graz

Hrnčiarová T (ed), Izakovičová Z, Pauditšová E, Krnáčová Z et al (2006) Krajinnoekologické podmienky Bratislavy. Veda, vydavateľstvo SAV, Bratislava

Janovicová K, Kubinská A, Javorčíková D (2003) Pečeňovky (Hepatophyta), rožteky (Anthoceratophyta) a machy (Bryophyta) na území Bratislavy (Slovensko). Liverworts (Hepatophyta), hornworts (Anthoceratophyta) and Mosses (Bryophyta) in the area of Bratislava (Slovakia). Bot ústav, Bratislava

Jarolímek I (1985) Syntaxonomický prehľad ruderálnych spoločenstiev Bratislavy. Biológia, Bratislava 40: 489–496

Jarolímek I (1994) Charakteristika pásmovitosti zástavby v Bratislave so zvláštnym zreteľom na hlavné typy ruderálnej vegetácie a jej stanovíšť. Zpr Čes Bot Společ, Praha, 29, Materiály 11: 47–55

Jehlík V (ed) (1998) Cizí expanzivní plevele České republiky a Slovenské republiky. Academia, Praha

Jursa M, Oťaheľová H (2005) Distribution of aquatic macrophytes in man-modified water-bodies of the Danube River in Bratislava region (Slovakia). Ekológia 24/4: 368–384

Lackovičová A (1978) Lišajníky Malých Karpát. Acta Ecol 6 (1977): 7–107

Lackovičová A, Pišút P, Guttová A (2008) Epiphytic lichens – bioindicators of air pollution in Bratislava (SW Slovakia). In: Kočárek P, Plášek V, Malachová K, Cimalová O. (eds) Environmental changes and biological assessment IV. Scripta Fac Rer Nat Univ Ostraviensis Nr. 186, Ostrava, p. 138–142

Letz R (2000) Flóra Bratislavy po dvoch storočiach od výjdenia Lumnitzerovho diela Flora Posoniensis. Bull Slov Bot Spoločn, Bratislava 22: 235–246

Lizoň P (2001) Červený zoznam húb Slovenska, 3. verzia (december 2001). In: Baláž D, Marhold K, Urban P (eds) Červený zoznam rastlín a živočíchov Slovenska. Ochr Prír 20 (Suppl.): 6–13

Lumnitzer S (1791) Flora Posoniensis exhibens plantas circa Posonium sponte crescentes secundum systema sexuale Linneanum digestas. Lipsiae

Marhold K, Hindák F (eds) (1998) Zoznam nižších a vyšších rastlín Slovenska. Checklist of non-vascular and vascular plants of Slovakia. Veda, vydavateľstvo SAV Bratislava

Marvan P, Heteša J, Hindák F, Hindáková A (2004) Phytoplankton of the Morava river (Czech Republic, Slovakia): past and present. – Oceanol and Hydrobiol Studies, Gdańsk 33/4: 42–60

Michalko J, Berta J, Magic D, Maglocký Š (1977) Ekologická valorizácia záujmového územia Bratislavy pre potreby jej urbanizačného rozvoja. Ms. Depon. Institute of Botany SAS, Bratislava

Ondrášek I (2002) Recentný výskyt niektorých vzácnych a ohrozených druhov cievnatých rastlín na juhozápadnom Slovensku. Bull Slov Bot Spoločn, Bratislava 24: 133–138

Oťaheľová H , Valachovič M (2006) Diversity of macrophytes in aquatic habitats of the Danube River (Bratislava region, Slovakia). Thaiszia 16: 27–40

Pantocsek J (1902) Adatok Pozsony város és vidékemoszat vinánhoz. (Adnationes phycologicae territorii Posoniensis). Verh. Vereins Natur-u. Heilk., Pressburg, Neue Folge 13: 67–71

Pišút I, Lackovičová A (1991) Flechtenindikation im Gebiet von Bratislava (S. W. Slowakei). VDI Berichte 901: 132–142

Reháčková T, Pauditšová E (2006) Vegetácia v urbánnom prostredí. Cicero, s. r.o. Bratislava

Ripková S, Ďuriška O (2009) The current knowledge of funga of the Devínska Kobyla Mountains. Acta Bot Univ Comen 44: 41–58

Uher B, Kováčik Ľ, Kučera P, Hindáková A, Pivko D (2005) Cyanobaktérie a riasy na kamenných substrátoch objektov kultúrno-historického významu v Bratislave. Bull Slov Bot Spoločn, Bratislava 27: 11–16

Uher B, Kováčik Ľ, Degma P, Vozárová A (2006) Distribúcia cyanobaktérií a rias na stavebnom kameni Presbytéria Dómu sv. Martina v Bratislave. – Bull Slov Bot Spoločn, Bratislava 28: 11–20

Zahlbruckner A (1894) Zur Flechtenflora des Pressburger Komitates. Verh. Vereins Natur-Heilk., Pressburg, Neue Folge 8 (1892/93): 19–73, 273–284

Zahlbruckner A (1899) Zur Flechtenflora des Pressburger Komitates. Verh. Vereins Natur-Heilk., Pressburg, Neue Folge 10 (1897/98): 17–29

Internet pages

(http://biodiversity-chm.eea.europa.eu/information/document/F1088156525/F1125582140). Accessed between September and November 2009.

http://www.bratislava.sk. Accessed between September and November 2009.

http://www.dunaj.broz.sk. Accessed between September and November 2009.

http://www.sopsr.sk. Accessed between September and November 2009.

Brussels

Sandrine Godefroid

Fig. 1 View of the city of Brussels (photo Rony Cnop)

Abstract Brussels, which occupies 161 km², and located in the centre of Belgium, is situated in an area of different biogeographical influences where species of Atlantic and medio-European origins grow together. The flora has taken advantage of this heterogeneity of mosaics by developing a higher diversity than that of the adjacent areas. A total of 730 plant species have been recorded in the city, which represents half of the Belgian flora. The bryoflora comprise 225 species, of which

Sandrine Godefroid (✉)
National Botanic Garden of Belgium, Domein van Bouchout, 1860 Meise, Belgium
e-mail: sandrine.godefroid@br.fgov.be

J.G. Kelcey and N. Müller (eds.), *Plants and Habitats of European Cities*,
DOI 10.1007/978-0-387-89684-7_5,

40 are hepatics and 185 are mosses; they represent about 33% of the Belgian bryoflora. 1306 species of fungi and 86 lichenised fungi have been recorded in the city so far. Nine important EU-habitats are present, mainly in three Special Areas of Conservation that cover a total of 1,900 ha (12% of the city). Current strategies for nature development are based on three different actions: (1) the increase in total surfaces of green spaces, (2) the increase in ecological quality of green spaces and (3) the development of ecological networks. The diversity of various taxonomic groups and habitats is particularly well monitored providing a biological database that has been integrated into and provides for the effective planning, design and management of the city's environment.

Natural Environment of the City

Brussels covers an area of 161 km^2 in the centre of Belgium; see Fig. 2. The city has developed in the relatively broad Senne valley, which has a mean width of 1,500 m; the base is 15 m a.s.l.; see Fig. 3. The river itself has a mean slope of 1.2‰,

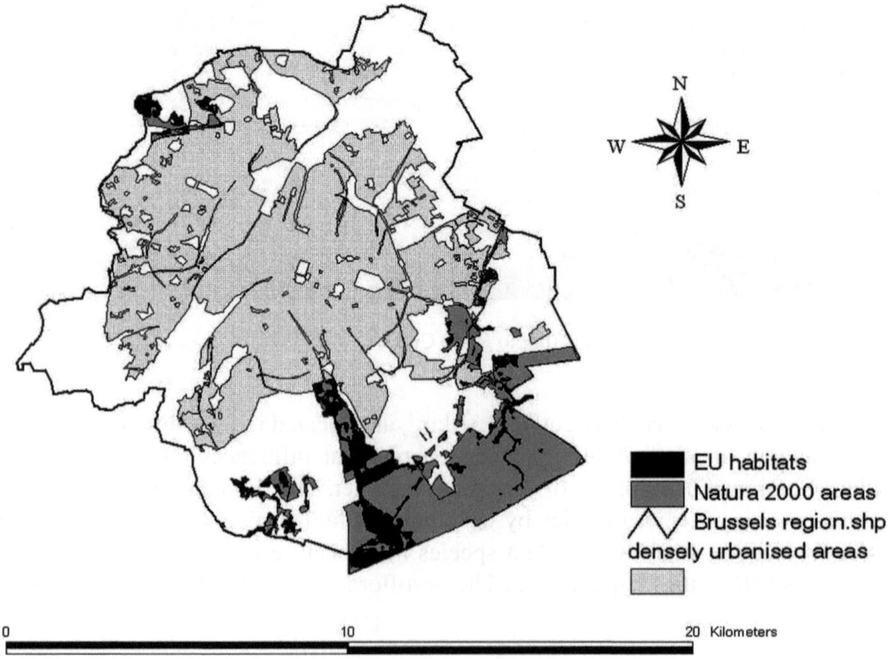

Fig. 2 Plan of Brussels

Fig. 3 Although the Senne river is culverted over most of its course in Brussels, it is still visible in the suburbs of the city

which means slow flows (3 m^3/sec), frequent inundations (in the past) and good conditions for wetlands and the creation of water bodies, reservoirs and ponds (Martens 1998).

The valley, which is aligned south-north, is asymmetric; in the east, there are steep slopes representing the first elevations of middle Belgium in contrast to the maritime lowland. Two tributaries of the Senne (the Maelbeek and the Woluwe) flow in deep and sandy valleys with the highest point about 120 m a.s.l. On the west, two other tributaries (the Pede and the Molenbeek) show a less marked course over a loamy basement with the highest point barely reaching 80 m a.s.l. (Martens 1998).

The solid geology of the Cambrian substrate is covered by successive layers of Tertiary clays, sands and sandstone followed by Quaternary loess. This alternation of permeable and impermeable layers results in perched water tables. Rivers have cut through these deposits resulting in the formation of springs. For most of their length, the watercourses cross Ypresian clay, which forms the base of the numerous ponds of the city (Saintenoy-Simon 1998). The water in and around Brussels is highly mineralised with a high pH of ca. 7.8 in the rivulet Woluwe and ca. 7.3 with 65.6 mg l^{-1} of calcium in the watertable under the Sonian Forest (IBGE 1995).

Soils are in general silty and sandy-silty. The valley bottoms comprise silty alluvium derived from erosion. Various wetlands, which have never been altered, still have interesting gley soils. The soil of the vast area of forest (it occupies 10%

of the city) has not been disturbed for 10,000 years. All these soils are prone to compaction and erosion due to recreation activities. They are also often disturbed by the deposition of rubble, rubbish and various other kinds of exogenous substrates. When assessed in terms of land use, most of the soils of the city are more or less degraded. The unaltered soils are only present in some extensive and largely undisturbed green areas. In the ancient agricultural areas, the upper soil layers have disappeared. So, we can observe shortened profiles in many areas re-colonised by spontaneous vegetation. In other cases, soils are profoundly altered (IBGE 1990).

Brussels is characterised by a temperate, mild and humid climate, resulting from two influences: the maritime climate of Flanders and the continental climate of the Ardennes. As a result, the winters are mild, there is little difference between the seasons and the rainfall is relatively evenly spread throughout the years. The mean annual temperature is 9.9°C; mean annual precipitation is 835 mm. The period without frost lasts 194 days. The vegetation period (temperature >10°C) is 172 days. In Brussels, the mean annual temperature has risen by 1.5°C in the last 40 years (Fig. 4), resulting in the "heat island effect" that is characteristic of cities.

Brussels has six distinctive features that differentiate it from other big cities (Table 1):

1. A substantial part of the biggest forest of North Belgium (1,660 ha out of 4,400 ha of the famous Sonian Forest).
2. About a hundred small woodlands and parks.
3. Numerous water bodies.
4. Various wastelands.
5. A canal and a harbour.
6. Numerous protected semi-natural areas.

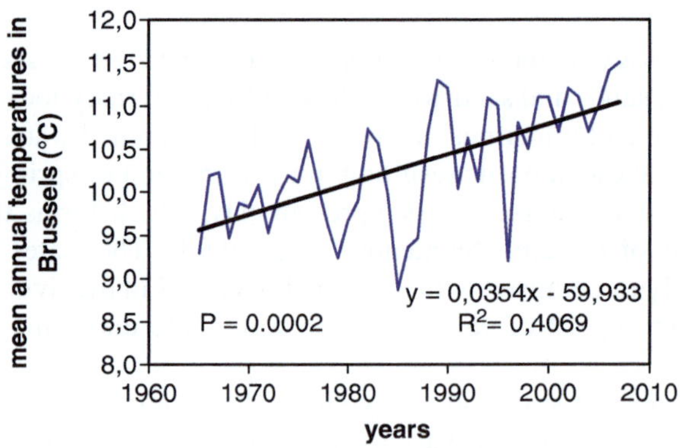

Fig. 4 Increase of the mean annual temperature in Brussels between 1965 and 2007. Data from the Royal Meteorological Institute

Table 1 Area of each land-use type in Brussels

Land use	ha	%
Standing water	94.36	0.57
Heathland	0.12	0.00
Grassland	645.16	3.91
Wetland	16.41	0.10
Quercus forest	433.15	2.62
Fagus forest	1,411.81	8.55
Alluvial forest	49.35	0.30
Other broad-leaved forests	77.72	0.47
Populus plantation	29.58	0.18
Conifer plantation	107.01	0.65
Shrub	284.38	1.72
Linear element (tree alignment, hedge, sunken road)	99.12	0.60
Wasteland	158.67	0.96
Arable land	337.84	2.05
Railway	331.13	2.01
Orchard	28.63	0.17
Park	1,162.76	7.05
Urbanised area	10,867.55	65.85
Road (outside urbanised areas)	368.37	2.23

Historical Development of the City

Settlements up to AD 1200

Before the establishment of the city, the territory of what is now the Brussels region has never undergone a massive human occupation. Since its earliest days, water has played an important role in the development of the city, which has been built around an island formed by two arms of the Senne River. The city was once called "Bruocsella", meaning "the home in the marsh". The first record of human presence was the finding of objects dating from about 7000 BC in the Sonian Forest. Excavations have discovered a 9 ha settlement from the "Michelsberg's culture", (3000–2200 BC). The Romans occupied the area in the sixth century AD. According to tradition, the city was named "Brussels" in AD 979 but there is no written or archaeological source to support this date.

AD 1200–1900

At the beginning of the twelfth century, trade became a main actor in western Europe. The commercial centres quickly became powerful cities, thanks to the rivers. Koln, Louvain, Ghent, Ypres, Antwerp and Bruges became centres of the textile trade. Thanks to its artisans and harbour, Brussels became an important trading crossroad (especially in clothes), resulting in the city becoming affluent during the next two centuries. The city expanded within Brabant, which was not in France but had been incorporated into the Germanic Empire in 923. In 1183, Brabant became a dukedom, and the first duke (Henry I (1190–1235), Count of Brussels and Louvain) enclosed Brussels with a wall, which was 4 km long. The wall also enclosed the island Saint-Géry and the first harbour of the Senne river, the place of the market which was to become the "Grand Place" of Brussels and the castle of Coudenberg.

The region was controlled by the Dukes of Burgundy from 1384 to 1477; then they lost power to the Habsburgs, who constructed the 28 km long "Channel of Willebroek", which enabled the region to develop even more. In 1555, Charles V surrendered to his son, Philip II of Spain; although the times were turbulent, the region remained under Spanish rule until 1713 when the Treaty of Utrecht, was signed to regulate the War of Spanish Succession. The Treaty gave the Spanish Netherlands (which included Belgium) to the Habsburgs, who continued to occupy the city and promote its wealth until the French Invasion of 1794. This was followed by the creation of the "United Kingdom of The Netherlands", which comprised Belgium and Luxembourg. On July 21, 1831, King Léopold became the first governor of the independent Belgium. Although Belgium remained neutral during Franco-Prussian war (1870–1871), there was tension between the Flemish people and those who spoke French – a linguistic division that still exists.

Brussels, particularly its green areas, carries the indelible trace of the reign of Léopold II, second King of the Belgians, rightfully nicknamed the king-masterbuilder. He created or preserved 7,500 ha of green areas (including 1,000 ha of public gardens) in the capital, which was practically devoid of them at the beginning of his reign. To accommodate the industrial expansion of Brussels and its increasing role as the capital city, Léopold devised a vast programme of urban development, with the following four priorities:

1. Cleaning up of the poor areas in the centre and improving communications between the poor and the upper districts of the city.
2. Construction of large access roads between the city and the surrounding countryside to allow the harmonious development of the outskirts.
3. Protection of landscapes and the creation of a belt of parks around Brussels.
4. Contributing to the prestige of the capital and its royal functions by building or enlarging public buildings.

AD 1900–1945

At the instigation of Victor Horta, superb Art Nouveau residences were built at the beginning of the twentieth century, by which time Brussels had become an important and successful cultural centre. In 1903, work started on the construction of a railway line to connect the north and south parts of the city; the work was to last for almost half a century. The railway was positioned on the hillside, between the low and the high parts of the city, where there are comparatively stable ground conditions and which would avoid the destruction of ancient monuments. However, without any debate, it passed through historic and densely populated quarters of the city, resulting in the loss of picturesque streets and about 1,200 houses.

Complex procedures and various technical problems delayed the work considerably so that by 1914, only a short section of the line had been completed. In 1918, changes resulted in national and local priorities, which included an attempt to complete the work quickly as a matter of prestige but at a time when there was little money available. Paradoxically, it was the crisis of the 1930s that put the completion of the project onto the agenda again. In 1935, a national office for the completion of the works was established but it was to take another 17 years before the inauguration, in 1952.

AD 1945–1960

The 1950s and 1960s constituted a particularly controversial period for Brussels and its architectural heritage. The list of ancient monuments that were destroyed in the first few decades after 1945 is impressive. During the same time, many urban monstrosities have been imposed on the city due to the reforming fever and property speculation – they have caused more damage to the urban fabric than the ravages of 1939–1945. The reconstruction of Brussels was based on "building high" linked to a separation of the four big functions: habitat, employment, circulation (traffic movement), and leisure and recreation. The considerable economic attraction and frenzy to build in the centre of the city was continued in the peripheral districts. By the end of the 1960s, all the development in Brussels was strictly determined by the implacable logic of property speculation. At first, the "reconstruction" of Brussels was associated with a Plan dating from before 1914, which involved (among other things) the construction of the north-south rail link described earlier. The highly populated historic quarters that were destroyed by the construction of the railway were replaced by a wide urban motorway. By 1955, most of the Senne River had been culverted throughout the city.

The organisation in 1958 of the first world exhibition was one of the big drivers of modernisation, aimed at giving a more contemporary aspect to the town centre. The road network was transformed, tunnels and viaducts were constructed, the main access roads were re-aligned and the first plans for a ring road were drawn. These huge infrastructure works, which were meant to guarantee accessibility to what was a temporary exhibition, raised deep criticisms. One of the strong criticisms was that the works had made Brussels a city orientated exclusively towards the car and inhospitable to and not concerned about its inhabitants. Coupled with "clean up campaigns" in the centre, it appears that the authorities would have favoured the exodus of the inhabitants from the centre towards the peripheral districts and the transformation of Brussels into a purely administrative city. Indeed, from 1958 onwards, Brussels was selected as head office of the future European Community, which resulted in the whole region of Brussels becoming a vast "built-up" zone.

AD 1960–2000

Since 1960, nature management in Brussels can be divided into three stages:

1960s to the late 1970s – the battle against nature. This period involved the suppression of most of the spontaneous vegetation, as the population wanted to control and "improve" nature. Most species now considered to be extinct disappeared during this disastrous period.

End of 1970s to the end of 1980s – new management perspectives were created, which were in a strict nature conservation mould, in a museological sense. However, it was not yet a holistic perception of the urban ecosystem because it concerned species conservation (for example, prohibiting hunting), the biotope conservation (for example, prohibiting the use of weed-killers on public land) and the site conservation (for example, creating nature reserves). The urbanisation processes that occurred during the period involved the loss of farmland and wasteland for industrial, commercial and residential developments and the construction of the ring road and large hospitals on the periphery. New parks were created within exiting green areas.

1989 to the present day – the perception of urban biological resources has become more all-embracing, going beyond the protection of species, biotopes and sites, even to the extent of integrating landscape management. The major goal is to improve inhabitants' living conditions. Therefore, at the present time, the integration of nature into urban planning does not only include the protection of plants and animals but also aims to create a high quality urban environment. Hence, after having denied the existence of spontaneous nature in the city and trying to destroy it (after having separated it from the rest by enclosing it in nature reserves), the present strategies for the inclusion of nature in urban development are based on different actions:

(a) Increasing the total area and spatial distribution of green spaces.
(b) Increasing the ecological quality of green spaces.
(c) The development of ecological networks.
(d) The management of invasive species.

There is also a trend to promote some green sites with a more "natural" aspect. Indeed, urban parks have often been created according to an ideal image governed by order, regularity, cleanliness and salubrity criteria, which were particularly unfavourable to the spontaneous flora. This objective can be carried out with the help of better ecological management practices such as mechanical methods (for example, mowing) instead of chemicals.

Brussels, which has a population of 1.1 million people, has a similar concentric structure to that found in other cities (De Bruyn and Lannoy 1991 in IBGE 1995). It comprises four concentric zones, varying in urbanisation from the dense core to the rural fringe:

1. Core dominated by commercial and administrative activities with a limited residential function.

2. Densely built-up districts, which were constructed in the last century.
3. Periphery, which is less densely built-up.
4. Suburbs which can be considered as the maximum (or peak) demographic growth zone. This part of the city still retains a relatively rural environment.

About half (8,500 ha) of the city is occupied by vegetation. These green spaces are unequally spread throughout the city, both quantitatively and qualitatively, from an average greenness of 10% in the centre (mainly urban parks), to 30% or up to 71% in the periphery districts where there is more semi-natural vegetation including forests and relict agriculture areas. These open and green spaces can be typified as follows: private gardens (32%), forests (20%), public parks and gardens (12%), large private domains (10%), wasteland (7%), agriculture areas (7%), recreation areas (4%), green spaces linked with road system (3%), railway verges (3%) and cemeteries (2%) (Gryseels 1998).

Changes of the Environment Due to City Growth

Drainage of Marshes and the Creation of Impermeable Surfaces

Among numerous reasons for the drainage of the marshes are the culverting and diverting of watercourses, especially those related to the disposal of surface-water run-off. This is exacerbated by the lowering of the water table for more than a century as water is extracted from the aquifers to meet the needs of the growing human population. The water table is now so low that most of the sources in Brussels have dried up. In addition, since 1875 the flows in the watercourses have been low. As in many other large cities, the problem of water flows is particularly marked in Brussels where the soil is almost entirely covered by impermeable materials, preventing the seepage of precipitation into the groundwater. This problem was not so acute in the past because the roads were unsurfaced and buildings were much less abundant. Furthermore, the numerous "ponds" located on watercourses acted as "detention basins". It is only after the ponds were filled in that flooding occurred (Société Royale Belge de Géographie 1991).

Pollution

Another problem linked to the construction of hard surfaces is pollution caused by road run-off. In the winter, high concentrations of salt (NaCl) are found in ponds and watercourses as a result of the spreading of de-icing salt on roads and footpaths. The use of other chemical compounds, for example, herbicides, insecticides and fungicides on streets, footpaths and other areas has an adverse impact on Brussels' wetlands. Eutrophication of the rivers and lakes is also a major problem; the addition of nitrates and phosphates results in excessive plant growth, which produces a very high oxygen demand as it decays.

Flora

Vascular Plants

As in many large European cities, the geography, ecology and landscape of Brussels have combined to produce many diversified areas. These features include the Senne river and its tributaries and the associated wetlands, which when drained become highly productive market gardens, silty soils that are favourable to cereal production and a diversity of materials including sands (of different types), sandstones, clays, calcareous outcrops – all with different uses. In additon, the city is adjacent to hills and valleys that are clothed by ancient forests. Brussels is situated in an area of different biogeographical influences where plants of Atlantic origins grow together with medio-European ones. These features have combined to provide a mosaic of habitats that support a flora that is much more diverse than that of the rather monotonous surrounding areas.

The city supports 730 plant species (half the Belgian flora), which represents an average of 120 species of wild plants per km^2. The city centre, because of its intense urbanisation, is characterised by few wild plants (on average 50 species km^2); the peripheral areas show an amazing floristic richness (between 200 and 300 species km^2); see Fig. 5.

Other zones contain 160 or 180 species km^2. Half of the city flora is considered to be extremely rare with frequencies between 0 and 9%; see Fig. 6. At the scale of the phytogeographical district, this percentage falls to 8%. Very few species are widespread; only 7% of them are common to very common.

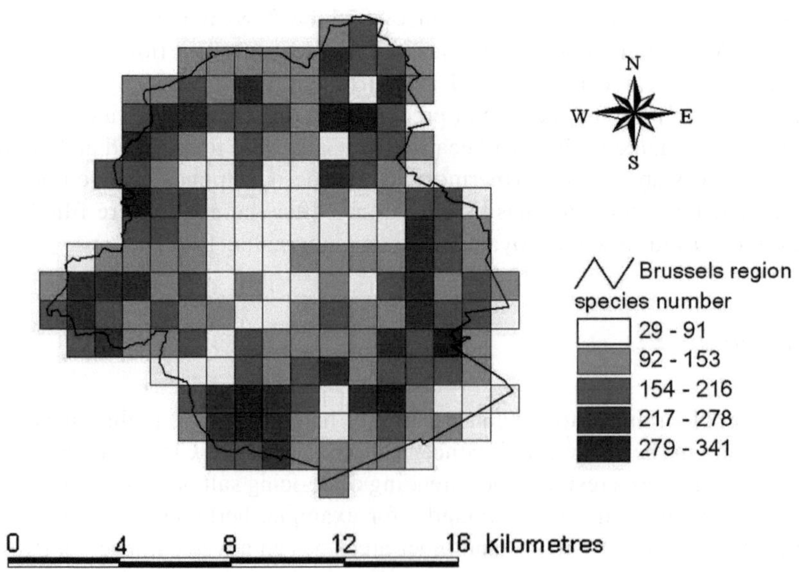

Fig. 5 Spatial distribution of plant species richness in Brussels. Each grid cell represents 1 km^2

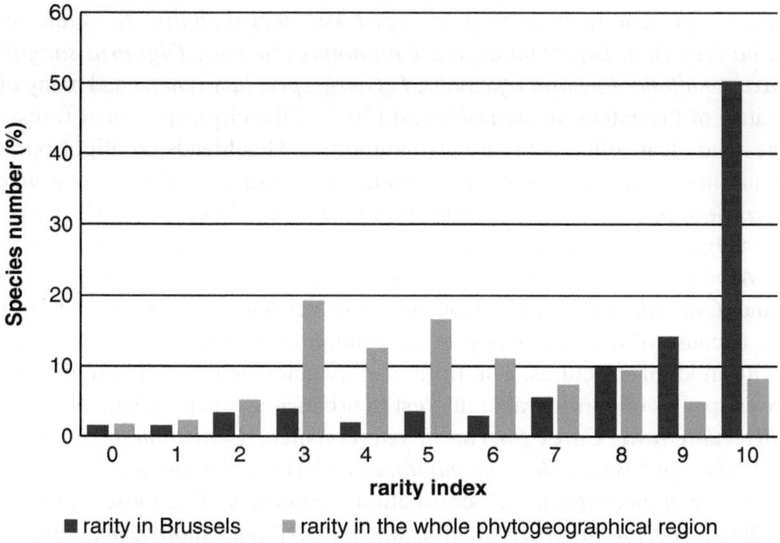

Fig. 6 Frequency distribution of the rarity index of Brussels flora, calculated for the city and for the phytogeographical district

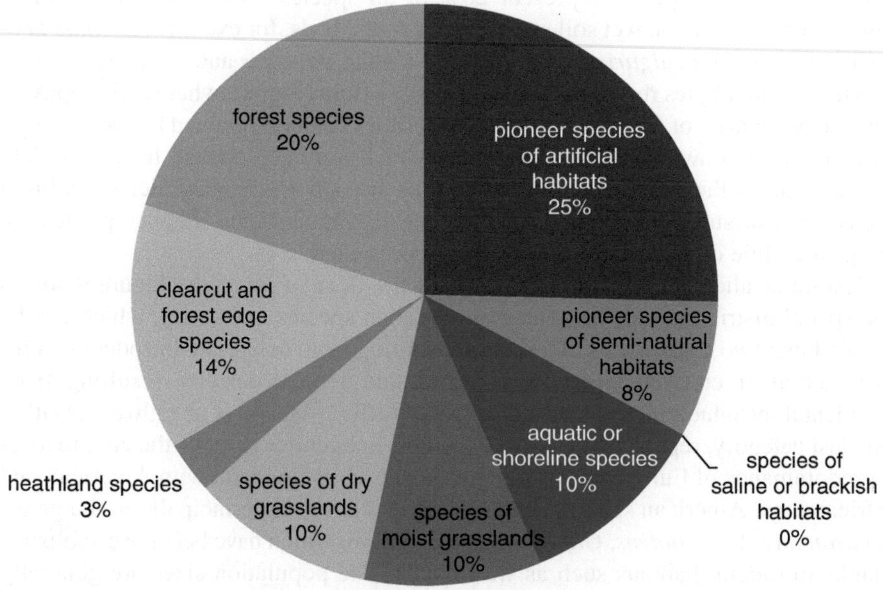

Fig. 7 Eco-sociological spectra of plant species recorded in Brussels

There is a predominance of pioneer species of artificial habitats, such as wastelands, road verges and fields (Fig. 7); these man-made and highly disturbed biotopes are very well represented in the city and support 25% of the urban flora.

The species present include *Ballota nigra* ssp. *meridionalis, Berteroa incana, Anisantha tectorum, Lepidium draba, Coronopus didymus, Digitaria sanguinalis, Diplotaxis tenuifolia, Fumaria officinalis, Legousia speculum-veneris* and many others.

Because of the extensive area of forest (10% of the city), species of forests, forest edges and clear-felled areas are also numerous. Heathlands on oligotrophic soil (which are not common in Brussels) contain species such as *Molinia caerulea* ssp. *caerulea, Galium saxatile, Carex pilulifera, Luzula multiflora, Vaccinium myrtillus, Calluna vulgaris* and *Melampyrum pratense*. The species found on walls and rocks include *Asplenium ruta-muraria, A. trichomanes, Erysimum cheiri, Pseudofumaria lutea* and *Cymbalaria muralis*. However, the vegetation of walls is now much reduced because of the destruction or restoration of the old walls or their replacement with modern structures. The flora of grasslands on dry soil is represented by numerous species of considerable interest in urban areas such as *Aira caryophyllea, Anthyllis vulneraria, Carex flacca, C. caryophyllea, Cerastium semidecandrum, Helictotrichon pubescens, Jasione montana,* and *Vulpia bromoides*.

Among the pioneer species of semi-natural habitats, we find those representative of the *Nanacyperion* such as *Centaurium pulchellum, Gnaphalium uliginosum, Hypericum humifusum, Isolepis setacea* and *Sagina apetala,* which are scarce and only occur in a small number of biotopes on open and humid soil, most often in humid post-cultural wastelands at the first stage of re-colonisation. Hydrophilous and hygrophilous species represent 20% of all species. Many of them occur in lawns and meadows on wet soil and reed and sedge beds, for example, *Lychnis flos-cuculi, Lysimachia vulgaris, Ajuga reptans, Caltha palustris* and *Carex disticha.*

Hemicryptophytes represent half of the city's flora (Fig. 8), whereas therophytes represent a quarter of the species. Such proportions can be explained by the presence of exposed land awaiting re-development, which has been unused for a sufficient period to allow the establishment of numerous therophytes. Some relict agricultural areas are also suitable for the establishment of annual plants. Phanerophytes are frequent, while other life forms are rarely represented.

Exotic or alien species represent 20% of the flora of Brussels. Figure 9 shows the spatial distribution of the percentage of alien species in the city, which can be divided into two categories: first, species resulting from deliberate introductions for ornamental or cultivation purposes (63%) and second, species resulting from accidental introductions, most often via road traffic, waterways or railways. Within this last category, representing 37% of the non-indigenous flora of the city, there is a predominance of European and Asiatic species in comparison with American and African taxa. American species present in Brussels belong principally to the genus *Amaranthus, Aster, Bidens, Galinsoga* and *Solidago,* which have become established thanks to ruderal habitats such as wastelands. The population sizes are generally low, except for *Solidago canadensis* and *S. gigantea,* whose appearance in a biotope is always followed by a rapid and spectacular increase.

The greatest threats to the indigenous vegetation are Asiatic species, particularly invasive species such as *Impatiens parviflora* (a native of Siberia and Turkestan), which has extensively colonised the field layer of numerous forest biotopes in Brussels. Other invasive Asiatic species include *Heracleum mantegazzianum* (Fig. 10), *Fallopia japonica* and *F. sachalinensis*.

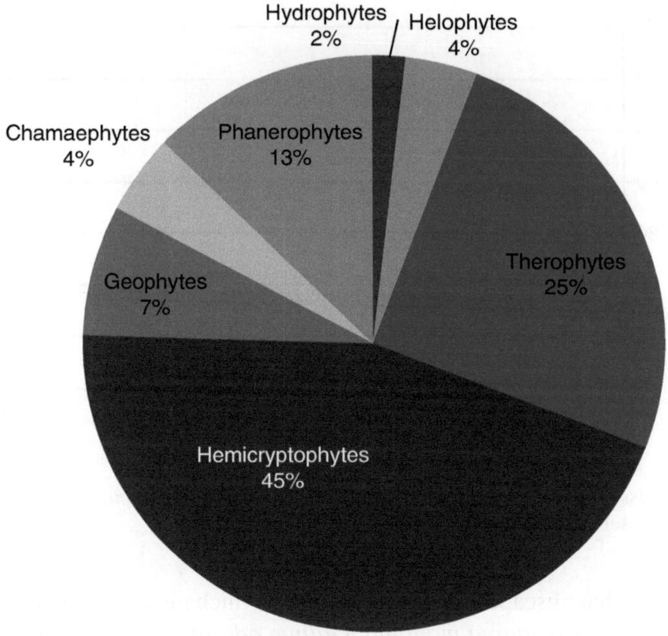

Fig. 8 Biological spectra of plant species recorded in Brussels

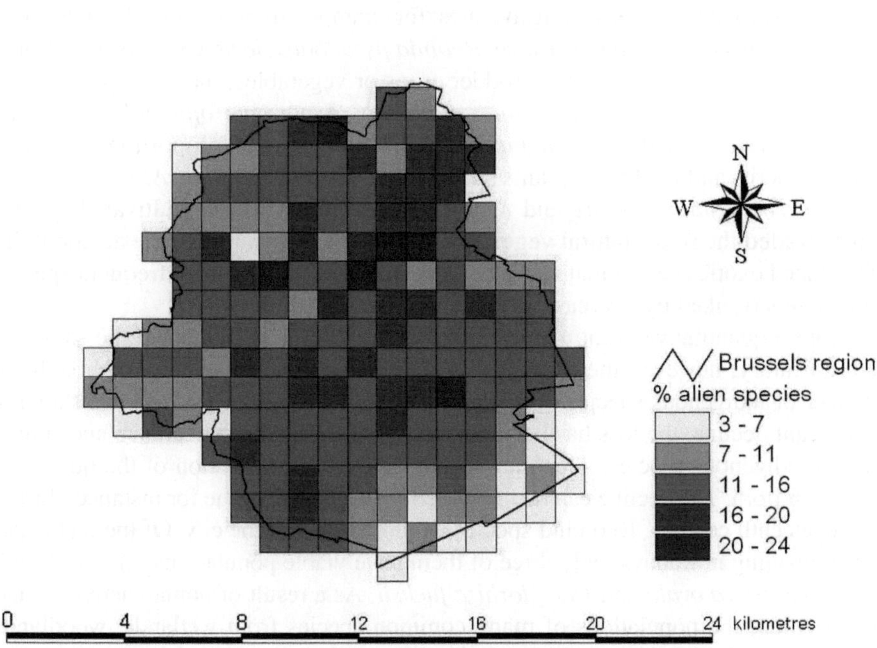

Fig. 9 Spatial distribution of the percentage of alien species in Brussels. Each grid cell represents 1 km²

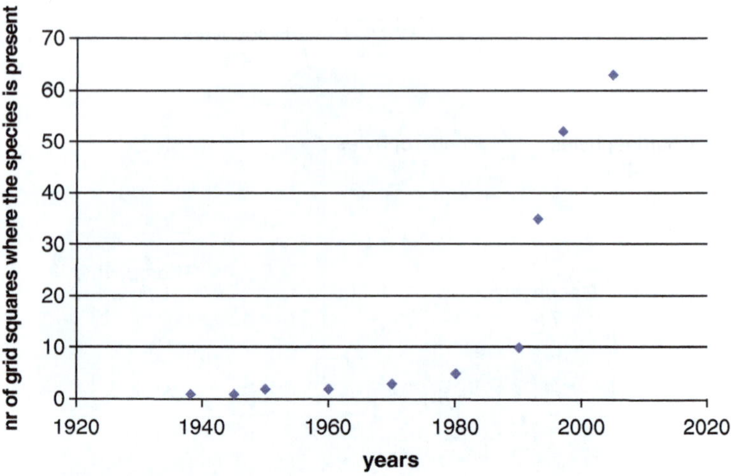

Fig. 10 Progression of the number of grid squares where *Heracleum mantegazzianum* has been recorded in Brussels

Fifty species have escaped from cultivation, of which most are ornamental gardens species, including *Centaurea montana, Galium odoratum, Impatiens glandulifera, Pseudofumaria lutea, Lathyrus latifolius, Lysimachia punctata, Muscari neglectum, Physalis alkekengi* var. *franchetii* and *Pentaglottis sempervirens*. Other species have self-seeded from trees along avenues, for example, *Acer platanoides, Castanea sativa, Juglans regia, Quercus rubra, Robinia pseudoacacia* and *Sorbus aria*. Some species were originally grown as fodder crops or vegetables; these species include *Medicago sativa* spp. *sativa, Rumex patientia, Asparagus officinalis, Brassica napus, B. oleracea* and *Trifolium incarnatum*. Finally, a small proportion comprises culinary herbs and medicinal plants such as *Artemisia absinthium, Allium schoenoprasum, Foeniculum vulgare* and *Mentha* x *piperita*. All these cultivated species have invaded the semi-natural vegetation of the city in contrast to the accidentally introduced exotic species that never become invasive. The 50 most frequent species in Brussels (ranked by decreasing frequency) are listed in Table 2.

From a quantitative point of view, the Brussels flora has remained the same for half a century; however, there has been a qualitative impoverishment. Although the number of indigenous species of wetlands, woodlands and arable land has shown a significant decline, the loss has been compensated for by the appearance and spread of non-indigenous species, which are responsible for the reduction of the quality of the urban flora. The decline can be illustrated by the Orchidaceae for instance: during the nineteenth century, 13 orchid species were recorded in the city. Of the eight species remaining nowadays; only three of them have viable populations, *Epipactis helleborine, Listera ovata* and *Dactylorhiza fuchsii*. As a result of human activities and urbanisation, the populations of many common species from wetlands, woodlands and arable land (Fig. 11) have declined since the 1970s, including *Matricaria recutita, Apera spica-venti, Spergula arvensis, Valeriana repens, Typha latifolia, Lythrum salicaria, Arum maculatum, Hyacinthoides non-scripta,* and *Adoxa moschatellina*.

Table 2 The 50 most frequent species in Brussels (ranked by decreasing frequency)

Species	Frequency (%)	Species	Frequency (%)
Cirsium arvense	98.94	*Galium aparine*	80.42
Poa annua	98.41	*Heracleum sphondylium*	80.42
Urtica dioica	98.41	*Sisymbrium officinale*	80.42
Plantago major ssp. *major*	97.88	***Conyza canadensis***	79.89
Stellaria media ssp. *media*	97.35	*Plantago lanceolata*	79.89
Sambucus nigra	96.83	*Achillea millefolium*	78.31
Acer pseudoplatanus	95.77	*Dactylis glomerata*	77.78
Trifolium repens	94.71	*Persicaria maculosa*	77.25
Ranunculus repens	94.18	*Sagina procumbens*	77.25
Artemisia vulgaris	92.06	*Sonchus asper*	76.72
Tussilago farfara	92.06	*Sonchus oleraceus*	76.72
Rumex obtusifolius ssp. *obtusifolius*	91.53	*Sorbus aucuparia*	76.72
Salix caprea	91.53	*Geum urbanum*	75.66
Betula pendula	88.89	*Prunus avium*	75.66
Capsella bursa-pastoris	87.30	*Cerastium fontanum ssp. vulgare*	75.13
Cirsium vulgare	87.30	*Geranium robertianum*	75.13
Senecio vulgaris	87.30	***Matricaria discoidea***	74.07
Bellis perennis	86.77	***Acer platanoides***	73.54
Lolium perenne	86.77	*Aegopodium podagraria*	73.54
Calystegia sepium	86.24	*Arrhenatherum elatius*	73.02
Hedera helix	85.71	*Arctium minus*	72.49
Equisetum arvense	80.95	***Buddleja davidii***	71.96
Fraxinus excelsior	80.95	*Crataegus monogyna*	71.96
Lapsana communis ssp. *communis*	80.95	*Chamerion angustifolium*	71.96
Polygonum aviculare agg.	80.95	***Fallopia japonica***	71.96

Bold = neophytes

The average decline in the populations, over 20 years, is around 35%. This represents a high rate for species that are considered to be rather common around Brussels. The main reasons for the decline is the drastic reduction of arable land, the impoverishment of forest areas because of intensive recreation, the natural dynamics of wetlands and uncontrolled urbanisation.

In parallel with the marked decline of indigenous species, there has been an increase in the number and abundance of exotic plants, introduced incidentally or deliberately, such as *Buddleja davidii, Heracleum mantegazzianum, Senecio inaequidens, Ceratochloa carinata* and *Fallopia japonica.* These species have become established, often in great quantity in various biotopes, where they replace indigenous species. The spatial distribution of the species regression percentage shows that the centre of the city has lost less species than the suburbs, meaning that population extinction is mainly occurring in the peripheral areas of the city (Fig. 12).

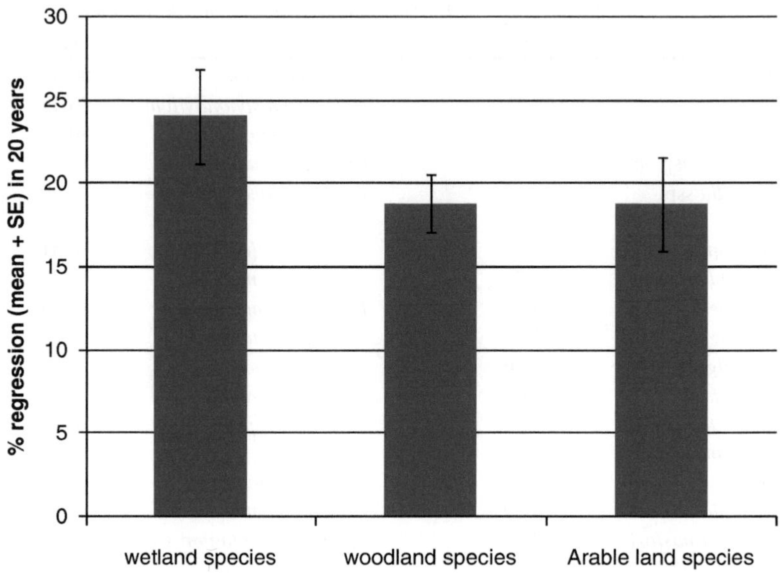

Fig. 11 Percentage regression of the most vulnerable species in Brussels between 1972 and 1994

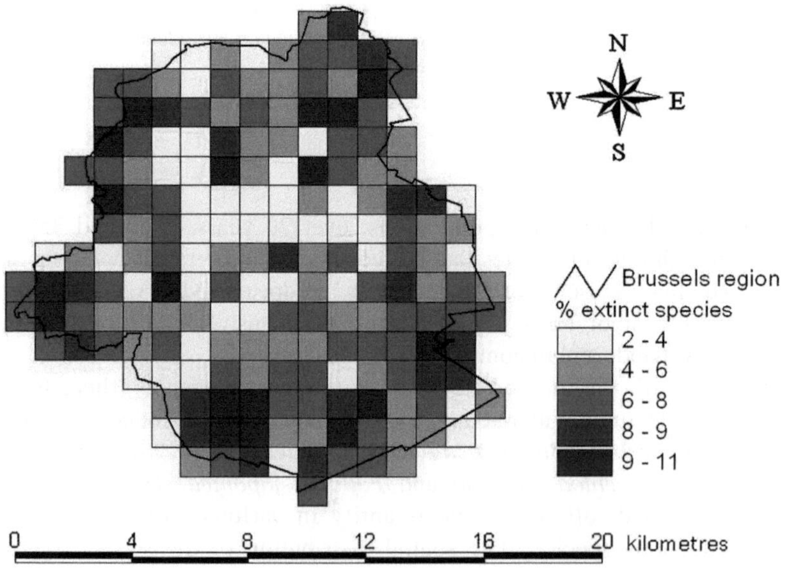

Fig. 12 Spatial distribution of species extinctions in the Brussels flora; each grid cell represents 1 km^2

Planted Trees and Shrubs

These days, there are about 90,000 trees along roads. They comprise 167 different taxa, including species, hybrids and numerous varieties, forms and cultivars (Geerinck 1998). Some species are only represented by one tree while there are many thousands of individuals of other taxa, for example, *Platanus* x *hispanica, Tilia* spp. and *Prunus serrulata.* Coniferous species are much more frequent in parks than along the roadsides.

Algae, Mosses and Liverworts

According to Vanderpoorten (1997), the bryophyte flora of Brussels comprises 225 species (40 hepatics and 185 musci), which represent about 33% of the Belgian bryophyte flora. The mean species richness is 34 species km^2 but it varies from less than 20 species km^2 to a maximum of 103 species km^2. Land use has a clear influence on the distribution of this species richness. Rich grid cells with more than 60 species km^2 occur in the southern part of the city, which include the largest forest of the area. The area is locally characterised by the presence of *Capylopgeia muelleriana, Campylopus flexuosus, C. pyriformis, Dicranum montanum, Dicranodontium denudatum, Leucobryum glaucum* and *Plagiothecium undulatum.* Some places in the forest area are particularly remarkable for the richness of their bryophytes, as a result of the presence of habitats such as calcareous slopes with characteristic species including *Plagiochila porelloides, Encalypta streptocarpa, Plagiothecium cavifolium* and *Fraxinus-Salix* woods including old *Sambucus* on which epiphytes such as *Orthotrichum obtusifolium* and *Cryphaea heteromalla* can be found. The southern part of Brussels also includes old sand quarries with silicicolous species including *Diplophyllum obtusifolium, Gymnocolea inflata, Lophozia bicrenata, Pogonatum nanum* and *Pohlia annotina.* In the northern part of the city, a complex of marshes, old *Salix* woodland and ponds possesses rich assemblages of epiphytes, for example, *Orthotrichum pallens, O. pulchellum, O. pumilum, Tortula papillosa,* and *T. virescens;* aquatic species, such as *Riccia fluitans* and pioneer species of pond margins including *Aphanorhegma patens* and *Physcomitrium pyriflorme.*

Some of the species recorded in Brussels are of great interest because of their rarity in Belgium; this is the case for *Calypogeia azurea, Rhynchostegiella curviseta, Riccia subbifurca, Tortula papillosa* and *T. virescens.* One species, *Ephemerum stellatum* (a European "Vulnerable" species), has its only Belgian population in Brussels.

Fungi (including Lichenised Fungi)

So far, Vanholen and De Kesel (2000) have recorded 1,306 species of fungi in the city, of which 58% are very rare and 19% are rare. Only a relatively small number of species are common (7%) or very common (2%), the most common species being

Hypoxylon fragiforme. The most common saprophytes are *Stereum hirsutum* and *Trametes versicolor;* the three most common mycorrhizal fungi are *Scleroderma citrinum, Laccaria amethystina* and *Russula ochroleuca*. At the present time, the most common of the "disease-causing" fungi are *Ganoderma lipsiense, Nectria cinnabarina* and *Armillaria mellea*.

Historical and recent observations have recorded a total of 86 lichen species in the city. A recent study (Vanholen and De Kesel 2000) recorded 21 species in 6 green spaces, the most common species being *Lepraria incana;* the rarest are *Chaenotheca ferruginea, Platismatia glauca* and *Usnea* cf. *subfloridana*. The most common epiphytic species recorded are relatively pollution-tolerant, which indicates that the air quality of Brussels could still be improved.

Habitats

Brussels is a green city with many different semi-natural vegetation types, including nine listed in Annex 1 of the EC Habitats Directive, 92/43; some of these occur in the three Special Areas of Conservation that have been designated, which cover in total 1,900 ha (12% of the surface of the city). Table 3 gives the extent and patchiness of the most characteristic habitats, which are described below.

Forests

(a) Limestone *Fagus* forest with *Mercuralis perennis* (EUH codes 9130 and 9150; Corine codes 41.13 and 41.16).

Phytosociology: *Asperulo-Fagetum; Carici-Fagetum.*
Topography: only on the slope of valleys.
Pedology: this unit is represented on the outcrops of calcareous sands.
Characteristic species:

- Canopy: *Fagus sylvatica, Acer pseudoplatanus, Carpinus betulus, Fraxinus excelsior, Prunus avium* and *Quercus robur.*
- Shrub layer: *Acer campestre, Cornus sanguinea, Euonymus europaeus, Ligustrum vulgare, Viburnum opulus, Acer platanoides, A. pseudoplatanus, Carpinus betulus, Corylus avellana, Crataegus monogyna, Fagus sylvatica, Fraxinus excelsior, Prunus avium, Ribes uva-crispa, Sambucus nigra, Tilia platyphyllos* and *Ulmus minor.*
- Herb layer: *Clematis vitalba, Melica uniflora, Mercuralis perennis, Paris quadrifolia, Phyteuma nigrum, Primula elatior, Tamus communis, Ajuga reptans, Anemone nemorosa, Arum maculatum, Brachypodium sylvaticum, Campanula trachelium, Carex flacca, Circaea lutetiana, Deschampsia*

Table 3 Synoptic table showing the importance of the main vegetation units in Brussels in terms of extent (area) and patchiness (nr. of polygons)

Vegetation unit	Nr. of polygons	Total area (m^2)	Min. patch area (m^2)	Max. patch area (m^2)	Mean. patch area (m^2)
Fagus forest with ferns*	80	3,199,895	83	371,993	39,999
Fagus forest with *Milium effusum,* ferns and *Rubus* sp.*	47	3,165,850	965	425,275	67,359
Young plantation: coppice under scattered full-grown trees with scarce ground vegetation (ferns)	61	1,599,093	689	121,078	26,215
Fagus forest with *Anemone nemorosa and Hyacinthoides non-scripta**	9	1,113,339	12,969	349,807	123,704
Mixed *Quercus* forest*	55	1,087,956	193	92,381	19,781
Conifer plantation	99	1,074,137	445	112,926	10,850
Fagus forest with *Deschampsia flexuosa*	34	693,781	162	106,304	20,405
Fagus or *Quercus* forest with *Pteridium aquilinum and Lonicera periclymenum*	21	668,241	991	107,617	31,821
Fagus forest with *Luzula sylvatica*	38	631,283	1,055	71,350	16,603
Mixed *Fraxinus-Quercus* forest*	40	500,126	585	66,643	12,503

(continued)

Table 3 (continued)

Vegetation unit	Nr. of polygons	Total area (m^2)	Min. patch area (m^2)	Max. patch area (m^2)	Mean. patch area (m^2)
Young close plantation without ground flora	46	401,782	883	51,986	8,734
Young plantation in an opening	46	397,315	630	58,647	8,637
Opening	24	329,537	710	107,120	13,731
Fagus forest with *Deschampsia flexuosa and Luzula sylvatica*	11	207,590	1,273	56,631	18,872
Quercus forest with *Hyacinthoides non-ncriyta**	6	172,540	5,430	53,392	28,757
Fraxinus forest with *Carex* and related communities*	6	132,090	14,080	37,493	22,015
Ponds*	13	114,991	1,109	39,680	8,645
Young plantation: coppice with scare ground vegetation (ferns)	4	110,530	4,904	41,597	27,633
Grasslands*	4	58,999	5,203	26,630	14,750
Fagus forest with *Leucobryum glaucum*	6	58,515	1,223	22,599	9002C753
Wetlands *sensu lato**	6	55,020	924	28,216	9,176
Quercus forest with *Vaccinium myrtillus* and *Deschampsia flexuosa*	5	48,152	4,206	16,696	9,630
Betula forest	5	46,527	992	35,095	9,305

Limestone beech forest with *Mercurialis perennis**	3	38,757	4,473	26,327	12,919
Fagus or *Quercus* forest with *Lamium galeobdolon*	7	38,564	1,683	10,129	5,509
Bare beech forest	6	36,516	1,270	14,334	6,086
Quercus forest with *Hyacinthoides non-scripta* and *Pteridium aquilinum**	3	23,174	1,892	17,615	7,725
Fagus forest with *Molinia caerulea* and *Calluna vulgaris**	3	14,948	1,828	7,660	4,983
Salix and *Alnus* groves	1	2,819	2,819	2,819	2,819

Items are ranked in order of decreasing total area. Habitat types listed in the Directive 92/43/EEC are marked by an asterisk *cespitosa, Geranium robertianum, Geum urbanum, Hedera helix, Lamiastrium galeobdolon, Polygonatum multiflorum, Potentilla sterilis, Sanicula europaea* and *Urtica dioica.*

(b) Mixed *Fraxinus-Quercus* forest (EUH code 9160; Corine code 41.24).

Phytosociology: *Primulo-Carpinetum.*
Topography: essentially localised on the upper flanks of the valleys; the unit is also sometimes found at the bottom of the slopes.
Pedology: this unit develops on rather deep, nutrient-rich and moist loamy soils, on deep and aerated colluvium and on deep well-drained alluvium.
Characteristic species:

- Canopy: *Fraxinus excelsior, Acer pseudoplatanus, Aesculus hippocastanum, Fagus sylvatica, Quercus petraea* and *Q. robur.*
- Shrub layer: *Acer pseudoplatanus, Fraxinus excelsior, Sambucus nigra, Aesculus hippocastanum, Betula pendula, Carpinus betulus, Fagus sylvatica,* and *Salix caprea.*
- Herb layer: very dense and rather diversified. The following species are dominant or frequent: *Cardamine flexuosa, Circaea lutetiana, Geum urbanum, Glechoma hederacea, Impatiens parviflora, Milium effusum, Urtica dioica,* seedlings of *Acer pseudoplatanus* and *Fraxinus excelsior*. Rather frequent: *Ajuga reptans, Lamiastrium galeobdolon, Ranunculus repens, Rubus* sp., *Veronica montana.* Rare: *Adoxa moschatellina, Anemone nemorosa, Arum maculatum, Athyrium filix-femina, Carex remota, C. sylvatica, Cirsium vulgare, Deschampsia cespitosa, Dryopteris carthusiana, D. dilatata, D. filix-mas, Impatiens noli-tangere, Lysimachia nemorum, Moehringia trinervia, Oxalis acetosella, Paris quadrifolia, Pteridium aquilinum, Scrophularia nodosa* and *Viola riviniana.*

(c) *Fagus* forest with *Anemone nemorosa* and *Hyacinthoides non-scripta* (EUH code 9120; Corine code 41.12).

Phytosociology: *Milio-Fagetum.*
Topography: essentially localised on plateaux.
Pedology: on loamy soils.
Characteristic species:

- Canopy: *Fagus sylvatica, Larix* sp. and *Quercus robur.*
- Shrub layer: Weak or none: *Betula pendula, Carpinus betulus, Fagus sylvatica* and *Sambucus nigra.*
- Herb layer: *Anemone nemorosa, Dryopteris dilatata, D. carthusiana, Hyacinthoides non-scripta, Rubus* sp. and *Milium effusum* are common. *Holcus mollis* may sometimes dominate. Monospecific covers of *Convallaria majalis, Lonicera periclymenum* and *Maianthemum bifolium* are locally present. Places where the soil has been compacted by logging activities are locally colonised by *Carex remota, Deschampsia cespitosa, Juncus effusus, Impatiens*

parviflora and *Persicaria hydropiper.* Species that occur sporadically: *Athyrium filix-femina, Circaea lutetiana, Dryopteris filix mas, Glechoma hederacea, Lamiastrium galeobdolon, Lysimachia nemorum, Luzula pilosa, Moehringia trinervia, Polygonatum multiflorum, Pteridium aquilinum* and *Ranunculus ficaria.*

(d) Mixed *Quercus* forest (EUH code 9160; Corine code 41.24).

Phytosociology: *Primulo-Carpinetum.*
Topography: very diversified: the unit is present at the bottom of valleys as well as on hillsides or plateaus.
Pedology: on nutrient-rich and deep loamy soils.
Characteristic species:

- Canopy: *Acer pseudoplatanus, Fraxinus excelsior, Quercus robur, Aesculus hippocastanum, Alnus glutinosa, A. incana, Betula pendula, Carpinus betulus, Castanea sativa, Fagus sylvatica, Salix caprea, Sorbus aucuparia* and *Tilia platyphyllos.*
- Shrub layer: *Acer pseudoplatanus, Carpinus betulus, Sambucus nigra, Corylus avellana, Fagus sylvatica, Fraxinus excelsior, Prunus avium, P. serotina, Sambucus racemosa, Sorbus aucuparia, Ulmus* sp. Variant on nutrient-rich to calcareous soils: *Cornus sanguinea, Ribes rubrum, R. uva-crispa* and *Symphoricarpos albus.*
- Herb layer: *Anemone nemorosa, Circaea lutetiana, Hedera helix, Milium effusum, Athyrium filix-femina, Dryopteris carthusiana, Geum urbanum, Glechoma hederacea, Lamiastrium galeobdolon, Lysimachia nummularia, Oxalis acetosella, Pteridium aquilinum, Scrophularia nodosa, Silene dioica* and *Veronica montana.* Variant on nutrient-rich to calcareous soils: *Adoxa moschatellina, Arum maculatum, Geranium robertianum, Paris quadrifolia, Polygonatum multiflorum, Primula elatior, Sanicula europaea* and *Tamus communis. Impatiens parviflora* (a locally invasive species) occurs frequently, as does *Carex remota* but to a lesser extent. Patches of *Galium aparine* and *Urtica dioica* are not uncommon.

(e) *Fagus* or *Quercus* forest with *Lamiastrium galeobdolon* (EUH code 9160; Corine code 41.24).

Phytosociology: *Primulo-Carpinetum lamietosum.*
Topography: in the bottom of the valleys.
Pedology: on moist and deep soils.
Characteristic species:

- Canopy: either *Fagus sylvatica* or *Quercus robur, Fagus sylvatica, Fraxinus excelsior* and *Populus alba.*
- Shrub layer: Either *Fagus sylvatica* or *Acer pseudoplatanus, Fagus sylvatica, Fraxinus excelsior* and *Populus alba.*

- Herb layer: *Circaea lutetiana, Lamiastrium galeobdolon, Milium effusum, Athyrium filix-femina, Carex remota, Dryopteris carthusiana, D. dilatata, Oxalis acetosella, Ranunculus ficaria* and *Urtica dioica.*

(f) *Qercus* forest with *Hyacinthoides non-scripta* (EUH code 9130; Corine code 41.13).

Phytosociology: *Endymio-Carpinetum typicum.*
Topography: on plateaux.
Pedology: on fairly deep loamy soils.
Characteristic species:

- Canopy: *Quercus robur, Acer pseudoplatanus, Prunus avium, P. serotina* and *Quercus petraea.*
- Shrub layer: *Acer pseudoplatanus, Corylus avellana, Sambucus nigra, Ilex aquifolium, Prunus avium, P. spinosa, Ribes rubrum* and *Sorbus aucuparia.*
- Herb layer: *Anemone nemorosa, Athyrium filix-femina, Hyacinthoides non-scripta, Rubus* sp., *Circaea lutetiana, Dryopteris dilatata, Hedera helix, Holcus mollis, Lamiastrium galeobdolon, Milium effusus* and *Lonicera periclymenum* and *Polygonatum multiflorum.* Variant on nutrient-rich and moist soils: *Deschampsia cespitosa, Festuca gigantea, Glechoma hederacea, Melica uniflora* and *Paris quadrifolia.*

(g) *Quercus* forest with *Hyacinthoides non-scripta* and *Pteridium aquilinum* (EUH code 9130; Corine code 41.13).

Phytosociology: *Endymio-Carpinetum holcetosum.*
Topography: on plateaux.
Pedology: on fairly deep loamy soils.
Characteristic species:

- Canopy: *Quercus petraea, Q. robur, Acer pseudoplatanus, Betula pendula* and *Prunus serotina.*
- Shrub layer: *Acer pseudoplatanus, Corylus avellana, Prunus serotina* and *Sorbus aucuparia.*
- Herb layer: *Hyacinthoides non-scripta, Pteridium aquilinum, Rubus* sp., *Dryopteris carthusiana, Lonicera periclymenum* and *Milium effusum.*
- Moss layer: *Polytrichum formosum.*

(h) *Fagus* forest with *Milium effusum,* ferns and *Rubus* sp. (EUH code 9120; Corine code 41.12).

Phytosociology: *Milio-Fagetum.*
Topography: mainly on plateaux.
Pedology: on loamy soils.
Characteristic species:

- Canopy: *Fagus sylvatica, Quercus petraea* and *Q. robur,*. Sometimes mixed with *Larix* sp.

- Shrub layer: Weak or none: *Acer pseudoplatanus, Fagus sylvatica, Sambucus ebulus* and *Sorbus aucuparia.*
- Herb layer: *Athyrium filix-femina, Dryopteris carthusiana, D. dilatata, Milium effusum, Rubus* sp., *Anemone nemorosa, Calamagrostis epigeios, Circaea lutetiana, Convallaria majalis, Deschampsia cespitosa, Epilobium montanum, Festuca gigantea, Galeopsis tetrahit, Holcus mollis, Lonicera periclymenum, Luzula pilosa, L. sylvatica, Oxalis acetosella, Poa nemoralis, Scrophularia nodosa* and *Teucrium scorodonia.* The following species occur where the soil has been compacted: *Carex remota, Juncus effusus, Persicaria hydropiper* and at the foot of *Fagus* trees, *Deschampsia cespitosa* with the moss *Leucobryum glaucum.*

(i) *Fagus* forest with ferns (EUH code 9120; Corine code 41.12).

Phytosociology: *Milio-Fagetum.*
Topography: in shady valleys, in deeply embanked ravines, on northern slopes and on plateaus.
Pedology: mainly developed on moist loamy soils.
Characteristic species:

- Canopy: *Fagus sylvatica, Quercus robur* and *Q. petraea* (*Quercus* is dominant at the bottom of valleys).
- Shrub layer: Weak or none: *Fagus sylvatica* (very rare), *Sambucus nigra, S. racemosa* and *Salix capraea.*
- Herb: *Dryopteris carthusiana, D.dilatata, Athyrium filix-femina, Oxalis acetosella* and *Rubus* sp. are frequent while *Carex pilulifera, Luzula pilosa, Milium effusum, Moehringia trinervia, Teucrium scorodonia* and *Pteridium aquilinum* are scattered. The following species occur where the soil has been compacted: *Carex remota, Deschampsia cespitosa, Impatiens parviflora, Juncus effusus* and *Persicaria hydropiper. Holcus mollis* may be locally dominant. In openings following clear-felling or windthrow, *Rubus* may form an almost impenetrable scrub.

(j) *Fagus* or *Quercus* forest with *Pteridium aquilinum* and *Lonicera periclymenum* (EUH and Corine codes – not listed).

Phytosociology: *Fago-Quercetum convallarietosum.*
Topography: on plateaux and changes of slope, sometimes in contact with conifer plantations.
Pedology: on well-drained acidic soils.
Characteristic species:

- Canopy: either *Fagus sylvatica* and *Betula pendula* or *Betula pendula, Quercus petraea, Q. robur, Q. rubra, Castanea sativa, Prunus serotina* and *Sorbus aucuparia,* sometimes mixed with conifers, such as *Pinus sylvestris* and *Larix* sp.
- Shrub layer: Either *Fagus sylvatica, Frangula alnus* and *Sorbus aucuparia* or *Frangula alnus, Prunus serotina, Sorbus aucuparia, Acer pseudoplatanus, Carpinus betulus* and *Fagus sylvatica.*

- Herb layer: *Lonicera periclymenum, Maianthemum bifolium, Oxalis acetosella, Pteridium aquilinum, Rubus* sp., *Teucrium scorodonia, Dryopteris carthusiana, D. dilatata, Hedera helix, Holcus mollis, Luzula pilosa, L. sylvatica* and *Milium effusum*. Patches of *Impatiens parviflora* and *Juncus effusus* occur locally. *Deschampsia flexuosa* is frequently present at the foot of beeches.
- Moss layer: *Polytrichum formosum.*

(k) *Quercus* forest with *Vaccinium myrtillus* and *Deschampsia flexuosa* (EUH and Corine codes – not listed).

Phytosociology: *Querco-Betuletum.*
Topography: on plateaux, on changes of slope and on hillside.
Pedology: on well-drained acidic loamy sands.
Characteristic species:

- Canopy: *Betula pendula, Quercus petraea, Q. robur, Castanea sativa* and *Fagus sylvatica,* sometimes *Larix* sp.
- Shrub layer: *Sorbus aucuparia, Fagus sylvatica, Frangula alnus, Prunus serotina, Quercus petraea* and *Q. robur.*
- Herb layer: *Deschampsia flexuosa, Galium saxatile, Molinia caerulea* ssp. *caerulea, Vaccinium myrtillus, Carex pilulifera, Dryopteris carthusiana, D. dilatata. Holcus mollis, Lonicera periclymenum, Luzula multiflora, Maianthemum bifolium, Oxalis acetosella, Pteridium aquilinum, Rubus* sp., *Teucrium scorodonia* and *Veronica officinalis.* Patches of *Impatiens parviflora* occur occasionally.
- Moss layer: *Leucobryum glaucum* and *Polytrichum formosum.*

(l) *Fagus* forest with *Deschampsia flexuosa* and *Luzula sylvatica* (EUH and Corine codes – not listed).

Phytosociology: *Fago-Quercetum* (intermediate variant between the *F.Q. convallarietosum* and the *F.Q. luzuletosum sylvaticae*).
Topography: on plateaux.
Pedology: on loamy sands, on sands with loess and on well-drained acidic soils.
Characteristic species:

- Canopy: *Fagus sylvatica.*
- Shrub layer: Poorly developed or absent; *Betula pendula, Fagus sylvatica, Sambucus nigra* and *S. racemosa.*
- Herb layer: *Deschampsia flexuosa, Luzula sylvatica, Carex pilulifera, C. ovalis, C. pallescens, Dryopteris carthusiana, D. dilatata, Chamerion angustifolium, Galium saxatile* (very rare), *Holcus mollis, Hypericum pulchrum, Luzula multiflora, L. pilosa, Rubus* sp. and *Urtica dioica.* The following species occur where the soil has been compacted: *Carex remota, Deschampsia cespitosa, Impatiens parviflora, Juncus effusus* and *Persicaria hydropiper. Lonicera periclymenum, Pteridium aquilinum* and *Teucrium scorodonia* are found occasionally.
- Moss layer: *Leucobryum glaucum* and *Polytrichum formosum.*

(m) *Fagus* forest with *Luzula sylvatica* (EUH and Corine codes– not listed).

Phytosociology: *Fago-Quercetum luzuletosum sylvaticae.*
Topography: developed on shady aspects of moderate to rather steep slopes in small valleys and ravines.
Pedology: on loamy sands, on sands with loess and on well-drained acidic soils.
Characteristic species:

- Canopy: *Fagus sylvatica.*
- Shrub layer: Poorly developed or absent; *Betula pendula, Fagus sylvatica, Sambucus nigra* and *S. racemosa.*
- Herb layer: *Luzula sylvatica* may have a continuous monospecific ground cover over huge areas; it may also form extended patches accompanied with the guild of the beech forest with *Deschampsia flexuosa* and *Luzula sylvatica* (unit l) with the following species: *Blechnum spicant, Calluna vulgaris, Cytisus scoparius, Molinia caerulea* and *Sorbus aucuparia.*

(n) *Fagus* forest with *Deschampsia flexuosa* (EUH and Corine codes – not listed).

Phytosociology: *Fago-Quercetum.*
Topography: developed over a large area on slightly sloping plateaux; localised fragments on ridges.
Pedology: on well-drained nutrient-poor acidic soils on loams, loamy-sands or loamy-stony substrates.
Characteristic species:

- Canopy: *Fagus sylvatica.*
- Shrub layer: Poorly developed or absent; *Betula pendula, Fagus sylvatica, Sambucus nigra* and *S. racemosa.*
- Herb layer: *Deschampsia flexuosa* may sometimes form a dense carpet covering extensive areas or may form tufts accompanied by species of the beech forest with *Deschampsia flexuosa* and *Luzula sylvatica*. Vegetation type 1, see previous page *D. flexuosa* may form a ring around old beech trunks in open communities.

(o) *Fagus* forest with *Leucobryum glaucum* (EUH and Corine codes – not listed).
Phytosociology: *Fago-Quercetum leucobryetosum.*
Topography: on slight slopes.
Pedology: on very dry and nutrient-poor soils (podzols) developed on sandy substrates, in outcrop areas of flint and in areas where the wood was formerly burned.
Characteristic species:

- Canopy: *Fagus sylvatica.*
- Shrub layer: Weak or none: *Betula pendula, Fagus sylvatica, Sambucus nigra* and *S. racemosa.*
- Herb layer: except for *Deschampsia flexuosa,* the herb layer is species-poor and looks like that of the "beech forest with *Deschampsia flexuosa*".
- Moss layer: *Leucobryum glaucum* may form, in some places within the beech forest with *Deschampsia flexuosa,* a multitude of small cushions giving a particular physiognomy to the community.

(p) *Fagus* forest with *Molinia caerulea* and *Calluna vulgaris* (EUH and Corine codes – not listed).

Phytosociology: *Fago-Quercetum molinietosum.*
Topography: on plateaux.
Pedology: on well-drained acidic soils.
Characteristic species:

- Canopy: *Fagus sylvatica.*
- Shrub layer: Poorly developed or absent; *Betula pendula, Fagus sylvatica* and *Sambucus racemosa.*
- Herb layer: *Calluna vulgaris, Deschampsia flexuosa, Molinia caerulea, Agrostis capillaris, Carex pilulifera, Dryopteris carthusiana, D. dilatata, Galium saxatile, Holcus mollis, Luzula multiflora, L. sylvatica, Rubus* sp. and *Vaccinium myrtillus*. Rarely: *Lonicera periclymenum, Oxalis acetosella* and *Teucrium scorodonia.*
- Moss layer: *Leucobryum glaucum* and *Polytrichum formosum.*

(q) *Betula* forest (EUH and Corine codes – not listed).

Phytosociology: *Querco-Betuletum typicum.*
Topography: variable.
Pedology: variable.
Characteristic species:

- Canopy: *Betula pendula* and *Fagus sylvatica,* sometimes mixed with conifers such as *Larix* sp.
- Shrub layer: *Betula pendula, Quercus petraea, Sorbus aucuparia, Cytisus scoparius* and *Fagus sylvatica.*
- Herb layer: *Dryopteris carthusiana, D. dilatata, Lonicera periclymenum, Luzula sylvatica* and *Milium effusum.* Variant on more acidic substrates: *Calluna vulgaris, Deschampsia flexuosa, Carex pilulifera, Chamerion angustifolium* and *Luzula sylvatica.*

(r) *Fraxinus* forest with *Carex* and related communities (EUH code 91EO; Corine codes 44.3; 44.2 and 44.13).

Phytosociology: *Carici-remotae Fraxinetum* Koch ex Faber 1936.
Topography: related to brook alluvium and seepage creeks and springs with a calcium-rich water.
Pedology: on soils with superficial and fluctuating watertable (0–20 cm), where calcareous sands are in contact with Ypresian clay.
Characteristic species:

- Canopy: *Fraxinus excelsior, Alnus glutinosa* and *Quercus robur.*
- Shrub layer: *Fraxinus excelsior, Acer pseudoplatanus, Alnus glutinosa, Carpinus betulus, Corylus avellana, Prunus padus, Salix caprea, Sambucus nigra* and *Viburnum opulus.*
- Herb layer: the characteristically dominant species are *Carex pendula, C. remota, Chrysosplenium oppositifolium* (*C. alternifolium* is much scarcer)

Fig. 13 Old *Salix* spp. characterise the *Salix* forest that is present in many wetland areas in Brussels

and *Equisetum telmateia.* Characteristically frequent species are *Carex strigosa, Impatiens noli-tangere, Lysimachia nemorum, L. nummularia, Rumex sanguineus* and *Veronica montana.* Rather frequent hydrophilous species are *Cardamine pratensis, Deschampsia cespitosa, Juncus effusus, Persicaria hydropiper* and *Ranunculus repens.* A variant develops on shady places, the species include *Athyrium filix-femina, Dryopteris carthusiana* and *D. dilatata,* which are frequent and *Oxalis acetosella,* which is scarce. *Urtica dioica* forms a ruderal variant.

(s) *Salix* and *Alnus* groves (Fig. 13) (EUH and Corine codes – not listed).

Phytosociology: *Salicion cinereae, Alnion glutinosae, Sambuco Salicion.*
Topography: situated at the bottom of humid valleys.
Pedology: on hydromorphic soils.
Characteristic species: the herbaceous vegetation is similar to that of reedbeds and tall-herb communities: *Carex acutiformis, C. pendula, Chrysosplenium oppositifolium, Petasites hybridus, Phragmites australis* and *Scrophularia auriculata.* A dry variant is dominated by *Urtica dioica.*

Wetlands

Wetlands *sensu lato* (Fig. 14), (EUH code 6430; Corine codes 37.7 and 37.8).
Phytosociology: many different units related to the *Phragmitetea*-class.
Topography: situated at the bottom of humid valleys.
Pedology: on hydromorphic soils.

Fig. 14 Various semi-natural marshlands are still present in the suburbs; they support a large variety of species belonging to the most diversified plant communities in Brussels

Characteristic species: wetland communities, which are rather diverse, include or are a combination of reedbed, flooded reedbed, tall-herb community with *Carex acutiformis,* tall-herb community with *C. riparia,* community with *Scirpus sylvaticus,* tall-herb community with *Scrophularia umbrosa* and *Sparganium erectum,* floodable community with *Mentha aquatica* and *Scrophularia umbrosa,* wetland with *Glyceria maxima,* tall-herb community with *Petasites officinalis* and ruderal tall-herb community with *Urtica dioica.*

Grasslands

Grasslands (Fig. 15) (EUH code 6510; Corine code 38.2).

Phytosociology: many different units related to the *Molinio-Arrhenatheretea*-class.

Topography: at the bottom of small valleys.

Pedology: on nutrient-rich moist, sometimes slightly flooded, soils.

Characteristic species: *Agrostis stolonifera, Ajuga reptans, Bellis perennis, Cardamine pratensis, Carex remota, Circaea lutetiana, Cirsium arvense, Galium palustre, Glechoma hederacea, Holcus lanatus, Hypericum hirsutum, Lycopus europaeus, Lysimachia nummularia, Mentha aquatica, Plantago major, Poa annua, P. trivialis, Persicaria hydropiper, Prunella vulgaris, Ranunculus repens, Rumex conglomeratus, R. obtusifolius,*

Fig. 15 Diversified and colourful grasslands still exist in Brussels. Most of them are mown each year to preserve their species diversity. Some are developing together with fruit trees

Stellaria uglinosa, Trifolium repens, Veronica chamaedrys, V. montana, V. serpyllifolia and *Urtica dioica.*

Ponds (EUH code 3150; Corine codes 22.13 x (22.41 or 22.421).

Phytosociology: many different units related to the *Potametea-* or the *Lemnetea-*classes.

Topography: at the bottom of valleys.

Pedology: on hydromorphic soils.

Comments: these man-made ponds must be cleaned out regularly to prevent the development of the hydrosere to marsh and then scrub and woodland. However, the vegetation is variable; some are almost never colonised by aquatic vegetation while others have an aquatic vegetation that is sometimes or nearly always luxuriant.

Characteristic species: *Lemna minor* is a frequent free-floating species. *Spirodela polyrhiza* and *Potamogeton natans* are rare. *Lemna trisulca* is not frequent. The submerged macrophytes include *Callitriche* sp., *Ceratophyllum demersum, Elodea canadensis, Myriophyllum spicatum, Potamogeton pectinatus* and *Ranunculus circinatus* (which is very rare). A vegetation of heliophilous tall herbs develops on sunny and non-trampled banks: *Carex acutiformis, C. pendula, C. pseudocyperus, Cirsium oleraceum, Epilobium hirsutum, Equisetum telmateia, Filipendula ulmaria, Iris pseudacorus, Lysimachia vulgaris, Lythrum salicaria* and *Scirpus sylvaticus. Salix* or *Alnus* groves are present on marshy soils. Fragments of the *Fraxinus* forest with *Carex* spp occur in seepage areas. Locally rare species include *Potamogeton lucens, P. pectinatus, Ranunculus circinatus, Sagittaria sagittifolia, Cyperus fuscus* and *Oenanthe aquatica.*

City Centre (Historical City, Business and Shopping Districts)

The highly urbanised city centre has very little green space and contains few species; old buildings sometimes harbour *Herniaria glabra, Dryopteris filix-mas, Buddleja davidii, Senecio viscosus* and *Chamerion angustifolium*. Cracks of old walls are a habitat for *Pseudofumaria lutea, Cymbalaria muralis, Parietaria judaica, Asplenium trichomanes, A. ruta-muraria* and even the rare *A. scolopendrium* and *Ceterach officinarum*.

Residential Areas

High Density Housing Areas (Block- and Ribbon-development, Multi-storey Buildings)

A recent study has shown the relative importance of high density residential development in influencing the plant species composition and richness in Brussels (Godefroid and Koedam 2007). Species that are typically found in high density housing areas are *Arctium minus, Chamerion angustifolium, Galinsoga ciliata, Hordeum murinum, Lolium perenne, Matricaria discoidea* and *Trifolium repens*.

Low Density Housing Areas (Terraced and Detached Housing Areas, Villas)

The species found in low density housing areas include *Pseudofumaria lutea, Fallopia convolvulus, Cymbalaria muralis, Euphorbia peplus, Hypochaeris radicata* and *Hyacinthoides non-scripta*.

Industrial Areas

Industrial areas are characterised by derelict land, which is sometimes polluted (from former industrial land uses). It includes areas with very little or a lot of green space and the characteristic species are *Vulpia myuros, Lepidium ruderale, Convolvulus arvensis, Eupatorium cannabinum, Chenopodium album, Anisantha sterilis, Matricaria inodora* and *Artemisia vulgaris*.

Transport Routes and Areas

Railways

Railway embankments are frequently invaded by bramble (*Rubus* spp.), *Clematis vitalba, Fallopia japonica* and various wooded plants. Among the trees, *Robinia pseudoacacia* has been frequently planted in order to stabilise the slopes. It is accompanied by spontaneous shrubs like *Cornus sanguinea, Rosa canina, Prunus spinosa, Crataegus monogyna, Salix capraea, Betula pendula* and *Sambucus nigra*. The herb layer varies according to the substrate, acidic grasslands occur where the railway crosses siliceous sand while a calcareous flora will develop on ballasts with a mix of species from grassland, waste grounds and clear-felled areas; the species include *Leucanthemum vulgare, Medicago lupulina, Agrimonia eupatoria, Artemisia vulgaris* and *Chamerion angustifolium.*

Roads

Most of the roads and pavements contain a poor flora, which comprise plants that can tolerate trampling without damage such as *Polygonum aviculare agg., Poa annua, Matricaria discoidea* and *Coronopus didymus*. The predominant vegetation of road verges is short, frequently mown grass, which includes *Lotus corniculatus, Leontodon autumnalis, Achillea millefolium* and the halophytes *Plantago coronopus* and *Puccinellia distans,* which have colonised and established as a consequence of the use of de-icing salt. *Heracleum mantegazzianum* and *Pastinaca sativa* occur frequently along the major peripheral roads and the major roads into the city.

Harbours

The harbour of Brussels has a modern infrastructure where relatively few plants grow, although some exotic species are frequent, such as *Fallopia japonica, Setaria viridis, S. verticillata, Buddleja davidii, Datura stramonium, Sorghum halepense, Echinochloa crus-galli* and *Panicum capillare*. Locally, a few species indicate some soil humidity, for example, *Eupatorium cannabinum, Bidens frondosa* and *Persicaria amphibia.*

Recreation Areas

Parks

The 100 parks, squares and gardens that occur in the Brussels region, were designed to provide beautiful landscape perspectives. Their botanical interest varies widely; for example, some parks contain outstanding collections of trees, shrubs or ornamental plants. There is a continuum between the monotonous parks comprising lawns surrounded by hedges (in which the flora is reduced to a few species) and the

Fig. 16 Typical view of an urban park in Brussels that is managed in a nature-friendly way. Spontaneous vegetation is left in areas with good potential for developing a species-rich community

hilly parks, with springs, streams and ponds and artificial rockeries, in which some parts are more or less wild (Fig. 16).

The most nature-friendly parks include beautiful forest fragments. Broadly speaking, the city's parks include the following:

(a) Lawns, which in the older parks in particular, sometimes support a wide variety of species, such as *Cardamine pratensis, Potentilla anserina, Prunella vulgaris, Veronica chamaedrys, Leucanthemum vulgare, Lysimachia nummularia* and *Galium verum,* etc.
(b) Groves and semi-natural woodland with an interesting flora, which contain beautiful trees: indigenous species (*Fagus sylvatica, Quercus* spp, *Carpinus betulus, Tilia* spp. and *Fraxinus excelsior*) and exotic trees (*Castanea sativa, Aesculus hippocastanum* and *Robinia pseudoacacia*). In most parks, these woods are rich in nitrophytes, for example, *Circaea lutetiana, Aegopodium podagraria, Alliaria petiolata, Arum maculatum* and *Impatiens parviflora.*
(c) Woodland areas including new woodland where pteridophytes are quite abundant, *Quercus-Carpinus* forests, acidic woodland with *Pteridium aquilinum, Luzula sylvatica* and *Holcus mollis, Salix* woodland, *Alnus* groves and relict of *Fraxinus excelsior* forests.
(d) Lakes and streams; running waters (Fig. 17) are sometimes saturated in $CaCO_3$ after passing through calcareous sands. The preciptation of lime-rich water results in the formation of small areas of travertine, which provide a substrate for bryophytes such as *Cratoneuron filicinum.*

Fig. 17 Various streams flow through the green areas of Brussels contributing to plant diversity of the city (here is an old park that has returned to wilderness)

Allotments

Allotments, which are still numerous in the Brussels region, are bordered by hedges and contain many cultivated flowers and ornamental shrubs. They support a flora that is typical of moist and eutrophic soils. The characteristic allotment flora comprises the following:

(a) Species of nutrient-rich and strongly anthropised environments: *Senecio vulgaris, Chenopodium album, Sonchus oleraceus, Stellaria media* and *Solanum nigrum.*
(b) Annual ruderals: *Conyza canadensis, Hordeum murinum, Sisymbrium officinale* and *Lactuca serriola.*
(c) Species of row crops: *Euphorbia helioscopia, E. peplus, Chenopodium polyspermum, Echinochloa crus-galli, Persicaria maculosa* and *Galinsoga ciliata.*

Many nitrophilous species are also present: *Urtica dioica, Galium aparine, Artemisia vulgaris, Ballota nigra* ssp. *foetida, Tanacetum vulgare,* as well as trampling-tolerant perennials (*Polygonum aviculare agg., Plantago major, Matricaria discoidea*) and small pioneer plants of moist soil (*Juncus bufonius, Gnaphalium uliginosum*) or of mudbanks rich in nitrates (*Ranunculus sceleratus, Rorippa palustris*). Some species of arable habitats are also present: *Papaver rhoeas, Sinapis arvensis, Matricaria recutita, Apera spica-venti* and *Vicia hirsuta.*

Cemeteries

Cemeteries contain an interesting and diversified flora, especially those that are not very well maintained. On sandy materials, there is a characteristic type of grassland with *Aira caryophyllea, Centaurium erythraea, Hypericum humifusum, Myosotis discolor, Spergularia rubra, Aphanes arvensis* and even the rare *Jasione montana* and *Anacamptis pyramidalis*. Climbing plants including *Humulus lupulus, Hedera helix, Bryonia dioica* and *Clematis vitalba* are invading the gravestones together with smaller plants like *Cymbalaria muralis, Pseudofumaria lutea* and *Asplenium ruta-muraria*. Many annual plants of arable and wasteland habitats are found on the areas of bare ground. In addition, many ornamental plants have been introduced and are more or less naturalised.

Open Land

Arable

Although densely built-up, the periphery of Brussels supports an active agricultural industry. The eutrophic soils used for the production of spring cereals are characterised by the *Polygono-Chenopodion* vegetation type, represented by *Anagallis arvensis* ssp. *arvensis, Spergula arvensis, Lamium amplexicaule, Raphanus raphanistrum, Fallopia convolvulus, Chenopodium album* and *Persicaria maculosa*. The *Panico-Setarion* vegetation types occur on the mesotrophic and siliceous soils; representative species include *Echinochloa crus-galli* and *Setaria viridis*. A group related to the *Aphanion* is found in fields; oats *(Avena)* or barley *(Hordeum)* on loess, silty or gravelly soil is locally represented by *Aphanes arvensis, Matricaria recutita* and *Legousia speculum-veneris*. Nanocyperion species, with *Isolepis setacea, Centaurium pulchellum, Juncus bufonius* and *Spergularia rubra* occur on the damper substrates.

Waste Grounds

The most characteristic biotope of the city occurs on the numerous areas of waste land. These sites are floristically very rich and contain many rare species. Several different plant communities can be distinguished according to the substrate, examples of which are as follows:

(a) Cinders with *Herniaria glabra* and *H. hirsuta*.
(b) Cobblestone tracks with *Eragrostis minor, Digitaria sanguinalis, D. ischaemum, Setaria viridis, S. pumila* and *Amaranthus albus*.
(c) Paving stones with *Sedum acre, S. album, S. rupestre, Saxifraga tridactylites* and *Scleranthus annuus;* ballasts with *Arenaria serpyllifolia, Anisantha tectorum, Amarhanthus retroflexus, Apera interrupta, Berteroa incana* and *Senecio inaequidens*.

(d) Communities with *Melilotus albus* and *Daucus carota*.
(e) Communities with *Medicago sativa* spp. *sativa* and *M. falcata*.
(f) Grassland with *Arrhenatherum elatius, Silene vulgaris,* etc.

The biotopes may also contain many ornamental species that have escaped from surrounding gardens, become established and are spreading rapidly. These species include *Rosa canina, Salix* spp., *Buddleja davidii* and various other introduced species. Many rare species have been recorded on waste grounds, such as *Aristolochia clematitis, Galega officinalis, Atropa belladonna, Rhinanthus minor* and some orchids like *Dactylorhiza fuchsii* and *Ophrys apifera*.

Water

Brussels contains several semi-natural wetlands, whose conservation has contributed to the high species richness of the city. The wetland flora accounts for 23% of the flora of Brussels, which is 171 species. The distribution of these species shows three different patterns:

(a) Wide distribution including at least partially the city centre – *Epilobium hirsutum, Salix alba* and *Calystegia sepium*.
(b) Distribution typically suburban, forming a peripheral belt and excluding the city centre – *Cirsium oleraceum, Phragmites australis, Lemna minor, Iris pseudacorus* and *Humulus lupulus*.
(c) Distribution strictly peri-urban but more fragmented (largely discontinuous peripheral belt) – *Equisetum telmateia, Caltha palustris, Carex strigosa,* and *Cardamine amara*.

These three distribution patterns illustrate a gradient of increasing sensitivity of the species to various disturbances that characterise urbanisation, drainage, eutrophication and pollution. The species frequency is inversely proportional to their degree of hygrophily.

Some species are rare in Brussels but still common outside the city, for example, *Crepis paludosa* and *Achillea ptarmica*. On the other hand, the city contains some species that are rare in the surroundings, for example, *Carex paniculata, C. strigosa, Eleocharis palustris, Equisetum sylvaticum, Leersia oryzoides, Catabrosa aquatica, Lemna trisulca, Myriophyllum spicatum, Potamogeton crispus* and *Sagittaria sagittifolia*. Most of these species only occur at one site and therefore may disappear in the near future.

Nature Conservation, Environmental Planning and Education

Compartmentalisation and fragmentation of the organisation of knowledge, which is reflected in the structures and ways in which both academic and government institutions work, has tended to lead to an incomplete understanding of complex human situations and to the formulation of unwise, strongly sectorial policies (Celecia 1997). This was particularly the case for Brussels before the creation of the Brussels Institute for Environment Management in 1989. Now, most of the environmental problems in the city are dealt with by this Institute, which centralises the information and has research programmes with scientific teams from universities. During the last 40 years, there have been three stages of nature conservation management. First (in the 1960s to 1970s), the battle against nature involved the suppression of most of the spontaneous vegetation elements as the population required a "controlled and improved" nature. Most of the species quoted as being extinct now disappeared during this disastrous period. Second (1970s to 1980s), new management perspectives were created; however, they were produced in a strict museological sense (that is, managing collections in a museum). However, it was not yet a holistic perception of the urban ecosystem because it concerned the conservation of species (for example, by prohibiting hunting), the biotope conservation (for example, by prohibiting the use of weed-killers on public land) and site conservation (for example, by creating nature reserves). Third, from 1989 to the present day, the perception of urban biological resources have become more all-embracing, going beyond the protection of species, biotopes and sites even to the extent of integrating landscape management. The major goal is to improve inhabitants' living conditions. Therefore, at the present time, the integration of nature into urban planning does not only involve the protection of plants and animals but aims to create a high quality of urban environment. Hence, after having denied the existence of spontaneous nature in the city and trying to destroy it (after having separated it from the rest by enclosing it in nature reserves), present strategies for the inclusion of nature in urban development are based on three different actions: (1) the increase in total surfaces of green spaces; (2) the increase in ecological quality of green spaces and (3) the development of ecological networks. At the present time, there is also a trend to promote some green sites with a more "natural" aspect.

Indeed, urban parks have often been created according to an ideal image governed by order, regularity, cleanliness and salubrity criteria, which were particularly unfavourable to the spontaneous flora. This objective can be carried out with the help of better ecological management practices such as mechanical methods (for example, mowing) instead of chemicals. The phytosociological data collected in some lawns in urban parks have emphasised the presence of an interesting flora within the dominant Poaceae. This natural potential is important in the urban context. It foresees a favourable evolution of these biotopes by mowing which may contribute to the regression of some undesirable taxa such as the Poaceae and to the development of meadows and marsh, whose status is much more precarious, as previously mentioned. More than 30 years ago, a "green" map of the city was produced (based on aerial photographs), which showed the typology of the urban space and the principal vegetation types. An updated version of the map was prepared in 1991, in order to study the changes in green space. Other studies were also undertaken at this time including a phytosociological analysis of 80 semi-natural sites, different surveys of animal species (birds, mammals, reptiles and amphibians), a complete inventory of the street trees, a fungi inventory, a survey of the corticolous macro-lichens and a bryological survey. As a consequence, the environmental planning, design and management of Brussels are based on an effective integration of biological data.

Closing Comments

The geological diversity, geomorphology and human influence has induced in Brussels the development of a whole variety of habitats that are characterised by a diverse flora. Some areas (nature reserves and semi-natural protected sites) have some stability, while others are often reworked and are highly dynamic. The intensive research conducted in recent years has helped to provide a comprehensive overview of the Brussels flora, which belongs to a combination of the Atlantic and central European/sub-Mediterranean floristic zones. The Atlantic element is illustrated by the presence of *Hyacinthoides non-scripta, Tamus communis, Narcissus pseudonarcissus, Equisetum telmateia, Pulicaria dysenterica, Fumaria capreolata, Ornithogalum umbellatum* ssp. *umbellatum* and *Ranunculus ficaria* ssp. *ficaria.* The (sub)central-European, (sub)mountain and (sub)Mediterranean plants include *Oreopteris limbosperma, Veronica montana, Poa chaixii, Phyteuma nigrum, P. spicatum, Euphorbia amygdaloides, Neottia nidus-avis* and *Ophrys apifera.*

We have seen that the political and economic development of Brussels (capital of Europe) has often lead to the destruction of valuable ecosystems and habitats. Although mentalities are changing since the creation of the Brussels Capital administrative region in 1989 and the Brussels Institute for Environment Management, it is still difficult to meet the challenge of conserving biological diversity in a city that is the capital of Europe and therefore, wants to sustain economic growth and to keep a certain prestige. In a recent case, the plant-monitoring programme resulted in a potential major disaster being avoided and the consequential saving of important biological resources.

Literature Cited

Celecia J, (1997) Urban ecology: biodiversity and contemporary stakes of inventories. J. Agric. Tradit. Bot. Appl. 39 (2): 241–263.

Geerinck D, (1998) Les arbres en ville. Richesse, diversité, situation. In : IBGE (ed). Qualité de l'Environnement et Biodiversité en Région de Bruxelles Capitale. Inventaire et suivi de la Flore et de la Faune. Documents de Travail de l'IRSNB n°93, pp. 87–99.

Godefroid S, Koedam N, (2007) Urban plant species patterns are highly driven by density and function of built-up areas. Land. Ecol. 22: 1227–1239.

Gryseels M, (1998) Natuur en groene ruimten in het Brussels Hoofdstedelijk Gewest. In : IBGE (ed). Qualité de l'Environnement et Biodiversité en Région de Bruxelles Capitale. Inventaire et suivi de la Flore et de la Faune. Documents de Travail de l'IRSNB n°93. Bruxelles.

IBGE, (1990) L'Environnement à Bruxelles. Un premier état de la situation. Les cahiers de l'IBGE n°2. Institut Bruxelles pour la Gestion de l'Environnement. Bruxelles.

IBGE, (1995) Rapport sur l'Etat de l'Environnement en Région de Bruxelles-Capitale. Les cahiers de l'IBGE n°9. Institut Bruxelles pour la Gestion de l'Environnement. Bruxelles.

Martens JM, (1998) Bruocsella, la cité de l'Iris jaune. In: IBGE (ed). Qualité de l'Environnement et Biodiversité en Région de Bruxelles Capitale. Inventaire et suivi de la Flore et de la Faune. Documents de Travail de l'IRSNB n°93, pp. 35–42.

Saintenoy-Simon J, (1998) Etude de la flore de la Région de Bruxelles-Capitale. In: IBGE (ed). Qualité de l'Environnement et Biodiversité en Région de Bruxelles Capitale. Inventaire et suivi de la Flore et de la Faune. Documents de Travail de l'IRSNB n°93, pp. 43–66.

Société Royale Belge de Géographie, (1991) Itinéraire des zones humides bruxelloises. Collection 'Hommes et Paysages' n°17, Société Royale Belge de Géographie.

Vanderpoorten A, (1997) A bryological survey of the Brussels Capital Region (Belgium). Scripta Botanica Belgica 14.

Vanholen B, De Kesel A, (2000) Inventarisatie en monitoring van de mycoflora en de lichenen van het Brussels Hoofdstedelijk Gewest. Jaarrapport 1999 Werkgroep mycologie. Nationale Plantentuin van België, Meise.

Bucharest

Marilena Onete and Mihaela Paucă-Comănescu

Fig. 1 Parliament Palace, situated in the centre of the city, the second largest building in the world after the Pentagon

Marilena Onete (✉)
Ecology, Taxonomy and Nature Conservation Centre, Institute of Biology, Romanian Academy, Spl. Independentei 296, Sector 6, 060031 Bucharest, Romania
e-mail: m_onete@yahoo.com; marilena.onete@ibiol

J.G. Kelcey and N. Müller (eds.), *Plants and Habitats of European Cities*,
DOI 10.1007/978-0-387-89684-7_6,

Abstract The green spaces in Bucharest have different ages and degrees of structural complexity (vegetational and architectural). They can be considered to be artificial ecosystems because they are created for aesthetic reasons by the planting of ornamental trees, shrubs and herbaceous species. The native vegetation has almost disappeared from the city having been replaced by the artificial ecosystems described. Some species that are planted in these artificial systems are species of the natural former forests of Romania. The city contains remnants of natural vegetation, which has adapted to the modified conditions of the urban environment. However, the authorities are ignoring the necessity of maintaining and increasing the green spaces of the city. Even Natural Law 749/2006 which regulates the administration of green spaces in the city is being systematically ignored by the authorities. The structural habitat diversity and the limited botanical surveys of vascular and non-vascular plants and lichenised fungi indicate that the city contains an extensive and interesting flora. However, the vegetation and the flora of the city need to be studied much more comprehensively, not least to ensure that the planning, design and management of the city as a whole and the green spaces in particular are based on sound scientific information.

Natural Environment

Location

Bucharest is situated in southern Romania in the south-eastern part of the European peninsula, with the city centre at latitude 44°26′07′76″ N and longitude 26°06′09,09″ E and between 55 and 95 m a.s.l. The city is located about 55 km north of the Danube River and 70–80 km south of the sub-Carpathian Mountains. It is about 200 km west of the Black Sea. Covering 238 km^2 and with a population of more than two million inhabitants, Bucharest is the largest city of the country with 10% of the Romanian population living in it; see Fig. 2.

Geology and Topography

The city lies in the middle of the Romanian Plain, which is the largest expanse of flat land in Romania. Consequently, there is very little variation in the rather monotonous topography. The city is situated in the transition between the Moesic Platform and the outer epicratonic Carpathian Avanfoss. The Platform is bounded to the north and west by the Carpathians, to the south by the Balkans and to the east/north-east by the Peceneaga-Camena geological fault. The Intramoesic fault and its related western faults are of major importance because the central part of the platform is mobile and a major centre for earthquakes and

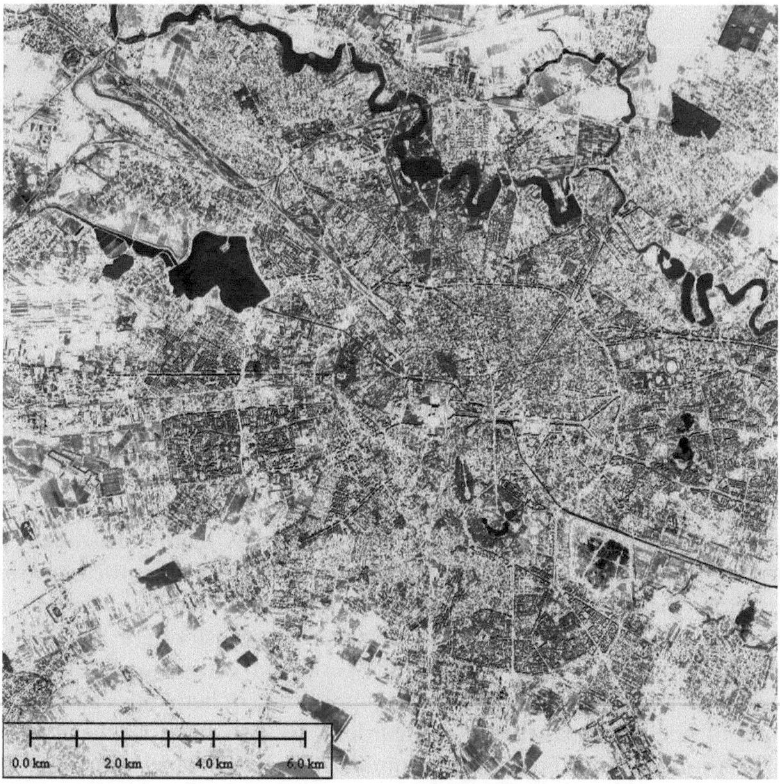

Fig. 2 Bucharest city map (1:15,000) © Urban Planning Centre

other seismic activity. Two earthquake types have been identified in the Bucharest area; they are classified according to the depth of the hypocentre (Lăcătuşu et al. 2008):

1. Around the Intramoesic Fault, the prevalent earthquakes have deep hypocentres with a magnitude of 4–5°.
2. Near Bucharest, the earthquakes have a lower magnitude (2–4°) and arise in the earth's crust.

The city overlies Upper Cretaceous limestone with strata of variable thickness and comprising limestone, marls and sandstone as well as clay and littoral-lacustrine sands overlying clay and coal, with siliclastic deposits in the upper layers that are finer above and coarser below.

The surface geology comprises sand and gravel (brought down by the Carpathian rivers) and loess (Lăcătuşu et al. 2008). The Colentina, Dâmboviţa and Argeş rivers have had a major influence on the topography of the land to the north of the city.

Soil

Using the Romanian System of Soils Taxonomy and the Food and Agriculture Organisation of the United Nation soil group, Lăcătuşu et al. (2008) have identified two soil types in Bucharest:

1. Soils of natural or mainly semi-natural origin – including soils from the peri-urban and locally from the semi-central areas. These soils belong to Protisol, Chernisol, Luvisol and Hidrisol classes and comprise alluviosols with clay texture, regosols with clay-sandy texture, chernozems with clay-loam texture, gleysol with clay-loamy texture and stagnosol with clay texture.
2. Soils of anthropogenic origin – belonging to the Protisol class with sandy-clay and clay-loamy texture developed on anthropogenic parental material, usually ca. 50 cm thick but thinner (30 cm) where the parental material is skeletal without horizons. These soils are predominant (48.5%) within the city, especially in the central area.

Clay-loams form 86.25% of the soils and are neutral to weakly alkaline with a moderate organic content (humus) and varying levels of total nitrogen, phosphorus and soluble potassium. All the soils have been disturbed to varying degrees. Some have been removed and replaced by imported soil from elsewhere in Romania, while others have been disturbed by construction and landscape works. The soils of agricultural land in the peripheral areas have been altered as the consequence of drainage works and the application of herbicides and fertilisers. The soils of housing areas, parks and gardens are derived from the forest soils typical of the large forests that once dominated the landscape of the area that is now Bucharest. The soil chemistry is likely to have been altered by air and water pollution.

Drainage

Bucharest is part of the hydrographical basins of Colentina River (inferior course) and Dâmboviţa River (medium course). The Dâmboviţa enters the north-west side of the city and almost immediately it is drained and flows into Lake Morii, from where it is canalised for the whole of its length leaving the city in the south-east. The river banks above the concrete re-enforcement are sown with grass mixtures and colonised by ruderal vegetation. Trees are very rare but still present near the river banks close to the streets.

Historically, the work of straightening and deepening the Dâmboviţa River began in 1880 (it was finished in 1883) in order to protect the town from flooding. The works resulted in the loss of the tributaries and islets. The straightening resulted in some of the meanders on the left or right side of the new channel being retained. After the riverbed was deepened, the meadows dried out, no more ponds were formed and the health of the city's population benefited. A new network of sewers was constructed to take domestic sewage and run-off to the river, into which it was discharged together with the oily residues from the Grozaveşti power-plant. Bridges were built across the river; by 1889 there were 12 such bridges, 7 of stone and 5 of iron.

Important works have been undertaken to increase the water supply, both potable and for industry. For example, potable water was provided from the Dâmboviţa (via water treatment works of Arcuda), from underground water and from deep wells. Industrial, public utility and irrigation water was brought in through the diversion channel from the Argeş River at Crivina, where a dam was built. Between 1956 and 1960, a new 1.5 m diameter water pipe was constructed from Crivina to the Arcuda. This increase in water consumption was matched by the increased size and capacity of the sewage system for household and industrial wastewater. Surface water run-off from the roads and other hard surfaces is discharged into some of the park lakes and Dâmboviţa River.

Climate

Situated in continental temperate climatic conditions, Bucharest is especially influenced by the eastern continental air masses and to a lesser extent by the western and southern air masses. The continental eastern influence results in large thermal amplitudes of up to 70°C between harsh winters and very hot summers. The western and southern influences explain the presence of long and warm autumns, some mild winter days and early and short springs. An Italian traveller in 1855 stated: "*the Bucharestian autumn is unusually long contrasting with a spring that here does not exist; it is the most beautiful season, not even Italy can offer days more lovely and cleaner, nights more pleasant and brighter.*"

In spite of the mild winter, sometimes the snow cover can be deep, for example, 4 m in February 1875; in January 1917 (during the German), the snow covered low houses in Obor and in February 1954, the city was covered with 2–7 m of snow for 6 weeks. The mean period of sub-zero temperatures lasts for 90–100 days, while during summer there are up to 46 "tropical days", when the temperature is >30°C.

Within this general climatic context, there are local thermal changes generated by the structure and activity of the city, which show differences between the typical climatic conditions in the centre and those on the periphery. In 1985, Croitoru and Târcob described various thermal differences between the centre of the city and the outskirts. They found that in the mornings the average daily temperature is reached by about 08:00 hour in the city and 1 hour later outside the city in Băneasa. The lowest diurnal mean temperature occurs in January and February: –21°C in the city and 1°C lower outside. The lowest monthly mean temperatures were recorded in January dropping to –2.4°C in the city and –3°C in the peripheral areas. The mean monthly temperature during July is 22.7°C in the city and 22.4°C outside. The dominant winds come from the east, followed by west, north-east and south-west winds. As with the temperature regime, the wind regime reveals differences between the built-up and surrounding areas.

Annual mean precipitation is 585 mm and falls on an average of 117 days per year. Recently there has been a marked decrease in the number of the days when precipitation occurs and an increase in the duration of precipitation on the days when it does rain (or snow). Differences in land use and the landscape have produced three micro-climates:

1. Central zone – influenced by urban buildings with extremely high temperature and a high frequency of atmospheric calm and nebulosity, the effects of which are mitigated by green spaces.
2. Industrial zones – as the consequence of air pollution there is a greater frequency of fog and heavy rain. The Dâmbovița valley has a unique character with high air humidity and wind currents and frequent fog, especially during autumn and winter.
3. Peripheral residential zones – the micro-climate, which is similar to that outside the city, is characterised by stronger winds and lower temperatures.

Air and Water Quality

Currently, the biggest problem for the Bucharest authorities is to ensure a healthy environment for the people in relation to identifying and controlling the multiple pollution sources that are seriously and adversely affecting the quality of the air, water and soil. The pollution sources are not only multiple but are combined in some areas, with emissions taking place at different levels in the atmosphere as a result of differing discharge heights. Bucharest is one of the most polluted cities in Romania because of vehicle exhaust emissions and discharges from power stations and industrial plants. However, the contribution of pollutants from industrial sources has decreased in the last few years as a result of reduced industrial activity and the enforcement of pollution control measures. With approximately 37% of motor vehicles being older than 8 years, they contribute greatly to the pollution levels; for example, 92% of the total atmospheric NO_X in the city centre is from motor vehicle exhausts. According to emission surveys, the second most important source of major pollutants is power stations, which produce 92% of the SO_2 pollution. The next important source is industry, which is the main contributor of particulate matter; 27% of particulate matter pollution is attributable to power stations with 26% being produced by motor vehicles.

The increased construction of residential areas with ten-storey apartment blocks results in a significant three-dimensional pollution impact (Gheorghe and Nica 2008). The buildings create obstacles that cause calm conditions, which occur at twice the frequency in the city as they do on the periphery. This situation is very harmful in relation to the concentration of air pollutants – in the calm areas they have long accumulation and short disperson times. Wind intensity is higher outside the city; long-term monitoring revealed that the city's surroundings (forests and lakes) have good air circulation patterns that are favourable to the maintenance of a relatively stable atmosphere and are capable of dispersing polluntants more quickly.

Another feature that contributes to the high level of pollution is the design of the main avenues and boulevards of the city, which are aligned in a star pattern with the centre in the heart of the city. These roads represent the main connections within the centre and between the centre and the peripheral areas and beyond. The rapid development and expansion of the city, coupled with the hugely increased traffic levels, have produced a serious decrease in air quality.

Bucharest is the only European capital that does not have major sewage treatment works; 85% of the treatment plants do not work properly. This is exacerbated by road and other run-off being discharged to the river and to some of the park lakes. As a result the water in the Dâmboviţa River is contaminated with phosphates, nitrates and ammonium, detergents and urea. The phosphate level is 30–40 times higher than the standard quality norms.

Historical Development

Pre-thirteenth Century

The first record of hominid occupation in the Bucharest area is from the Palaeolithic period (about 15,000 BC). Archaeological evidence (many copper and gold objects) proved that the area continued to be inhabited at the beginning of the Neolithic period and during the Bronze and Iron Age periods.

Many hypotheses have surrounded the question of the identity of these early inhabitants. What is more certain is that Herodotus wrote about the Geto-Dacian population at the end of the sixth century BC. In addition to hunting and fishing, they practised agriculture (both crop production and animal husbandry), viticulture and apiculture (from the Bronze Age). The Roman domination started during the time of Decebal in the first century AD. There is a lot of evidence that in the second and third centuries there were Roman settlements in the Bucharest area. Following the withdrawal of the Roman administration and army, the Dacian-Roman population (peasants, artisans and poorer people) remained in the forest area.

Place names have been important in the study of the history of the city. For example, the names of watercourse and rivers (for example, Ialomiţa, Dâmboviţa, Prahova, Mostiştea, Neajlov, Snagov, Colentina and Ilfov) indicate the existence of the Slavic people in the sixth century. Other names indicate the presence of ancient Turkish people and Tatars from the ninth to the thirteenth centuries, the final period of migration.

Thirteenth to Eighteenth Centuries

The presence of a settlement (a "citadel on the Dâmboviţa") called Bucharest was established by 1368. Bucharest probably existed as a village before the foundation of the Romanian Provinces in the thirteenth century. During the time of the Black Prince (Negru Vodă), in the fourteenth century, a fortress tower was built and in AD 1459 a charter refers to fortifications to provide a defensive outpost against the Turkish invasion of Wallachia.

The capital was transferred from Târgovişte to Bucharest. Despite its eventful history, Bucharest continued to expand economically and in terms of human population. At the end of the sixteenth century, the city had developed to occupy both sides of the Dâmboviţa River. Short periods of peace and prosperity alternated with difficult years, which included bloodshed, plunder, fire, hunger and epidemics. This cycle of events was characteristic of Bucharest, which for many centuries was ruled at various times by Greeks (Phanariot), followed by Ottomans and then Russians. During these times, Bucharest became an important political, administrative, commercial, economic and cultural centre.

Nineteenth Century

The modernisation process which started in 1806–1812 was "in full swing" by the middle of the nineteenth century. In 1829, General Kisselef acceded to power as the Tsarist governor, setting up theatres and services for architecture and medicine, draining pools and marshes and establishing city boundaries. The streets were paved with stones from the rivers, the sides of the boulevards were planted with *Tilia* spp. and lighting was provided along the roads. Between the late 1830s and 1840s, the city suffered considerable damage, for example, by an earthquake in 1838. In 1839, the Dâmboviţa flooded three quarters of the city while in 1847 another flood affected the entire commercial centre up to 2.0 km from the river; in the same year a fire destroyed many houses in the suburbs.

During the mid-1800s and at a time of Romanian prosperity and unity as well as the development of language, literature and politics, the idea of Romanian nationality began to crystallise. In 1848, a revolution occurred but it was short-lived, being suppressed by Turkish and Russian troops who entered Bucharest, each of them occupying one half of the city. Nevertheless, the revolution was influential in the formation of modern Romania. The Austrians made a pact with the Turks resulting in their influence over various aspects of the city and its life. In 1860, roads were systematically paved with hand polished stones, new boulevards were constructed (for example, Academiei Boulavard) and the number of houses with stone walls increased.

Following the unification of the provinces in 1862, Bucharest became the sole capital of a united Romania. In 1869, the first railway in Romania was built between Bucharest and Giurgiu followed by the first horse-drawn tram concession in 1871. In addition, more roads were built and many public and private buildings were constructed. These activities coupled with an increase in foreign investment resulted in substantial industrial expansion. During this period, the area and population of the city grew continuously with new districts appearing in the peripheral areas and replacing the old slums in the central area. The city also grew in height, for example, the construction of houses with two floors and multi-storey hotels. The intervention and use of re-inforced concrete techniques made even higher buildings possible. By 1899 Bucharest had 282,071 inhabitants.

1900–1945

From 1902, the number of new factories started to increase and 4 years later the surfacing of the roads with asphalt began. New buildings and districts continued to be built on wasteland and on the dry land (resulting from the draining of wetlands) to replace the former slums. The population of the city reached 331,307 in 1910 and by 1914 it occupied an area of 5,614 ha. The First World War started in 1914 and in November and December of that year the first German troops occupied Bucharest. On 1 December 1918, full sovereignty was gained over the territories inhabited by Romanian people and the national unitary state was established. Between 1918 and 1945, Bucharest experienced extraordinary economic growth; industrialisation progressed rapidly, resulting in the city becoming an important industrial centre for heavy industry.

New industries evolved, for example, chemical, electro-technical, aeroplane manufacture and the production of vegetable oils; these resulted in the considerable expansion of the city and its population. The 1939 census recorded a population of 870,000 inhabitants in an area of 7,800 ha. Districts were established beyond the city limits; new suburban settlements were established, which later tended to merge into the city. This is the period when the high-rise "blocks" of flats first appeared – along the large boulevards. The 1940 earthquake caused some of the blocks to collapse and resulted in later blocks being more solidly built.

In 1941, Bucharest was bombed by the Soviet airforce with considerable damage being caused. In June 1944, Romania concluded an armistice with the USSR, USA and UK. German aircraft based north of Bucharest (at Otopeni and Buzău – now international airports) systematically raided the city. After August 1944, the history of Bucharest is that of the transition to communism. The city's expansion continued, especially along the main access routes. New complexes of dwellings emerged and the city limit was pushed further out, reaching the orbital railway in the north and north-west.

1945–1990

In December 1947, the People's Republic of Romania was created, leading 2 years later to the abdication of King Mihai. In 1950, the number of administrative divisions of the city, which occupied ca. 256 km^2, was increased from four to eight. These changes were accompanied by others, for example, the introduction of trolleybuses in 1949, a transport system that had no rails, was quiet and did not pollute the atmosphere. The district central heating system was introduced in 1958; by the end of 1978, the length of the pipelines had reached 346 km. New housing districts for workers started to be built in 1957, reaching a peak in 1961. The housing districts comprised multi-storey blocks of flats generally located towards the city outskirts. Smaller residential complexes consisting of houses of one or two storeys were constructed in many areas of the city.

By 1965, the Communist Party was thoroughly established with Nicolae Ceauşescu as Secretary General of the Party; the country was re-named "The Socialist Republic of Romania". In February 1968, the Great National Assembly decided the administrative organisation of the country, with the capital being Bucharest municipality, the city itself was organised with eight sectors (reduced to six in 1979). In 1974, Ceauşescu was elected President of the country.

Works for the underground railway started in 1975, the same year as flooding affected the city, though not as severely as it affected the surrounding countryside. On 4 March 1977, the city was subject to a major earthquake, measuring 7.2 on the Richter scale. It resulted in substantial damage to buildings and 1,570 casualties. Following the earthquake, many buildings were restored and new buildings constructed, with modified foundations to enable them to better withstand seismic activity. In order to alleviate the many problems arising from the rapid expansion of the city's population, multi-storey (four and ten) blocks of flats were built in each district.

In 1980 the construction of the "House of the People" began resulting in the destruction of many old houses and churches typical of Bucharest. Work continued until 1997 when it became the Palace of the Parliament. The entire surrounding district was "urbanised". Set on an artificial hill, the Palace of the Parliament, which is 84 m high (12 floors above and as many below ground), dominates the landscape. It is the world's second largest structure in terms of built area (330,000 m^2) after the Pentagon, see Fig. 1.

1990 to Present

The Revolution of 16–25 December 1989 put a violent end to communist rule. In 1991, a new democratic Constitution was approved by a national referendum; however, the different reforms implemented by the successive Governments have fallen below expectations. Romania had become impoverished during the Ceauşescu era, especially in the 1980s. Thus, the transition began from a low base, much lower than in the other former communist states.

After 1990, many residents of Bucharest left the town to build villas in the surrounding areas, and large residential projects were developed for the middle class. The construction of blocks of flats expanded both in Bucharest and its surroundings. However, a decade later, some people returned to the city because of the increasingly heavy traffic and the continued long delays in undertaking and completing infrastructure works. In the lead up to the accession of Romania to the European Union on 1 January 2007, huge rises in land prices drove a second wave of residential expansion in the areas surrounding the city. Developers used so-called empty green areas, clearing them (of forest and farmland) and constructing apartment blocks and villas. There are now proposals for two more highways that will relieve traffic in the localities around Bucharest; that is, the widening of the present ring road and the construction of a second, outer one.

Flora

Foreign visitors who came to Bucharest between 1870 and 1905 spoke of the cheerful character of "*the city submerged in vegetation*". However, in recent years, it has lost many green spaces as more and more land has been developed. The total amount of public green space in the city is about 3,000 ha (comprising forests, parks, public gardens, squares and the "gardens" adjacent to the blocks of flats), which represents about 15% of the whole administrative territory.

Two forests (Băneasa-Tunari and Cernica – Pustnicu) occur in the Romanian Plain 10–15 km from Bucharest; both are adjacent to the Colentina river. The forest Băneasa-Tunari is now within the new luxury district of Tunari and therefore subject to considerable deleterious human activities, including the clear felling of forests and the destruction of herbaceous vegetation by fire (caused by barbeques), trampling and collecting. Some of the satellite settlements of Bucharest contain small areas of forest, including Mogoşoaia, Pantelimon and Pasărea.

In the southern part of the city, there are small forests framed by the Argeş, Dâmboviţa, Câlnic, Sabar and Ciorogârla rivers. The rivers, forest and lakes that surround the city are major contributors to the beauty of the landscape. The lakes, which are adjacent to the forests, have the same name, for example, Cernica and Pustnicu, Tunari and Băneasa. The Buftea forest is separated from the Colentina River by a chain of small lakes. Comana forest is adjacent to the Neajlov River in the north.

The development of the city and its satellites has resulted in massive forest clearance. Following the clearance, there was a considerable expansion in agricultural production; now most of this land is unused and consequently it has attracted residential and retail developers, among others. Small lakes and wetlands have been drained, the size of the forests patches has been reduced and as a result there have been significant changes in the landscape diversity, which have induced changes in the local climate.

There is a shortage of published information about the flora of the city, for example – there is no "Bucharest Flora". All the published papers (Morariu 1943; Sanda and Popescu 1971; Popescu et al. 1971; Nedelcu et al. 1972a, b; Spiridon 1973; Anastasiu 1994) describe studies made in the peripheral areas of the city or on sites with naturalised and ruderal vegetation within it. The only overview of the flora is restricted to trees (Prodan 1922). Inventories of the plants of three central parks (Cişmigiu, Izvor and Unirii) and adjacent forests (Băneasa and Baloteşti) have been carried out recently (Onete and Paucă-Comănescu 2008a, b), following the urban-rural gradient to establish the main bio-indicators of air pollution. The studies were carried out as part of a project to enable people to become more aware of the health problems associated with air pollution and urban ecosystems degradation.

Angiosperms

Up to now 680 species (530 herbaceous and 150 woody) have been recorded in the city and the adjacent forests. The 50 most common/abundant herbaceous taxa are listed in Table 1.

Ornamental annual species have been planted in most of the parks and green spaces; for example, *Viola* x *wittrockiana, Calendula officinalis. Lolium perenne, Poa pratensis* and *Festuca pratensis* have been sown to create lawns. In the parks, the herb layer under the trees adjacent to the boundary fences and where there is intense trampling is poorly developed with much bare ground. However, it is rich in ruderal species when compared with the sown areas where the dominant species are few and derived from seed-mixtures used by the parks administrators.

During summer, the soil becomes dry, cracked, open and largely devoid of vegetation resulting in it supporting few species. Large populations of *Polygonum aviculare* agg. have colonised the green spaces of the larger streets; there are no or rarely any other accompanying species. *Portulaca oleracea* is abundant and often dominant in areas with compacted, very dry and cracked soil, where the grass species have vanished or are scarce. In the autumn, when the environmental conditions are more suitable for plant growth, late flowering species that mainly spread vegetatively, occur in the areas of bare soil that are created during summer; they include *Achillea millefolium, Agrostis stolonifera, Cynodon dactylon, Dactylis glomerata, Polygonum aviculare* agg., *Trifolium pratense, T. repens* and *Potentilla reptans.*

It appears that plants of some species change their phenology; for instance, *Fragaria viridis* usually flowers in May and June but it has been found in fruit in November. The perennial grass *Phleum pratense* and *Setaria viridis* normally flower in June to August but has been found to flower in November. Following mowing and a very hot summer, the vegetation cover is dominated by species such as *Cichorium intybus* and *Medicago sativa* ssp. *sativa.* In the autumn, a complex of abiotic factors provides the plants with optimal conditions for growth. In most of the parks, "hidden" less trampled areas among trees on nutrient-poor soils can be found to contain mainly ruderal species such as *Arctium minus, Atriplex patula, Brassica oleracea, Capsella bursa-pastoris, Chelidonium majus, Cirsium vulgare, Conyza canadensis, Lamium album, Malva neglecta* and *Solanum nigrum.*

Table 1 The 50 most frequent herbaceous taxa in the city

Taxon
Abutilon theophrasti
Achillea millefolium
Aegilops cylindrica
Alliaria petiolata
Amaranthus retroflexus
Arctium minus
Atriplex patula
Ballota nigra
Bellis perennis
Berteroa incana
Capsella bursa-pastoris
Carduus acanthoides
Carduus nutans
Chelidonium majus
Chenopodium glaucum
Chenopodium opulifolium
Cichorium intybus
Conium maculatum
Cirsium arvense
Convolvulus arvensis
Conyza canadensis
Daucus carota
Datura stramonium
Erigeron annuus
Galium aparine
Geum urbanum
Lactuca serriola
Lamium album
Lamium amplexicaule
Leucanthemum vulgare
Malva sylvestris
Oxalis corniculata
Plantago lanceolata
Plantago major
Polygonum aviculare agg.
Portulaca oleracea
Sclerochloa dura
Descurainia Sophia
Sisymbrium loeselii
Prunella vulgaris
Potentilla inclinata
Rorippa pyrenaica
Rumex crispus
Rumex obtusifolius
Stellaria media
Silene latifolia
Taraxacum officinale
Urtica dioica
Veronica arvensis
Viola arvensis

The soil-moisture levels in the parks are lower than in the forests, the reason being that in the parks, the asphalt or concrete paths without vegetation and surrounding high circulated boulevards increase the albedo and therefore the ambient temperature which in turn increase the water evaporation from the soil.

Trees and Shrubs

The city is situated in the *Quercus* forest zone of Romania; the surrounding forests are mainly *Quercus* forests in which the main canopy species are *Quercus cerris* and *Q. frainetto, Acer campestre* and *Carpinus betulus; Ulmus minor* and *Acer tataricum* occur quite frequently. The dominant shrub layer species are *Crataegus monogyna, Ligustrum vulgare, Clematis vitalba* and *Cornus mas*. The forests close to Bucharest (and unfortunately even those far from the city) contain an abundance of ruderal species as a result of continual human intervention and disturbance (Sanda and Popescu 1971).

Plant associations with *Quercus cerris, Q. robur, Carpinus betulus* and *Tilia tomentosa* are abundant in Băneasa forests; they include *Acer campestre, A. platanoides, Carpinus betulus, Fraxinus excelsior, Tilia cordata* and *Quercus robur* in the canopy; the shrub layer includes *Cornus mas, Crataegus monogyna* and *Ligustrum vulgare*. The ground is dominated by carpets of *Hedera helix*. Young, vigorous individuals of *Tilia tomentosa* indicate that the natural regeneration of the forest is good.

The present forests, which are relicts of the former forest, are characterised by species adapted to xerophylous conditions. *Quercus cerris* and *Q. frainetto* establish in areas with compact soil that are very dry, especially at the end of the summer/beginning of the autumn. In areas with more loose soil, on wetter depressions, the optimal conditions for survival are for *Carpinus betulus, Quercus robur, Q. pedunculiflora, Fraxinus, Ulmus* and *Tilia* species.

The natural tree and shrub vegetation of the city has mainly disappeared, being replaced by non-native species from China, Japan, America and other countries. There are still some remnants of the natural vegetation, which has adapted to the modified environmental conditions of the city, including nutrient status (eutrophic or oligotrophic), changes in soil pH and increased temperature and pollution. Most are non-native ornamental species such as *Albizia julibrissin, Amygdalus communis, Berberis julianae, Buxus sempervirens, Catalpa bignonioides, Celtis australis, Chamaecyparis lawsoniana, Euphorbia pulcherrima, Forsythia europaea, Fraxinus americana, Ginkgo biloba, Hibiscus syriacus, Ilex aquifolium, Mahonia aquifolium, Parthenocissus tricuspidata, Paulownia tomentosa, Platanus* x *hispanica, Rhododendron sinogrande, Rhus hirta, Salix babylonica, Sophora japonica, Spiraea alba, Symphoricarpos albus, Tamarix tetrandra, Tilia rubra, Viburnum rhytidophyllum* and *Wisteria sinensis*. Saplings of *Ailanthus altissima* are abundant along the sides of many streets – it has become an invasive species in recent years.

The native species include *Acer platanoides, A. pseudoplatanus, A. tataricum, Berberis vulgaris, Betula pendula, Carpinus betulus, Cerasus mahaleb, Cornus mas, C. sanguinea, Corylus avellana, Crataegus monogyna, Fraxinus excelsior, Hedera helix, Ligustrum vulgare, Populus alba, P. tremula, Quercus cerris, Q. robur, Rosa canina, Rubus caesius, Salix alba, S. fragilis, Sambucus nigra, Tilia cordata, T. platyphyllos, T. tomentosa, Ulmus glabra* and *U. minor.*

Other species planted in the city green spaces for aesthetic and amenity reasons, including *Acer negundo, Aesculus carnea, A. hippocastanum, Juglans regia* and *Buxus sempervirens, Mahonia aquifolium, Hibiscus syriacus, Symphoricarpos albus* and *Viburnum rhytidophyllum* tend to naturalise in the Botanic Garden (Anastasiu 1994). However, in the parks and other green spaces of the city spontaneous regeneration does not occur (or does so only very rarely) because of the removal of propagules by management operations.

Herbaceous Species (Including Grasses and Sedges)

Studies of the herbs, grasses and sedges of the surroundings of Bucharest were carried out in the 1970s (see Popescu et al. 1971 and Nedelcu et al. 1972a, b); however, it was only in 2006–2008 that investigations of the city's flora were undertaken (Onete and Paucă-Comănescu 2008a, b). The city's green spaces support relatively large populations of species such as *Agrostis stolonifera, Bromus arvensis, Cynodon dactylon, Dactylis glomerata, Carex divulsa* ssp. *divulsa, C. divulsa* ssp. *leersii, C. praecox, Festuca pratensis, Lolium perenne, Setaria pumila* and *S. viridis.* There are large populations of *Lolium perenne, Trifolium hybridum, T. pratense,* and *T. repens* and *Lotus corniculatus* in the parks, especially during the autumn. In the parks and other green spaces with large areas between the trees (for example, Izvor Park), there are large populations of *Medicago sativa* ssp. *falcata, M. lupulina, M. minima* and *M. sativa.*

The wetlands in the city and their margins contain small populations of species such as *Alisma plantago-aquatica, Epilobium hirsutum, Galium palustre, Mentha aquatica, Myosoton aquaticum, Persicaria hydropiper, P. mite, Typha angustifolia,* and *Phragmites australis.* Ruderal vegetation is dominant on the banks.

Most of the lakes in the city have concrete banks; consequently, *Typha* spp. and *Phragmites* find it very difficult to colonise and establish. However, they can survive and spread in places where the water table is high. In Izvor Park, a small wetland was created unwittingly a few years ago where these two species appeared spontaneously and survived until 2008 when they were removed by the park managers. In the natural lakes without bank re-inforcement, both species form belts in shallow water and help with the bio-purification of the polluted water.

Aquatic Plants

Although the aquatic plants of the surroundings of Bucharest were studied more than three decades ago (Nedelcu et al. 1972a, b), it appears that there are no recent or relatively recent studies within the city. The well-developed drainage network of the Romanian Plain and the surroundings of Bucharest have numerous and diverse still waters and other wetlands. In the depressions, because of spring floods, the water stagnates over a period of time, sometimes for the entire year, creating conditions for assemblages of aquatic plants. *Lemna minor, L. trisulca* and *Utricularia vulgaris* occur in the still water bodies within and close to the city with *Hydrocharis morsus-ranae* occurring near the margins. Other hydrophytes that occur in this habitat include *Nuphar lutea, Nymphaea alba, Nymphoides peltata* and *Potamogeton natans,* together with the helophytes *Phragmites* and *Typha* spp. The submerged zone is represented by *Ceratophyllum demersum, Potamogeton lucens, Myriophyllum spicatum* and *M. verticillatum.* During late spring and the beginning of summer, when the water in Dâmbovița canal flows slowly, submerged vegetation develops and algal blooms occur.

Gymnosperms

Coniferous and mixed habitats do not regenerate naturally in the *Quercus* forest zone. However, many coniferous taxa have been planted throughout the city together with other arborescent species that are adapted to the city environment (Prodan 1922; Onete and Paucă-Comănescu 2008a, b). The species include *Chamaecyparis lawsoniana, Larix decidua, Picea abies, P. glauca, P. pungens, Pinus nigra, P. strobus, P. sylvestris* and *Pseudotsuga glauca.* The trees may have their natural shapes or be stunted according to the place where they grow. The Botanical Garden contains specimen trees of *Ginkgo biloba* while *Juniperus horizontalis, J. virginiana, Taxus baccata, Thuja occidentalis* and *T. orientalis* have been planted in most of the parks, creating a structural complexity within them and increasing the plant species biodiversity.

Pteridophytes

Ferns have not been recorded in the city indicating the need for more investigations of their presence and distribution. It is known that *Marsilea quadrifolia, Salvinia natans,* and *Azolla filiculoides* occur in wetlands in areas surrounding Bucharest (Nedelcu et al. 1972a, b).

Algae

The Dâmbovița River and most of the lakes are subject to phytoplankton blooms during the spring to autumn. Some of the lakes are also subject to excessive growths of filamentous algae.

Bryophytes

A recent review of the liverworts of Bucharest by Gomoiu and Ştefănuţ (2008) and The Hornwort and Liverwort Atlas of Romania (Ştefănuţ 2008) reported the occurrence of the following species: *Conocephalum salebrosum,* which was found for the first time in the Botanical Garden in 2005 and *Frullania dilatata* found in the Botanical Garden in 1982 with only male plants being found in 2005 in Pantelimon, Lebăda Forest. The following species, which were found in Bucharest in the early twentieth century have not been found in recent times: *Lunularia cruciata* (found in the Botanical Garden greenhouses in 1915 – it is an introduced species), *Metzgeria furcata* (last found in 1978), *Porella platyphylla* (found in the Botanical Garden in 1982), *Radula complanata* (found in Băneasa Forest in 1941 and 1995), *Riccia cavernosa* (found on the Dâmbovița bank at Grozăveşti in 1900, 1901 and 1915) and *Riccia fluitans* found in Ciurel and Grozăveşti in 1900, 1901 and 1915. The habitats suitable for *Riccia cavernosa* and *Riccia fluitans* were destroyed by the re-inforcement of the Dâmbovița River banks.

Marchantia polymorpha was found in Cotroceni, on the wall of the Botanical Institute and the Botanical Garden in 1896, 1909 and 1915; the Artillery school in 1901 and 1915; Ştirbei Vodă Street in 1903 and 1922; Botanical Garden in 1960, 1961 and 1982 and on the roof of the building of the Rector of the Polytechnic University; ssp. *ruderalis* (male only) in Cişmigiu Park in 2004 and 2005.

Fungi (including Lichenised Fungi)

As a result of habitat fragmentation, reduction in the size of the forests, soil compaction, air pollution and climate change, there has been a noticeable decrease in the number of macrofungi, and mycorrhizal fungi of the city. The inventory of fungi species in the central parks of Bucharest (Mogîldea 2008) includes *Agaricus campestris* var. *campestris, Boletus rubellus, Coprinus comatus, Coprinellus disseminatus, C. micaceus, Fistulina hepatica, Fomes fomentarius, Ganoderma applanatum, G. lucidum, Laetiporus sulphureus, Marasmius rotula, Phellinus pomaceus, Pluteus cervinus, Polyporus squamosus, Schizophyllum commune, Scleroderma verrucosum, Stereum hirsutum, Trametes versicolor* and *Volvariella bombycina.*

The mycocoenoses composition of the green spaces is the result of the interaction between many biotic and abiotic variables. Human activities held in the parks such as trampling, mowing, the removal of dead wood and leaf litter, the introduction of exotic trees and shrubs and changes in herbaceous plant populations have had a negative impact on the fungi.

Lichenised Fungi

Lichens from the central parks of Bucharest have been listed by Gomoiu and Ştefănuţ (2008) including *Candelaria concolor, Hypogymnia physodes, Parmelia saxatilis, P. sulcata, Xanthoria parietina* and species from genus *Lecanora* and *Melanelia.* Because of their sensitivity to air pollutants, studies of the lichens of the common species found in the central parks and the peripheral forests were carried out between 2005 and 2008. Chemical analyses showed that the heavy metal content of the park species was higher than those in the peripheral areas.

Habitats

With one exception there are no natural or semi-natural habitats in Bucharest; they only occur in the forests around the city. The exception is Băneasa forest, which comprises small patches of some natural and semi-natural habitats; see Doniţă et al. (2005). The natural and semi-natural habitats around the city comprise patches of forest that vary in size but are generally small. The forests occur as "islands" in a sea of arable land and without any connection between them.

Forests

Deciduous

Mixed Danubian forests with *Quercus robur* and *Tilia tomentosa* with *Scutellaria altissima* (EUH code R4147; Corine code, there is no corresponding Romanian plant community).

The canopy of these forest types includes *Quercus robur* or/and *Q. cerris, Fraxinus excelsior, F. angustifolia, Tilia cordata, T. tomentosa, Acer platanoides, A. campestre,* etc. The shrub layer includes species such as *Cornus mas, C. sanguinea, Corylus avellana, Crataegus monogyna, Euonymus europaeus, Ligustrum vulgare* and *Viburnum lantana*. The herbaceous vegetation growing on basic or weakly acidic, eutrophic soils comprises *Arum orientale, Carex sylvatica, C. pillosa, Cardamine bulbifera, Euphorbia amygdaloides, Geranium robertianum, Ghechoma hirsuta, Mercuralis perennis* and *Pulmonaria officinalis.* The characteristic species *Scutellaria altissima* has been not found in Băneasa Forest.

Coniferous and Mixed Forests

Coniferous and mixed habitats do not regenerate in the *Quercus* forests zone of Romania.

Scrub

Ponto-Panonic scrubs with *Prunus spinosa* and *Crataegus monogyna* (EUH code R3122; Corine Code 31.8B3).

This vegetation type occurs on different soils generally with a high soil-moisture deficit during the summer months. The species associations, which are typical of clear-felled *Quercus* forests, include *Prunus spinosa* and *Crataegus monogyna* together with *Rubus caesius, Rosa canina, Euonymus europaeus, Pyrus pyraster, Ligustrum vulgare, Humulus lupulus, Clematis vitalba, Prunus fruticans, Cornus mas, Vicia tenuifolia, Bromopsis inermis, Origanum vulgare, Festuca valesiaca, Agrimonia eupatoria, Dactylis glomerata* and *Teucrium chamaedrys*.

Grassland

Dry

West-Pontic grasslands with *Poa bulbosa, Artemisia austriaca, Cynodon dactylon* and *Poa angustifolia* (EUH code R3420; Corine code, there is no corresponding Romanian plant community).

This habitat comprises degraded vegetation of floodplain pastures. The characteristic species *Poa bulbosa* can be found only in spring when the watertable is high; in the summer, the plants are overwhelmed by species such as *Artemisia austriaca, Cynodon dactylon, Lolium perenne, Lotus corniculatus, Medicago lupulina, M. minima* and *Poa angustifolia.*

Damp (= Semi Dry)

Danubian-Panonic grasslands with *Agrostis stolonifera* (EUH code R3715; Corine code, there is no corresponding Romanian plant community).

This habitat occurs on alluvial and sandy-clay soils. The characteristic species, *Agrostis stolonifera,* occurs together with *Alopercurus pratensis, Poa pratensis, P. trivialis, Daucus carota, Medicago lupulina, Potentilla reptans* and *Lotus corniculatus.*

Danubian-Pontic grasslands with *Poa pratensis, Festuca pratensis* and *Alopercurus pratensis* (EUH code R3716; Corine code, there is no corresponding Romanian plant community).

The most representative species of this habitat are *Poa pratensis, Festuca pratensis, Dactylis glomerata, Elytrigia repens, Trifolium pratense, Lotus corniculatus, Potentilla reptans, Holcus lanatus, Agrostis capillaris, Taraxacum officinale, Leucanthemum vulgare* and *Stellaria graminea.*

Wet

Wet grassland occurs in small patches in the city and on the margins of lakes in Herăstrău, Bordei, Colentina, and Mogoşoaia, where the edges are more natural and without concrete re-inforcement.

Danubian communities with *Typha angustifolia* and *T. latifolia* (EUH code R5305; Corine code 53.13).

This plant community occurs on clay soils (sometimes slightly salty), the phytocoenses comprise *Typha angustifolia* and *T. latifolia* together with *Phramites australis, Alisma plantago-aquatica, Lythrum salicaria, Lysimachia vulgaris, Solanum dulcamara, Persicaria hydropiper, Galium palustre, Mentha aquatica, Lemna minor, Myriophyllum spicatum* and *Ceratophyllum demersum.*

Danubian communities with *Phragmites australis* and *Schenoplectus lacustris* (EUH code R5309; Corine Code 53.1).

The phytocoenosis of this plant community is dominated by *Phragmites australis,* growing on soils rich in organic matter. The other species present include *Mentha aquatica, Typha angustifolia* and *T. latifolia, Alisma plantago-aquatica, Lemna minor, Schoenoplectus lacustris, Stachys palustris* and *Hydrocharis morsus-ranae.*

Ponto-Danibian communities with *Bidens tripartita, Echinochloa crus-galli* and *Persicaria hydropiper* (EC Habitat code R5312; Corine code 24.52).

The most representative plant species of this habitat are *Echinochloa crus-galli, Bidens tripartita, B. frondosa, Persicaria hydropiper, Symphytum officinale, Rumex palustris, Veronica anagalis-aquatica* and *Rorippa austriaca.*

Open

Anthropogenic communities that occur along transport routes (EC Habitat Code R8701; Corine code, there is no corresponding Romanian plant community).

The species that occur in this community type include *Cephalaria transsilvanica, Leonurus marrubiastrum, Nepeta cataria* and *Marrubium vulgare, Artemisia vulgaris, Elytrigia repens, Ballota nigra, Cirsium arvense, Cynodon dactylon, Lepidium draba, Capsella bursa-pastoris, Conium maculatum, Geum urbanum, Glechoma hederacea* and *Verbena officinalis.*

Anthropogenic communities occurring in abandoned fields, around blocks of flats and on clay soils with low water content (EUH code R8704; Corine code, there is no corresponding Romanian plant community).

The characteristic species of this habitat include *Polygonum aviculare* agg., *Lolium perenne, Schlerochloa dura, Plantago major, Poa annua, Lepidium ruderale, Matricaria perforata, Lepidium ruderale, Malva pussila, Sagina procumbens, Amaranthus crispus* and *Euclidium syriacum.*

Aquatic

The aquatic plants are removed from the park lakes during summer for amenity reasons.

Danubinan communities with *Lemna minor, L. trisulca, Spirodela polyrhiza* and *Wolffia arrhiza* (EUH code R2202; Corine Code 22.411).

The plant communities of these eutrophic water bodies are poorly developed. The submerged vegetation is predominantly *Lemna trisulca,* which occurs among stands of *Phragmites* or *Typha.* Other submerged species include *Myriophyllum spicatum, Ceratophyllum demersum, Potamogeton crispus* and *P. pectinatus.* The floating-leafed zone (when it occurs) is dominated by *Lemna minor. Alisma plantago-aquatica* can also be found on the margins of the lakes.

Danubian communities with *Nymphaea alba, Trapa natans, Nuphar lutea* and *Potamogeton natans* (EUH code R2207; Corine code 22.43).

The charactersitic species of basins with still or slow running water include *Nymphaea alba, Trapa natans, Nuphar lutea* and *Potamogeton natans* together with *Ceratophyllum demersum, Elodea canadensis, Myriophyllum spicatum, Lemna minor* and *L. trisulca. Trapa natans* appears only in the lakes from the surrounding forests.

City Centre

The city centre is typically densely developed (roads, shops, offices, hotels and apartments). However, there are some large parks and other areas of open space around the centre; see Fig. 3. Within the centre, there are small areas of mature trees, including *Aesculus hippocastanum, Ailanthus altissima, Albizia julibrissin,*

Fig. 3 Part of the city centre seen from Unirii Park © Marilena Onete

Catalpa bignonioides, Paulownia tomentosa, Platanus x *hispanica, Quercus* spp., *Tilia* spp., *Picea* spp. and *Thuja* spp.

Most of the mature *Platanus* x *hispanica* trees along the road sides have been cut down because of the damage that falling branches have caused to cars parked along the streets. However, a large number of young trees of this species have been planted along the boulevards and in the parks. The lawns, which include such species as *Lolium perenne, Agrostis* spp., *Festuca* spp. and *Carex* spp., are being colonised by ruderal species, for example, species from the genera *Cirsium, Atriplex, Amaranthus, Arctium, Artemisia, Lepidium, Rumex, Sisymbrium* and *Sonchus. Polygonum aviculare* agg. is dominant in the open spaces. Cracks in the pavements and similar areas contain such plant species as *Aegilops cylindrica, Lepidium ruderale, Matricaria chamomilla, Polygonum aviculare* agg. and *Sclerochloa dura.*

Residential Areas

(a) Low Density

The gardens (including courtyards) vary in size from small (about 250 m^2) to large (about 500 m^2); the most frequent are about 250 m^2 while some may have only one plant – a mature tree; see Fig. 4.

Fig. 4 Vegetation within a low density housing area © Florin Bodescu

The structure and species composition of the gardens are also variable, some (mainly the older houses) are dominated by trees (often fruit trees) and shrubs such as *Cerasus avium, C. vulgaris, Morus alba, Prunus cerasifera* varieties, *P. domestica* or other *Prunus* species, *Persica vulgaris, Armeniaca vulgaris* and *Ribes* spp. Most of the large gardens have a substantial cover of *Vitis vinifera* to provide shade and grapes. Vegetable gardens are to be found in some of the old gardens. The climbers *Parthenocissus tricuspidata, Wisteria sinensis, Vitis vinifera* and *Hedera helix* cover the walls of the old houses.

Newly created large gardens generally comprise a large lawn with several horticultural varieties of native and non-native taxa including *Picea* spp., *Pinus* spp., *Juniperus* spp., *Magnolia* spp., *Forsythia europaea, Hibiscus syriacus, Syringa vulgaris* and *Tamarix tetrandra.* Ornamental herbaceous species from the genera *Begonia, Allium, Tagetes, Cosmos, Rudbeckia, Dahlia, Gaillardia, Santolina, Lilium, Tulipa* and *Papaver* are often planted in these gardens, together with the native species *Convallaria majalis,* which in such situations is being used as an "ornamental" species.

(b) High Density

The high density developments that virtually surround the city are bordered by narrow (2–4 m wide) "gardens" and contain small squares of "green space".

Both areas contain (often predominantly) mature to over-mature trees including *Sophora japonica, Ailanthus altissima,* and species within the genera *Tilia, Populus, Aesculus* and *Fraxinus;* see Fig. 5. The trees often form a dense canopy resulting in little light reaching the ground and therefore the absence of shrub and herb layer vegetation. In many cases, the canopy is in contact with the building causing the apartments to be heavily shaded. Many of the trees are suffering considerable dieback/canker probably exacerbated by age and air pollution. Most of them have been or are being cut down as a precaution against shedding branches or being blown over by strong winds.

The green spaces are bordered by hedges comprising species such as *Hibiscus* spp., *Symphoricarpos albus, Ligustrum vulgare* and *Taxus baccata.* The green space between the hedges and the buildings containing sparse sown or spontaneous vegetation dominated by *Cirsium* spp., *Chelidonium majus, Convolvulus arvensis, Geum urbanum, Plantago lanceolata, P. major, Polygonum aviculare* agg., *Sonchus arvensis, Rumex crispus* and *Taraxacum officinale.* In the grassland areas, the dominant species are *Poa annua, Festuca pratensis, Cynodon dactylon, Bromus commutatus, B. sterilis, Dactylis glomerata, Elytrigia repens* and *Setaria viridis* together with *Galium aparine, Convolvulus arvensis, Crepis foetida, Erigeron annuus, Taraxacum officinale* and *Malva neglecta.*

In recent times, some of the "squares" within the high-rise blocks of flats have been converted to car parks and play areas while "low-rise" offices and shops have been built in some of the larger open spaces resulting in the clearance of the vegetation.

Fig. 5 Large parts of the city are occupied by high density blocks of flats © Marilena Onete

Employment Areas

(a) Commercial

Most of the commercial areas are located in parts of the city previously occupied by old buildings or brownfield sites that have been colonised by ruderal vegetation. These "features" have been replaced by shopping centres and new office blocks. The car parks of the shopping centres have been planted with widely spaced trees including *Platanus* x *hispanica* and *Tilia* spp. – most of them are suffering from drought caused by excessive heat and large areas of tarmac or other hard surfaces. Small areas adjacent to new car parks and buildings have been planted with ornamental shrubs whose shapes and foliage are attractive to the human eye.

(b) Industrial

During the communist period, the industrial areas were established on the outskirts of the city but as the city has expanded, these areas have been incorporated into it. However, some industrial areas (mainly derelict) still occur in the central parts, for example, close to the Parliament Palace and in the Obor, Progresul, Semănătoarea-Grozăveşti zones. These areas were re-located to the periphery and the land was used for the construction of new residential districts. The urban wastelands and green network of the city have been described by Culescu and Tudora (2007), Culescu (2008). However, while the vegetation is a mixture of native, spontaneous and alien species, details of the plant communities have yet to be studied; see Fig. 6.

Fig. 6 Typical industrial area © ROMSTAL

(c) Educational

Located in the Cotroceni district near to Cotroceni Palace (the residency of the President) is the 17.5 ha University Botanic Garden "Dimitrie Brândză". The garden, which includes 4,000 m^2 of greenhouses, contains over 10,000 plant species. Founded by Carol Davila in 1860, the first Botanic garden was situated near to the Medical Faculty. The garden was re-located in 1884 by Dimitrie Brândză (a well-known Romanian botanist) and Fuchs (a Belgian landscape architect). Following the construction of the greenhouses and stocking with plants, the garden was inaugurated in 1891. Since then, the garden has been seriously affected by many events including major floods in 1891, the First World War (when it was used by the German army) and the Second World War (it was bombed by Anglo-American troops in April 1944). In some parks and green spaces (for example, Cişmigiu and the Romanian Academy), the species of the trees and shrubs are labelled in order to help people recognise them.

Transport Routes

Railways

The first railway line was constructed in 1869; it was soon followed by a network of others. At one time, the city had nine railway stations; five have subsequently disappeared. The vegetation on the embankments and cuttings is mainly tall grassland with some trees (*Populus* spp.) and shrubs (for example, *Rosa canina, Rubus caesius*); see Fig. 7. *Ailanthus altissima* scrub of about 20–30 cm is dominant on some embankments,

Fig. 7 Railway into the city, photo taken near by the main railway station Gara de Nord © Marilena Onete

where it occurs with *Populus tremula.* Young individuals of *Lycopersicon esculentum* and *Clematis vitalba* can also be found. The herbaceous species present include *Artemisia annua, Chenopodium album, Tragopogon pratensis, Crepis foetida, C. biennis, Plantago lanceolata, P. media, Taraxacum officinale, Daucus carota, Portulaca oleracea, Conyza canadensis, Poa angustifolia, P. annua, Festuca pratensis* and *Cynodon dactylon.*

Roads

With the exception of some of the main boulevards, the road verges are generally devoid of vegetation. Several of the main boulevards, adjacent to parks, are lined with semi-mature/mature trees, for example, *Aesculus hippocastanum, Platanus* x *hispanica, Populus nigra Italica, Quercus robur* and *Q. rubra.* The central reservations are mainly short (mown) neutral grassland dominated by grass species such as *Cynodon dactylon, Setaria viridis, Trifolium repens, Polygonum aviculare* agg., and *Convolvulus arvensis*. Beds of ornamental flowers (including *Viola* spp., *Calendula* spp., *Rosa* spp., *Leucanthemum* spp., etc.) have been established in some of the central reservations.

Airports

The runways and taxiways of Băneasa Airport, which lies within the city boundary, are set in extensive areas of grassland, which have not been investigated botanically.

Recreational Areas

Parks

Bucharest has about 36 parks. Some are more like small gardens while others are more complex (for example, Herăstrău, Carol, Alexandru Ioan Cuza and National). The largest park comprises a contiguous series of individual parks associated with the lakes on the north side of the city. Over time, new districts appeared and the older ones were completed; parks appeared and disappeared, new boulevards and streets were opened and many blocks of flats and houses were built, surrounded by "green space". In 1906, some waste ground was devoted to public gardens (Ioanid Park) or private gardens (Filipescu Park).

In 1843, the German horticulturist Meyer and his assistant Hörer created a park adjacent to Kisselef Boulevard using trees and shrubs, which were brought from Italy. Foreign travellers wrote that on both sides of the Boulevard there was a public garden, one of the most beautiful gardens in Europe. At the same time Meyer established the Cişmigiu Garden, a lovely park in the city centre. The lake that existed was drained and the fields were converted to lawns and shrubberies and planted

Fig. 8 Cişmigiu Park situated in the city centre © Marilena Onete

with many tree species. So by 1846, the people were able to talk about a garden that extended from the city centre. Foreign travellers were impressed by the beauty of the gardens, not only of the splendid gardens of the nobleman but also of the public gardens, stating that no other city, except Paris, could compare and that Cişmigiu excelled everything that Germany could offer. Bucharest was a wonderful, interesting city of striking contrasts (Fig. 8).

During the communist period, new parks were built, particularly on the sites of former dumps, for example, 23 August, Tineretului and Circului. The existing parks were enlarged (Herăstrău Park by 50 ha) and private parks and gardens were opened to the public. This increased green areas (parks, gardens, "squares of open space" many of which were densely planted with trees and shrubs), made the city healthier and more beautiful, and restored its previous state as a city covered in gardens. In 1977, intense plantation work aimed to increase the green area (2,570 ha) to 20 m^2/inhabitant. The green spaces surrounding the city were also improved, with the string of lakes on the Colentina River maintained and enlarged, together with the new lakes at Plumbuita and Pantelimon.

The species diversity in the parks is higher than that of the surrounding areas, mainly because of the planting of many species (especially ornamentals) by the park managers for aesthetic reasons. The trees and shrubs of the parks include alien species such as *Quercus* spp., *Juniperus* spp., *Aesculus hippocastanum, A. carnea, Acer negundo, Albizia julibrissin, Ailanthus altissima, Catalpa bignonioides, Celtis australis, Paulownia tomentosa, Picea glauca, Platanus* x *hispanica, Buxus sempervirens, Forsythia europaea, Rhus hirta, Robinia pseudoacacia, Tamarix tetrandra* and *Taxus baccata* together with native species: *A. platanoides, A. pseudoplatanus, A. tataricum, Betula pendula, Fraxinus excelsior, Pinus sylvestris, Quercus cerris, Q. robur, Tilia cordata, T. tomentosa, Ulmus minor, Cornus sanguinea, Crataegus monogyna, Hedera helix, Ligustrum vulgare, Rosa canina* and *Rubus caesius.* The distribution of some trees and shrubs in some of the smaller parks (for example, Cişmigiu, Izvor and Unirii) is kept under strict control.

The herbaceous species present in the parks comprise mainly ruderal species, which are generally common in the city but none are characteristic of a particular park. The species include *Achillea millefolium, Agrostis stolonifera, Alliaria petiolata, Brassica elongata, B. oleracea, Bromus arvensis, Capsella bursa-pastoris, Chenopodium album, Cichorium intybus, Convolvulus arvensis, Conyza canadensis, Cynodon dactylon, Dactylis glomerata, Daucus carota, Erigeron annuus, Euphorbia cyparissias, Festuca pratensis, F. rubra, Galium aparine, Geranium pusillum, Geum urbanum, Glechoma hederacea, Lamium album, L. amplexicaule, Linaria vulgaris, Lolium perenne, Lotus corniculatus, Lysimachia nummularia, Malva sylvestris, M. neglecta, Medicago sativa* ssp. *falcata, M. lupulina, M. sativa, Myosotis arvensis, Oxalis corniculata, O. stricta, Phytolacca Americana, Plantago lanceolata, P. media, Poa angustifolia, P. annua, P. pratensis, Polygonum aviculare* agg., *Potentilla reptans, Prunella vulgaris, Rumex crispus, Setaria pumila, S. viridis, Silene latifolia, Solanum nigrum, Stellaria media, Taraxacum officinale, Trifolium pratense, T. repens* and *Urtica dioica.* It is probable that further surveys will discover some interesting and possibly rare species, especially in the less well-used areas of the parks.

Cemeteries

The earliest cemeteries were located outside the city, but as the city expanded they became incorporated within it. Most of the trees and shrubs are ornamental and include *Jasminum officinale, Paulownia tomentosa, Philadelphus coronarius, Syringa vulgaris, Pinus* spp., *Picea* spp. and *Thuja* spp. Some herbaceous species "creep" into this protected space where they are accompanied by exotic or native flowering plants, for example, species of *Viola, Calendula* and *Vinca.*

Open Land

Agricultural

The peripheral areas of and land adjacent to the city comprise relatively large areas that are used for the production of cereals and other crops including *Zea mays, Triticum aestivum, Secale cereale, Hordeum vulgare* and *Helianthus annuus.* In communist times, the land was used by the State (via Co-operatives) for intensive agriculture (monocultures). Following the 1989 revolution, the land was returned to the former private owners in small units. Some of the landowners have continued to manage the land for agricultural purposes (although sometimes for different crops) while others have not and left the land fallow. Currently, much of the arable and fallow land is being used for residential, commercial and industrial developments.

Land Awaiting Re-development

As described earlier, following the 1989 revolution proposed developments and a considerable number (in various stages) of constructions were abandoned. This resulted in many brownfield sites and partly constructed buildings (and their surroundings). These sites became colonised by species that are typical of open, eutrophic conditions such as *Amaranthus retroflexus, Arctium tomentosum, Artemisia annua, Chelidonium majus, Conium maculatum, Datura stramonium, Echium vulgare, Medicago sativa ssp. falcata, Sambucus ebulus, Xanthium spinosum* and *X. strumarium.* As a result of increased economic development in the 5–10 years, many of the abandoned developments have been re-started and new developments commenced on the abandoned sites – both resulting in the consequential loss of the vegetation.

Aquatic

Moving Water

Although it is slightly sinuous, most of the River Dâmbovița was straightened during the canalisation process; the banks are steep and made of concrete, see Fig. 9. Grass species such as *Lolium perenne, Agrostis* spp., *Festuca* spp. and *Carex* spp. have been sown on the soil between the top of the concrete banks and the boundary fence. This area which also contains ruderal species is being colonised by *Ailanthus altissima.* Between the fences and asphalt roads, *Tilia* ssp. dominates the landscape. In the lawn, many ruderal species grow, as in parks, more young trees of *Ailanthus altissimo* are appearing. The river does not support any floating-leafed or emergent vegetation but can support submerged vegetation. The other river system originates to the north-west of the city. The Crevedia and Colentina rivers flow into a series of sinuous lakes, which continue into, through and out of Bucharest. On the river banks can be found trees such as *Acer campestre, Alnus glutinosa, Fraxinus angustifolia, Populus alba, P. nigra, Salix alba, S. fragilis, Tilia tomentosa, T. cordata, Ulmus laevis* and *U. minor* and shrubs such as *Cornus sanguinea, Euonymus europaeus, Frangula alnus, Salix cinerea, Sambucus nigra, Viburnum opulus,* etc. dominating *Rubus caesius* and *Galium aparine* in the lower layer. Herbaceous vegetation from the rivers and also some lakes margins comprise *Iris pseudacorus,* different *Carex* species, *Caltha palustris* and many other species.

Still Water

Reservoirs

Lake Morii is a balancing lake or impoundment reservoir constructed to protect the city against flooding. The lake and its margins are also used for amenity purposes. The reservoir has a margin of about 7 km and a surface area of 246 ha. The water

Fig. 9 Dâmbovița River as it "flows" through the city below Lacul Morii © Marilena Onete

is contained within 15 m high bunds. In the northern part of the lake, there is an artificial island with a wharf.

Plantago lanceolata, Poa pratensis, Polygonum aviculare agg. and *Portulaca oleracea* grow from the cracks in the concrete banks while *Persicaria hydropiper* and *Phragmites australis* occur on the shoreline where soil has accumulated over the concrete banks. The herbaceous vegetation above the concrete banks includes *Achillea millefolium, Amaranthus retroflexus, Artemisia annua, Atriplex oblongifolia, Berteroa incana, Bromopsis inermis, B. sterilis, B. tectorum, Lepidium draba, Carduus nutans, Centaurea* spp., *Cephalaria transylvanica, Chenopodium album, Cirsium arvense, C. vulgare, Convolvulus arvensis, Coronilla varia, Crepis foetida, Cynodon dactylon, Dactylis glomerata, Daucus carota, Descurainia sophia, Erigeron annuus, Conyza canadensis, Eryngium campestre, Festuca pratensis, Hordeum murinum, Lactuca serriola, Lathyrus tuberosus, Lolium perenne, Lotus corniculatus, Malva neglecta, M. sylvestris, Medicago sativa* ssp. *falcata* , *M. lupulina, Melilotus albus, M. officinalis, Plantago lanceolata, Poa pratensis, Polygonum aviculare* agg., *Portulaca oleracea, Potentilla argentea, P. reptans, Sambucus ebulus, Setaria italica, S. verticillata, S. viridis, Silene latifolia, Sonchus arvensis, S. oleraceus, Symphytum officinale, Tragopogon pratensis, Trifolium pratense, T. repens, Xanthium spinosum* and *X. strumarium.* Trees and shrubs such as *Gleditschia triachantos, Morus alba, Prunus cerasifera, P. spinosa, Rosa canina,* and *Rubus caesius* occur but only rarely.

Ornamental and Recreational Lakes

Most of the parks and many other open spaces contain lakes of varying sizes and depths. The bank material and slopes are also variable. The water is eutrophic. However, the bed and margins are usually made of concrete. Submerged vegetation

Fig. 10 Lake in the middle of Cişmigiu Park © Marilena Onete

is represented by *Myriophyllum* species, which is removed in the summer to allow people to swim safely. Very small amounts of floating-leafed vegetation occur in some lakes, for example, *Nymphaea alba* and *Nymphoides peltata.* Emergent/marginal vegetation is also scarce with localised occurrences of *Typha latifolia* and *Phragmites australis.* Some of the lake margins are shaded by trees, including *Salix alba, S. babylonica sensu lato* and *S. fragilis,* which have been pollarded; see Fig. 10. However in some areas, because of neglect, the trees have not been pollarded for many years and are now starting to collapse into the water. Blooms of phytoplankton and in some cases excessive growths of filamentous algae may become a big problem in the summer and autumn.

Nature Conservation, Environmental Planning and Education

Nature conservation is not the main priority of the city municipality. The area of the city's green spaces has decreased in recent years. Some have been sold to the private sector for the development of hotels and commercial centres. The nature reserves, which are not well managed, are in the forests outside the city and not within it. Nature trails are provided in every nature reserve but people are not yet able to use them properly.

Table 2 lists the only taxa in Bucharest that are protected or are of nature conservation concern; of these only one (*Galanthus nivalis*) is listed in Annex V of the EC Habitats Directive (regarding plant species of Community interest whose taking in the wild and exploitation may be subject to management measures). Annex V is

Table 2 Taxa occurring in Bucharest that are protected or are of nature conservation concern

Taxon	EC habitats directive	National legislation	IUCN threat categories	Red lists
Crocus flavus	–	–	Vulnerable	+
Corylus colurna	–	–	Vulnerable/rare	+
Galanthus nivalis	Annex V	Annex 5A	Not threatened	+

included in Annex 5A of the Romanian Government's Urgency Ordinance 57/2007. The Red List of the Higher Plants of Romania (Oltean et al. 1994) has been prepared using the Directive 92/43/CEE, WCMC – Conservation Status Listing from 1992 and also different articles of the Romanian scientists about Romanian flora.

Galanthus nivalis and *Crocus flavus* are considered by Ciocârlan (2000) to be frequent in the wild in Romania but are often planted in the city's parks. *Corylus colurna,* which is a rare species in the wild, is also planted in the city as part of landscaping schemes. Some protected plant species grow wild in Romania and have the status "natural monument" but are planted in the city as ornamental species and/or for hedges; they include *Taxus baccata* and *Ilex aquifolium,* which is a sub-endemic Tertiary relict that is "Vulnerable/Rare" in the Romanian Red List.

Bucharest contains some individual tree species that are protected because of their beauty, rarity, age or size, for example, *Corylus colurna* and *Platanus* x *hispanica* in Cişmigiu Park and *Ginko biloba* in the botanical garden.

Closing Comments

The biggest current problem for the Bucharest authorities is to ensure a healthy environment for the people (in relation to identifying and controlling the multiple pollution sources that are seriously affecting the quality of the air, water and soil). However, the authorities are ignoring the necessity of maintaining and increasing the green spaces of the city. Even Natural Law 749/2006 which regulates the administration of green spaces in the city is being systematically ignored by the authorities. Large new development projects are converting green spaces into retail and entertainment areas for people, for example, the creation of a swimming pool with sand beaches and spas (Sănătate Prin Apa). Some non-governmental organisations fight for the preservation and enlargement of the green spaces in the city but not with much success.

The green spaces in Bucharest have different degrees of structural complexity (vegetational and architectural). They can be considered to be artificial ecosystems because they are created for aesthetic reasons by the planting of ornamental trees, shrubs and herbaceous species. The native vegetation has almost disappeared from the city, having been replaced by the artificial ecosystems described. However, some the species that are planted in these artificial systems are species of the natural former forests that occurred in the sylvo-steppic zone of the country. However, the city contains some remnants of natural vegetation, which have adapted to the modified conditions of the urban environment, for example, the application of fertilisers and herbicides or the converse – the lack of the nutrients, changes in soil pH, increased temperature and pollution.

The vegetation and flora of Bucharest needs to be studied much more comprehensively in order to ensure that the planning, design and management of the city as a whole and the green spaces in particular are based on sound scientific information.

Literature Cited

Anastasiu P (1994) Naturalized plants in Bucharest city. Acta Bot. Hort. Buc., 135–137

Ciocârlan V (2000) Ilustrated flora of Romania. Pteridophyta et Spermatophyta. Editura Ceres, Bucharest

Culescu M (2008) Bucharest industrial wasteland – between ecology and development, 3 rd Conference of the Competence Network Urban Ecology "Urban Biodiversity and Design – Implementing the Convention of Biological diversity in towns and cities", Book of abstracts, pag. 56

Culescu M, Tudora I (2007) Post-industrial Bucharest and re-deffinition of a new urban landscape, ACUM 2 (Art, Urban Comunity, Mobilization) – Public space and social re-insertion of a artistic and architectural project. Editura Universitara "ION MINCU", Bucharest

Doniță N, Popescu A, Paucă-Comănescu M, Mihăilescu S, Biriş IA (2005) The habitats from Romania. Editura Tehnică, Bucharest

Gheorghe R, Nica M (2008) The capital city of Romania. In: Onete M, Ion M (ed) Planning environmental management. Methodology for assisting municipality, Editura Ars Docendi, Bucharest

Gomoiu I, Ştefănuț S (2008), Lichen and mosses as bioindicator of air pollution in Bucharest. In: Onete M (ed) Species monitoring in the Central Parks of Bucharest. Editura Ars Docendi, Bucharest

Lăcătuşu R, Anastasiu N, Popescu M, Enciu P (2008) The geo-atlas of Bucharest City (Geoatlasul municipiului Bucharest). Editura Estfalia, Bucharest

Mogîldea DE (2008), Macrofungi in urban ecosystem. In: Onete M (ed) Species monitoring in the Central Parks of Bucharest. Editura Ars Docendi, Bucharest

Morariu I (1943) Antropophile plant associations from Bucharest surrounding with observations on their distribution in the country and especially in Transilvania. Bul. Grad. Bot. Cluj. XXII (3–4):131–224

Nedelcu GA, Popescu A, Sanda V (1972a) Coenological researches upon the helophyts from Bucharest surroundings. St. Cerc. Biol. Seria Botanica, 24(1):3–8

Nedelcu GA, Popescu A, Sanda V (1972b) Researches upon the sociology of macrophytes from Bucharest surroundings. Hidrobiologia, 13:189–198

Oltean M, Negrean G, Popescu A, Roman N, Dihoru G, Sanda V, Mihăilescu S (1994) The red list of higher plants from Romania. Studii, Sinteze, Documentații de Ecologie. Institutul de Biologie. Academia Română, I:5–52

Onete M, Paucă-Comănescu M (2008a) Synanthropic flora from Bucharest central parks, Romanian Ecological Society Conference. Book of abstracts. Editura Ars Docendi, 179–181.

Onete M, Paucă-Comănescu M (2008b) Heavy metal content assessment in plants from Bucharest. In: Onete M (ed) Species monitoring in the Central Parks of Bucharest. Editura Ars Docendi, Bucharest

Popescu A. Sanda V, Ionescu Al (1971) Researches upon herbaceous vegetation from Bucharest surroundings. St. Cerc. Biol. Seria Botanica, 23(1):47–55

Prodan I (1922) An overview upon arborescente flora from Bucharest. Revista padurilor, 397–404

Sanda V, Popescu A (1971) Phytocoenological researches in the forests from Bucharest surroundings. St. Cerc. Biol. Seria Botanica, 23(2):125–142

Spiridon L (1973) The ruderal vegetation from Bcuharest city surroundings. Analele Univ Bucharest. Biol Veget, XXII: 129–132

Ştefănuţ S (2008) The Hornwort and Liverwort Atlas of Romania. Edit. Ars Docendi. Bucharest

London

Michael J. Crawley

Fig. 1 Tower Bridge looking west along the River Thames, with The Tower of London behind, to the right (Photo Arpat Ozgul)

Abstract The native flora of London has declined inexorably since the Roman occupation, as natural habitats were built-over or dumped upon. The species richness of surviving fragments of semi-natural vegetation has declined under the interacting effects of drainage, trampling, acidification, eutrophication, dog-fouling, harvesting and botanical collection. However, the decline in native species has been matched by a steady increase in the number of alien species. London's important

Michael J. Crawley (✉)
Imperial College London, Silwood Park, Ascot, Berkshire SL5 7PY, England
e-mail: m.crawley@imperial.ac.uk

J.G. Kelcey and N. Müller (eds.), *Plants and Habitats of European Cities*,
DOI 10.1007/978-0-387-89684-7_7,

alien species evolved in far-flung corners of the globe (for example, *Buddleja davidii* in China, *Conyza sumatrensis* in South America, *Epilobium ciliatum* in North America and *Crassula helmsii* in New Zealand), yet they come together to form strange new plant communities. The replacement of native by alien flora might be the despair of conservationists, but the dynamics of extinction and invasion are endlessly fascinating to those of us who describe ourselves as urban botanists.

Natural Environment

London lies on the Greenwich Meridian (zero degrees longitude) at roughly 51° 30′ North latitude, just upstream of the point at which the River Thames flows into its estuary, see Fig. 2.

The area has an exceptionally long history of human occupation, and some of Britain's earliest signs of human settlement (about 400,000 years ago) come from Swanscombe on the south shore of the estuary. Swanscombe is one of the richest of British Palaeolithic sites, and also the home of the oldest known English human

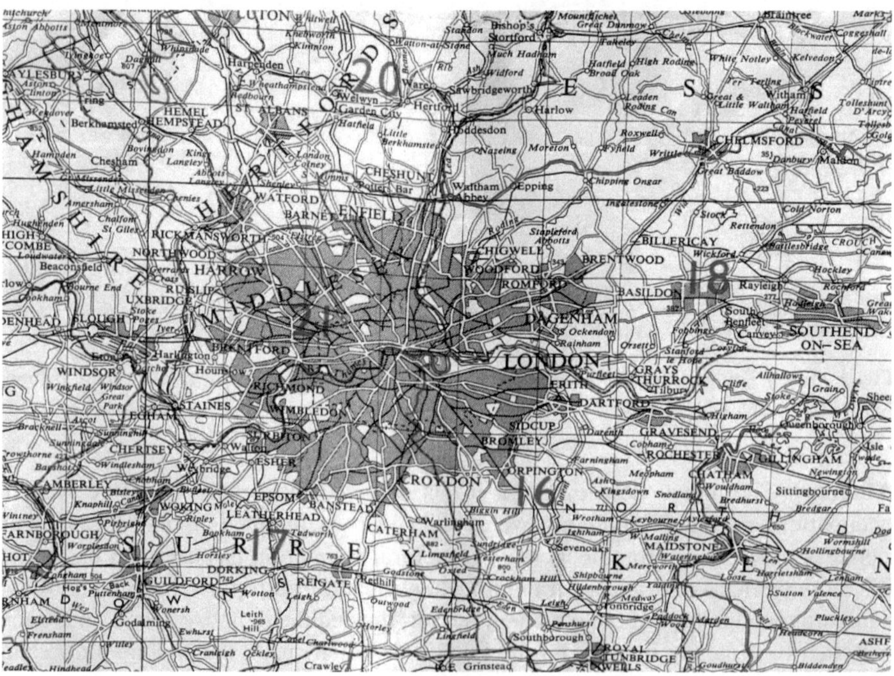

Fig. 2 London and surrounding countryside. Numbers represent "vice-counties" which are the traditional spatial recording units for British natural history. North of the River Thames, London covers all of Middlesex (v.c. 21) and the western parts of Essex (v.c. 18), while south of the Thames, London occupies the northern part of Surrey (v.c. 17) and the north-western part of Kent (v.c. 16)

remains (Stringer 2006), but between about 400,000 and 100,000 years ago there was a dramatically decreasing signal of human presence (hand-axes and flake tools) in the terraces of the River Thames, as successive glacial advances pushed human populations further south. There were *Rangifer tarandus* and *Mammuthus primigenius* in the cold periods, and *Coelodonta antiquitatis, Palaeodoxodon antiquus, Bos taurus primigenius, Equus caballus, Cervus elephus, Canis lupus* and *Trogontherium cuvieri* in the warmer periods. These animals roamed woodland dominated by *Quercus robur, Carpinus betulus, Alnus glutinosa, Corylus avellana* and *Salix* spp. with grasslands bordering the river.

At the peak of the Anglian glaciation, some 450,000 years ago, the River Thames flowed west to east along the snout of a massive ice sheet substantially to the north of its current course, through St Albans towards Clacton and Colchester. At this time, Britain was a peninsula of Western Europe, joined to the Eurasian continent by a wide land bridge running from East Anglia to the Isle of White, whose spine was the chalk ridge between what is now south-east England and France. After the ice had retreated, the Hoxnian interglacial saw the creation of a staircase of terraces of sands, gravels and silts, on which London is built. At the end of the last Ice Age, 10,000–15,000 years ago, the terraces were clothed with *Betula nana, Thalictrum alpinum, Oxyria digyna* and other typical arctic plants, but many currently familiar urban weeds were already present at that time, including *Polygonum aviculare* agg., *Rumex crispus* and *Urtica dioica*. Later, an all but impenetrable waterlogged forest of *Alnus glutinosa* and *Quercus robur* restricted human habitation, and vast swamps, up to 5 km wide in places, flanked the River Thames from Brentford downstream to the Isle of Dogs. Despite the long history of early human occupation, fire and flint tools were relatively feeble instruments of ecological change, and it was not until the Roman period that substantial areas of woodland began to be felled for fuel. The clearance of dense *Alnus, Salix* and *Quercus* woodland was continued by Saxons and Danes, so that by the Norman Conquest in 1066, the bulk of London's woodland had gone.

Geology

The city lies in a shallow bowl of chalk that outcrops in the Chiltern Hills to the north and in the North Downs to the south. This had enormous significance for the development of London, because the aquifer provided an abundant and reliable water supply for the inhabitants. On top of the chalk, in the bottom of the bowl, lie the younger, post-Cretaceous strata. These Eocene sands and clays comprise a rich loamy soil of considerable agricultural and horticultural importance, especially in the south-eastern (Kent) quarter. Next come the Blackheath Pebble Beds, which give rise to a less fertile soil, forming the Commons of Hayes, Keston, Chislehurt and Plumstead which were dominated by *Betula pendula, Pinus sylvestris, Ulex europeus, Cytisus scoparius,* and *Calluna vulgaris*. Next comes the extensive deposit of the London Clay, varying in depth in different parts of the city from 20 to 130 m and lying beneath most of the conurbation. The London Clay proved to be better for making bricks than for growing crops, and London's famous yellow-brown bricks, the so-called London Stock, were created in countless kilns and brickfields all over the city. In due course, the abandoned clay pits formed habitats for scores of interesting plants, either growing on the rubbish that was typically dumped into them, or in and around the lake that often formed in the bottom of the abandoned brick pit. Above the London Clay are scattered outcrops of sands and gravels. The Bagshot Sands, for instance, cap several of the local hills, as at Highgate and Hampstead, and these supported fragments of lowland heath (H2 *Calluna vulgaris-Ulex minor* heath (Rodwell 1991b). On top of the solid geology lie the various alluvial deposits laid down by the river since the Ice Ages, with the various terraces tracing the history of the river's course.

Climate

By the standards of southern England, the climate of London is drier, calmer and sunnier than the average. The urban heat-island effect adds an extra 2 or 3 degrees to mean air temperatures at most times of the year compared with the surrounding countryside. Frosts are much less frequent in the built-up areas, where heat is stored in the great mass of brick, concrete and asphalt. The average annual rainfall is about 650 mm, but the range is substantial, from 800 mm at Biggin Hill to just 530 mm at river level near Rainham. The growing season for plants (when mean daily air temperature exceeds 5.6°C) is about 3 weeks longer in the centre of town (281 days) than on the outskirts at Wisley in Surrey (259 days). Surface temperatures on the man-made substrates in central London can be very high and lethal to many native plant species. London is relatively sheltered from the prevailing south-westerly winds but occasional storms (as in October 1987 and January 1990) cause severe structural damage, and can uproot mature trees in great numbers. Flooding is a perennial risk for the capital, especially when easterly winds coincide with high spring tides in the estuary. The Thames Barrier was opened in 1982 to exclude these tidal surges, but predicted increases in mean sea level and increased frequency and severity of storms have led to the development of plans to build a more sophisticated replacement barrier in the near future.

Air pollution has been a recurrent theme of London life since the capital began to function by burning coal imported by sea from Newcastle, towards the end of the Medieval period. The London smog (a literary and literal combination of smoke and fog) was legendary, and had severe consequences for the health of the human inhabitants as well as for plant life. In 1937, the air pollution amounted to 125 tonnes of solids deposited per square kilometre per year, of which 17 tonnes was of sulphate and 2 tonnes was of tar (Larsen et al. 2004). The London "pea-soupers" earned the capital the nickname of "The Smoke" and the Great Smog of 5–9 December 1952 killed thousands of people and debilitated thousands more.

The leaves of vascular plants were coated by soot by the end of summer. We learn that although *Corylus avellana* was common, "*nuts are not formed in the neighbourhood of populated, and therefore smoky, districts*". One of the most dramatic botanical consequences of London's air pollution was the virtual elimination of lichens and bryophytes. The acidic effects of sulphur dioxide pollution from coal burning turned London into a "lichen desert", and of the 130 species recorded in Epping Forest in 1800, just 28 lichen species survived in 1975. At the peak of sulphur dioxide pollution in 1970, Jack Laundon, a lichenologist from the Natural History Museum, recorded only 9 species growing on trees within 16 km from Charing Cross (Laundon 1970). Since then air quality has improved considerably and lichens are returning to London's parks and gardens (Rose and Hawsworth 1981). For instance, there were only 11 lichen species in the Chelsea Physic Garden in the 1970s, but the flora had recovered to 38 species by 2004 (when 11 of the lichens were found on wooden benches; Ann Allen and Peter Jones, 2005, unpublished).

Human and Botanical History

Despite the long history of human occupation, the first proper town to develop here was probably Roman London (*Londinium*). After the invasion of AD 43, the Romans built the first city on the twin hills on either side of the Walbrook stream on the Taplow Terrace, on the banks of the navigable Thames. They built the first wooden bridge across the river at the point where the River Fleet flowed into the River Thames, which was probably the lowest part of the river at which it could be bridged, and close to the tidal limit at the time (the land was perhaps 5 m higher than it is today).

The town rapidly became an important and thriving port, and *Londinium* was already a notable commercial centre when it was sacked by Boadicea in AD 61. After this, the Romans built a wall around the town, enclosing 134 ha. The walled town served as the provincial capital of Roman Britain from then onwards. Inside the wall, people became the dominant ecological force but outside the city wall, apart from tree-clearance for fuel, the countryside was still little affected by human activity. The town became the hub of an extraordinary network of Roman roads, radiating out from London like the spokes of a wheel, to the most far-flung corners of England, bringing food to the population of London from the rich corn and cattle lands of Essex, Kent and Hertfordshire and taking soldiers to the provinces. The Roman roads also facilitated the movement of plants, both intentionally and unintentionally, all over the country. The Romans brought with them many plant species new to Britain, including at least 50 new plant foods, mostly fruits, herbs and vegetables.

These introductions represented a major diversification of the plant component of the British diet at this time, adding important nutrients, variety of flavours, ways of expressing cultural identity, as well as social status (van der Veen et al. 2008).

The list of vegetables introduced to Britain includes *Allium sativum, A.ascalonicum, A. porrum, Brassica oleracea* var. *capitata, B.rapa, Raphanus sativus, Apium graveolens* var. *dulce, Pisum sativum* and *Asparagus officinalis* ssp. *officinalis* along with herbs like *Rosmarinus officinalis, Thymus* spp. *Laurus nobilis, Ocimum basilicum* and *Mentha spicata*. They also introduced herbs that were used in brewing and for medicinal purposes, and instigated new farming practices and more productive grains, so that bread became a much more important part of the London diet. Among the trees, *Juglans regia* and *Castanea sativa* were introduced (it is not clear whether *Acer pseudoplatanus* was a Roman or a later introduction). They also introduced a wider variety of cultivated fruit trees, including apples (as opposed to *Malus sylvestris*), *Prunus domestica* ssp. *insititia, Vitis vinifera, Morus nigra* and cherries (*Prunus* spp.).

There was a period when the Romans prohibited the establishment of vineyards outside Italy, in order to safeguard the Italian wine trade, but in the third century the emperor Probus granted permission to Britain, Spain and Gaul to re-establish *Vitis vinifera*. Romans were among the first to use gardens as a place of relaxation and decoration, and we can assume that they introduced many new species to London as garden plants. They were also responsible for the introduction of vegetables-turned-weeds like *Aegopodium podagraria, Sinapis alba* and *Smyrnium olusatrum* but we shall never know the full list of plant species that were deliberately or unintentionally introduced by the Romans and which arrived with subsequent waves of immigration. The population of Roman London probably peaked at or about 50,000 people, a figure that was not surpassed until after 1500.

After the Romans left in ca. AD 400 and before the Saxons took possession, nature is likely to have reclaimed much of the site of *Londinium*, and from the seventh to the eleventh centuries there was much open ground within the Roman city walls. London disappeared from history for almost 200 years, although it probably remained as a major Saxon settlement and business centre among the Roman ruins. As early as the seventh century, however, it was Britain's biggest settlement, and by ca. 730 the Venerable Bede was writing of London as "*a trading centre for many nations who visit it by land and sea*", no doubt increasing the variety of alien plant species growing there. London was pillaged when England was invaded by the Danes in 851, but King Alfred re-captured the town in 886 and the commercial prosperity of London revived. The walls were restored and the quays rebuilt.

William the Conqueror established his power base at Windsor Castle, 25 km upstream, but London developed rapidly. In 1067 work started on the Tower of London and the first stone bridge across the Thames was completed in 1176. Westminster had royal palaces and great abbeys, but at the time of the Doomsday Book in 1086 London still contained more than 500 ha of arable land and substantial pasturage, while the surrounding county of Middlesex had enough woodland to provide pannage for 20,000 swine. It took many years before the new London broke out from the limits defined by its Roman Wall, but once out, there was no stopping its growth.

By 1268, during the reign of Henry III, the city was "*among the most noble and celebrated cities of the world ... is one of the most renowned, possessing above all others abundant wealth, excessive commerce, great grandeur and magnificence*" (Fitzstephen twelfth century). Its development as a seat of government in the thirteenth and fourteenth centuries, its rise as a world centre of trade in the sixteenth century, the development of its financial institutions in the late seventeenth century, and its role as a colonial then imperial capital in the eighteenth and nineteenth centuries, all contributed to its phenomenal expansion.

Until the Great Fire of London in 1666, the town consisted of close-packed wooden buildings, and the desolation in the aftermath of the Great Fire saw a dramatic increase in the abundance of *Sisymbrium irio* (see below). The city grew by building layer upon layer over the dung heaps and rubbish tips of the previous generation, in conditions of squalor and filth that we associate these days with shanty towns in the third world. In these malodorous and generally offensive circumstances, thrived a particular guild of vascular plants vividly invoked by Thomas Johnson in the second edition of *Gerard's Herbal* (1636). Of *Descurania sophia* he writes that it "*groweth almost everywhere in ruins of old buildings by high waies and in filthie obscure base places*". In the Colne district of Buckinghamshire, immense quantities of street sweepings and household refuse were brought from the capital in barges and deposited on the banks of the canal or as land-fill in adjacent brick-yards. There was a major refuse site near what became Kings Cross station, which was home to dustmen and cinder-sifters as well as more general scavengers who lived on the refuse of the city.

In addition to the parks that were kept open for Royal hawking (hunting with raptors), deer (Cervidae) hunting, hare (*Lepus euopaeus*) coursing and fox (*Vulpes vulpes*) hunting (for example, Regents Park in 1562), London became home to several great gardens. There were Royal parks like Hampton Court and gentlemen's gardens, of which the most celebrated were Syon House (seat of the Duke of Northumberland, opposite Kew Gardens), Osterly (home of the Child family) and Cannons (where the pretentious James Brydges, 1st Duke of Chandos lived; the poet Alexander Pope's considered this to be an ideal example of archetypical bad taste and mis-spent wealth). Chelsea Physic Garden was founded in 1673 by the Worshipful Society of Apothecaries as a collection of medicinal plants. The garden's celebrated Cedars of Lebanon (*Cedrus libani*) were planted in about 1683; the last came down in 1903. Chelsea Physic Garden may have been the source of several early garden escapes like *Cymbalaria muralis* but there is no solid evidence for the precise origin of this plant, nor indeed for most of London's alien plants. The Royal Botanic Gardens at Kew evolved from modest beginnings in 1759 and were extended in 1841. Kew now houses one of the world's largest collections of living plants with 40,000 taxa (including roughly 10% of the world total of seed plants and ferns), while the scientific collections in the Herbarium comprise about seven million specimens. Several species are alleged to have escaped from Kew Gardens but most of these cases are fanciful. London has no equivalent of Oxford Botanic Garden's famous escapee, *Senecio squalidus,* which went from the garden's walls in the early nineteenth century to colonise Britain along the ballast of the newly created railway lines (Druce 1886).

The only reasonably well-documented escapes from Kew Gardens are *Galinsoga parviflora* (escaped in 1796 and now a common weed of cultivated and bare ground) and *Ceratochloa carinata* (escaped in 1919, now widespread by the River Thames from Chiswick to Twickenham).

From the start, London was ringed by market gardens producing food and flowers for the urban population: *Brassica oleracea* var. *capitata, Brassica oleracea* var. *botrytis* (cauliflower and broccoli), *Lactuca sativa, Daucus carota* ssp. *sativus, Apium graveolens* var. *dulce, Allium cepa, Rheum* x *hybridum, Asparagus officinalis* ssp. *officinalis* were grown, along with orchards of *Malus domestica* varieties, *Prunus avium* and *P. domestica* ssp. *domestica* while the main soft fruits were *Ribes uva-crispa, R. rubrum, R. nigrum, Rubus idaeus, Rubus loganobaccus, Rubus fruticosus* agg and *Fragaria ananassa.* Of the land devoted to growing cut flowers, most was given over to *Narcissus* varieties and other bulbs, while the land under glass, mainly in the Lea Valley was used for the production of *Lycopersicon esculentum.* Much of the richest market garden ground (for example, the fertile brick-earths of Middlesex and South Essex) was lost under suburban sprawl, as vegetable-growing was driven steadily outwards by the radial spread of London.

At the end of the eighteenth century Acts of Parliament allowed rich landowners to take into private ownership large areas of previously common land including what was known at the time as "waste", and these newly enclosed lands were turned over to cultivation. Subsequently, much of this land was sold to speculative house-builders as the period of urban sprawl began at the end of the nineteenth century. Before 1800 the increase in London's population had been driven by immigration because, prior to that date, more people died in the capital than were born there. By the turn of the nineteenth century, only 10% of the capital's population lived within the jurisdiction of The City. The population of London grew from roughly 850,000 in 1800 to 3.5 million in 1900 and 7.5 million in 2000.

By 1800 the port of London had become the busiest in the world: "*Britain has drawn upon the universe for her supplies and in turn sends her products to the outposts of the world until her commercial prosperity is one of the wonders of the age*" (Druce in Hayward and Druce 1919). The Port of London handled about 40% of the country's total imports including foodstuffs, and vast quantities of raw materials for the wide variety of industries in the London area. The ships often had animals on board, along with hay to feed them. Seeds were swept out of the holds and dumped on the pier, while other seeds arrived among packing materials, and others attached to imported hides and wool. Ballast dumped from arriving vessels built up in great heaps on the shore and formed islands in the shallows of the river. The ballast contained countless seeds and fragments of plants from all over the world. Some of these survived the vicissitudes of their long journey to flower under London skies on these piles of foreign soil. Before the docks were built, cargo boats were moored side by side across much of the width of the river, and their cargoes were unloaded into small boats then rowed upstream to the wharf at which they were to be landed.

This system was the cause of much theft and dishonest handling of goods, and Parliament decided to create a major system of docks to accommodate the cargo ships more efficiently. In 1802 the West India Docks were opened, followed in 1806 by East India Docks, Royal Victoria Docks (1855) and Royal Albert Docks (1876). The buildings associated with these docks were among the largest ever built to that date. "*The docks are impossible to describe*" wrote Verlaine in 1872 (in Carter 1971). "*They are unbelievable! Tyre and Carthage all rolled into one.*" Subsequently, the Regent's canal was built to link the docks to the rest of England, meeting the Grand Union Canal at Paddington Basin. The dock-sides, wharves and warehouses became favourite hunting grounds for London botanists (Curtis 1775–1798; de Crespigny 1877).

The period of commercial dominance of the canals was short-lived. The arrival of the railways caused a revolution for the landscape of London, as vast brick viaducts brought railway lines into the heart of the city from all points of the compass, and imposing terminus buildings advertised the wealth and importance of the various railway companies. Euston station opened in 1837, Waterloo in 1848, Kings Cross in 1852, Victoria in 1860 and St Pancras in 1868. These were not just stations as we now understand them, but included coal stores, granaries and stabling for large numbers of horses. In 1863 the first Underground Railway system in the world was started, and the associated works included the stone-work embanking of substantial lengths of the Thames (for example, construction of the Victoria Embankment) further hastening the demise of the riverside flora. Along with the railroad tracks themselves, came vast areas of sidings and engine sheds, all providing habitat for plant species adapted to freely draining, nutrient-poor, well-lit substrates including cinder, slag, clinker and ballast. Species like *Chaenorhinum minus* and *Senecio viscosus* that had previously been great rarities were able to prosper, and the network of railway lines provided highly effective corridors for plant dispersal (most famously, perhaps, for *Senecio squalidus,* above). Huge areas of flat, open ground were used as marshalling yards. At Feltham, for instance, much of the Southern Region freight for London destinations was sorted out, and transfer traffic to the Midland and Eastern Regions was marshalled into trains. After the yard was abandoned in 1969, the area became a *de facto* wilderness, now dominated by an extensive woodland of *Betula pendula* with an understorey rich in colourful alien species like *Potentilla recta* and *Galega officinalis*. The Great Exhibition in South Kensington in 1862 was the source of many new botanical records for London, presumably resulting from the combination of disturbance from the building work, introduction of species and the presence at the Exhibition of unprecedented numbers of experienced plant hunters from all over the world.

The sprawl of suburbs developed as the railways allowed the evolution of daily commuting as a way of life. Even as recently as 1920, much of the area of Greater London was semi-rural, but by 1970 most of it was highly urbanised. There was a massive increase in the number of gardens in the newly developed suburbs, with a consequent increase in plants that were either escapes or outcasts from horticulture.

Fly-tipping of garden waste on roadsides, railway cuttings and waste ground led to great increases in geophytes like *Narcissus* and *Hyacinthoides* cultivars, along with vigorous rhizomatous perennials like *Aster* and *Solidago* that tend to outgrow their space in the garden.

John Claudius Loudon (1829) had been appalled at the prospect of the loss of Hampstead Heath to urban development. This was still a vast, uncultivated region of fields and woodlands, freely enjoyed by the whole community as a "green lung". Loudon led a chorus of protest, and published an elaborate plan for the whole capital as a series of concentric circles in which belts of "city" alternated with "rural" belts, creating "breathing spaces for the Metropolis". His integration of public transport, mains supplies of water and gas, irrigation canals and drainage, in a pattern that made sense in terms of sanitation, economics and aesthetics was a model of its type (Ponte 1991). The Green Belt dates from an idea proposed in 1934 and followed eventually by legislation in 1955. The aim was to check the unrestricted sprawl of large built-up areas, to prevent neighbouring towns from merging into one another. This would assist in safeguarding the countryside from encroachment, preserve the setting and special character of historic towns, and assist in urban regeneration by encouraging the recycling of derelict and other urban land. Once an area of land had been defined as green belt, it provided opportunities for access to the open countryside for the urban population for outdoor sport and recreation. It retained attractive landscapes close to where people live, secured nature conservation interests and retained land in agricultural, forestry and related uses (Anon 1995).

Bombing and the associated fires (between September 1940 and May 1941) created extensive areas of open ground and rubble in previously built-up areas, especially in The City and the East End. In 1943, Salisbury recorded 126 species on bombed ruins in formerly plant-free areas. In contrast to the aftermath of the Great Fire in 1666, there was no great resurgence of *Sisymbrium irio* after the Blitz but its place was taken by other *Sisymbrium* species, mostly *S. orientale* but also *S. loeselii* and S. *altissimum. Senecio squalidus* and *S. viscosus* both did well on the extensive waste ground created by bombing; both had been late additions to London's flora (1867 and 1838, respectively). *Galinsoga parviflora* was found on 14% of bombed sites, while *Solanum nigrum* and *S. dulcamara* were both locally abundant. Ponds formed in bomb craters were rapidly colonised by *Sparganium erectum, Alisma plantago-aquatica, Juncus articulatus, J. inflexus, J. conglomeratus* and *J. effusus. Chamerion angustifolium,* increased from being a rare garden-escape at the time of Curtis' *Flora Londinensis* (1775–1798) to being one of the commonest species, dominating large areas of derelict land in the capital after the Second World War.

Habitats

The wholly built-up areas are buried beneath stone, brick and asphalt, and have no remaining bare soil and hence no extensive vegetation. Even here, however, there are epiphytic plants on the stonework (escaped ornamental ferns, and brick-dwelling woody plants like *Buddleja davidii,* there are weeds in the sediment collected in roof gutters, for example *Poa annua* and *Stellaria media* and trample-resistant, bryophytes such as *Bryum argenteum* in joints between the paving slabs).

Semi-natural habitats represented in the London Area include bogs, marshes, heaths, various kinds of grassland (chalk grasslands (CG2) on the Downs, mesotrophic grasslands (MG5) on the clays, and acidic grasslands (U1) on the Bagshot Sands, woods (mostly W5 or W10), river banks and salt marshes. The codes are from the National Vegetation Classification (see Rodwell 1991a, b, 1992, 2000). Details of the floras of these places can be found in the local County Floras of Hertfordshire (Dony 1967), Essex (Jermyn 1974), Kent (Philp 1982), Surrey (Salmon 1931; Lousley 1976; Leslie 1987), Berkshire (Crawley 2005) and Buckinghamshire (Druce 1926; Maycock and Woods 2005). It is the man-made habitats that are the focus of this account.

The classic urban habitat for plants is the vertical brick wall. Railway bridges, viaducts and brick-lined cuttings are the preferred substrate for native ferns like *Asplenium ruta-muraria* with less common *A. trichomanes* and *A. adiantum-nigrum.* Damp brickwork beneath a broken down-pipe will often have *Phyllitis scolopendrium* and *Geranium robertianum* along with the ubiquitous *Cymbalaria muralis.* Garden escapes like *Campanula poscharskyana* are widespread and locally abundant, but *Asarina procumbens* is still relatively uncommon. *Parietaria judaica* is the most likely species to be found on damp steps or in stair-wells. One of London's characteristic wall plants these days is *Pseudofumaria lutea,* which was a rare garden escape in 1909, but is now frequent and locally abundant on brick walls of buildings, boundaries and bridges.

London rubbish dumps have long been the haunt of botanists searching for rare alien plants, and the tips at Dagenham, Grays, Tilbury, Hackney Marsh and Yiewsley were the most celebrated sites. Melville and Smith found 250 adventive species in 1928 including a dense tall forest of *Heracleum mantegazzianum* with *Rumex patientia,* extensive dense coverage of *Solanum nigum* and *Chenopodium rubrum* with *Consolida orientalis, Brassica juncea, Raphanus sativus, Lepidium sativum, Camelina sativa, Beta vulgaris* ssp. *vulgaris, Cucumis sativus, Cucurbita pepo, Ricinus communis, Fagopyrum esculentum, Cannabis sativa, Xanthium strumarium, X. spinosum, Guizotia abyssinica, Calendula officinalis, Centaurea diluta, Allium cepa, Lolium temulentum, Secale cereale, Triticum aestivum, Hordeum vulgare, H. distichon, Avena sativa, Polypogon monspeliensis* and *Panicum miliaceum.* These days, *Silybum marianum* is restricted to waste tips and soil piles, typically as a garden outcast.

Urban lawns are often the only green habitats in central London. Here are found interesting native species like *Trifolium micranthum, T. subterraneum, Ornithopus perpusillus, Montia fontana* ssp. *chondrosperma* and *Hypericum humifusum* along with abundant aliens like *Oxalis exilis* and *Veronica filiformis*. The latter species is a pest of lawns in several parts of London, escaped from its original use as an ornamental plant for rockeries. *Veronica filiformis* does not set seed regularly in Britain, and is distributed by fragments of shoot, either on footwear, in imperfectly composted grass clippings, or attached to mowing machines, tractors and other equipment moved from one lawn to another.

Parks and Cemeteries

London's parks range from Richmond Park which extends to almost 1,000 ha and includes a National Nature Reserve and several Sites of Special Scientific Interest, down to tiny pockets of closely mown lawn surrounded by towering buildings, see Figs. 3–5. Most parks are formally managed with large areas of turf, scattered trees and small plots of ornamental shrubbery, but wildlife-friendly areas are being created under more enlightened management. The closely mown grasses (*Agrostis capillaris, Festuca rubra, Holcus lanatus, Lolium perenne* and *Poa pratensis*) are often accompanied by little more than *Bellis perennis, Ranunculus repens, Taraxacum* spp. and *Plantago lanceolata,* typically impoverished by eutrophication as a result of dog-fouling. Less polluted lawns have *Achillea millefolium, Trifolium repens, Crepis capillaris* and *Hypochaeris radicata,* while the most species-rich support *T. micranthum, T. arvense, T. striatum, Ornithopus perpusillus* and *Spergularia rubra,* with less frequent grasses and sedges including *Koelaria macrantha, Poa humilis* and *Carex muricata* ssp. *lamprocarpa.* Neglected areas typically develop into a tall-grass community dominated by *Arrhenatherum elatius, Dactylis glomerata, Elytrigia repens* or *Lolium perenne* (MG1; Rodwell 1992) but with some botanical interest occasionally provided by *Centaurea nigra, Senecio erucifolius, Lotus pedunculatus, Scrophularia nodosa, Lathyrus pratensis* or *Vicia cracca.*

The older trees are a splendid feature of many of London's parks and cemeteries. There are fine veteran individuals of *Quercus robur* and stately avenues of *Tilia* x *europaea* (a few of which, as in Bushy Park, support abundant *Viscum album*). Graveyards often have ancient *Taxus baccata*, along with other native trees like *Ilex aquifolium, Betula pendula, Corylus avellana* and *Fraxinus excelsior*. Most of the

Fig. 3 View from Greenwich Park towards Canary Wharf (Photo Daniel Kräher)

Fig. 4 Rembrandt Gardens in Westminster (Photo Daniel Kräher)

parkland trees, however, are planted exotics. Among the conifers are many fine elderly *Cedrus libani* and *Sequoiadendron giganteum,* with younger plantings of X *Cupressocyparis leylandii* and numerous cultivars of *Chamaecyparis lawsoniana.* The most frequent broadleaf tree by far is *Platanus* x *hispanica,* but there are frequent *Populus* x *canadensis, P. nigra* "Italica", *Robinia pseudoacacia, Aesculus hippocastanum, Tilia cordata* and *Quercus ilex.* Rarer trees include *Fagus sylvatica* var. *purpurea, Prunus dulcis, Acer platanoides, Araucaria araucana, Quercus cerris* (quite commonly self-sown), *Carpinus betulus, Juglans regia, Quercus* x *pseudosuber, Catalpa bignonioides, Prunus cersifera* "Pissardii", *Sorbus intermedia* and *Ficus carica,* often with fruits.

In 1832 Parliament passed a bill encouraging the establishment of private cemeteries outside the city of London, and later passed a bill to completely close all inner London churchyards. Over the next decade, seven cemeteries were set up, comprising what are now affectionately known as the "Magnificent Seven": Kensal Green (in 1832), West Norwood (1837), Highgate (1839), Abney Park, Nunhead and Brompton (all 1840) and Tower Hamlets (1841). Kensal Green Cemetery, for instance, is of great historic and wildlife interest, incorporating several formerly widespread habitats of the countryside, now lost from most of the city. Regionally rare and uncommon plants grow in the rich neutral grassland between gravestones, including *Achillea ptarmica, Carex divulsa* ssp. *divulsa, Persicaria bistorta, Sanguisorba officinalis, Silaum silaus* and *Valeriana officinalis.* In the north-west of the cemetery, dry grassland includes *Clinopodium vulgare* and *Leontodon hispidus,* both scarce in north London. The more elaborate tombs are habitat for ferns, with *Phyllitis scolopendrium, Asplenium adiantum-nigrum* and *A. ruta-muraria,* occasionally with rarer ferns like *A. trichomanes* and *Ceterach officinarum* (Table 1), while gravelled grave-tops are often dominated by one or more of *Sedum acre, S. album, S. spurium* or *S. rupestre.*

Fig. 5 Panorama view across London from the The Royal Obeservatory in Greenwich Park (centre of the view: The Old Royal Naval College with Canary Wharf behind, to the left: Central London) (Photo Daniel Kräher)

Some cemeteries contain fragments of ancient, unimproved grassland, the most frequent of which are acid grasslands, typically dominated by *Festuca rubra, Rumex acetosella, Pilosella officinarum* and *Luzula campestris,* with *Hypochaeris radicata, Vicia tetrasperma, Lotus corniculatus, Leontodon autumnalis, Leucanthemum vulgare, Stellaria graminea, Vicia sativa* ssp. *nigra* and *Veronica chamaedrys,* along with less frequent species including *Potentilla erecta, Hypericum humifusum, Campanula rotundifolia, Carex muricata* ssp.*lamprocarpa, Salvia verbenaca, Galium verum, Ranunculus bulbosus* and *Carex divulsa* ssp. *divulsa.* More fertile ground might support *Allium vineale, Filipendula ulmaria, Prunella vulgaris, Geranium molle, Lysimachia nummularia, Agrimonia eupatoria* or *Carex spicata.* Drier parts of some cemeteries have *Festuca ovina, Aira praecox, Geranium pusillum, Vulpia bromoides* or *Aira caryophyllea,* while moister parts often support large populations of *Cardamine pratensis.* More calcareous cemetery grasslands have *Pimpinella saxifraga, Sanguisorba minor, Anacamptis pyramidalis, Briza media, Primula veris, Carex flacca, C. divulsa* ssp. *divulsa, Galium mollugo* and *Saxifraga granulata* with *Silene vulgaris, Erigeron acer* and *Verbascum nigrum* on calcareous graves.

Shady parts of the graveyard sometimes have woodland plants such as *Viola riviniana, Festuca gigantea, Circaea lutetiana, Primula vulgaris, Anemone nemorosa, Geum urbanum,* and *Mercuralis perennis,* while damp, unkempt corners often support *Pseudofumaria lutea, Pentaglottis sempervirens, Anchusa officinalis, Chelidonioum majus, Parietaria judaica, Lapsana communis, Linaria purpurea, Medicago lupulina, Stachys sylvatica* and *Mycelis muralis.*

Table 1 Ferns found growing on walls in London (data from Fred Rumsey and John Edgington up to 2008). This mixture of natives and alien species escaped from window boxes or plant pots creates a highly distinctive urban community, benefiting from the capital's heat island and the abundance of damp brickwork, stonework and cement

Adiantaceae	Dryopteridaceae
Adiantum capillus-veneris	*Cyrtomium caryotideum*[a]
Adiantum raddianum[a]	*Cyrtomium falcatum*[a]
Pellaea falcata[a]	*Cyrtomium fortunei*[a]
Pellaea rotundifolia[a]	*Cyrtomium macrophyllum*[a]
	Dryopteris cycadina[a]
Pteridaceae[a]	*Dryopteris dilatata*
Pteris cretica[a]	*Dryopteris filix-mas*
Pteris multifida[a]	*Polystichum aculeatum*
Pteris nipponica[a]	*Polystichum polyblepharum*[a]
Pteris tremula[a]	*Polystichum setiferum*
Pteris vittata[a]	*Polystichum tsus-simense*[a]
Polypodiaceae	Woodsiaceae
Polypodium interjectum	*Athyrium filix-femina*
Polypodium vulgare	*Gymnocarpium robertianum*
	Cystopteris alpina[b]
Dennstaedtiaceae	*Cystopteris fragilis*
Pteridium aquilinum ssp. *aquilinum*	
	Blechnaceae
Aspleniaceae	*Doodia australis*[a]
Asplenium adiantum-nigrum ssp *adiantum nigrum* var. *silesiacum*	
Asplenium fontanum[b]	
Asplenium marinum	
Asplenium ruta-muraria ssp. *ruta-muraria*	
Asplenium scolopendrium ssp. *scolopendrium*	
Asplenium trichomanes ssp. *quadrivalens*	
Asplenium viride	
Ceterach officinalis	

[a]Taxon not native to the British Isles and Ireland
[b]Taxon believed to be extinct in the British Isles and Ireland

Three of the capital's most extensive man-made plant habitats (motorway verges, railway cuttings and canal banks) are compared in Table 2.

Motorway verges are free from human trespass but their botanical interest is often limited by eutrophication. They tend to be dominated by bulky dicotyledons, such as *Conium maculatum, Dipsacus fullonum, Picris echioides, Brassica napus, Hirschfeldia incana* and *Heracleum mantegazzianu. Galega officinalis* is locally abundant, forming drifts of white, pale pink or bluish-mauve flowers, and *Securigera varia* is occasional in similar places. Note that *Bunias orientalis,* so abundant in this habitat in central Europe, is still an uncommon plant in London. Motorways and larger trunk roads that are heavily salted for de-icing in winter support a fascinating community of seaside

Table 2 The floras of selected man-made habitats in London in 2008

Motorway	(%)	Railway	(%)	Canal bank	(%)
(a) Distinctive species					
Dipsacus fullonum	80	*Betula pendula*	52	*Glyceria maxima*	57
Conium maculatum	55	*Pteridium aquilinum*	34	*Alnus glutinosa*	50
Glechoma hederacea	52	*Ligustrum ovalifolium*	33	*Oenanthe crocata*	43
Plantago coronopus	51	*Clematis vitalba*	32	*Phragmites australis*	29
Ulex europaeus	49	*Fallopia baldschuanica*	19	*Carex riparia*	25
Cirsium arvense	44	*Prunus laurocerasus*	19		
Cochlearia danica	41	*Quercus ilex*	18		
Bellis perennis	38	*Pyracantha coccinea*	14		
Spergularia marina	26	*Chamerion angustifolium*	12		
Leucanthemum vulgare	26	*Fallopia japonica*	12		
Medicago lupulina	23	*Solidago canadensis*	12		
Puccinellia distans	22	*Lapsana communis*	11		
		Quercus cerris	8		
		Aster x *versicolor*	8		
(b) Widespread species					
Artemisia vulgaris	82	*Buddleja davidii*	89	*Dactylis glomerata*	98
Dactylis glomerata	74	*Hedera helix*	72	*Hedera helix*	93
Picris echioides	71	*Acer pseudoplatanus*	63	*Lamium album*	93
Arrhenatherum elatius	69	*Conyza sumatrensis*	57	*Sambucus nigra*	93
Fraxinus excelsior	69	*Quercus robur*	52	*Arrhenatherum elatius*	86
Plantago lanceolata	68	*Arrhenatherum elatius*	44	*Galium aparine*	86
Galium aparine	66	*Fraxinus excelsior*	42	*Acer pseudoplatanus*	79
Acer pseudoplatanus	56	*Dactylis glomerata*	40	*Anthriscus sylvestris*	79
Anthriscus sylvestris	56	*Poa annua*	26	*Buddleja davidii*	79
Senecio jacobaea	56	*Galium aparine*	25	*Conyza sumatrensis*	79
Urtica dioica	44	*Urtica dioica*	25	*Ranunculus repens*	79
Heracleum sphondylium	44	*Artemisia vulgaris*	24	*Fraxinus excelsior*	71

Quercus robur	44	*Senecio squalidus*	19	*Plantago lanceolata*	64
Achillea millefolium	41	*Sonchus oleraceus*	18	*Salix cinerea* ssp. *oleifolia*	64
Ranunculus repens	39	*Stellaria media*	16	*Urtica dioica*	64
Malva sylvestris	39	*Sambucus nigra*	14	*Artemisia vulgaris*	57
Lamium album	38	*Anthriscus sylvestris*	14	*Heracleum sphondylium*	57
Conyza sumatrensis	38	*Ilex aquifolium*	14	*Poa annua*	50
Poa annua	38	*Lamium album*	13	*Sonchus oleraceus*	50
Sonchus oleraceus	34	*Salix cinerea* ssp. *oleifolia*	13	*Senecio jacobaea*	50
Stellaria media	34	*Corylus avellana*	13	*Lamium purpureum*	50
Cirsium vulgare	34	*Senecio jacobaea*	12	*Cirsium vulgare*	50
Senecio squalidus	33	*Chamerion angustifolium*	12	*Malva sylvestris*	50
Senecio vulgaris	31	*Plantago lanceolata*	9	*Hirschfeldia incana*	43
Hedera helix	30	*Ulex europaeus*	9	*Picris echioides*	43
Cardamine hirsuta	30	*Heracleum sphondylium*	8	*Quercus robur*	43
Buddleja davidii	28	*Lamium purpureum*	8	*Cardamine hirsuta*	43
Hirschfeldia incana	25	*Picris echioides*	8	*Stellaria media*	36
Ilex aquifolium	25	*Hirschfeldia incana*	7	*Senecio vulgaris*	36
Sambucus nigra	23	*Cardamine hirsuta*	7	*Sisymbrium officinale*	36
Corylus avellana	23	*Cirsium vulgare*	7	*Ilex aquifolium*	29
Potentilla reptans	21	*Sisymbrium officinale*	7	*Achillea millefolium*	29

Species from motorway verge, railway and canal bank in 100 m^{-2} samples from 350 habitats across London were used to calculate percentage occurrence (%), then to select the 50 most frequent species from each community. The upper rows (a) show the species that were distinctive to each habitat (that is they are in the top 50 species for that habitat but not for the other two habitats, although these species may occur at lower frequencies in the other habitats). More of the distinctive railway species are aliens than in the other two communities. The lower rows (b) show the frequencies of widespread species that appeared in the top 50 for all three plant communities (for example, *Artemisia vulgaris* was in 82% of motorway samples, 24% of railway samples and 57% of canal bank samples), illustrating the substantial overall similarity of the three floras plants in the central reservation and on the bare ground that forms a strip about 30 cm wide along the edge of the kerb. These species, known as "salt adventives", include grasses like *Puccinellia distans* and *Echinochloa crus-gali,* and herbs, for example, *Cochlearia danica, Plantago coronopus* and *Spergularia marina.* Less abundant members of this community include *Cynodon dactylon, Hordeum murinum, Juncus gerardii, Parapholis strigosa, Puccinellia fasciculata, P. rupestris, Atriplex littoralis, Lepidium ruderale, Sagina maritima* and *Armeria maritima.* The North American alien *Hordeum jubatum* is not a seaside plant in its native country, but it is spreading rapidly in this salty roadside habitat. *Bassia scoparia* is a summer bedding plant first seen naturalised in Yorkshire in 1978 on Hull docks and its approach-roads, which subsequently spread along the motorway system, and it is now abundant on salted roadsides, especially on the M25 motorway and in south-east London.

The railway embankments and cuttings are much older, of course, than those of the motorways (mid-nineteenth century rather than late twentieth century), and are consequently more heterogeneous in their floras. The track-bed itself is typically kept scrupulously clear of plant life but where herbicide has not been applied too recently you will find *Chaenorhinum minus, Conyza sumatrensis, C. canadensis* and *C. floribunda* along with *Senecio squalidus* growing on the ballast. In the absence of intervention, succession would soon lead to the creation of *Buddleja-Conyza* scrub (Table 3).

On the bank beside the tracks are bulky native species belonging to MG1 *Arrhenatherum elatius* (Rodwell 1992) like *Anthriscus sylvestris, Heracleum sphondylium, Artemisia vulgaris* with aliens including *Fallopia japonica, Solidago canadensis, Alcea rosea, Lysimachia punctata, Leucanthemum* x *superbum* and *Colutea arborescens*. Bulbs thrown out from adjacent gardens grace the banks in spring, with *Narcissus* "Ice Follies" and *N.* "Golden Harvest", *Hyacinthoides* x *hispanica* (in various shades from white through pink to pale blue), *Galanthus elwesii* and *G. plicatus* in addition to the common *G. nivalis* "Flore Pleno", and the ubiquitous grape hyacinth *Muscari armeniacum.* In late summer, the most conspicuous flowers of London's railway banks are the various kinds of *Aster.* The hybrid *A.* x *versicolor* is the commonest taxon, but *A.* x *salignus, A. laevis, A. lanceolatus, A. novae-angliae* and *A. novae-belgii* have been reliably recorded. The trackside fence is often festooned with massive plants of *Fallopia baldschuanica,* draped with the bright pinkish-purple flowers of *Lathyrus latifolius* or hung with the fruits of *Humulus lupulus.*

The canals were built during the relatively brief period between the industrial revolution in the mid-eighteenth century and the railway revolution of the early nineteenth century. These days, London's canals form a system of linear nature reserves reaching right into the heart of the city, and the former tow-paths allow access to numerous fascinating wetlands. The downside of this open access is that copious dog faeces leads to a uniformly monotonous flora dominated by a small number of nitrogen-loving species. The striking feature of the comparison in Table 2 is the basic similarity of the three floras, probably as a result of convergence resulting from eutrophication and acidification from atmospheric inputs, coupled with exclusion of grazing animals. In the twenty-first century, however, London's most distinctive and characteristic plant community is *Buddleja-Conyza* scrub (Table 3). This community is not yet characterised under the National Vegetation Classification (NVC, Rodwell 2000), but is often extremely species-rich, with alien species usually making up more than 80% of the plant list.

Table 3 *Buddleja-Conyza* scrub community. This vegetation type is not described in the NVC (Rodwell 1991, 2000) but is a highly distinctive plant community in London, and is particularly rich in alien species. A composite list from a variety of sites in London in 2004 to 2008 is given below. Constant species: *Buddleja davidii, Conyza sumatrensis, Artemisia vulgaris, Epilobium ciliatum, Picris echioides, Sambucus nigra, Senecio squalidus.* Physiognomy: this is a heterogeneous scrub vegetation with thickets of *Buddleja*, often with sapling trees of *Acer pseudoplatanus, Salix cinerea, S. caprea* and *Sambucus nigra,* interspersed with open clearings dominated by *Conyza sumatrensis* (locally replaced by *C. floribunda*) and other herbs like *Hirschfeldia incana, Picris echioides, Oenothera glazioviana*, and *Senecio squalidus*. The perennial ground flora of the clearings is variable, but typically patchy with clonal plants like *Glechoma hederacea, Fragaria vesca, Prunella vulgaris, Hypericum perforatum* or *Sedum acre.* Habitat: *Buddleja-Conyza* scrub is characteristic of sunny sites on flat, freely drained ground, often on cinder, ballast or building rubble, but also in basements of terrace houses and abandoned gardens throughout London. Zonation and succession: the community colonises bare ground within 1 or 2 years of soil disturbance and can persist for at least 20 years. Typically it gives way to deciduous woodland if left undisturbed for longer periods. The dominant trees of older communities include *Acer pseudoplatanus, Salix* spp., *Betula pendula* and a variety of woody neophytes. Affinities: plant communities of open vegetation (e.g. OV23 *Lolium perenne-Dactylis* glomerata community and OV42 *Cymbalaria muralis* community of brick walls; Rodwell 2000) often have similar species lists but differ in relative abundance. *Buddleja-Conyza* scrub is usually species-rich, particularly in alien species, because at least early in succession levels of competition are very low and many species are able to recruit from seed. Plant species with wind-dispersed seeds predominate. Species composition of *Buddleja-Conyza* scrub community by growth form is as follows

Trees: *Acer pseudoplatanus, A. campestre, A. platanoides, Aesculus hippocastanum, Betula pendula, Carpinus betulus, Castanea sativa, Chamaecyparis lawsoniana, Fagus sylvatica, Fraxinus excelsior, Malus domestica, Populus tremula, Quercus cerris, Q. robur, Robinia pseudoacacia, Sorbus aucuparia, Taxus baccata, Tilia* x *europaea, Ulmus procera* (o), *Quercus ilex, Sorbus intermedia, Thuja plicata* (r)

Shrubs: *Buddleja davidii* (d), *Sambucus nigra, Vinca major* (ld), *Berberis darwinii, B.* x *stenophylla, B. thunbergii, Brachyglottis (Senecio)* "Sunshine," *Cornus alba, Corylus avellana, Cotoneaster simonsii, Crataegus monogyna, Hypericum calycinum, H.* x *inodorum* "Elstead," *Ilex aquifolium, Lavandula* x *intermedia, Leycesteria formosa, Lonicera pileata, Prunus laurocerasus, Pyracantha coccinea, Rosa rugosa, Symphoricarpos* x *chenaultii* "Hancock," *Viburnum tinus, Vinca minor* (f), *Cytisus scoparius, Elaeagnus* x *ebbingei, E. pungens* "Maculata," *Escallonia macrantha, Lonicera nitida, Osmanthus heterophyllus, Pachysandra terminalis, Potentilla fruticosa, Prunus lusitanica, Ribes sanguineum, R. nigrum, Rosmarinus officinalis, Salix caprea, S. cinerea, Sambucus racemosa, Spartium junceum, Symphoricarpos albus, Syringa vulgaris, Ulex europaeus, Viburnum davidii, V. rhytidophyllum* (o), *Ficus carica, Genista hispanica, G. monspessulana* (r)

Grasses: *Aira praecox, Agrostis capillaris, A. gigantea, A. stolonifera, Anisantha sterilis, Anthoxanthum odoratum, Arrhenatherum elatius, Bromus hordeaceus, Dactylis glomerata, Elytrigia repens, Festuca rubra* ssp. *commutata, megastachys* and *rubra, Holcus lanatus, H. mollis, Hordeum murinum, Lolium perenne, Poa annua, P. pratensis, P. trivialis, Vulpia bromoides, V. myuros* (f), *Carex pendula, Phalaris arundinacea* "Picta," *P. canariensis, Polypogon viridis* (o), *Briza maxima, B. minima* (r)

(continued)

Table 3 (continued)

Herbs: *Artemisia vulgaris, Calystegia sepium, C. sylvatica, Chamerion angustifolium, Conyza sumatrensis, C. floribunda, Fallopia japonica, F. baldschuanica, Pentaglottis sempervirens, Rubus fruticosus* agg., *Urtica dioica* (locally dominant), *Galium aparine, Pteridium aquilinum, Sisymbrium officinale, Solanum dulcamara* (a), *Agrimonia eupatoria, Arctium minus, A. nemorosum, A. lappa, Ballota nigra, Brassica napus, Capsella bursa-pastoris, Cardamine hirsuta, Centaurium erythraea, Cerastium fontanum, C. glomeratum, Conyza canadensis, Crepis vesicaria* ssp. *taraxacifolia, Diplotaxis tenuifolia, Dipsacus fullonum, Epilobium ciliatum, E. parviflorum, Eupatorium cannabinum, Glechoma hederacea, Hirschfeldia incana, Hypericum perforatum, Lactuca serriola, Lamium album, L. purpureum, Lapsana communis, Leucanthemum vulgare, Lotus corniculatus, Malva sylvestris, Matricaria recutita, M. discoidea, Oenothera glazioviana, Oxalis exilis, Papaver dubium, P. somniferum, Picris echioides, Plantago coronopus, Polygonum aviculare* agg., *Prunella vulgaris, Ranunculus repens, Reseda luteola, Rumex acetosella, R. acetosa, R. crispus, R. obtusifolius, Sagina procumbens, Sedum acre, Senecio jacobaea, S. squalidus, Sonchus arvensis, S. asper, S. oleraceus, Stachys sylvatica, Stellaria graminea, S. holostea, S. media, Symphytum officinale, S.* x *uplandicum, Trifolium dubium, T. hybridum, T. repens, T. pratense, Tripleurospermum inodorum, Verbascum nigrum, V. thapsus, Vicia hirsuta, V. sativa* ssp. *segetalis, V. tetrasperma* (f), *Aegopodium podagraria, Alliaria petiolata, Anagallis arvensis, Anthriscus sylvestris, Aquilegia vulgaris, Armoracia rusticana, Bryonia dioica, Cardamine flexuosa, C. pratensis, Carduus crispus, Centaurea nigra, Cerastium tomentosum, Chelidonium majus, Chenopodium album, Cicerbita macrophylla, Cichorium intybus, Cirsium arvense, C. palustre, C. vulgare, Clematis vitalba, Conopodium majus, Convallaria majalis, Digitalis purpurea, Diplotaxis muralis, Dryopteris filix-mas, Duchesnea indica, Epilobium hirsutum, E. montanum, Epipactis helleborine, Equisetum arvense, Erigeron acer, Erodium cicutarium, Euphorbia lathyris, E. peplus, Fragaria* x *ananassa, Galega officinalis, Geranium dissectum, G. molle, G. endressii, G. pyrenaicum, G. robertianum, Geum urbanum, Hedera helix, Heracleum sphondylium, Humulus lupulus, Hyacinthoides non-scripta, Hypochaeris radicata, Lactuca virosa, Lobelia erinus, Lonicera periclymenum, Lunaria annua, Lysimachia nemorum, L. punctata, Mahonia aquifolium, Malva neglecta, Mentha* x *villosa, Mercurialis annua, Moehringia trinervia, Myosotis arvensis, M. discolor, M. sylvatica, Onopordum acanthium, Ornithogalum angustifolium, Oxalis corniculata, Pastinaca sativa, Persicaria amplexicaulis, Picris hieracioides, Plantago major, P. lanceolata, Polygonum arenastrum, Primula vulgaris, Ranunculus acris, R. bulbosus, R. ficaria, R. flammula, Reseda lutea, Rumex sanguineus, R. cristatus, Sanguisorba minor, Senecio erucifolius, S. inaequidens, S. viscosus, S. vulgaris, Sinapis arvensis, Tamus communis, Tanacetum parthenium, T. vulgare, Tragopogon pratensis* ssp. *minor, Trifolium arvense, T. campestre, T. micranthum, Verbascum virgatum, V. blattaria, Verbena officinalis, Veronica arvensis, V. chamaedrys, V. hederifolia, Vicia cracca* (o), *Conyza bonariensis, Dipsacus laciniatus, Lotus glaber, Rumex palustris, R. pulcher, R. maritimus, Sisymbrium orientale, Trifolium medium, Verbascum densiflorum, V. lychnitis, V. phlomoides, V. speciosum* (r)

(*d*) dominant, (*a*) abundant, (*f*) frequent, (*o*) occasional, (*r*) rare

Flora

In 1869, Trimen and Dyer reported that Middlesex (essentially the area of London north of the Thames) had 768 native plant species and 91 naturalised alien species, representing an alien contribution to the total flora of about 11%. By 1975, Kent had documented 756 native species (a reduction of 12 species) but 953 alien species in the same area, taking the alien contribution up to 57% of the total flora. This apparent massive increase in aliens is almost entirely due to the adoption of a much broader concept of which alien species were worth recording, along with a substantial increase in recording effort by London's botanists. It is unlikely that the number of alien species growing in the capital had increased substantially between 1869 and 1975, and many of Kent's 953 species had their first records before 1869. The 32 Boroughs of Greater London were created in 1965, and along with surrounding largely suburbanised countryside in Hertfordshire, Essex, Kent, Surrey and Buckinghamshire up to a radius of 32 km from St Paul's Cathedral (3,424 km^2), formed the area covered by Rodney Burton's *Flora of the London Area* published in 1983. This used data collected by the members of the London Natural History Society between 1965 and 1976, and mapped the distribution of 2055 taxa (including hybrids) at a scale of 2×2 km in 856 "tetrads". Subsequently, plant records were collected by members of the Botanical Society of the British Isles as part of the Atlas 2000 project that led to publication of the *The New Atlas of the British and Irish Flora* (Preston et al. 2002). More recently, updates of the distribution maps have been incorporated on the BSBI web site (http://www.bsbi.org.uk), and the current (2008) figure for the flora of the London area is 2,381 species and hybrids, of which 56% are aliens, a figure very close to Kent's estimate of the proportion of aliens from Middlesex in 1975 (the data for 2008 and 2009 can be obtained from the author on request at m.crawley@imperial.ac.uk).

The most conspicuous of London's plants are its street trees. Much the commonest of these is *Platanus* x *hispanica*. Of the parent species, *P. orientalis* was introduced to Britain in 1562, and *P. occidentalis* in 1636, but little grown. The hybrid probably originated in Spain or southern France and was first grown in London at Barnes in 1680; the famous Berkley Square planes were planted in 1789, but the first record of natural regeneration was in 1944, and several distinctive clones are frequently found in London. The abundance of *Aesculus hippocastanum* as a street tree in London only became obvious in recent years, after the trees began to be attacked by the alien leaf miner *Cameraria ohridella*. The insect was first noticed in London at Wimbledon in July 2002, now *Aesculus* canopies all over London are completely brown by the middle of summer, while all the other street trees like *Ailanthus altissima, Catalpa bignonoides, Robinia pseudoacacia, Liquidamber styraciflua, Liriodendron tulipifera,* various *Tilia* spp. and *Taxodium* are still bright green.

Unusually for a crucifer, *Lepidium draba* is a rhizomatous, patch-forming perennial which forms dense monocultures in waste places, railway banks and dry roadsides. It can be a pernicious and persistent weed. According to botanical folklore, its alternative English name, Thanet Cress, derives from the fact that it was introduced to Britain in mattress-stuffing brought back by fever-stricken troops returning from the ill-fated expedition to the island of Walcheren in 1809 (106 men died in action fighting the French, but 4,000 died from disease). The troops landed at Ramsgate in Kent and their mattresses were subsequently sold to a farmer on the Isle of Thanet, who ploughed in their contents for manure. Needless to say, there is no evidence to support this story, and the truth is likely to be more prosaic. The plant appears to have been introduced at several docks in England and Wales in the early years of the nineteenth century. Although the first record from Kent was from 1829, it was recorded from Swansea docks in 1802, some 7 years before the Walcheren invasion.

Maritime plants were introduced to London with seaside turf used on golf courses (*Glaux maritima, Plantago maritima*) as on Wimbledon Common, and others were found occasionally in turfing contractors' yards. *Ammophila arenaria* was planted in bunkers on Northwood golf course in 1913. *Gaultheria shallon* is locally invasive in woodland on acid sands. *Hypericum* x *inodorum* "Elstead" is a garden cultivar of the hybrid *H. androsaemum* x *H. hircinum* which is very frequent and locally abundant on waste ground, and a frequent component of the *Buddleja-Conyza* community (Table 3). *Sedum rupestre* is the commonest stonecrop of walls and roofs, often found growing with *Campanula poscharskyana*. The most common urban brambles on wasteland tend to be *Rubus armeniacus* (with its chalky-white under-leaves and deliciously sweet fruits) and *R. laciniatus,* (with its deeply dissected, dark green leaves), both of which are bird-sown from allotments and gardens. The genus *Oxalis* is conspicuously represented in London by *O. articulata, O. corniculata, O. exilis* and *O. stricta* and less commonly by *O. debilis, O. dillenii, O. incarnata* and *O. latifolia.* The three *Impatiens* species, *I. capensis, I glandulifera* and *I. parviflora,* are found all over the capital, but *I. balfourii* is local and rare. *Valerianella carinata* is now as widespread as *V. locusta,* but *V. dentata* is no more than occasional, and *V. eriocarpa* and *V. rimosa* are both rare.

The increasing alien *Myriophyllum aquaticum* joins native species like *M. alterniflorum* (uncommon) and *M. spicatum* (locally abundant) in still water of flooded gravel pits or slow rivers, but *M. verticillatum* is rare. The invasive alien *Crassula helmsii* occurs on ponds all over London, often covering the entire water surface in a dense blanket of shoots.

Oenothera glazioviana is common throughout the capital and is typically more abundant than either *O. biennis* or *O. cambrica* with which it often grows in swarms of impossibly complicated variants and intermediate forms. It is thought that the species *O. glazioviana* arose by hybridisation of two alien taxa in Europe, either on waste ground or in a garden, before being transported back to the New World, where its parents originated, by the horticulture trade (Carter and Co. were selling the plant in England in 1860). The plant was found by Hugo de Vries in a potato field near Hilversum in Holland in 1886, and he made the study of its various forms his life-long work (the model system for his "mutation theory" of evolution (Dietrich, Wagner and Raven 1997).

Smyrnium perfoliatum is much more common in London than elsewhere in Britain, but there is no clear documentation that it escaped from Kew Gardens, although it is certainly abundant there and often weeded out (it may have been thrown over the wall). It is also a weed in Chelsea Physic Garden, which may be the source of escapes in Battersea. The typically seaside alien *Smyrnium olusatrum* is locally abundant on roadside banks and by canals throughout the capital. London is also a good hunting ground for alien *Solanum* species with *S. chenopodioides, S. diflorum, S. laciniatum, S. nigrum* ssp. *schultesii, S. physalifolium, S. rostratum* and *S. sarachoides* all having been found in recent years. In addition to the ubiquitous *Verbascum thapsus,* there are *V. blattaria, V. densiflorum, V. lychnitis, V. phlomoides, V. speciosum* and *V. virgatum. Leycesteria formosa* is increasingly common as a bird-sown alien on open stony ground, often growing with *Centranthus ruber* (the latter occasionally with white flowers).

The distribution of *Viscum album* within London is extremely curious. It is most common with *Tilia* x *europaea* as a host and ancient parkland as a habitat. But even within these constraints it is extremely local. For example, it is abundant in Bushy Park and Hampton Court, but it is absent from Kew Gardens and Syon Park only a few miles down the river, and is not to be found in most of London's parks despite the abundance of apparently suitable host trees.

Extinctions

Mourning the loss of treasured native plant species has been a recurrent lament in the writings of London's botanists. Here is Salmon, writing in 1931: "*the list of Commons etc. now lost to the botanist – by being either enclosed or built over – is melancholy reading.*" In similar vein, the "*acquisition of Hampstead Heath for public use has now practically extinguished its interest as a botanical area.*" One of London's most celebrated yet controversial extinctions is *Maianthemum bifolium.*

It had gone from Ken Wood on Hampstead Heath by 1924, and despite being reintroduced in 1933 it never re-established. In fact, it may never have been native, since it was not described until 1813, and other rarities like *Polygonatum verticillatum* and *Scilla verna* were known at this date to have been planted at Caen Wood House, nearby. Extinctions of definite native species include *Dianthus armeria* which was locally common in Gerard's time (1597), but has not been seen since 1680. Several plant species were picked to extinction by local people; *Convallaria majalis* was formerly abundant on Hampstead Heath, while *Primula vulgaris* and various ferns were seriously depleted by collectors. Some individuals were infamous for their depredations of rare plants. For instance, Miles of Cowley, a parson near Uxbridge "*is orchis mad, takes them all up, leaves none to seed, so extirpates all wherever he comes, which is cruel, and deserves chastisement*" (Irvine 1838).

Bogs and swamps within London went early. Out of 28 species in the Leg of Mutton Pond on Hampstead Heath only five remained in 1912 (*Hottonia palustris, Carex ovalis, Nardus stricta, Molinia caerulea* ssp. *caerulea* and *Salix repens*), while *Menyanthes trifoliata* declined from common to extinct, and aquatics like *Butomus umbellatus, Valeriana dioica, Sagitaria sagittifolia* and *Schoenoplectus triqueter* were still found on the Surrey shore of the River Thames in the mid-nineteenth century.

A fascinating insight into the flora of London at the beginning of the nineteenth century is afforded by Cooper's *Flora Metropolitana* (1836) which consists of species lists from floral rambles in 130 famous locations like Barnes Common, Battersea Fields, Epping Forest, Erith, Hackney, Hampstead Heath, Highgate, Hyde Park and Wimbledon Common. Forty years later, de Crespigny wrote *A New London Flora* (1877) with species lists from many of the same locations. For instance, Hounslow Heath had been a famous botanising site "*upon which many rare plants grew formerly*" including *Gentiana pneumonanthe, Damasomium alisma, Lythrum hyssopifolium, Pilularia globulifera* and *Lycopodiella inundata.* By 1877 however, it was fenced off and converted into a military parade ground and firing range. During World War I, the site was used as a military airfield, but by the 1930s it was privately owned by Fairey Aviation Company and used for aircraft assembly and testing. Then the nearby hamlet of Heath Row was demolished to make way for the civilian airport that bears its name, which opened on 31 May 1946.

Of course, the extinction of a plant species is always an hypothesis rather than a fact, and it is possible that a species thought to be lost may be rediscovered or reintroduced. The conditional nature of extinctions is illustrated vividly by the flora of London. Of the 58 native species listed as extinct by Trimen and Dyer in 1869, a remarkable 34 were subsequently rediscovered. Of the 129 extinctions listed by Kent (1975), 25 have been rediscovered, but a further 25 species have been lost (Kent 2000). This kind of turnover has probably been a feature of urban floras since the beginning of human settlement, although the rate of change is likely to have varied with fluctuations in trade, sanitation, horticultural taste, and building technology.

Decreasers

Sadly, there were no attempts to make quantitative estimates of plant abundance in London in the nineteenth century, but from Trimen and Dyer's account of the flora of Middlesex in 1869 the sad demise of charismatic native species is evident on every page. In contrast, if we compare their assessment of the abundance of alien species with the assessment one would give for the same alien species in 2008, it is striking how often the plants are *more* abundant now than then (of course, we do not know that "rare" in 1869 meant the same thing as "rare" in 2009).

In 1597, this is what Gerard wrote about *Diplotaxis tenuifolia: "most brick and stone walls about London covered with it,"* and although it is still occasional in east London in chalk pits and on cinder railway sidings, it has almost certainly declined substantially. *Rumex palustris* in marshes and damp waste ground was probably more plentiful by the Thames in London than in any other comparable area in Britain. *Sisymbrium irio* (London rocket) was abundant "*almost everywhere in the suburbs of London*" (Merrett 1666) and "*in 1667 and 1668, after the City was burnt it grew very abundantly on the ruins round St Paul's cathedral*" (Ray 1690), yet no specimens had been seen since 1832, and neither Trimen nor Dyer (1869) had ever met with the plant themselves. It is currently staging a modest comeback.

It is not surprising that species of Chenopodiaceae and Polygonaceae associated with dung heaps and middens are much declined, but the greatly reduced frequency of *Chenopodium bonus-henricus* is probably exacerbated by the fact that few people these days grow the plant as a vegetable crop. The decline of many of London's species no doubt reflects the fact that they are no longer introduced in large numbers, either unintentionally (for example, as a result of better cleaning of seed crops) or intentionally because of changes in horticultural fashion (as with many of the formerly common medicinal plants like *Leonurus cardiaca, Marrubium vulgare, Hyocyamus niger, Atropa belladonna, Artemisia absinthium* and *Inula helenium*).

Increasers

The aliens arrived in London by countless routes, including grain imported for milling or brewing which contained seeds of cornfield weeds from the country of origin (these were sifted out and either thrown onto piles outside the mill, or sold as chicken food), contaminants of agricultural seed, commercial bird seed, as burrs stuck to the sacking of bales used for transporting soft or loose cargos, among packaging materials used for moving fragile goods, in consignments of cotton, wool or hides, in clots of dung attached to fleeces, in plant-pots or root-balls of imported trees and shrubs, in foreign hay containing the seeds of pasture weeds, among sand and gravel of discarded ships' ballast, and so forth. Druce (in Hayward and Druce 1919) waxes lyrical about the "*botanist who goes down to the sea in ships and visits many lands has his imagination stimulated again and again in after years by the sight at home of a lonely alien visitor that speaks of steppe and prairie, of bluff and nullah far across the main.*"

He saw alien plants as "*the epitome of centuries of the world's discovery, invention, exploration and conquest over the realm of nature.*" It is clear that most of the introduced species were casuals: they relied on repeated introduction, and did not form self-replacing populations. Some, however, have become completely naturalised, as did the South American *Coronopus didymus,* a species that was first noted on ballast heaps then spread to cultivated ground (as, for instance, in Chelsea Physic Garden, Hyde Park and the grounds of Buckingham Palace; Kent 1975).

Perhaps the most rapid and effective spread was the invasion during the first half of the twentieth century by *Matricaria discoidea:* first recorded in 1871, within 30 years it had spread to occupy almost all of London, and now it is found on paths, gateways and allotments in every tetrad within 20 miles of St Paul's. The North American willow-herb, *Epilobium ciliatum,* was first found in Britain in 1891 and showed a spectacularly rapid increase in distribution and abundance. For the last 25 years it has been the most frequent member of the genus on waste and cultivated ground. It is often a weed in horticultural nurseries, and transport of rosettes in flowerpots no doubt increased the rate of introduction of the plant into gardens across London. *Senecio squalidus,* which originates from bare volcanic pumice in mediterranean Sicily is pre-adapted to cope with high temperatures and dry conditions of railway ballast (see above).

Another species exhibiting a sudden increase in distribution and abundance is *Chamerion angustifolium.* In Trimen and Dyer (1869) the species was described as "*Gravelly banks and woods, rare*". Known "*from only a few scattered localities (in Middlesex) until the end of the nineteenth century when it began its spectacular spread which accelerated in the first thirty years (of the twentieth century). It is now widespread and very common in all parts of Middlesex*" (Kent 1975). It is still not clear whether alien genotypes imported from America were the cause of this sudden change in ecological behaviour.

Lactuca serriola is ubiquitous today on disturbed waste ground, but was formerly rare. Burton (1983) writes that "*an alien race appeared in London (about 1930), increased rapidly around the developing suburbs, and is now abundant in many parts of our area, especially where motorway works and the like have recently bared the ground.*" *Hirschfeldia incana* was relatively local and scarce in the 1970s but is now widespread and abundant on dry waste ground in full sun, on motorway embankments, railways and canal banks (see Table 2). However, much the most spectacular of the increasing plants in London in recent years have been *Buddleja davidii* and *Conyza sumatrensis*. These have become so common that they now form their own distinctive plant community (Table 3). *Buddleja davidii* is much planted in parks and gardens as a "Butterfly Bush" and spreads freely by abundant, wind-borne seeds. It is native to China, where it is found in thickets on mountain slopes between 800 and 3,000 m a.s.l., from Gansu to Zhejiang. Seedlings, saplings and mature plants are now common all over London, on railways, waste ground, chalk pits, gravel workings and high up on old walls, chimney stacks, etc. It was first recorded as regenerating in London in 1927, but inspection of old railway photographs shows very little evidence of the plant until recently and it appears that the plant has been abundant on railways only since about 1980.

Its economic effects are not yet felt in full, and the damage inflicted by the roots of *Buddleja* growing through brick walls has yet to be widely appreciated. The effects of *Fallopia japonica* in breaking up road surfaces and undermining the foundations of new-build houses are much better known. There are substantial direct costs to property developers working on brownfield sites and former waste ground, because they need to guarantee that the weed has been effectively controlled before they can hand over the new buildings. This involves considerable expense in clearance, herbicide treatment and insurance against future claims for damages (Child et al. 1999). In Britain, we are extremely fortunate that our problem plants are so few, given the almost total disregard for quarantine or containment of imported plants shown by our predecessors.

Conyza sumatrensis has been naturalised in London since 1983 (see *BSBI News* 65: 34–38), and has been locally abundant on urban wasteland in the capital since 1990. These days, it is locally dominant on waste ground, railway ballast, abandoned pavements and other open spaces throughout the capital. The other recent arrival, *C. floribunda* is less frequent but can be locally abundant, while *C. bonariensis* is still highly localised in its distribution within London (B. Wurzell, 2008, personal communication).

Most Recent Increasers

Climate warming, coupled with the effects of the urban heat island, have created conditions in which many new plant species are capable of surviving the winter, growing to flowering size and setting seed. For instance, various *Citrus* spp. and *Persea americana* are now able to ripen their fruits in London. Likewise, there are now many more species of ferns found on walls in London than in former years (Table 1). Many of the most conspicuous newly established species are woody members of the family Rosacea, commonly grown in gardens and escaping from them via fruit-feeding birds. Among the trees, *Prunus serotina* is locally frequent on acid, heathy ground to the south of London (but nowhere near as invasive as it is on the outskirts of Paris), *Amelanchier lamarkii, Crataegus persimilis* "Prunifolia" and *Sorbus intermedia* are commonly self-sown along with numerous shrubby species of *Cotoneaster,* most abundantly *C. simonsii,* but with frequent sightings of *C. dielsianus, C. franchetii, C. horizontalis, C. integrifolius, C. lacteus, C. salicifolius* and *C.* x *watereri* and occasional records of dozens more species from this huge genus.

A native species showing rapid increase in the early years of the twenty-first century is *Iris foetidissima,* once uncommon, but now locally frequent in rough grass in partial shade, and *Duchesnea indica* now widespread and increasingly abundant all over the capital. Of the alien trees, *Ailanthus altissima* seedlings are increasingly abundant, and thickets are formed by suckering roots around existing planted trees. *Robinia pseudoacacia* also forms thickets around the base of planted trees and establishes from seed all over London. Both of these trees are highly invasive in Continental Europe.

The South African *Senecio inaequidens* is a serious invader of pasture land in other parts of the world, (for example, in South America). It is naturalised in London, and spreading rapidly by seed, especially in the East End, but it is not yet invasive of established vegetation. Seedling recruitment is restricted to open sites like gravel or cinder, where competition from perennial plants is minimal. However, the plant can persist for many years once established, even in what has become relatively dense grassland. *Urtica membranacea* is the most recent addition to the flora of London, first found in 2007. The spread of *Ludwigia grandiflora* has only been noted in the last few years, but the plant is currently being actively persecuted by The Department of Environment, Food and Rural Affairs, with teams of workmen spraying glyphosate herbicide on infested waterways.

Closing Comments

The native flora of London has declined inexorably since the Roman occupation, as natural habitats were built-over or dumped upon, and the species-richness of surviving fragments of semi-natural vegetation declined under the interacting effects of drainage, trampling, acidification, eutrophication, dog-fouling, harvesting and botanical collection. However, the decline in native species has been matched by a steady increase in the number of alien species. London's important alien species evolved in far-flung corners of the globe (for example, *Buddleja davidii* in China, *Conyza sumatrensis* in South America, *Epilobium ciliatum* in North America and *Crassula helmsii* in New Zealand), yet they come together to form strange new plant communities (Table 3). The replacement of native by alien flora might be the despair of conservationists, but the dynamics of extinction and invasion are endlessly fascinating to those of us who describe ourselves as urban botanists.

Literature Cited

Anon (1995) Planning Policy Guidance Note 2. HMSO, London

Burton RM (1983) Flora of the London Area. London Natural History Society, London.

Carter AE (1971) Paul Verlaine. Twayne Publishers, New York.

Child, L., M. Wade, and S. Harrison (1999) Strategic invasive plant management, linking policy and practice: A case study of Fallopia japonica in Swansea, South Wales (United Kingdom). Pages 291–301 in G. Brundu, editor. Plant Invasions: Species Ecology and Ecosystem Management. Backhuys, Leijden.

Cooper D (1836) Flora Metropolitana. S. Highley, London.

Crawley MJ (2005) The Flora of Berkshire. Brambleby Books, Harpenden.

Curtis W (1775–1798) Flora Londinensis. London

de Crespigny EC (1877) A New London Flora or Handbook to the Botanical Localities of the Metropolitan Districts. Hardwicke and Bogue, London.

Dietrich W, Wagner WL, Raven PH (1997) Systematics of *Oenothera* Section *Oenothera* Subsection *Oenothera* (Onagraceae). *Systematic Botany Monographs*, 50: 1–234.

Dony JG (1967) Flora of Hertfordshire. Hitchin Urban District Council, Hitchin.

Druce GC (1886) The Flora of Oxfordshire. Parker & Co., Oxford.

Druce GC(1926) The Flora of Buckinghamshire. T. Buncle & Co., Arbroath.

Fitzstephen W (12th Century) *Descriptio Londinae.*

Gerard J (1597) The Herball or Generall Historie of Plantes. Bollifant, London.

Hayward I, Druce GC (1919) *The Adventive flora of Tweedside* T. Buncle & Co., Arbroath.

Irvine A (1838) The London Flora. Smith, Elder and Co., London.

Jermyn ST (1974) Flora of Essex. Essex Naturalist's Trust, Colchester.

Johnson T (1636) Gerard's Herball. Second edition, revised and enlarged. Dover, New York.

Kent DH (1975) The Historical Flora of Middlesex. The Ray Society, London.

Kent DH (2000) Supplement to the The Historical Flora of Middlesex. Ray Society, London.

Larsen RS, Bell JNB, James PW, Chimonides PJ, Rumsey FJ, Tremper A, Purvis OW (2004) Lichen and bryophyte distribution on oak in London in relation to air pollution and bark acidity. In 5th Symposium of the International Association for Lichenology, pp. 332–340, Tartu, ESTONIA.

Laundon JR (1970) London's lichens. London Naturalist 49: 20–69.

Leslie AC (1987) Flora of Surrey: Supplement and Checklist. A.C. & P. Leslie, Guildford.

Loudon JC (1829) Hortus Britannicus. London.

Lousley JE (1976) Flora of Surrey. David & Charles, Newton Abbot.

Maycock R, Woods (2005) A Checklist of the Plants of Buckinghamshire (including Milton Keynes and Slough). Milton Keynes Natural History Society, Milton Keynes.

Melville R, Smith RL (1928) Adventive Flora of the Metropolis. 1. Report of the Botanical Exchange Club of the British Isles 8:444–454.

Merrett C (1666) Pinax Rerum Naturalium Britannicarum, continens Vegetabilia, animalia et Fossilia in hac Insula reperta inchoatus, etc., London.

Philp EG (1982) Atlas of the Kent Flora. The Kent Field Club, West Malling.

Ponte A (1991) Public parks in Great Britain and the United States: from a 'Spirit of the Place' to a 'Spirit of Civilization'. In Mosser M, Teyssot G (eds) The History of Garden Design. Thames and Hudson, London.

Preston CD, Pearman DA, Dines TD (2002) New Atlas of the British and Irish Flora. Oxford University Press, Oxford.

Ray J (1690) Synopsis Methodica Stirpium Brittanicarum, London.

Rodwell J (1991a) British Plant Communities: Volume 1. Woodlands and Scrub. Cambridge University Press, Cambridge.

Rodwell J (1991b) British Plant Communities: Volume 2. Mires and Heaths. Cambridge University Press, Cambridge.

Rodwell J (1992) British Plant Communities: Volume 3. Grasslands and Montane Communities. Cambridge University Press, Cambridge.

Rodwell J (2000) British Plant Communities: Volume 5. Maritime and Weed Communities. Cambridge University Press, Cambridge.

Rose CI, Hawksworth DL (1981) Lichen recolonization of London's cleaner air. Nature 289:289–292.

Salmon CES (1931) Flora of Surrey. G. Bell & Sons, Ltd., London.

Salisbury EJ (1943) The flora of bombed areas. Nature 151:462–466.

Stringer C (2006) Homo Britannicus. Allen Lane, London.

Trimen H, Dyer WTT (1869) Flora of Middlesex. Robert Hardwicke, London.

Veen M van der, Livarda A, Hill A (2008) New plant foods in Roman Britain – dispersal and social access. Environmental Archaeology 13: 11–36.

Annex 50 most frequent plant species in the urbanised area of London

Acer pseudoplatanus
Aegopodium podagraria
Artemisia vulgaris
Atriplex prostrata
Ballota nigra
Buddleja davidii
Calystegia silvatica
Cardamine hirsuta
Cirsium arvense
Cirsium vulgare
Cochlearia danica
Conium maculatum
Conyza floribunda
Conyza sumatrensis
Dactylis glomerata
Diplotaxis tenuifolia
Dipsacus fullonum
Epilobium ciliatum
Fallopia baldschuanica
Fallopia japonica
Fraxinus excelsior
Galega officinalis
Galium aparine
Geranium robertianum
Geum urbanum
Glechoma hederacea
Hedera helix
Hirschfeldia incana
Hordeum murinum
Lactuca serriola
Lamium album
Malva sylvestris
Medicago lupulina
Mercurialis annua
Pentaglottis sempervirens
Picris echioides
Plantago coronopus
Plantago lanceolata
Poa annua
Polygonum aviculare agg.
Potentilla reptans
Pseudofumaria lutea
Quercus robur
Sagina procumbens
Sambucus nigra
Senecio squalidus
Senecio vulgaris
Solanum dulcamara
Sonchus oleraceus
Stellaria media agg.

Bold = neophytes

Maastricht

Eduard J. Weeda

Fig. 1 Maastricht has a number of ecological parks that are periodically grazed by a herd of sheep, which now and again migrate through the city from one park to another (Photo C. Frissen-Moors)

Abstract Maastricht, the birthplace of the Euro, is the southernmost city in The Netherlands. Thanks to its favourable position in the transition from a hilly loess and limestone area to a river valley, the city supports a rich flora and fauna. The earliest botanical records are from excavations of Neolithic and Iron Age sites in which the remains of cereals and weed species were found. A greater variety of

Eduard J. Weeda (✉)
Alterra Wageningen UR, P.O. Box 47, 6700AA Wageningen, The Netherlands
e-mail: ejweeda@hotmail.com

J.G. Kelcey and N. Müller (eds.), *Plants and Habitats of European Cities*,
DOI 10.1007/978-0-387-89684-7_8, © Springer Science+Business Media, LLC 2011

food plants were found in the excavations of Roman sites. In recent times, 721 vascular plants (17% are neophytes) have been recorded in the city, of which 461 were recorded in the central area and 260 only in the outskirts. The St. Pietersberg (a limestone hill) and three man-made habitats (the Medieval walls, the fortifications and the limestone revetments of the river Jeker) are of special botanical interest. The old walls are the most botanically valuable urban habitat in the city, especially for vascular species and lichens, whilst the fortifications are a remarkable element in terms of bryophytes and grassland species. A substantial amount of effort is taken to preserve the ecological value of the old walls and fortifications and to enlarge the area of the flowering plants, which also encourage bees and butterflies.

Natural Environment

Maastricht (50°51′ North, 5°40′ East) is the southern-most town of The Netherlands and the capital of the Province of Limburg (Ramakers 2005). The western fringe of the built-up area partially coincides with the Belgian-Dutch frontier (Fig. 2). The southern part of Limburg extends into the hilly parts of central Europe, thus making an exception to the general lowland character of The Netherlands.

The city is situated on both sides of the River Meuse, which flows into The Netherlands from Belgium a few kilometers to the south. The river valley, which is only 3 km wide and deeply incised south of Maastricht, widens into a "bowl-shaped" area that is twice as wide and in which most of Maastricht has been built. The bottom of the valley lies about 45 m a.s.l., while the built-up area generally extends to 60 m a.s.l. The city is surrounded by arable land and orchards with wooded slopes within 1–2 km of the town's eastern and southern edges.

The south-side of the town borders the St. Pietersberg, a hill originally 117 m high; large parts of it have been excavated for limestone for the manufacture of cement (Van Schaïk 1983). The hill is the type locality of the Maastrichtien, a limestone division of the Upper Cretaceous. The small river Jeker (a tributary of the Meuse) lies to the west of the St. Pietersberg. On reaching the historical centre of Maastricht, the Jeker splits into two branches, one flowing adjacent to the fourteenth century town-wall, the other crossing the town between the thirteenth and fourteenth century walls (Fig. 2).

As a rain-fed river, the Meuse is subject to large changes in velocity and flow, which in some years result in substantial rises in water level, especially in autumn and early spring. The northern part of the municipality contains two small old villages named Borgharen and Itteren, which have been built on slightly elevated sites within a low-lying area between the Meuse and a parallel shipping canal – the Julianakanaal. During periods of extreme flows in 1993 and 1995 most of this land was flooded twice in the winter, which underlines its unsuitability for urban development.

Most of the urban area as well as the surrounding agricultural belt are situated on colluvial soils with a large amount of loess. Nevertheless, limestone plays an important part in the urban ecology of Maastricht, since many organisms are dependent on various kinds of historical buildings that are constructed from it.

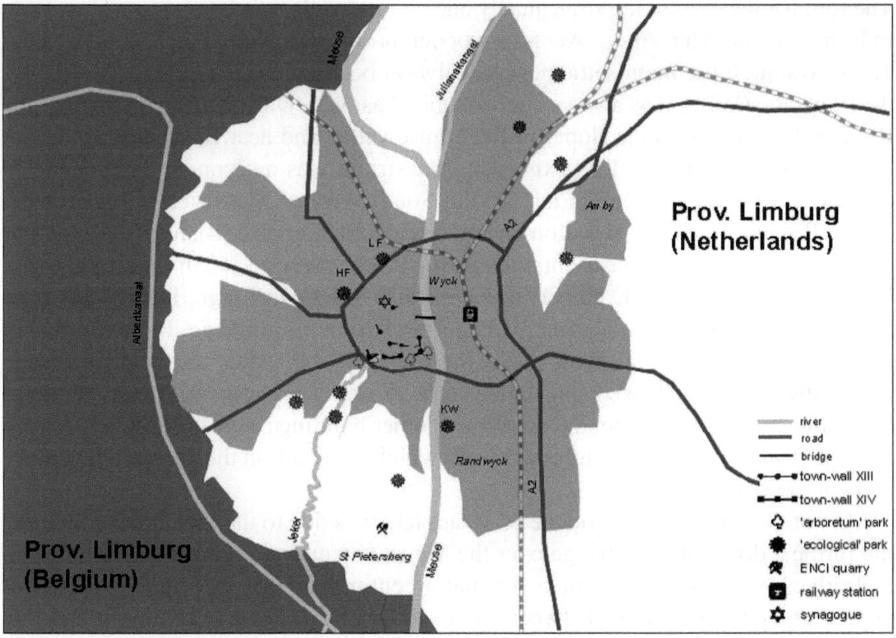

Fig. 2 Map of Maastricht. *HF* Hoge Fronten; *KW* Kleine Weerd; *LF* Lage Fronten. Town-wall XIII/XIV = thirteenth/fourteenth century town-wall (Map composed by H.P.J. Huiskes)

The Maastricht region belongs to the maritime part of the temperate zone of Europe. Within The Netherlands it has relatively warm summers, as might be expected from its southern position. It also used to be one of the driest parts of the country (annual rainfall less than 700 mm/year) but in the last few decades the amount of precipitation has been increasing. Over the period 1971 to 2000 the mean temperature in Maastricht was 9.8°C, with a mean minimum of 6.0°C and a mean maximum of 13.7°C. The warmest months are July and August with 17.7 (min. 13.0, max. 22.5)°C and 17.6 (min. 12.8, max. 22.7)°C, respectively. The coldest month is January with 2.6 (min. – 0.1, max. 5.0)°C. The annual rainfall amounts to 740 mm, the annual sunshine to 1,480 hours (data from Koninklijk Nederlands Meteorologisch Instituut, De Bilt).

Historical Development of the Town

Maastricht is one of the oldest towns in The Netherlands, dating back to the Roman era. There is ample archaeological evidence to show that the city has been continually inhabited from the first century AD onwards, which is exceptional in The Netherlands. Like many old towns, it originated as a settlement relating to a ford across the river.

The ford formed part of the strategically and commercially important route from Köln to Boulogne-sur-Mer. About AD 50 a wooden bridge was built, which survived until 1275. Although the main settlement has always been on the left side of the Meuse, the right side (Wyck) has always been inhabited as well. During their occupation, the Romans built villas on the slopes of the Meuse valley and near Borgharen.

In the first centuries of its existence, Maastricht was a secondary settlement to Tongeren (in present Belgium and 15 km South West of Maastricht). Towards the end of Roman times, the situation changed when St. Gervase (Bishop of Tongeren) established his seat at Maastricht, thus giving it the "prestige" of an Episcopal town. In the eighth century, St. Hubert transferred the chair to Liege, but by this time Maastricht had become a commercial centre of some importance. In the thirteenth century, a town-wall was built enclosing about 36 ha. After the wooden bridge across the Meuse collapsed (presumed to be during a procession), it was replaced by a stone bridge. Subsequently, houses and other buildings were constructed outside the wall necessitating a second town-wall, which was built in the fourteenth century and enclosed 114 ha.

Part of the strategic importance of Maastricht was due to the fact that it contained the furthest down-stream bridge over the Meuse; meandering impeded the building of bridges over its lower course for many centuries. Time and again Maastricht proved an apple of discord between rulers and nations. For a long time it had two masters, the Bishop of Liege and the Duke of Brabant. At the beginning of the Dutch uprising against Spain, it became a garrison-town. When the soldiers joined the revolt in 1579, it was besieged by Spanish soldiers. In 1632, Maastricht was captured by the Dutch army and became the furthest central European outpost of the Dutch Republic. Its strategic position required the construction of vast fortifications (Bonnemayer 1986). The town was temporarily captured by the French on three occasions; the last occupation was from 1795 to 1814. Their best-known conquest is a fossil, the skull of the marine reptile *Mosasaurus giganteus,* found in 1770 in limestone at St. Pietersberg (Bardet and Jagt 1996; see also Dortangs et al. 2002).

After the Viennese Congress in 1814 when The Netherlands and Belgium were united into one kingdom, Maastricht became the capital of the Province that was given the name Limburg. When the Belgians revolted against the unification in 1830, the presence of garrison enabled Maastricht to be retained by the Dutch side. As a result, Limburg was divided into a western (Belgian) and an eastern (Dutch) part; the latter retained Maastricht as its (very eccentric) capital. At first, its isolated position proved disadvantageous, because it seemed to nullify the efforts of the Dutch–Belgian king to make Maastricht a vital link between the industry of Liege and the Dutch harbours. During the nineteenth century, a great variety of factories were founded, ceramic and glass works became famous. Railways offered a new connection with the rest of The Netherlands and with the larger towns in the neighbouring parts of Belgium and Germany, like Liege and Aachen. The fortifications lost their function and were largely demolished to create boulevards, new housing and industry. Nevertheless, the number of inhabitants only grew slowly. Whereas Maastricht had about 20,000 inhabitants by the mid-seventeenth century, this figure increased to about 30,000 by the end of the nineteenth century. Many of them were labourers living in small and unhealthy lodgings.

In 1920, the area of the municipality was enlarged from 4.5 to 33 km² by the annexation of several surrounding villages. The eastern side of this area contained low-lying land that functioned as "flood storage" for the Meuse in periods of high flows; in 1935 this function was taken over by the Julianakanaal, a shipping canal that was built in the Meuse valley. It served to replace the meandering Meuse north of Borgharen, which could not be canalised because it formed the Belgian–Dutch border. This released the flood storage area for house building. To meet the increasing amount of traffic, a new bridge over the Meuse was built 250 m north of the old bridge (which was renovated) but both bridges were destroyed for strategic reasons in 1940. The older bridge was rebuilt in 1947; the younger not until 1960.

In 1926, the ENCI (Eerste Nederlandsche Cement Industrie) was founded for large-scale limestone extraction from the St. Pietersberg for cement production. This resulted in a significant increase in employment in Maastricht; consequently the most special piece of nature near the town became a large quarry surrounded by steeply sloping slides. Shortly after 1945, there was a 33% shortage of houses, the highest figure of any Dutch town at that time. From 1954 onwards, an area of new, partially self-supporting quarters ("parishes") was built. In the inner town, many of the old houses were demolished and replaced by new ones whereas the large monumental buildings were restored, where it was possible to do so. The area on the northern side of the town became an industrial district and a harbour. As a part of a ring-road around the old town centre, two new bridges were constructed. In 1970, the municipality was again enlarged by the annexation of neighbouring villages, resulting in 70% increase in its area.

At present, the municipality of Maastricht covers 60 km², about half of which forms a coherent urban area. In 2006, the city had its maximum number of inhabitants (120,000); since that year a slight decline has been observed. Apart from the low-lying Borgharen and Itteren, almost all of the former villages within the municipality have been incorporated into the town area; only Amby on the north-east side remains somewhat separate from the rest of the town. An extension of the built-up area to the north-west has been planned (Belvédère). The town has attracted almost all kinds of regional functions including governmental, educational, cultural, medical, industrial and commercial. It is the major shopping centre for southern Limburg and adjacent areas of Belgium. The problems associated with the highway from Amsterdam to the Mediterranean region, which crosses the eastern half of the built-up area, require urgent solutions, including the construction of a tunnel to ameliorate the negative effects on the air quality and the local traffic (Klasberg et al. 2009).

A new function for Maastricht was created in 1975 by the foundation of a university. The medical faculty is located in a new quarter on the south-side of the town (Randwyck), while other faculties occupy historical buildings in the inner town. Another addition to the town's profile is inspired by the European Union: Maastricht is establishing itself as a centre of European co-operation. The Maastricht-Aachen airport has been established near Beek, north-east of the city. It is possible to say that the euro was born in Maastricht in 1992, when the Treaty of Maastricht included the decision to introduce a common European currency.

Changes of the Environment Due to Town Growth

The formation of urban soils started in Roman times, when the Romans constructed transport routes using pebbles, thus creating "high ways" (the literal sense of being constructed at a higher level). A striking feature in the trace of the Roman way across the Meuse is the zigzag bend of the way on the western bank, suggesting that a marshy piece of land had to be avoided. This is confirmed by archaeological records of a number of marsh plants on the fringe of the Roman settlement.

Because of its low elevation in the Meuse Valley, ever since its foundation Maastricht has been threatened by flooding during the winter months. For a long time, building of new town quarters had to be preceded by raising the ground level by importing material from the surrounding area. In the south-eastern parts, such a land-fill could be dispensed with, because the basic parts of the former fortifications already offered a sufficiently high-lying surface for house building.

Apart from being an important industrial town itself, Maastricht is flanked by larger industrial areas at 10–20 km on its northern, eastern and southern sides. To the north and east, a former coal-mining belt extends from Sittard over Heerlen to Kerkrade, where it links up with the Aachen agglomeration. To the south, the industry around Liege occupies a stretch of several tens of kilometers of the Meuse valley. In addition to the regional air pollution caused by industry and agriculture, there is also considerable traffic-generated pollution, notably along the motorway crossing the eastern half of Maastricht. This poor air quality is mirrored by the scanty epiphytic lichen and moss flora in the town.

Besides the Medieval town walls, two man-made habitats have a special interest for plants. One is found on the ramparts of the fortifications, notably retaining walls covered with small amounts of earth. Here, several rare moss species have been observed. The other is formed by waterside revetments constructed from limestone blocks.

Flora

Archaeological Aspects

There is a lot of archaeological information about the present urban area of Maastricht, see Table 1 (data from the archaeobotanical database RADAR, kindly provided by O. Brinkkemper).

As the result of a large number of excavations, the Roman period is especially well documented (Bakels and Dijkman 2000). Many of the finds are charred fruits, which allow plants to be identified to different taxonomic levels (family, genus, species or even cultivar). Only those taxa identified to species level are given in Table 1; they are listed according to the habitat from which the species most likely originated. However, some species that are classified as arable weeds may also occur in grassland, for example, *Odontites vernus, Ranunculus sardous* and *Rumex acetosella* but in arable land their fruits have a better chance of being transported to human settlements and conserved there.

The first vascular plants documented from the Maastricht area are cereals and arable weeds, whose records date back to the Neolithic (fifth millennium BC), when the central European loess belt was taken into cultivation. They relate to an underground cereal silo within the present quarter of Randwyck. Besides *Hordeum vulgare,* some kinds of *Triticum* spp. and the "semi-cereal" *Bromus secalinus,* a number of weeds were present, including *Chenopodium album, Fallopia convolvulus, Lapsana communis, Rumex acetosella, Persicaria lapathifolia* and *Vicia hirsuta.* Presumably the fruits of several of these plants were eaten by people, as was the case with *Corylus avellana,* which was also found in the silo. An Iron Age find in ca. 500 BC, also in Randwyck, identified *Panicum miliaceum* as a new cereal and *Echinochloa crus-galli* and *Galium aparine* as additional weed species. Several other useful plants, for example, *Hyoscyamus niger* and *Malva sylvestris,* are known (from other excavations) to have arrived in the Low Countries well before the Romans.

Excavations in the present town centre have resulted in the finding of a much larger spectrum of cultivated plants and weeds from the Roman period. In addition to cereals an interesting variety of vegetables and condiments were cultivated, while berries and nuts of various shrubs and trees were eaten. Several of these fruits will have been gathered in the wild, like those of *Sambucus nigra, Rubus fruticosus* agg. and *Crataegus* spp. However, the Romans also cultivated a variety of fruit-bearing trees and shrubs, for example, *Prunus avium, Pyrus communis* and *Malus domestica* (Bakels and Jacomet 2003). *Prunus avium* is now considered to be a member of the wild woodland flora of the region, but presumably it was introduced by the Romans since there are no Dutch archaeobotanical records of it before the Roman times. The St. Pietersberg was among the first sites in The Netherlands on which vineyards were established while fruit production still continues on the outskirts of Maastricht.

Some 50 species of arable weeds have been documented from Roman times, most of them being obligatory or facultative cornfield weeds, at least in central and western Europe. Only a few of them may have originated from the wild vegetation, for example, *Cirsium arvense* from river banks. Most cornfield weeds are considered to be archaeophytes. In the few cases that the same species occurs both in arable fields and on calcareous rocks, like *Arenaria serpyllifolia, Medicago lupulina* and *Silene vulgaris,* it is not clear which kind of habitat is likely to be "original" or oldest in this region. Examples of ruderal plants that were most likely deliberately introduced by the Romans include *Reseda luteola, Nepeta cataria* and *Sambucus ebulus:* the former is a dye plant; the other two are medicinal herbs.

Although it is easy to classify a considerable number of plants as archaeophytes, for many other species it is difficult to decide whether they are old human introductions or belong to the spontaneous flora. A number of ruderal weeds,

Table 1 Archaeological records of crops and other plant species from the present urban area of Maastricht

	Neolithic	Iron Age	Roman Era	Early Middle Age	High and Late Middle Age	Modern History	Recent (≥1,800)	Recent (≥1,800)
	5000–4000 BC	ca. 500 BC	AD 0–500	AD 500–1000	AD 1000–1500	1500 –1950	Maastricht	S. Limburg
1. Cereals								
Triticum dicoccon	x	x	x				-	-
Triticum monococcum	x		x				-	-
Hordeum vulgare	x	x	x		x	x	-	!
Panicum miliaceum		x	x	x		x	!	!
Triticum aestivum	cf		x	x	x	x	-	!
Triticum spelta			x	x	x	x	-	-
Secale cereale			x	x	x	x	!	!
Avena sativa					x		!	!
Fagopyrum esculentum					x	x	!	!
Oryza sativa[a]					x	x	-	-
2. Vegetables & condiments								
Anethum graveolens			x		x	x	!	!
Apium graveolens			x				-	p
Beta vulgaris			x				!	!
Brassica rapa			x		x		!	!
Coriandrum sativum			x	x	x	x	-	p
Lens culinaris			x	x			-	-
Linum usitatissimum			x	x	x	x	!	!
Papaver somniferum			x		x	x	!	!
Pisum sativum			x	x			-	p

Satureja hortensis		x			x	-	p
Vicia faba		x			x	-	p
Brassica nigra				x	x	!	!
Carum carvi				x	x	!	!
Portulaca oleracea (*sensu lato*)				x	x	!	!
3. Fruit plants (wild or cultivated)							
Corylus avellana	x	x		x	x	!	!
Prunus spinosa	x	x	x			!	!
Cornus mas		x			x	!	!
Crataegus laevigata		x				!	!
Crataegus monogyna		x				!	!
Ficus carica[a]		x		x	x	!	!
Fragaria vesca		x				!	!
Juglans regia		x	x	x	x	!	!
Prunus avium		x	x	x	x	!	!
Prunus domestica		x	x	x	x	!	!
Pyrus communis		x		x	x	(!)	!
Rubus fruticosus agg.		x	x	x	x	!	!
Rubus idaeus		x	x	x	x	!	!
Sambucus nigra		x	x	x	x	!	!
Vitis vinifera		x	x	x	x	!	!
Malus domestica			x	x	x	!	!
Mespilus germanica				x	x	(!)	!
Morus nigra				x	x	-	-
Ribes rubrum				x	x	!	!
Rubus caesius				x	x	!	!
Vaccinium myrtillus				x	x	-	!

(continued)

Table 1 (continued)

	Neolithic	Iron Age	Roman Era	Early Middle Age	High and Late Middle Age	Modern History	Recent (≥1,800)	Recent (≥1,800)
	5000–4000 BC	ca. 500 BC	AD 0–500	AD 500–1000	AD 1000–1500	1500 –1950	Maastricht	S. Limburg
4. Medicinal herbs, dyes & other useful plants								
Daucus carota			x	x		x	!	!
Hyoscyamus niger			x	x		x	!	!
Hypericum perforatum			x				!	!
Malva sylvestris			x	x			!	!
Nepeta cataria			x				†	!
Reseda luteola			x	x	x	x	!	!
Rhamnus cathartica			x				-	!
Sambucus ebulus			x	x	x	x	!	!
Verbena officinalis			x		x	x	!	!
Cannabis sativa					x	x	!	!
Humulus lupulus					x	x	!	!
5. Arable weeds								
Chenopodium album	x	x	x		x	x	!	!
Fallopia convolvulus	x	x	x		x	x	!	!
Lapsana communis	x		x		x	x	!	!
Persicaria lapathifolia	x		x		x	x	!	!
Rumex acetosella	x		x	x	x	x	!	!
Vicia hirsuta	x		x	x	x	x	!	!
Bromus secalinus	cf		cf	cf		cf	-	†
Echinochloa crus-galli		x	x	x		x	!	!
Galium aparine		x	x	x	x		!	!
Aethusa cynapium			x		x		!	!
Agrostemma githago			x	x	x	x	†	!
Amaranthus blitum			x				!	!

Anagallis arvensis	x			x	!	!
Anthemis arvensis	x	x		x	†	!
Anthemis cotula	x	x	x	x	†	!
Aphanes arvensis	x				!	!
Arenaria serpyllifolia	x			x	!	!
Capsella bursa-pastoris	x			x	!	!
Centaurea cyanus	x		x	x	!	!
Chenopodium ficifolium	x			x	!	!
Chenopodium polyspermum	x		x	x	!	!
Cirsium arvense	x				!	!
Cuscuta epilinum	x				†	†
Euphorbia helioscopia	x	x			!	!
Fumaria officinalis	x		x	x	!	!
Galium spurium	x				†	†
Legousia speculum-veneris	x		x	x	!	!
Lithospermum arvense	x				†	!
Medicago lupulina	x	x		x	!	!
Nigella arvensis	x				-	(†)
Odontites vernus (sensu lato)	x				!	!
Orlaya grandiflora	x	x	x		(?)	†
Papaver argemone	x	x	x	x	!	!
Persicaria maculosa	x	x	x	x	!	!
Polygonum aviculare agg.	x	x	x	x	!	!
Ranunculus sardous	x	x		x	!	!
Raphanus raphanistrum	x		x	x	!	!
Silene vulgaris	x			x	!	!

(continued)

Table 1 (continued)

	Neolithic	Iron Age	Roman Era	Early Middle Age	High and Late Middle Age	Modern History	Recent (≥1,800)	Recent (≥1,800)
	5000–4000 BC	ca. 500 BC	AD 0–500	AD 500–1000	AD 1000–1500	1500 –1950	Maastricht	S. Limburg
Solanum nigrum			x	x	x	x	!	!
Sonchus asper			x		x	x	!	!
Spergula arvensis			x	x		x	!	!
Stachys annua			x				(?)	(†)
Stellaria media			x	x	x	x	!	!
Thlaspi arvense			x	x	x	x	!	!
Urtica urens			x	x	x	x	!	!
Valerianella dentata			x		x	x	(!)	!
Valerianella rimosa			x		x		-	†
Vicia sativa (sensu lato)			x	x	x		!	!
Vicia tetrasperma			x	x			!	!
Apera spica-venti				x			!	!
Chenopodium hybridum				x	x		!	!
Sinapis arvensis				x	x	x	!	!
Arnoseris minima					x	x	-	!
Scleranthus annuus					x	x	!	!
Setaria verticillata					x		!	!
Sonchus arvensis					x		!	!
6. Other ruderal plants								
Carduus crispus			x				!	!
Conium maculatum			x			x	!	!
Plantago major			x		x	x	!	!
Reseda lutea			x				!	!
Rumex obtusifolius			x	x	x		!	!
Urtica dioica			x			x	!	!
Alliaria petiolata				x			!	!

Artemisia vulgaris		x			!	!
Poa annua		x	x		!	!
Torilis japonica		x			!	!
7. Grassland plants						
Carex ovalis	x				!	!
Glechoma hederacea	x				!	!
Leucanthemum vulgare	x			x	!	!
Linum catharticum	x				!	!
Picris hieracioides	x				!	!
Plantago lanceolata	x	x	x	x	!	!
Potentilla anserina	x				!	!
Prunella vulgaris	x	x	x	x	!	!
Ranunculus repens	x			x	!	!
Stellaria graminea	x				!	!
Taraxacum officinale agg.	x			x	!	!
Potentilla erecta	cf	x			-	!
Achillea millefolium		x			!	!
Knautia arvensis		x	x	x	!	!
Phleum pratense		x			!	!
Centaurea jacea (sensu lato)		cf	x		!	!
Calluna vulgaris			x	x	!	!
Carex hirta			x		!	!
Leontodon autumnalis			x		!	!
8. Pioneers of wet soil						
Cyperus fuscus	x				-	!
Juncus bufonius	x				!	!
Persicaria hydropiper	x	x	x	x	!	!

(continued)

Table 1 (continued)

	Neolithic	Iron Age	Roman Era	Early Middle Age	High and Late Middle Age	Modern History	Recent (≥1,800)	Recent (≥1,800)
	5000–4000 BC	ca. 500 BC	AD 0–500	AD 500–1000	AD 1000–1500	1500 –1950	Maastricht	S. Limburg
Ranunculus sceleratus			x				!	!
Tripleurospermum maritimum				x		x	!	!
Bidens tripartita					x		!	!
9. Marsh and waterside plants								
Alisma plantago-aquatica			x		x		!	!
Carex disticha			x				!	!
Carex riparia			x				!	!
Cirsium palustre			x				!	!
Eleocharis palustris			x	x	x		!	!
Filipendula ulmaria			x				!	!
Galium palustre			x		x		!	!
Glyceria fluitans			x				!	!
Glyceria maxima			x				!	!
Juncus articulatus			x				!	!
Juncus subnodulosus[b]			x				-	-
Lychnis flos-cuculi			x		x		!	!
Lycopus europaeus			x		x		!	!
Oenanthe aquatica			x				!	!
Oenanthe fistulosa			x				!	!
Ranunculus flammula			x	x	x		!	!
Schoenoplectus lacustris (sensu lato)			x				!	!
Stachys palustris			x				!	!
Stellaria palustris			x			x	†	!

Berula erecta			x		(!)	!
Lythrum salicaria			x		!	!
Solanum dulcamara			x		!	!
Caltha palustris				x	!	!
Carex viridula (sensu lato)				x	-	!
Carex otrubae				x	!	!
Carex panicea				x	-	!
Phragmites australis				x	!	!
10. Miscellaneous						
Viscum album	x				!	!
Carex muricata (sensu lato)		x			!	!
Carex remota				x	!	!

x = identified to species level; cf = identified to species-group level, probably concerning the indicated species. The last two columns indicate the presence of species in the last two centuries in the Maastricht town area and in the whole region of southern Limburg; ! = present 1990–2006; (!) = still present in 1980–1989; † = (locally) extinct; - = no records from 1800 onwards; (?) = old record(s) from Maastricht without further specification; (†) = old record(s) from border regions of southern Limburg with Belgium and/or Germany; p (crops) = presumably still cultivated in the region, but no data available

[a]It is very unlikely that *Oryza sativa* and *Ficus carica* would have occurred as complete fruiting plants; most probably their fruits were brought from the Mediterranean region

[b]Twentieth century records of *Juncus subnodulosus* in southern Limburg are spurious

like *Carduus crispus* and *Rumex obtusifolius,* may also occur in natural habitats like riverbanks and riverine woods. It is still more difficult to determine which taxa in the grassland flora are archaeophytes and which are native species. It is necessary to bear in mind that archaeological finds always relate to human occupation and activities. Moreover, the plant finds are interpreted according to our image of the habitats of the species, which is derived from the present situation. Therefore, our designation of archaeophytes is in danger of circular arguments. For this reason no attempt has been made to estimate the number of archaeophytes in the Maastricht flora.

Interestingly enough, some samples of charred (carbonised) seeds from the Roman period also contain a number of marsh and wet grassland plants, which are not expected to occur in an urban or arable context. Such mixtures have been found in several excavation sites in or very close to the historical centre of Maastricht. There are two plausible interpretations – first, refuse has been deposited on marshy land; second, in addition to various food plants for human consumption, hay from the surrounding area has been stored in the settlement. As is shown by Table 1, the finding of remains of marsh plants in excavations indicate that they were present well into the Middle Age. Because the main interest of these finds is historical ecology rather than the history of the flora, their interpretation will be considered in more detail later.

Botanical History

There has been considerable interest in the flora of Maastricht and its surroundings since the beginning of the nineteenth century. About eighty grassland and woodland plants and arable weeds have vanished since that time, mainly as the result of changes in land use and overall habitat destruction. In the case of some orchids, collecting has contributed to the loss. On the other hand, many newcomers have been "imported" via the Meuse, railways, roads and industrial activities.

Seven hundred and twenty-one native or established vascular species (about half the Dutch vascular flora) have been observed within the urban area of Maastricht between 1990 and 2006. This number has been calculated on the basis of species lists of 36 grid-cells of 1 km^2 that were surveyed in that period (data kindly provided by the Natuurhistorisch Genootschap in Limburg, Roermond). These cells cover almost all of the built-up area of the city. Of course such a figure is only an approximation, since the progress of establishment of new species is in full swing and their "membership" of the flora can only be judged after a number of years. There are three grid-cells in which more than 400 species were recorded; one of these grid-cells includes an outer-meadow (Kleine Weerd), the second one contains a large portion of the fortifications (Hoge Fronten) and in the third one a large quarry and remnants of the hill (St. Pietersberg) are situated.

The 50 species found most frequently in Maastricht are listed in Table 2. They occur in at least 32 of the 36 km^2 grid-cells. A largely similar list could have been composed in any other Dutch town of some importance. Only two or three of the

Table 2 The 50 most frequent herbaceous vascular plants in Maastricht, arranged according to their main habitats

1. Plants of trampled areas	4. Grassland plants
Lolium perenne	*Achillea millefolium*
Plantago major ssp. *major*	*Arrhenatherum elatius*
Poa annua	*Bellis perennis*
Polygonum aviculare agg.	*Dactylis glomerata*
	Daucus carota
2. Ruderal plants	*Heracleum sphondylium*
Anisantha sterilis	*Persicaria amphibia*
Artemisia vulgaris	*Plantago lanceolata*
Carduus crispus	*Potentilla anserine*
Cirsium arvense	*Ranunculus repens*
Cirsium vulgare	*Taraxacum officinale* agg.
Convolvulus arvensis	*Trifolium pratense*
Conyza canadensis	*Trifolium repens*
Crepis capillaris	
Elytrigia repens	5. Wood-fringe plants
Lactuca serriola	*Alliaria petiolata*
Rumex obtusifolius	*Bryonia dioica*
Senecio inaequidens[a]	*Calystegia sepium*
Tanacetum vulgare	*Galium aparine*
Tussilago farfara	*Glechoma hederacea*
	Lapsana communis
3. Arable weeds	*Sambucus nigra*
Capsella bursa-pastoris	*Urtica dioica*
Cardamine hirsuta	
Chenopodium album	
Lamium purpureum	
Matricaria recutita	
Mercurialis annua	
Papaver rhoeas	
Persicaria maculosa	
Senecio vulgaris	
Sonchus asper	
Sonchus oleraceus	
Stellaria media	

[a]The listed species occur in at least 32 of 36 grid-cells of 1 km^{-2} covering the urban area
Bold = neophytes

species listed are not so common on a country-wide scale – *Bryonia dioica, Mercurialis annua* and perhaps *Papaver rhoeas*. In The Netherlands (where non-calcareous soils are predominant), the three species occur mainly in the relatively lime-rich areas, while *Mercurialis* also reaches its northern limit in this country.

Of the 721 vascular plant species, 461 (64%) occur in the town centre of Maastricht. This figure is based on inventories of five grid-cells covering the core of the urban area on both sides of the Meuse and including the medieval town walls, most of the railway land, 1 km of the Jeker and 2 km of the Meuse but excluding most of the seventeenth to nineteenth century fortifications, which contain many elements of the "rural" flora. The most significant part of the town centre flora is the flora of the walls. Within the city *Cystopteris fragilis, Erysimum cheiri, Arabis sagittata* and *Pseudofumaria alba* are confined to the town centre, while *Hieracium speluncarum* (a hawkweed species from the *H. sect. Amplexicaulia*) has its main occurrence there and is among the most striking plants when flowering.

Of the species recorded between 1990 and 2006, 260 species were only recorded in the outer parts of Maastricht; many of them are roadside plants. In some cases their occurrence might be due to planting or sowing, for example, *Prunus spinosa, Lolium multiflorum* and *Vicia sativa* ssp. *sativa*. Other frequent plants in grassy habitats in the outskirts of the city are *Alopercurus pratensis, Cichorium intybus* and the neophytes *Trifolium hybridum* and *Crepis vesicaria* ssp. *taraxacifolia*. Remarkably enough, some archaeophytes are also characteristic of roadsides in the outer parts of Maastricht, while being absent from the town centre, for example, *Conium maculatum* and *Armoracia rusticana*. The same holds true for some annual grasses, like *Apera spica-venti* and *Avena fatua,* which are better known as cornfield weeds. Besides these roadside plants, there are a number of marsh and wet grassland plants, which characterise the outskirts of Maastricht with regard to the more central parts, for example, *Rumex conglomeratus, Eleocharis palustris* and *Myosotis scorpioides*. The same applies to woodland plants like *Ribes rubrum, Ajuga reptans* and *Adoxa moschatellina*.

In the "rural" parts of the municipality, the St. Pietersberg remains a site of interest in spite of the loss of many species such as *Spiranthes spiralis, Gentianella campestris* and *Vincetoxicum hirundinaria*. Examples of plants still having their main or only recent Dutch occurrence on this hill are *Helianthemum nummularium,* which is found in remnant limestone grassland, *Orobanche hederae* in woodlands on the eastern slope and *Crepis foetida* and *Centaurea calcitrapa* in ruderal habitats.

Both the urban area of Maastricht and the St. Pietersberg are rich in apomictic "microspecies" of the genus *Hieracium*. A major factor in favour of this diversity is the great variety of stony, more or less stable, sunny to partially shaded habitats, whose substrate might be short- or long-weathered, basic to acid and poor or rich in humus. Within the town the fortifications are among the richer sites, notably the Lage Fronten, which not only contain old walls but also disused railway land (Verschoor and Egelmeers 2006). Table 1 lists only the sections not the microspecies, which are still being studied. The same approach has been adopted with other critical taxa such as *Taraxacum* and *Rubus*.

Neophytes

Neophytes make up 17% of the total vascular flora in Maastricht, both in the whole urban area and in the town centre (Table 3). However, in the centre the share of neophytes per square kilometre grid-cell is higher (14%) than in the town area as a whole (11%).

Table 3 Comparison of species numbers in the whole urban area and in the town centre and the proportion of neophytes in the town's flora

	Total urban area		Town centre	
N of grid-cells (1 km²)	36		5	
	No.	%	No.	% (w.r.t. town centre)
Total N of vascular plant species	721	*100*	461	*100*
Woody plants (>1 m high)	74	*10*	54	*12*
Herbaceous plants (incl. dwarf shrubs)	647	*90*	407	*88*
Neophytes	120	*17*	80	*17*
Woody neophytes	18	*2.5*	15	*3.3*
Herbaceous neophytes	102	*14*	65	*14*

While the total number of woody species is considerably higher in the outer parts of the city, the proportion of woody neophytes in the centre is higher. This might be due to the multitude of stony habitats in the historical core of the town, which results in high density of potential germination sites. In addition, there are a large number of exotic trees in the parks on the southern fringe of the town centre, but no spontaneous offspring of these tree species outside the parks has been reported so far.

The 43 most frequent neophytes in Maastricht are listed in Table 4.

These species have been recorded in at least eight out of the 36 km²grid-cells taken as a base for determining the frequency of species in the urban area. The majority of them are characteristic of man-made or highly disturbed habitats, their most "natural" habitat being riversides. Their occurrence in various kinds of vegetation will be discussed later. The most noteworthy neophyte in the town is *Geranium rotundifolium,* which reached its northern limit in Maastricht in the early twentieth century and is only slowly establishing itself further to the north. A characteristic neophyte of the Meuse valley is *Sisymbrium austriacum* ssp. *chrysanthum,* originally a Pyrenean plant, which was imported with wool to the factories of Verviers in eastern Belgium and has been conveyed from there into the Meuse valley by the river Vesdre. A similar pathway has been taken by *Senecio inaequidens,* a species from Southern Africa that became established in the Meuse valley in Limburg about 1942 from where it colonised other areas, and becoming a common species in large parts of The Netherlands from ca.1975 onwards – as it has in adjacent countries.

Frequent woody newcomers include *Buddleja davidii, Robinia pseudoacacia* and *Acer platanoides. Buddleja,* which is planted in gardens to attract butterflies, frequently establishes itself far from where it is planted. On stony grounds it may form dense scrub, especially in the less well-used parts of railway land. Another favourite habitat of this species is the "joint" between buildings and adjacent pavements. *Robinia* and *Acer* mainly colonise and become established on talus close to where they have been planted.

Table 4 The 43 most widespread neophytes in the urban area of Maastricht, arranged according to their main habitats

	frequency class	occurrence in town centre
1. Herbaceous plants (incl. dwarf shrubs) tread plants		
Matricaria discoidea	V	+
Eragrostis pilosa	III	+
Eragrostis minor	II	+
2. Plants of ruderal habitats (in wider sense)		
Senecio inaequidens	V	+
Conyza canadensis	V	+
Epilobium ciliatum	IV	+
Fallopia japonica	IV	+
Erigeron annuus	IV	+
Solidago gigantea	IV	+
Amaranthus retroflexus	III	+
Lolium multiflorum	III	-
Oenothera glazioviana	III	+
Geranium rotundifolium	III	+
Solanum nigrum ssp. *schultesii*	III	+
Heracleum mantegazzianum	III	+
Solidago canadensis	II	+
Oenothera biennis	II	+
Helianthus tuberosus	II	+
Brassica napus	II	+
Datura stramonium	II	+
Oenothera parviflora	II	+
Potentilla indica	II	+
3. Arable weeds (mainly garden weeds)		
Galinsoga quadriradiata	V	+
Veronica persica	V	+
Oxalis stricta	III	+
Oxalis corniculata	III	+
Galinsoga parviflora	III	+
Coronopus didymus	II	+
4. Wall plants		
Cymbalaria muralis	II	+
Pseudofumaria lutea	II	+
5. Grassland plants		
Veronica filiformis	V	+
Medicago sativa	IV	+
Trifolium hybridum	III	-
Geranium pyrenaicum	II	+
Crepis vesicaria ssp. *taraxacifolia*	II	---

(continued)

Table 4 (continued)

	frequency class	occurrence in town centre
Pilosella praealta (sensu lato) (incl. P. bauhinii)	II	+
6. Riverside plants		
Bidens frondosa	III	+
Impatiens glandulifera	II	+
Acorus calamus	II	+
7. Woody plants (>1 m high)		
Buddleja davidii	IV	+
Robinia pseudoacacia	III	+
Acer platanoides	III	+
Mahonia aquifolium	II	+

The listed species occur in at least 8 of 36 of 1 km² grid-cells covering the urban area
Frequency classes: V=29–36, IV=22–28, III=15–21, II=8–14 grid-cells. Occurrence in town centre: +=observed in town centre 1990–2006; - -=observed in town centre 1980–1989, but not later-on; ---=not observed in town centre. The town centre is represented by five grid-cells of 1 km²

Invasive Species

Some ubiquitous native species have considerable capacity to crowd out other plants. In the case of *Urtica dioica,* hypertrophy is the main stimulating factor, for example, on the margins of the Jeker. *Arrhenatherum elatius* will become dominant after the abandonment of mowing or grazing, as was demonstrated in the fortifications before grassland management was restored. *Cirsium arvense* takes an intermediate position between these two species. In woodlands, *Hedera helix* is by far the strongest competitor in the herb layer; in this respect Maastricht is similar to southern Limburg. The dominance of *Hedera helix* can be reversed or prevented by sheep grazing but this is only possible in fenced nature parks.

The most aggressive of the invasive species is the neophyte *Fallopia japonica.* In the fortifications (Hoge Fronten, see later) it has not only colonised parts of the grassland vegetation, but it is also undermining the walls. For these reasons it is controlled by chemical means to preserve the historical and biological interest of the fortifications. Similar problems may be caused by the even more robust *Fallopia sachalinensis,* which has been observed in a few sites in the city.

Other neophytes that exhibit strong invasive capacity are *Solidago canadensis* and *Solidago gigantea.* Large areas of the former have become established in the moats of the fortifications. *Solidago gigantea* appears especially well-adapted to riverine conditions. Both *Solidago* species are avoided by cattle so that they are unaffected by a moderate grazing regime; consequently if vulnerable vegetation types are threatened by them it is necessary to control them by regular mowing. They have some importance as a nectar source for bees and other insects that fly in summer and autumn although the plants they replace, such as *Vicia sepium* and *Origanum vulgare* often have a more specific interest for insects.

Planted Trees

Tilia x *europaea* has been the most common tree along the roads of Maastricht for at least the last three centuries. Other frequently planted trees in the town are *Platanus* x *hispanica, Acer pseudoplatanus* and *Fagus sylvatica. Populus* x *canadensis* mainly occurs on the margins of the urban area. Locally, for example in Amby, *Viscum album* can be seen in the tops of *Populus* spp. but this parasite is far more numerous in rural parts of the Meuse valley north of the town.

A chain of parks on the south side of the town centre has been laid out as an arboretum containing some 160 tree species and cultivars (Graatsma and De Jong 1995). Among the striking features of these parks are the dark, evergreen groups of *Taxus baccata.*

Mosses and Liverworts

From a bryological point of view the fortifications are the most remarkable element in the urban area; they support several thermophilous species that are on the northern edge of their range. These mosses thrive on old wall parts covered with a thin layer of earth. In 1995 *Reboulia hemisphaerica* was recorded in the Hoge Fronten – a thallous liverwort, which had been found several times in southern Limburg in the nineteenth century but had not been observed for 120 years. At the same site many acrocarpous mosses were found, including *Pterygoneurum ovatum* – the only Dutch record since 1980 – as well as *Encalypta vulgaris* and *Microbryum curvicollum* (Nieuwkoop 1995).

In 2005 *Funaria pulchella* was observed at Fort St. Pieter, the first Dutch record for this mainly Mediterranean species which is apparently spreading northwards. In Belgium this moss has only been found at three sites – all of them since 1991 (Van Melick 2005).

Unlike the town itself the adjacent hilly area of southern Limburg has a rich epiphytic moss flora, including *Pylaisia polyantha, Porella platyphylla, Leucodon sciuroides, Anomodon viticulosus,* and a number of Orthotrichaceae. These epiphytes are found on trees and shrubs with rich and subneutral barks like *Populus* x *canadensis, Salix alba, S. cinerea, Fraxinus excelsior* and *Sambucus nigra.* Most good sites for epiphytic mosses are situated in brook valleys, suggesting that the moist atmosphere in these valleys counterbalances regional air pollution. Even the Bunderbos, which is only 2 km from the fringe of the urban area, contains many epiphytic mosses.

Fungi (including Lichenised Fungi)

Lichenised Fungi

As for lichens, old town-walls and fortifications are again the main sites of interest within Maastricht (Aptroot and Van Herk 1999). In 1998 *Caloplaca crenulatella* was observed as a new species for The Netherlands – it was found on the cement of the fourteenth century town-wall. *Leptogium schraderi* and *Vezdaea aestivalis* were found on the brickwork of the same wall. Remarkable finds in the Hoge Fronten include *Lecidella anomaloides* and *Caloplaca albolutescens,* also growing on bricks.

The richest natural location for epilithic lichens in The Netherlands, the Bemelerberg, is situated only 2 km east of the border of the urban area. Here *Toninia sedifolia, Catapyrenium squamulosum, Squamarina cartilaginea, Fulgensia fulgens* and *Psora decipiens* have been found. The first two have also been recorded from the St. Pietersberg.

Very few records of fungi are known from the town – mycologists' interest being focused on the woods and calcareous grasslands in the neighbouring rural area. Three fungi are worth mentioning (P.H. Kelderman, 2009, personal communication). First, the calciphilous gasteromycete *Tulostoma brumale,* which was recorded on the upper side of the fourteenth century town-wall; elsewhere in South Limburg it is a rare fungus of mossy chalk grassland. Second and found at the same site is the endangered agaric *Omphalina rickenii.* Third was the ascomycete *Geopora sumneriana,* which has been observed under *Cedrus,* which was planted as an ornamental tree. It is an example of a fungus that was found to be less rare than supposed after its substrate (decaying *Cedrus* needles) was inspected more often in the appropriate season (spring).

Habitats

Aquatic Vegetation

The most frequent aquatics within the town area are *Potamogeton pectinatus, Nuphar lutea, Lemna minor, Ceratophyllum demersum* and *Myriophyllum spicatum.* All of these species are either characteristic or tolerant of eutrophic, hard water (that is rich in bicarbonate). At first sight this seems self-evident because of the town's position in a river valley bordering eutrophic (loess) and calcareous (limestone) deposits. Moreover much of the surrounding area has been heavily manured for agricultural use; as a result the run-off water is rich in nutrients. However, in the past, some aquatics of less eutrophic or less hard water also occurred in or near Maastricht. *Hottonia palustris,* which once occurred in pools at several locations, has now become very rare. At the beginning of the twentieth century *Ranunculus hederaceus* was observed near Amby, but this small western European aquatic has not been found in southern Limburg for 70 years.

At present, *Ranunculus fluitans* and *Potamogeton nodosus* are the most interesting aquatics in Maastricht. Both are riverine species characteristic of European Union Habitat 3260 "Water courses of plain to montane levels with the *Ranunculus fluitans* and *Callitricho-Batrachion* vegetation." *Ranunculus fluitans* is found both in the Jeker and the Meuse; after a period of decline it has recently recovered thanks to a reduction in water pollution. During the last few decades, *Potamogeton nodosus* has gradually extended its distribution in both the Julianakanaal and the Meuse. As a thermophilous species it has probably benefited from a rise in water temperature.

Marsh and Waterside Vegetation

As a result of archaeological excavations, 27 species of marsh and waterside plants have been recorded from Roman and/or Mediaeval Maastricht (Table 1). Twenty-three of them are still present. *Berula erecta,* which was still present in the 1980s, may be found again in the newly created pools or cleaner watercourses. The three species that have vanished – *Stellaria palustris, Juncus subnodulosus* and *Carex panicea* – are indicative of mesotrophic marshes. Presumably they occurred in the hay fields in the Jeker valley. Interestingly enough, in the nineteenth century marsh plants like *Apium repens, Blysmus compressus* and *Triglochin palustris* were found in peaty marshland along the Jeker near Maastricht (Graatsma 2003). These three marsh plants used to occur both in coastal regions and in the interior of The Netherlands but became largely confined to coastal parts in the course of the twentieth century. In the last few decades they have re-appeared in some interior regions thanks to the restoration of wetlands. Currently, only few mesotraphent marsh plants occur in Maastricht, for example, *Ranunculus flammula, Carex disticha* and *Cirsium palustre*. "Wet nature development" on the fringes of the town might increase the chances for mesotrophic habitats; for example, it is hoped that an ecological project that is being implemented in the Jeker valley will provide opportunities for some mesotraphent species to re-establish themselves (De Mars and Verhulst 2005).

Nowadays the most frequent waterside plants in the city are *Stachys palustris, Iris pseudacorus, Phragmites australis, Scrophularia auriculata, Rumex hydrolapathum, Lythrum salicaria, Mentha aquatica, Glyceria maxima, Typha latifolia* and *Filipendula ulmaria*. In addition, *Acorus calamus, Sagittaria sagittifolia* and *Schoenoplectus lacustris* occur at several sites. Like the aquatics these waterside species are indicative or tolerant of eutrophic, hard water. Some marsh umbellifers whose occurrence within southern Limburg was confined to the Meuse valley – *Oenanthe aquatica, Oenanthe fistulosa* and *Sium latifolium* – have become very rare or extinct.

A comparatively large number of waterside plants are found near the motorway west of the former village of Amby, where an estate and two ecological parks containing some ditches, ponds and partly wet pastures are located.

The Winter Channel and Adjacent Areas of the Meuse

Within the contour of the urban area the Meuse is largely contained within high walls with narrow bank strips. Whereas some indigenous or archaeophytic species such as *Cuscuta europaea* and *Saponaria officinalis* are characteristic of river banks too, this habitat is especially known for its richness of neophytes including *Senecio inaequidens, Solidago gigantea, Helianthus tuberosus, Aster lanceolatus, Bidens frondosa, Erigeron annuus* and *Impatiens glandulifera*. For several of these plants the riverside has functioned as a stepping stone from where they have invaded other habitats. Recent acquisitions to the riverside flora in the Maastricht region include

Angelica archangelica and *Lepidium latifolium*. Both are somewhat salt tolerant but it is not clear whether their appearance along the Meuse is correlated with an increase in its salinity.

Within the urban framework of southern Maastricht, there is an outer meadow on the eastern bank of the Meuse, called Kleine Weerd, which was taken out of agricultural use in 1994 and transferred to nature conservancy organisations. Since then, it has been managed as a pasture grazed by Galloway cattle and Konik horses all year without the use of fertilisers. A mosaic of rough and grassy vegetation has developed. While common species like *Arrhenatherum elatius, Elytrigia repens, Cirsium arvense* and *Urtica dioica* soon became prominent, a number of less common plants have appeared, for example, *Ajuga reptans, Cruciata laevipes, Origanum vulgare, Inula conyzae* and even *Astragalus glycyphyllos* (Lejeune 2002).

Little or nothing remains of the pioneer vegetation that occurred in those parts of the river bed that become dry in the summer months. The rarest of these species (and the first to be lost) is *Sisymbrium supinum,* which only occurred before 1850. While it is now an "EC Habitat Directive species" threatened throughout its range, its ecological requirements are not well understood. Other examples of lost species are *Mentha pulegium, Inula britannica, Pulicaria vulgaris* and *Limosella aquatica.* Re-inforcement of the river banks seems to be the main cause for the disappearance of their former habitat. However, the present absence of the latter three is somewhat surprising because over the last few decades they are recovering quite well in other parts of the Dutch riverine system.

On the eastern side of the regular floodplain there was an 8 km long low-lying strip, which had an overflow function up to 1935 and then became part of the town area (see p. 241). At the beginning of the twentieth century it still contained damp meadows with rare species like *Orchis ustulata, O. coriophora* and *Galium glaucum.* To be honest, it must be admitted that they had already vanished by ca. 1920 due to the intensification of agricultural management (De Wever 1913). Subsequently almost the whole of the valley bottom has been added to the urban area, causing a continuous decline of meadow plants like *Colchicum autumnale, Primula veris* and *Saxifraga granulata,* which used to be rather common in the region. About 4 km outside the northern limit of the municipality some remnants of wet Meuse meadows have been preserved.

Woodland Vegetation

There are now only small remnants of woodland left in the urban area; nevertheless the number of woodland plant species is considerable, which mirrors the general richness of the wood flora of the region. Again the area west of Amby (Severen) proves to be comparatively rich in comparison with the rest of the urban area.

The most frequent herbaceous wood plants in the town area are *Geum urbanum, Ficaria verna, Silene dioica, Epipactis helleborine, Dryopteris filix-mas, Scrophularia nodosa, Viola odorata, Stachys sylvatica, Arum maculatum, Poa nemoralis,*

Brachypodium sylvaticum and *Lamiastrum galeobdolon*. However, the latter is found far more as a garden escape with strongly variegated leaves (ssp. *argentatum*) than as the native taxon. The rather high frequency of *Dryopteris filix-mas* and *Brachypodium sylvaticum* is partly due to their occurrence on moist or shaded walls, where they find a secondary habitat. Some woodland plants have settled in groves along the motorway, notably *Epipactis helleborine* and *Poa nemoralis*. An example of a woodland species invading small woodlands with comparative ease is *Allium ursinum*.

A few woodland species that are considered native to southern Limburg as a whole only occur in Maastricht as garden escapes; these include *Aquilegia vulgaris, Campanula persicifolia, Digitalis purpurea* and *Pulmonaria officinalis*. Although "cultivated" flower colours (pink-flowered *Aquilegia,* white-flowered *Campanula* and *Digitalis*) may be used as a criterion for determining introductions, it is often impossible to distinguish native and introduced populations on purely morphological features (unlike *Lamiastrum galeobdolon* ssp. *argentatum*). Nevertheless, it should be kept in mind that urban populations of these plants lack the "naturalness value" of those in some rural parts of southern Limburg. A similar situation occurs with some woody species, for example, *Cornus mas*.

While most woodland plants occurring in Maastricht are rather common in southern Limburg, several woods just outside the municipality are rich in rare wood plants. Some examples may be quoted which are situated within 3 km from the boundary of the urban area. The Bunderbos north-east of Maastricht contains a number of wells accompanied by spring-wood mosses such as *Hookeria lucens* and *Trichocolea tomentella*. To the east-north-east the Dellen contains the acidophytic, oreophilous ferns *Phegopteris connectilis* and *Oreopteris limbosperma*. To the south-east the Savelsbos has *Euphorbia amygdaloides* (at the north-eastern limit of its area), *Neottia nidus-avis* and *Anemone ranunculoides*.

Grassland and "Rock" Vegetation in the Fortifications

The greatest number of rare grassland and pioneer species within Maastricht are found on the fortifications: the Bossche Fronten (divided into Hoge) on the north-west side of the historical town centre, and Fort St. Pieter on the south-side. Most of the 36 Red List grassland species and pioneers of stony ground (Table 5) have their largest or only populations in these parts of the town. The Fronten owe much of their fame among nature-lovers and ecologists to the lizard *Podarcis muralis,* which has its northern-most station here (Frissen-Moors and Tilmans 2009). The restoration of the walls of the Hoge Fronten caused the lizard population to decline dramatically but thanks to the persistent actions of herpetologists the Hoge Fronten was given protected status (Kruyntjens 1994). Legal protection of the Lage Fronten is still awaited.

The fortifications qualify as semi-natural habitats in spite of their man-made form. They do so because of the very long period over which the grassland has developed, their role as a refuge for native grassland species and the similarities they have with riverine grassland.

A prominent vegetation type in the Hoge Fronten is the *Arrhenatheretum elatioris,* which covers fairly large coherent surfaces both in the bastions and the moats. Such sites of this plant community have become quite rare in The Netherlands, most examples of the *Arrhenatheretum* covering narrow strips on dykes and waysides. Well-developed examples of this association represent European Union Habitat 6520 "Lowland hay meadows". They have a great aesthetic value thanks to their rich variety of attractive flowering plants, including many Fabaceae and Asteraceae. Characteristic species include *Trisetum flavescens, Crepis biennis, Knautia arvensis, Daucus carota* and *Vicia sepium.* In addition the *Arrhenatheretum* in the Fronten contains a number of species which have their main habitat in calcareous grasslands (at least within southern Limburg), for example, *Sanguisorba minor, Anthyllis vulneraria, Plantago media, Leontodon hispidus, Linum catharticum* and *Carex flacca.* Still more special is the isolated occurrence of *Veronica austriaca* ssp. *teucrium,* which elsewhere in The Netherlands is virtually confined to the Rhine system. These and other slender or short-living species, such as *Centaurium erythraea* and *Saxifraga granulata,* can only maintain themselves as long as the vegetation does not become dense or tall. In a bastion grassland of the Lage Fronten *Muscari comosum,* formerly a plant of loess fields has one of its last surviving populations in the region.

Another flowery, though much rougher vegetation type occurs mainly in the moats; it is characterised by co-dominance of *Origanum vulgare* and *Rubus caesius.* It is crossed by cattle paths, where *Odontites vernus* ssp. *serotinus* takes its chance.

Near the end of the twentieth century, after a long period of neglect, most of the moats and the bastions were covered by rough and highly productive vegetation for the greater part. From 1992 onwards a mixed management regime was introduced involving periodic grazing by sheep as well as mowing. Thanks to this management most of the diversity in structure and species composition has been restored. Nevertheless the invasiveness of *Solidago canadensis* and *Fallopia japonica* is still a major problem.

It is striking that characteristic species of the *Arrhenatheretum* occupy a large proportion of the wall vegetation of the Fronten, while specific wall plants such as *Cymbalaria muralis* only play a minor part. This phenomenon is repeatedly seen on slightly inclining, earth-backed walls. In fact, the walls may have functioned as a refuge for slender or short-living grassland species (for example, *Sanguisorba minor* and *Veronica austriaca* ssp. *teucrium*) whose occurrence in the grassy places was temporarily endangered, either by over-grazing and trampling or by neglect resulting in dominance of tall herbs and grasses. Among the species having their main occurrence on the walls are *Campanula rotundifolia, Echium vulgare* and *Inula conyzae.*

The diversity of plant species as well as the stability of the plant populations and the variety of bee nesting sites (including holes in walls) is among the factors that contribute to the rich insect fauna. Many pollen or nectar collecting insects benefit from the opulent occurrence of representatives of Fabaceae and *Asteraceae,* as well as *Knautia arvensis* and many other plant species. The entomological value of the Hoge Fronten is illustrated by the presence of oligo- or monolectic

Table 5 Species in the National Red List that occur in the urban area of Maastricht, arranged by their main habitats

	Red List category	Frequency class in Maastricht		Red List category	Frequency class in Maastricht
1. Ruderal plants			5. Grassland plants		
Hyoscyamus niger	VU	I	*Allium oleraceum*	VU	I
Mentha suaveolens	EN	I	*Anacamptis pyramidalis*	NT	I
Verbena officinalis	NT	IV	*Anthyllis vulneraria*	VU	I
			Botrychium lunaria	VU	I
2. Arable weeds			*Campanula rapunculus*	VU	I
Aphanes arvensis	EN	I	*Carlina vulgaris*	VU	I
Centaurea cyanus	NT	I	*Centaurea scabiosa*	VU	I
Consolida regalis	CR	I	*Colchicum autumnale*	EN	I
Kickxia elatine	VU	I	*Cuscuta epithymum*	VU	I
Legousia speculum-veneris	CR	I	*Cynosurus cristatus*	NT	I
Sherardia arvensis	VU	I	*Dianthus armeria*	EN	I
Silene noctiflora	EN	I	*Genista tinctoria*	EN	I
Torilis arvensis	EN	I	*Knautia arvensis*	NT	III
			Linum catharticum	VU	I
3. Wall plants			*Nardus stricta*	NT	I
Arabis sagittata	EN	I	*Odontites vernus* ssp. *serotinus*	NT	II
Erysimum cheiri	CR	I	*Ononis repens* ssp. *spinosa*	NT	I
Cystopteris fragilis	EN	I	*Orobanche minor*	EN	I
Hieracium speluncarum[a]	NT	I	*Plantago media*	VU	II
Rumex scutatus	NT	I	*Polygala vulgaris*	NT	I
			Primula veris	VU	I
4. Pioneers of stony ground			*Rhinanthus alectorolophus*	VU	II
Cerastium pumilum	NT	I	*Rhinanthus minor*	NT	I
Clinopodium acinos	VU	I	*Salvia pratensis*	VU	I
Minuartia hybrida	CR	I	*Saxifraga granulata*	EN	I
Sedum rupestre	EN	I	*Scabiosa columbaria*	EN	I
			Sedum sexangulare	VU	I
			Tetragonolobus maritimus[a]	NT	I
			Thymus pulegioides	VU	I
			Trisetum flavescens	NT	II
			Valerianella carinata	NT	I
			Veronica austriaca ssp. *teucrium*	EN	I

(continued)

Table 5 (continued)

	Red List category	Frequency class in Maastricht		Red List category	Frequency class in Maastricht
			6. Wood-fringe plants		
			Agrimonia eupatoria	NT	II
			Arabis glabra	EN	I
			Atropa belladonna	NT	I
			Clinopodium menthifolium[a]	EN	I
			Clinopodium vulgare	VU	I
			Cruciata laevipes	VU	I
			Fragaria vesca	NT	I
			Hieracium murorum sensu lato	VU	I
			Hypericum hirsutum	VU	I
			Sambucus ebulus	EN	I

Red List categories: *CR* critically endangered, *EN* endangered, *VU* vulnerable, *NT* near threatened. Frequency classes based on presence in 36 grid-cells of 1 km^2; IV = 22–28, III = 15–21, II = 8–14, I = 1–7 grid-cells
[a]Neophytes

bees such as *Andrena lathyri* (forages on *Vicia sepium*), *Melitta tricincta* (forages on *Odontites vernus*), *Chelostoma distinctum* (forages on *Campanula rotundifolia*) and *Hoplites adunca* (forages on *Echium vulgare*) (I.P. Raemakers, 2009, personal communication).

Fort St. Pieter is a refuge for a number of plants that once inhabited the St. Pietersberg, for example, *Silene nutans,* which has always been a rare species in southern Limburg. On some projecting parts of the fortress, where the stone is covered by a shallow layer of earth, a durable pioneer vegetation occurs with *Koeleria macrantha, Thymus pulegioides, Sedum album, Cerastium pumilum, Arenaria serpyllifolia* ssp. *leptoclados* and the lichen *Cladonia humilis*. This vegetation is the more remarkable for the occurrence of *Funaria pulchella* (which is new to the Dutch moss flora) and one tiny specimen of *Botrychium lunaria*. Its floristic composition corresponds with European Union Habitat 6110 "Rupicolous calcareous or basophilic grasslands of the *Alysso-Sedion albi*" but at present it is only found on an artificial rocky substratum.

Old Walls

Nowadays old walls are the most valuable urban habitat within Maastricht, see Fig. 3. The number of plant species and vegetation types characteristic of walls is greater than anywhere else in The Netherlands.

Fig. 3 Nowadays old walls are among the most valuable urban habitats of Maastricht, supporting, for example, a large population of *Erysimum cheiri* (Photo H.J. Buurman)

Some wall plants, like *Poa compressa* and *Saxifraga tridactylites,* are found on limestone rocks in southern Limburg as well, so they may have been derived from the flora of the region. The same holds true for *Arabis sagittata,* a very rare plant in The Netherlands. In Maastricht it occurs on the synagogue and near the former Capuchin monastery on the north-western fringe of the historical town centre. All three plants are found both on the sides and flat tops of walls, while *Poa compressa* and *Saxifraga tridactylites* have found another kind of urban habitat on railway embankments.

In addition to these three flowering plants, some ferns are the only mural vascular plants that have their origin in the natural vegetation of the same region. They used to occur on shaded loess faces and rocks, but have vanished or become very rare in this habitat. In the urban area of Maastricht, *Asplenium trichomanes* and *Asplenium scolopendrium* are rather widespread, while *Cystopteris fragilis* is characteristic of the limestone walls along the Jeker. Before 1990 *Asplenium adiantum-nigrum* and *Gymnocarpium robertianum* were also known from Maastricht, but they have not been seen in the urban area since that time. In 1986, the first Dutch record of *Asplenium septentrionale* was made in the city, but it grew on imported stone in the harbour and has not established. By far the most common fern on walls is *Asplenium ruta-muraria,* a species that is only sporadically found on natural rocks in the region. Recently *Asplenium scolopendrium* has been found to be increasing both on the sheltered walls and on the sides of wooded valleys in southern Limburg. This may be partly due to mild winters and partly to the dispersion of spores from gardens, where is often used as an ornamental plant.

Indeed, many characteristic wall plants owe their present distribution to introductions for ornamental purposes. One of the most conspicuous examples is the well-known, fragrant *Erysimum cheiri,* which is one of the oldest ornamental plants in Europe. It is characteristic of very old walls with rough, not quite perpendicular surfaces. In Maastricht it is confined to remnants of the oldest (thirteenth century) town wall. In the Netherlands this species, which is on the north-eastern edge of its area, has become extremely rare.

About AD 1600 *Cymbalaria muralis* and *Pseudofumaria lutea* were introduced as ornamentals from southern Europe. The substrate requirements of both are considerably less critical than that of *Erysimum cheiri;* consequently they have a much wider distribution both in The Netherlands and within the Maastricht area. A rare garden escape in southern Limburg is *Pseudofumaria alba,* which has established itself very locally on walls; in Maastricht only one site is known at present.

Another neophyte in the mural vegetation of Maastricht is *Hieracium speluncarum,* which was first observed in 1876. It is not clear whether it was introduced deliberately. In Maastricht the species may be associated with *Erysimum cheiri* or *Arabis sagittata* but it also plays a role in other kinds of wall vegetation, be it on vertical or on horizontal surfaces. It is found both in the historical town centre, in the fortifications and on quay walls. In the course of the twentieth century it has spread to some other locations in southern Limburg.

Although an inconspicuous plant, *Parietaria officinalis* must nevertheless be classified as a garden escape. After being planted in the botanical garden of the Museum of Natural History in 1914, it has colonised various sites in Maastricht. It is locally abundant on the limestone walls along the Jeker; among its frequent companions are *Cystopteris fragilis, Scrophularia auriculata* and the liverwort *Conocephalum conicum. Parietaria officinalis* may also thrive in the joints between building façades and pavements. The related *Parietaria judaica* is an even more recent acquisition to the town's flora. First observed in 1943, it was no doubt brought down from Belgium by the Meuse; the quay walls along the river are still its main habitat in Maastricht.

Railway Verges and Goods Yards

Railway verges support a number of biennials and short-living perennials that attract attention because of the richly colourful flowers, for instance *Echium vulgare, Verbascum nigrum, Verbascum thapsus, Reseda lutea* and *Campanula rapunculus*. Less obvious species, which grow in abundance along railways are *Vulpia myuros, Viola arvensis* as well as the C_4 plants *Digitaria sanguinalis, Setaria viridis* and *Amaranthus retroflexus*. In comparison with other (C_3) plants, C_4 plants have a faster mechanism of carbon fixation that works at temperatures above 12°C only. The three quoted examples and many other C_4 plants of central and western Europe are annuals germinating in late spring, which thanks to their short life cycles are able to benefit from the open conditions created by the use of herbicides.

The vegetation of the main station area was thoroughly investigated in the 1980s but this terrain is now largely inaccessible. Among the rare species found at that time was *Lythrum hyssopifolium,* which is worth mentioning because it reaches its northern limit in The Netherlands. It is mainly observed as a pioneer of the unstable conditions created by pools that dry out during the summer, especially along the lower course of rivers; but in Maastricht a population has persisted for a number of years on gravel.

Another area of railway land that supports a rich flora is the former goods-station of Boschpoort, which borders the fortifications of Lage Fronten. Because it is now disused *Betula pendula* and shrubs like *Rosa canina, Rubus* spp. and *Buddleja davidii* are colonising while low-growing pioneers like *Arabis arenosa* and *Erigeron acer* have decreased in recent years. A number of *Hieracium* and *Pilosella* spp. have survived so far. In 2007 part of the scrub was cleared, which has resulted in the re-establishment of the herbaceous vegetation.

Motorway Verges and Noise Attenuation Mounds

A highly varied grassland and ruderal vegetation occurs on the motorway verges adjacent to the south-eastern quarters of the town (Harle 2008). Thanks to mowing the motorway verges have a moderate nutrient status enabling low-growing plants like *Trifolium campestre, Centaurium erythraea* and *Luzula campestris* to grow side by side with more robust species like *Campanula rapunculus, Inula conyzae, Agrimonia eupatoria* and *Origanum vulgare*. A recent acquisition is *Blackstonia perfoliata,* a southern calciphilous plant, which is colonising both Limburg and the coastal area of The Netherlands. In Limburg road verges serve as the route of expansion of other species, for example, *Orobanche minor,* which is another striking species.

A recently developed habitat offering new chances for grassland and ruderal species is formed by the noise attenuation mounds along motorways, which were formed in the south-eastern parts of Maastricht in 1998 using soil with rather low nutrient content. The vegetation is still "rough" and rich in robust species in the genera *Verbascum, Melilotus, Dipsacus, Helianthus* and *Solidago* but also slender plants like *Sherardia arvensis, Geranium rotundifolium* and *Carex spicata* have found their way to this new habitat. Regrettably highly invasive *Fallopia japonica* has settled too and is expanding in several places.

As the result of salting the roads in winter, some halophytes have colonised the road sides and the central reservations. By far the most common of these species in Maastricht is *Cochlearia danica,* with *Plantago coronopus* and *Puccinellia distans* ssp. *distans* being found in a few places.

Changing Fate of Field and Garden Weeds

A well-documented and striking change in the flora concerns arable weeds. Few other groups so sharply mirror changes in human soil use and techniques. Generally speaking weeds showing classical types of ecological mimicry (that is likeliness to a crop in essential features such as stem height and fruit size) have been eradicated by modern seed cleaning techniques. Nowadays a successful form of ecological mimicry is resistance against the same chemicals as crop species – like "weedy" grasses between cereals and Chenopodiaceae between beets. Moreover the use of herbicides has enabled some weeds (notably summer germinators, including C_4 plants) to invade urban areas.

Some cornfield weeds documented from the Roman era survived in the surroundings of Maastricht up to the nineteenth century but have vanished since, namely, *Orlaya grandiflora, Stachys annua* and *Nigella arvensis* (Table 1). The same applies to the specific flax-field weed *Cuscuta epilinum.* In the course of the twentieth century many more arable weeds have disappeared or become very rare in southern Limburg, for instance *Agrostemma githago, Lithospermum arvense* and *Ranunculus arvensis* – three cornfield weeds whose fruit features do not offer an effective "survival trick" anymore. Several other cornfield weeds now mainly survive in alternative habitats; they include *Papaver rhoeas* and *Avena fatua* in disturbed verges; *Papaver dubium, Veronica hederifolia* and *Valerianella locusta* in old walls and other stony sites; *Chaenorhinum minus* and *Setaria viridis* along railways and *Vicia hirsuta* in grassland.

Nevertheless the Maastricht area still contains a long list of arable weeds, 12 of which are in the "Top 50" of most frequent species – see Table 2. Another 24 species are common, being recorded in more than half of the 36 km² grid-cells; examples are *Euphorbia helioscopia, Galinsoga quadriradiata, Veronica persica, Geranium dissectum, Anagallis arvensis* ssp. *arvensis, Fumaria officinalis, Lamium amplexicaule* and *Raphanus raphanistrum.* Most of these weeds are either generalists as to the type of crop or they have their optimum between root-crops. Some of them, for instance *Chenopodium album* and *Fallopia convolvulus,* belong to the oldest documented members of the Maastricht flora (Table 1).

It is striking that most neophytes among arable weeds are more often found as garden weeds (both in vegetable and ornamental gardens) than in arable land. This applies to *Galinsoga quadriradiata, G. parviflora, Oxalis stricta, O. corniculata, Coronopus didymus, Veronica peregrina* and *Claytonia perfoliata.* Only *Veronica persica* is also a common weed in arable fields.

Now and again rare weeds of calcareous fields are still recorded in the town area for example *Kickxia elatine, Legousia speculum-veneris* and *Torilis arvensis* all of which have been observed after 2000 and probably have germinated from the seedbank. They show that the possibilities for a more exquisite weed flora in fields are not exhausted at present.

Nature Conservation, Environmental Planning and Education

Red Data Species

No less than 65 species of the Dutch Red List of vascular plants have been recorded in Maastricht since 1990 (Table 4).

The most frequent of these is the ruderal species *Verbena officinalis*. About two-third of the Red List species are grassland or wood-fringe plants, of which *Knautia arvensis, Odontites vernus* ssp. *serotinus, Trisetum flavescens, Agrimonia eupatoria, Plantago media* and *Rhinanthus alectorolophus* are the most widespread. The remnants of the fortifications are a local hot spot of botanical diversity. The presence of Red Data Species in relation to habitats is discussed in the next section.

The position of Maastricht on the southern edge of The Netherlands implies that the Dutch flora is not an appropriate basis for the evaluation of the town's flora and vegetation, which is better judged on central European criteria. This marginal position is one of the reasons why the Province of Limburg has compiled a list of threatened species for the Province (Cortenraad and Mulder 1998). If the data from 1990 to 2006 surveys are evaluated according to this provincial list, the urban area of Maastricht would contain 218 species whose survival in the region is threatened to some degree. Species that are proportionately more common in Maastricht than in the remaining parts of southern Limburg include *Inula conyzae, Geranium rotundifolium, Sedum album, Carex otrubae, Odontites vernus* ssp. *serotinus* and *Veronica polita*. The first three are characteristic of sunny, stony sites that benefit from the abundance of old walls and gravelly spots in the town. *Carex otrubae* and *Odontites vernus* are found in moist, grassy, clayey habitats in valleys, surviving rather well in the Meuse valley within Maastricht. *Veronica polita* is an arable weed that apparently maintains itself better in vegetable gardens and disturbed verges then in arable fields nowadays.

NATURA 2000 Areas

Rural parts of the municipality of Maastricht comprise two elements of the European Union Directive Network NATURA 2000. To the south of the town parts of the St. Pietersberg and the Jeker valley belong to a Natura 2000 site that has its counterparts in the Flemish and Walloon parts of Belgium (Schaminée and Janssen 2009). On the Belgian side of the St. Pietersberg a considerably larger area of limestone grassland has been preserved than in the part within The Netherlands. On both sides of the frontier these grasslands are mown and periodically grazed by wandering herds of the local “limestone” race of sheep. Recently the St. Pietersberg has regained its place among the best butterfly areas in The Netherlands.

North of Maastricht on the Dutch side of the Grensmaas (the Meuse section serving as frontier with Belgium) is a 50 km long, ribbon-shaped NATURA 2000 area, extending from Borgharen northwards to Thorn in central Limburg; the southernmost 7 km belongs to the municipality of Maastricht. Because of its position at the transition between the hills and the lowland it has a special position among Dutch rivers, being rather shallow, containing gravel banks and being contained within high grounds. Moreover its frontier status has safeguarded it from such radical canalisations as other rivers have been subject to. Nowadays a combination of water catchment, gravel production and nature functions is pursued.

Two further NATURA 2000 areas lie on the eastern slope of the Meuse valley, only a few kilometers distance from Maastricht. To the south-east the Savelsbos is among the most species-rich slope-woods of the region; moreover it contains a small but valuable piece of limestone grassland. To the north-east the Bunder and Elsloërbos are equally rich but wetter ravine woods with calcareous and non-calcareous springs as well as some wet grasslands.

Nature Parks

Maastricht has a number of nature parks in the transition belt between urban and rural areas. To the west of the Meuse, the Jeker valley and the old fortifications form the backbone of the “green infrastructure”. To the east of the Meuse a similar function is performed by rural enclaves in the built-area near the former village of Amby.

In addition to the Hoge and Lage Fronten and parts of the Jeker valley, several woodlands, grasslands and ponds on the north-eastern side of Maastricht (near Amby) have an urban nature function. One of these grasslands is grazed all year by Galloway cattle and Konik horses, while another is periodically grazed by sheep (for example, the Hoge Fronten). In order to maintain a varied structure with sufficient proportion of open vegetation, cutting of scrub and mowing are also carried out.

The objectives of the nature parks include a combination of ecological, nature conservation, educational and recreational functions. Outdoor lessons are provided for school children and several nature parks have been adopted by schools. Public and private organisations and volunteers co-operate in the regional “Centre for Nature and Environment Education” (CNME Maastricht and region). The Centre is responsible for the management of most nature parks in or near the urban area.

Most of the nature parks are open for informal recreation and some provide playgrounds for children. Panels at the entrances give information about terrain conditions and management. In parts of the Hoge Fronten the access is limited in favour of the Wall Lizard population, and during periods of sheep grazing some nature parks are closed to the public.

Legal Protection of Wall plants and of the Wall Lizard

In 1991 rare wall plants were added to the list of protected plants in The Netherlands. This protection applies to specific wall ferns (excluding *Asplenium ruta-muraria*), *Catapodium rigidum, Arabis sagittata, Erysimum cheiri, Hieracium speluncarum, Parietaria judaica* and *Pseudofumaria lutea*. Except for some very rare ferns all these wall plants occur in Maastricht. If a wall on which any of these species are growing needs to be reconstructed, measures must be taken to prohibit their local extinction. In this way many old walls in the town centre of Maastricht enjoy botanical protection by law. In the fortifications (Hoge and Lage Fronten) protected wall plants play a less important role but here the presence of the Wall Lizard is a legal argument for their protection.

Closing Comments

Thanks to its favourable position in the transition from a hilly loess and limestone area to a river valley, Maastricht has excellent opportunities to support a rich flora and fauna. A substantial amount of effort is taken to preserve the ecological value of the old walls and fortifications and to enlarge the area of flowering plants, which also encourage bees and butterflies. It would be a great contribution to the botanical interest of the area if special attention was given to the restoration of two biotopes that occur in the immediate vicinity of the city: (a) mesotrophic wet habitats and (b) the cereal fields, with their characteristic weed flora. In view of a number of archaeological botanical records it is worthwhile trying to restore mesotrophic conditions near Amby and on the flanks of the Jeker valley, at least so far as the groundwater quality and supply allow. If the arable fields on the slopes surrounding Maastricht are taken out of intensive agricultural use, they might be converted into cereal and weed fields by way of contributing to the restoration of old-fashioned agricultural landscapes.

Further Reading

Aptroot A, Van Herk CM (1999) Korstmossen in Limburg, voorjaarsweekend 1998. Buxbaumiella 49: 14–26

Bakels C, Dijkman W (2000) Maastricht in the first millennium *A.D.* – the archaeological evidence. Archaeologica Mosana II, Maastricht

Bakels C, Jacomet S (2003) Access to luxury foods in Central Europe during the Roman period: the archaeobotanical evidence. World Archaeology 34: 542–557

Bardet N, Jagt JWM (1996) Mosasaurus hoffmanni, le «Grand Animal fossile des Carrières de Maestricht»: deux siècles d'histoire. Bulletin du Muséum national d'Histoire naturelle 18(4): 569–593

Bonnemayer JJAM (1986) De Bossche Fronten. Cultuurhistorie en natuurhistorie hand in hand. Natuurhistorisch Maandblad 75: 4–9

Cortenraad J, Mulder T (1998) Actualisering van de lijst van bedreigde planten in Limburg. Natuurhistorisch Maandblad 87: 161–170

De Mars H, Verhulst H (2005) Een eeuw verdroging in het Jekerdal. Natuurhistorisch Maandblad 94: 227–231

De Wever A (1913) Lijst der wildgroeiende planten in Z.-Limburg III. Jaarboek van het Natuurhistorisch Genootschap in Limburg 1913: 1–73

Dortangs RW, Schulp AS, Mulder EWA, Jagt JWM, Peeters HHG, De Graaf DT (2002) A large new mosasaur from the Upper Cretaceous of The Netherlands. Netherlands Journal of Geosciences 81: 1–8

Frissen-Moors CMM, Tilmans RAM (2009) Muurhagedis Podarcis muralis. In: Creemers RCM, Van Delft JJCW (eds.) De amfibieën en reptielen van Nederland. Nederlandse Fauna 9. Naturalis, EIS, Leiden

Graatsma BG (2003) De flora van de omstreken van Maastricht in de 19[e] eeuw. Stichting Natuurpublicaties Limburg, Maastricht

Graatsma B, De Jong T (1995) 'Onbekommerd' in de voetsporen van Jac. P. Thijsse. CNME, Maatricht

Harle N (2008) Floristische rijkdom tussen stad en land. Natuurhistorisch Maandblad 97: 213–221, 225–231

Klasberg M, Bindels E, Breitkopf S (2009) Meerwaarde van de landschappelijke benadering van het project A2 Maastricht voor de waarde van groen en natuur. Groen 20/10: 20–27

Kruyntjens B (1994) De Hoge Fronten: restauratie, consolidatie en beheer in 1992 en 1993. Natuurhistorisch Maandblad 83: 154–163

Lejeune M (2002) De vegetatie van de Kleine Weerd (Maastricht) 1996–2000. Ruigte en co … of is er meer aan de hand? Natuurhistorisch Maandblad 91: 160–169

Nieuwkoop JAW (1996) Reboulia hemispherica L. Raddi terug in Nederland, met opmerkingen over de mosflora van de Hoge Fronten. Buxbaumiella 40: 37–40

Ramakers E (2005) Historische atlas van Maastricht. 2000 jaar aan Maas en Jeker. SUN, Amsterdam

Schaminée JHJ, Janssen JAM (2009) Europese Natuur in Nederland. Natura 2000-gebieden van Hoog Nederland. KNNV Uitgeverij, Zeist

Van Melick HMH (2005) Funaria pulchella H. Philib. (Gaaf krulmos) nieuw voor Nederland. Buxbaumiella 73: 21–25

Van Schaïk DC (1983) De Sint Pietersberg. Met een aanvullend gedeelte van 1938-1983. EF & EF, Thorn

Verschoor G, Egelmeers J (2006) De flora van de Lage fronten. Natuurhistorisch Maandblad 95: 217–224

Milton Keynes

John G. Kelcey

Fig. 1 Central Milton Keynes

Abstract In terms of the age and scale of the urban developments described in the other chapters, Milton Keynes is new and small (occupying 9,000 ha with a target population of 250,000). It differs in its origin from most of those cities, which have grown (organically, as architects say) from a cluster of buildings on a trade route, being imposed on a predominantly rural area as the consequence of a Government decision. There was very little published botanical information available about the city before the start of development in 1971. Because 80% of the

John G. Kelcey (✉)
Čečkovice 14, Bor u Tachova, 348 02, Czech Republic
e-mail: johnkelcey@hotmail.com

J.G. Kelcey and N. Müller (eds.), *Plants and Habitats of European Cities*,
DOI 10.1007/978-0-387-89684-7_9, © Springer Science+Business Media, LLC 2011

area was in intensive agricultural production the flora was "impoverished", mainly comprising species of disturbed soil (arable land) and those that are characteristic of agriculturally improved grassland on damp, meso- to eutrophic soils. The areas of greatest botanical interest were at the extremes of the age range and separated by 1,000 years or so old woodland and young clay workings. Most of the habitats present have been created since 1971 and therefore comprise plant communities/associations that are unique and not found in the countryside or probably in any other urban area. They provide exciting opportunities for botanists to study the relationship between species that would not otherwise occur together and the dynamics of new plant communities.

Introduction

This chapter differs from all the others because it describes the procedures and processes involved in the conversion of 9,000 ha (90 km^2) of countryside into an urban area and the consequences for the habitats and the flora.

The British "Garden City Movement" was started by Ebenezer Howard in the closing years of the nineteenth century. It was to become the New Towns Movement of which Milton Keynes is the biggest and one of the last of the 28 New Towns that were completed or started in Britain between 1946 and 1968. The New Towns were unique in being imposed on an area by the Government and an anacronism in terms of democracy. They were not accountable in terms of planning and related matters and obtained such planning and other permissions as they needed from the Government. "The City of Trees", as Milton Keynes was to be called (for publicity reasons), is unique among the New Towns because it was located in an area where there was no substantial pre-existing urban development. The start of development resulted in physical characteristics and the habitats and flora changing rapidly especially between 1972 and 1983, after which the type and rate of change continued but at a slower rate.

Natural Environment

Location

Milton Keynes is located in the Midland Plain about half way between London and Birmingham, at the geographical co-ordinates 52°02′ 31″ North: 00°45′ 30″ West. Most of the eastern boundary is formed by the M1 motorway, see Fig. 2. The northern boundary is formed by the River Ouse. There are no significant physical southern and western boundaries.

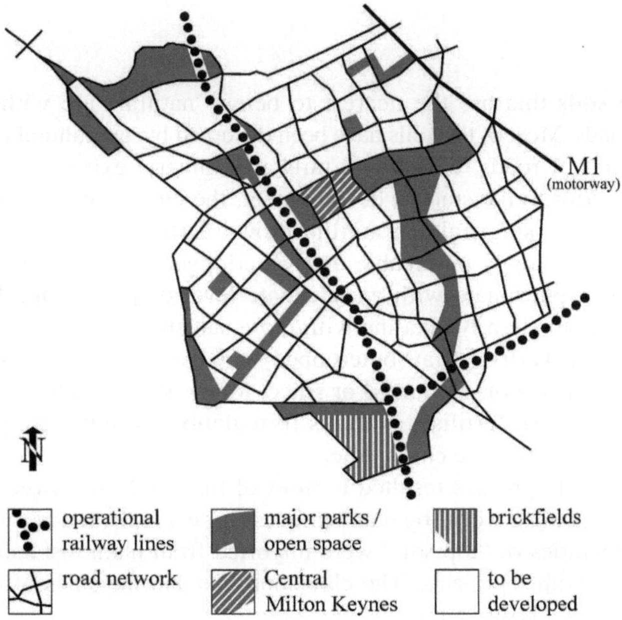

Fig. 2 Diagrammatic development proposals ca.1970

Topography

The general area in which Milton Keynes is located is relatively flat. The city itself has a gently undulating topography formed by two broad and shallow valleys separated by a central ridge, which is "cut" by small re-entrant valleys. The northern boundary is formed by the River Ouse. There are about 60 m between the lowest and highest points of the city. Spoil from excavations was used to create artificial hills to add interest to the form of the physical landscape.

Geology

Most of the surface area of the city comprises shrinkable clays. The solid geology is predominantly Oxford Clay, which extends for considerable distances to the east and west. Jurassic limestone outcrops on the southern edge of the Ouse Valley. The drift geology comprises mainly a variable thickness of calcareous Boulder Clay, Head and glacial lake deposits. First and second river terrace gravels and alluvium occupy the valleys. Greensand and chalk outcrop to the south with limestone and ironstone to the north.

Soil

Probably the soils that are the nearest to being "natural" are within the three largest woodlands. Most of the soils have been disturbed by agricultural management, the construction of roads, canals and railways, mineral extraction and "urban" development. Prior to the start of development, the city comprised four types of clay soil, including stagnogleyic argillic brown earths and pelo-alluvial gleys, both of which are slowly permeable and waterlogged in winter. The soils were generally low in phosphate with an adequate level of potassium. The pH was generally neutral to slightly alkaline with more calcareous soils occurring where the limestone and Oxford Clay outcropped or were close to the surface. Very localised acidic conditions occurred or since. Many of the fields were subject to annual applications of fertiliser (various formulations of nitrogen, phosphorous and potassium) and in some cases lime.

The development process resulted in most of the soil being severely disturbed with large areas being covered by hard surfaces (for example, roads and buildings). Substantial quantities of "top soil" were imported from unknown sources for use in landscape and other projects. The clay spoil tips and the sides of the clay pits comprised exposed Oxford Clay.

No known systematic or other soil surveys are known to occurred before or since the start of the development.

Agriculture and Agricultural Land Quality

Virtually, all of the agricultural land was Grade 3 (average) on the national agricultural land classification, with localised areas of Grade 2 and larger areas of Grade 4. Most of the land was used for sheep and cattle grazing and the production of wheat and barley.

Drainage

Virtually, all the land comprises highly impermeable clays and consequently the land drainage was very poor. The natural drainage of the area is divided about equally between two catchments, which are drained by two "main" watercourses (River Ouzel and the "Loughton Brook," which has different names along its length) and their tributaries. The water flows northwards into the River Ouse, which eventually flows into the North Sea. Many of the agricultural fields were under-drained and bordered by ditches. Other surface run-off was and probably still is discharged to the canal and a disused gravel pit.

The River Ouse flows through the city for about 12 km at an average flow of 2.5 cumecs (=2.5 $m^{-3} s^{-1}$). The next largest river is the Ouzel of which 8 km is within the city. In the late 1960s the average flow was about 1.6 cumecs. The third largest is the Loughton Brook of which about 12 km is within the city; the "bank full" channel width varies throughout its length. The Simpson and Broughton Books (about 4 and 3 km long, respectively) are tributaries of the River Ouzel. At the time of designation, the Ouse and Ouzel rivers were subject to organic pollution. As described later, the construction of the city resulted in the River Ouzel and the Loughton Brook being subject to considerable changes including bank re-enforcement and re-alignment.

From the land drainage point of view, the location of the city on impervious clays at the top of a major catchment is undesirable. It would have been preferable (from a drainage perspective) to build it on the Greensand to the south, which would have allowed soakaways to be used. However, as will be described, it was necessary to dispose of the run-off by positive drainage to a series of balancing lakes. A new sewage works was built to treat foul sewage, which was then discharged to maturation lagoons outside the city.

Climate

The annual mean precipitation varies between 508 and 635 mm. The annual mean temperature (reduced to sea level) is 10°C; the highest (28°C) occurs between June and August while the lowest (0.5°C) occurs in January/February. The prevailing wind is from the south-west; the velocity is generally 30 km h^{-1}, rarely exceeding 70 km h^{-1}. The average maximum amount (9 h) of sunshine occurs in June, the lowest (1 h) in December. There are no extremes in the micro-climate.

It was recognised that the development would influence air movements, local temperatures and the patterns of cold air drainage and frost pockets. However, there are no known data to indicate how the development has influenced the city's climate – for example, whether a "heat island" has established. As a matter of principle, it is likely that the conversion of the rural environment to an urban one will have had some effect on the micro-climate.

Air and Water Quality

Prior to the designation of Milton Keynes some of the established urban developments were subject to Smoke Control Orders. Smoke, ozone and sulphur dioxide levels were at acceptable levels in the 1970s. Such air pollution as there was mainly originated from long distance transport. A lichen survey undertaken in the early 1980s indicated that the air quality of the city was five, based on a scale of 0–10, where 10 was no pollution; 5 is equivalent to a winter SO_2 level of 60 $\mu g\ m^{-3}$.

The brick-making process in the south-west of the city resulted in the discharge of fluorine, which can be a major pollutant and could be a major problem to people and plants because the prevailing south-westerly winds would blow the discharge across the city. Assessments made in the mid-/late 1970s using lichens and bryophytes showed that there was a very low level of local aerial contamination but only sufficient to affect plant species that are sensitive to fluorine. Brick production eventually stopped in the mid- to late 1990s, resulting in the cessation of fluorine emissions.

Although there was no problem with heavy metal contamination from motor vehicles, allotments were not located close to what was to become the grid of major roads.

Up to the construction of a new sewage works in the mid-1970s, the upper sections of the Rivers Ouse and Ouzel were substantially polluted by discharges from the "combined" sewage works that existed prior to that time. The new sewage works produced a strong localised smell, which the landscape architects tried to mask by the dense planting of aromatic varieties of *Populus* spp., for example, *P. trichocarpa.* Urban run-off and the application of fertilizers also contributed to the pollution loading of various still and moving water bodies.

History

10,000 BC to AD 1000

The first recorded hominid activity occurred during the Palaeolithic when most of Britain was covered by deciduous forest. Human activity continued throughout the Mesolithic with the first permanent settlement occurring in the Neolithic. The first significant settlement took place about 1720 ± 80 BC (Bronze Age). The early activities and subsequent settlements were mainly associated with the river valleys, but in the Bronze Age, there was some movement from the lighter sands and gravels of the valleys to the heavier clay soils of the higher ground. Seven substantial settlements were built between 700 and 100 BC. *Quercus, Crataegus* and *Corylus avellana* were present in the second century BC. In the Late Iron Age the area was settled by people from northern France. The Romans who invaded Britain in the mid first century AD constructed a major road from London to North Wales; the section through Milton Keynes remained as part of the national highway network until the mid-twentieth century. During Roman times, the city area contained two to four villas and 26 "native" settlements with a population of 900–1,300 people – a small town occupying 7.5 ha with a population of 1,200 people was built just outside of what is now Milton Keynes.

The Romans developed an agricultural economy, with cereals including *Triticum* spp. (Emmer and Spelt wheats), *Hordeum* sp, *Daucus carota, Brassica oleracea, Pastinaca sativa, Coriandrum sativum* and fruit trees including species/varieties of the genera *Pyrus, Malus, Castanea, Morus* and *Juglans*. In addition, it is likely that there would have been fodder plants for the cattle, goats and horses and plants for producing dyes and for the making of textiles. The animal remains indicate that areas of woodland were present during this time.

The Saxons, who replaced the Romans, inherited a landscape that had been managed for agriculture for 1,000 years. The Saxons abandoned large areas of agricultural land, which was eventually colonised by scrub and then woodland. Nine settlements were established during the Late Saxon/Early Medieval times, including all the villages that existed in 1969 (and still do).

AD 1000 to AD 1500

The Medieval landscape of Milton Keynes comprised 20 settlements set in the "open field system" of agriculture. A large area of parkland was recorded in the southern part of the city in the fourteenth century. There were five woodlands (of which three still exist); coverts (small woodlands) were planted to encourage game for hunting. Several fishponds (one of 4.8 ha) and moats were created. The late twentieth century road network was established during this period. Several monastic houses were established in the twelfth century; in the fifteenth/sixteenth centuries many of the villages were totally or partially abandoned – some of the depopulation being caused by land enclosures.

1500–1967

Much of the land was enclosed in the 1500s mainly for sheep farming with more large-scale enclosures occurring in the seventeenth to nineteenth centuries. Hedges and/fences (with or without ditches) were used to demarcate field boundaries. In the eighteenth century two woodlands were recorded; neither of them existed in 1970. It appears that there was very little change until the major transport infrastructure was constructed through the city during the late eighteenth/early nineteenth centuries. During this period 21 km of canal and the main railway line from London to North Wales were constructed through what is now Milton Keynes. The development of the railway industry resulted in the creation of a new town and the expansion of an existing town. By the mid-nineteenth century the city area comprised 16 scattered settlements – four small "towns", eight villages and five hamlets. The large-scale extraction of clay for brick-making started in 1894.

In 1962, following assessments of population growth and housing need the Local Planning Authority envisaged the need to accommodate 250,000 people by the end of the century. In 1967, and following Government Inquiries to examine the need for the creation of a new town, the Government approved the designation and

Fig. 3 Typical rural landscape in and adjacent to Milton Keynes before the start of development

construction of Milton Keynes and established Milton Keynes Development Corporation (a Government Agency) to build it. About this time the population of "Milton Keynes" was 39,050, of which 36, 810 lived in the four small towns. This gave a population density of just over four people/ha (Fig. 3).

1967–1971 (Planning and Design)

The overall planning and design of the city occurred between about 1967 and 1971 and is contained in The Plan for Milton Keynes, which was published in 1970. The Plan provided for the development of housing, employment and recreation for 250,000 people, which included the existing population, the incoming population (mainly from London) and the natural growth they produced. The Master Plan anticipated that the development would start in 1971 and be completed in 1991, so increasing the population from 40,000 to 250,000 in 20 years (an average of ca. 10,000 people/year). This would involve converting mainly agricultural land to housing, employment, associated infrastructure and parks at an average rate of ca. 400 ha (4.0 km^2) per year. The total cost of the development was estimated at £700 million (£7.5 billion at 2009 prices), 70% being provided by the public sector and 30% by the private sector. The Plan set out six major objectives:

(a) Opportunity and freedom of choice.
(b) Easy movement and access.
(c) Balance and variety.
(d) The creation of an attractive city.
(e) Public awareness and participation.
(f) Efficient and imaginative use of resources.

The fundamental design was based on a network of roads aligned east/west and north/south to create blocks of land about 1.0 km^2, known as grid squares, see Fig. 2. Most of the squares were allocated for housing, others for employment and still others as parks and open spaces. The city centre was to occupy ca. 160 ha in the centre of the Designated Area. In addition to the broad policies contained in the Master Plan detailed design "policies" were devised for various elements of the city; however, by the late 1970s most of them had been abandoned or changed significantly. The land uses in 1970, as proposed by the Master Plan in 1970 and as revised by the Development Corporation in 1990 are given in Table 1.

In 1992, the Development Corporation published its estimated end date land budget, which is given in Table 2. The "end date" is not given. The average density of dwellings at that time was 27 dwellings/ha. It is likely that post 1992 projections will show a decrease in the area of open space and an increase in the area of development.

Table 1 Land uses of Milton Keynes designated area, actual in 1970, as proposed in the Master Plan (1970) and as revised by the Development Corporation in 1990

Year	1970		1990	
	Proposed in the Master Plan		Development Corp. Revision	
Source	ha	%	ha	%
Housing (gross)	4,700	53	3,690	42
Parks and other open space	1,160	13	1,970	22
Roads and railways	1,000	11.3	1,250	14
Industry, shops, offices	800	9	1,220	14
Reserve Land	297	4	Not available	
Social/education	297	4}		
Centres/health	260	3}	475	5
Brickfields	240	2.7	275	3

Table 2 Land budget for the completed city. Development corporation 1992

Land use	ha	%
Gross residential (including local centres and local open space	3,690	41.5
Gross employment	1,060	12.0
Central Milton Keynes (city centre)	160	2.0
Other (facilities services and utilities outside the city)	475	5.5
Brickfields	275	3.0
Roads and reservations	1,250	14.0
Parks and open spaces	1,970	22.0

The overall landscape policy for the city can best be described as "amenity forestry". The Master Plan committed the Development Corporation to protecting the woodlands and as many trees and hedgerows as possible together with a programme of extensive tree planting. Trees became the landscape/environmental icon resulting in Milton Keynes calling itself "The City of Trees". Now and most appropriately there is a tree cathedral. A guiding concept was the continuity of green space throughout the city so that, for example, it would be possible for a person to drive a car through the city without realising they had passed through an urban area.

Except for general tree and shrub planting and the provision of local open space and play areas there were no specific policies for the landscaping of residential areas. In order to reduce journey time to and from work the employment areas were distributed throughout the city and close to residential areas. There was no overall landscape strategy for these areas although some were located adjacent or close to parks. The Corporation "designed" the tree and shrub planting and maintained it. The schools were designed and built by the County Council; consequently the Corporation had no influence over the design of the buildings or the landscaping.

The Master Plan provided for a network of major roads with a total length of about 160 km and a major new national road. The "grid roads" (as the network was called) were bordered on each side by a 4.5 m wide grass verge and a landscape reservation (in reality tree plantations) that varied between ca. 16 m and ca. 36 m wide. This would provide 1.3 km^2 of grassland (excluding the central reservations of the dual carriageways) and 7.2 km^2 of woodland, which together would occupy about 20% of the city. Landscape design policies were developed for footpaths/cycleway network and the canal. The former, which involved the planting of five species of *Tilia* with an understorey of shrubs on both sides of the "path" was never implemented. It was replaced by shrub planting, most of which was removed at the request of the police for safety reasons. As will be described later, only a few hundred meters (at most) of the canal landscaping policy were implemented.

A hierarchy of parks was established; local parks to be provided at 0.6 ha/1,000 people – usually between 1 and 2 ha per residential grid square, district parks to serve 15,000 people and 1,160 ha of linear parks, which are mainly the floodplains of the river valley – because the land could not be built on. The three largest woodlands were also included in the linear parks. Subsequently the area of the linear parks was extended to 1,650 ha. The Master Plan also proposed the provision of sports grounds at the ratio of 1.0 ha/1,000 population and five golf courses (a total of about 250–300 ha – however, only two have been built). Allotments (or leisure gardens) were to be provided at 0.5 ha/1,000 population.

The Plan was prepared without virtually any ecological information or consideration of the contribution it could make to the detailed planning, design and management of the city, not only in terms of nature conservation but also the wider role of ecology in urban development. The only reference to nature conservation is one short paragraph, which states the obvious, namely that the development will modify or destroy many wildlife habitats in and on the edge of the city and that the Nature Conservancy (now Natural England) has drawn attention to some important (but unidentified) sites which should be evaluated and safeguarded.

1971 to Present (Detailed Design and Implementation)

The detailed planning, design and construction, which started in 1971 had to cope with many administrative changes of Government, radical changes of Government policy (not least the change from public to private finance/housing provision) and the re-organisation of local government and water/drainage authorities. For example, in the beginning the Corporation was its own drainage authority, but having gone through two national re-organisations, it is now the responsibility of two – one a Government Agency and the other a private company.

In 1992, the Development Corporation published a Planning Manual, which makes only two passing references to ecology and nature conservation and no references to biodiversity, sustainability, climate change, global warming or the use of natural resources. In 1996, Milton Keynes Council sought "*ex post facto*" to establish a wildlife corridor system throughout its administrative area (of which the city was a part) although such a network had been established (by default) in the 1971 Master Plan.

On 31 March 1992, the Corporation was incorporated into another Government Agency, which was charged with disposing of the Corporation's assets and responsibilities. Some were transferred to the Borough Council or to Trusts. Prior to it being abolished the Corporation established the Milton Keynes Parks Trust, a charitable Limited Company (and therefore publicly unaccountable) formed to manage the larger parks and open spaces. The objective was to prevent the Local Authority having responsibility for managing the parks and open spaces that the Corporation had established.

Prior to its dissolution the Corporation had obtained ownership of a substantial area of land and had obtained planning permission for much of it from the Government. The Corporation and its successor also provided much of the basic infrastructure, which formed the structure for much of the subsequent development. Consequently, although the local authority took over many of the functions of the Corporation, its role as the planning authority for Milton Keynes remained small.

Since the mid-1970s residential and other developments have occurred in the small towns and villages that surround Milton Keynes, including areas adjacent to it. As a result, the urban area now extends beyond the city boundary.

Flora

Introduction

The botanical history of Milton Keynes is poor, the last flora, which included the area of what is now Milton Keynes was published in 1926. There were no further publications until 1975, when the Milton Keynes Natural History Society published a small number of papers in its very short-lived Journal. There was nothing more until 2000, when the Society published a check list of plants recorded in "Milton

Keynes" between 1987 and 1999. The list, which is derived from a 2×2 km grid survey, relates to the administrative area of Milton Keynes Council (the local authority), which covers 309 km^2 (including the 90 km^2 of Milton Keynes city). With the exception of two small towns and a few villages, the 219 km^2 outside the city is rural. In 2005 Maycock and Wood published a check list of the flora of Buckinghamshire (the county that includes Milton Keynes), which makes *ad hoc* references to the presence of species in the city. The nomenclature follows that in the publications listed in Further Reading.

In August 1972, the Development Corporation appointed an ecologist with a very wide and impossible brief, part of which included undertaking botanical surveys and advising on planning, design and management issues. The first botanical surveys were undertaken in 1973 and continued annually until the 1983 field season. Further, *ad hoc* surveys were undertaken between 1984 and 1992. The results of no more than 50% of the surveys were made publicly available in a series of mainly simplistic reports known as "Ecological Studies in Milton Keynes", which included not only botanical studies – 85 were published between 1973 and 1983 and 35 between 1983 and 1992. The Development Corporation, its successors and the relevant Government Departments have stated that they do not intend to fund the detailed and comprehensive analysis of the data obtained between 1973 and 1992.

In 1992, responsibility for the management of most of the parks and larger open spaces was transferred to a charitable Trust, the Milton Keynes Parks Trust. The occasional botanical surveys carried out by the Trust are mainly restricted to angiosperms of the old and new woodlands. To date the Parks Trust has published 35 reports, not all of a botanical nature.

Geographical Flora

In late 1972, consideration was given to mapping the distribution of selected plant and animal taxa on a 2 km grid and to undertaking a habitat or vegetation mapping exercise. Both projects were abandoned because the start of large-scale development earlier that year and the major changes that would continue in the ensuing years would be too fast (given the resources available) making the exercise impossible. As an aside, it is interesting to compare this situation with that of archaeology, where the Corporation employed two senior archaeologists, several assistants and many (may be up to 100 temporary) staff to undertake "rescue archaeological" work.

Ecological Flora

In the circumstances the most appropriate approach was to establish an "ecological flora", which was to be achieved by undertaking botanical surveys of one or more sites or habitats.

Species Check Lists

In 1979, a check list of the plant species that had been recorded in the city between 1968 and 1978 (inclusive) was compiled using information from *ad hoc* records and ecological surveys undertaken up to 1978. For various reasons, the exercise was unsatisfactory and unreliable.

For reasons that are described later, neither the 2000 nor the 2005 publications are especially helpful in understanding the structure and distribution of the flora of Milton Keynes. In addition, because Milton Keynes is adjacent to two counties and close to another, Buckinghamshire is not an appropriate geographical context in which to assess the flora of the city.

The lists were compiled for different geographical units of different sizes using different methods and different amounts of effort. Nevertheless, some of the quantitative information in relation to bryophytes is interesting, for example, 104 moss and 17 liverwort species were recorded in the city between 1968/78 compared with 122 and 13 in 1987/99 recorded in the Milton Keynes Council area, which is x2.5 larger than the city.

It is not possible to list the fifty most frequent angiosperm species because since 1972 the effects of development will have caused the list and the sequence within it to change on at least an annual basis. Certainly, such a list compiled in 1971 would be substantially different in content and frequency than one prepared in 1981, 1991, 2001 and 2011. However, it is reasonable to assert that the lists would comprise a relatively high proportion of ephemeral/ruderal species that are characteristic of open disturbed ground such as *Plantago major, Convolvulus arvensis, Polygonum aviculare* agg., *Lepidium draba, Papaver* spp., *Chenopodium* spp., *Capsella bursa-pastoris, Kickxia* spp., *Lactuca serriola* and *Fumaria* spp. and herbs and grasses that are typical of dry to damp neutral, mesotrophic soils; for example, *Trifolium pratense, Cirsium arvense, C. vulgare, Artemisia vulgaris, Arrhenatherum elatius, Dactylis glomerata* and cultivars/strains of *Lolium perenne, Poa pratensis, Agrostis capillaris, Trifolium repens* and other species used in horticultural or agricultural seed mixes.

Taxonomic Divisions

Algae

No information about the algal flora was known before 1973, when surveys of the existing and new lakes started – no studies have been made of the terrestrial algae. The surveys carried out during the following 10 years showed that:

1. *Chara/Nitella* "meadows" occurred extensively on the beds of the clay pits and some of the newly constructed balancing lakes.
2. The more common of the filamentous algae were those associated with eutrophic conditions, the most abundant being *Entermorpha intestinalis* and *Cladophora* spp.
3. Phytoplankton was the most studied.

One hundred and sixty-three taxa in ten Classes have been recorded. Bacillariophyceae (59), Chlorophyceae (54), Cyanophyceae (16), Euglenophyceae (6), Cryptophyceae (6), Xanthophyceae (3), Chrysophyceae (3), Rhodophycaea (2) and Dinophyceae (2). Because of the difficulties in identification many taxa were only recorded to the generic level and recorded as "sp. or spp." Therefore, the number of taxa of and below "species" rank was likely to be substantially higher suggesting that the total number of taxa is likely to exceed 300.

Bryophytes

Prior to 1973 there were few records of the bryophytes in the city. The bryophyte surveys undertaken after 1973 were restricted to specific habitats – churchyards, two disused brick tips and pits, two woodlands, the disused railway, selected road verges and the canal; they are also referred to in other surveys. However, there is insufficient information to provide a general description of the bryophyte flora. The number of bryophyte species recorded in different habitats is given in Table 3.

Pteridophytes

The city contains sparse populations of only 12 pteridophyte species, a situation that is likely to have existed for decades if not centuries. *Equisetum arvense* was and probably still is the most common of the Equisetaceae. The next most frequent species is *E. palustre* with *E. telmateia* and *E. fluviatile* occurring rarely. *Asplenium trichomanes* and *Polypodium vulgare* occurred in walls, the former at only one site. *Pteridium aquilinum* is dominant over a relatively large area of a railway cutting but very rare elsewhere. *Dryopteris dilatata, D. filix-mas* and *Athyrum felix-femina* mainly occurred in the three major woodlands. *Ophioglossum vulgatum* occurred rarely. *Polystichum aculeatum* is only known from one locality.

Table 3 Number of bryophyte species recorded in different habitats

Woodlands	43
Brickfields	31
Disused railway line	52
Road verges	28
Canal (terrestrial and aquatic habitats)	49
Churchyards	58

Gymnosperms

Between 1936 and 1940 a small part of the largest woodland was partly clear felled and replanted, in part, with *Picea abies*. In the 1970s the woodland was underplanted with a variety of conifers including *Larix* spp., *Pseudotsuga menziesii, Thuja plicata, Chamaecyparis lawsoniana, Picea abies* and *Tsuga heterophylla.* From 1959 to 1965 one of the other woodlands was partly felled and planted with *Picea abies, Chamaecyparis lawsoniana* and *Tsuga heterophylla.* Gymnosperms also occur sporadically in some of the older, established parks and grounds of former large houses, for example, *Pinus sylvestris, Ginkgo biloba, Sequoiadendron gigantum, Metasequoia glyptostroboides, Cedrus libani, Larix* spp. and *Taxus baccata.* X *Cupressocyparis leylandii* was and continues to be common hedge species in residential areas.

Despite the Corporation's initial reluctance to plant conifers in landscape and forestry schemes because "evergreens" are alien to the lowland landscape of England, in later years many taxa were to pass through the Corporation's nurseries; they include *Cedrus* (5 taxa), *Pinus* (15), *Picea* (5), *Chamaecyparis* (16) and *Juniperus* (13). *Taxodium distichum* has been planted in the new parks.

Angiosperms

Prior to the start of extensive tree planting in the 1972/73 planting season by far the greatest number of mature and semi-mature trees and tree species occurred in the hedgerows, the most abundant taxon being *Ulmus procera.* The next most frequent hedgerow trees were *Fraxinus excelsior*>*Quercus robur*>*Salix* spp.>*Acer campestre* they are also the four most frequent tree species in the larger woodlands. The first two were suffering considerable crown die back – a feature of both species in the countryside as a whole.

The situation in respect of *Ulmus* changed rapidly because over the following few years all the mature and semi-mature *Ulmus* (*procera* and other taxa) were killed by *Ophiostoma novo-ulmi.* However, most of the plants that were not destroyed by the development survived as suckers. This loss had implications for design and construction because the constraints imposed by the mature trees (and their roots, etc.) no longer existed. The only two veteran trees in the city were also *Ulmus procera;* despite injections to try to save them they both eventually succumbed to the fungus.

Pollarded *Salix fragilis* line both banks of the River Ouse and occurred sporadically elsewhere. Because many had not been pollarded since the 1950s and 1960s the crowns had become too heavy resulting in the stems (which were hollow) collapsing. In order to prevent further loss a re-pollarding programme was instituted in the mid-1970s. The crowns of the trees often supported "epiphytes" such as seedlings and saplings of other tree and some shrub species, including *Rubus fruticosus* agg.

In the early 1970s and in order to exploit the opportunities to increase the population of the nationally rare taxon *Populus nigra* ssp. *betulifolia* (which did not occur in the city but is likely to have done so), cuttings were taken from a female tree in an adjacent town and male cuttings from trees relatively close to the city. The cuttings were propagated by the Corporation and planted out in the ratio one female to five male plants.

In order to provide wood for the production of "Milton Keynes Cricket Bats" *Salix alba* var. *caerulea* was planted in the Ouse Valley.

Between its first tree planting season in 1972/73 and its last in 1991, the Development Corporation planted several hundreds of thousands if not millions of trees and shrubs of about 1,800 taxa (according to the Corporation's nursery records). However, for reasons that are considered later the list is difficult to interpret. Some of the most frequently planted genera of tall trees include *Acer* (43 taxa), *Prunus* (64), *Sorbus* (35) and *Populus* (31). The most frequently planted small tree/shrub genera include *Malus* (48 taxa), *Salix* (45 – including some tall tree taxa and probably many hybrids), *Cotoneaster* (48) and *Ilex* (18). It is misleading to read too much into the number of taxa used whether in total or within a genera. While the total species-richness may be high, the relative abundances may tell a different story.

As with the trees, most of the shrubs occurred in the hedgerows and woodlands with secondary abundances on disused clay workings and railway embankments. The most abundant species on neutral soils included *Crateagus monogyna, Prunus spinosa, Sambucus nigra* and *Salix* spp. The species found on calcareous soils were similar to those that occurred on neutral soils plus the calcicole species *Euonymus europaeus, Viburnum lantana, V. opulus, Rhamnus cathartica* and *Ligustrum vulgare.* The acidic species *Cytisus scoparius* and *Ulex europaeus* occurred occasionally, regardless of soil type.

Both *Crataegus monogyna* and *C. laevigata* occur in the city, the latter mainly in the three old woodlands and occasionally or rarely in some of the hedgerows. Most plants were recorded as *C. monogyna* and although no detailed surveys were carried out, general observations indicated that although *C. monogyna* was much more abundant than *C. laevigata;* the most abundant taxon by far was the hybrid *C.*x *media.*

The gradual removal of most of the field hedges and areas of scrub coupled with the landscaping resulted in a radical change in the relative abundance of shrub species. The native species of the previously existing habitats were "replaced" with non-native genera of which the following were the most frequent, *Berberis* (33 taxa), *Euonymus* (20), *Viburnum* (24), *Rosa* (201), *Cytisus* (22), *Escallonia* (17) and *Syringa* (20) being densely planted in housing and employment landscape schemes.

In 1972, at the start of the development the most abundant herbs and grasses were those that are typical of arable and other disturbed ground conditions and pastures on mesotrophic, dry-damp soils. The species of disturbed, open soils (excluding mineral workings) included *Polygonum aviculare* agg., *Plantago major, Poa annua, Convolvulus arvensis* and *Aethusa cynapium.* The grassland species included *Lolium perenne, Poa pratensis, P. trivialis, Dactylis glomerata, Arrhenatherum elatius, Ranunculus acris, R. repens, R. bulbosus, Plantago lanceolata, Lotus corniculatus, Bellis perennis, Taraxacum officinale* agg. and *Trifolium dubium.*

The next most abundant species were those of calcareous soils, which occurred on the Oxford Clay brick workings and where Jurrasic limestone was near the surface or outcropped. The species in these areas included *Ophrys apifera, Briza media, Bromus erecta, Blackstonia perfoliata, Trifolium campestre, Lotus glaber* and *Cirsium acaule.*

Although most of the arable land and pasture were developed, relatively large areas of the original pasture (generally dry to damp mesotrophic grassland) have been retained mainly within the parks. In addition many new areas of mesotrophic grassland have been created, for example, on the road verges, the lawns of houses, open spaces and in the wet/dry balancing lakes.

The development or reclamation of the disused brick works and the planting of some of the wider road verges with trees has resulted in the loss of most of the calcareous flora.

Although no statutorily protected plant species or species of nature conservation concern occur in the city, some species of local interest were found post-1972; however, it is likely that they were present for a long time before then. They are:

1. *Ulmus* spp.
   The elm flora of Milton Keynes and District was of considerable botanical interest and importance. The city lies on the transition between three elm zones:
   (a) *Ulmus procera,* which extends from the Thames Valley north to north Buckinghamshire.
   (b) Old forest edge of north Buckinghamshire – of particular interest in relation to *U. minor.*
   (c) Grassland area, where *Ulmus* is rare.
   In addition to the dominance of *U. procera,* the elm flora included *U. glabra, U. plotii* and many clones of *Ulmus minor* (a particularly special clone). In 1973, a survey was undertaken of the elm flora of 36 villages in and adjacent to Milton Keynes. Cuttings of the *U. minor* clones were taken and propagated in the Corporation's tree nursery for planting in the city when the ravages of Dutch Elm Disease had passed. Cuttings were also taken to the Jodrell Laboratory of the University of Manchester.
2. *Glyceria fluitans* x *declinata* – recorded in the wild for the first time (in Britain) in one of the woodlands. Some plants were transplanted to similar situations in the largest woodland.
3. *Brachypodium pinnatum, sensu stricto* – recorded on a railway cutting close to what is now the main railway station.
4. *Carex nigra* – a nationally common taxon but very rare in and around Milton Keynes. It was found (with the next species) in the only acidic marsh in the city. A major road was re-aligned to ensure its survival.
5. *Eriophorum angustifolium* – a nationally common taxon but very rare in and around Milton Keynes. It was found in the only acidic marsh in the city. A major road was re-aligned to ensure its survival.
6. *Carduus tenuiflorus* – a species more typically found in coastal areas was recorded on the bank of the River Ouse.
7. *Bolboschoenus maritimus* – a relatively large stand was found in a disused gravel pit from where some plants were moved to other water bodies in the city.

8. *Ophrys apifera* – mainly occurred on exposed Oxford Clay resulting from clay extraction. The populations varied annually within and between sites, from only a few plants in some years to tens of thousands in others.
9. *Cladium mariscus* – was transplanted from a National Nature Reserve in eastern England and established on the margin of a balancing lake.
10. *Downingia elegans* – one of several alien species that appeared in areas of newly sown grassland suggesting that seeds of the species were in the seed mix and that it comprised seed from non-native sources. Despite what was said to be tight controls (for example, by specifying seed of British provenance) over the contractor it was always impossible to ascertain the sources of the seed parents. It is one of many species that is likely to have been introduced in grass seed mixes, for example, *Ranunculus marginatus* introduced into a nature reserve via a wildflower seed mix.

Non-Native Trees

A major botanical effect of Milton Keynes has been the substantial increase in species-richness, partly as the result of the creation of new habitat types but more significantly the planting of a large number of non-native taxa. Most, probably all, of the native species that were planted in and after the 1971/72 planting season are likely to have non-native genomes. This raises the issue of whether "non-native" should relate to the species or the genome. It also raises the issues of the geographical area in which a taxon is native and at what period in history. For example, *Pinus sylvestris* is a native British species that has been widely planted and grows spontantaneously on sandy soils. It is most unlikely to be a native species of Milton Keynes. It is also reasonable to speculate that some of the other native British species that occur in Milton Keynes only do so because of human activities.

Although it is probable that most of the plants have been propagated by cloning, it is reasonable to assume that the introduction of non-native genomes has, nevertheless, increased the genetic diversity although it may have resulted in a reduction in the native genetic diversity depending on what that was and the extent of breeding between native and non-native plants of the same taxon.

Of the ca. 1,800 taxa that passed through the Corporation's nurseries, most were non-native woody species. Only 71 taxa are native to Britain; under natural conditions many of these would have been rare and 18 are unlikely to have occurred at all.

This figure does not take into account inaccurate taxonomy, for example, wrongly labelled taxa and hybrids. It also excludes taxa that did not pass through the nursery administration, for example, those used in non-Development Corporation landscape schemes and substitutions made by landscape architects and/or contractors.

The Corporation's inventory does not include geophytes (for example, the many cultivars/varieties of *Narcissus, Crocus* and *Galanthus*), the species (and strains) in grass seed mixes and the aquatic plants excavated from water bodies outside Milton Keynes. In addition there are the "horticultural" species (trees, shrubs, herbaceous plants, fruit and vegetables) grown in the established and new gardens and allotments.

Fungi (including Lichenised Fungi)

Fungi

Prior to about 1970, there was no known information about the fungi of the city except for the occurrence of *Nectria galligena, Microsphaera alphitoides* and *Armillaria mellea* in the larger woodlands. *Ophiostoma novo-ulmi,* the fungus that causes symptoms called "Dutch Elm Disease" was rampant from the early 1970s until most of the mature and semi-mature *Ulmus* spp. had died. After that *Ulmus* became visibly less common but continued to colonise and establish as suckers, which grew about 5–6 m and a stem diameter of about 15 cm before being re-infected.

A survey of the macro-fungi in 1983 recorded 212 species, which is probably less than 50% of the likely number of macro-fungi that occur in the city. One hundred and forty-two species were Agricales; 29 were Aphyllophorales, 6 Gastromycetes and 35 were a mixture of Basidiomycetes, Ascomycetes and Myxomycetes. *Coprinus comatus* was the most frequent species followed by species such as *Hebeloma crustuliniforme* and *Pholiota ochrochlora.* The rare species *Hericium ramosum* was found only at a parkland site. The major fungal habitats were the deciduous woodlands, hedgerows and grasslands.

Lichenised Fungi

Nothing was known about the lichen flora prior to the start of development. Ninety-one species of lichen were found in 1981/82. The survey found that the distribution of lichens was significantly influenced by air quality and habitat degradation. Most of the saxicolous taxa were found on old walls and on gravestones while most of the corticolous species mainly occurred on mature, well-lit trees. The new substrates created by development operations supported few species; those present included *Lecanora dispera, L. muralis* and *Candelariella aurella.* The number of species found on different substrates is given in Table 4.

Table 4 Number of lichen species found on different substrates

Deciduous trees	32
Coniferous trees	2
Fence posts	7
Soil	4
Mortar and cement	38
Asbestos	19
Bricks	11
Sandstone	17

The more abundant species include *Lecanora conizaeoides* (which occurred throughout the city, especially in the residential and industrial areas), *Parmelia saxatilis* and *Hypogymnia physodes* (abundant in the less populated areas), *Candelariella vitellina, Lecanora dispersa, Calcoplaca aurantia, Rinodina subexigua, Verrucaria muralis, V. nigrescens, Xanthoria aureola* and *X. parietina.* Three species, *Cladonia digitata, C. ochrochlora* and *Rinodina exigua,* are rare in the region. Four species are non-native having been introduced for retail purposes for floral displays and other uses.

Habitats

The habitats can be divided into three categories. First, those that existed prior to 1971 and continue to do so in their original or only slightly changed form. Second, those habitats that existed in 1971 and which have been substantially changed and/or reduced in size or lost altogether and third, habitats that have been deliberately or inadvertently created as the result of development operations.

Woodlands: Large

The city contains three "large" woods, which are old (originating from the eleventh century) or even primary – in the case of the largest. Part of the larger wood was clear felled in 1936–1940 and planted with *Fraxinus excelsior* and *Picea abies*. In the late 1960s, the woodlands were reported to have been neglected for decades, although between 1959 and 1965 one of them was partly clear felled and planted with *Picea abies, Chamaecyparis lawsoniana* and *Tsuga heterophylla.*

The number of species of plants and fungi found in the woodlands is listed in Table 5.

The number of vascular plants has been recorded several times between the mid-1970s and mid-1990s. The figures in Table 5 are the lowest and highest

Table 5 Number of species of plants and fungi found in the woodlands

	Linford (40 ha)	Howe Park (25 ha)	Shenley (25 ha)
Vascular cryptogams	184–210 (270)	115–174 (218)	132–163 (176)
Bryophytes	33	36	No information
Algae	No information	6 (Genera)	No information
Fungi	114 {46}	46 {11}	No information
Lichens	29	23	13

Key: the number in (..) is the accumulated total. The number in {..} is the number of species found only in that woodland numbers recorded in the survey. The total number from all the surveys is given in brackets.

The soils are likely to be the most natural soils in the city. They vary from free draining loam to waterlogged gleys and include clay loams and mull gley and drier acid soil. In general terms the soil pH was 7.0. The main woodland type (as classified according to the British National Vegetation Classification) is *Fraxinus excelsior, Acer campestre* and *Mercuralis perennis* woodland. The most abundant canopy species are *Fraxinus excelsior>Populus tremula> Acer campestre> Quercus robur>Carpinus betulus. Malus sylvestris* occurs in the sub-canopy. The most frequent shrub species were *Prunus spinosa* and *Corylus avellana* with *Crateagus monogyna, C.laevigata, Ligustrum vulgare, Viburnum lantana, V. opulus, Salix* spp. and *Sambucus nigra*. The herb layer species include *Rubus fruticosus* agg., *Luzula pilosa, Geum urbanum, Glechoma hederacea, Urtica dioica, Anenome nemorosa, Milium effusum, Mercuralis perennis, Carex sylvatica,C.remota, Melica uniflora, Primula vulgaris, Dactylorhiza fuchsii, Ranunculus auricomus, Poa trivialis, P. pratensis, Festuca gigantea, Circaea lutetiana, Hyacinthoides non-scripta, Arum maculatum, Galium odoratum, Lamiatsrum galeobdolon, Viola riviniana, Veronica chamaedrys* and *Listera ovata.*

The wetter areas are dominated by *Populus tremula* with *Salix* spp., with a field layer comprising such species as *Carex strigosa, C. remota, C. pendula, C. paniculata, Deschampsia cespitosa, Filipendula ulmaria, Lychnis flos-cuculi* and *Ranunuculus ficaria.* The rarer species (in the context of Milton Keynes) were *Carex strigosa, Epipactis helleborine, Orchis mascula, Paris quadrifolia, Platanthera chlorantha, Ranunculus auricomus, Stachys officinalis, Lathyrus sylvestris, Sanicula europaea* and *Glyceria fluitans* x *declinata* (the rarest species found in the city).

The more common bryophyte species found included such species as *Acrocladium cuspidatum, Atrichum undulatum, Dicranella heteromalla, Fissidens bryoides, F. taxifolius, Pohlia nutans, Bryum capillare, Funaria hygrometrica, Mnium hornum, Thamnium alopercurum, Plagiothecium denticulatum, Lophocolea bidentata, L. cuspidatum* and *Pellia* spp. The fungi included *Microsphaera alphitoides, Armillaria mellea, Ophiostoma novo-ulmi, Xylaria hypoxylon, Daldinia concentrica, Mycena inclinata* and many other species. The lichen species included such species as *Lecanora conizaeoides, Lepraria incana, Cladonia chlorophaea, C. frimbriata, Hypogymnia phycodes, H. tubulosa, Cetraria glauca, Evernia prunastri* and *Parmalina sulcata.*

In 1972, the Corporation started a woodland “improvement” programme, which involved the restoration of the woodland for timber production, visual amenity and recreation. The latter involved re-establishing the ride (path/track) system, the creation of new rides (some ca. 15 m wide) and the construction of surfaced paths and a trim trail. The ditch systems were restored and over-deepened to improve drainage and tree growth. In order to make the woods more visually attractive and to increase their diversity they were planted with many non-native species including *Alnus incana, Amelanchier lamarkii, Eucalyptus gunnii, Larix* spp. *Nothofagus* spp,

Pseudotsuga menziesii, Quercus cerris, Q.rubra, Q frainetto, Thuja plicata, Chamaecyparis lawsoniana, Fagus sylvatica 'Purpurea', *Picea abies* and *Tsuga heterophylla.*

Woodlands: Small

At the time of its designation the city contained 31 small "woodlands" varying in size from 0.008 to 2.0 ha, 15 being about 0.5 ha. The total number of tree and shrub species found was 37, the number in each wood varying from 3 to 14. Thirteen of the 28 tree species were non-native including *Quercus cerris* and *Pinus sylvestris.* The most frequent tree species was *Quercus robur* followed by *Fraxinus excelsior.* The most frequent shrub species was *Crataegus monogyna;* other shrub species included *Viburnum lantana, Corylus avellana* and *Sambucus nigra.*

The number of herb layer species per woodland varied from 7 to 37, the total for all the small woodlands was 115, the five most frequent being *Urtica dioica>Poa trivialis>Stachys sylvatica>Galium aparine>Rubus fruticosus* agg. Ten locally uncommon species were found and eight species that Natural England considers to be indicators of ancient woodland.

Woodlands: New

Starting in 1972 woodlands (or more appropriately tree belts) of variable size, width, length and species composition were planted in visually "sensitive" areas (for example, sections of the city boundary) 10–15 years before any development was to occur in the general area. The tree species included native species such as *Fraxinus excelsior, Betula pendula, Salix fragilis,* and many non-native species for example *Acer platanoides, A. pseuoplatanus, Aesculus hippocastanum, Quercus rubra, Alnus incana, A.cordata, Larix decidua* and native and non-native *Populus* spp. The small trees and shrub layer species included *Sorbus aucuparia, Acer campestre, Crataegus monogyna, Corylus avellana, Prunus spinosa, Ligustrum vulgare, Salix* spp. and *Cornus sanguinea* and non-native species, for example, *Rosa mollis.* The herb layer varied in accordance with the management, which was usually the intensive use of a variety of herbicides. In general terms, the herb layer was dominated by tall species of neutral, nutrient-rich soils such as *Anthriscus sylvestris, Urtica dioica, Aegopodium podograria, Cirsium vulgare, C.arvense, Galium aparine, Elytrigia repens* and *Arrhenatherum elatius.*

In 1991, a deciduous woodland of about 11 ha was created on permanent pasture on the west side of the city. The pasture was bordered on three sides by a hedgerow and on the fourth by an ancient trackway. A mature hedgerow crosses the site. The canopy species comprised *Quercus robur* (8,000 plants), *Betula pendula* (4,000) *Carpinus betulus* (4,000). A survey undertaken at the time recorded 39 "field layer" species. The following year a seed mix comprising 20 species of wildflowers was sown. Two years later the number of field layer species had increased to 136. However, a survey undertaken in 2001 found four tree, 18 shrub layer and 122 herb layer species.

Scrub

Scrub habitats comprising native species were rare before 1972; they became more extensive for a few years after that as former agricultural land awaited development and then declined as more and more development occurred. The shrub species present were primarily *Crataegus monogyna/C.*x *media, Prunus spinosa, Rosa* spp., *Sambucus nigra, Rubus fruticosus* agg. and *R. caesius*. The scrub on the calcareous soils (for example, disused limestone quarries) contained such species as *Ligustrum vulgare, Acer campestre, Euonymus europaeus* and *Rhamnus cathartica.* Scrub habitats were considered to be an indication of neglect/dereliction and therefore undesirable. Consequently, strenuous efforts were made to remove them. An "interesting" area of scrub was formed by an extensive stand of *Fallopia sachalinensis,* which was eventually destroyed by a development.

Despite this "policy", about 0.25 ha of scrub habitat was created on an area of mown grassland using 2000 plants of nine native species, *Crataegus monogyna, Prunus spinosa, Cornus sanguinea, Viburnum lantana, Rosa* spp., *Sambucus nigra, Rubus fruticosus* agg., *Ligustrum vulgare* and *Rhamnus cathartica* in the ratio 45:20:10:5:5:5:5:1:1. The planting positions were generated from a table of random numbers. The spontaneous herb layer included such species as *Dactylis glomerata, Festuca rubra, Clinopodium vulgare, Silene dioica, Pimpinella saxifraga, Dactylorhiza fuchsiii, Primula veris* and *Hyacinthoides non-scripta.* On seeing this area from the train, the Corporation's Chairman threatened to have the scrub removed because it gave a "wrong image" of the city. On being told that the shrubs had been planted by children, he reluctantly changed his mind.

An experiment to thicken a hedge by natural colonisation rather than planting failed. Although a ca. 10 m wide strip of pasture adjacent to the hedge was fenced off, the anticipated colonisation did not occur.

Despite the opposition to scrub habitats comprising native taxa, the Corporation planted many shrubberies (*de facto* scrub) of varying size throughout the city (including housing and commercial areas) using many non-native species, as described previously.

Hedgerows

Hedgerows comprise linear scrub and tall grassland and usually a ditch. They vary in their age, species composition and management. Some originated from the primary woodland or were planted to form the boundaries of parishes or ancient track ways. The fields were enclosed between the sixteenth and nineteenth centuries, mainly the eighteenth – usually by hedges. In the nineteenth century they were planted to form the boundaries of roads, canals and railways. The management varied from unmanaged to severe annual cutting.

The retention and protection of hedges in developments were considered very important – as sources of established "mature" trees and shrubs rather than as hedgerows. Consequently a "hedgerow survey" was carried out by the Corporation's Forestry Officer in 1969. The survey was a visual one with the objective of determining the visually "dominant" tree and shrub species present to provide information to architects, landscape architects and engineers for incorporation into their designs. At the time of the survey, the dominant hedgerow tree was *Ulmus procera* and the next most frequent species were *Fraxinus excelsior* and *Quercus robur*. The most abundant shrubs were *Crataegus monogyna* and *Prunus spinosa.* Despite this policy and future considerations, very few (probably <0.1%) hedgerows or the individual components were retained.

As the result of research by Max Hooper (of what has become Natural England) in the 1960s/70s, the protection of "ancient hedgerows" gained considerable political weight and importance. Hooper devised a method for dating hedges based on the formula (99 x No. species in a 30-yard section) – 16 = the age of a hedge in hundreds of years. However, he cautioned that the method may not work in all areas, especially where multi-species hedges were planted and therefore before it could be applied to an area local correlation, studies were required. The political weight filtered down the Corporation and as a result detailed hedgerow studies were carried out in 1977 and 1979. The first study concluded that there were discrepancies of 200 years between known and predicted dates. The second study concluded that Hooper's equation was applicable to hedges on parish boundaries and bordering ancient trackways and that the presence of *Acer campestre* and *Corylus avellana* indicated that the hedge was probably of woodland origin. The survey found that the presence of *Acer campestre* in a hedge indicates that it may be >300 years old. *Cornus sanguinea* was rarely found in a hedge with less than five species suggesting that its presence may indicate an age of >500 years.

Twenty-eight tree and shrub species were recorded; the most common tree species were *Ulmus procera* (mainly as suckers), followed by *Fraxinus excelsior* and *Acer campestre. Crataegus monogyna* was the dominant shrub followed by *Prunus spinosa, Sambucus nigra* and *Rosa* agg. The tree and shrub species with the lowest frequency included *Corylus avellana, Rhamnus cathartica* and *Cornus sanguinea.* Seventy species of herbs and grasses were recorded, the ten most common being *Rubus fruticosus* agg > *Urtica dioica* > *Galium aparine* > *Cirsium vulgare* > *Anthriscus sylvestris* > *Cirsium arvense* > *Glechoma hederacea* > *Hedera helix* > *Hercaleum sphondylium* > *Tamus communis.*

Grassland

Although quite a lot of the grassland is developed on "ridge and furrow" and is therefore structurally quite old, the vegetation is young, most being established in the 1940s for cattle, sheep and/or horse grazing. Although not surveyed until after 1972, the grasslands appear to have been agricultural leys, which were created by the sowing of agricultural grass seed mixes dominated by strains of *Lolium perenne, L. temulentum* with other grass species and *Trifolium repens*. The grasslands were subsequently treated with fertilisers and herbicides and limed occasionally. The fields were also under-drained.

One hundred and fifty-nine angiosperms and 13 bryophyte species have been recorded in the agricultural grasslands as a whole. The more abundant angiosperms include *Lolium perenne, Alopercurus myosuroides, Festuca rubra, Phleum pratense, Dactylis glomerata, Holcus lanatus, Agrostis stolonifera, A. capillaris, Cynosorus cristatus, Trifolium repens, Taraxacum officinale, Rumex acetosa, Cirsium arvense, Plantago lanceolata, P. major* and *Achillea millefolium*. The less frequently occurring species included *Picris echioides, Lotus corniculatus, Hordeum secalinum, Leontodon autumnalis* and *Trifolium pratense*.

Some relatively small areas of the mesotrophic ridge and furrow grassland escaped the worst excess of agricultural management and retained a "semi-natural" vegetation, of which the following are some of the characteristic species, *Festuca rubra, Holcus lanatus, Poa pratensis, Bromus hordeaceus, Trisetum flavescens, Hordeum secalinum, Helictotrichon pubescens, Briza media, Achillea ptarmica, Cruciata laevipes, Phleum bertolonii, Tragopogon pratensis, Primula veris, Conopodium majus, Vicia sativa, Sanguisorba minor, S. officinalis, Lathyrus pratensis, Thalictrum flavum, Galium verum, Leucanthemum vulgare, Pimpinella saxafraga* and *Luzula multiflora*. The areas with wetter soils contained species such as *Alopercurus geniculatus, Carex hirta, Ranunuculus ficaria* and *Filipendula ulmaria, Epilobium hirsutum, Deschampsia cespitosa, Cirsium palustre, Stachys officinalis* and the pteridophyte *Ophiloglossum vulgatum*. Most of these areas were ploughed to remove an impediment to development.

Mesotrophic soils of some of the village greens and other areas of open space were drier, less nutrient-rich and supported larger populations of greater variety of species than occurred in the agricultural grasslands. The more abundant species in these areas included *Alopercurus pratensis, Trisetum flavescens, Cynosorus cristatus, Hordeum secalinum, Anthoxanthum odoratum, Pimpinella saxifraga, Lotus corniculatus, Trifolium pratense, Cerastium fontanum, Prunella vulgaris* and *Lathyrus pratensis*.

The calcareous grassland mainly occurred on the disused clay workings, the cutting slopes of the railway lines (operational and disused) and on the land reservations for pre-1972 road-widening schemes. The more abundant species included *Brachypodium pinnatum, Bromus erecta, Clinopodium vulgare, Cirsium acaule, Agrimonia eupatoria, Carex flacca, Daucus carota, Linum catharticum, Centaurium erythraea* and *Plantago media*.

Small patches of "acidic" grassland occurred sporadically. The species present included *Holcus mollis, Festuca ovina, Jasione montana, Rumex acetosella, Pteridium aquilinum* and *Succisa pratensis*.

Most of the agricultural grassland was converted to other land uses, mainly housing, and employment. Large areas were used for the construction of balancing lakes and Central Milton Keynes and for mineral extraction. In addition, some areas of botanically interesting grassland were destroyed by landscaping (actually tree planting). However, some of the mesotrophic grassland was retained as part of the linear park system, especially in the river valleys. In addition, new mesotrophic grassland was created as the result of construction of the wet/dry balancing lakes, creation of parks and open spaces, golf courses and road verges.

Land was taken out of agricultural use on an "as needed" basis, though because of changes in development programmes quite large areas remained fallow from a few months to about 2 years. Because of the lack of management the vegetation became temporarily dominated by species that are typical of tall grassland/herbs on dry to damp mesotrophic soils.

Although cattle and sheep grazing were retained during the construction of the city, the area gradually reduced until it was restricted to the major areas of open spaces such as parts of the river valleys and the wet/dry balancing lakes. However, even this eventually gave way to mowing and horse grazing. No studies have been carried out to determine whether the post-1972 changes have resulted in significant changes in the plant species to be found in the different grassland types. General observations suggest that although there has been substantial reduction in the extent of grassland since 1972, most (if not all) of the species remain.

Marsh

The city contains only one area of marsh, which occupies 1.3 ha adjacent to one of the watercourses. The substrate comprises organic-rich alluvial clay with thin seams of peat. The vegetation contained 131 species, which are predominantly species associated with waterlogged eutrophic soils, such as *Epilobium hirsutum, Angelica sylvestris, Carex disticha, Equisetum fluviatile, Galium palustre, Lotus pedunculatus, Filipendula ulmaria, Pulicaria dysenterica, Stachys officinalis, Deschampsia cespitosa* and *Cirsium palustre* with an "understorey" comprising lower growing species such as *Glechoma hederacea, Agrostis stolonifera* and *Caltha palustris.* The marsh contained two species of interest, namely, *Carex nigra* and *Eriophorum angustifolium.* Although both species are nationally common, the former is uncommon in the general area while the latter is rare in the county. As a consequence the proposed road that would have destroyed the marsh was re-aligned to avoid it.

Arable

Prior to 1972, the area of arable land was similar to that of the agricultural grasslands. The cereal crops comprised mainly varieties of *Triticum* with some *Hordeum.* The ruderal/ephemeral understorey species included, for example, *Polygonum aviculare* agg., *Convolvulus arvensis, Stellaria media, Coronopus squamatus, Solanum nigrum, Fallopia convolvulus, Kickxia elatine, K. spuria* and

Veronica persica. The fields were bordered by a 1–1.5 m wide "verge" of tall mestrophic grassland comprising species such as *Arrhenatherum elatius, Dactylis glomerata, Cirsium arvense, Senecio jacobaea, Lapsana communis, Conium maculatum, Solanum dulcamara, Chaerophyllum temulum, Urtica dioica* and *Hercaleum sphondylium.*

As development progressed, more and more land was taken out of agricultural production and used mainly for the construction of housing, employment areas, city centre, roads, educational establishments and the associated landscape. This habitat is now absent or virtually absent from the city. The species found in it are now rare and confined to other disturbed/open habitats.

Disturbed/Open Habitats

Prior to 1972, the disturbed/open habitats were restricted to mineral workings, arable fields and the road verges adjacent to the carriageway. From 1972 onwards, these habitats were to become the most extensive and dynamic. During most of the construction phase, there were extensive areas of unmanaged, open/disturbed ground, comprising one or a mixture of arable land taken out of agricultural use and awaiting development and exposed glacial deposits and Oxford Clay. The habitat was temporary, lasting for varying times between 3 months and about 2 years, with a mean of about 6–12 months. The flora "moved around" as construction works were completed and new sites started. Open/disturbed habitats also occurred in the woodlands associated with the construction of footpaths and ditches and in association with the construction of surfaced footpaths in the parks.

The newly exposed land was quickly colonised by many species that were previously relatively uncommon in the city. Because of the lack of competition, they were able to establish large populations, for example, the total area covered by *Conium maculatum* occupied several hectares. The other characteristic species included *Lactuca serriola, Papaver rhoeas, P. dubium, P. somniferum, Oenanthera* sp., *Galeopsis speciosa, G. tetrahit, Conyza canadensis, Cichorium intybus, Arabidopsis thaliana, Galium aparine, Thlapsi arvense, Malva neglecta, Anagallis arvensis, Odontites vernus, Pastinaca sativa, Reseda luteola, R. lutea, Tussilago farfara, Urtica urens, Chenopodium album, Atriplex* spp. and *Verbascum thapsus.* The lower growing species included *Valerianella locusta, Coronopus squamatus, Kickxia elatine, K. spuria, Vulpia myuros* and *Senecio viscosus.*

Ponds

All the ponds and lakes are man-made, the earliest dating from Medieval times. It is probable that the city area contained about 200 ponds prior to its designation although nothing is known about them.

As land was taken out of agricultural use for development, the ponds were filled in because they were considered a safety hazard. However, one or two adventurous planners and architects were able to circumvent the activities of the landscape managers and retain and enhance a small number of ponds in housing and shopping areas. Some ponds were retained because they were protected for archaeological reasons. The ornamental water features built around sculptures were designed and maintained to prevent plant growth.

A survey in 1984 identified 105 ponds within the city; however, it is likely that this is only 50–75% of the number present in 1971. Ninety-five of the ponds were subject to detailed biological surveys. The most abundant of the 15 submerged species found were *Elodea canadensis, Callitriche* spp. and *Ranunculus aquatilis* agg. together with two bryophytes – *Drepanocladus aduncus* and *Fontinalis antipyretica.* The floating-leafed zone contained eight species of which one or a combination of *Callitriche* spp., *Glyceria fluitans, Lemna minor* and/or *Potamogeton natans* were the most frequent. The ponds supported 50 emergent/marginal taxa, the ten most fequent being *Solanum dulcamara, Alisma plantago-aquatica, Ranunculus scleratus, Veronica beccabunga, Rorippa nasturtium-aquaticum, Glyceria declinata, Myosotis scorpioides, Mentha aquatica, Apium nodiflorum* and *Eleocharis palustris.*

One of the top ten ponds in terms of nature conservation value was a pond in a shopping centre that an architect had saved from destruction by the Forestry and Conservation Department and restored that as an architectural feature.

Moving Water

As described earlier the city contains several watercourses. The most abundant submerged and floating-leafed were *Potamogeton pectinatus, Ranunculus fluitans,* and *Nuphar lutea* and *Sagittaria sagittifolia,* respectively. The taller marginal/emergent species included *Butomus umbellatus, Sparganium erectum, Schoenoplectus lacustris, Glyceria maxima, Rorippa amphibium, R. nasturtium-aquaticum, Mentha aquatica, Juncus effusus, J. inflexus, Veronica beccabunga, Myosotis scorpioides, Carex riparia,* and *Apium nodiflorum* with an understorey comprising such species as *Caltha palustris, Galium palustre, Lychnis flos-cuculi* and *Ranunculus flammula.* The bankside vegetation comprised species that are typical of damp, neutral nutrient-rich soils, for example, *Eupatorium cannabinum, Elytrigia repens, Conium maculatum, Phalaris arundinacea, Deschampsia cespitosa, Picris echioides, Chenopodium* spp., *Tanacetum vulgare, Dipsacus fullonum, Arrhenatherum elatius, Epilobium hirsutum, Anthriscus sylvestris, Epilobium hirsutum, Scrophularia auriculata* and the parastic species *Cuscuta europaea. Clematis viticella* has been found close to the River Ouzel.

Between 1971 and 1983 about 4.5 km of the River Ouzel were re-aligned and the banks re-enforced as part of the construction of two large off-stream balancing lakes. Most of the course of the Loughton Brook was changed, the south-western section was widened in places as part of the linear park design to form "swales", sections have been "lost" by the construction of "on stream" balancing lakes while other sections have been straightened and re-enforced. The banks, margins and beds of the altered sections were colonised by plants from the unaltered sections and elsewhere.

Housing Areas

The nineteenth century housing density is relatively high while the density of the residential areas built in the twentieth century is relatively low, with only one or two high-rise residential blocks being built. No surveys of the gardens were carried out but general observations in 1972 indicated that most houses had front and/or rear gardens bordered by walls, fences or hedges, comprising, for example, *Ligustrum ovalifolium* or x *Cupressocyparis leylandii.* Most of the gardens comprise a lawn bordered by shrubs or flowerbeds and with the occasional tree, for example, *Fagus syvlatica* 'Purpurea', or *Prunus* spp. together with fruit trees including varieties of *Malus* and *Pyrus.* The lawns are dominated by grass species, such as *Lolium perenne, Poa* spp., *Festuca* spp. and *Agrostis* spp. with some dicotyledons including *Taraxacum officinale, Trifolium dubium* and *Bellis perennis*. The shrubs are mainly of non-native genera or species, for example, species and/or varieties of *Forsythia, Viburnum, Berberis, Pyracantha, Buddleja, Syringa vulgaris, Laburnum, Mahonia, Ribes* and *Rubus.* The borders contain *Erysimum cheiri, Aubretia deltoidea, Paeonia officinalis, Leucanthemum* x *superbum, Iberis umbellata* and varieties of *Delphinium, Iris, Antirrihinum, Viola* x *wittrockiana, Primula, Tagetes* and *Lupinus.*

The Master Plan proposed 80,000 dwellings on 3,700 ha (41% of the city), which gives an average plot size of about 460 m^2. The 1971–1992 average housing density was 27 dwellings/ha, with few (if any) residential developments of more than two to three storeys. However, the density ranged from five dwellings/ha (plots) to 65 dwellings/ha for rental housing. The lower cost housing was 30–45 dwellings/ha; higher cost housing density was 10–20 dwellings/ha.

The Corporation adopted a policy of retaining control over the landscaping and maintenance of the rental housing areas, which up to the late 1970s formed the majority of the new housing (Fig. 4).

The tenants were not consulted about the design and choice of species used in "their" front gardens. The landscaping varied within and between residential developments and between adjacent dwellings. In some situations, lawns were established using commercial seed mixes comprising strains of *Lolium perenne, Festuca* spp., *Poa* spp., *Agrostis* spp. and *Trifolium repens* with an individual small tree or shrub, for example, *Rhus.* Other gardens were densely planted with monocultures of low growing shrub species such as *Hypericum* spp., *Berberis* spp., *Lavandula* varieties or ground cover species such as *Cotoneaster horizontalis, Vinca* sp. and *Hedera helix.* Soon after planting many of the plants were found on the roadsides in and around the city while others were removed when the Government allowed tenants the right to buy their houses.

Fig. 4 Rental housing scheme

The verges of the service roads, squares and other incidental open space within housing areas were sown with a commercial grass seed mix (similar to that described in the previous paragraph) and planted with well-spaced trees such as *Tilia platyphyllos, T. cordata, Acer platanoides, A. saccharinum, Platanus acerifolia* and *Sorbus* spp. with groups of shrubs, for example, *Berberis verruculosa, B. julianae, Symphoricarpus* spp., *Lonicera pileata, Prunus laurocerasus* and *Cotoneaster conspicuus* 'Decora' in some places.

The landscaping of one of the early private housing schemes comprised *Rhus typhina* and *Cytisus* x *praecox* with *Parthenocissus quinquefolia* on the sides of the single storey houses. One of the earliest rental housing schemes was planted with *Symphoricarpus* x *chenaultii, Lonicera pileata, Prunus laurocerasus* 'Zabeliana', *Rosa* spp., *Platanus acerifolia, Tilia cordata* and *Acer saccharinum.*

Other species occurring in the new housing areas include *Cotinus coggygria, Cotoneaster* x *intermedia, C. wateri, Aralia chinensis, Eleagnus umbellata, Galanthus elwesii* and *G. woronowii, Parthenocissus inserta, Sorbaria sorbifolia, Euonymus fortunei, Lathyrus odoratus, Lonicera henryi* and *Stachys annua, Carthamus lanatus, Brassica rapa* ssp. *oleifera* (probably a relict of previous agricultural production), *Arbutus unedo, Vicia parviflora, Stachys annua, Cardamine bulbifera, Scorzonera hispanica, Sorbaria sorbifolia, Euonymus fortunei, Festuca brevipila, Erysimum* x *marshallii* and *Cytisus battanieri.*

In the late 1970s/early 1980s, the Government abandoned its public housing programme. Consequently, all future housing was built by private developers. This had two significant effects, first, there was substantially less landscaping and the owners of the houses were responsible for their own small front gardens (where they existed). The front gardens usually comprised a small lawn with one or two trees – *Salix babylonica* (or *S.* x *sepulcralis* or *S.* x *pendulina*) being a particular favourite.

The structure and species composition of rear gardens was always a matter for the tenant or owner. In general terms, they comprised a lawn with a small number of shrubs (*Chaenomeles speciosa, Forsythia* spp.) and annual bedding plants such as *Viola* x *wittrockiana, Primula* spp. and *Salvia* spp.

The domestic lawns and the verges of the service roads were eventually colonised by species typical of dry to damp mestrophic soils, such as *Plantago lanceolata, Ranunculus repens, Bellis perennis, Leontodon autumnalis, Trifolium dubium, Taraxacum officinale* agg., *Trifolium repens, T. pratense, Plantago major, P. lanceolata, Cerastium fontanum, Geranium dissectum, G. molle, Ranunculus repens, Prunella vulgaris* and *Hypochaeris radicata.*

Employment Areas

The few industrial and commercial areas that existed at the time of designation were stark and largely devoid of vegetation. As with the housing areas the Corporation initially planned, designed and maintained most of the employment areas. The landscape comprised a combination of earth shaping and planting. Initially the ground was sown with a commercial seed mix and subsequently planted with trees and/or shrubs at densities varying from open to very dense. The tree and shrub species included *Tilia cordata, Acer platanoides* and *Prunus spinosa*. The species found in the grassland were similar to those found in the lawns and verges of the housing areas, which in turn reflected the species composition of the pastures.

A commercial centre including offices and retail in the north-west of city was built on the site of a former orchard. Many of the fruit trees, mainly *Malus,* were retained. The development was then landscaped with species that included climbers such as *Lonicera japonica, Fallopia baldschuanica, Parthenocissus tricuspidata* and *Wisteria sinensis.* The shrub species included *Prunus subhirtella* 'Autumnalis', *Cytisus* x *praecox, Viburnum bodnantense* 'Dawn', *Cotoneaster salicifolius* 'Afterglow', *Cortaderia selloana* and *Hypericum calycinum.*

Central Milton Keynes

Central Milton Keynes (city centre), which occupies about 2.0–2.5 km, is built on the highest point of Milton Keynes, which was formerly occupied by arable land and pasture. It comprises three major areas – the shopping buildings, commercial areas and central area housing.

The landscape of the main glass shopping building comprises two elements, inside and outside. The internal landscaping comprised 47 planting beds, each 10.8 m long and 1.8 m wide. One hundred and fifty taxa, dominated by non-native species (including tropical species), were planted. They included *Fatsia japonica, Sparmannia africana, Pinus radiata, Phoenix* spp., *Ficus* sp., *Yucca elephantipes, Acacia* spp., *Photinia fraseri* 'Red Robin', *Trachycarpus fortunei, Amomum cardamom, Passiflora caerulea* and *Vitis vinifera* 'Leo Millot'. The boulevards that border the city centre comprise avenues of *Platanus* x *hispanica* and *Aesculus hippocastanum.*

Other species that occur in Central Milton Keynes include *Crocus chrysanthus* x *biflorus, Lonicera japonica, Cotoneaster salicifolius, Pachysandra terminalis, Eryngium campestre, Euphorbia characias, Rhodotypos scandens, Sisyrinchium striatum* and *Rhododendron scandens.*

Walls and Similar Structures

Old limestone walls, farm buildings, bridges and similar structures (most of which have been retained within the development) support a variety of lichens including *Cladonia* spp., *Caloplaca aurantiaca, C. hepplana, Physcia caesia, Verrucaria muralis, V. nigrescens, Xanthoria aureola and X. parietina.* The vascular cryptogams include *Poa compressa, Sedum spurium, Alyssum saxatile, Erophila verna, Polypodium vulgare, Valerianella locusta* and *Catapodium rigidum.*

Churchyards

At the time of its designation, Milton Keynes contained 14 churchyards (including small cemeteries). Since then a "cathedral" has been built in Central Milton Keynes and a tree cathedral was planted in one of the linear parks. A crematorium was built in the 1970s. The number of bryophyte species found in the churchyards varied from 11 to 22, the more common species included *Brachythecium rutabulum, Bryum argenteum, B. capillare, Camptothecium sericeum, Ceratodon purpureus, Grimmia pulvinata, Hypnum cupressiforme* var. *resupinatum, Tortula muralis, Dicranum scoparium, Eurhynchium conferatum* and *E. praelongum.* The number of lichen species found varied from 10 to 46 – one site contained less than 12 species while two contained more than 37 species, the species included *Caloplaca heppiana, Acarospora fuscata, Candelariella vitellina, Lecidea sulphurea, L. tumida, L. scraba* and *Paramelia saxatilis.* Some of the less common tree species such as *Juglans nigra* and *Ginkgo biloba* occur in churchyards. The herbaceous species include *Lotus corniculatus, Primula veris, Saxifraga granulata, Pimpinella saxifrage* and some rare or uncommon species such as *Ornithogalum nutans.*

Mineral Workings (Clay)

In the late nineteenth century, it was discovered that the calcareous Oxford Clay, which contains high proportion of organic matter, could be used to make large quantities of cheap bricks. However, the main expansion of brick production in Milton Keynes did not start until the late 1920s. At the time the city was designated it contained two operational brickworks (the larger occupying 3% – 275 ha of the city) and three abandoned exploratory pits and their associated spoil tips. The Corporation bought two of the exploratory pits/tips and the smaller of the two operational workings, which was closed, the buildings were demolished but the flooded pits were retained.

These mineral workings provided the greatest areas of botanical and ecological interest of any of the city's terrestrial and aquatic habitats. The topography was variable with varying sizes of hummocks and hollows with different amplitudes and different degrees of slope and aspect. The terrestrial vegetation comprised a continuum (between and within workings) of calcareous habitats from bare clay to dense scrub/embryonic deciduous woodland. The number of angiosperm taxa recorded at one of the disused tips was 160 (comprising 13 trees, 11 shrubs and 136 herbs and grasses); another contained 170 taxa. In both cases, the herbs and grasses were under-recorded, probably of the order of 25%.

The old disused tips supported a mixture of open to dense scrub and "grassland". The scrub comprised young trees such as *Ulmus procera, Fagus sylavtica, Acer campestre* and *Quercus robur* (which is as much a "colonising" as a "mature canopy" species). The shrubs were mainly *Rosa canina, R. arvensis, Rubus caesius* and *Corylus avellana* with some *Sambucus nigra, Ulex europaeus* and *Cornus sanguinea.* The grassland component varied from species characteristic of nutrient-rich soils to those of open, calcareous clay. The former included species such as *Rumex obtusifolius, Urtica dioica, Torilis japonica, Elytrigia repens, Solanum dulcamara, Knautia arvensis, Ranunculus repens* and *Leucanthemum vulgare* together with the relict populations of the calcareous flora of the recently tipped or excavated material.

The characteristic species of the open calcareous clay included *Cirsium acaule, Erigeron acer, Carex flacca, Linum catharticum, Briza media, Dactylorhiza fuchsii* (and some hybrid swarms), *Carlina vulgaris, Lotus glaber, Anthyllis vulneraria, Trifolium campestre, Ononis repens, O. spinosa, O. repens* x *spinosa* and *Euphrasia* agg. Relatively large areas were colonised by *Tussilago farfara.* The population size of *Ophrys apifera* varied from year to year and place to place. In some years thousands of spikes would appear on all the sites, in other years the number of flower spikes would just about manage double figures while in still other years some sites would have a large number and some none or only a small number. Interestingly, virtually all of the relatively large stands of *Phragmites autralis* in and adjacent to the city occurred on the slopes of the larger of the operational tips.

The pits were conical or pyramidal with a surface area of about 1–4 ha and a depth of 7–27 m. They were steep sided and consequently the marginal vegetation was narrow. The light penetration (measured by Secchi disc) was 5–10 m, the pH was 8 and the water had relatively high conductivity. The phytoplankton of two of the pits was dominated by Bacillariophyceae with low densities of Chlorophycaea and Cyanophycaea. The most abundant species in one pit was *Dinobryon* sp. with *Ceratium hirundinella.* The submerged zone varied within and between the pits; *Chara* spp. were dominant on the substrate of most of the pits. *Elodea canadensis, Myriophyllum spicatum* and *Lagarosiphon crispa* formed dense submerged "mats" in some of the pits.

The other submerged species include *Lemna triscula* and *Potamogeton lucens*. The floating-leafed vegetation was restricted to *Persicaria amphibia* and *Potamogeton natans* – with stems up to 5 m long. The marginal vegetation, which was low growing and sparse, included *Typha angustifolia, Phragmites australis, Alisma plantago-aquatica, Juncus effusus, J. inflexus, Schoenoplectus lacustris, Carex acutiformis, C. acuta, C. otrubae, Eleocharis palustris, Sparganium erectum* and *Mentha aquatica*. When some of the *Typha* and *Phragmites* were transplanted to nutrient-rich soil on the edges of the balancing lakes the plants grew from 1.0 to 1.5 m up to 3.0 m+ high.

The bryophytes found on the clay tips included *Fissidens taxifolius, Tortula muralis, Barbula fallax, B. hornschuchiana, Dicranella varia, Bryum bicolour, Campylium protensum, Brachythecium rutabulum* and *Riccardia pinguis. Campylium polgamum, Acrocladium cuspidatum* and *Drepanocladus aduncus* occurred on the water's edge of the pits. The last two also occurred in the water and wet hollows.

From the mid- to late 1970s, all except one of the disused brick workings were gradually destroyed. As described, one of the exploratory tips was used as a landfill site and eventually converted to an outdoor events arena – most of the area was sown with a commercial grass seed mix mainly comprising strains of *Lolium perenne*. The steep back slope was planted with *Salix* spp.

The company who owned the operational brickworks wished to rationalise the largest of the brick works and in so doing offered what is now Natural England an agreement to secure the largest and botanically most interesting of the tips and one of the pits and its associated tip. Natural England refused the offer and the areas were destroyed and restored.

In the early 1980s the other exploratory pit was incorporated into a balancing lake. The pits associated with the smaller of the brickworks (which was closed in 1971) remained until the early 1980s when one was used for the disposal of spoil and other material. The operation resulted in total loss of the aquatic and terrestrial (bank side) vegetation both of which were of significant botanical interest. When the filling was completed the surface was to be restored to playing fields. However, as a result of the threat of legal action, the area was restored to "hummock/hollow topography" with small ponds occupying the hollows and trees and shrubs being planted on the hummocks. Grassland habitats were also established. The area now contains 223 species (angiosperms). The other pit was retained and unaffected by the landfill activities.

Mineral Workings (Gravel)

Several years before the designation of Milton Keynes, a 6.5 ha abandoned gravel working had been converted into a balancing lake to take run-off from what was then a new industrial area. In 1975, the lake was enlarged to 9.0 ha to increase its balancing capacity. The maximum water depth is about 2.0 m deep, the pH 8 and the light penetration (secchi disc) 0.4 m. At that time, the aquatic vegetation was virtually absent except for a small area that was dominated by *Potamogeton lucens* with an abundance of *Bolboschoenus maritimus* and *Typha latifolia* on the margins.

In 1978, *Phragmites autralis* and *Nymphaea alba* were planted around the margins and in the lake, respectively. The latter was successful, the former failed. The total phytoplankton concentration was found to be typical of that of northern temperate lakes, having large spring and small summer bloom. At least 50 taxa were recorded, the dominant being species of *Chlamydomas, Ankistrodesmus, Scendesmus, Rhodomonas, Merismopedia, Oscillatoria* and *Nitzshia.*

In 1977, the Government started the construction of a national road across a section of the floodplain of the Ouse valley Park, which the Corporation had informally designated a "nature reserve" because of its value for breeding waders. The contractor asked the Corporation for permission to extract gravel from the park; the Corporation agreed subject to very stringent conditions including the protection of the bird breeding area, which was excluded from the contractor's licence. In addition the contractor was required to restore the excavated area to a "wildlife plan" that was designed and supervised by Corporation officers. The restoration included the creation of damp grassland and five interconnecting lakes, which varied in size from 0.55 to 2.8 ha. In addition 7,000 trees and shrubs were planted including *Alnus glutinosa, Quercus robur, Prunus avium, Salix* spp. and *Crataegus monogyna.* Following completion of the restoration the nature reserve became know as the "Wildlife Conservation Area".

An abandoned dry working area on terrace gravels comprised a mixture of dry to wet soils and open/disturbed conditions. The species present included *Deschampsia cespitosa, Elytrigia repens, Agrostis stolonifera, Achillea millefolium* and an abudance of *Conium maculatum* and *Dipsacus fullonum.* The damp/wet areas supported such species as *Lythrum salicaria, Senecio aquaticus, Juncus inflexus* and *Typha latifolia.*

Canal

The city contains a 21 km section of a canal that was opened in 1801 and is 1.5 m deep. The "canal" comprises a complex of habitats, which includes the water, a towing path and a boundary hedge. Although virtually all commercial use of the canal had stopped many years previously before the city was designated, it was still well used by pleasure boats and for fishing. In 1971, the Corporation devised a Canalside Landscape Policy the main components being the removal of the existing hedge, the construction of a 3 m surfaced broad walk with a row of *Populus nigra'* 'Italica' on each side and a ca. 5 m wide belt of scrub on the "outer side" to replace the lost hedgerow. Only a few hundred meters of the Policy was implemented before it was abandoned.

The four most abundant phytoplankton species found in the canal as a whole were *Ankistrodesmus falcatus, Chlorella* sp., *Cyclotella/Stephanodiscus* and *Scenedesmus quadricauda. Chara/Nitella* spp. occurred frequently on the substrate.

The macrophytic flora was typical of slow moving/still eutrophic waters of southern England. A survey carried out in 1975 discovered that 246 of the 550 plant species recorded in Milton Keynes (at that time) were associated with the canal.

The submerged zone was dominated by *Elodea canadensis* except in the northern section, which is dominated by *Potamogeton pectinatus* and the bryophytes, *Fontinalis antipyretica* and *Drapenocladus aduncus.* The other submerged species present were *Ceratophyllum demersum, Potamogeton pusillus, P. trichoides* and *P. compressus* – the last two being rare in eastern and central England. The floating-leafed zone comprised mainly *Callitriche* spp., *Nuphar lutea, Potamogton natans* and *Sagittaria sagittifolia,* while the more frequent of the marginal species included *Juncus effusus, Sparganium erectum, Rumex hydrolapathum, Butomus umbellatus, Acorus calamus, Alisma planto-aquatica, Carex riparia, Eupatorium cannabinum, Filipendula ulmaria, Epilobium hirsutum* and *Schoenoplectus lacustris.*

The terrestrial bryophytes recorded on the concrete edges and elswhere included *Fissidens adianthoides, Dicronoweisia cirrata, Tortula laevipila, Grimmia apocarpa, Orthotrichum affine, Amblystegium serpens* and *Cratoneuron filicinum.*

Railway Land

The main railway from London to Wales and parts of Scotland runs south to north through Milton Keynes. The line was opened in 1838; subsequently branch lines were built, with two being closed in the 1960s.

Operational

The main operational line is carried on a series of embankments and passes through a series of cuttings. The latter is mainly in Oxford Clay, and it is assumed that the former comprise fill material derived from the cuttings and/or imported. The absence of steam trains (and therefore accidental fires) has resulted in the cuttings and embankments being colonised by tree and shrub species such as *Fraxinus excelsior, Acer pseudoplatanus, Ulmus procera, Salix* spp., *Viburnum lantana, Rosa* spp., *Prunus spinosa, Crateagus monogyna* and *Rubus fruticosus* agg. The grassland component of the embankments comprises species characteristic of tall grassland on dry neutral soil, such as *Elytrigia repens, Arrhenatherum elatius, Lolium perenne, Melilotus officinalis* and *Cirsium arvense*. The herbaceous vegetation of the cuttings is characteristic of calcareous clay "soil"; the species present include *Trisetum flavescens, Carlina vulgaris, Primula veris, Briza media, Ononis repens, Hieracium* spp. and *Daucus carota.* A survey in 1979 found that 103 species occurred on both the west- and east-facing sides with 37 species occurring on the east-facing side but not on the west and 28 on the west-facing side but not on the east. A large population of *Brachypodium pinnatum sensu stricto* was found on one cutting slope. An extensive area of *Pteridium aquilinum* occurs on a railway cutting. The ballast and ash of the permanent way and similar areas included *Arenaria serpyllifolia, Lepidium heterophyllum, Chaenorhinum minus, Campanula rotundifolia,*

Linaria repens and *L. vulgaris*. A substantial section of the railway cutting in the centre of the city was used for the construction of the city centre railway station.

Disused

The closure of one of the branch lines in 1960 has resulted in most of the verges being colonised by trees, shrubs and species of tall grassland. Two hundred and sixty-nine angiosperm species were recorded along the line in 1974. The more abundant trees and shrubs included *Fraxinus excelsior, Acer pseudoplatanus, Rosa* spp., *Prunus spinosa, Crateagus monogyna, Sambucus nigra, Rubus caesius* and *R. fruticosus*. The other woody species present included *Ulmus glabra, Acer campestre, Viburnum lantana, V. opulus, Rhamnus cathartica, Salix caprea* and *S. cinerea*. The tall grassland is characterised by species of dry, unmanaged neutral to calcareous soils such as *Dactylis glomerata, Arrhenatherum elatius, Poa trivialis, Holcus lanatus, Cynosurus cristatus, Agrimonia eupatoria, Conium maculatum, Cirsium arvense, Conopodium majus, Astragulus glycyphyllos, Silaum silaus, Carex spicata, Hypericum perforatum, Dipsacus fullonum, Centaurea nigra, C. scabiosa* and *Vicia cracca*. The track and its margins supported lower growing species and species of open conditions such as *Lotus corniculatus, Juncus bufonius, Erodium cicutarium, Galium verum, Festuca rubra, Capsella bursa-pastoris, Convolvulus arvensis* and *Arabidopsis thaliana*.

The line contained 51 species of bryophyte, which included *Dicranella varia, Mnium longirostrum, Bryum capillare, Hypnum cuppresiforme, Plagiothecium curvifolium* and *Camptothecium lutsecens*.

In the mid-1970s, a 3 m wide cycleway/footpath was constructed along the disused railway line resulting in the removal of virtually all of the ballast ash and its and the replacement with tarmac resulting in the loss of about 13 species and a marked reduction in the frequency of others, particularly species such as *Chaenorhinum minus, Vulpia myuros, Astragalus glycyphyllos, Cirsium eriophorum* and *Lathyrus sylvestris* (the last two being rare species in Milton Keynes).

Road Verges

Many of the roads that existed outside the urban areas in 1970 had existed for several centuries, including the main London to North Wales Trunk Road, which was built by the Romans and improved by successive generations. The roads are bounded on each side by a grass verge (of varying widths) and a hedge. The vegetation of the verges of the minor roads is typical of that found on dry to damp, mesotrophic soils. The major road through the city had a wide verge on one side or the other; the wider verges contained a mixture of scrub and calcareous grassland.

The roadside hedgerows contained relatively few trees; those that did occur included *Fraxinus excelsior, Ulmus* spp (as suckers), *Acer campestre* and *Quercus robur.* The dominant shrub was *Crateagus monogyna/C.*x *media;* the other abundant shrubs included *Prunus spinosa* and *Rosa* agg. with *Viburnum lantana, V. opulus, Cornus sanguinea, Ligustrum vulgare* and *Rhamnus cathartica* occurring on the more calcareous soils. Several climbing species occurred such as *Tamus communis, Clematis vitalba, Bryonia dioica* and *Hedera helix.* The herbs and grasses present included *Elytrigia repens, Arrhenatherum elatius, Cirsium arvense, Dactylis glomerata, Hercaleum sphondylium, Achillea millefolium, Anthriscus sylvestris, Trifolium repens, T. pratense, Centaurea nigra, Lolium perenne, Potentilla reptans* and *Plantago lanceolata.*

The wider verges of the major road support such species as *Blackstonia perfoliata, Centaurium erythraea, Erigeron acer, Campanula latifolia, Dactylorhiza fuschsii* (and hybrids), *Primula veris* and *Ophrys apifera.* Damp soils, especially adjacent ditches, contained species such as *Pulicaria dysenterica, Rumex hydrolapathum* and *Angelica sylvestris. Orchis mascula* was recorded in an area of tall dense suckers of *Ulmus procera.*

As a result of spray and vehicle encroachment, the open conditions about 0.5–0.75 m wide adjacent to the carriageway support ruderal/ephemeral species, for example, *Convolvulus arvensis, Plantago major, Matricaria discoidea, Chenopodium spp., Stellaria media, Poa annua, Polygonum aviculare* agg., *Coronopus squamatus, Matricaria discoidea* and the halophyte *Puccinellia distans.*

The bryophytes found on the road verges that existed prior to 1972 included *Fissidens bryoides, F. taxifolius, Hypnum cuppresiforme, Plagiothecium succulentum, Isoptergium elegans, Leska polycarpa, Bryum caespiticium* and *Trichostomum sinuosum.*

Most of the minor roads have become footpaths and cycleways, with little change to the vegetation. However, the botanically interesting land reservations for the widening of the major trunk road were largely destroyed by landscape schemes intended to screen the city from the adjacent rural area and to provide a visual boundary.

Road Verges: New Major Roads

The Grid Road Landscape policy which was devised in 1971 by the Landscape, and Forestry and Conservation Departments, comprised four objectives:

1. Achieve a visual character consistent with the vegetation of lowland Britain (in essence "no conifers").
2. Create species zones with sharp comprehensible boundaries.
3. Achieve a planting mixture capable of responding to the full-range situations that might arise.
4. Enable plant requirements to be forecast for years in advance.

The city was divided into six "tree zones", each zone was to be represented by a dominant canopy, sub-canopy and shrub species. The species were to comprise 70% of the planting at each level, the remaining 30% to be determined by the landscape designer. For example, 70% of the forest tree, small tree and shrub species in the central zone would be *Aesculus hippocastanum, Taxus baccata* and *Prunus laurocerasus,* respectively. In the eastern zone, the species would be *Quercus cerris, Prunus avium* and *Prunus cerasifera.* The zones were not defined on the basis of prevailing ecological conditions; the species were not chosen for their suitability for those conditions. Of the total of 18 species, seven are non-native and 11 are native (one was totally unsuitable and another was rare as a native taxon – it was more likely that a cultivar was the intended taxon). However, the policy was not implemented; the choice of species to be planted along the grid roads was left to the discretion of the individual landscape designer. The tree species included *Fraxinus excelsior, Prunus avium, Robinia pseudoacacia, Acer palantoides, A. pseudoplatanus, A. saccharinum, Tilia* x *europaea, Salix fragilis, Alnus glutinosa, A. incana, Aesculus hippocastanum, Betula pendula, Populus* spp., *Taxus baccata* and *Thuja plicata.* The lower canopy species planted included *Malus sylvestris, Sorbus aucuparia, Acer campestre* and *Cotoneaster* spp. The shrub layer species included *Prunus spinosa, Buddleja davidii, Salix caprea, Prunus laurocerasus, Symphoricarpos albus, Sambucus nigra, Ligustrum vulgare, Elaeagnus* x *ebbingei* and *Viburnum opulus.*

Immediately after construction, the verges were sown with commercial grass seed mixes similar to those used in employment and other areas. It is likely that the seed comprised more species than was "written on the packet". The landscape areas were subsequently planted with trees, the remaining grassland was colonised by many herbs and grasses. In later years, the central reservation and some of the cuttings were planted with geophytes, especially species and varities of *Narcissus* and *Crocus* (Fig. 5).

Fig. 5 Aerial view of a section of a grid road showing the landscape reservations in 1973

Attempts to establish a field layer were hampered by the dense planting and application of fertilisers and herbicides, which were aimed at fast tree growth to obtain canopy closure as soon as possible. Nevertheless it proved possible to circumvent this policy and plant wildflowers in localised pockets. The species, which were transplanted individually or as turves from two of the largest woodlands, included *Primula vulgaris, Silene dioica, Mercuralis perennis, Sanicula europaea, Paris quadrifolia, Hypericum hirsutum. Listera ovata* and *Hyacinthoides non-scripta*. Subsequently seed and plants were used, resulting in the establishment of 30 species, of which the most successful were *Silene dioica* and *Geum urbanum; Primula vulgaris* and *Hyacinthoides non-scripta* were of intermediate success.

Hebeloma spp., *Lactarius glyciosmus, L. torminosus* and *L. turpis* were abundant in the areas where *Betula* had been planted. The other fungal species recorded in the plantations included *Paxillus involutus, Naucaria eschariodes* and *Stropharia coronilla.*

Parks

At the time of the designation of Milton Keynes, the area of parks and other public open space was restricted to small areas within the three towns, and village greens. The local parks comprised frequently mown grassland comprising species within the genera *Festuca, Lolium, Poa* and *Agrostis* with herbs such as *Bellis perennis, Plantago lanceolata, Hypochaeris radicata, Taraxacum officinale, Trifolium repens* and *T. dubium.*

Most of the public space was formed by the Linear Parks, which were based on the river valleys because the land could not be built on. The design and use of the parks comprised three elements – "strings" (for example, paths, cycleways) running through a "setting" (for example, sports grounds, woodland and grazing land, usually grassland) and connecting beads (or nodes comprising play parks, local parks, picnic areas and similar features). This "design" appears to have had no noticeable effect on the plants to be found in the park, save for tree and shrub planting.

The species to be found in the linear parks are mainly the mesotrophic grassland species described in the grasslands section (above). The loss of grassland to development has been exacerbated by the further reductions resulting from tree planting, although it is unlikely there has been any loss of species.

The fungi recorded in the grasslands included such species as *Coprinus comatus, C. atramentarius, C. plicatilis, C. niveus, Agaricus campestris, Bolbitius vitellinus, Pholiota ochrochlora* and *Lepiota rhacodes.*

A herb garden was established in the grounds of a "field centre". The garden, which contains at least 64 species, is divided into four sections. The first contains plants used for dyes, for example, *Isatis tinctoria, Allium cepa, Genista tinctoria.* The second group is concerned with medicinal species including *Atropa belladonna, Helleborus niger, Glechoma hederacea* and *Ruta graveolens.* The third group comprises medicinal herbs, for instance *Achillea millefolium, Aegopodium podagraria, Alchemilla mollis* and *Aphanes arvensis.* The final group are utility herbs such as *Carum carvi, Dipsacus fullonum, Filipendula ulmaria* and *Myrrhis odorata.*

Sports Grounds, Playing Fields and Golf Courses

The vegetation of the sports grounds and playing fields is dominated by grass species that are typical of neutral, nutrient-rich soils, including *Lolium perenne, Cynosorus cristatus, Agrostis capillaris, A. stolonifera, Holcus lanatus, Plantago lanceolata, Ranunculus repens* and *Trifolium* spp.

The city contains two golf courses; one was designed and built by the Corporation about 1976. The design included the deliberate creation of several different habitat types using native species. It was therefore surprising that a survey undertaken in 1996 recorded only 69 angiosperms, although the subsequent management of the course may not have been sympathetic to the survival of the habitats that were created and their component flora. Most of the species recorded in 1996 were characteristic of meso- and eutrophic soils, the taller species included *Holcus lanatus, Chamerion angustifolium, Rumex obtusifolius,* low growing species *Geranium dissectum, Primula veris, P. vulgaris, Carex divulsa* ssp. *divulsa, Vicia sativa* and species of open conditions such as *Poa annua, Cardamine flexuosa* and *Veronica agrestis.*

Allotments

Allotments, mainly for the growing of vegetables, existed prior to the start of the development. Some were well managed while others were partly disused. The development of the city resulted in considerable demand for more allotments, which the Corporation provided. As with the established allotments, the new ones were mainly used for growing vegetables such as *Phaseolus coccineus, Vicia faba, Brassica oleracea* var. *capitata, B. oleracea* var. *botrytis, B. oleracea* var. *gemmifera, Allium cepa, A. schoenoprasum, A . porrum, Daucus carota* ssp. *sativus, Solanum tuberosum* varieties, *Beta vulgaris* ssp. *vulgaris, Lactuca sativa* varieties, *Raphanus sativus, Petroselinum crispum, Fragaria* x *ananassa, Rubus idaeus* and occasionally flowers such as *Lathyrus odoratus*. The allotment plots were generally well managed including the application of fertilisers and herbicides. Arable "weeds" such as *Stellaria media, Senecio vulgaris, Elytrigia repens* and *Taraxacum* agg. occur within the plots while the marginal footpaths contained species such as *Plantago major, Poa annua, Polygonum aviculare* agg., *Arrhenatherum elatius* and *Dactylis glomerata.*

Balancing Lakes (Wet)

The seven wet lakes, which vary in size from ca. 0.5 to 68 ha, were built between 1972 and 1983 – three were on and four were off stream (Fig. 6).

Initally, the lakes were "colonised" by phytoplankton and filamentous algae that were in the watercourses. The dominant/abundant phytoplankton species change from lake to lake and monthly within a lake. Six years after being flooded, one of

Fig. 6 Aerial view of a wet balancing lake

the lakes contained 47 taxa, dominated by *Rhodomonas* spp, another contained 53 taxa of which the most frequent taxa were *Anabaena flos-aquae, Oscillatoria* sp. and *Chlamydomonas* sp. with *Navicula* sp. being the most abundant diatom. Filamentous algae within the genera *Cladophora, Rhizoclonium, Ulothrix, Microspora, Tribonema, Oscillatoria, Zygonema* and *Entermoprha intestinals* established soon after flooding; as with the phytoplankton, other taxa followed, including *Chara/Nitella* which formed extensive beds on the substrate.

Shortly after their completion the lakes were planted with submerged, floating-leafed and marginal species while other species colonised naturally. The submerged species planted were mainly *Potamogeton perfoliatus,* and *P. crispus. Elodea canadensis, Lemna triscula, Ranunculus aquatilis* agg. (including *R. circinatus), Potamogeton pectinatus, Myriophyllum spicatum,* which colonised naturally and became dominant. The floating-leafed species *Sagittaria sagittifolia, Potamogeton natans, Callitriche* sp. *Nymphaea alba* and *Persicaria amphibia* were planted but virtually all of the plants died.

The marginal species that were planted were obtained from several sources (usually within the city and usually as rhizomes. The species included *Typha latifolia, T. angustifolium, Glyceria maxima* (only planted in the water-skiing lake), *Sparganium erectum, Bidens tripartita, Carex pseudocyperus, Phragmites australis, Juncus effusus, J. inflexus, Iris pseudacorus, Caltha palustris, Veronica beccabunga, Lythrum salicaria, Schoenoplectus lacustris, Phalaris arundinacea, Carex riparia, Butomus umbellatus, Mentha aquatica* and one plant of *Cladium mariscus.* The planting was augmented by the natural colonisation by some of the species that had been and some that had not been planted including *Typha latifolia, Carex riparia, Glyceria fluitans, Epilobium hirsutum, Ranunculus scleratus, Alisma plantago-aquatica, Myosotis scorpioides, Eleocharis palustris, Lycopus europaeus, Eupatorium cannabinum, Apium nodiflorum, Cotula coronopifolia, Potentilla norvegica, Lythrum hyssopifolium* and *Limnanthes douglasii.*

Wet/dry Balancing Lakes

The wet/dry balancing lakes were constructed by the modification of a valley. Following construction, 15 cm of topsoil was spread over the slopes and dam walls and sown with a commercial grass seed mix at 35 $g\,m^{-2}$. Once the grassland had become established, it was grazed by sheep and/or horses. The open/disturbed conditions of the valley bottoms were colonised by ruderal ephemeral species such as *Polygonum aviculare* agg., *Coronopus squamatus, Plantago major* and *Poa annua.* The more abundant species include *Lolium perenne* (the dominant species), *Festuca rubra, Agrostis stolonifera, A. capillaris, Picris echioides, Dactylis glomerata, Hordeum secalinum, Phleum pratense, Cirsium vulgare, C. arvense, Rumex crispus* and *Elytrigia repens. Holcus lanatus, Potentilla reptans, Trifolium pratense, T. dubium, Prunella vulgaris, Taraxacum officinale, Lotus corniculatus, Achillea millefolium, Plantago lanceolata, Crepis* sp., *Geranium molle, Taraxacum officinale* agg. together with species characteristic of damp and/or open conditions occurring in the most frequently flooded areas, for example, *Deschampsia cespitosa, Alopercurus geniculatus, Juncus inflexus, J. effusus* and *Ranunculus repens.*

Nature Conservation, Environmental Planning and Education

The statement in The Plan for Milton Keynes states that the University College London would undertake a 20 year ecological monitoring programme; it was not implemented. The Corporation instituted what turned out to be a 10 year *ad hoc* monitoring programme mainly using ground surveys augmented by remote sensing (using a combination of vertical and oblique monochrome, normal colour and false-colour infra-red imagery) as well as ground photography. Some of the surveys have been subject to varying levels of analysis while most have not been analysed at all, nor have they been subject to any comprehensive analysis that provides a detailed picture of what happens to the habitats and the flora and fauna of an area that has been converted from mainly agricultural use to totally urban.

In the1970s two exercises were undertaken to determine whether the application of a broad-scale nature conservation evaluation would have had any influence on the planning and design of the city. One method used a "manual" determination, the other used remote sensing. It was concluded that the scale was too broad and unsuitable for the detailed assessment needed. Consequently an *ex post facto* site/habitat evalution was carried out using data obtained from the various site, habitat and species surveys undertaken up to that time. This exercise concluded that had such an assessment been carried out before and used in the preparation of the Master Plan, most if not all of the areas of natural history value could have been retained intact or with minimum change.

Arguably the most important issues to emerge were the creation of new habitats/vegetation types/plant associations and the translocation of species and plant communities that would be destroyed – both were very strongly objected to and resisted by the predecessors of Natural England and voluntary nature conservation organisations.

A study was also undertaken to determine the value of remote sensing in the monitoring of the wet balancing lakes, using three emulsions – monochrome, normal colour and false-colour infra-red. It was concluded that the best results were obtained from the application of all three but if for some reason only one can be obtained to used.

No statutorily protected plant species or species identified as being of national nature conservation concern (that is, listed in the IUCN or national Red Data Books or the Berne Convention) were recorded before or after the start of development. At the time of designation, there were no statutorily protected sites in the city. Natural England did not consider any site in or close to the city had sufficient scientific merit to be seriously considered for Notification as a "Site of Special Scientific Interest". However, in 1994, one of the three larger woodlands within the city and a 3.7 ha of grassland adjacent to the western boundary were Notified as "Sites of Special Scientific Interest". The same year the local authority designated the restored brickworks site and the adjacent pit as a Local Nature Reserve.

In 1981, the Corporation gave the Wildlife Conservation Area (described earlier) together with a substantial endownment to ensure its proper management as a Wildlife Trust.

The Corporation initiated an ambitious environmental education programme primarily aimed at children. The programme started in 1974 with the formation of a "Junior Conservation Corps", with the objective of involving children of all ages in caring for their environment. By 1980, the scheme involved 2,350 children in 65 schools. The activities included the collection and sowing of tree, shrub and wildflower seed, bulb planting (ca. 40,000 bulbs/year), transplanting species from sites that were to be developed, planting of aquatic plants in new ponds, the creation and planting of insect gardens and other habitat creation works in school grounds, establishment and maintenance of tree nurseries, and collecting rubbish. A major habitat creation excercise in the grounds of one of the village schools involved the children in creating, planting and maintaining a variety of habitats including woodland, shrubberies, different types of grassland and two ponds. Twenty-seven tree and shrub species and 16 aquatic species were planted and a wildflower seed mix was sown.

Guided walks for adults and/or children started in 1976. Adult and junior environmental trails were created in some of the parks and woodlands. The junior trails were in the form of workbooks with opportunities to draw/colour in plants and animals. Educational packs relating to the plants and animals of the city were devised and made available to schools and other organisations concerned with children. Six copies of a natural history slide pack (with notes) was prepared and made available to schools. Because of the lack of support from the local education authority and the winding down of the Corporation, education programmes were terminated in the mid-1980s.

Closing Comments

In terms of the age and scale of the other urban developments described in other chapters of this book, Milton Keynes is new and small. It differs in its origin from most of those cities, which have grown (organically, as architects say) from a cluster of buildings on a trade route. Milton Keynes was imposed on a predominantly rural area as the consequence of a decision by a Socialist Government in pursuit of "paradise" – in the same way that Peter the Great founded St. Petersburg and subsequent politicians founded Brazilia. Milton Keynes was to be built at high speed – 20 years from beginning to end not that British New Towns have an end; like other urban areas they continue to expand, only the speed changes.

There was very little published botanical information available about the city before the start of development in 1971. Because 80% of the area was in intensive agricultural production the flora was impoverished, mainly comprising species of disturbed soil (arable land) and those that are characterisitic of agriculturally improved grassland on damp, meso- to eutrophic soils. The areas of greatest botanical interest were at the extremes of the age range and separated by 1,000 years or so – old woodland and young clay workings. The former have been retained, although in altered state while the latter have been virtually destroyed.

However, the most important aspect was not the protection of what existed because except as described in the last paragraph there was not much worth protecting but in the opportunities for the creation of habitats and plant communities – most by design and some by default. Some were created by the modification of existing habitats while others, and by far the most extensive, were created *de novo*. The latter included various types of woodland, scrub, grassland and aquatic habitats including those associated with gardens and landscape schemes adjacent to and within buildings (commercial and retail). This situation provided an opportunity (that was not fully exploited) to "take ecosystems apart" and "reconstruct them", – although most of it was done "in the dark".

The development of Milton Keynes has resulted in an increase in the plant biodiversity of the area by 300–400%, most of which are non-native taxa. However, most of the plant material (even the native taxa) came from outside Britain (let alone England), where it was probably propagated by cloning. This gives rise to an interesting dichotomy. On the one hand the importation and planting of large numbers of plants of non-native species from sources mainly from outside Britain has substantially increased the pre-existing species and genetic diversity, on the other hand the genetic diversity of the imported taxa is likely to be very restricted. The same principle applies to the genetic diversity of the imported native taxa. The consequences raise questions of endless botanical fascination and debate.

Virtually all of these habitats have plant communities/associations that are unique and therefore are not found in the countryside or probably in any other urban area. They provide exciting opportunities for botanists to study the relationship between species that would not otherwise occur together and the dynamics of new plant communities. Most of the existing lowland plant communities are the result of human influence on whatever the natural vegetation was. Consequently are the new communities any "better" or "worse" or any more or less desirable than what they have replaced?

Further Reading

Bendixson T, Platt J (1992) Milton Keynes Image and Reality. Granta Editions, Cambridge

Cox J (1994) Woodland Islands – a Review of the Issues Arising from The Isolation of Ancient Semi-natural Woodlands within the Urban Matrix, with Special Reference to Milton Keynes. Hampshire and Isle of Wight Wildlife Trust, Curdridge, Hampshire

Croft RA Mynard DC (1993) The Changing Landscape of Milton Keynes. Buckinghamshire Archaeological Society, Buckingham

Maycock R, Woods A (2005) A Checklist of the Plants of Buckinhamshire (including Milton Keynes and Slough). Milton Keynes Natural History Society. Milton Keynes

Milton Keynes Development Corporation (1970) The Plan for Milton Keynes. Milton Keynes Development Corporation, Milton Keynes

Milton Keynes Development Corporation (1974 to 1990) Ecological Studies in Milton Keynes Volumes 1–120. Milton Keynes Development Corporation, Milton Keynes

Milton Keynes Development Corporation (1992) The Milton Keynes Planning Manual. Milton Keynes Development Corporation, Milton Keynes

Milton Keynes Natural History Society (2000) Milton Keynes More than Concrete Cows – Real Animals and Plants too. Milton Keynes Natural History Society, Milton Keynes

Milton Keynes Natural History Society (1975–1977) Journals 1–4. Milton Keynes Natural History Society, Milton Keynes

Milton Keynes Parks Trust (1993 to 2008) Ecological Studies in Milton Keynes Volumes 121 to 153. Milton Keynes Parks Trust, Milton Keynes

Walker D (1981) The Archicture and Planning of Milton Keynes. Architectural Press, London

Moscow

Alexander Shvetsov

Fig. 1 The Kremlin

Abstract The city, as part of the cultural landscape, is a mobile and highly dynamic system, the changes of which are related to a variety of economic and social factors. Periods of stabilisation alternate with periods of fast changes and development of the city. Vegetative cover as an element of urban environment is directly or indirectly affected by all these factors and phenomena. Species of natural communities are of special concern. The reduction in the population size and frequency of many native species can adversely affect their survival. It is possible that many rare species will

Alexander Shvetsov (✉)
Main Botanical Garden Russian Academy of Sciences, Botanicheskaya St., 4, 127276 Moscow, Russia
e-mail: floramoscow@mail.ru

J.G. Kelcey and N. Müller (eds.), *Plants and Habitats of European Cities*, 321
DOI 10.1007/978-0-387-89684-7_10, © Springer Science+Business Media, LLC 2011

disappear from the city, primarily those of wet and dry meadows, forest edges, meadows and marsh communities. There is also another danger, which is generally overlooked – the loss of native genotypes and their replacement by alien genotypes of the same taxon. These negative processes can also affect non-native species, for example, plants of old parks and "relics of cultivation" such as *Chenopodium bonus-henricus, Luzula luzuloides, Phyteuma spicatum* and *Poa chaixii.*

Natural Environment of the City

Moscow is located at 55°45′ north and 37°37′ east, in the centre of the Russian Plain and in the watershed of the rivers Volga and Oka. Moscow, the capital of the Russian Federation, is the largest city of the country and the most northern of the largest cities in the world. In 1996, it occupied 994 km^2 of which 878 km^2 are within the boundaries of the Moscow ring road (MKAD), which is 109 km long. Within the ring road, the city has an oval shape; 42 km north/south and 35 km west/east. With the new area around the MKAD, these figures are likely to increase to 38 and 52 km, respectively (see Fig. 2). In 1995, the population was 8.8 million (Moscow: Encyclopedia 1997). In the central part of the city, the developed surface is 7,128 $m^{-2}ha^{-1}$, which is reduced to 4,460 $m^{-2}ha^{-1}$ in the peripheral areas. By 1997, the average numbers of floors per apartment block was five.

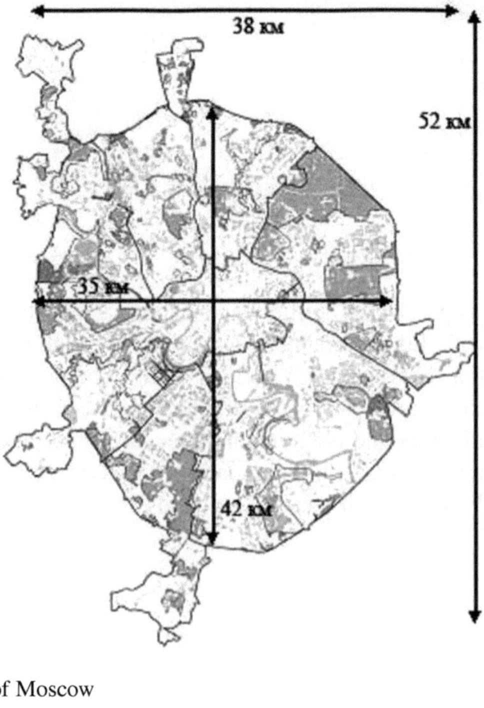

Fig. 2 The territory of Moscow

Geology

The territory of Moscow is located in the Russian Platform and is founded on crystalline schists formed by orogenesis during the Archaean and Proterozoic times. The formation is overlain by 1,650 m of the Riphean, Vendian, Cambrian, Devonian, Carboniferous, Jurassic, Cretaceous and Quaternary sedimentary rocks. The Carboniferous deposits are extensive and comprises mainly limestone up to 350 m deep. Prior to the construction of the Moskva – Volga channel, Coal Measures outcropped in the river bank above the city centre; at present it is below the level of the river. The Carboniferous deposits are overlain by Jurassic sands and clays up to 80 m thick, which in turn are covered by 75 m of Cretaceous sands and clays. Outcrops of Jurassic and Cretaceous deposits occur in the valley of the Moskva River and some of its tributaries above and below the city centre.

Virtually, all of the surface geology of the city comprises glacial deposits 40–60 m thick resulting from three glacial periods of the early to mid-Pleistocene. The deposits include moraines, fluvio-glacial sandy sub-moraines and lacustrine-glacial formations. Alluvial deposits are represented by sand and clay. Hundreds of years of human activities have resulted in the formation of technogenic deposits (cultural layer) (0.5) 4–6 (20) m deep which occupy about 70% of the city's territory, including the whole of the city centre. The thickest deposits are characteristic of landfill sites of old quarries, river valleys, ravines and swamps. In some areas of the city, a modern technogenic sediment deposit is growing at an estimated 10 cm/year.

Changes in hydro-geological conditions resulted in a rise in the level of the groundwater, which is most widespread in the northern and eastern parts of the city, where a high watertable occupies 40% of the territory and relates mostly to the watershed spaces of moraine and fluvio-glacial plains.

Topography

The relief of Moscow is formed by Quaternary glaciations and by river erosion (see Fig. 4).

The variation in the topography of the city is due primarily to its location at the crossroads of three large geomorphological regions. The northern part of the city is located in the Smolensk-Moscow Upland, which comprises a Middle Quaternary plain 165–185 m a.s.l. with a fluvio-glacial at a lower level (155–175 m a.s.l).

The southern part of the city is occupied by the Moskvoretsko-Oksky erosion plain which has a maximum height of 255 m a.s.l. and is 130–140 m above the river bank. The area is characterised by extensive dissections that form ravines and small river valleys. The Mescherskaya lake-glacial lowland in the eastern part of the city is a Middle Quaternary outwash plain with a maximum height of 140–160 m a.s.l. The outwash plain is characterised by a hollow and undissected relief with rare moraine butte; almost half of the area (43%) has no run-off.

The valley complex of the Moskva river and its major tributaries (Yauza, Setun and Skhodnya) occupies a third of the total area of the city which is divided into four levels: the flood plain and three terraces. The greatest area is occupied by the third terrace, which is 30–35 m above river level and occurs throughout the city. The floodplain of the Moskva river no longer exists – it was flooded when the canal was built.

Human impact on the terrain started with the first settlement and the emergence of agriculture and the formation of burial mounds. Some of these "micro-changes" in the topography still occur in the city. As a result of human economic activity the natural terrain changed simultaneously in two directions, first lowering, such as levelling elevations and second, elevations such as the accumulation of the cultural layer. As a result of these processes, the relief of the city was levelling out more and more resulting in the loss of the natural micro-relief. The intensity of human activity and its effect on the natural terrain has continued to increase over the millennia.

In the process of development of the territory a large number of small rivers, ravines, swamps and ponds have been culverted; especially significant are the landfill areas, which are 3–8 m deep. A large number of elevations were created during the construction of highways and railroads, some of the latter reaching 13–15 m. In some cases the relief was lowered to reduce the vertical alignment of transport routes, for example, some of the railway cuttings are 12–17 m deep and 200 m long. In the central and long inhabited parts of the city the terrain has changed to a greater extent than in the suburbs. Nevertheless even in the dense and older parts of the city the relief of the former natural terrain is easily visible.

Climate

Moscow is located in a zone of moderate continental climate, whose character is significantly stronger relative to other major European cities. The largest annual amplitude in temperature is 28°C. The dominant air mass is air of moderate latitudes, generally coming from the Atlantic Ocean with the Arctic air from the north and north-east and tropical air from southern Europe also occurring during the year. The average monthly temperature and rainfall data are given in Table 1.

The average annual temperature between 1901 and 2000 was 4.53°C. However, further analysis shows a gradual warming, the data for 1961 to 1990 and 1991 to 2000 show an average annual temperature of 5.02 and 5.75°C, respectively (Isaev et al. 2002). There are 141 frost-free days, the frosts generally start about 29 September and end about 10 May, with severe frosts occurring between 24 November and 10 March. "Permanent" snow cover occurs around 26 November and finally melts about 11 April – by the end of the winter the depth of the snow is (on average) 30–35 cm.

The length of the growing season (number of days with an average daily temperature of ≥5°C) is 175 days (from 18 April to 11 October). The average temperature in July is about 18.1°C. During the incursions of the cold air from the Arctic in June, the temperature might drop to −2 to −4°C. Long-term observations, which started in 1880, show that there has been a gradual warming in the northern

Table 1 The average monthly temperature and rainfall

Month	I	II	III	IV	V	VI	VII	VIII	IX	X	XI	XII	Average annual
Temperature (°C)													
1901–2000	−9.3	−8.6	−3.4	5.1	12.4	16.7	18.4	16.6	10.9	4.4	−2	−6.8	4.53
1961–1990	−9.4	−7.7	−2.2	5.8	13.3	16.8	18.4	16.7	11.1	4.9	−1.4	−6.2	5.02
1991–2000	−5.9	−5.9	−1	7.3	12.7	17.7	18.8	16.8	11.1	5.6	−2.5	−5.5	5.75
Precipitation (mm)													
1901–2000	40	34	36	41	52	75	85	78	61	58	50	47	658
1961–1990	43	36	35	44	50	77	91	78	62	57	56	55	684
1991–2000	56	42	41	35	43	90	73	80	65	74	53	51	702

hemisphere, which reached a peak about 1930. In the period from 1950 to 1970 cooling was observed, which is now believed to have been replaced by new global warming. The average annual precipitation for 1901 to 2000 was 658 mm, of which about 70% occurs in the warm season (April-October). As with the temperature, the average annual precipitation shows a slight increase, for example, in 1961 to 2000 it was 684 mm and in 1991 to 2000 it was 702 mm.

The influence of a city, especially a large one, on the climate and atmospheric phenomena is extremely strong. Large cities transform natural climatic conditions, creating a local urban climate. Microclimatic differences within the city are due to the diversity of urban landscapes and a combination of local conditions, both natural and artificial. The general trends of climate changes in cities can be summarised as increasing temperature, precipitation and air pollution and decreasing humidity and solar radiation.

Increasing air temperature is caused by a number of factors for example the temperature of the stone walls and pavements is, on average, 7–10°C higher than the air temperature. The temperature measurements in Moscow in summer time showed that the temperature of a red brick wall reached 41°C, asphalt pavement registered 45°C, very hot iron roofs showed temperatures of more than 50°C while the surface temperature of lawns was 25°C. Given the large area of heat retaining/reflecting material, the magnitude of the thermal effects on the air temperature is substantial. The albedo of snow has declined by two or more times as the result of pollution – the albedo of freshly fallen snow is 85%, while that of polluted and melting snow is 40%. As a result, snow melts one or two weeks earlier in the city than in the surrounding area. At the same time there are factors and conditions which lead to the lowering of the air temperature, for example, woody vegetation reduces the temperature by 1–2°C (sometimes 3.5–4.0°C). Vegetation has high albedo, obscures the soil and reduces soil temperatures and evaporation. Soil covered with vegetation, receives, on average, 18% less heat than bare soil. Reservoirs in the warm weather also contribute to cooling the air, but the extent of their influence is small. The air temperature is temporarily reduced (by 1.2–5.0°C) at its interface with lawns and streets by irrigation.

In the areas of the city with high buildings, narrow streets and enclosed courtyards, the daytime air temperature is slightly lower due to shading than in the adjacent broad streets. In large cities, the factors and conditions that cause increases and decreases in the air temperature interact with each other, the overall effect being an increase in the temperature of the air, soil, surface water and groundwater between the periphery of the city and the centre. These factors may be of paramount importance in relation to the effect that global warming may have on the city's environment and its flora.

The difference between the mean annual air temperature and the length of frost-free period in the central and outlying areas of the city is 1.4°C and 25 days longer, respectively.

Some researchers suggest a zonal model of the temperature distribution within the city, while others offer a sectoral model of the meso-climatic division of the city. The latter identifies three sectors – Central, West and East. The central part is the warmest and least ventilated area of the city, the average temperature is 5.0–7.0°C, which is 2–4°C higher than the western suburbs. The western sector is the coldest and most windy part of the city, with an average annual temperature of 3.0–5.0°C, which exceeds the temperature of the western suburbs by no more than 2°C. This area has the cleanest air – the prevailing winds come from the forests adjacent to the west side of the city. The eastern sector has intermediate meteorological conditions between the central and western sectors, average annual temperature being 4.0–6.0°C. The prevailing winds bring the polluted air from the entire city into this area.

Surface Water

The Moskva River crosses the city from north-west to south-east for almost 80 km (Fig. 4).

Virtually all of its banks have been re-enforced with stone revetments. Between 1933 and 1937 major "improvement" works were carried out, including the construction of a channel to connect the river to the Volga River (which raised the level through the city centre by 6 m) and the construction of a 4 km^2 reservoir to regulate the flow and provide potable water. At present, persistent ice does not form on any parts of the river and in some places (in the centre and in the southern part of the city) ice does not form at all. The average water temperature of the river is highest in the central part of the city and at outfalls from sewage works, these conditions allow adventive thermophilic organisms to exist in such places and migratory waterbirds over winter.

The city has a total of more than 100 watercourses, but many of them have been culverted or dammed to create reservoirs. Twenty-five rivers have not been culverted and 18 have only been partially culverted. The total length of "exposed" moving water (excluding the Moskva) is 187 km; the longest are Yauza (29 km within the city), Likhoborka (18 km), Gorodnya (16 km) and Setun (13 km). The large number of oxbow lakes in the Moskva floodplain has been destroyed. The city contains many artificial lakes, mainly in the outskirts and an abundance of ponds formed in the beds of the river, some of which have existed from the seventeenth and eighteenth centuries. The water bodies occupy a total of 800 ha with 5.5% of the total number being 5.0 ha or more, the largest is the Borisovsky (86 ha), which was constructed in the sixteenth century.

The hydrographic network has been subject to the direct and indirect impacts of human activities, from which it continues to suffer. The direct impacts include the construction of new water bodies (canals, reservoirs and ponds), reconstruction of the rivers (dredging, enlargement, flow regulation and the re-enforcement of the banks) and infilling of lakes, ponds and marshes. The indirect impacts include water pollution from sewage and urban run-off and reduction in water depth as a result of artificially lowering the underground water.

Soils

The soil cover of modern Moscow is quite varied and complex in structure and origin. This is due to a combination of natural and anthropogenic factors. These include the extent of the variation in the topography (which creates differences in drainage and humidity), the diversity of the geological structure and the nature of underlying rocks, differences in soil age and the duration of soil-forming processes and the size and composition of the cultural layer (or the ground on which soil is formed).

The soils of the city, as well as the vegetation, are represented by those with natural and semi-natural differences, modified and transformed soil formations (preserved mainly in the peripheral zone), soils of various cultural origins and artificial soils (emerging in the context of anthropogenic landscape).

The natural soils are represented by Soddy-podzolic soils, which are characteristic of the area and climate conditions. Among them are the forest soils that have passed through the stages of agricultural use and reforestation after cropping or fire. To some extent, these factor have probably affected almost all of the remaining natural areas. In addition, there are several types of swamp and alluvial meadow soils.

From the regional point of view, the natural soil cover refers to three soil areas (coinciding with the geomorphological areas):

1. The Northern and Western (Smolensk-Moscow Upland) region with sod-heavy and medium-podzolic soils.
2. Southern (Teplostanskaya Upland) – with a predominance of sod-light-podzolic loam soils.
3. East (Meschersky lowland) – with sod-podzolic soils of light mechanical composition in combination with peat and peaty-gley soils.

Modified versions of the natural undisturbed soil combine the middle and lower part of the profile and anthropogenically disturbed upper layers.

Human economic activity has resulted in the formation of new soil types, which have two primary origins. First, those that develop spontaneously (mostly without direct human intervention) on the bulk substrate and disturbed ground (mounds, wasteland and built area) where there has been a long period of artificial waterlogging. Second, those which originate from active and purposeful management (cultural soils of urban parks, ornamental and vegetable gardens and agricultural production). Of course, these soils develop under similar and often identical influences as natural soils, for example, the climatic conditions.

According to soil scientists (Stroganova 1998), the main difference between anthropogenic and natural soils is that the former are formed on loose, disturbed ground and on the cultural layer, which contain construction and household waste in the upper layer and are characterised by a high degree of contamination with heavy metals (often exceeding background levels by 5–100 times) and petroleum products. These anthropogenic soils have distinct physical, chemical and mechanical properties (low moisture, high compaction, large particle sizes and nutrient levels that may vary from very low to very high). The application of de-icing salt (NaCl) results in urban run-off being polluted during snow melts. The sodium level can be as much as 115–130 mg l^{-1} with chloride levels of up to 75 mg l^{-1}, which are two orders of magnitude higher than the levels found in typical soils. As a general rule, urban soils also have a high content of nitrogen, phosphorus and potassium compared to typical soils. Thus, P_2O_5 content in urban soils varies from 5 to

150 mg/100 g of soil while the sod-podzolic soils of the Moscow region have from 5 to 15 mg/100 g of soil. Typically, the amount of organic matter in urban soils is also higher than in typical natural soils; for example, the content of humus in the old part of town is 8–12% (the average is 4–6%).

Urban soils are characterised by a fairly high degree of variability, spatially and in structure, for example, mechanical and physico-chemical composition. In contrast to the natural soil in which soil-geochemical processes are quite slow and gradual, the urban soils are exposed to the severe disturbance and alteration by highly intensive processes, often resulting in the death of one soil system and the emergence of a new one. Thus, the form of existence of urban soils consists of repeated disturbance, re-mixing, introduction of additional materials (including chemicals) and the consequential rejuvenation of the soil profile.

Contamination has become the reason for the city's "replacing" soil policies. Probably, Moscow is unique in applying such a measure on a very large scale. The top soil and the vegetation it contains are removed to a depth of up to several tens of centimeters and replaced with so-called nutrient-rich soil, which comprises a mixture of peat, humus and silt from irrigated fields. From the 1990s these activities have been carried out over a large part of the city, the soil being replaced not only in the built-up parts of the city but also in the forests, floodplains, and even on the slopes of ravines. In 2000 alone over 300,000 m^{-3} of soil was brought into the city for this purpose (Frolov and Antonenko 2002).

The Natural Landscapes

Moscow is located at the junction of three physiographical provinces (Fig. 3), Moscow (II), Moskvoretskaya-Oksky (III) and Meschersky (IV), in which different authors have highlighted a varying number of landscapes and natural-territorial complexes of lower rank. Province 1 lies outside the city boundary.

A characteristic feature of the landscape structure of the modern Moscow is based on the fact that the majority of indigenous landscapes almost converge in the central part of the city (Fig. 4), their main characteristic differing strongly from each other (Nizovtsev and Shurkina 1997).

The landscapes on the right bank of the Moskva River (zones 1–3 in Fig. 4) are located on the watershed of morainic plains and have a maximum height of 254 m a.s.l. They are highly dissected and well drained (see Fig. 4). The indigenous forest of these landscapes comprises broad-leaved species in which *Tilia cordata* and *Quercus robur* are dominant with some *Picea abies.* The herb layer includes *Anemone ranunculoides, Asarum europaeum, Carex pilosa, Lamiastrum galeobdolon, Mercuralis perennis* and *Pulmonaria obscura.*

The landscapes of the north-east and east of the city (zones 6–8 in Fig. 4) comprise fluvio-glacial flat plains with altitudes of 140–150 m a.s.l. formed by fluvio-glacial sands on the moraine and fluvio-glacial loam on sand. The soils are poorly drained, medium-podzolic and gleys. The landscape contains a large area of

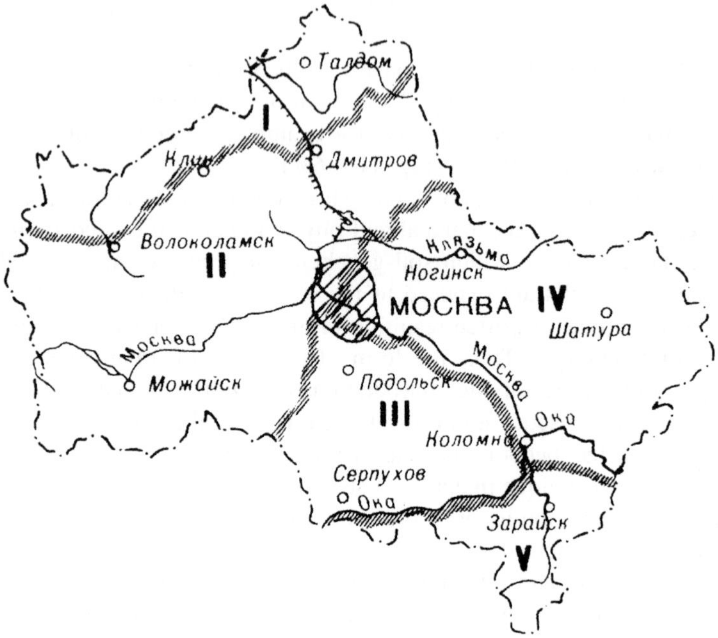

Fig. 3 Division of the territory of Moscow region

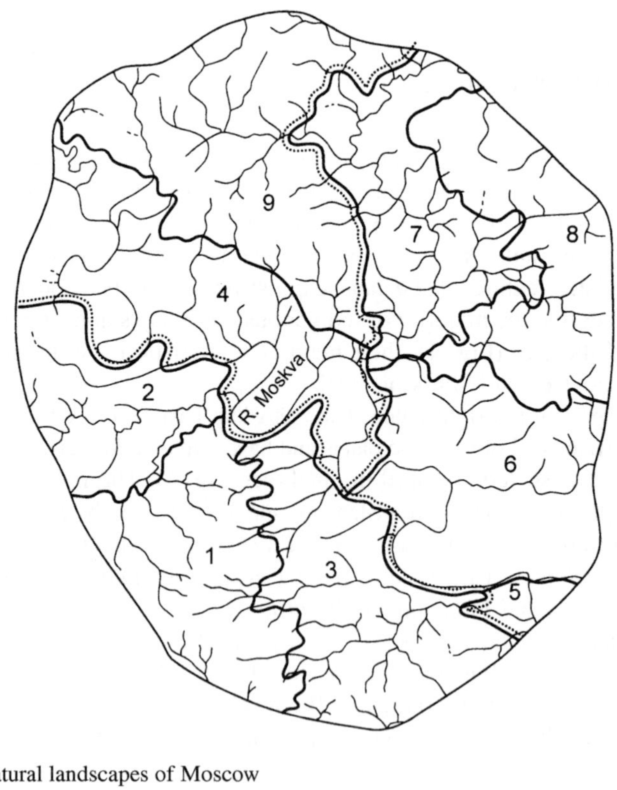

Fig. 4 The natural landscapes of Moscow

wet soils. The indigenous forest consists of coniferous and mixed forests with a predominance of *Pinus sylvestris* and *Picea abies* often on damp soils. The area is characterised by the extensive distribution and high abundance of *Vaccinium myrtillus* and *Vaccinium vitis-idaea*. The northern and eastern parts of the city are also characterised (in part) by *Sphagnum* bogs some of which still survive.

The landscape of the northern part of the city (zone 9, Fig. 4) is intermediate between the southern and eastern landscapes described above. The Valley landscape of the Moskva river (zone 4, Fig. 4) includes its floodplain and the river terraces in which soils of light mechanical composition prevail. There is high degree of variation in the soil-water regime. *Pinus sylvestris* forests dominate but there are areas with a predominance of deciduous species.

Historical Development

From ancient times, Moscow benefited from many auspicious events in its territorial and population development, in particular, from its advantageous geographical location in the centre of the Russian plain as well as diversity of natural conditions. Through the Moskva River it was possible to reach the main waterways of the Volga, Oka and Klyasma Rivers, which facilitated the spread of the population both in the latitudinal and longitudinal directions. The numerous tributaries of the Moskva River made it easy to move through the territory and to colonise the diverse landscapes close to and distant from Moscow. The natural environment played an important role in the emergence of the city. The earliest archaeological evidence of human occupation of the modern city dates from the Mesolithic (about 7,000–9,000 years ago).

About 2,500 years ago Iron Age settlers with their established agricultural practices started to colonise the area that is now occupied by the city. The use of iron substantially increased the production potential of the people and drastically increased their impact on the surrounding environment. Within the limits of the modern city there are ten Iron Age hill forts. The Iron Age settlements have a peculiar feature – their duration of almost a thousand years; they first appeared in the eighth and ninth centuries BC and continued to the seventh and eighth centuries AD. The human culture of the Moscow Basin at that time has been named the "Dyakovskoe" after Dyakovo, which was the first hill fort to be investigated.

Judging by the composition of the pollen preserved in the cultural layer of the Dyakovo Hill Fort, the extent of arable and pasture land around the fort at the beginning of the first millennium AD was continually expanding and shrinking. Probably at that time secondary vegetation started to replace the primary forest – evidence suggests that significant areas of open space had been created by forest clearance within a radius of several kilometers around the settlements. The analysis of burnt grain found in the cultural layer of the Dyakovo Hill Fort shows that the main agricultural plants were *Panicum miliaceum, Hordeum vulgare* and *Triticum aestivum.*

It is estimated that the total population living within the limits of modern Moscow during the Iron Age reached several tens of thousands – resulting in substantial changes to the natural landscape. Slavic tribes began colonising the Moscow region in the tenth century. By the late eleventh/early twelfth centuries, there had been a major increase in the number of settlements in the middle section of the Moskva River. The southern part of the modern city, located on the right bank of the Moskva River, was especially intensively populated. The well-drained soils provided excellent conditions for ploughing. One hundred and nineteen tombs (barrows) and five settlements were found along a 2 km section of the valley of the Yazvenka River.

Natural conditions in the northern and eastern parts of the modern city were less favourable for agriculture; therefore these areas remained sparsely populated even in the eleventh to thirteenth centuries. The unfavourable agricultural conditions contributed to the survival and preservation of large forest tracts (now the parks of Izmailovo, Kuzminki and Losiniy Ostrov).

Consequently, the territory of the modern city has been populated by people from the earliest times. The natural components of the landscape, especially plant associations, underwent significant changes and transformations during these early times. New natural-anthropogenic and anthropogenic complexes arose or began to be formed as the consequence of developing economic activities. There is no doubt that these changes, to one extent or another, are reflected in the past and present composition of the flora. The waves of the new geographically and ethnographically different human communities populating the territory were the main source of the origin of new species (cultivated, random and accompanying human culture). Probably, the most significant changes occurred in the period of mass colonisation of the region by the Slavs.

History of the City's Formation and Development

The first reference to "Moscow" dates from 1147. In the initial period of its history the city was a fortress on the Borovitsky Hill (25 m above the level of the Moskva River), representing the cape between the two rivers – Moskva and Neglinnaya. The ancient fortress had an irregular oval shape, occupied more than 3 ha and was bound by a ditch and a 7 m high earth rampart. The first stonewall around the Kremlin was built in 1366 to 1367 and existed for more than 120 years. In 1490, the Milanese master Pietro Antonio Solari began the construction of the new Kremlin wall, which still exists today.

During the sixteenth century, the city grew very quickly – it comprised a coalescence of four "cities" each surrounded by its own wall – the Kremlin (with a perimeter of about 2 km), Kitai-Gorod (enclosed by a stone wall in 1535 to 1538), Bely Gorod (enclosed by a brick wall in 1585 to 1591 and with a perimeter of about 7 km) and Skorodom (an early 1590s wooden fortress wall on the earthen ramparts with a perimeter of about 16 km). Towards the end of the century, the city's territory reached 1,878 ha (Fig. 5).

In the middle of the eighteenth century, a sequential expansion of the city occurred and the earth ramparts became its boundary (Kamer-Kollezhsky val),

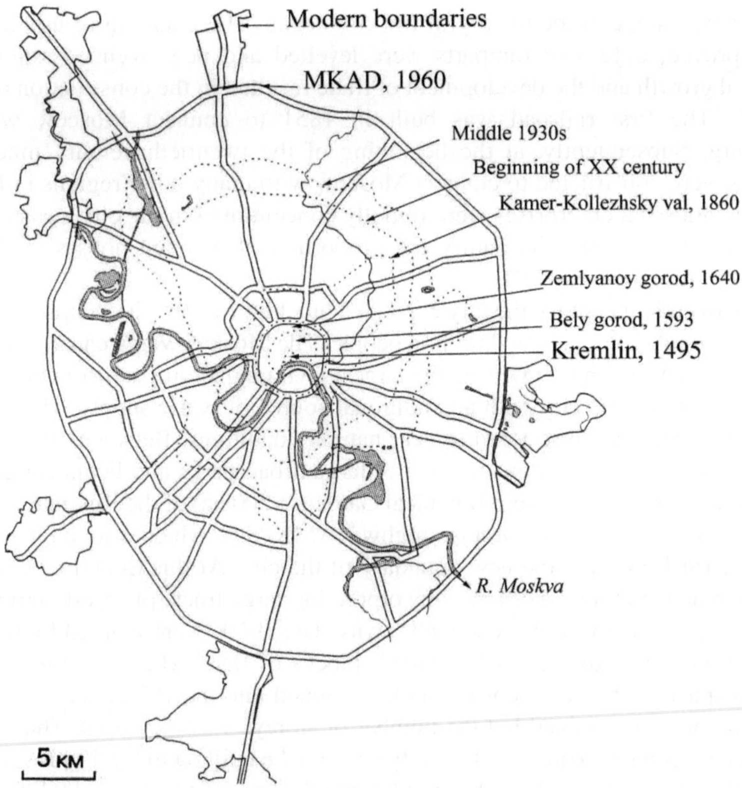

Fig. 5 Territorial growth of the city of Moscow

which was more than 32 km long. Towards the end of the century, Moscow occupied 63 km^2 and supported a population of between 150,000 and 200,000 people.

Unlike other European cities, wooden houses were predominant in Moscow at that time, for example, in 1796 they comprised 78% of all structures. The proportion of stone buildings decreased from 95% in the historical centre to 8% in the outlying areas. The ornamental and vegetable gardens were located near the houses. Many open areas remained in the city, including ravines, swamps, valleys, meadow and pastures. At the end of the eighteenth century such plots comprised a third of the city. The first avenues appeared in the city at the end of the eighteenth century. About 1756 P. Demidov established Moscow's first botanical garden, which is described by P.S. Pallas (Pallas 1781). Large out-of-town estates with regular landscaped parks became an outstanding characteristic of eighteenth century Moscow – some of them are now located within the city's boundaries.

In the nineteenth century, the growth and transformation of the urban territory became more intense; however, a fire in 1812 destroyed about 70% of the apartment buildings. In September 1812, Moscow was captured and occupied by the French army. Following the re-capture of the city from the French the following month, a new stage in the history of the city began. The planning of urban streets was improved, defensive ramparts were levelled and new avenues established. Industrial growth and the development of trade resulted in the construction of a rail network. The first railroad was built in 1851 to connect Moscow with St. Petersburg. Subsequently, at the beginning of the twentieth century nine more railroads were constructed to connect Moscow with many other regions in Russia. The new industrial enterprises were initially concentrated in the city's eastern sector. During the nineteenth century the proportion of wooden houses in the city remained quite high – at 54%.

In the twentieth century the city expanded quickly; by 1917 it occupied 242 km^2 and had a population of about 2 million people. The Moskva-Volga canal, which was built between 1932 and 1937, was the largest civil engineering construction of that decade, providing the city with an additional source of water supply. The flows in the Moskva River were restored to near-natural conditions. Between 1930 and the 1950s, Moscow witnessed the creation of large urban parks and botanical gardens, the largest of which is the Main Botanical Garden (>300 ha) of the Russian Academy of Sciences. The Moscow circular highway (MKAD), which was built between 1959 and 1961, became the new boundary of the city. At this time the territory of Moscow reached almost 879 km^2 – incorporating large tracts of forest, agricultural land, rural populations and even small towns. The 1960s were marked by the mass construction of predominantly five-storey blocks of flats, which are free standing and well spaced. This mass construction continued into the 1970s; the "free-standing" principle was retained but the number of storeys was increased. The last vast and still unoccupied territories of the city began to be built on in the 1990s stretching beyond the limits of the MKAD and enlarging the area of the city to 994 km^2.

In summary, throughout its history the area of the city, the proportion of "stone" buildings and their height have constantly increased. In spite of the circular character of its administrative boundaries and transport arteries, sectoral tendencies have always been well expressed. The city has spread by separate tongue-shaped sectors whose direction has changed in different periods. The individual sectors of the city are essentially different in their architecture and social structure. The process of development of new areas and their incorporation into the city in all periods of its expansion has resulted in the destruction of the natural drainage systems (river, ponds, and swamp) and forests. The increase in the population and maximisation of building in natural areas has been accompanied by an increase in the anthropogenic pressure, which ultimately leads to the worsening and gradual degradation of the quality of the landscape and its natural components.

Flora

The flora of the city comprises 1647 vascular plant species, related to 640 genera and 136 families. The overwhelming majority of families, genera and species (98%) are from the Division Magnoliophyta. The flora is dominated by two families the Asteraceae and Poaceae, which together comprise 22% of all the species (see Table 2). A noticeable increase in the role played by some families rich in adventive species (for example, Brassicaceae, Fabaceae, Rosaceae and Chenopodiaceae) is a characteristic feature of the flora of cities and anthropogenic landscapes, in comparison with regional rural flora. This situation is also observed in the flora of Moscow, in which Brassicaceae occupies third place in the number of species. At the same time, the proportion of other families is reduced in comparison to the regional flora (for example, Caryophyllaceae and Lamiaceae).

The genera with the largest number of species are *Carex* (47 species), *Hieracium* (25), *Polygonum sensu lato* (22) and *Veronica* (21). Perennial non-woody species are predominant in the flora, comprising about 56% of its composition (Table 3) while the proportion of annual species is approximately only half – 27%. In total, non-woody species account for about 90% of the city's flora with trees and shrubs accounting for only about 9%.

Dwarf shrubs include species such as *Andromeda polifolia, Arctostaphylos uva-ursi, Calluna vulgaris, Chamaedaphne calyculata, Ledum palustre, Linnaea borealis, Vaccinium oxycoccus, V. myrtillus, V. uliginum, V. vitis-idaea* and *Vinca minor*. The dwarf sub-shrubs include species such as *Chimaphila umbellata, Thymus marschallianus, T. ovatus, T.* x *loevyanus* and *T. serpyllum;* sub-shrubs include *Artemesia abrotanum, A. pontica* and *Pachysandra terminalis*.

Table 2 Families of the flora with the largest number of species

Family	Species number	% of all species	Family	Species number	% of all species
Asteraceae	192	11.7	Apiaceae	52	3.2
Poaceae	169	10.3	Caryophyllaceae	52	3.2
Brassicaceae	104	6.3	Ranunculaceae	46	2.8
Rosaceae	99	6	Chenopodiaceae	39	2.4
Fabaceae	74	4.5	Polygonaceae	38	2.3
Cyperaceae	69	4.2	Boraginaceae	31	1.9
Scrophulariaceae	63	3.8	Salicaceae	27	1.6
Lamiaceae	53	3.2	Orchidaceae	22	1.3

Table 3 The "life forms" of the flora

Life form	Species number	% of all species
Tree	76	4.6
Shrub	70	4.3
Dwarf shrub	11	0.7
Dwarf sub-shrub	5	0.3
Sub-shrub	3	0.2
Herbaceous	1,482	90
Perennial	929	56.4
Biennial	103	6.3
Annual	450	27.3

Table 4 Taxonomic composition of the main factions

	Native species	Xenophytes	Ergasiophytes
Family number	103	53	83
Genera number	358	224	256
Species number	823	440	384

Alien plant species play a significant role in the composition of the flora, accounting for half of it – 824 species. Analysis has shown a similar situation in a whole series of indices not only between groups of native and alien species but also between two basic factions of the alien, namely xenophytes and ergasiophytes. The number of families, genera and species in the native faction is higher than in every alien one (see Table 4). Among alien plants the families and genera variety is higher in the ergasiophytes (83 families, 256 genera) but xenophytes are the highest in terms of number of species (440).

The species of Asteraceae, Poaceae and Cyperaceae families hold the leading position in the native faction, comprising a total of 25% of the native flora of the city. The family of Poaceae leads among xenophytes and together with Asteraceae, Brassicaceae and Chenopodiaceae, it accounts for a half of this faction (Table 5 and Fig. 6).

Most of the species of ergasiophytes are within the Asteraceae, just as in the native faction. It is characteristic that the role of the Rosaceae is large – it occupies the second position in terms of the number of Ergasiophyte species. In total, the species of three leading families (Asteraceae, Rosaceae and Poaceae) comprise 27% of the ergasiophytes faction. The genera with the largest number of species are *Carex* (47 species), *Hieracium* (25), *Polygonum sensu lato* (22) and *Veronica* (21).

There are also differences in the composition of the largest genera. So far as native species are concerned, the largest genera are *Carex* (43 species), *Hieracium* (24) and *Alchemilla* (19); among the Xenophytes it is *Artemisia* (12), *Bromus sensu lato* (12), *Polygonum sensu lato* (9) and in the ergasiophytes faction – *Acer* (8), *Allium* (8) and *Populus* (8).

The proportion (72.3%) of life forms in the native flora are predominately perennial non-woody species, the proportion of annual and biennial species is more than three times lower. By comparison, tree and shrub species are represented by a virtually identical number of species – in total they represent about 5% of the composition of the native flora. The relationship between the number of perennials and the number of annuals and biennials is reversed in the xenophyte faction where the largest number of species are annuals and biennials, which comprise more than 60% of the species composition of the faction. In the composition of ergasiophytes, as in the native flora, the proportion of perennial non-woody species is the largest (about 45%). The characteristic feature of ergasiophytes is the high variety of woody plants, which comprise more than 25% of this faction.

When assessed in terms of naturalisation, the unstable, short-lived ephemerophytes predominate, comprising 53% of all alien species. They occur in the city as

Table 5 Composition of the largest families of the basic factions

Native species	Species number	%	Xenophytes	Species number	%	Ergasiophytes	Species number	%
Asteraceae	86	10.4	Poaceae	80	18.2	Asteraceae	42	11
Poaceae	64	7.8	Asteraceae	64	14.5	Rosaceae	37	9.6
Cyperaceae	59	7.2	Brassicaceae	57	13	Poaceae	25	6.5
Rosaceae	48	5.8	Chenopodiaceae	27	6.1	Fabaceae	19	4.9
Scrophulariaceae	38	4.6	Fabaceae	24	5.5	Brassicaceae	17	4.4
Caryophyllaceae	35	4.3	Scrophulariaceae	15	3.4	Apiaceae	12	3.1
Fabaceae	31	3.8	Rosaceae	14	3.2	Salicaceae	12	3.1
Ranunculaceae	31	3.8	Lamiaceae	14	3.2	Scrophulariaceae	10	2.6
Brassicaceae	30	3.6	Polygonaceae	14	3.2	Lamiaceae	10	2.6
Lamiaceae	29	3.5	Caryophyllaceae	12	2.7	Ranunculaceae	10	2.6
Apiaceae	29	3.5	Apiaceae	11	2.5	Solanaceae	10	2.6

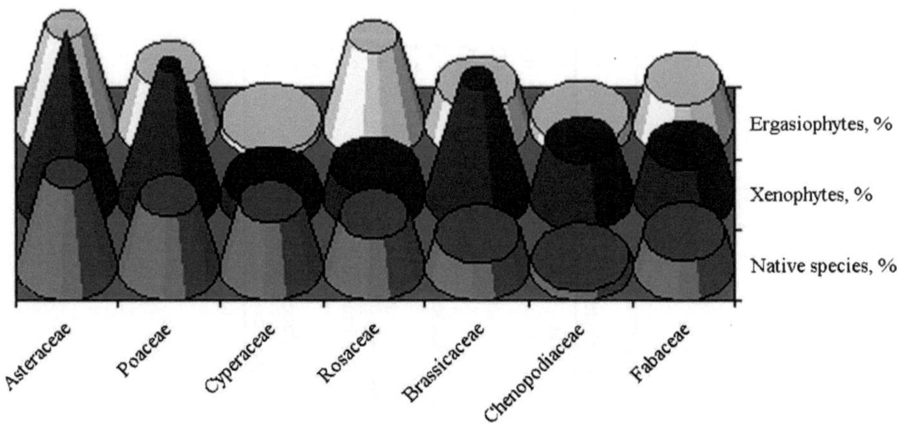

Fig. 6 The leading families of the main factions of the flora of the city of Moscow

the result of unintentional introduction by people. Twenty-four percent of the species that occur in Moscow are coloniophytes. The large number of alien species indicates that a substantial proportion of the city's flora is made up of unstable components. Coloniophytes are quite vulnerable to extinction because they have limited propagation potential and therefore frequently exist only because of their ability to reproduce vegetatively. The percentage of ephemerophytes is almost identical to that of xenophytes – 52% and 54%, respectively. It should be noted that epekophytes predominate among xenophytes, comprising 17% of the faction while the percentage of alien species that establish and propagate in natural and semi-natural habitats account for 4%. In the ergasiophytes, the proportions are reversed; agriophytes comprise 14% and epekophytes about 11%.

This indicates that xenophytes have mainly colonised and established in man-made habitats, whereas "fugitives from the culture" – plants cultivated by man (for example, species such as *Acer negundo, Echinocystis lobata, Heracleum sosnowskyi, Hippophae rhamnoides, Impatiens glandulifera, I. parviflora, Fallopia japonica, Solidago gigantean, Veronica filiformis, Vinca minor* and *Viola odorata*) are the main species that are colonising and spreading into natural and semi-natural habitats. Annual and biennial species are predominant (52%) among the spreading plants – epekophytes and agriophytes in the faction of xenophytes. On the other hand, in the faction of ergasiophytes, the perennial non-woody plants predominate (70%), and the proportion of woody species is also large – 20%.

The plants found in the city are represented by populations of species of native and alien origin, including *Achillea ptarmica* 'Pleno Flore' and *Phalaris arundinacea*. The former are encountered only in natural habitats, the latter in disturbed and man-made habitats. Sometimes, plants of introduced populations establish more quickly than native ones, for example, *Circaea lutetiana* has spread widely within the main botanical garden of the Russian Academy of Science. Other forest species such as *Anemone nemorosa, Campanula latifolia, C. trachelium,* and *Corydalis solida* have also spread from cultivated populations. It is likely that the species that have spread in the botanical gardens have done so from samples obtained in different geographical origin, that is, they have mixed gene pools – this is assumed in relation to the populations of *Circaea lutetiana*.

Culture plays an exceptionally important role as a source of the alien faction of flora. For example, even in the last 20 years *Adenocaulon adhaerescens, Cardamine leucantha, Corydalis ochotensis, Hylomecon vernalis, Meehania urticifolia* and *Schizopepon bryoniifolius* have spread from botanical collections into forest habitats. Some species are intentionally introduced by people into natural and semi-natural habitats, for example, *Lunaria rediviva* was planted in the urban forests for the purpose "of repatriation". This plant established rapidly and now seriously competes with local forest species. *Eichhornia crassipes* has been introduced into ponds and watercourses for the purpose of decreasing pollution. However, in the summer it spreads and forms dense "mats" in some ponds. The introduction of such a species is believed to be causing serious damage to natural vegetation.

The sources of the propagules of alien species changes over time, for example, in recent years horticultural activities including the importation of soil for lawns and gardens have become the most significant source of alien plants, especially the seeds of ruderal species such as *Amaranthus blitum, Cardamine hirsuta, Claytonia perfoliata, Euphorbia peplus, Anchusa arvensis, Oxalis repens, Sideritis montana, Silene dichotoma, Sinapis alba, Stachys annua* and *Veronica peregrina*.

In the last 150–170 years, 104 plant species known from herbarium collections have disappeared from the modern city. The representative families include Ranunculaceae (9 species), Orchidaceae (8 species), Cyperaceae (8 species), Scrophulariaceae (7 species) and Poaceae (6 species). Among the species that have disappeared 70% are aboriginal plants mainly those of wetlands, including marshes, wooded swamps, wet meadows, aquatic and riversides (33 species). Nineteen forest and forest edge species have become extinct in Moscow together with 19 species of short grassland, moist meadows and margins of meadows.

Several of the aboriginal species that have disappeared were only known from a single record finding; they have not been re-found either within the city or its suburbs. Some of the species do not now occur in the Moscow region or are known only from very few sites. They include *Arctostaphylos uva-ursi, Botrychium virginianum, Carex dioica, C. hartmanii, C. panicea, Cypripedium guttatum, Diplazium sibiricum, Elatine alsinastrum, Eleocharis quinqueflora, Equisetum scirpoides, Eriophorum gracile, Liparis loeselii, Melampyrum cristatum, Ophioglossum vulgatum, Pedicularis sceptrum-carolinum, Peucedanum oreoselinum* and *Ranunculus polyphyllus*. A number of species that no longer occur in the city are also declining in Moscow and adjacent regions, for example, *Coeloglossum viride, Cypripedium calceolus, C. guttatum, Gladiolus imbricatus, Hottonia palustris, Orchis militaris* and *Pulsatilla patens*.

Thirty percent of the species that have disappeared are aliens, most of which were only known from a single finding; they include *Anagallis arvensis* ssp. *foemina, Avena strigosa, Centaurea arenaria, C. trichocephala, Ceratocephala testiculata, Chaerophyllum hirsutum, Cirsium serrulatum, Meniocus linifolius, Onosma tinctoria, Scrophularia vernalis, Sherardia arvensis, Stipagrostis plumosa, Agrostemma githago, Androsace maxima* and *Senecio vernalis*, and were more widely distributed. The history of *Ranunculus bulbosus* is interesting, because it existed at one site in the city for a hundred years and then after 1962 it was no longer found.

Cultivated Flora of the City

The cultivation of plants is an integral part of human civilisation; therefore it will not be an exaggeration to say that the plants in the city have been cultivated over the entire duration of its existence. The main purposes of the cultivation of plants are aesthetics, cognitive (ornamental plants, development of botanical gardens) and utilitarian (cultivation of food, fodder and industrial crops). The cultivation and growing of plants is closely associated with practical human activities; therefore, it occurs in the sphere of influence of a whole series of social factors as a result of which it is subject to dynamic changes. There is an opinion that a plant growing in "antiquity" was pursued only for utilitarian purposes and that it was only considerably later that it acquired an aesthetic and cognitive purpose. It is desirable to think that this may not be so; human beings have always been attracted to beauty and knowledge. Like the spontaneous flora, the composition of the cultivated flora depends on natural and climatic factors. The historical and cultural idiosyncrasy of the country or region gives some individual features to it. Simultaneously, the cultivated flora is influenced by globalisation, which ensures the similarity of its composition in different countries and cities.

The composition of cultivated flora changes with time but the data about its composition in the past centuries is very scarce and this is an obstacle for carrying out detailed analyses of it. There are some well known historical facts that Tsar Aleksey Mikhaylovich (1629–1676) energetically organised the pharmaceutical vegetable gardens in Moscow, for which he ordered plants to be sent from different regions of Russia. The Tsar is also known for his experiments to increase the variety of cultural plants, for example, *Vitis vinifera, Gossypium* spp. and *Morus* sp. There are records mentioning a "Fir tree" (*Abies sibirica*) and "Sibirica" pine (*Pinus sibirica*). The composition of the cultivated flora in that period was apparently noticeably different from that of today. The plant names are not always reliable. Aromatic herbs, for example, are frequently mentioned, including *Artemesia abrotanum, Hyssopus officinalis, Tanacetum balsamita, T. vulgare, Ruta graveolens* and *Salvia* sp. indicating a preference for species with such an attribute.

The development of horticulture in the eighteenth century coincided with the growth of cultural, scientific and economic connections with Europe and was accompanied by a sharp increase in the diversity of cultivated plants that were introduced into Moscow. During this period the first botanical gardens appeared, the largest of which was the garden of P. Demidov. According to the description of Pallas there were 2,224 plant species in the garden (Pallas 1781). By the end of the eighteenth century the city had 1,523 gardens, 156 greenhouses and 376 vegetable gardens (Moscow in the descriptions 1997).

Some plants, which were cultivated at that time, became naturalised and nowadays are encountered in the old homestead parks. They include *Aquilegia vulgaris, Arrhenatherum elatius, Bellis perennis, Luzula luzuloides, Phyteuma spicatum,*

Poa chaixii, Rhamnus cathartica, Sorbaria sorbifolia, Trisetum flavescens, Vinca minor, Viola odorata and *Hieracium* spp.

In the eighteenth century, the homestead parks (especially those that had been landscaped) were dominated by a small number of woody species, particularly *Tilia* spp. In the nineteenth century, this approach changed and it became fashionable to create parks with large numbers of woody species. During the subsequent years, this tendency became the basic strategy for landscape planting, for example, the creation of the parks in the city between 1930 and 1950 involved the planting of 100–200 different woody species.

In the middle of the 1980s (during the Soviet period), landscaping was mainly characterised by a "forest" approach, which involved the planting of large quantities of woody plants. The creation of lawns and flower gardens had low value and priority. The variety and availability of ornamental non-woody plants were very poor.

Between 1990 and 2000 the availability of the variety of plants underwent significant revolutionary changes. The variety of ornamental species substantially increased, mainly due to the import of seeds and living plants. The nature of landscape planting also changed. As described, in the previous decade priority was given to "forest" planting. Starting in the 1990s and still continuing is the increase of lawns and ornamental plants as the structural elements in landscape design. Flowerbeds and lawns became an indispensable element of private gardens and landscapes associated with offices and in residential districts. A new consumer demand has developed based on the "rarity" of plants and their "novelty" in terms of cultivated/horticultural varieties. Large private collections of plants have appeared – comparable in their value and function to the botanical gardens.

Significant changes have also occurred in the composition of cultivated species. In particular, the proportion of monocotyledonous plants has increased because of the beautiful flowers of the Iridaceae and Liliaceae and the infloresences of the Cyperaceae and Poaceae. In previous times, species of the Cyperaceae were practically unused as ornamental plants while in lawns the Poaceae was represented by a single species, *Phalaris arundinacea*. Now, they represent more than 5% of the plant species and varieties that are being used for decorative purposes.

At present, the composition of the cultivated flora is not stable, it changes over years. Currently, it is going through the formation phase, as a result it can only be estimated, at or about 1,200 non-woody species (of 115 families) in cultivation in the city and its suburbs. The largest number of species (30%) is in the Asteraceae, Liliaceae and Ranunculaceae. Of the 1,200 species, only a little more than 300 species are commonly found in the city, of which the 20 most frequent taxa are *Ageratum houstonianum, Alcea rosea, Aquilegia vulgaris, Astilbe x arendsii, Begonia semperflorens, Brunnera sibirica, Calendula officinalis, Hemerocallis fulva, Iris germanica, Narcissus poeticus, Petunia x hybrida, Phlox paniculata, Rudbeckia laciniata, Saponaria officinalis, Sedum spectabile, Solidago gigantean, Symphytum caucasicum, Tagetes patula, Tulipa gesneriana* and *Viola x wittrockiana*. The most frequently grown native species is *Convallaria majalis*.

Species composition and the relationship between annual and perennial species are noticeably variable in the basic types of new plantings. The urban flowerbeds generally contain only a small number of taxa, based mainly on annuals such as *Ageratum houstonianum, Begonia semperflorens, Petunia x hybrida* and *Tagetes patula.* Some perennial plants are grown as annuals, for example, *Tulipa gesneriana.* In recent years it has become fashionable to plant some species that were previously only grown inside; they include *Chlorophytum comosum, Coleus blumei,* and *Pelargonium zonale.* In general private gardens and flowerbeds are considerably richer in species with perennial plants predominating.

There are more than 370 woody species in the city, of these about 170 species are encountered most frequently in the urban area, including *Acer ginnala, Acer negundo, Acer platanoides, Aesculus hippocastanum, Betula pendula, Cerasus vulgaris, Fraxinus pennsylvanica, Malus domestica, Prunus avium, Salix alba, S. fragilis, Tilia cordata, T. platyphyllos, Ulmus laevis* and *Populus* spp. The shrub species include *Caragana arborescens, Cotoneaster lucidus, Lonicera tatarica, Physocarpus opulifolius, Spiraea chamaedryfolia, Cornus alba, Symphoricarpos rivularis, Syringa josikaea, S. vulgaris, Crataegus* spp. and *Rosa* spp. Of the coniferous species *Larix decidua, Picea pungens* and *Thuja occidentalis* are encountered more frequently than others. *Parthenocissus inserta* is the only climbing species found in the city. Evergreen "leafy" trees and shrubs are rare, with *Mahonia aquifolium* being encountered more frequently than others.

Bryophytes

At present, the composition of mosses and liverworts of Moscow has not been fully studied. According to the preliminary data, 188 species have been recorded in the city, comprising 1 species of Anthocerothae, 26 Hepaticae and 161 Musci (Abramova and Ignatov 2004). The species-richness of mosses is greater in the large forest areas on the outskirts of the city than within it, for example, 136 species were recorded in Kuntsevo and 112 species in the Losiniy Ostrov. The greatest diversity has been recorded in the *Betula* forests, in which *Atrichum undulatum, Plagiomnium cuspidatum, Oxyrrhynchium hians, Cirriphyllum piliferum,* among other species, are widespread. *Rhodobryum roseum, Ptilium crista-castrensis, Brachythecium albicans, Dicranum scoparium* and *Sphagnum centrale* are encountered in different types of *Pinus* forests. *Plagiomnium undulatum, Rhytidiadelphus triquetrus, Hylocomium splendens* and *Plagiochila porelloides* are common in different types of *Picea* forests. In Sphagnum bogs, *Sphagnum angustifolium, S. balticum, S. fimbriatum, S. riparium, S. girgensohnii* and *S. squarrosum* are widespread. The urban habitats are populated by only a few species, including *Pylaisia polyantha, Bryum argenteum, Ceratodon purpureus, Funaria hygrometrica, Barbula unguiculata, Brachythecium salebrosum, Amblystegium serpens, Marchantia polymorpha, Didymodon fallax* and *Schistidium apocarpum.*

Fungi (including Lichenised Fungi)

Lichenised Fungi

At present 87 species of lichen have been recorded in the city, the most frequent being *Physcia stellaris, Phaeophyscia orbicularis, Parmelia sulcata, Scoliciosporum chlorococcum, Hypogymnia physodes, Cladonia coniocraea* and *C. fimbriata* (Biazrov 2004).

The variety of lichens in the central part of the city (within the limits of the Sadovoye Kol'tso (Moscow inner ring road)) is minimal, only two species being encountered – *Phaeophyscia orbicularis* and *Parmelia sulcata.* A much larger number of species occur in the outskirts of the city, especially in the preserved large forest areas – 65 in Losiniy Ostrov, 38 in the Serebryaniy Bor and 36 in Kuntsevo. The natural forest areas are important refuges for lichens; for example, 35 species were recorded in only one of them, for example, *Buellia punctata, Caloplaca decipiens, Candelariella xanthostigma, Cladonia botrytis, Evernia mesomorpha, Lecanora muralis, Peltigera canina, Ramalina farinacea, Usnea hirta* and *Xanthoria fallax.*

Habitats

The plant species diversity of the different habitats change within rather wide limits (see Tables 6 and 7); however, the maximum diversity both for natural (forest habitats – 392 species) and urban habitats (waste ground – 408 species; railway land – 423 species) is similar.

Table 6 Taxonometric diversity according to species-richness

Habitats	Number of species	Number of families
Buildings of the1930–1950	172	36
Buildings of the1970s	194	43
Fallow land	200	43
New lawns	217	40
Buildings of the 1960s	218	47
Wasteland	218	37
Buildings of the1990s	226	38
Grassland	243	46
City centre	254	43
Buildings of the old city	283	45
Aquatic	312	65
Woodland	392	75
Waste ground	408	63
Railway land	423	62

Wasteland = brownfield and similar sites. Waste ground = rubbish dumps and similar areas

Table 7 Taxonomic diversity according (approximately) to age and naturalness of the habitat

Habitats	Number of species	Number of families
Buildings of the old City	283	45
Railway land	423	62
Buildings of the1930s–1950s	172	36
Buildings of the 1960s	218	47
Buildings of the1970s	194	43
Buildings of the1990s	226	38
City centre	254	43
Waste ground	408	63
Woodlands	392	75
Grassland	243	46
New lawns	217	40
Fallow land	200	43

Waste ground = rubbish dumps and similar areas

It is important to emphasise the difference in the character and origin of the species diversity of these habitats. The waste ground and railway vegetation is characterised by casual, unstable species, and the structure of the plant communities changes practically every year. The proportion of "unstable" alien species (ephemerophytes) in the flora of these habitats exceeds 25% while in the forest habitats it only accounts for 4%.

The ratio of annual to perennial herbaceous plants is a rather useful parameter in analysing the urban flora. In the young habitats (new lawns, waste grounds and fallow), the ratio of these groups of plants is almost the same although one slightly exceeds the other in places. On railway land, the older buildings (the old city, city centre buildings of the 1930–1950s) and on the more recent buildings, the prevalence of perennials is insignificant. In contrast, the number of perennials associated with the buildings of the 1960s and 1970s is twice that of the older buildings. In the natural habitats (forests, meadows and aquatic habitats), on fallow land and the slopes of railway cuttings the number of perennial species exceeds annuals by four times. The high proportion of annuals is characteristic of unstable, temporary habitats and for the initial stages in the formation of the vegetative cover. On the other hand, the high proportion of annuals in the conditionally stable vegetation types associated with old buildings is the result of constant anthropogenic pressure, which interferes with the succession of the plant communities.

The ratio of native to alien species in the flora of different types of habitats also varies (Table 8). The greatest number of alien plants occurs on railway land and waste ground. In these habitats the alien species comprise 50% of the waste ground flora or even exceeds the number of native species on railway land. The percentage of alien species associated with the various types of apartment blocks is approximately twice as low as that of native species. In the flora of natural habitats native species prevail, in general terms the presence of alien species is insignificant ranging from 5% in meadow habitats to 15% in aquatic and forest habitats.

Ergasiophytes prevail in the flora of the majority of habitats (Table 8). In the areas of the "newer" buildings (that is, 1960s), forest habitats and on fallow land

Table 8 Percentage of alien species in different habitats

Habitats	Alien species (% of all species)	Xenophytes (% of all species)	Ergasiophytes (% of all species)
Railway land	51	30	21
Waste ground	50	24	26
Buildings of the old city	34	14	19
Buildings of the 1970s	28	8	20
Wasteland	28	17	11
Buildings of the 1930s–1950s	28	13	15
New lawns	26	14	12
Buildings of the 1960s	26	9	17
Buildings of the 1990s	22	11	11
Buildings of the city centre	21	9	11
Fallow land	20	5	15
Woodland	15	3	12
Aquatic	13	7	6
Grassland	6	2	3

Wasteland = brownfield and similar sites. Waste ground = rubbish dumps and similar areas

the number of ergasiophytes is twice to four times higher than xenophytes, which are generally only abundant in the flora of railway land and to some extent unstable, younger habitats (for example, wasteland and new lawns). It was found that the prevalence of xenophytes in the areas of the "newest" buildings (1980s to 1990s) is temporary; in due course ergasiophytes will become abundant.

Within the administrative borders of the city the large area of natural landscape have been preserved and have become a refuge for a wide diversity of natural vegetation types and their flora. The vegetation of such sites reflects the characteristic features of the natural landscape of the city – powerful anthropogenic pressure could not destroy geographical heterogeneity of the urban territory (Gutnikov and Shvetsov 2004).

The highest diversity of natural flora and vegetation occurs in the peripheral zone of the city, including the two largest forests "Losiny Ostrov" (about 2,910 ha) and Bitza Forest (about 2,230 ha). Areas of natural habitats in the inner parts of the city are smaller and considerably less diverse than those in the periphery. The greatest number (18) of the "Moscow Region Red Book Species" occurs in the Moscvoretzk-Skhodnia landscape zone (zone 4 in Fig. 4), probably as a result of the high habitat diversity that occurs in this area.

The ecological character of rare native species is a good indicator of the natural diversity of different city landscapes. Characteristic species of the landscapes of the southern part of the city (the right bank of the Moskva river), (landscape zones 1–3 in Fig. 4) are broad-leaved forest, meso-xerophytic meadows and forest edges, which include such species as *Corydalis cava, C. marschalliana, Gentiana cruciata, Koeleria cristata, Lathyrus niger, Omphalodes scorpioides, Polystichum braunii, Senecio erucifolius* and *Thymus marschallianus*. The species of *Pinus* and wet forests are characteristic of the western part of the city (landscape zone 4, Fig. 4), the species present include *Chimaphila umbellata, Corallorhiza trifida,*

Goodyera repens, Linnaea borealis, Malaxis monophyllos and *Pyrola media.* In the northern part (landscape zones 7-8 in Fig. 4) the wet meadows and bogs contain *Andromeda polifolia, Carex chordorrhiza, C. lasiocarpa, Dactylorhiza longifolia, Drosera rotundifolia, Ledum palustre, Montia fontana, Salix myrtilloides* and *Scheuchzeria palustris*.

Forests

Moscow is located in the mixed coniferous/broad-leaved forest zone. The forest vegetation comprises a mosaic of nemoral and boreal species. The main factors determining the features of the habitat are one or a combination of the geological structure, relief and anthropogenic factors. Broad-leaved forests prevail in the southern part of the city (Fig. 7). The canopy species include *Acer platanoides, Quercus robur, Tilia cordata,* with some *Fraxinus excelsior, Ulmus glabra* and *U. laevis*. The shrub layer species include *Corylus avellana, Euonymus verrucosa* and *Lonicera xylosteum*. Such species as *Actaea spicata, Aegopodium podagraria, Anemone ranunculoides, Asarum europaeum, Campanula latifolia, Carex pilosa, Corydalis solida, Ficaria verna, Gagea lutea, Lamiastrum galeobdolon, Lathyrus vernus, Mercurialis perennis, Poa nemoralis, Polygonatum multiflorum, Pulmonaria*

Fig. 7 Tsaritsyno, *Tilia cordata* old forest

obscura, Ranunculus cassubicus, Stellaria holostea and *Viola mirabilis* are typically found in the herb layer. The rare species in such habitats include *Corydalis cava, C. intermedia, C. marschalliana, Lathyrus niger, Omphalodes scorpioides* and *Polystichum braunii.*

The largest sites where conifers (*Picea abies* and *Pinus sylvestris*) predominate occur in the western and eastern parts of the city. These are habitats that contain an abundance of *Vaccinium myrtillus* and *Vaccinium vitis-idaea.* Many rare species occur in the conifer forest communities including *Carex echinata, C. vaginata, Chimaphila umbellata, Corallorhiza trifida, Goodyera repens, Hypopitys monotropa, Linnaea borealis, Lycopodium annotinum, Malaxis monophyllos* and *Pyrola media. Alnus glutinosa, A. incana, Salix alba* and *S. fragilis* communities occur on the banks of the rivers and estuaries. *Betula pendula* is widespread throughout the city occurring in all forest communities or forms independent stands, which appear in fallows and areas of forest that have been felled. These "*Betula* forests" are being restored to *Tilia cordata* or *Picea abies* forests.

The percentage of alien plants in the flora of the forest habitats is insignificant, reaching only an average of 15%. The most abundant (12%) of the alien species are ergasiophytes; the percentage of xenophytes is much lower (about 3%). Another important feature is the significant proportion (30%) of alien tree and shrub species, the most widespread being *Acer negundo, Amelanchier spicata, Fraxinus pennsylvanica, Malus domestica, Sambucus racemosa, Sorbaria sorbifolia* and *Cornus alba.* Among herbaceous plants the most common is *Impatiens parviflora;* other less common species include *Aquilegia vulgaris, Epilobium pseudorubescens, Geum macrophyllum, Impatiens glandulifera, Lilium martagon, Scilla siberica, Telekia speciosa, Vinca minor* and *Viola odorata.*

Some species of forest habitats have the capability of colonising, for example, stone walls and fence posts support *Athyrium filix-femina, Dryopteris carthusiana* and *Gymnocarpium dryopteris.* Spring ephemerals such as *Anemone ranunculoides, Corydalis solida, Ficaria verna* and *Gagea minima* actively colonise city squares, small urban parks, hedgerows and even the old city. In the last 10–20 years *Epipactis helleborine* has colonised the city. Some species show a dual reaction to urban conditions – they are unable to colonise urban habitats directly but when introduced they are able to become established, for example, *Anemone nemorosa, Asarum europaeum, Campanula latifolia* and *Matteuccia struthiopteris.*

The prospects for the preservation of the forest communities in the city are good. The basic problems are natural successions as a result of which the variety of species is decreasing and the direct destruction of forest habitats by human activities. Broad-leaved forest communities and a mosaic of nemoral species are becoming established in the city. At the same time, communities of the *Pinus sylvestris* forests are deteriorating because the species is not regenerating. In addition the structure of the community is made complicated by the growth of *Sorbus aucuparia* and some other species in the sub-canopy layers. As a whole, practically all urban and semi-natural habitats have a tendency for successive development towards formation of forest communities.

Meadow Flora and Meadow Habitats

The species composition of meadow habitats is rather variable; the main factors determining the species composition are one or a combination of the geological, relief, origin of the habitat, nutrient-richness of the substrate and the groundwater regime.

High species diversity is a characteristic of meso-xerophytic meadows formed on the sandy slopes of river valleys (Fig. 8). Not only do they contain meadow species but also plant species of open woodland and the woodland edge, which indicate that the land could have been previously occupied by forest. The species include *Campanula persicifolia, Corydalis solida, Helichrysum arenarium, Pteridium aquilinum* and *Vaccinium vitis-idaea*. The characteristic grassland species are *Agrimonia eupatoria, Anthemis tinctoria, Campanula bononiensis, Campanula glomerata, C. rotundifolia, Carex praecox, Centaurea scabiosa, Dianthus fischeri, Eryngium planum, Fragaria viridis, Galium verum, Inula salicina, Lathyrus sylvestris, Leontodon hispidus, Medicago sativa ssp. falcata, Origanum vulgare, Poa angustifolia, Primula veris, Ranunculus polyanthemos, Rumex thyrsiflorus, Senecio jacobaea, Seseli libanotis, Silene nutans, Solidago virgaurea, Thalictrum minus, Tragopogon orientalis, Trifolium montanum, Verbascum nigrum, Veronica teucrium, Viola hirta* and *Viscaria viscosa* The drier sites contain *Filipendula vulgaris, Koeleria cristata, K. delavignei, Phleum phleoides* and *Thymus marschallianus.* The short-grass meadows on wet, nutrient-poor soils contain *Agrostis capillaris, Alchemilla* spp., *Anthoxanthum odoratum, Briza media, Carex pallescens, Deschampsia cespitosa, Festuca rubra, Potentilla erecta, Ranunculus acris* and *Succisa pratensis*. *Botrychium lunaria, Nardus stricta* and some other rare species are to be found occasionally in such meadows. The following

Fig. 8 Kolomenskoye, grassland and Moskva River

species are more abundant in other mesophytic meadows: *Achillea millefolium, Elytrigia repens, Alopercurus pratensis, Centaurea jacea, Dactylis glomerata, Festuca pratensis, F.rubra, Geranium pratense, Hieracium umbellatum, Lathyrus pratensis, Leucanthemum vulgare, Phleum pratense, Pimpinella saxifraga, Poa pratensis, Prunella vulgaris, Ranunculus acris, Solidago virgaurea, Stellaria graminea, Trifolium hybridum, T. medium, T. pratense, Vicia cracca* and *V. sepium.*

The proportion of alien species in the flora of meadow habitats is small and amounts to an average of only 5%. The two most common alien species are *Arrhenatherum elatius* and *Trisetum flavescens*.

The survival of meadow communities is only possible under certain conditions, which are difficult to maintain in a large city. In addition to the direct destruction of meadow communities by human activity, they suffer from the expansion of *Calamagrostis epigeios* and trees such as *Betula pendula, Populus tremula* and *Salix* spp. which change the structure of the herb communities and cause a decrease in their diversity.

Aquatic Habitats and their flora

The ecological spectrum of aquatic habitats and therefore species diversity decreases from the suburbs to the city centre. In the city centre, the diversity is at its lowest, there is practically no place for marginal plants because the rivers have stone embankments. The predominant aquatic species are the hydrophytes *Ceratophyllum demersum, Potamogeton lucens, P. pectinatus* and *P. perfoliatus.*

The variety of aquatic habitats is largest in the peripheral areas, for example, the floodplain marshes, old river beds, shallow waters, river banks, brooks and springs. The most abundant species include *Phragmites australis, Acorus calamus, Butomus umbellatus, Calamagrostis canescens, Calla palustris, Carex cespitosa, C. rostrata, C. vesicaria, Equisetum fluviatile, Glyceria maxima, Phalaris arundinacea, Scirpus lacustris, Sparganium erectum, Typha angustifolia* and *T. latifolia.* Communities of *Scolochloa festucacea* are only found in the western part of the city. The communities of free-floating, submerged and rooted hydrophytes are also variable in their species composition, which includes such species as *Ranunculus circinatus, Ceratophyllum demersum, Hydrocharis morsus-ranae, Lemna minor, L. trisulca, Myriophyllum spicatum, M. verticillatum, Nuphar lutea, Nymphaea candida, Potamogeton lucens, P. natans, P. pectinatus, P. perfoliatus, Spirodela polyrhiza, Stratiotes aloides, Utricularia vulgaris* and *Zannichellia palustris*. Occasionally occurring species include *Nuphar pumila, Potamogeton praelongus, P. trichoides, Sparganium natans* and *Utricularia minor.*

The number (19) of alien species in the aquatic habitats is insignificant. Although some of them only have a local distribution, the coenotic role of each of them is, as a rule, significant. They are dominant or in abundance in the reservoirs in which they occur. The species include *Alisma gramineum* ssp. *gramineum, Cabomba caroliniana, Ceratophyllum submersum, Elodea canadensis* (the most abundant of all), *Phragmites altissimus, Vallisneria spiralis, Wolffia arrhiza* and *Zizania latifolia*. The majority of the alien species just mentioned are heat-loving plants, whose distribution in Moscow can be interpreted as an indication to the changes in the temperature of urban reservoirs.

In recent years, the city administration has been trying to combat water pollution with the help of the tropical species *Eichhornia crassipes* and *Pistia stratiotes,* which grow rapidly in some reservoirs in the summer. In the autumn, after the first frost, it dies without causing a major oxygen demand. Such experiments with non-native species represent serious ecological threat to the natural communities.

Water plants are a dynamic component of urban flora, for example, *Nuphar lutea* and *Nymphaea candida* were rather rare in the 1980s but during the 1990s they spread rapidly in the city reservoirs and are now found even in the Moskva river in the city centre. In the first half of the twentieth century, *Zannichellia palustris* was considered a rare plant in the city (it was known only from a herbarium of the first quarter of the nineteenth century), now it is widespread throughout the city's reservoirs.

What are the prospects for this group of plants? The number of aquatic habitats (for example, ponds, rivers, and bogs), especially of natural origin is declining. Regulation of the drainage system (for example, an increase or reduction in flow, velocity and seasonal dynamics) has resulted in the changing of conditions in reservoirs and the construction of stone or concrete embankments have resulted in the disappearance of "natural" banks and the associated species. Current engineering construction methods provide stone or concrete embankments, which exclude the opportunity for marginal and bank-side habitats and their component species.

At the same time, habitats of an anthropogenic origin (for example, ditches along roads and railway lines, temporary reservoirs and balancing lakes/impoundment reservoirs) are occupied by some hydrophyte and marginal species such as *Alisma plantago-aquatica, Callitriche cophocarpa, C. palustris, Catabrosa aquatica, Cyperus fuscus, Glyceria fluitans, Lemna minor, Limosella aquatica, Potamogeton* spp., *Scutellaria galericulata, Typha latifolia* and *Zannichellia palustris*.

Some *Sphagnum* bogs have been preserved on the outskirts of the city, which are now the only places in the city to see *Andromeda polifolia, Carex chordorrhiza, C. limosa, Chamaedaphne calyculata, Drosera rotundifolia, Eriophorum vaginatum, Ledum palustre, Vaccinium oxycoccus, Salix myrtilloides, Scheuchzeria palustris* and *Vaccinium uliginosum*. Many bogs have disappeared during the development of the city. The reason is not only in their direct destruction human activities; a bog is a complex ecosystem (especially hydrologically) that probably cannot exist for a long time in a large modern city.

Buildings

Buildings are the fundamental component of urban habitats – they make it what it is; they differ in age, architectural design, materials and their location in relation to the city centre. An important characteristic is the housing density. The city centre (on the border of Zemlyanoy Gorod) and the old city (on the border of Kamer-Kollezhsky Val) are the oldest areas of the city and have the highest housing density

(Fig. 9). The age of the buildings and the amount of developed land (hard surfaces) decreases towards the periphery of the city.

The diversity of the flora is higher in the old areas of the city and in the buildings of the 1960s to 1990s (Table 6, Fig. 10). The percentage of alien species is higher in the old city (34% of flora of any given habitat); in the other types of development the proportion is less than a third of the species found in these habitats (Table 8). Ergasiophytes are the most numerous of the alien species, the proportion reaching a maximum in the buildings of the 1960s and 1970s.

Species resistant to trampling, for example, *Capsella bursa-pastoris, Tripleurospermum inodorum, Plantago major, Poa annua, Polygonum aviculare* agg., *Potentilla anserina, Taraxacum officinale* and *Trifolium repens* form an important part of the flora associated with buildings. Along fences and in less trampled sites, tall grass and herbs are abundant, including *Elytrigia repens, Anthriscus sylvestris, Arctium tomentosum, Artemisia vulgaris, Bromopsis inermis, Calamagrostis epigeios, Carduus crispus, Cirsium arvense, Leonurus quinquelobatus, Sonchus arvensis,* and *Urtica dioica.* The species that occur in shady and nutrient-rich conditions include *Aegopodium podagraria, Anthriscus sylvestris, Chelidonium majus, Geum urbanum, Glechoma hederacea, Impatiens parviflora, Lamium album, Lapsana communis, Rumex obtusifolius, Stellaria media* and *Urtica dioica.* In addition to the alien species listed above the most widely distributed are *Acer negundo, Conyza canadensis, Fallopia japonica, Fraxinus pennsylvanica, Galinsoga ciliata, G. parviflora, Helianthus subcanescens, Solidago gigantea* and *Symphytum caucasicum.*

Fig. 9 Sretenka Street, in the city centre

Fig. 10 Maryino district, buildings of the1980s

On the gradient between the periphery and the city centre there is a marked increase in the frequency and abundance of some species of which *Aethusa cynapium, Chenopodium hybridum* and *Elsholtzia ciliata* are the most typical. Their frequency in the city centre and the old city is higher than in the buildings of the 1960s to 1990s.

The significant part of the city that was built in the 1960s and 1970s is located between the boundary of the middle city of the 1930s and the MKAD (Fig. 5). The buildings comprise five storey (and more) free-standing blocks of flats. The characteristic feature of these habitats is dense forest stands between the blocks, which has been formed by planting and spontaneous growth of forest vegetation. Their structure is rather variable and includes practically all the native tree and shrub species as well as many introduced species; the species occurring in these areas include *Acer negundo, A. platanoides, Aesculus hippocastanum, Betula pendula, Caragana arborescens, Cerasus vulgaris, Fraxinus pennsylvanica, Malus domestica, Prunus avium, Quercus robur, Ribes nigrum, Spiraea chamaedryfolia, Cornus alba, Symphoricarpos rivularis, Syringa vulgaris, Tilia cordata, Tilia platyphyllos, Ulmus laevis* and species in the genera *Crataegus, Populus* and *Rosa*. Coniferous trees, which include *Larix decidua, Picea abies, P. pungens, Pinus sylvestris* and *Thuja occidentalis,* are rare. Ecological conditions in these plantings are similar to forest communities (light exposure, illuminance, humidity, the structure and nutrient-richness of the soil and leaf litter). As a result the plant communities formed are those in which shade-tolerant nitrophilous and forest species (native and alien) play a significant role – *Aegopodium podagraria, Alliaria petiolata, Festuca gigantea, Geum urbanum, Impatiens parviflora, Lamium album, Lapsana communis, Lysimachia vulgaris, Ranunculus cassubicus, Urtica dioica* and *Viola odorata*. The presence of ephemerals such as *Anemone nemorosa, A. ranunculoides, Corydalis solida, Ficaria verna, Gagea minima, Scilla siberica* and other forest species (for example, *Brachypodium sylvaticum, Carex sylvatica, Poa nemoralis* and *Pulmonaria obscura*) is also characteristic. The housing areas of the 1990s differ by having larger areas of lawn and fewer trees.

Railways

The city is crossed by eleven main railway lines connected by the "Circular Railway". In total, they occupy about 3% of the urban area. The land occupied by the railway network represents a set of separate habitat types, each differing appreciably in their ecological condition – as indicated by the structure of the vegetation and its species composition.

The substrate of railway land comprises loose material such as sand and gravel, which form an open, pioneer habitat, which is maintained in a condition – "of constant disturbance". This is the basic ecological factor that is favourable to the reproduction and existence of various species of different origin, primarily segetal, ruderal, and alien. Such conditions, when competition of coenotically active species is weakened, are favourable for a lot of native plants.

The track is characterised by a high level of species diversity, the source of which is the constant "drifting" of plants. High diversity is also promoted by heterogeneity of the prevailing ecological conditions at different sites, including the mineral structure of the substrate (for example, sand and tarmac), and the intensity of use and the character of the transported cargoes (disused lines, passenger, cargo and places for unloading cars). Four hundred and twenty-three plant species belonging to 62 families have been recorded on railway land. Annual and biennial species (46%) are the most numerous with the proportion of perennial herbs being only slightly lower at 42%; trees make up about 12%. A characteristic feature of the flora is the high proportion of alien species – 51%, which is greater than any other habitats in the city. Xenophytes account for 59% of the alien flora. Another important characteristic of the alien flora of the railway land is the high percentage (48%) of unstable species (ephemerophytes).

The structure of the "track" flora is very dynamic, at some sites the change of structure happens practically annually. Although the periods of significant change are characteristic of the railway flora as a whole they can be interspersed (at a particular site) with periods when the vegetation becomes stabilised and the number of species decreases. Species with high (more than 40%) frequency make up only about 15% of the flora. They include species that are common in the city such as *Artemisia vulgaris, Calamagrostis epigeios, Chenopodium album, Conyza canadensis, Impatiens parviflora, Lepidium ruderale, Matricaria perforata, Fallopia convolvulus, Sonchus oleraceus* and *Tanacetum vulgare.* Some ephemerophytes, for example, *Ambrosia artemisiifolia, Helianthus annuus, Panicum miliaceum, Secale cereale* and *Triticum aestivum* also show high frequency which can be explained by the constancy of their drift. Among species mainly associated with railway land (they occur in it more often than in others habitats) are *Amaranthus albus, A. blitoides, Artemisia sieversiana, Atriplex tatarica, Anisantha tectorum, Diplotaxis muralis, Kochia scoparia, Salsola collina* and *Sisymbrium wolgense.*

The lower level of use of the railway tracks results in the rapid colonisation of the ballast by grassland and woody species. Most of the colonising species are native with alien species playing a relatively minor role. *Acer negundo* is the species that is able to colonise this habitat the most rapidly.

Flora of the cuttings and embankments differ from the flora of the track. The structure of the flora changes in the sequence, track – embankment slope – cutting slope. There is a reduction in the species diversity of species (423 OR 243 OR 214), the percentage of alien species (51% OR 30% OR 14%), ephemerophytes (45% OR 17% OR 10%), while the proportion of perennial herbaceous species increase in the order (42% OR 72% OR 78%). The vegetation of the cutting and embankment slopes is more stable and permanent than that of the track. In a typical case the abundant species included *Achillea millefolium, Elytrigia repens, Bromopsis inermis, Calamagrostis epigeios, Carex praecox, Dactylis glomerata, Festuca pratensis, Galium mollugo, Hieracium umbellatum, Medicago sativa ssp. falcata, Potentilla argentea, Solidago virgaurea* and *Vicia cracca.* In addition characteristic southern, thermophilous species also occur, for example, *Elytrigia intermedia, Centaurea scabiosa, Echinops sphaerocephalus, Koeleria cristata, Salvia nemorosa* and *S. verticillata.* On neglected sites the more abundant species include *Arctium tomentosum, Artemisia vulgaris, Pastinaca sativa, Solidago gigantea* and *Urtica dioica.* In areas shaded by trees the dominant communities include *Chelidonium majus, Impatiens parviflora* and *Lamium album.*

Track-side ditches provide suitable conditions for a wide spectrum of moisture-loving plants, among them are *Alisma plantago-aquatica, Calla palustris, Carex rostrata, Cicuta virosa, Eleocharis palustris, Epilobium palustre, Equisetum fluviatile, Iris pseudacorus, Lemna minor, Rumex aquaticus, Typha latifolia* and *Utricularia vulgaris.*

The importance of the railway as “donors” that is, a primary source of distribution of species to neighbouring land is not as great as it is believed to be. Because of the special environmental conditions (relating to heat and light) and their weak competitive ability, plants that are specific to railways are usually limited in their distribution by those factors and consequently they are unable to colonise and become established on neighbouring land. In the city, railways can act as local refuges of the native flora, especially some rare species such as *Dactylorhiza* spp.

Recently Abandoned Garden Areas

In some of the settlements that have been incorporated into the city (small towns and villages) only examples of the rather large and long abandoned domestic gardens remain. Despite the small area occupied by this habitat, its importance in the modern city is great. *Chenopodium bonus-henricus* is almost exclusively confined to these places. The abandoned gardens are the centres of existence of many native forest species, for example, *Anemone ranunculoides, Athyrium filix-femina, Campanula latifolia, C. trachelium, Carex sylvatica, Corydalis solida, Dryopteris carthusiana, D. filix-mas, Ficaria verna, Pulmonaria obscura, Ranunculus cassubicus* and many naturalised species including *Allium ursinum, Anemone nemorosa, Aquilegia vulgaris, Hepatica nobilis, Lilium martagon, Narcissus poeticus, Scilla siberica, Tulipa gesneriana, Vinca minor* and *Viola odorata.* A relatively large number of veteran trees, both native and introduced occur in these former gardens; the species include *Fraxinus excelsior, Phellodendron amurense, Pinus sibirica, Quercus robur, Salix alba, Tilia cordata* and *Ulmus laevis.* The abandoned gardens have become centres for the natural restoration of vegetation; therefore they are of considerable scientific interest, for example, for studying the processes of reversion and formation of plant communities and changes in the plant species associations.

Cemeteries

There are 59 cemeteries in Moscow, with a total area of 1,000 ha (Moscow 1997). The physiognomic image of the cemeteries is defined by forest vegetation. The large number of forest trees (planted and spontaneous) creates a shaded environment. The plantings are variable in their structure and species composition. In the old urban cemeteries (eighteenth and nineteenth centuries), broad-leaved trees are generally dominant, including *Acer platanoides, Tilia cordata* and *Ulmus laevis* while the abundance of old trees of *Populus alba* is characteristic. In the planting of the cemeteries of the 1950s, the abundance of *Populus* spp., *Acer negundo, Betula pendula, Fraxinus pennsylvanica* was at its most popular. In the former village cemeteries, the most abundant species is *Betula pendula* with some *Tilia cordata, Ulmus laevis* and even *Syringa vulgaris* (on occasions). Conifers such as *Larix decidua, Picea abies, P. pungens, Pinus sylvestris* and *Thuja occidentalis* do not, as a rule, play an important role, being represented by individual trees. The predominant shrubs are alien species such as *Caragana arborescens, Spiraea chamaedryfolia, Syringa vulgaris, Philadelphus* spp. and *Rosa* varieties. The presence of native species, for example, *Corylus avellana, Euonymus verrucosa* and *Lonicera xylosteum* is insignificant.

In floristic terms, a relatively high level of diversity is characteristic of this habitat, 426 species representing 83 families have been recorded in it. Perennial herbs are the most numerous species comprising about 62% of the flora. Among the spontaneous species shade-tolerant ruderal, nitrophilous species are the most common, including *Aegopodium podagraria, Anthriscus sylvestris, Arctium tomentosum, Artemisia vulgaris, Chelidonium majus, Geum urbanum, Glechoma hederacea, Impatiens parviflora, Lamium album, Lapsana communis, Leonurus quinquelobatus, Lysimachia nummularia, Rumex obtusifolius* and *Urtica dioica* as well as some forest and forest ruderal plants including *Alliaria petiolata, Cardamine impatiens, Poa nemoralis* and *Ranunculus cassubicus.* There is also a large abundance of species that are characteristic of compact soil, for instance *Plantago major, Poa annua,* etc. The proportion of xenophytes in the floral structure of cemeteries is insignificant and accounts for only about 3%, which is related to the shady conditions of the habitats.

The characteristics feature of the habitat is a high diversity of cultivated plants concentrated into a rather small area. Cemeteries act as centres of cultivation for various plant species from which spontaneous species spread. In the older cemeteries cultivated species predominate (comprising about 60% of the species found). The most widespread of the cultivated species include *Aquilegia vulgaris, Bellis perennis, Dianthus barbatus, Fragaria* x *ananassa, Hemerocallis fulva, Hesperis matronalis, Iris germanica, Narcissus poeticus* and *Pyrethrum parthenium.* Some species occur in this habitat more often than in other city sites, for example, *Aconitum napellus, Brunnera sibirica, Dicentra formosa, D. spectabilis, Euphorbia cyparissias, Geum quellyon, Vinca minor* and *Viola odorata,* which are typical species of cemeteries. The frequent use of native plants, especially forest species, is another characteristic feature of this habitat, the most common being *Asarum europaeum, Athyrium filix-femina, Convallaria majalis, Dryopteris filix-mas* and *Pulmonaria obscura.* An important feature of the cemetery flora is the large number of ephemerals, the most common native species are *Anemone ranunculoides, Corydalis solida, Ficaria verna, Gagea lutea* and *G. minima;* the more common of the alien species includes *Galanthus nivalis, Leucojum vernum* and *Scilla siberica.*

Wasteland

Wastelands are unused sites with exposed and disturbed substrates. Herbaceous species are dominant, tree species being represented by seedlings, saplings or individual plants, which do not form continuous shade. Although these habitats appear all over the city they occur more typically in the peripheral zone and new housing areas. In the city centre where the number of buildings and hard surfaces are the most extensive, wastelands (in their typical form) are rare and insignificant, occurring only in relation to earthworks and the construction of new buildings.

Annual and biennial plants are the most frequent comprising 51% of the species found; the percentage of perennials is a little lower, at 45%, while the number of tree species is insignificant (4%). Alien species comprise about 28% of the floral structure of wastelands, of which most are xenophytes. The nucleus of the flora comprises species that are typical of the early colonisation stage of exposed, disturbed soil, such as *Elytrigia repens, Amaranthus retroflexus, Arctium tomentosum, Artemisia vulgaris, Atriplex sagittata, Bidens tripartita, Capsella bursa-pastoris, Chenopodium album, C. glaucum, Cirsium arvense, Echinochloa crus-galli, Conyza canadensis, Erysimum cheiranthoides, Lactuca serriola, Lepidium ruderale, Matricaria perforata, Medicago lupulina, Persicaria lapathifolium, Rorippa palustris, Solanum nigrum, Sonchus arvensis, Tanacetum vulgare* and *Trifolium pratense.* In a typical situation, these also contain an abundance of *Atriplex sagittata, Chenopodium album, Matricaria perforata* and some other annuals. The presence of higrophilous plants, such as *Bidens tripartita, Rumex maritimus, Ranunculus sceleratus, Juncus bufonius* and *Gnaphalium uliginosum,* is also characteristic. In depressions with poor drainage, pools appear and become colonised by species

Fig. 11 Wasteland – city "desert" with *Corispermum* species

that include *Typha latifolia, Alisma plantago-aquatica* and *Bulboschoenus maritimus.* The high abundance of *Atriplex* spp. and *Chenopodium* spp. species such as *Puccinellia distans* and a number of plants listed above indicate a degree of mineralisation and salinisation of wastelands soil. *Acer negundo* is the most common wasteland tree species.

In the city areas where sand or clayey sand prevail, there is an increase in the frequency and abundance of such species as *Berteroa incana, Crepis tectorum, Lactuca serriola, Lepidium densiflorum, Potentilla supina, Sisymbrium altissimum* and *S. loeselii* and other species that prefer a "warmer" and lighter mechanical soil structure. The classification of wasteland vegetation is difficult because it is very unstable with significant changes occurring in its structure and species composition even in the second year of its existence (Fig. 11).

Fallow Land

Fallow land communities are formed in abandoned kitchen gardens, former arable land, in association with earthworks, on spoil heaps/mounds with different soil structures including building and industrial waste (for example, slag heaps), on waste grounds and other places with no tree shade. Fallow land is relatively common in the peripheral areas but virtually absent from the older areas of dense housing, where there is little space. Perennial herbaceous species are the most numerous (66%)

while the percentage of annual and biennial species is appreciably lower (22%); tree species comprise 12% of species recorded in this habitats. Alien species account for about 20% of the flora, of which ergasiophytes are appreciably the most numerous. The nucleus of the flora comprises such species as *Achillea millefolium, Elytrigia repens, Arctium tomentosum, Artemisia vulgaris, Centaurea jacea, Dactylis glomerata, Lathyrus pratensis, Poa compressa, Rumex crispus, Trifolium hybridum, T. medium, T. pratense, Tussilago farfara* and *Vicia cracca*. The physiognomy of these communities is defined first of all by *Calamagrostis epigeios* and some other species with high abundance of *Bromopsis inermis, Pastinaca sativa* and *Tanacetum vulgare*. As for alien species, great profusion of *Helianthus subcanescens* and *Solidago gigantea* is characteristic. Tree species are represented by individual or small groups of trees, the most common being *Acer negundo, Betula pendula, Populus tremula* and *Salix caprea*.

Waste Ground

The waste ground habitat includes rubbish-tips and similar sites. Together with railway land within city limits, waste ground is the main habitat for the occurrence of alien species. Four hundred and eight species representing 63 families were recorded in this habitat; annual and biennial species are the most numerous comprising 51% of the total flora. The proportion of alien species is as large as the proportion found in association with railway land, amounting to 50% of the flora. Unlike railway land ergasiophytes prevail on waste grounds and account for 52% of the alien flora. The proportion of ephemerophytes is higher than on railway land, accounting for about 55% of the alien species.

The predominant vegetation comprises nitrophilous species of exposed and damaged soil, for example, *Elytrigia repens, Amaranthus retroflexus, Arctium tomentosum, Artemisia vulgaris, Atriplex sagittata, A. patula, Bidens tripartita, Cannabis sativa, Capsella bursa-pastoris, Chenopodium album, C. glaucum, C. rubrum, Cirsium arvense, Echinochloa crus-galli, Conyza canadensis, Erysimum cheiranthoides, Lactuca serriola, Lepidium ruderale, Matricaria perforata, Persicaria lapathifolium, Puccinellia distans, Rorippa palustris, Solanum nigrum, Sonchus arvensis* and *Tanacetum vulgare*. A characteristic feature of the waste ground flora is a high percentage of cultivated (by origin) plants which account for about 23% of all the species recorded. Practically all food plants are represented; among them is a high frequency of *Anethum graveolens, Armeniaca vulgaris, Citrullus lanatus, Coriandrum sativum, Cucurbita pepo, Helianthus annuus, Lycopersicon esculentum, Malus domestica, Melo sativus, Solanum tuberosum* varieties and *Vitis vinifera*. Ornamental plants are widely represented with plenty of species such as *Calendula officinalis, Cosmos bipinnatus, Echinocystis lobata* and *Malva mauritiana*. A typical plant of waste ground is *Phalaris canariensis* – a basic component of bird food. The species composition of this habitat is very unstable because it depends substantially on the incoming material. In addition many of the species are simply incapable of reproduction or existing for a long time given the site conditions. The termination of waste ground operations results in the rapid colonisation of fast-growing ruderal species such as *Artemisia vulgaris* and to the establishment of fallow land communities. At present this habitat is disappearing from the city's territory.

New Lawns

In recent years, considerable efforts have been made to create new lawns throughout the city. Consequently, this habitat has become one of the most important in the city. Two hundred and seventeen plant species representing 38 families have been found. Annual and biennial plants prevail, accounting for 55% of the flora of this habitat. Alien plants make up about 26% of the flora of which the most numerous are xenophytes, which comprise 55% of the alien species.

The lawns are created by the sowing of commercial seed mixes that may include *Dactylis glomerata, Festuca arundinacea, F. pratensis, F. rubra, Lolium multiflorum, L. perenne* and *Poa pratensis.* The flora of the young lawns is similar in its species composition to the flora of the wastelands. The more abundant species are those that are characteristic of exposed, damaged soil such as *Elytrigia repens, Amaranthus retroflexus, Artemisia vulgaris, Bidens tripartita, Capsella bursa-pastoris, Chenopodium album, C. glaucum, C. rubrum, Cirsium arvense, Echinochloa crus-galli, Erysimum cheiranthoides, Lepidium ruderale, Matricaria perforata, Myosoton aquaticum, Persicaria lapathifolium, Raphanus raphanistrum, Rorippa palustris, Senecio vulgaris, Solanum nigrum, Sonchus oleraceus, Tanacetum vulgare* and *Thlaspi arvense.* The greatest influence in the formation of the species composition of the flora is the soil, in which a wide range of plant species is brought into the city. The species include those typical of damp habitats, for example, *Bidens tripartita, Oenanthe aquatica, Rumex maritimus, Ranunculus sceleratus* and *Typha latifolia.* Alien and ruderal species including *Amaranthus blitum, Ambrosia artemisiifolia, Atriplex sagittata, Bidens frondosa, Cannabis sativa, Cyclachaena xanthiifolia, Rumex stenophyllus* and *Stachys annua* are also very characteristic. For some species lawns have become the main habitat in the city, including *Amaranthus blitum, Bidens frondosa, Cannabis sativa* and *Rumex stenophyllus*. Therefore by making lawns human activity is disseminating a lot of plant species throughout the city.

Lawns of Old Parks

The typical species of the meadow lawns of the old parks include *Arrhenatherum elatius* and *Trisetum flavescens*. *Bellis perennis* is also a characteristic species – judging from herbarium specimens that have occurred in some parks from the middle of the nineteenth century. Characteristic plants of the shaded lawns in the old parks are *Luzula luzuloides, Poa chaixii* and *Hieracium* spp.

Nature Conservation, Environmental Planning and Education

Protected and Rare Plant Species

Fifty species of those registered in the Red Book of the Moscow region (2008) occur in the territory of the modern city (Table 9). The threat categories are defined in Appendix VIII of this book.

Critically Endangered/Endangered

The Endangered threat category contains three species – *Carex disticha, Dactylorhiza longifolia* and *Equisetum variegatum.*

Vulnerable

The species listed in the Vulnerable category include *Corallorhiza trifida, Epipactis palustris, Montia fontana, Nuphar pumila, Platanthera chlorantha, Polystichum braunii* and *Pyrola media.*

Rare

Most of the species occur in the "Rare" category include *Chimaphila umbellata, Conioselinum tataricum, Corydalis cava, Corydalis marschalliana, Gentiana cruciata, Goodyera repens, Lathyrus niger, Malaxis monophyllos, Omphalodes scorpioides, Salix myrtilloides* and *Utricularia minor.*

A study of the flora of the city was originally proposed in 1914 by a group of students and their teachers. However, its realisation was prevented by the 1914–1918 war. During the following years, the flora of Moscow as well as of other cities of the Soviet Union was not a subject of specific research. It was not until 1980 that the author (of this chapter) was able to begin systematic studies of the Moscow flora (Shvetsov 1997). Sometime later studies were undertaken of the lichens and mosses in the parks.

Table 9 Species listed in the Red Book of the Moscow region that occur in Moscow

Threat categories	Number of species
1. Endangered (EN, CR)	3
2. Vulnerable (VU)	17
3. Rare (R)	26
4. Data deficient (DD)	4
Total	50

In approximately the same years the attitude towards natural territories changed. Initially they were considered to be recreational areas and it was believed that the basic function of vegetation was its ability to improve the quality of the urban environment. Currently, great attention is paid to the preservation of natural diversity in particular. But the nature of the care for the natural sites in general remains the same. Considerable efforts are aimed at creating lawns and flowerbeds within the natural areas and a large number of trees and shrubs. The embankments of ponds and rivers are re-enforced by stone or wood, which create more comfortable conditions for people, but they are not always favourable for the natural communities.

Natural territories are now actively used for ecological education. The basic role in this activity is provided by botanical gardens, museums, nature reserves (Kolomenskoye, Kuzminki and Tsaritsyno) and natural parks (Losiniy Ostrov). There are also a number of popular guidebooks about the animals and plants of the city's parks. Natural territories have also become places of research for school children and students. The city authorities hold annual competitions for the best flowerbeds and their exhibitions of floral design. These events besides being a means of education also encourage citizens to make greater use of ornamental plants.

Closing Comments

The city, as part of a cultural landscape, is a rather mobile and highly dynamic system, the changes of which are related to a variety of economic and social factors. Periods of stabilisation alternate with periods of fast changes and development of the city. Vegetative cover as an element of urban environment directly or indirectly is affected by all these factors and phenomena. Various changes in the municipal economy and urban infrastructure, economic ties and human values affect the composition and dynamics of the flora. Therefore it is very difficult to make a forecast of possible changes of the botanical composition of the city in future years.

In the modern city there is an obvious tendency towards the reduction of ecological variety. On the one hand, it is related to the direct influence on natural complexes, such as the destruction of sites that are "inconvenient" for the city, for example, bogs, ravines and valleys of small rivers, drainage of meadows, woods and bogs and the construction of quays along the rivers. On the other hand, there are the indirect effects of human activity, the most important being the process of changes in the economic systems and land use. This process is closely related to the successional dynamics of communities which leads to the reduction in the number of plants and ultimately their disappearance.

Species of natural communities are of special concern. The reduction in the population size and frequency of many native species can adversely affect their survival. It is possible that a lot of rare species will disappear from the city, primarily those of wet and dry meadows, forest edges, meadows and marsh communities. There is also another danger, which is generally overlooked – the loss of native genotypes and their replacement by alien genotypes of the same species. These negative processes can also affect alien species, for example, plants of old parks and "relics of cultivation" such as *Chenopodium bonus-henricus, Luzula luzuloides, Phyteuma spicatum* and *Poa chaixii*. In the process of reconstruction of the railways, many colonophytes have disappeared.

However, an optimistic forecast can be based on the ability of native species to adapt to adverse environmental conditions and on their ability to occupy some urban habitats.

Literature Cited

Abramova M.I., Ignatov M.S. (2004) Mosses in conditions of recreational use of forests. In: Influence of recreation on forest ecosystems and their components. ONTI PNC of RAN, Pushino. P. 177–214.

Biazrov L.G. (2004) Lichens of Moscow recreational forest plantings In: Influence of recreation on forest ecosystems and their components. ONTI PNC of RAN, Pushino. P. 149–176.

Frolov T.Y., Antonenko S.L. (2002) Use of imported vegetative soil for Moscow gardening in 2000. *Ecology of the large city* 6: 158–162.

Gutnikov A.V., Shvetsov A.N. (2004) Landscape indication of valuable natural objects in the area of Moscow. Bulletin of the Main Botanical Garden 187: 50–70.

Isaev A.A., Gutnikov V.A., Sherstiukov B.G. (2002) Scientific – applied directory on Moscow climate. Air temperature, precipitation (by months and per year), data on heating period (1879–2000). MSU, Moscow. 160 p.

Moscow. Encyclopedia (1997). Big Russian Encyclopedia, Moscow. 976p.

Moscow in the descriptions of 18th century (1997). Yanus-K, Moscow. 320 p.

Nizovtsev V.A., Shurkina E.A. (1997) Landscape preconditions of the foundation of Moscow. In: History of study, application and protection of Moscow and Moscow region natural resources. Yanus-K, Moscow. P. 26–34.

Pallas P.S. (1781) Enumeratio plantarum quae in Horto viri illustris atque excell. D-ni Procopii a Demidof. Consiliari status actualis, et orphanotrophaei moscuensis summi benefactoris, Moscuae vigent, recensente P.S. Pallas, Academico Petropolitano. Academy of sciences, St. Petersburg. 163 p.

Red Book of the Moscow region (2008). KMK Scientific Press Ltd., Moscow. 828 p.

Shvetsov A.N. (1997) The synopsis of the flora in the area of Moscow. *Bulletin of the Main Botanical Garden* 174: 47–57.

Stroganova M.N. (1998) Urban soil: genesis, systematic and ecological significance (Moscow as an example). Ph.D. Thesis of the Moscow State University.

Poznań

Bogdan Jackowiak

Fig. 1 Town hall of Poznań

Abstract Poznań is a typical large central European city in terms of its spatial structure and intensive urbanisation. It has been subject to systematic botanical research for almost 200 years, consequently it is possible to follow changes in the city's flora over a long period. The floristic studies indicate, amongst other things, that cities are areas where strong species selection takes place. The loss of indigenous elements and the spread of cosmopolitan plants lead to the uniformity of urban flora, at least within the

Bogdan Jackowiak (✉)
Department of Plant Taxonomy, Faculty of Biology, Adam Mickiewicz University,
Umultowska 89, 61-614 Poznań, Poland
e-mail: bogjack@amu.edu.pl

J.G. Kelcey and N. Müller (eds.), *Plants and Habitats of European Cities*,
DOI 10.1007/978-0-387-89684-7_11, © Springer Science+Business Media, LLC 2011

same biogeographical region. Degradation of habitats and direct human impact on isolated populations of plants are not the only causes of species extinction, a crucial role is played by invasive alien species. Generally, there is weak competitiveness in urban environment, which promotes the establishment and spread of invasive species. As a consequence the city is a source of expansive species. This is evidenced in the history of colonisation of several alien species that occur on a massive scale in the Wielkopolski National Park, which is close to the city. Knowledge of the mechanisms of changes in urban flora and its role in modifying the vegetation cover outside the administrative boundaries of a city suggests that it is advisable to undertake specific actions in terms of environmental management.

Natural Environment of the City

Poznań is located at the geographical co-ordinates 16°48′ to 17°06′ North longitude and 52°18′ to 52°29′ east latitude. The city, which is within the Wielkopolska Lakeland macro-region, contains three meso-regions: the Poznań Lakeland (western part), the Gniezno Lakeland (the eastern districts) and the Poznań Ravine of the Warta River, which is a narrow strip that divides the other two. The natural relief was formed in the Poznań phase of the Baltic glaciation. It is lowland that lies approximately 80–100 m a.s.l. Most of the city occupies flat moraine upland (Fig. 2).

Since the earliest times in its development, the axis of Poznań has been the ravine of the Warta River, which connects two wide pro-glacial stream valleys, the Warta-Odra and Toruń-Eberswald. The natural landscape of the town was significantly influenced by glacial lakes that flowed into the Warta River.

The hydrographic network of the town includes two lakes: Lake Kierskie (309.2 ha) and Lake Strzeszyńskie (32 ha). Additionally, Lake Swarzędzkie lies adjacent to the north-western boundary of the city. Reservoirs also play an important role, for example, the Malta reservoir (ca. 70 ha) in the valley of the Cybina river and Rusałka (ca. 50 ha) located in the valley of the Bogdanka River. In addition, there is an extensive complex of clay pits in the valley of the Strumień Junikowski (Fig. 2).

About half of Poznań is situated mainly on Quaternary sands and gravels. The lower moraine clay accounts for 34%, upper moraine clay for 14% and loams for only 1%. These deposits overlie mottled clays of the Pliocene, which only outcrop occasionally in the city (Bartkowski 1981). The predominant soils are podsols, pseudo-podsols and lixiviated brown earths which cover the moraine upland and the outwash plain. Only a small area of the town is occupied by phaeozems and degraded phaeozems. Alluvial soils and low peaty soils occur in the valleys of the Warta River and the smaller rivers and streams.

In terms of the climatic regionalisation of Poland, Poznań belongs to the "middle district", which comprises the eastern part of the Wielkopolska region and the western part of Mazovia. The significant feature of this district is that it has the lowest annual rainfall in Poland, <550 mm (Fig. 3). Poznań is situated in the zone of the so-called western circulation with the influence of the continental climate.

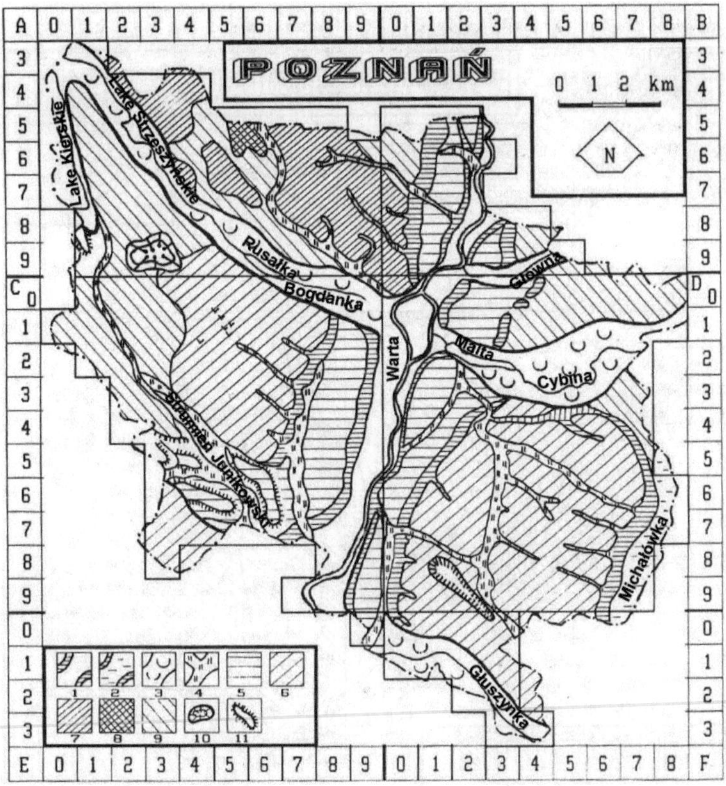

Fig. 2 Geomorphological and hydrographical layout of Poznań (After Bartkowski 1981). Key: (1) the bottom of the valley, (2) glacial troughs with a strongly levelled bottom, (3) glacial troughs weakly transformed, (4) small valleys on the moraine upland, (5) non-flooded terraces, (6) the flat moraine upland, (7) moraine undulating upland, (8) hilly moraine upland, (9) sandur, (10) kame hill, (11) esker ridge

Fig. 3 Gaussen-Walter climatic diagram for Poznań (data from 1950–1980) (After Jackowiak 1993)

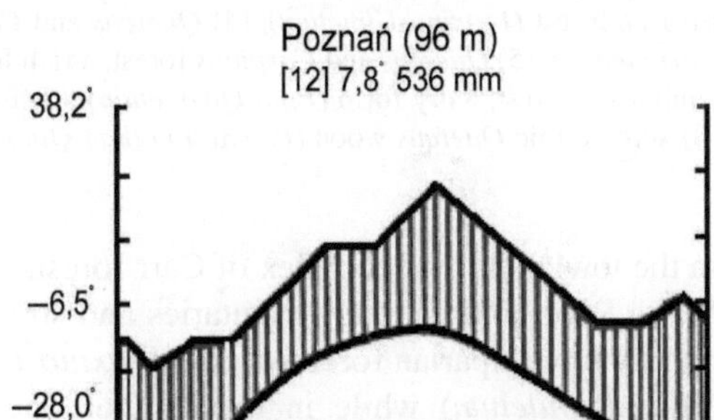

The dominant type of natural vegetation is the *Galio silvatici-Carpinetum* forest (Bartkowski 1981), which formerly occupied the vast areas of the moraine upland and non-inundated terraces of the Warta River valley (Fig. 4).

A smaller area was covered by *Leucobryo-Pinetum* forest (on poor habitats of the outwash plain) and *Pino-Quercetum* forest, which occurred on the podsol soils

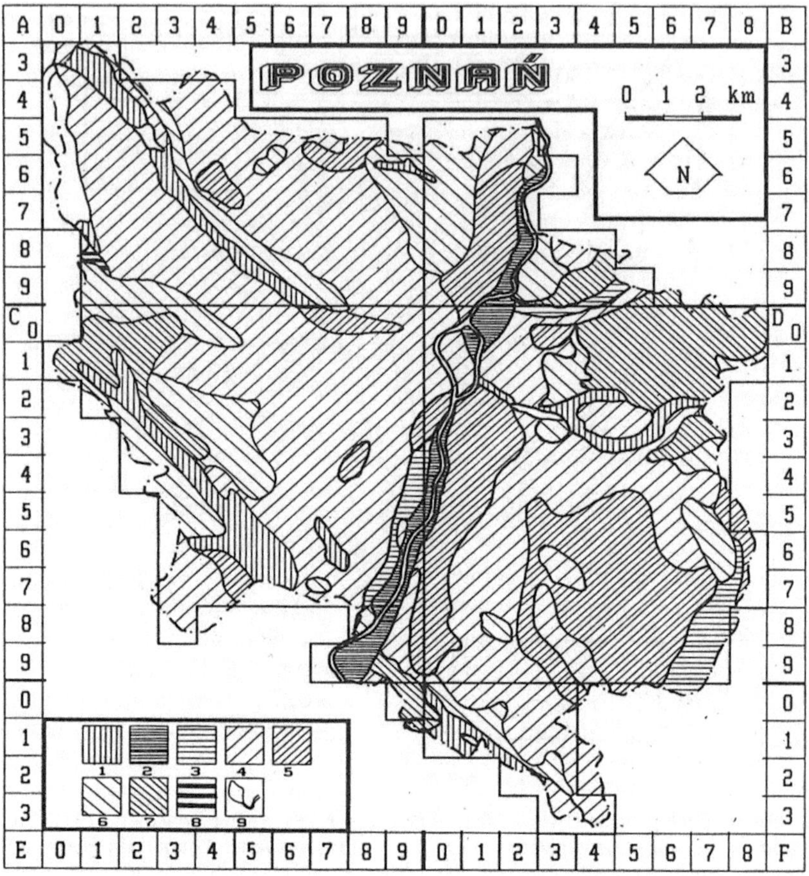

Fig. 4 Potential natural vegetation (After Wojterski et al. 1981). Key: (1) *Fraxinus-Alnus* riparian forest *(Circaeo-Alnetum),* (2) *Salix-Populus* wood *(Salicetum albae-fragilis),* (3) *Fraxinus-Ulmus* riparian forest *(Fraxino-Ulmetum),* (4) *Quercus* and *Carpinus* forest, a poor form *(Galio silvatici-Carpinetum),* (5) *Quercus* and *Carpinus* forest, a rich form *(Galio silvatici-Carpinetum),* (6) mixed coniferous forest, a dry form *(Pino-Quercetum),* (7) fresh coniferous forest *(Leucobryo-Pinetum),* (8) xerothermic *Quercus* wood *(Potentillo albae-Quercetum),* (9) watercourses and reservoirs

in the lowlands. The complex of Carr forests was associated with the valleys of the Warta River and its many tributaries and streams. In the larger valleys, there were mainly three riparian forests types (*Fraxino-Ulmetum, Salicetum albae-fragilis* and *Salici-Populetum*) while in the smaller ones the forests were primarily of the *Circaeo-Alnetum* type. As an addition to the forest communities, there were scattered *Carici elongatae-Alnetum* and *Potentillo albae- Quercetum* forests. In the geobotanical division of Poland, Poznań belongs to the Poznań-Gniezno district, included in the Wielkopolska-Kujawy Lowland (Szafer 1972); however, it is outside the natural range of such important forest-forming trees as *Fagus sylvatica, Picea abies, Abies alba, Taxus baccata* and *Larix polonica.*

Historical Development of the City

Poznań is one of the oldest and largest towns in Poland. Currently it is the economic and cultural centre of Wielkopolska, a region located in western Poland. At least four important periods of its history can be identified (Jackowiak 1990).

AD 1253

The area of the present town was inhabited as early as the Middle Paleolithic Age (10,000–7,000 years BC) when agriculture started to develop (Gąsiorowski 1973). The early form of the town was a fortified settlement, which was probably established at the turn of the eighth and ninth century AD on Ostrów Tumski – a sandy island formed at the confluence of Warta and Cybina rivers (Fig. 5).

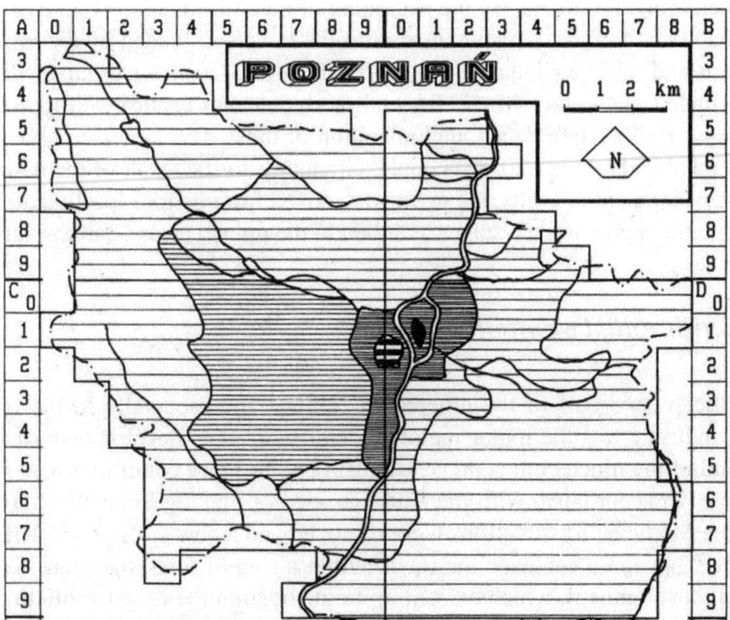

Fig. 5 Spatial development of Poznań (After Jackowiak 1993). Key: (1) the borough of Ostrów Tumski at the beginning of the eleventh century, (2) left-bank town (founded in 1253), (3–7) areas incorporated into the city before: 1896 (3), 1900 (4), 1939 (5), 1945 (6), 1980 (7)

Foundation and Medieval Development of the Town

Poznań was founded in 1253 on the left bank of the Warta River, which was to determine the western direction of the city's development until the mid-1900s (Bartkowski 1981). Soon after its foundation, the city, which originally occupied 21 ha, was fortified. In subsequent decades, the area inside the city walls was slowly built up. From the fourteenth to the sixteenth century, Poznań was included in the important east–west trade route and the North–South trade route from Cracow to the Baltic Sea (Gąsiorowski 1973). At the same time, the settlement outside the city walls was progressing rapidly; the process was called "colonisation" under German Law (Zajchowska 1977).

Seventeenth and Eighteenth Centuries

During this period, further settlement occurred with housing developments being built on both banks of the Warta River. In addition, the surrounding rural areas were also being intensively developed. By the beginning of this period, a kind of a city agglomeration had been formed, which consisted of the town surrounded by the town walls, settlements and suburbs situated along the fortification as well as a group of 12 smaller towns founded at the turn of the fifteenth and sixteenth centuries (Topolski 1973; Zajchowska 1977). The Poznań agglomeration of those days occupied about 338 ha and was inhabited by over 20,000 people. Gradually, as the result of increasing trade and the development of crafts, the town was moved from its ford location adjacent to the river to the higher terrace stages and then to the upland areas (Bartkowski 1981).

Nineteenth and Twentieth Centuries

During the second half of the nineteenth century and especially in the twentieth century, industry was the major factor in determining the development of the city. At the end of the nineteenth century, political and military conditions also played a major role. In accordance with the Prussian idea of making Poznań a "stronghold town", the Medieval fortifications were removed and a new system of fortifications was built. This again enclosed the town within a ring of forts, therefore restricting its spatial development, which existed up to the beginning of the twentieth century. At this time the destroyed fortifications were replaced by green belts (Bartkowski 1981). The first phase of the town's development in the twentieth century was still focused in a western direction. It was only in the second half of the century that the eastern moraine upland on the right bank of the Warta River started to be intensively developed for housing. Soon afterwards, residential development commenced, with similar intensity, in the northern areas of the city.

In the time of the botanical surveys from 1980 to 1985, the city occupied 228.6 km^2 and had 575,100 inhabitants. At present, Poznań covers 262 km^2 and has 565,000 inhabitants. The population density is 2,158 people per km^2. The highest number of residents was recorded in 1990 (almost 600,000).

Changes to the Natural Environment Accompanying the City's Growth

Spatial Structure of the Contemporary Town

Anthropogenic elements have played a more important role in the shaping of modern Poznań than its natural features. Despite Poznań being distinguished by a high proportion of green space, recent trends have resulted in the conversion of this and agricultural land to roads and residential and industrial developments (Bartkowski 1981). Currently, the built-up area covers 41% of the city area, 22% comprises arable land and 27% green areas (including approximately 13% of forest). The remainder is covered by reservoirs and wasteland. In terms of the dominant forms of land use, seven spatial complexes can be distinguished (Fig. 6).

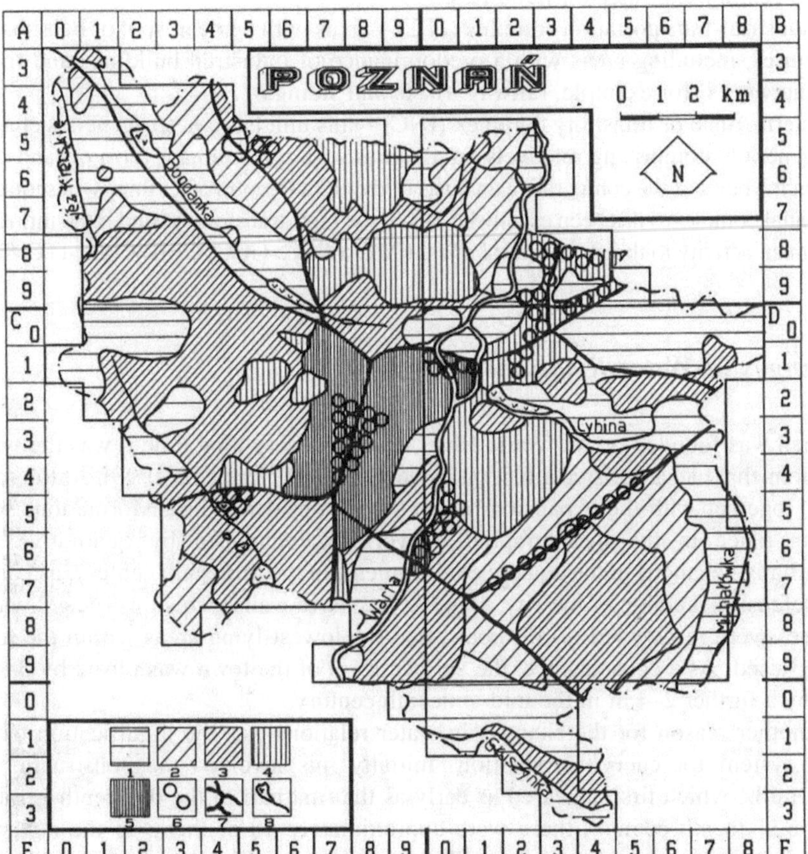

Fig. 6 Division of the town on the basis of urban space use (After Jackowiak 1993). Key: (l) forest-meadow complex, (2) agricultural complex, (3) garden-cottage complex, (4) low density block housing complex, (5) high density tenement housing complex, (6) industrial-transportation complex, (7) railway tracks, (8) water network

Forest-meadow complex (FMC) – mainly the "green wedges" along main rivers, larger tree stands (for example, the grounds of the former forts in the Citadel Park) as well as vast, extensively utilised land around the airports in Ławica and Krzesiny.

Agricultural complex (AGC) – comprises areas within the city boundary that are predominantly arable land and characterised by scattered, rural-type buildings.

Garden-cottage complex (GCC) – includes areas with predominately garden plots usually associated with arable land. A characteristic feature of this complex is the presence of low, exclusive residential developments.

Complex of block building (well spaced) (CBB) – groups of flats constructed from prefabricated concrete, often located on former arable land, rarely including small fragments of the garden-cottage complex.

Complex of tenement-houses (compact) building (CCB) – comprises the densely built-up medieval centre of the town and the districts with the predominant nineteenth century and interwar buildings.

Industrial-transportation complex (ITC) – areas with a linear, spot or belt-shaped structures, including areas with a predominance of industrial buildings and transport networks (for example, railway tracks and sidings).

Intermediate or transitory complex (INC) – this unit has a heterogeneous character, generally comprising relatively small areas with no dominant form of land use.

On the basis of the comparative analysis of the flora (among other things), a sequence of spatial complexes has been established from the most transformed under the influence of human activity to the most natural ones: FMC >AGC>GCC>CBB>CCB>ITC>INC.

Changes in Water Relations

Poznań was founded on the Warta River, and for a very long time it was the water relations that had a decisive effect on its development (Kaniecki 1993). At the same time, together with the expansion of the town, considerable transformations were taking place in the hydrographic network, accompanied by qualitative and quantitative changes in the aquatic environment. In the thirteenth century, as part of the defensive function of the city, massive defensive walls were constructed, watercourses were re-directed to fill moats, and the lowest-lying areas within the town were raised. As a consequence, the surface level of the town was raised by 3–4 m and by a further 2–4 m in the mid-sixteenth century.

Another reason for the changes in water relations was the modification of the river system for energy production. Initially this involved the construction of watermills, which first appeared as early as the first half of the thirteenth century; by the sixteenth century, there were approximately 40 of them. In the next two centuries, the construction of watermills was accelerated. The operation of watermills was associated with the construction of canals and weirs (that divided the river channels) as well as the regular dredging of the smaller and medium-sized rivers. The construction of watermills contributed also to the changes in water relations in adjacent areas. Damming the rivers resulted in the raising of the watertable in these areas so that it was very close to the surface while the construction of drainage ditches in wetlands had the opposite effect.

The decisive role of water in the life of Poznań continued until the early nineteenth century when it was decided to turn the city into a fortress resulting in it being again enclosed within a ring of defensive walls and moats. Numerous sluices were constructed to facilitate the flooding of the lowest-lying parts of the city and preventing access to those areas. Towards the end of the nineteenth century, as a consequence of the substantial increase in the population, old moats, tributaries of the Warta river, ponds and wetlands were filled in and the land used for building. In some places, the embankments were more than 5 or even 10 m high.

At the turn of the nineteenth and twentieth centuries, the sewage system was built, reducing the number of watercourses into which the municipal and suburban sewage was discharged. At that time, works were initiated to protect Poznań against flooding by the Warta – the works were completed in the 1920s. The city walls were demolished and the areas surrounding them drained and filled.

The present hydrographical system of Poznań is artificial – it is regulated and comprises watercourses with banks that are frequently reinforced with concrete slabs. A major problem is the discharge of urban run-off directly to watercourses. This has led to the depletion of the first level underground water resources and, consequently, the drying out of the soil and the loss of aeration. The construction of increasingly higher buildings and the increasing depth of their foundations have further exacerbated hydrological problems.

In addition to moving waters, there are still waters of natural origin (lakes, oxbow lakes) and artificial reservoirs (detention reservoirs, swimming pools, pits of former mineral workings, recreational water bodies, storage reservoirs, fish ponds, storm water tanks, park ponds, former moats and mill ponds). Most of them are eutrophic and saprophytic. The quality of the water in the on-stream reservoirs is to a large degree dependent on that of the "feeder" watercourses. These waters contain two types of contaminants: (i) biogenic substances (phosphates (as PO_4), total phosphorous, ammoniacal nitrogen (as NH_4) and total nitrogen), which influence the oxygen balance and (ii) substances accumulated in living organisms and exhibiting potential toxicity (heavy metals and polycyclic aromatic hydrocarbons).

Changes in Soil Conditions

Changes in water relations, particularly changes in land use related to development, are primarily associated with soil transformation. Currently, over 40% of the area of Poznań is occupied by urban soils. The soils have a markedly shortened profile in which natural and anthropogenic materials are mixed and where different types of liquid and gaseous substances are found. Within the city, soils of this type gradually replace hortisols and soils which have been in agricultural use for centuries.

The former occur occasionally in the built-up areas in the form of "islands" surrounded by urban soils. Unfortunately, they are exposed to adverse impacts leading to their degradation (Breś 2000), for example, a high heavy metal content, which has been confirmed not only by the analyses of the soils but also vegetables sold in market and grown in gardens. In over 60% of the soil samples, an elevated lead content was found, 20% of the samples were contaminated with cadmium and 10% with zinc. At the same time, it was observed that a considerable proportion of the soils contained insufficient amounts of available manganese and copper for proper plant growth and development. The heavy metals contamination is strongly correlated with the distance from roads.

Soil transformation in the built-up areas, particularly along roadsides, is also related to an increase in salinity and an associated decrease in plant vigour, especially in roadside trees. A characteristic feature of soils formed as a result of construction of high-rise apartment blocks in large housing districts is their markedly elevated content of calcium compounds and a high soil pH level (pH >7).

Changes in Climatic Conditions

Although the first regular meteorological measurements in Poznań were taken as early as 1848, comprehensive data have only become available in the last 50 years (Woś 2005). Three notable changes have been observed in recent years:

(a) During the last 10–20 years of the twentieth century, the number of hours of sunshine exceeded the annual average (1,597) by over 300. This increase was correlated with a decreasing number of sunless days as well as an increase in the number of cloudless days.
(b) From the early 1970s to 2000 (especially 1988–2000), the mean annual temperature was higher than the 50 year mean of 8.2°C. At the same time, an increase was observed in the number of warm days, that is, the days on which air temperature within the 24 h period was higher than 0°C, while the number of frosty days decreased.
(c) Less evident trends related to precipitation. Both the total annual precipitation (mean for the last 50 years was 516 mm) and the total number of days with precipitation (mean for the last 50 years was 158 mm) are characterised by a relatively large variation. Generally speaking, recent years have seen an upward trend in the values of both parameters.

Flora

Poznań has been subject to geobotanical studies since the first half of the nineteenth century, the most thoroughly studied group being vascular plants. The first reference was in 1828; the list of flowering plants that was published 8 years later comprised

722 species, including 234 horticultural and about 100 feral species. A comprehensive picture of changes taking place in the flora of Poznań in the nineteenth and twentieth centuries is provided by synthetic studies published in 1850, 1896, 1951 and 1990 (Jackowiak 1990). Despite these investigations and the published research papers, there are still no synthetic studies of other taxonomic groups: algae, bryophytes, fungi and lichens.

Vascular Plants

Sources and Data Collection Methods

In the investigations conducted in the 1980s, a total of 8,000 source data was used, including 4,500 herbarium records collected in the Poznań herbarium (POZ). The method used in the contemporary studies of the flora of Poznań involved mapping the species found in 1 km^2 grid, which is based on the national mapping grid of 10 km^2 (Zając 1978). Mapping was conducted in a total of 243 squares, including 186 plots entirely located within the area of Poznań and 57 adjacent squares (Jackowiak 1993). The species composition of these squares is the sum of information collected in the field in the form of floristic and ecological relevés. In the course of relevé collection, both the phenological aspect and habitat variation occurring in the basic squares were taken into consideration. The degree of habitat preservation was assessed on a six-point hemeroby scale (Sukopp 1972). Together with the localisation of relevés in the grid squares, their position in the distinguished spatial utilisation complexes was determined. Thus, it was possible to determine the variation of flora depending on the form and intensity of land use.

Geographical-Historical Structure and Dynamics of Flora in the Nineteenth and Twentieth Centuries

During the period 1828 to 1990, a total of 1,299 spontaneously occurring species of vascular plants were found within the administrative area of Poznań (Table 1). Over 64% of the flora comprises native species, of which almost 27% are apophytes. The predominant alien species are ephemerophytes, which appear transitionally and originate either from local cultivation or were brought in with transport from other, sometimes very distant geographical regions. The proportion of archaeophytes in the flora of Poznań is 9.5%, which is higher than the proportion of permanently naturalised neophytes. In the built-up zone of the city there were 815 species, which is 62.7% of the total flora of the city, the majority (57.1%) of which were alien species.

The importance of individual groups of plants changed in the course of the nineteenth and twentieth centuries in a very characteristic manner. The intensification of urban development resulted in a decreasing number of native species (on average by 6% in 50 years) and archaeophytes (11% in the first half of the

Table 1 Geographical-historical differentiation of the flora of Poznań in 1828–1990 and 1980–1990

	Total flora (1828–1990)		Present-day flora (1980–1990)	
	Number of species	%	Number of species	%
Native species, including:	834	64.2	701	69.4
Non-synanthropic species	484	37.3	352	34.8
Native synanthropic species (apophytes)	350	26.9	349	34.6
Alien species (anthropophytes), including	465	35.8	309	30.6
Archaeophytes	124	9.5	102	10.1
Neophytes	109	8.4	105	10.4
Casual alien species	232	17.9	102	10.1
Total number of species	1,299	100.0	1,010	100.0

Table 2 Changes in the proportion of native and aliens species that permanently established in Poznań between 1850 and 1990

Years		1850	1896	1951	1990
Native species	NS	834	784	735	701
	%	82.4	81.6	78.9	77.2
Archaeophytes	NS	124	118	104	102
	%	12.3	12.2	11.2	11.2
Neophytes	NS	54	58	92	102
	%	5.3	7.0	9.9	11.5
In total	NS	1012	960	931	908
	%	100.0	100.0	100.0	100.0

NS number of species

twentieth century), on the other hand, there was an almost twofold increase in the number of neophytes (Table 2). These phenomena eventually resulted in a fundamental transformation of the species composition of the flora.

Dynamics of the Native Flora

Probably, as late as the first half of the nineteenth century, in the areas now occupied by Poznań, most habitats typical of the natural landscape of central Wielkopolska were present, although in many of them human activity had made an impact. In the mid-1800s there were at least 834 native species within the city, see Table 2. The structure of the flora was consistent with the pattern formed by natural processes and climatic and habitat conditions under which the city was developing. This meant that the most numerous group of plants comprised herbaceous perennials, the proportion of which was many times higher than that of annual species or woody plants (trees, shrubs and low shrubs).

Within the next 150 years, which were important in relation to the development of the city, the number of native plants was reduced by at least 101 species and the forecast for the foreseeable future is even more pessimistic. Based on a detailed analysis of resources of the native species populations conducted in the late 1980s, it was found that 125 native species were directly endangered and approximately 126 species were threatened with extinction. The former concerns taxa that are found in only one or two sites in the city, while the latter relates to species observed in 10–20 localities and often in isolated populations. The number of native species that have disappeared or were adversely affected by the development of the city is definitely higher than that of plants resistant to changing environmental conditions and which can still be found in at least some districts of Poznań (Table 3).

The extinct and endangered species of Poznań include almost all of the life forms, from shrubs to annuals. The most numerous groups are the cryptophytes and hydrophytes. The urbanisation processes result in the decline (to varying degrees) of representatives of most habitats. Species have disappeared as a result of the drainage of peat-bogs, swampy and wet meadows as well as the reinforcement of river banks, regulation of rivers and streams and the construction of reservoirs. As a consequence of the destruction or degradation of these habitats, the following species (among others) have become extinct (**) or are directly endangered (*) in the city area: aquatic species, *Potamogeton compressus***, *P. obtusifolius**, *Najas marina**, *Zannichellia palustris***, *Utricularia vulgaris** and *U. minor**; peat-bog species, *Ledum palustre***, *Andromeda polifolia***, *Drosera anglica***, *D. rotundifolia*** and *Vaccinium oxycoccos**; peat-bog and meadow species, *Pinguicula vulgaris***, *Orchis militaris***, *O. coriophora***, *O. morio*** and *O. ustulata*** as well as species of wet riparian forests, *Circaea lutetiana***, *Astrantia major***, *Sanicula europaea***, *Mercuralis perennis** and *Daphne mezereum**.

The increasing urbanisation of the city is also a threat to the continued existence of species of coniferous forests, not only those that are naturally rare but also those that are common in many other parts of Wielkopolska. The threat arises not only as a consequence of coniferous forests being built on but also as a result of chemical changes (alkalisation) occurring in soils of the areas in which *Pinus* stands were

Table 3 Dynamic trends of native and established alien plants (anthropophytes) in Poznań in the nineteenth and twentieth centuries ($n = 1{,}067$ species)

Species receding from the area of the city		Species spreading in the area of the city	
Threat category	Number of species	Spread category	Number of species
Extinct	124	Strongly expansive throughout the city	50
Directly endangered	125	Expansive in certain areas	63
Strongly endangered	126	Moderately expansive	154
Threatened with extinction	193	Stable or spreading, but not expansive	232
In total	568	In total	499

consciously preserved as an element of the urban green system. Among the extinct and endangered coniferous forest species of Poznań are such species as *Vaccinium vitis-idaea**, *Monotropa hypopitys***, *Chimaphila umbellata***, *Nardus stricta**, *Lycopodium clavatum**, *L. annotinum*** and *Diphasiastrum alpinum**.

In addition, numerous species of xerothermic swards that occur most frequently on valley slopes or on elevations of different type (for example, eskers and kames) are also threatened by severe urbanisation pressures. In the area now occupied by Poznań, localities of vegetation of this type were naturally very rare, consequently, the development or degradation of them results in a substantial increase in the threat to many of the component species, for example, *Campanula bononiensis**, *C. sibirica***, *Origanum vulgare**, *Eryngium planum**, *Lithospermum officinale** and *Cirsium acaule**.

The survival in the city of species found in broad-leaved forests is also threatened. For example, scarce fragments of the *Quercus-Carpinus* forest (which were widely distributed in the area of the present-day Poznań) have been preserved within the city limits but they have turned out to be insufficient to save many species, for instance, *Actaea spicata***, *Viola mirabilis**, *Neottia nidus-avis***, *Lathraea squamaria***, *Impatiens noli-tangere**, *Platanthera chlorantha***, *Equisetum sylvaticum*** and *Phyteuma spicatum** that are now extinct or facing a direct threat of extinction.

Spontaneous vegetation is not completely defenceless against human activity, since even under the most extreme environmental conditions of inner city or industrial districts many interesting species can be found. A considerable proportion of these species (including herbaceous perennials) originate from forest communities (less frequently coniferous forests), meadows and xerophilous swards as well as riparian scrub. In this way, urbanisation results in native wild plants being subject to a specific selection process. In the same ecological groups, apart from more numerous receding species, there are plants that are resistant to anthropogenic pressure. Those that have become "liberated" from this pressure are able to expand their distribution range and population size. However, such species are in the minority. Among native plants, considerable adaptability to urban conditions is found in no more than one in four species. As a result, the flora of the city becomes more homogenous and monotonous compared with undeveloped areas. Species best adapted to urban conditions include (among others) well-known herbaceous plants such as *Chenopodium album, C. glaucum, C. rubrum, Cirsium arvense, C. vulgare, Rumex acetosella* and *R. crispus,* as well as grasses including *Dactylis glomerata, Poa annua, Lolium perenne, Elytrigia repens* and *Calamagrostis epigejos.* The tree and shrub species include: *Acer platanoides, A. pseudoplatanus, Rubus caesius, Sambucus nigra* and *Salix purpurea.*

Thanks to the colonising abilities of vascular plants, goods yards, railway tracks, embankments, waste dumps, unsightly municipal landfills, gaps in walls and pavements are not completely free of the vegetation cover. The micro-climatic and aesthetic value of these plants, although obvious, still remains underestimated. This is manifested not only in the pejorative term "weeds", but also in their persistent control, even in places where there is no such need.

Dynamics of Alien Species

The first botanical traces of human activity in the area of the present-day city include Neolithic remains of cereals. However, the intensification of the "transformation" of the flora can be related primarily to two events: (a) the foundation of a fortified town at the turn of the eighth and ninth centuries and (b) the intensive development of agricultural settlements following the foundation of the town on the left bank of the Warta in 1253. These coincide with the acceleration and extension of deforestation in central Wielkopolska, related to the establishment of settlement under German law. Although at present we know little of the dynamics of native plant species in that period, more is known about the alien species, which enriched the flora of the city. They arrived at different periods from different and sometimes distant regions of the world.

Archaeobotanical and phytogeographical studies indicate that until the end of the fifteenth century, at least 140 alien species of vascular plants became established in Poland, of which as many as 124 species were observed in Poznań as late as the nineteenth and twentieth centuries (Tables 1 and 2). Such a high proportion of alien species established before the end of the fifteenth century (the so-called archaeophytes) in the flora of the city is related to the long history of agricultural settlement in the region and the central role of Poznań in the development of Poland. Over 70% of archaeophytes came from areas that were the cradle of European agriculture – the Mediterranean and Irano-turanian regions. It is reasonable to assume that most of these species were brought to the area of Poznań with crops imported from those regions. They are commonly called "weeds", because they are found in cultivated fields and gardens against human wishes as well as in the so-called ruderal habitats associated with human settlements. The archaeophytes found in Poznań include, among others:

(a) Species known in central Europe since the Neolithic (*Sinapis arvensis, Agrostemma githago, Avena fatua, Fallopia convolvulus* and *Bromus secalinus*).
(b) Plants brought in during the Hallstatt period (*Anagallis arvensis, Hyoscyamus niger, Malva neglecta* and *Thlaspi arvense*).
(c) The grass, *Echinochloa crus-galli,* brought to Europe during the Roman period.
(d) Species known since the Middle Ages (*Consolida regalis, Onopordum acanthium, Urtica urens, Anthemis arvensis, Verbena officinalis* and *Euphorbia helioscopia*).

Archaeophytes include common species such as those found mainly in fields of cereals, for example, *Centaurea cyanus, Papaver argemone, P. dubium, P. rhoeas, Apera spica-venti* and *Matricaria recutita;* weeds infesting fields of cereals and root crops, like *Tripleurospermum inodorum;* those common in root crops and gardens-*Setaria pumila* and *S. viridis,* and species widely distributed in ruderal habitats such as *Hordeum murinum.*

The influx of alien species did not cease at the end of the fifteenth century. On the contrary, the establishment of transport routes between the European and American continents as well as an intensifying development of urban civilisation in Europe initiated the next wave of expansion of plant species, this time from North America, which was discovered by Christopher Columbus in 1492. This time constitutes not only a turning point on the basis of how we distinguish older arrivals (archaeophytes) from species established in the later period (neophytes) but also a turning point in the direction of spread of vascular plants. Since the beginning of the sixteenth century, 109 species have become established in the area of Poznań, of which 25% originated from North America and 20% from Asia. Most of the remaining 55% are species from southern and south-eastern Europe. Only a few species originated from the South and Central Americas. Neophytes found in the flora of Poznań include

(a) Species known in Poland since the sixteenth century: *Marrubium vulgare* from southern Europe and south-western Asia and *Mercurialis annua* from south-western Europe; *Acorus calamus* – recorded in Poland since the seventeenth century and originating from southern and central Asia.
(b) Species that have become established in Poland since the eighteenth century: *Robinia pseudoacacia* and *Conyza canadensis,* both from North America.
(c) Species observed since the nineteenth century: *Senecio vernalis* from south-eastern Europe and western Asia, *Galinsoga parviflora* from central America and *Elodea canadensis* and *Solidago canadensis* from North America.
(d) Species brought in to Poland in the twentieth century – *Galinsoga quadriradiata* from central America and *Vicia grandiflora* from southern Europe and south-western Asia.

The better known neophytes in the flora of Poznań include common trees, *Acer negundo, Quercus rubra* and *Aesculus hippocastanus;* commonly distributed shrubs, *Prunus serotina* and *Lycium barbarum;* species established in broad-leaved forests, *Impatiens parviflora,* and common ruderal plants, *Matricaria discoidea, Sisymbrium loeselii* and *S. altissimum.*

The process of establishment of alien species in the area of Poznań has not been completed. This view is erected on at last two premises: first, the group of neophytes established in Poland comprises over 250 species (more than twice the number found in Poznań) and second, within the last 150 years, 230 alien species have established temporarily; it is likely that at least some will eventually naturalise. This forecast is confirmed by observations since 1990 which show that some thermophilous species that did not previously occur spontaneously have established and spread rapidly in many parts of the city, for example, *Ailanthus altissima, Elaeagnus angustifolius* and *Juglans regia.*

The Fifty Most Frequent Species

The occurrence of species in Poznań covers virtually the entire range of variability, from 1 to 242 localities. Species occurring most frequently have an important effect on the physiognomy of plant cover in the city (Tables 3 and 4). The 50 most

Table 4 Fifty most frequent species in Poznań

Species	Fr	Status	RG	Species	Fr	Status	RG
Achillea millefolium	100	Ap	H	*Bromus hordeaceus*	71	Ap	T
Artemisia vulgaris	90	Ap	Ch	*Apera spica-venti*	70	Ar	T
Taraxacum officinale	90	Ap	H	*Conyza canadensis*	70	N	TH
Convolvulus arvensis	86	Ar	GHli	*Bromopsis inermis*	70	Ap	H
Cirsium arvense	86	Ap	G	*Tanacetum vulgare*	70	Ap	H
Urtica dioica	86	Ap	H	*Galium mollugo*	69	Ap	H
Chenopodium album	85	Ap	T	*Berteroa incana*	68	Ap	HT
Elytrigia repens	84	Ap	G	*Euphorbia cyparissias*	67	Ap	HG
Dactylis glomerata	82	Ap	H	*Arrhenatherum elatius*	67	Ap	H
Polygonum aviculare agg.	81	Ap	T	*Matricaria discoidea*	67	N	T
Artemisia campestris	81	Ap	Ch	*Sisymbrium altissimum*	66	N	TH
Plantago major	79	Ap	H	*Rumex crispus*	65	Ap	H
Capsella bursa-pastoris	78	Ar	T	*Rubus caesius*	65	Ap	ChN
Rumex acetosa	77	Ap	H	*Lactuca serriola*	65	Ar	H
Trifolium repens	77	Ap	CH	*Silene latifolia*	65	Ap	T
Tripleurospermum inodorum	76	Ar	TH	*Trifolium pratense*	65	Ap	H
Poa annua	75	Ap	TH	*Rumex acetosella*	65	Ap	GH
Stellaria media	75	Ap	T	*Tussilago farfara*	64	Ap	G
Plantago lanceolata	75	Ap	H	*Sambucus nigra*	63	Ap	N
Equisetum arvense	74	Ap	G	*Sisymbrium loeselii*	63	N	HT
Lolium perenne	74	Ap	H	*Calamagrostis epigejos*	63	Ap	G
Medicago lupulina	72	Ap	TH	*Galium aparine*	63	Ap	T
Daucus carota	72	Ap	H	*Pimpinella saxifraga*	63	Ap	H
Fallopia convolvulus	72	Ar	T	*Ballota nigra*	63	Ar	CH

Key: *Fr* frequency of occurrence in %; Status: *Ap* apophytes, *Ar* archaeophytes, *N* neophytes; *RG* Raunkier's group: *C* herbaceous chamaephytes, *Ch* lignified chamaephytes, *H* hemicryptophytes, *G* geophytes, *T* therophytes, *li* lianas commonly occurring (those with a frequency of >60%) are listed in Table 4. The number of apophytes is predominant (74%), followed by archaeophytes (16%) and finally neophytes (10%). Among the dominant species the greatest number are hemicryptophytes and therophytes, with a combined proportion of more than 70%. Other life forms are represented by only a few species. Among the woody perennials the two most frequent are *Sambucus nigra* and *Rubus caesius*.

Planted Trees and Shrubs

The city is characterised by an exceptionally rich dendroflora. This primarily comprises all trees and shrubs planted in the municipal forests, arboreta, botanical gardens, parks, cemeteries, domestic gardens, residential districts and along roadsides. In the Poznań arboretum there are over 900 species, including cultivars and varieties.

Pinus sylvestris is the dominant species in urban forests, where it is frequently accompanied by *Betula pendula* and *Populus tremula* and much less often by *Quercus robur, Q. petraea, Tilia cordata, Fagus sylvatica Acer platanoides, A. pseudoplatanus* and *A. campestre*. Among the alien species, those planted most frequently in municipal forests are *Robinia pseudoacacia, Q. rubra, Prunus serotina, Syringa vulgaris* and *Symphoricarpos albus*. Along the banks of the Warta and its tributaries as well as water bodies, *Acer negundo* and *Cornus sericea* were introduced together with different cultivars of *Populus* (including *Populus* x *canadensis* and *P. nigra* 'Italica').

Many of these species are also found in municipal parks, usually as cultivars, for example, *Acer platanoides* 'Schwedleri' and *Acer pseudoplatanus* 'Atropurpureum'. Frequently occurring species in the older parks include *Platanus* x *acerifolia* and *P.* x *hispanica* with *Aesculus hippocastanum* being the most common. The latter species is at the same time a permanent component of roadsides in the city and the suburban streets. A characteristic element of the city's cemeteries are various species and varieties of conifers such as *Thuja orientalis, T. occidentalis* and *T. plicata*. The dominant component of the dendroflora of the fortifications is *Robinia pseudoacacia*.

Until recently, the trees in the housing districts were predominantly representatives of the "Section" of the Balsam Poplars (*Populus* spp.), which are being removed because they pose a threat to buildings and are a hazard to people. Apart from that, these districts are planted with many varieties of exotic trees and shrubs from the genera: *Amorpha, Hippophae, Cornus, Elaeagnus, Forsythia, Lonicera, Spiraea, Rosa, Malus, Prunus, Pyrus, Crataegus, Colutea, Caragana, Laburnum, Berberis, Rhus, Ailanthus* and many coniferous tree and shrub species.

Designed landscapes which are rich and varied in terms of their species composition are the primary source of diaspores of the spontaneous dendroflora. Overall, the spontaneous dendroflora contains about 160 taxa, of which the most widely distributed are *Sambucus nigra, Rubus caesius, Acer platanoides, A. negundo, A. pseudoplatanus, Prunus cerasifera, Betula pendula, Fraxinus excelsior, Juglans regia* and *Ulmus laevis* (Winiecka-Nowak 2009, unpublished).

Algae

The first mention of algae in the reservoirs of Poznań dates from 1860 and is concerned with diatoms. Now, it is known that the numerous and diverse water bodies comprise a large diversity of algae from the Chlorophyceae, Cyanobacteria (Cyanophyta), Euglenophyceae Cryptophyceae, Dinophyceae and Chrysophyceae (Szeląg-Wasielewska 2002). The precise number of taxa is difficult to establish mainly because of methodological problems, the very small dimensions and rarity of some of the taxa and the difficulties of identification.

Taxa within the Cyanobacteria are commonly found in all water bodies investigated so far, with each water sample containing from several to about 20 species. They live all year long, while in the summer months, in fertile or contaminated water bodies, they form “blooms”, which result in turbidity and change the colour of the water. Filamentous species are the most frequent: *Aphanizomenon flos-aquae, Anabaena flos-aquae, Plantktothrix agardhii, Pseudanabaena limnetica,* as well as a colonial form: *Microcystis aeruginosa.* A massive bloom of the Cyanobacteria results in a substantial reduction in the dissolved oxygen which results in the death of many animal groups. These algae also produce substances, which are toxic to other aquatic organisms and human beings.

Euglenids, which are considered to be an indicator of organic pollution, are more typically found in small water bodies that are rich in organic matter, including small ponds and puddles, where they sometimes appear in large quantities. In the area of Poznań, representatives of such genera as *Euglena, Phacus, Trachelomonas* and *Colacium* are found most frequently. Their proportion in the species composition of phycoflora in individual water bodies is typically low. A relatively wide range of aquatic habitats is inhabited by unicellular algae, usually Cryptophyceae – primarily from two genera: *Cryptomonas* (*C. erosa, C. marssonii, C. obovata, C. ovata* and *C. reflexa*) and *Rhodomonas* (*R. lacustris* and *R. lens*); colourless forms of the Cryptophyceae from the genera *Cryptaulax* and *Chilomonas* also occur.

An important component of the city phycoflora is the Dinoflagellates (Order – Peridiniales). This group is represented by several dozen of species, mostly belonging to three genera: *Ceratium, Gymnodinium* and *Peridinium,* of which the most frequent (often on a massive scale) are *Ceratium hirundinella* and *Peridinium cinctum.* In contrast, *Diplopsalis acuta* is rare. The water bodies of Poznań also contain species in the Classes Chrysophyceae, Bacillariophyceae and Xanthophyceae. Particularly numerous are species of diatoms, which may comprise as much as 50% of the phycoflora of reservoirs; the most frequently reported species are from the genera: *Asterionella, Fragilaria* and *Diatoma.* Species of the genera *Cycotella* and *Stephanodiscus* sometimes occur on a massive scale. Among the Chrysophyceae, representatives of the genera: *Chrysococcus, Kephyrion* and *Dinobryon* are commonly found while among the least recognised Xanthophyceae, species are reported from the genera: *Goniochloris, Ophiocytium, Tetraedriella* and *Tribonema.*

In terms of the number of species, the richest family in the phycoflora of the water bodies of Poznań is the Chlorophyceae. The most frequently found are species from the Orders Chlorococcales, Volvocales, Ulotrichales, Conjugales and Charales. Numerous species from the genera: *Scenedesmus, Chlorella, Monoraphidium, Pediastrum* and *Oocystis* occur abundantly. Worthy of note is the presence of the three genera of the Order Charales, which are *Chara, Nitella* and *Nitellopsis*.

Bryophytes

The first references to the bryophytes of Poznań are from 1894. Until the end of the 1970s, studies of the bryophytes focused on natural habitats. Afterwards they included anthropogenic habitats such as tram and railway tracks, ornamental gardens, municipal parks, cemeteries, and fortifications. Although a comprehensive study of this group of plants is still lacking, to date a total of 138 species and two varieties of Musci as well as eight Hepaticae species have been identified (Rusińska 2002).

Unfortunately, at present, information on the occurrence of some of them is only of historical importance. The group of extremely rare species in Poland, which became extinct in Poznań as a result of the city's development, comprises *Paludella squarrosa* (a glacial relict), *Drepanocladus sendtnerii, Dicranella varia, Pleuridium subulatum, Pottia davalliana, Campyliadelphus elodes, Helodium blandowii, Sphagnum subnitens* and *Porella platyphylla* (Rusińska 2002).

According to the Rusińska, species that are rarely recorded in the region, Poland or even Europe still occur outside the city centre; they include species usually associated with mountainous areas, for example, *Brachythecium populeum, B. reflexum, Hypnum pallescens* and *Anomodon viticulosus* as well as other species such as *Buxbaumia aphylla, Isothecium alopecuroides, Plagiothecium latebricola, Fontinalis antipyretica* and *Fissidens adianthoides*.

The most common mosses in the densely built-up areas include: *Bryum argenteum, Streblotrichum convulatum, Funaria hygrometrica* and *Ceratodon purpureus* and among the Hepaticae, *Marchantia polymorpha. Tortula muralis* is frequently found on the weathered mortar of urban walls, on fences and concrete posts together with the ubiquitous mosses: *Amblystegium serpens, Brachythecium rutabulum, Hypnum cupressiforme* and less often *Grimmia pulvinata, Schistidium apocarpum* and *Orthotrichum anomalum*. Lawns in the city centre contain such species as: *Eurhynchium hians, Rhytidiadelphus squarrosus, Brachythecium albicans* and *Plagiomnium undulatum*. The walls of forts in the former Poznań citadel support a unique bryoflora primarily comprising calciphilous species: *Bryoerythrophyllum recurvirostrum, Camptothecium lutescens, Cratoneuron filicinum, Didymodon fallax, Distichium capillaceum, Encalypta streptocarpa, E. vulgaris* and *Philonotis fontana*, as well as a mountain species – *Leskella nervosa*.

Recent studies by Fudali (2005) discovered the presence of 71 species of Musci and six Hepaticae species in 21 parks and ten cemeteries of Poznań. Worthy of note are the forest species: *Atrichum undulatum, Aulacomnium androgynum, Brachythecium salebrosum, B. velutinum, Cirriphyllum piliferum, Plagiomnium cuspidatum* and *Pohlia nutans*.

Fungi (including Lichenised Fungi)

Fungi

The first mycological information from the area of Poznań comes from the nineteenth century. Much attention was devoted to this highly diverse group of organisms in the 1930s and again at the turn of the twentieth and twenty-first centuries. Despite limited data, it has already been determined that mycobiota of Poznań comprise over 550 species belonging to the macromycetes, including the fungi that produce large or relatively large fruiting bodies (Celka 2002).

Poznań contains numerous species that are strictly protected in Poland; they include *Morchella esculenta, M. conica, M. elata, Langermania gigantea, Phallus impudicus, P. hadriani* and *Sparassis crispa*.

Information about some species is only of historical importance now. For instance, among six species from the genus *Geastrum*, two species (*G. melanocephalum* and *G. striatum*) have not been recorded in recent years. On the other hand, as late as 2000, *Volvariella surrecta*, a species parasitising the fruiting bodies of other fungi (*Lepista nebularis* and *Melanoleuca brevipes*) has been identified for the first time in Poland.

Fungi which in the urban environment commonly produce fruiting bodies from spring to autumn include *Coprinus atramentarius* and *C. comatus*, which are frequently found on road verges and in Squares; *C. disseminatus* and *C. micaceus* that grow in large numbers on rotting tree trunks. *Auricularia auricula-judae* bears fruiting bodies on live or dead specimens of *Sambucus nigra* while *Letiporus sulphureus* grows on broad-leaved trees in parks, gardens and along streets. Specimens of this species have been found on *Robinia pseudoacacia*, an alien species in the flora of Poland. A characteristic element of the urban arboreal mycobiota of Poznań is also *Polyporus squamosus*, which produces fruiting bodies that may be >50 cm in diameter.

Different "agarics", particularly *Agaricus arvensis, A. campestris, A. xanthodermus* and *A. bitorquis*, occur in lawns, parks, shrubberies, gardens and near dumps. A specific group of coprophilous fungi (found on animal excrement) is represented by *Bolbitius vitellinus*.

This considerable richness of terrestrial and litter fungi is related to the well-preserved municipal forests. Most of the edible fungi occur outside the urban area together with a large group of poisonous fungi which also appear in municipal parks, particularly in parks with well-developed, old stands of trees.

Lichenised Fungi

The first information about lichens in Poznań is from the 1920s. For a long time such studies were extensive rather than intensive in character. A comprehensive list of historical data as well as an analysis of the current status of lichen biotas was prepared in 1999 (Kepel 2002). At least 227 lichen species were found, of which 178 were recorded towards the end of the twentieth century. This is the richest biota of lichens of all the large cities of Poland.

However, the lichen biota is subject to strong transformations involving, among others, the local extinction of numerous species. Among the 48 species which are no longer known to occur in Poznań are not only species that are naturally rare but also species that were previously found relatively often, for example: *Pleurosticta acetabulum, Ramalina fraxinea, Usnea hirta, Lecanora carpinea, Candelaria concolor* and *Lecidella elaechroma*. Declining species, although still observed in the city, include, among others: *Candelariella xanthostigma, Lecanora argentata, Physcia stellaris* and *Pseudevernia furfuracea*. Species that are particularly threatened with extinction are large lichens with "bushy" or leafy thalli, growing on tree bark. Species with a crust-like thallus, particularly those growing in the wild on calciferous rocks, are much better established in the city, occupying walls, concrete posts and steps. This group is represented in Poznań by *Aspicilia calcarea, Candelariella aurella, Sarcogyne regularis, Lecidella stigmatea, Lecanora albescens, L. dispersa, L. muralis, Caloplaca holocarpa* and *C. citrina*. In recent years, some species have considerably extended their local range, a good example is *Lecanora conizaeoides,* a species with a powdery thallus, which as late as the mid-twentieth century was not reported in Poznań; now it occurs on almost all trees except for those in the immediate city centre.

Changes in the species composition and distribution of lichens mentioned here are determined by the permanently increasing air pollution. Maps prepared on the basis of their distribution make it possible to distinguish seven air pollution zones, with the central zone being a "total lichen desert".

Habitats

Spatial Variation

The city comprises a high diversity of ecological systems, in which natural and anthropogenic factors overlap. This is also reflected in the spatial distribution of habitats and in the structure of local ranges of vascular plants. An indication of the scale of diversity can be obtained comparing the species-richness of the flora. The difference between the least and the most species-rich is more than eightfold: from approximately 40 to >340 species. On an average, in Poznań there are over 150 species of wild vascular species per 1 km^2. The spatial distribution of habitats and the flora is, on the one hand, determined by the hydrographic system, which controls the direction of development of the vegetation and on the other hand, by the form and intensity of building activities. Although on a small scale, habitats very often form a kind of a mosaic in the city, four distinct floristic and habitat zones can be distinguished.

Zone 1

Zone 1 occupies more than 37% of the city; it covers very strongly transformed areas, dominated by tall, high density housing (tenements) and blocks of flats. It is the zone

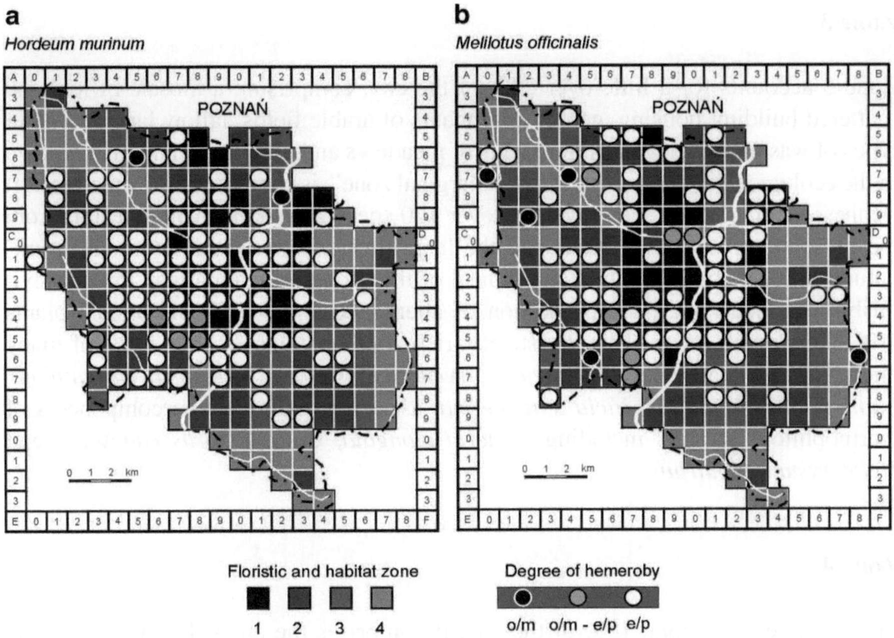

Fig. 7 Distribution maps of the selected species characteristic for the particular floristic and habitat zones (7a *Hordeum murinum*, 7b *Melilotus officinalis*). Key: o – oligo-, m – mezo-, e – eu-, p – polyhemeroby habitats

characterised by the lowest number of vascular plants (on average 126 species km^2), the highest proportion of alien species (36.8%) and a high frequency of annual plants (31.9%). In terms of ecology, the flora is distinguished by: (i) the dominance of species forming short-lived, pioneer ruderal communities, fragmentarily developed associations of garden weeds and root crops and (ii) a relatively high proportion of thermophilous plants preferring soils rich in soluble nitrogen compounds. Species, which mainly occur in this zone, include *Aethusa cynapium* (a representative of the native flora), the archaeophytes *Hordeum murinum* (Fig. 7a) and *Urtica urens* and the neophytes *Bunias orientalis* and *Galinsoga quadriradiata*.

Zone 2

Zone 2 covers approximately 25% of the city area and overlaps to a considerable extent with the area of low but relatively dense housing mixed in places with tall buildings. On average, there are 40 species more per 1 km^2 than in Zone 1, while the proportions of alien plants (24.4%) and annual species (23.4%) are markedly lower. The predominant native species include *Melilotus officinalis* (Fig. 7b), *Coronilla varia, Medicago sativa* ssp. *falcata* and *Silene vulgaris*. The noreworthy archaeophytes *Digitaria ischaemum, Malva neglecta, Setaria pumila* and *S. viridis* occur in large numbers. Neophytes are represented by species such as *Epilobium ciliatum*.

Zone 3

Zone 3 accounts for a little over 22% of the city, comprising a mosaic of low and scattered building housing, gardens, remnants of arable fields, fallow land and other types of wasteland, as well as fragments of meadows and grassland. This area, known in the ecological literature as the "inner marginal zone", is characterised by the greatest richness of plant species (on average, over 170 species per km^2), which results from the presence of natural habitats and the large diversity of synanthropic and semi-natural habitats, both of which are typical of this Zone. In comparison to the previously described zones, the proportion of alien species (19.7%) and annual plants (19.2%) is much lower. This floristically rich zone contains an abundance of many alien species, particularly *Anthemis arvensis* (Fig. 8a), *Artemisia absynthium, Lamium purpureum* and *Vicia tetrasperma,* as well as many native components of thermophilous swards, including *Armeria elongata, Corynephorus canescens* and *Helichrysum arenarium.*

Zone 4

Zone 4 covers less than 15% of the city. It comprises the areas that have been the least transformed by urbanisation. In the literature, it is frequently referred to as the "outer marginal zone". It has an atypical shape because it includes not only the areas

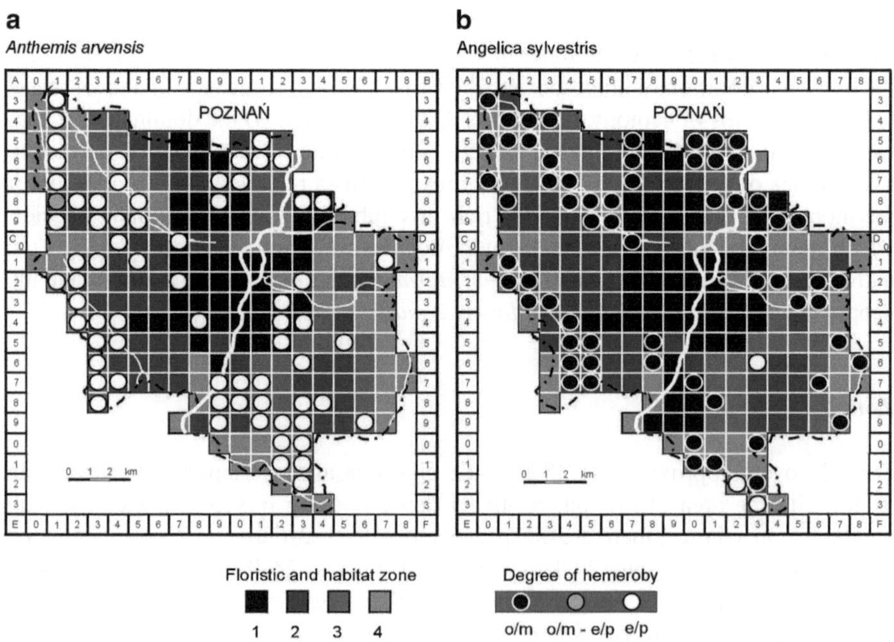

Fig. 8 Distribution maps of selected species (8a *Anthemis arvensis*, 8b *Angelica sylvestris*) characteristic for the particular floristic and habitat zones. Key: o – oligo-, m – mezo-, e – eu-, p – polyhemeroby habitats

adjacent to the city boundary but also fragments of the urban green space system extending towards the city centre and based on the valleys of the Warta and its tributaries. The average area in this zone is floristically poorer with 140 species per km^2 compared with 170 in Zone 3 and 166 in Zone 2. However, quantity is not always equivalent to quality. Although the outer marginal zone contains slightly fewer species than Zones 2 and 3, its flora is definitely better preserved, and the least transformed by human activity. This is demonstrated by (among other things) the lower proportion of annual plants (15.4%) and the proportion of alien species (13.2%). This means that in this part of the city only about 10% of the species are of alien origin, while in Zone 1 it is about 30%. Consequently, the outer marginal zone is a refuge of the most valuable components of the native forest, marshland, meadow and waterside flora, which provide visible evidence of the natural history of Poznań. This is not restricted to extremely rare plants (which are considered to be the treasures of the city flora) but also includes the native species that are relatively common outside the city's. Species such as *Caltha plaustris, Angelica sylvestris* (Fig. 8b), *Cirsium oleraceum, Myosotis scorpioides, Corylus avellana* and *Carpinus betulus* are mainly found in the outer Zone and less frequently in the inner Zone.

Important European Union Habitats

On the basis of the classification in the "Interpretation Manual of European Union Habitats" (July 2007), a total of 11 important European habitats have been identified in Poznań, including one that is completely degraded. The asterisk indicates habitats of priority significance.

1. Inland dunes with open *Corynephorus* and *Agrostis* grasslands (EUH code 2330; Corine code 64.1).

   Small areas with relatively well-developed grassland vegetation are found in open, sandy outwash areas in the north-eastern and north-western parts of the city. The most frequently found diagnostic species include *Corynephorus canescens, Agrostis capillaris* and *Spergula morisonii* while *Teesdalia nudicaulis* is observed less often and in much smaller numbers.

2. Xeric sand (inland) calcareous grasslands (*Koelerion glaucae*) (EUH code 6120*; Corine code 34.12).

   A rare habitat that occurs in the north-eastern and western parts of the city; the swards are only relatively well-developed in few places, the characteristic species include *Koeleria glauca, K. macrantha, Astragalus arenarius, Arabis arenosa, Carex ligerica* (very rarely), *C. praecox, Dianthus deltoides, Gypsophila fastigiata, Helichrysum arenarium, Herniaria glabra, Petrorhagia prolifera* and *Silene chlorantha.*

3. Semi-natural dry grasslands and scrubland facies on calcareous substrates (*Festuco-Brometalia*) (EUH code 6210; Corine code 34.31).

Until recently this habitat was found on the steep slopes of the Warta valley, in the northern part of the city (Fig. 2); however, it disappeared as a result of the natural succession of thermophilous scrub. Advantageous conditions for the formation of *Festuco-Brometalia* grasslands were created in several of the city forts, constituting an element of an incompletely realised ring-type system of green space. On the slopes of fortification embankments, rich in calcium compounds, multi-species grasslands are formed comprising such species as *Anthyllis vulneraria, Arabis hirsuta, Brachypodium pinnatum, Bromopsis erecta, Bromopsis inermis* ssp. *inermis, Campanula glomerata, Carex caryophyllea, Centaurea scabiosa, Dianthus carthusianorum, Leontodon hispidus, Medicago sativa* ssp. *falcata, Polygala comosa, Primula veris, Sanguisorba minor* and *Veronica austriaca* ssp. *teucrium*.

4. Natural eutrophic lakes with *Hydrocharition*-type vegetation (EUH code 3150; Corine code 22.13).

The best preserved habitat of this type is found in ponds occurring in the Bogdanka valley and in the Strumień Junikowski valley, which are left-bank tributaries of the Warta (Fig. 2). The vegetation of these water bodies comprises such species as: *Lemna minor, L. trisulca, Spirodela polyrhiza, Hydrocharis morsus-ranae, Stratiotes aloides, Utricularia australis* (very rarely) and *U. vulgaris*. As late as the mid-twentieth century, well-developed communities of pondweeds were also observed, including such species as: *Potamogeton lucens, P. praelongus, P. zizii* and *P. perfoliatus*. In recent times, most habitats of species from the genus *Potamogeton* have been completely or partly destroyed.

5. Watercourses in the plains with the *Ranunculus fluitans* and *Callitricho-Batrachion* (EUH code 3260; Corine code 24.4).

A habitat with a similar distribution pattern as the previous one but observed slightly more often. In addition to the areas mentioned in 4 (above), it is also found in the Warta valley and the valleys of its tributaries on the right bank (namely, Główna, Cybina and Głuszynka). The habitat also occurs in the valley of the Michałówka, along the eastern boundary of the city (Fig. 2). The vegetation formed in watercourses includes *Ranunculus trichophyllus* (most frequently), *R. aquatilis, Myriophyllum spicatum, M. verticillatum, Callitriche cophocarpa* and *Berula erecta*.

6. Rivers with muddy banks with *Chenopodion rubri* p.p. and *Bidention* p.p. vegetation (EUH code: 3270; Corine code 24.52).

A frequent habitat but relatively poorly developed in the Warta valley. The embankments of the river and their intensive use are not conducive to the occurrence of muddy habitats. However, periodically, after a reduction in water level very good conditions for the development of plant communities belonging to the *Chenopodion* and *Bidention* associations occur on the bottom of the artificial lake "Malta", created in the Cybina valley (Fig. 2). The primary components of this type of vegetation are *Bidens tripartita, B. cernua, B. frondosa* (an increasingly common neophyte), *Chenopodium rubrum, Xanthium riparium* and *Persicaria lapathifolium.*

7. *Molinia* meadows on calcareous, peaty or clayey-silt-laden soils (*Molinion caeruleae*) (EUH code: 6410; Corine code 37.31).

   A very rare habitat found in the Bogdanka valley, in the north-western part of the city and in the Michałówka valley, along the eastern boundary (Fig. 2). Fragmentarily developed meadow communities with *Molinia caerulea* are also found in other parts of the city. This species is most frequently accompanied by *Galium uliginosum, Crepis paludosa, Luzula multiflora, Juncus conglomeratus, Inula britannica, Lotus pedunculatus, Potentilla erecta, Sanguisorba officinalis, Serratula tinctoria, Selinum carvifolia* and *Viola palustris.* In the best preserved phytocoenoses, there are several plants that are rare throughout the region, for example, *Carex pallescens, Dianthus superbus, Inula salicina, Ophioglossum vulgatum, Silaum silaus, Tetragonolobus maritimus* and *Viola persicifolia.*

8. *Galio-Carpinetum* oak-hornbeam forests (EUH code 9170; Corine code 41.261).

   As it is shown on the map of potential natural vegetation, Poznań lies in the zone of dominance of this habitat type (Fig. 4), which previously occupied the vast areas of the moraine plateau (Fig. 3). In the period of the development and expansion of agriculture, it was mainly this habitat type that was cleared and converted to agricultural use which in turn was and is the land that is the most suitable for urban development. Currently, forests of this type are rare, although the species of plants originating from them constitute an important element of the city's flora. Relatively well-developed fragments of this forest type are still preserved on the non-flooded terraces in the Warta valley, the slopes of the Cybina valley and in the zone of the end moraine in the northern part of the city. The stand most often comprises *Quercus robur* and *Carpinus betulus,* less frequently – *Tilia cordata, Acer platanoides* and *A. pseudoplatanus*. The shrub layer is usually formed by *Corylus avellana, Cornus sanguinea* and *Rhamnus cathartica.* The herb layer includes *Anemone nemorosa, A. ranunculoides, Gagea lutea, Ranunculus ficaria, Lamiastrum galeobdolon, Hepatica nobilis* (very rarely), *Milium effusum, Polygonatum multiflorum* (very rarely) and *Pulmonaria obscura.*

9. Old acidophilous oak woods with *Quercus robur* on sandy plains (EUH code: 9190; Corine code 41.51).

   A very rare type of habitat in Poznań. Small areas have been identified on the flattened area in the end moraine zone, in the northern part of the city, and above the edge of the Cybina valley, in the eastern part of the city (Fig. 2). Floristically relatively poor, the forest is formed by *Quercus petraea* (dominant) and *Q. robur* with *Betula pendula* and *Pinus sylvestris* occurring less frequently. The shrub layer includes *Sorbus aucuparia* and *Frangula alnus* with such species as *Deschampsia flexuosa, Maianthemum bifolium, Melampyrum pratense, Festuca ovina* and *Pteridium aquilinum* in the herb layer.

10. Alluvial forests with *Alnus glutinosa* and *Fraxinus excelsior* (*Alno-Padion, Salicion albae*) (EUH code 91E0*; Corine code 44.3).

    This well-understood group of forest habitats, which comprises several distinct types, is associated with the river valleys of the city. Fragments of *Salix-Populus* and *Fraxinus-Ulmus* forests have been preserved in the valley areas of the larger rivers such as the Warta. In the southern section of the valley, the *Salix-Populus* forests occur on the low inundation terrace with *Fraxinus-Ulmus* above. *Fraxinus-Alnus* forests are found in the valley of the Warta tributaries (the Cybina, Główna, Głuszyna, Bogdanka and Strumień Junikowski) (Fig. 2). These forests differ not only in terms of the occupied habitats, but also the floristic composition:

    (a) *Salix-Populus* forests; canopy – *Salix alba, S. fragilis, Populus alba* and *P. nigra;* shrub layer – *S. purpurea, S. triandra, S. viminalis* and a herb layer – *Calystegia sepium, Humulus lupulus, Fallopia dumetorum, Solanum dulcamara* (climber) and *Bidens cernua, B. tripartita, B. frondosa, Persicaria hydropiper* and *P. minus.*
    (b) *Fraxinus-Ulmus* forests – the most important role is played by *Fraxinus excelsior, Ulmus laevis, Quercus robur* (canopy), *Cornus sanguinea, Prunus padus, Corylus avellana, Sambucus nigra* (shrub layer) and *Adoxa moschatellina, Anemone nemorosa, A. ranunculoides, Ranunculus ficaria* and *Gagea lutea* (herb layer).
    (c) *Fraxinus-Alnus* forest is formed by *Alnus glutinosa* (dominant), *Fraxinus excelsior, Betula pubescens* and *B. pendula* (canopy), *Frangula alnus, Prunus padus, Ribes spicatum* (shrub layer), *Humulus lupulus, Solanum dulcamara* (climber) and *Carex acutiformis, Iris pseudacorus, Filipendula ulmaria, Deschampsia caespitosa, Lysimachia vulgaris, Urtica dioica* and *Athyrium filix-femina* (herb layer).

11. Euro-Siberian steppe woods with *Quercus* spp. (EUH code: 91I0; Corine code 41.7A).

    Potential habitats of this type were identified in Poznań on the steep shores of Lake Kierskie, in the north-western part of the city (Fig. 3). As late as the second half of the twentieth century, there remained fragments of thermophilous scrub and grassland vegetation. They have been subsequently lost as the consequence of development.

Urban Forests

State, municipal and private forests cover an area of approximately 3,330 ha. The largest complexes are located in the north-western, north-eastern and eastern parts of the city where 50–60-year-old monocultures of *Pinus sylvestris* predominate. This species was routinely planted in habitats of the *Quercus-Carpinus* forest or mixed coniferous forest, which have been planted on land previously used for agriculture. In addition to *Pinus sylvestris* (which is dominant), *Betula pendula* and *Populus tremula* are commonly found in these forests. In the shrub layer, *Prunus serotina* occurs on a large scale and *Sorbus aucuparia* has also been frequently planted. The herb layer is predominately grasses, including *Festuca ovina, Calamagrostis epigeios* and *Deschampsia cespitosa.*

Urban forests have an important bioclimatic role as well as providing important leisure and recreational facilities for the city's inhabitants; as a consequence, the forests are subject to considerable public pressure. However, this provides a significant protective function for forest and non-forest communities of greater natural value by diverting public pressure from them.

In the vast complexes of *Pinus,* monocultures scarce and *Quercus* (*Q. robur* and *Q. sesilis*), *Carpinus, Fagus* and *Tilia* forests can be found. Relatively natural forest communities occur along the watercourses and on the shores of water reservoirs. In addition to the European Union Habitat forest types described above, the city contains small fragments of *Alnus* Carr (*Ribeso nigri-Alnetum*), which occurs in swampy habitats where the watertable is close to the surface.

Non-forest Habitats

Outside the built-up areas but within the city limits, there are well-preserved non-forest habitats of natural and semi-natural character. They are mainly associated with lakes and ponds of different origin as well as watercourses. The best preserved communities of non-forest vegetation are found in the Bogdanka valley, constituting the western fragment of the wedge-shaped green open space system which includes two large lakes Kierskie and Strzeszyńskie as well as the artificial lake Rusałka and numerous peat pits. A considerable variation in the vegetation of water and wetland habitats occurs in the valley of the Strumień Junikowski, a western tributary of the Warta. Within the valley, there are about 20 water bodies of different sizes, created as the result of the extraction of clay. Different types of meadow habitats remain in the Michałówka valley, running along the eastern boundary of the city, while in the Głuszynka valley (in the south-eastern area of Poznań), in addition to the meadows there are also xerophilous swards and thermophilous forest edge communities and scrub. In the Warta valley, despite its considerable transformation, different types of fragmented aquatic, waterside and rush communities occur.

Aquatic Habitats (Abbreviations – names of the researchers who described the plant association)

Riccio fluitantis-Lemnion trisulcae (R. Tx. et A. Schwabe 1974 in R. Tx. 1974)

Communities with *Lemna minor* and *L. trisulca* are found relatively commonly in ponds, ditches and peat pits, forming clusters on the surface of standing or almost stagnant waters.

Bidentetalia tripartitii (Br.-Bl. et R. Tx. 1943)

Larger areas of habitats with species from the genera *Bidens* and *Persicaria* (*Polygonum*) (*Bidention tripartitii* Nordh. 1940), as well as *Chenopodium* and *Atriplex* (*Chenopodion fluviatile* R. Tx. 1960), are located mainly in the Warta valley, while smaller phytocenoses are commonly found along banks of all watercourses and water bodies. Only some of them meet the criteria defined in the EC Habitat Directive.

Isoëto-Nanojuncetea (Br.-Bl. et R. Tx. 1943)

Temporarily wet habitats that are found less and less often in the Warta valley are occupied by the plant communities that include *Juncus bufonius, Eleocharis acicularis* and *Cyperus fuscus.* The vegetation type is disappearing as a consequence of the regulation of the river and the re-inforcement of the banks with concrete.

Charetalia (Sauer 1937)

Ephemeral communities of stoneworts (*Chara* and *Nitella*) that occur in peat pits located in the Bogdanka valley and other water bodies.

Potamion (Koch 1926 em. Oberd. 1957 and *Nympheion* Oberd. 1953)

A group of habitats and plant communities, which are highly diverse and, with few exceptions, strongly endangered as a result of regulation of watercourse and changes to the water chemistry. The most frequent community, which occurs in still, eutrophic water, is dominated by *Ceratophyllum demersum.* Communities occurring in isolated, shallow waters comprise such species as *Myriophyllum spicatum, Hydrocharis morsus-ranae* and *Stratiotes aloides.* Communities with *Nuphar lutea* and much less often with *Nymphaea alba* also occur as well as typically small phytocenoses with *Hottonia palustris.*

Wetlands

Phragmitetea (R. Tx. et Prsg. 1942)

Reed habitats, which are highly diverse, are relatively common in the river valleys of Poznań. Relatively well-developed phytocoenoses of typical reeds (*Phragmition* Koch 1926) are found on the shores of the lakes Kierskie, Strzeszyńskie and Rusałka and in the littoral zone of ponds in the valleys of the Cybina and Strumień Junikowski. Habitats of this type are much less well preserved in the Warta valley. The reed communities include the following species: *Hippuris vulgaris, Schoenoplectus lacustris, Typha angustifolia* and *T. latifolia, Sagittaria sagittifolia, Sparganium emersum, S. erectum, Eleocharis palustris, Equisetum fluviatile, Phragmites australis, Acorus calamus* and *Glyceria maxima,* as well as *Oenanthe aquatica* together with species from the genus *Rorippa.* The *Bulboscheonus maritimus* habitat occurs in the Warta valley but only rarely. The occurrence in the city of communities with an abundance of tall species of the Cyperaceae is noteworthy; they are generally well-developed and some of them cover large areas. The communities include species such as *Carex pseudocyperus, C. riparia, C. acutiformis, C. paniculata, C. rostrata, C. elata, C. appropinquata, C. acuta,*

C. vesicaria, C. vulpina and *C. disticha*. Within the city limits, there is also a calciferous habitat with *Cladium mariscus* – a habitat that is rarely found in the whole region. Other phytocoenoses with *Phalaris arundinacea* and *Iris pseudacorus* are commonly recorded in Wielkopolska.

Wetlands with low-growing species are represented by four plant communities, distinguished by: *Glyceria fluitans, G. notata, Leersia oryzoides* and very rarely *Rorippa nasturtium-aquaticum*.

Grasslands

In addition to meadows described above from the *Molinion* association (and mentioned in Annex 1 of the EC Habitat Directive), several highly diverse meadow and pasture communities occur within the city, for example:

(a) Low swards on periodically flooded sites (*Agropyro-Rumicion crispi* Nordh. 1940 em. R. Tx. 1950).
(b) Perennial herb communities developing in unmown marshy sites and comprising tall perennials (*Filipendulion ulmariae* Segal 1966).
(c) Eutrophic wet meadows occupying drained and well-fertilised locations (*Calthion palustris* R. Tx. 1936 em. Oberd. 1957).
(d) Boggy meadows occupying small, local depressions (*Scirpetum silvatici* Ralski 1931), including the *Juncus subnodulosus* habitats, which are very rare in the Wielkopolska region.
(e) Grazing meadow habitats, in which *Juncus effusus* is abundant.
(f) Intensively cultivated and managed meadows (*Alopercurus pratensis* Pass. 1964).
(g) Meadows formed on fertile, not very wet and non-boggy sites (*Arrhenatherum elatius* (Br.-Bl. 1925) Koch 1926).

Carex/Bryophyte Fens

Habitats of meadow bogs and fens rich in mosses (*Scheuchzerio-Caricetea nigrae* (Nordh. 1937) R. Tx. 1937) constitute unique systems within the administrative boundaries of the city with elements of the flora that are rare in Poznań and elsewhere in Poland. The habitats include fragmented fen with such species as *Carex lasiocarpa* and *C. diandra*, as well as communities of poor fens with *Carex curta* and *Agrostis canina*, which have been identified in the Bogdanka valley. Until recently the same complex contained a unique community of *Schoenus ferrugineus*.

Thermophilous Fringe Communities and Scrub

Two types of plant communities are found in the ecotone between the forest and grassland communities. First, scrub, comprising a belt of shrubs adjacent to the forest (*Rhamno-Prunetea* – Rivas Goday et Garb. 1961), in which the most frequent species is *Prunus spinosa.* Second, a narrow strip of thermophilous and photophilous perennial herbs (*Trifolio-Geranietea sanguinei* Müll. 1962) such as *Trifolium medium, T. alpestre* (rarely), *Agrimonia eupatoria* and *Vicia dumetorum.*

Residential Areas

As already mentioned, built-up areas located within the city limits include three main spatial complexes: (i) high density tenement housing, (ii) low density apartment blocks, and (iii) low density detached houses. They differ substantially in terms of the density and height of buildings, the area of paved surfaces (unavailable for higher plants), physical and chemical properties of soil and climatic conditions. These result in the diversification of the vascular flora, which is manifested, among other things, in the total number of species, the number of species found in areas comparable in size as well as species structure.

The flora of the tenement housing complex comprises about 310 species; the number of species per km^2 is frequently <100. More than 39.7% of the flora comprises alien species of which 18.7% are archaeophytes and 13.2% are neophytes; the remainder is composed of casual alien plants. A total of 377 species has been recorded in the area covered by the low density apartment blocks – 100–180 species per km^2. Approximately 39.8% of the species are aliens; as with the tenement housing complexes, the proportion of archaeophytes (19.4%) exceeds that of neophytes (12.5%). The flora of the low, detached housing complex is much richer. In this area, a total of 589 species were recorded and there were usually 180–220 taxa per km^2, while alien species comprise almost 37.2% of the flora. Predominance of archaeophytes (14.9%) over neophytes (12.4%) was slight, while ephemeral species accounted for a high proportion of the flora (9.9%).

Interesting findings were reported in the recently completed studies on the dendroflora of these complexes (Winiecka-Nowak 2009, unpublished). They not only confirmed the claim that with an increase in the density of building development the number of tree and shrub species decreases but also showed that the decrease mainly affected fruit-bearing species. In the low building (scattered) complex, the number of fruit-bearing phanerophytes was observed to be twice as high (on average about 40 species in the survey area) than in the complexes of high density and apartment blocks, where there was an average of about 20 species.

Transport Routes

Poznań, which is located halfway between Berlin and Warszawa, is a very important centre of railway and road transport. The role of the city as a centre for air transport has been increasing while river transport is of little importance.

Railway Land

The main railway station, with the related extensive railway track system, is located virtually in the city centre. In the eastern part of the city, there is a freight railway station that occupies an equally large area. The length of the railway tracks within the city is approximately 125 km.

Road Verges

The southern part of Poznań is crossed by 13 km of the motorway A2, which was constructed several years ago and which replaced the international road (E8) that runs through the northern part of the city. The main public transport is provided by high density tramway and bus networks. In this respect, the use of rail transport is of very limited extent. The Poznań road infrastructure comprises 1,039 km of roads and 460 engineering structures. Approximately 75% of roads are hard surfaced. The primary transport system of the city comprises 204 roads, which cover 45.9% of the total area of the city streets. The length of municipal transport routes is 982 km, of which 78% are bus routes. In 1997, a tram system, unique on the national scale, was opened in the city, the so-called Poznań Fast Tram. A special route of 6.1 km links the city centre with the Piątkowo housing district. Over 25 million passengers use the Poznań Fast Tram every year.

Airport

A small airport Ławica is located in the western part of Poznań, serving passenger airlines (over one million people annually), while in the south-eastern part of the city there is a military airport. A small river port, now rarely used, is located on the Warta.

Transport routes have a huge impact on the vegetation cover of the city, which is demonstrated both by landscaped areas and spontaneous vegetation. According to a survey in the mid-1990s, there were over 26,020 roadside trees under the supervision of the city authorities. Hedges, with an estimated total length of 40 km, are also associated with transportation routes. Road and railway verges serve as migration corridors for numerous plant species. For example, *Puccinellia distans* (a facultative halophyte of native origin) and *Bunias orientalis* (a neophyte originating from south-eastern Europe and western Asia) as well as many other alien species have spread rapidly throughout the city. Altogether, in industrial and transport areas approximately 430 species have been recorded to date, including as many as 105 neophytes and almost 70 archaeophytes. The more frequent road verge species include: *Poa annua, Lolium perenne, Matricaria discoidea, Polygonum aviculare* agg. and *Plantago major*. The railway land contains such species as *Sisymbrium altissimum, S. loeselli, Amaranthus retroflexus, Sedum acre* and *Artemisia vulgaris.*

Parks

In Poznań, there are approximately 40 parks of different sizes, covering a total of more than 440 ha. The greatest number of parks is located within a radius of about 6 km from the city centre (Mierzejewska 2001); however, they are mostly small. The larger parks include the green wedges that are linked to the forest areas. The biggest park (over 100 ha), the Cytadela (the central part of the urban fortifications) occurs in the city centre. The park is characterised not only by a highly diverse landscape design but also a rich spontaneous flora (over 400 species of vascular plants). It contains a large group of forest and meadow plants as well as several lithophytic species, which have found a suitable habitat in the gaps of the walls of the fortress. A unique group of parks includes the Botanical Garden of the Adam Mickiewicz University (17.2 ha) and the Dendrological Garden of the Poznań University of Life Sciences (4 ha). They are characterised by an exceptionally rich flora. In the former, over 7,000 taxa have been collected, while in the latter, there are approximately 900 species, varieties and cultivars of trees and shrubs. The park greenery is also associated with the centrally located old Zoological Garden (5.2 ha), while the new Zoological Garden, covering an area of 117 ha, has been established in the vast forest complex in the eastern part of the city.

Cemeteries

There are 19 cemeteries in the city, covering a total of almost 232 ha, of which almost 94% accounts for the two biggest municipal cemeteries, Junikowo (94 ha) and Miłostowo (98 ha). Nine much smaller parish cemeteries are of historical value. Apart from their primary function, cemeteries (whether large, small, old or recent including those established in the early twentieth century and with a designed landscape) are of considerable botanical importance. They characteristically combine the value of the formal landscape with the richness of the spontaneous flora. On the one hand, in the context of the background of the city's built-up areas old cemeteries form islands of forest and scrub flora, while, on the other hand, they offer a habitat for the epiphytic biota, particularly for bryophytes and lichens. Cemeteries are almost the only places within the city, where a sandstone substrate is found. There is also a large variety of slabs, gravestones and monuments made from different types of other crystalline rocks (less often calcareous rocks) and processed materials (for example, terrazzo and concrete, etc.). Preliminary studies of the vascular flora of three old Poznań cemeteries, with a total area of 13 ha, showed the presence of over 220 taxa, including 181 spontaneously occurring species. Almost 40% species are of alien origin (37 archaeophytes and 29 neophytes). Although in the flora segetal (20.5%) and ruderal species (15.4%) predominate, the proportion of forest and scrub species (19.8%) is also considerable.

Allotment and Domestic Gardens

In Poznań there are almost 100 complexes of allotment gardens and family gardens plots, occupying a total area of over 825 ha. They are scattered in a haphazard manner throughout the city with a large cluster in the western districts. Allotment gardens represent a unique type of habitat; very often they have been established on the poorest agricultural land or long-term wasteland. Until recently, they were predominantly used for growing vegetables and fruits for the immediate needs of their owners. As a consequence of the application of different types of cultivation measures, the owners improved the productivity of the soil, which also had some side-effects on the species composition of the spontaneous flora. Species that prefer arid, sandy habitats with low soil pH (for example, *Agrostis capillaris*) were replaced by species of higher trophic and soil moisture requirements, including nitrophilous plants. The characteristic flora in this type of habitat includes *Euphorbia helioscopia, E. peplus, Galinsoga parviflora, G. quadriradiata, Urtica urens, Solidàgo canadensis* and *S. gigantea.*

During the last decade there has been an increasing trend for allotment gardens to be used more for recreational purposes than food production. As a result the growing of vegetables and fruits has been replaced by cultivated lawns and ornamental plants of more and more exotic species. This change obviously has an effect on the species composition of the spontaneous flora. The previously dominant root crop species have been replaced by plants characteristic of trampled areas and grasslands.

Preliminary studies of the spontaneous flora of one of the bigger complexes of family allotment gardens recorded the presence of over 350 species, of which over 30% were alien species. Among the latter, archaeophytes predominated (50 species) associated with arable land on which the allotment complexes had been established and with habitats accompanying human settlements. As could have been predicted, a considerable proportion of the flora comprises plants that have spread from cultivated areas (26 species). Many native species have managed to survive in this habitats for many years, for example, *Dianthus superbus* and *Sanguisorba officinalis* as well as aquatic and waterside plants associated with ditches that were retained in the complex, *Caltha palustris, Cardamine amara* and *Symphytum officinale,* together with components of swards typical of sandy soils (*Corynephorus canescens, Spergula morisonii, Filago minima* and *F. arvensis*).

Habitats of a similar character are associated with gardens established around detached family houses. Currently, Poznań contains over 100 housing estates covering almost 4,000 ha, which is about 15% of the city area. In total, the different types of garden cover over 4,700 ha which is many times the area of the parks. Multi-species composition of the vegetation of domestic gardens, in which coniferous trees and shrubs as well as perennials predominate, is primarily for ornamental purposes. Spontaneous plants are generally eradicated from these habitats. Nevertheless several native and alien species are able to establish, particularly *Humulus lupulus* (climbing on fences), *Glechoma hederacea* (commonly found in lawns) as well as two species appearing under the canopy of trees that have been mulched with bark, *Oxalis corniculata* and *Cardamine hirsuta.*

Squares and Small Green Areas

An important element of urban green space is associated with Squares and small green areas. They usually occupy from several hundreds of square meters to about 2–3 ha. This group comprises: (a) small green areas adjacent to industrial plants and shopping and sports facilities, children's playgrounds and school gardens (jointly amounting to about 85 ha), (b) green areas in the precincts of hospitals and housing estates (approximately 460 ha), (c) small green areas adjacent to transport routes (lawns separating traffic lanes, green areas around tram terminus and bus stations, and roundabouts) with a total area of approximately 150 ha, and (d) other small recreational areas (86.9 ha). The purpose of these landscape areas varies and includes ornamental, recreational, and protective. The vegetation, which depends on the landscape objective, may include one or a combination of lawns, flowerbeds and small green areas with planted trees and shrubs of different species. The spontaneous flora depends on the age, design and maintenance of the area. It usually comprises annual and biennial plants that are commonly found in cities, very often of alien origin. Examples of some quite interesting species are *Amaranthus blitoides* and *A. blitum* var. *ascendens,* which occur in the city centre flowerbeds and *Stachys annua,* which is found among clusters of shrubs planted on the calcium-rich soils of new housing estates.

Agricultural Areas

Over 33% of the city (8,655 ha) comprises agricultural land, which, as a rule, are former suburban villages that have been incorporated into the administrative boundaries of Poznań. Some of them occupy large areas that still remain outside the areas of intensive development. In terms of the flora, they are the poorest within the entire city and frequently sustain no more than 80 species per km^2. Within the cereal crops, there are such species as *Apera spica-venti* (found in large quantities in many fields), *Centaurea cyanus, Papaver argemone, P. rhoeas, P. dubium, Matricaria recutita, Tripleurospermum inodorum,* several species from the genus *Veronica* and less and less frequently *Agrostemma githago*. The land used for growing root crops contains *Elytrigia repens, Avena fatua, Amaranthus retroflexus, A. chlorostachys* (rarely) and *Stellaria media.*

As the spatial development of the city progresses, the arable fields become fragmented and surrounded by buildings. This leads to an increase in the number of species, since the group of plants closely connected with cultivated fields is supplemented by species accompanying human settlements. This process is sometimes referred to as the "ruderalisation of segetal flora".

Wasteland

Throughout the city there are scattered areas which are periodically excluded from conscious, purposeful use. They are highly diverse in terms of size, duration of the lack of use, the history preceding their disuse and in their location in the urban space. This is reflected in the vegetation structure and species composition of their flora. Irrespective of this diversification, these areas serve an important biotic function. The areas of this type are difficult to estimate in terms of their size because they are not subject to detailed surveys. Most frequently, they are included in the category of land management of the area to which they are related. In the suburban zone, they comprise mainly fallow land. Most frequently there is an administrative decision approving the change of use from arable agriculture to non-agricultural, especially for built development or some other type of investment project. Although cultivation practices are no longer continued in such areas, some segetal weeds continue to grow in this habitat. However, gradually the dominant annual plants are replaced by perennials followed by the appearance of tree seedlings. The species composition of the habitats depends primarily on the character of the substrate.

Wastelands also form a considerable proportion of the land in industrial and transport complexes. The soil in such areas is generally very strongly transformed both in terms of its mechanical and chemical properties. This results in these habitats being colonised mainly by species that are resistant to environmental stress. A characteristic feature of the flora of industrial wastelands is the high proportion of neophytes.

Areas excluded from management are also found within residential districts. For example, former agricultural land that is allocated to large investment projects that will take many years to start and complete. Wasteland of this type becomes overgrown by typical segetal weeds. In the high density tenement-type housing, the occurrence of small patches of wastelands is, as a rule, transitory in character and more associated with the re-development or the allocation of areas previously serving other functions (for example, recreational) than new building development.

Nature Conservation, Environmental Planning and Education

Principles and directions of environmental management

Following national legislation, the primary directions of environmental planning, design and management are contained in "*Study on the conditions and directions of spatial management in the city of Poznań*" (2008) which covers the entire city and provides the basis for local spatial development plans. These plans are prepared at a relatively large scale (most frequently 1:5,000 and 1:10,000) and are the basis for deciding the location of all types of developments.

The "Study..." cited above stresses the primary importance of the wedge-ring system of the urban green space that forms the historically developed structure within which the way of management should be subordinate to the protection of natural values and resources. The primary task is to enhance the biological function of this system through the following measures:

(a) Preservation and enhancement of the continuity of the system, including the attempts to integrate scattered fragments of greenery, protect existing and realise new connections with the surrounding areas.
(b) Recommendations for the incorporation of the most valuable fragments of the system into the emerging local spatial development plans in order to enhance their protection.
(c) Establishment of nature protection in the areas of special natural values, preceded by the verification of their natural value.
(d) Preparation and realisation of transport connections, overhead and underground pipelines and cables in a manner that ensures the protection of the landscape, as well as minimising the environmental impact of investment projects.

The "Study..." is particularly focused on the valuable natural areas, including not only protected on the basis of the national legislation "*Act on Nature Protection*" but also those of local importance, connected with the specific character of the urbanised area. The "Study..." defines the boundaries of the valuable natural areas that are recommended for the different forms of nature protection and states that building development will be excluded from these areas, with the single exception of public investment projects of trans-regional importance being permitted.

An important provision of the "Study..." is the protection and development of the municipal forest areas, which constitute a refuge for biological and landscape diversity and provide leisure and recreation facilities for the inhabitants of the city. It is assumed that the forested area is going to increase and forest stands will be re-structured so that they are more related to site conditions. The need for the closer integration of municipal forests with the regional system of state forests is also stressed.

An essential goal of the environmental policy of the city is to protect and manage the parks. The objectives are to maintain their natural functions and to enhance their value for leisure and recreation. The number and area of municipal parks are assumed to increase, as they are to be established first of all in new housing districts, which have fewer parks than the central districts.

The most crucial goal connected with agricultural areas, which are potential investment sites, is to preserve the spatial order when planning new investment projects. The agricultural function is to be limited to those plots of land that are incorporated into the wedge-ring system of green areas. The environmental policy of the city authorities assumes the support for actions aimed at the preservation and establishment of new allotment garden complexes in suburban areas, treating them as an element that will supplement the system of urban green space.

Red List Species

The Red List of vascular plants of Poznań comprises 568 species which accounts for 53.2% of the permanent flora of the city. This is similar to the level of threat in other cities of central Europe (for example, West Berlin and Warsaw), where it is >50%. The Red List contains 200 of the species that are endangered and threatened in the Wielkopolska region and 39 taxa that have currently disappeared from Poland. An especially important group is the 18 species that were declared endangered on the national and regional scales (Table 5) and at the same time (the first half of the 1990s) they were observed in the area of Poznań, unfortunately, recent studies have confirmed that only some of them are still present.

On the other hand, the threat to vascular plants in Poznań is approximately 20% higher than the average level of decline in the regional floras of central Europe. For example, the Red List of vascular plants of the Wielkopolska region comprises about 30% of the flora (Żukowski and Jackowiak 1995). A characteristic feature of the Poznań flora is the considerable proportion of alien species in the list of endangered species. In Poznań this group comprises 40 species, which is about 7% of plants that are declining as the result of human activity. In the group of endangered anthropophytes, archaeophytes (34 species) predominate over neophytes (6 species). The highest number of alien species (23) occurs in the extinct category.

Among the unique species, the flora of Poznań includes species listed in Annex II of the EC Habitat Directive: *Angelica palustris* – observed occasionally since the first half of the nineteenth century, *Liparis loeselii* – reported for the last time in the mid-1980s and *Pulsatilla patens,* which became extinct in the middle of the nineteenth century.

More than 60 species in the Red List of large-fructification fungi endangered in Poland have been found in the city, they include *Amanita strobiliformis, Coltricia*

Table 5 A list of endangered and threatened species of Poznań, the Wielkopolska region and Poland (as of 1995)

	Threat category in:		
Name of species	Poland	Wielkopolska	Poznań
Alisma gramineum	R	V	I[a]
Angelica palustris	E	E	E[a]
Carex ligerica	R	V	E[a]
Cnidium dubium	V	V	E[a]
Corrigiola litoralis	V	V	E
Dactylorhiza maculata	V	V	E[a]
Dianthus superbus	V	V	R[a]
Drosera rotundifolia	R	V	E
Epipactis palustris	V	V	V[a]
Gentiana pneumonanthe	E	E	E
Gentianella uliginosa	V	V	E
Juncus subnodulosus	V	V	E[a]
Lathyrus palustris	V	V	V[a]
Liparis loeselii	V	E	E
Schoenus ferrugineus	V	E	E
Scorzonera purpurea	R	E	E
Veronica catenata	I	V	I
Viola persicifolia	V	E	E

Threat categories *E* endangered, *V* vulnerable, *R* rare, *I* indeterminate
[a]Species observed in Poznań after 2000

cinnamomea, Gyromitra gigas and *Macrotyphula fistulosa*. The city also contains an abundance of nationally rare or endangered lichen species, 43 species are legally protected, while as many as 51 species are listed in the "*Red list of the lichens in Poland*".

Areas of Natural Value

The following protected areas occur within the city boundaries:

1. Three areas of European importance, included in the Natura 2000 ecological network, of which one area is located entirely within the city limits and two are partially within it.
2. Reserves established on the basis of the national "Act on Nature Protection".
3. Many other areas of natural history value, of which several are protected in the form of "Ecological Areas".
4. Many single legally protected objects (trees, avenues and boulders).

Natura 2000 Sites

The list of the Natura 2000 areas, prepared by the Minister of the Environment, following the EC Habitats Directive, includes two habitat refuges and one bird refuge established within the city limits of Poznań.

1. Special Area of Conservation PLH300001 Biedrusko (9,641.7 ha) is located north of Poznań, mostly on a military training ground. In that area, 18 habitats from (Annex 1) of the Directive have been identified. Within the city limits, the areas of particular value include fragments of riverside carr, which is well preserved in the former manor house park at Radojewo.
2. Special Area of Conservation PLH300005 Fortifications in Poznań (137.39 ha) comprises a complex of nineteenth-century fortress facilities, the Citadel and three bunkers (a total of 22 objects), constituting the basis for the ring system of municipal greenery. They are wintering grounds for large colonies of bats.
3. Special Protection Area PLB300013 the Samica River Valley (2,390.98 ha) covers the upper and middle course of the Samica River, which is a left-bank tributary of the Warta. A small fragment of the bird refuge is located in the north-western outskirts of Poznań.

Nature Reserves

According to the Polish legal regulations, a nature reserve covers areas preserved in a natural or slightly altered condition, ecosystems, refuges and natural habitats, as well as habitats of plants, animals and fungi and formations and components of inanimate nature, exhibiting special natural, scientific, cultural or landscape values.

Such areas are established as nature reserves on the basis of an Ordinance of the Provincial Governor. The Ordinance defines the name of the reserve, its location and boundaries and a protection zone (if one was established), the goals of protection as well as the kind, type and sub-type of the reserve, and its supervising body.

1. "Żurawiniec" mire reserve (1.47 ha) was established in 1959 to protect its transitional bog and rare species of plants, such as *Drosera rotundifolia, Vaccinium oxycoccos* and several species from the genus *Sphagnum*. As the result of the impact of urbanisation, particularly the progressive drying out of the area and direct anthropogenic pressures of the local population, the most important values of the reserve have been lost. After extensive discussion, it was decided to maintain the status of the reserve, mainly for historical reasons.
2. Landscape reserve "Meteoryt Morasko" (54.54 ha) was established in 1971 in order to protect the landscape of the end terminal moraine as well as the sites where meteorites had fallen. In the reserve there are craters, partly filled with water. They are located among *Quercus-Carpinus* and *Ulmus-Fraxinus* forests of great natural value. In the large forest complex, the habitats of alder (*Alnus*) carr and acid *Quercus* forest have also been preserved.

Ecological Areas

One of the forms of nature protection established in the Polish law is the "Ecological Area". These areas are remnants of ecosystems worthy of protection because of their importance for the preservation of biological diversity. They are natural water reservoirs, small water bodies located in fields and forests, clusters of trees and shrubs, swamps, peat-bogs, dunes, phytocenoses of uncultivated vegetation, oxbow lakes, rock outcrops, escarpments, gravel-bars, natural habitats as well as localities of rare or protected species of plants, animals and fungi, their refuges and breeding grounds or seasonal (for example, wintering) grounds.

A network of 23 Ecological Areas was created in the city in 1994. It was proposed to establish another 24 Areas in subsequent years. Unfortunately, in 2004, as a result of changes to the "Act on Nature Protection" Ecological Areas became no longer legally valid. However, this does not mean that they have lost their natural value. Some of these sites, at present called "Areas of Natural Value", are given the status of Ecological Areas within the framework of work conducted on the local spatial management plans. The large number indicates that irrespective of the legal problems, there still remains many places of significant value for the preservation of natural biodiversity in the area of Poznań.

Nature Monuments

In Polish law, nature monuments are single formations of an animate and inanimate nature or their clusters of special natural, scientific, cultural, historical or landscape value that exhibit individual features, which distinguish them from other formations. Most frequently they are veteran trees, shrubs of native or alien species, springs, waterfalls, karst springs, rocks, ravines, erratic boulders and caves. There are 25 natural monuments in the city, most of them are trees: 12 single specimens and ten groups and avenues, comprising almost 1,000 trees. Additionally three erratic boulders have been established as nature monuments.

Closing Comments

Poznań is a large city in central Europe, whose history dates back many centuries. It is typical of European cities in terms of its spatial structure and intensive urbanisation during the last 20 years. It has been subject to systematic research in life sciences for almost 200 years, which facilitates periodic reviews, summaries and syntheses. Consequently it has been possible to follow changes in its flora in the period of accelerating industrialisation and fundamental cultural and social changes. Apart from many detailed observations, it is possible to reach several conclusions of a general nature in terms of the functioning of plants under urbanisation stress and the role of the city in the transformation of the flora in a wider spatial perspective.

The botanical studies of the city indicate, among other things, that cities are areas where strong species selection takes place. According to the general rule proposed by J.B. Faliński (2000), presented in the theory on synanthropisation, many native, stenotopic elements are replaced by eurytopic native or even alien species. The loss of indigenous elements and the spread of cosmopolitan plants lead to the uniformity of urban flora, at least within the same biogeographical region. The frequent extinction of species in urban areas is of catastrophic dimensions. Degradation of habitats and direct human impact on isolated populations of plants are not the only causes of species extinction. A factor which forms a particularly crucial role is the spread of invasive alien species. Poznań, like many other European cities, lies on a large river and is an important centre of rail and car transport. Conditions found in such habitats are especially lacking in competitiveness, which promotes the establishment and spread of invasive species. The first phase of the process occurs primarily in built-up areas followed by the spreading of species outside the urbanised zones. Consequently the city becomes, as it were, a source of expansive species. This is evidenced in the history of colonisation of several alien species that are currently found on a massive scale in the Wielkopolski National Park, located near Poznań.

Knowledge of the mechanisms of changes in urban flora and its role in modifying the vegetation cover outside the administrative boundaries of a city suggests that it is advisable to undertake specific actions in terms of environmental management. In addition to the standard methods of environmental management that are currently being used, it is also necessary to develop and apply new methods. It is of particular importance in view of the fact that despite huge transformations in urban areas there remain features of considerable natural value, even on the European scale.

Literature Cited

Bartkowski T (1981) Transurbacje miast Wielkopolski i niektóre zagadnienia przestrzenno-planistyczne ich rozwoju oraz zastosowanie do niektórych metod fizjografii urbanistycznej. Wyd. Nauk. UAM, ser. Geografia, 22: 1–228

Breś W (2000) Assessment of soil contamination by zinc, manganese, copper and iron on the area of Poznań. Rocz. AR Poznań CCCXXIII, Ogrodnictwo 31(1): 211–215

Celka D (2002) Królestwo grzybów. Kronika Miasta Poznania, 3: 70–80

Fudali E (2005) Bryophyte species diversity and ecology in the parks and cemeteries of selected Polish cites. Agricultural University of Wrocław, Wrocław

Gąsiorowski E (1973) Poznań średniowieczny (X-XV w.). In: Topolski J (ed) Poznań – zarys dziejów. Wyd. Pozn. Poznań

Faliński JB (2000) The Interpretation of Contemporary Vegetation Transformations on the Basis of the Theories of Synanthropisation and Syndynamics. In: Jackowiak B, Żukowski W (eds) Mechanisms of Anthropogenic Changes of the Plant Cover. Publications of the Department of Plant Taxonomie, AMU, 10: 9–30

Jackowiak B (1990) Antropogeniczne przemiany flory roślin naczyniowych Poznania. Wyd. Nauk. UAM, seria Biologia, 42: 1–232

Jackowiak B (1993) Atlas of the distribution of vascular plants in Poznań. Publications of the Department of Plant Taxonomy, AMU, 2: 1–409

Kaniecki A (1993) Dzieje miasta wodą pisane. I. Przemiany rzeźby i sieci wodnej. Wyd. Aquarius. Poznań

Kepel A (2002) Niedostrzegane porosty. Kronika Miasta Poznania, 3: 81–90

Mierzejewska L (2001) Tereny zielone w strukturze przestrzennej Poznania. Wyd. PTPN, Poznań

Rusińska A (2002) Wszędobylskie mchy i wątrobowce. Kronika Miasta Poznania, 3: 39–46

Sukopp H (1972) Wandel von Flora und Vegetation in Mitteleuropa unter dem Einfluss des Menschen. Ber. ü Landwirtschaft, 50 (1): 112–139

Szafer W (1972) Szata roślinna Polski Niżowej. In: Szafer W, Zarzycki K (eds) Szata roślinna Polski, 2, PWN, Warszawa

Szeląg-Wasielewska E (2002) Glony naszych wód. Kronika Miasta Poznania, 3: 25–38.

Topolski J (1973) W czasach nowożytnych (XVI-XVIII w.). In: Topolski J (ed) Poznań – zarys dziejów. Wyd. Pozn. Poznań

Wojterski T, Wojterska H, Wojterska M (1981) Mapa potencjalnej roślinności naturalnej środkowej Wielkopolski. Bad. Fizjogr. nad Polską Zach., 24: 143–163

Woś A (2005) ABC meteorologii. Wyd. Nauk. UAM, Poznań

Zając A (1978) Atlas of distribution of vascular plants on Poland (ATPOL). Taxon, 27(5/6): 481–484

Zajchowska S (1977) Rozwój sieci osadniczej Poznania i najbliższego zaplecza w średniowieczu. In: Błaszczyk W (ed) Początki i rozwój starego miasta w Poznaniu w świetle badań archeologicznych i urbanistyczno-architektonicznych. Bibl. Fontes Archeolog. Posnaniensis, 3: 37–65

Żukowski W, Jackowiak B (1995) List of endangered and threatened plants in Western Pomerania and Wielkopolska (Great Poland). In: Żukowski W, Jackowiak B (eds) Endangered and threatened vascular plants of Western Pomerania and Wielkopolska. Publications of the Department of Plant Taxonomy, AMU, 3: 10–96

St. Petersburg

Maria Ignatieva, Galina Konechnaya, and Glenn Stewart

Fig. 1 St. Petersburg – the "Venice of the North". The Peter and Paul Fortress was the first structure built in St. Petersburg (1703) and is one of the main symbols of the City

Abstract Peter the Great initiated a gigantic experiment to change an inhospitable natural wetland landscape into a major city and port by the construction of drainage canals and buildings, the spreading of fertile soil and the planting of millions of broad-leaved trees. During Soviet times the city was surrounded by high-rise apartment blocks. The city is now a UNESCO World Heritage Site. The historic parks and gardens have the highest plant biodiversity of all the urban biotopes in

Maria Ignatieva (✉)
Swedish University of Agricultural Sciences, Division of Landscape Architecture,
Department of Urban and Rural Development, P.O. Box 7012, SE-750 07 Uppsala, Sweden
e-mail: maria.ignatieva@slu.se

J.G. Kelcey and N. Müller (eds.), *Plants and Habitats of European Cities*,
DOI 10.1007/978-0-387-89684-7_12,

the central areas although the most common biotopes are associated with walls, buildings and embankments. There are six protected areas (2,150 ha) comprising unique native habitats and which contain the most rare species and "pure" undisturbed examples of natural habitats. The recent shift to a market economy and the consequential increase in air pollution has seen a decrease in lichen biodiversity and degradation of urban soils. The shift has also resulted in the sub-urbanisation of the city caused by a change in emphasis from public green space in the Soviet era to the large private gardens of affluent people. St. Petersburg is following international trends in landscape design; all the plant material for new public and private sectors is sourced from 'western' nurseries and based mainly on non-native, fashionable "global" taxa, as a consequence the urban flora is also becoming standardised.

Natural Environment of the City

St. Petersburg is the second largest city in the Russian Federation and Europe's fourth largest city with a population of 4.8 million. It is located in the north-west of Russia on the Neva River delta on the eastern coast of the Gulf of Finland (hereafter referred to as the "Gulf") in the Baltic Sea at 59°57′ north and 30°19′ east. The city is situated along the shores of the Neva Bay of the Gulf and islands of the river delta. Because of its geographical location, the city is famous for its "White Nights" which begin at the end of May and last for at least 50 days. The longest day is June 21 (18 h 45 min); the shortest day is December 22 (5 h 52 min).

Climate

The climate of St. Petersburg is humid continental of the cool summer sub-type due to the influence of the Baltic Sea and the frequent invasion of warm air masses causing thawing of the ice, which is typical of maritime climates. The average annual temperature is 4.3°C. The average winter temperature is −7.8°C (January and February) and the average summer temperature is 17.8°C. The mean duration of the frost-free period is 157 days. St. Petersburg is in a zone of high precipitation with an annual average of 550–600 mm. Air humidity is 78% on average (Pokrovskaya and Bychkova 1967). During the winter months, the average depth of snow cover is not more than 33 cm. Soil moisture is usually high because of the lower evapo-transpiration rate causing the cool climate. The prevailing winds are from the west, north-west and south-west.

Water occupies 7% of the surface area of the city, which is among the highest of most cities worldwide and the highest in Russia. There are 308 waterways (totalling 217 km) and 108 reservoirs, lakes, ponds and artificial pools (>1 ha), which occupy a total of 2,087 ha. The short and powerful Neva river, the city's main waterway, has a total length of 74 km, of which 32 km is in the city. As the only outlet from

Lake Ladoga, the largest lake in Europe, it carries an enormous volume of water, which is discharged into the Gulf. The average width of the Neva within the city is 400–600 m and the average depth is 8–11 m. The high waves generated by tornadoes in the Baltic Sea have caused numerous floods – approximately 290 since the city was built, the most devastating was in AD 1777 (3.21 m a.s.l), AD 1824 (4.21 m) and AD 1924 (3.8 m).

Topography

The elevation of St. Petersburg ranges from the sea level to 175.9 m at the summit of Orekhovaya Hill. Most of the city is no higher than 4 m a.s.l. The city centre is situated in the river delta. There are several distinctive natural sites within the city and the adjacent areas; they are defined as "natural formations, typical of certain processes and phenomena and as a rule unique within the surrounding scenery". They can be of geological, geomorphological, botanical, or hydrological interest. The *Pinus* forests of the hills in the Ozerki – Shuvalovo – Jukki – Osinovaya Roshcha – Koltushi area are easily visible from the city and have become a favourite area for private residential development. The sand dunes, with *Pinus* forests, that stretch along the Gulf coast from Sestroretsk to Zelenogorsk, are favourite recreational zones. Quite steep slopes of the upland terraces (around Pargolovo – Lanskaya – Udelnaya area) have many green areas (including remnants of native forests). The residual fluvio-glacial Suzdalskie Lakes are the most popular public swimming lakes. Since the eighteenth century, the Duderhof Hills, which are characterised by the sharp relief of undulating moraines have been a "Mecca" for botanists because of their unique forests with a species-rich herb layer. Lake Dudergofskoe (Duderhof) has artesian springs (about 30 near Villozi). Other areas of topographical interest include the sand bars and spits of Elagin and Krestovsky Islands and the hilly and moraine lake-hollow relief north of Zelenogorsk. Therefore, it is not surprising that many of the protected areas (with unique flora and vegetation) in St. Petersburg include these distinctive natural sites.

The St. Petersburg area comprises 18 regions covering 1,439 km^2 including St. Petersburg itself (605 km^2), nine suburban towns and 21 municipal settlements. The greatest distance between the western and eastern borders of the city is 60 km and between the north and south boundaries it is approximately 60 km. The central districts of the city are very densely built (80% of the entire city area).

Soils

Natural topsoils and subsoils have retained their original composition only outside the city boundary. The soil cover of St. Petersburg has a characteristic multi-layered texture conditioned by the climate, parent rocks, drainage, micro-climate and vegetation.

There are also soils of various mechanical compositions: medium and light loamy moraine soils and glacio-lacustrian loam soils; clay sand and sandy soils, as well as sandy soils on loam glacio-lacustrian bedding and alluvial sands. Inclusions of different sized boulders are common. As for the origin, most of the soils are podzols, developing under climatic coniferous forests. The lowlands are generally characterised by poor drainage and sometimes a high watertable. This creates conditions for the development of swamp-like soils, which were common in the area prior to the development of the city. Most of the soils in the areas adjacent to St. Petersburg provide favourable conditions for forests. However, the natural soil fertility is quite low for good agricultural production and in most cases the soils in agricultural areas require drainage and the addition of fertilisers.

Vegetation

St. Petersburg is situated in the Taiga vegetation zone (Southern Taiga sub-zone), which is characterised by the dominance of coniferous forests. The Southern Taiga sub-zone is differentiated from other Taiga sub-zones by the presence of small amounts of broad-leaved species such as *Quercus robur, Tilia cordata* and *Fraxinus excelsior*. The most common forest type in pre-St. Petersburg was primary *Picea abies* forest. The typical herb layer of this forest type included *Oxalis acetosella, Vaccinium myrtillus* and *Trientalis europaea* and several bryophytes such as *Pleurozium shreberi, Hylocomium splendens* and *Dicranum polysetum.* At higher altitudes the forests are predominantly *Pinus sylvestris* and mixed *Pinus-Picea.* There are also quite a few birch forests, *Betula pubescens* and *Betula pendula* with some *Populus tremula,* which have replaced *Picea* forests after felling. Because of high humidity wet forests with *Pinus sylvestris* and *Betula* spp. and bog mosses (*Sphagnum* spp.) below often replaced the original *Picea* forest. Prior to the construction of St. Petersburg, the lowlands were covered by two types of bog. First, raised *Sphagnum* peat-bogs with *Eriophorum vaginatum, Vaccinium uliginosum, Betula nana, Vaccinium oxycoccus, Ledum palustre* and slow growing stunted *Pinus.* Second, upland eutrophic swamps with a high watertable containing *Alnus glutinosa, Betula* spp. and a variety of grass and moss species.

The islands of the Neva delta were mostly covered by woodland and scrub comprising species such as *Alnus incana, Salix phylicifolia, S. caprea, S. cinerea* and *S. myrsinifolia*. The names of some of the islands, for example, "Birch Island" and "Willow Island" (Elagin Island) reflect perfectly the character of the original vegetation over which the new Russian capital was built. The natural vegetation also included *Typha latifolia, Phragmites australis* and *Carex* spp., which grew along the coast of the Gulf. The vegetation of sandy dunes of the Gulf comprised mainly *Leymus arenarius, Lathyrus japonicus, Honckenya peploides, Festuca arenaria* and *Carex arenaria.*

All the surviving Swedish topographical maps show that swamps and wet forests covered more than half of the area that was to become the Russian capital. Since the foundation of the city, the "swampy" nature of the St. Petersburg landscape became the most important historical as well as poetical theme. The city "*growing from the swamp*" with associated fog and dampness is the most powerful image of St. Petersburg – a very "St. Petersburg myth". This land was inhospitable and unsuitable for human occupation. All 300 years of the city's history have been a "conquest" over nature and a struggle to make natural landscapes much more healthy and friendly for people. The bogs were drained by the construction of canals, the tonnes of excavated material being spread over the adjacent land. Many St. Petersburg streets reflect their "wet" origin, for example, Bog Street (Goryshina 2003).

Historical Development of the City

Before Peter the Great

Before the foundation of St. Petersburg, the area had many settlements and fortifications, which were known to the Slavs as 'Izhorsk' and to Swedish settlers as 'Ingermanland'. Artefacts from the eighth century have been found in the area, which prior to the twelfth century was part of Novgorod State. When the area was under Swedish control in the seventeenth century, there were several settlements slightly upriver, in what is now eastern St. Petersburg with some large estates downstream, near the Fontanka River (Dubyago 1963). Interestingly, in spite of its swampy nature, most of the land on which the city is now located was under cultivation prior to its founding in 1703, although there were a few permanent structures in the low-lying areas (Goryshina 2003).

Peter the Great (1703–1725)

St. Petersburg was founded by the Russian Tsar Peter the Great in 1703. The decision to build a new capital was based on his wish "*to cut the window to Europe*" and include Russia in the political, economic and cultural life of Europe. Peter the Great (referred to as "Tsar Peter", or "the Tsar" hereafter) started a unique, large-scale experiment to change the natural landscape by applying newly devised principles of European urban planning design, architecture and art. Tsar Peter was deliberately defying nature. It took an enormous amount of money, the heroic efforts and lives of thousands of people to drain the land, straighten the rivers, re-enforce the river banks and build canals so that the land could support large buildings and other structures. Tens of thousands of wooden piles of *Quercus* were imported into the city from forests to the south-east. The canal sides were constructed of rock imported from islands in the Gulf. Nevertheless, frequent floods brought destruction and disease. Most materials for building, as well as food supplies, had to be imported from elsewhere in Russia (Ageeva 1999). The rationale and formal planning structure that was applied to the building of the city also incorporated some natural configuration of the existing waterways. The Baroque principles of urban design can be seen in the huge heroic scale of the open spaces and waterfronts as well as in the radial and grid structure of the streets, axis-perspectives and the logical system of views, and visual dominance. The Tsar's desire was to create a city of perfect proportions and straight lines (Ignatieva 2005). So it was that in a very short time this unfriendly swampy environment was transformed into a magnificent European capital – the *"New Amsterdam, Venice of the North, New Rome"* (Fig. 1). Another of the Tsar's aspirations was to impose on this "inhospitable environment" an urban and suburban landscape based on traditional European parks and dominated by more "cheerful" deciduous trees such as species in the genera *Quercus, Tilia, Acer* and *Ulmus.*

In 1713 (only a few years after the foundation of the city), the capital of the Russian Empire was moved from old Moscow to St. Petersburg. From the beginning the new capital contrasted dramatically with traditional Russian urban forms. Older Russian cities tended to follow the natural land forms and to face inward, shutting out the broad expanses all around, while the new city ignored the lie of the land, looked outward, and celebrated space (Ageeva 1999).

The development resulted in the destruction of most of the natural vegetation, the forests disappeared first; many trees were used in the construction of buildings, bridges and embankments as well as for firewood. Paradoxically, the Tsar cared about protecting and even preserving some original forests. For example, he initiated a special statutory forestry protection policy that prohibited the destruction of trees next to large rivers. This policy applied particularly to species of *Quercus, Larix, Pinus, Acer* and *Ulmus.*

In the early days of the city's history, animals such as *Canis lupus, Vulpes vulpes* and *Lepus europaeus* were very common. After Tsar Peter's time, most of the groves and forests unfortunately disappeared under the buildings of the dramatically expanding city. During the first half of the eighteenth century, the vegetation of St. Petersburg comprised a few remnant forests and bogs, meadows, wastelands and cultivated parks, gardens, orchards, and vegetable gardens. At least three urban "meadows" are mentioned in eighteenth century documents, including the publicly accessible "Tsar Meadow" next to Tsar Peter's Summer Palace. The meadow probably consisted of some wetland grasses, such as *Agrostis stolonifera* and *Deschampsia cespitosa* and weedy herbaceous species such as *Plantago major, Potentilla anserina* and *Taraxacum officinale* (Goryshina 2003). Early chronicles also mentioned a lot of "wastelands" (pustirei). The first botanical records describe the typical plants to be found in such urban meadows and wastelands. By the middle of the eighteenth century, lawns of most private parks and gardens included typical grasses such as *Poa pratensis* and *Agrostis gigantea.*

Industrial Revolution of the 1860s to the early Twentieth Century

The emancipation of the serfs in 1861 by the progressive Russian Tsar Alexander II resulted in the rapid industrial development of the city. Former peasants migrated to the Russian capital so that by 1900 the population had reached 1.25 million – compared with 0.5 million in 1865. This surpassed the population of Moscow (1 million) and transformed St. Petersburg into one of the largest industrial centres of the Old World. By the beginning of the twentieth century, the famous "strong and beautiful" unified urban landscape of St. Petersburg was complete. The construction of subsequent architectural neoclassic and Russian Modern – Art Nouveau buildings during the nineteenth and the beginning of the twentieth centuries did not destroy the unity of the architectural and urban design but enhanced it by adding more variety and character to the heritage landscapes.

In 1914, as a result of the beginning of World War I and the "Germanic" character of the original name, St. Petersburg was renamed Petrograd. The name of the city was changed to Leningrad in 1924 as the Communists' memorial to Lenin's activity and his death that year. The city was re-named St. Petersburg in 1991, after the collapse of the Soviet Union. Although the city reverted to the original name, the area in which it is situated is still called the "Leningrad Region" (=Leningradskaya Oblast is a component of the Russian Federation). In 2002 the surrounding Leningrad Region occupied 84,500 km^2 and had a population 1,669,205.

Soviet Era (1917–1991)

The most dramatic periods in St. Petersburg's history are related to the major political and economic changes that have occurred in Russia, including the first Russian revolution (1905–1907), followed by World War I (1914–1918), the February and October Bolshevik Revolutions of 1917 (which put an end to the Russian monarchy and signalled the beginning of the Soviet Socialist era) and its eventual collapse 74 years later when it was replaced by the capitalist system.

Leningrad was named in Soviet propaganda as "*the cradle of the revolution*" and "*the city of three revolutions*". The city's population decreased dramatically during this "bloody" time including thousands of Red Terror (1918) and Civil War victims (1918–1923). By 1920 Leningrad's population was a third of what it had been in 1915.

The change in ideology and the economy resulted in creating a very distinct Soviet architectural, design and landscape language. After the Revolution of 1917, all land and many private houses were nationalised. A special Decree ("About Protection of Nature Monuments, Gardens and Parks", 1921) declared that all the parks of the Tsar were "national property". St. Petersburg's famous palaces (The Hermitage, Russian Museum) and parks (Peterhof, Pavlovsk and Tsarskoye Selo) were converted to and opened as national museums and parks. A new social class, the proletariat (the factory workers), needed new types of green areas. The first years of the Soviet State were characterised by a process of rapid urbanisation and searching for new forms of landscape design that could satisfy new architectural and urban planning.

A new type of "socialist culture" – parks of culture and rest (park kulturi i otdicha) emerged in the late 1920s. They were new types of public parks with the objective of being a "complex of culture" with multi-functional programmes including sport,

Fig. 2 The Mars Field – an example of Soviet landscape design dedicated to the victims of the October Revolution. *Syringa vulgaris* is used as a major species

culture and the political education of the community (Bogovaya and Fursova 1988) (Fig. 2). Memorial public gardens, parks, squares and cemeteries were dedicated to the heroes and leaders of the revolution: Karl Marx, Friedrich Engels, Vladimir Lenin and later Joseph Stalin.

The 1920s to 1930 are famous for the reconstruction of the poor outskirts into regularly planned districts. St. Petersburg Constructivism provided interesting and progressive examples of Socialist architecture (Kurbatov 2008).

During World War II, Leningrad was besieged by German forces; the Siege, which started on September 8, 1941, was to last until January 27, 1944, a total of 900 days. Some 800,000 of the city's 3,000,000 inhabitants are estimated to have perished from hunger, cold or killed by warfare. This "Great Patriotic War" resulted in the creation of other types of Soviet landscapes, for example, Victory Parks, Memorial Complexes and Memorial Cemeteries.

Fortunately for St. Petersburg, Soviet developments between 1917 and 1991 did not impinge on the historical centre. As will be described, that was left to the "new elite" of post-1991. The central district had only some local changes and reconstruction, especially after the World War II. Russian Constructivism, Soviet Neoclassicism ("Stalin's Ampir") and the Russian version of Modernism became the leading architectural styles for most apartment blocks in the new districts.

The era of new residential development (the "Residential Neighbourhoods" with standard functional apartment blocks) began on the outskirts in the late 1940s. Many families finally had the opportunity to move from the kommunalkas (communal apartments set up by the Soviet Regime in the houses it nationalised) to separate apartments. Citizens and government environmental organisations planted the areas adjacent to the apartments with fast-growing species such as *Populus x berolinensis* (and other hybrid *Populus*), *Betula* spp. and *Salix* spp.

The period of "Developed Socialism" (1960s–1980s) saw new types of Soviet Parks dedicated to anniversaries of the Soviet State and Soviet Leaders, for example, a park to celebrate the 100th anniversary of the birth of Lenin and another to commemorate 30 years of the Komsomol Organisation. They were mostly multifunctional recreational parks built in new city districts and planted with a greater variety of native and introduced tree species.

Another very important post-war aspect of Leningrad's architecture and landscape architecture is the development of one of the best restoration schools. Most suburban historical parks and gardens such as Peterhof, Pavlovsky Park, Tsarskoye Selo, Gatchina parks and palaces were completely destroyed and many buildings were badly damaged as a result of the Siege and other military activity. Based on detailed historical research, they have been carefully and completely restored, see Fig. 3.

During the Soviet period a great deal of attention was given to the management of the forest remnants within the city and the suburbs. An impressive system of green belts consisting of forest parks was created around most Russian cities, for example, the St. Petersburg Green Belt covers 142,500 ha.

St. Petersburg is famous for its numerous heritage landscapes. The historic centre and related groups of monuments were designated a UNESCO World Heritage Site in 1991. The city is among the very few around the world where the unique urban

Fig. 3 Tsarskoye Selo – one of the best examples of historic garden restoration

planning structure (mostly palaces and buildings of the Baroque and pure Neoclassical styles), parks and gardens and numerous canals and bridges have survived. The central part of St. Petersburg has 700 architectural monuments of the eighteenth and nineteenth centuries and numerous historical parks under government protection.

Changes in the City Environment

Perestroika: 1991 to Present

Changes in the political and economic situation after the collapse of the communist system in 1991 resulted in the rapid expansion of corruption and the growth of criminal power; St. Petersburg was no exception. For example, the unstable situation led to serious difficulties and failures in the protection of the city's St. Petersburg's unique heritage landscapes (Ignatieva 2005). Economic difficulties and instability of the "perestroika" period resulted in a decline of the population from 5,023,500 in 1989, to 4,628,000 in 2001 and to 4,568,047 in 2008. Since 2000 when Vladimir Putin, the former St. Petersburger, was elected as President (subsequently Prime Minister) of Russia, the city has experienced and is experiencing a period of relative stability and positive development. Nevertheless, in the modern era of westernisation and globalisation, St. Petersburg is struggling to preserve the image of a unique World Heritage City.

Because of the instability and corruption in the post-perestroika period, eight monuments of cultural heritage significance have been lost since 2005. There are another 1,317 objects that are rapidly deteriorating (Ignatieva 2005). Unfortunately, in addition there are also numerous negative examples of spontaneous illegal construction of buildings in the historic centre. Many of the green courtyards and children's playgrounds as well as parts of gardens and parks (which are so important in the dense built environment of St. Petersburg) are being converted to private car parks, new houses and garages. The statistics are very sad: in 2003 the green areas of the city covered 11,970 ha; in 2006 it was only 10,535 ha, a loss of about 10% in 3 years.

The main problems of the Nevskaya Bay (eastern part of the Gulf) are the deteriorating water quality (eutrophication and the disappearance of hydrobionts and valuable fish species) and development proposals. Part of the Bay is being reclaimed for residential development and there have even been plans to fill it in for the construction of an airport.

Urban Climate

On warm clear days, the temperature in the city centre is 5°C higher than in the outskirts (Pokrovskaya and Bychkova 1967). This increase is attributable to the urban heat-island effect, which is caused by industrial discharges and heating systems within the city. The frost-free period in the centre is 25 days longer than in Lesnoy District. The city's buildings have a considerable influence on the wind conditions, for example, the mean monthly wind speeds in the central districts are 1–1.5 $m\,s^{-1}$ lower than on the shore of the Gulf. Because of the higher temperature many warm-loving plants can be grown in urban areas much more easily, compared with the surrounding natural landscapes. Additionally, most flowerbeds need less protection in winter in the central urban districts than they did in the 1960s to 1980s.

Tsar Peter was an initiator of urban phenology, for example, he ordered the collection of special data on the seasonal development of trees with the objective of discovering the most suitable places for successfully growing garden plants in the urban environment. Twentieth-century phenology studies of different species of *Acer* in urban areas have shown that in the warmer central areas of cities, bud opening, leaf growth and flowering can be as much as 10 days earlier than in the cooler outer areas (Bulygin and Firsov 1983).

Dramatic increases in the number of motor vehicles in St. Petersburg have contributed to increased air pollution (O_3, NO_x and SO_2). According to Bulygin and Firsov (1995), native coniferous species are especially sensitive to air pollution and are not surviving in the city.

Urban Soils

The urban soils of St. Petersburg differ from the natural zonal soils by the presence of a diagnostic soil horizon called "urbik". "Urbik" is the upper 5 cm of a mixed soil layer that includes many different materials of anthropogenic origin, for example, building rubble.

There are four groups of urban soils in St. Petersburg (Babikov and Melnichuk 2003):

1. Naturally undisturbed (found in urban forests).
2. Naturally disturbed (a typical "urbik" layer overlying an undisturbed lower soil profile).
3. Anthropogenic, deeply disturbed soils (urbanozems).
4. Technogenic surface soils (urbanotechnozems).

Four soil districts have been identified in the city. The central district is characterised by anthropogenic, deeply disturbed soils (urbanozems), which are formed on a deep cultural layer (mainly historical gardens). The upper part of the soil profile is represented by an "urbik" horizon. The second group is the "northern natural-anthropogenic" sandy soils where the geological formation is represented by post-glacial deposits (lake and sea sands) as, for example, in some fragments of forest remnants – "Sosnovka". The third group comprise the natural-anthropogenic (disturbed natural and reclaimed) soils of the Kirovsky Islands district. The natural geological foundations of these islands are lake-glacial and post-glacial deposits, which are relatively resistant to erosion. The fourth group is natural non-disturbed loamy soils such as in parks created from natural forest ecosystems, for example, in Park Alexandrino and Park "Pine Forest Glade". The geological foundation of this district is Cambrian clay overlain by glacial boulders.

Because St. Petersburg is located in a zone with a high watertable, from the beginning the areas for future construction and planting have had to be drained. In the time of economic and political instability, many land drainage systems in the parks were abandoned, which resulted in gleying and the development of spontaneous wetland plant communities. As a consequence of the establishment of swampy environments, many trees in the old parks became stressed and were eventually replaced during restoration works.

Another general tendency in urban soils of the city is the process of alkalisation and the increase in alkaliphile micro-organisms, for example, the bacterium *Azotobacter*. There is generally change in the soil reaction from acid (in the forest parks within fragments of native plant communities) to the alkaline soils associated with street tree plantings. For example, in "Sosnovka" the *Azotobacter* bacterium level was 2%, in the urban park of the Forest Technical Academy 10.5% and in the urban streets 15.25% (Babikov and Melnichuk 2003). The urban soils are also characterised by a high level of compactness and pollution by the salt products that are widely used for de-icing roads and pathways.

Flora

Vascular Plants

Tsar Peter had a passionate interest in natural history and encouraged his court scientists to research the existing flora of the new capital. The Tsar himself was one of the "first collectors of the herbarium". At least 11 scientists investigated the St. Petersburg flora during the eighteenth century. The most important work was the "Flora of Ingermanland" by Gorter, which is written in Latin and mentions at least 300 species. This was followed in 1801 and 1802 by a large monograph by Sobolevsky entitled "Sanktpeterburgskaya Flora"; his work lists 678 species of angiosperms and 362 non-vascular plants (Goryshina 2003).

With the development of the city, new plants arrived mostly with stock, hay and grain from the southern and western regions of Russia. For example, *Beckmannia eruciformis* came in hay from meadows in central Russia. Compared with many other ports around the globe, in the eighteenth and probably even during the nineteenth centuries, St. Petersburg was not a significant point of entry for adventive plants. In contrast to other Russian and most European cities, the imported goods were mainly fabrics and luxuries for the court and the nobility with some consumables such as sugar, wine and spices (Goryshina 2003). However, the development of a railway network in the second part of the nineteenth century played an important role in the arrival and distribution of spontaneous adventive plants in the city.

Most introduced decorative plants were used in the private parks and gardens of the Tsar and the nobility. In the beginning, new species arrived from foreign nurseries but very soon local nurseries were also established. Each garden style, such as French formal designs, the Picturesque or Gardenesque styles resulted in the planting of "fashionable" plants. The establishment of formal gardens involved the mass planting of broad-leaved European species of *Quercus, Tilia, Fraxinus, Ulmus and Acer*, and *Aesculus hippocastanum* and "parterre" species such as *Buxus sempervirens* and *Taxus baccata.* However, these fashionable plants could not cope with the rigorous St. Petersburg climate and were soon replaced by indigenous species such as *Vaccinium vitis-idaea* and *Juniperus communis* (Ignatieva 1982).

With developing interest in horticulture in the second part of the nineteenth century, new non-native plants were "promoted" and used in most of the public parks and gardens. Some of these "escaped" cultivation and grew elsewhere in the city. The Imperial Botanical Garden was an important source of "garden escapees", for example, during the nineteenth and twentieth centuries, 28 herbaceous species "escaped" and became widely established in urban habitats (Ignatieva and Konechnaya 1996). Another large park with a rich botanical collection of non-native trees, shrubs and herbaceous species was the Park of the Forest Technical Academy, from which 14 species "escaped".

The Leningrad botanist Nekrasova gave a good historical overview of the St. Petersburg urban flora at the end of the eighteenth and the beginning of the nineteenth centuries (Nekrasova 1959). The ruderal flora of the city was analysed for the first time by Yuri Gusev (1968).

1,855 species of vascular plants, excluding taxa below species rank and including native and non-native, species occur within the present administrative boundaries of the city. The species belong to 144 families, the most represented being the Asteraceae (195 species, comprising 67 cultivated non-natives and 128 indigenous and adventives), followed by the Rosaceae (157 species; 90 cultivated non-natives and 67 indigenous and adventives) and the Poaceae (154 species; 9 cultivated non-natives and 145 indigenous and adventives (Tsvelev 2000)). Interestingly, in an earlier flora of the Leningrad region (which excluded cultivated species of urban habitats) the Poaceae dominated with Asteraceae second and Rosaceae only in sixth place (Minyaev et al. 1981). In this publication the flora was more representative of the typical boreal indigenous flora and the distribution of the more representative families – Poaceae, Asteraceae, Cyperaceae, Brassicaceae and Fabaceae.

Within the boundaries of the Ladoga-Ilmen floristic region (in which St. Petersburg is located), there are 90 species of indigenous woody plants (comprising 45 genera in 26 families); 86 species are angiosperms and only four species of gymnosperms (among them are the most important forest trees, *Picea abies* and *Pinus sylvestris*). Of the woody plants the most common are representatives of the Salicaceae (two genera and 22 species), Rosaceae (seven genera and 15 species) and Ericaceae (seven genera and 10 species). Three woody species, *Betula pendula, B. pubescens* and *Vaccinium myrtillus,* are the most representative and commonly observed (Bulygin and Firsov 1995).

Two hundred and fourty-nine adventive species (14% of the total flora) have been recorded (commonly along railway tracks), they include: *Amaranthus retroflexus, Crepis tectorum, Conyza canadensis, Lactuca serriola, Xanthium albinum, Impatiens parviflora, Brassica napus* ssp. *campestris, Sisymbrium wolgense, Salsola tragus, Oenothera rubricaulis* and *Anisantha tectorum.*

Twenty species have completely naturalised in natural and semi-natural plant communities in the city; they are *Acer negundo, Heracleum sosnowskyi, Aster salignus, Conyza canadensis Galinsoga ciliata, G. parviflora, Helianthus tuberosus, Impatiens glandulifera, I. parviflora, Elodea canadensis, Lupinus polyphyllus, Epilobium adenocaulon, Oenothera rubricaulis, Festuca arundinacea, Amelanchier spicata, Aronia melanocarpa, Geum macrophyllum, Rosa rugosa* and *Chaenorhinum minus.*

Invasive plants species can be found not only in anthropogenic biotopes but also in natural and semi-natural plant communities (Konechnaya, personal communication) where, in some places, they are replacing indigenous species. There are 20 invasive plants in the St. Petersburg flora, the most aggressive being *Heracleum sosnowskyi,* which appeared *en masse* in forest margins and abandoned fields where it completely destroyed the indigenous herbaceous communities. This species is also abundant in urban habitats such, as roadside and grassland. Another invasive plant, *Aster salignus,* has colonised most of the coast of the Gulf and is spreading along urban rivers and waterways.

Impatiens glandulifera can be found along waterways and in the indigenous wet *Alnus incana* forests. *Impatiens parviflora* is very common in all urban parks, woodlands and indigenous *Alnus glutinosa* forests. Initially, *Helianthus tuberosus* only spread close to areas of cultivation (for example, in dacha gardens) but today it can be found in the indigenous coastal reed vegetation. *Echinocystis lobata,* which is a very common invasive species in other Russian regions, has only just begun invading the coastal and river banks. *Elodea canadensis* is one of the most common aquatic plants in all the still water bodies. *Conyza canadensis, Oenothera rubricaulis* and *Lupinus polyphyllus* are very common in dry, especially sandy, habitats. The American species *Amelanchier spicata* (which is successfully distributed by birds) is very common in most types of woody biotopes; it has recently become established as a shrub-layer species in *Pinus* forests. *Aronia melanocarpa,* which is also distributed by birds, can also be found in fragments of indigenous forest in the city. *Rosa rugosa* (one of the most popular non-native plants for use in hedges and shrubberies) is only found along the beaches of the Gulf. *Festuca arundinacea,* which was introduced as pasture grass, is now one of the most common plants in all urban habitats.

Bryophytes

There are three species of Anthocerotae, 127 Hepaticae and 416 species of Musci in the Leningrad Region (Biodiversity of the Leningrad Region 1999).

The typical bryophytes of the coniferous forests are the "green" mosses such as *Pleurozium shreberi, Hylocomium splendens* and *Dicranum polysetum,* which can be seen in the remnants of Taiga forests (for example, the forest parks such as "Sosnovka"). In swampy areas of forest remnants, *Polytrichum commune* and different species of *Sphagnum* (*S. angustifolium* and *S. flexuosum*) are very common. Mosses have been rarely researched in urban areas of St. Petersburg. The most comprehensive studies of the bryophyte flora have been undertaken in the different habitats of Elagin Island Park (Volkova et al. 2007). For example, in dry meadows dominated by *Festuca rubra,* there is a very well-established moss layer of *Rhytidiadelphus squarrosus, Plagiomnium ellipticum* and *Cirriphyllum piliferum* (80–90% cover). *Lolium perenne* lawns have patches of *Brachythecium rivulare, Pohlia wahlenbergii, Ceratodon purpureus, Funaria hygrometrica, Marchantia polymorpha* ssp. *ruderalis, Aneura pinguis* and *Pellia endiviifolia.* On the grassy slopes next to the Bolshaya Nevka river, the well-established plant communities contain a well-developed moss community that includes *Plagiomnium ellipticum, P. undulatum* and *Rhytidiadelphus squarrosus*. Some urban wastelands contain species such as *Polytrichum piliferum, Ceratodon purpureus* and *Bryum argenteum* (Doroshina, pers comm).

Fungi (including Lichenised Fungi)

Lichenised Fungi

Over 270 years, 87 scientific works have been published where lichens were mentioned.

More recently the lichens of the city have been very well researched by Malysheva (2003). Of the 284 species of lichens originally found within St. Petersburg, 131 species are now extinct. The most representative genera are *Cladonia* (21 species), *Lecanora* (13), *Ramalina* (8), *Melanelia* (7), *Bryoria* (7), *Physcia* (5), *Physconia* (5), and *Caloplaca* (5), which mainly occur in parks and gardens. In this sense, the St. Petersburg lichen flora is very similar to other cities in the European part of Russia in the Taiga zone (Malysheva 2003). Most lichens (74.5%) were found in historic parks, which probably reflect the stability of their habitats and their similarity to natural habitats. As for the substrate, most urban lichens colonise tree bark (74.5%). Quite a few lichens (22.9%) have established on the granite embankments, monuments, building basements and on boulders. Another quite common habitat is rotting wood, such as old fences and posts. Because of trampling and other activities, there are very few ground lichens (13.1%). The number of lichen species in suburban districts is much higher that in the densely built centre. In the city, only one or two species are commonly found whereas 68 species are frequent in the suburbs. However, there are also some "lichen islands" within the city centre, which have unusual assemblages and a high number of species. There is a very distinctive group of lichens (9.8%) that colonise new artificial substrates, for example, cement, concrete, brick and plastic. The group includes such widespread species as *Lecanora dispersa, L. crenulata, Caloplaca holocarpa* and *Verrucaria muralis.* Nitrophilous species, such as *Scoliciosporum chlorococcum, Physcia dubia, P. tenella* and *Phaeophyscia orbicularis* are abundant on the bark of trees.

Of the existing 153 species, 31 are "endangered", 75 belong to a group whose occurrence has not changed since records began and 47 are species recorded for the first time between 1990 and 2002. One hundred and six of the 153 species were recorded in the city before 1930. The genera that have completely disappeared from urban areas include *Leptogium, Lobaria, Nephroma* and *Stereocaulon.* Epiphytes (78 species) were the first to become extinct followed by terrestrial lichens (39 species). Arctic-alpine species such as *Flavocetraria nivalis* and *Shaerophorus fragilis* have also completely disappeared from the urban lichen flora, the most significant being the boreal species, which comprised 25% of all lichen species. Other species in genera such as *Cladonia, Peltigera* and *Ramalina* have declined considerably.

Atmospheric pollution has influenced the changes in the lichen flora. Of the lichen "forms", the crustaceous lichens are dominant (43.8%), followed by foliose (28.8%) and fruticose (27.4%), which are directly related to habitats affected by pollution. Lichens with small contact surfaces are able to survive better in polluted areas; however, there are morphological anomalies such as dis-pigmentation of thallomes and a decrease in size. The lichen indicator map of air pollution was used successfully in an assessment of the quality of the city's environment (Fig. 4). Using the Index of Atmospheric Purity (IAP), five pollution zones were suggested: I – the zone of intense pollution, IAP = 0–5, covers 9% of the city area, II – zone of significant pollution, IAP = 6–10, covers 29%, III – zone of moderate pollution, IAP = 11–15, covers 22%, IV – zone of average pollution, IAP = 16–20 (15%), V – zone of mild pollution, IAP > 21 (25%).

The lichen indicator map shows that there is a mosaic of polluted zones, mainly industrial areas that first appeared in the city from the middle of the nineteenth century. The factories were located not in one particular industrial zone but distributed throughout the city districts, including the centre and in suburbs. St. Petersburg has quite acute air pollution problems with up to 60% of the city experiencing high pollution stress.

Habitats

The dynamic of urban vegetation reflects St. Petersburg's history. In the eighteenth century, there were quite a few remnants of native plant communities in the central area of the city, for example, next to the Kazan Church, *Alnus glutinosa* forest and the *Picea* forests of Vasilevsky Island. The latter includes species such as *Picea abies, Prunus avium, Sorbus aucuparia, Ribes nigrum, Viburnum opulus* and *Vaccinium myrtillus* with a typical Taiga herb layer of *Oxalis acetosella, Pyrola rotundifolia* and *Trientalis europaea.* Next to Alexandro-Nevsky Laura there were

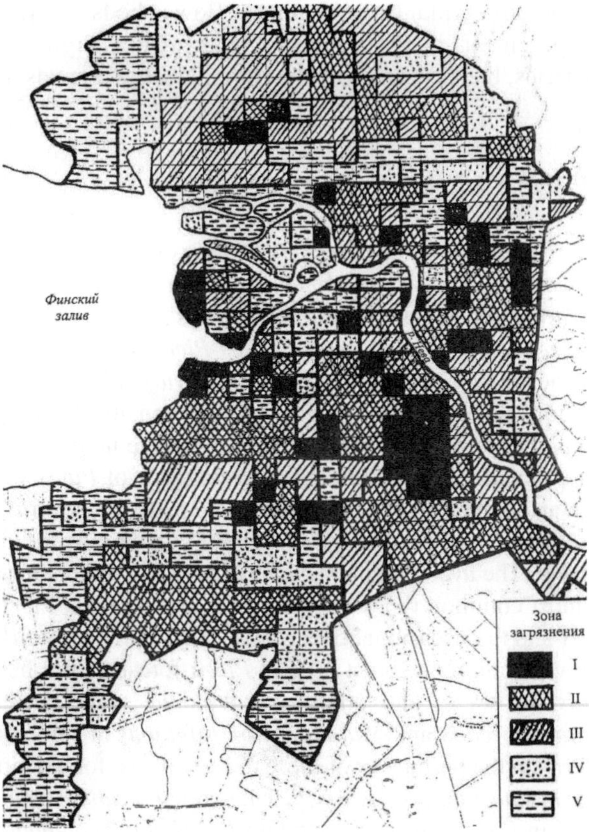

Fig. 4 The lichen indicator map of St. Petersburg (from Malysheva 2003). Pollution zones: (1) the zone of intense pollution. (2) zone of significant pollution. (3) zone of moderate pollution. (4) zone of average pollution. (5) zone of mild pollution

typical *Sphagnum* bogs with *Betula nana, Ledum palustre* and even *Drosera rotundifolia,* which were first discovered by early botanists (Nekrasova 1959). City residents even collected *Vaccinium oxycoccus* from the city's *Sphagnum* bogs during the eighteenth century. As a result of development and expansion, most of the native forest and wetland vegetation had disappeared from the city centre by the end of the eighteenth century.

At present, the wide range of natural habitats (forests, wetlands, bogs and *Salix* scrub) can only be found in indigenous remnants of the protected areas, in fragments of the "greenbelt" and in some public parks and gardens that were based on native vegetation (Pavlovsky Park). There are also semi-natural habitats such as modified indigenous plant communities in forest parks. Urban habitats are located within old districts (eighteenth to nineteenth and the beginning of the twentieth centuries) and in new residential areas of the twentieth and the beginning of the twenty-first centuries. The habitats include parks and gardens, plantings in residential areas, street plantings, hedges, flower-beds, wastelands, granite embankments, pavement cracks, railway tracks, road margins, private garden "dachas", cemeteries, the city port, canals, rivers, lakes and ponds.

Natural Habitats

In 2003, approximately 28,000 hectares (20% of the city) were covered by different fragments of native Taiga forests, which mainly occurred in the northern suburbs and in a few heavily urbanised zones (forest parks such as "Sosnovka"). The forests were predominantly *Picea* or *Pinus* or mixed *Pinus-Picea* with characteristic species dominating the herb layer, for example, *Calluna vulgaris, Vaccinium myrtillus, V. vitis-idaea* and *Lerchenfeldia flexuosa.* Quite often the *Pinus* forests contain *Juniperus communis* and *Sorbus aucuparia.* The average number of species in this plant community is about 34. Forests with a dominance of *Pinus* with some *Picea* and *Betula* have an abundance of *Vaccinium vitis-idaea, V. myrtillus, Melampyrum pratense, Pleurozium shreberi* and other "green" mosses in the herb and moss layers, respectively. The average number of species found in this forest type is 26.

One of the most common types of native remnant forest is *Picea* forest with a green moss layer and the occasional occurrence of *Vaccinium myrtillus* and *V. vitis-idaea.* Another very common forest type is *Picea* forest with a herb layer dominated by *Oxalis acetosella* and other typical boreal plants such as *Trientalis europaea, Linnaea borealis* and *Maianthemum bifolium.* In some cases, there are also different ferns in the herb layer of these *Picea* forests, for example, *Dryopteris carthusiana, D. expansa* and *Athyrium filix-femina; Pteridium aquilinum* is more commonly found in *Pinus* forests. *Picea* with *Polytrichum commune* can be found in those sections of forest with impeded drainage. In swampy areas there are also *Picea* forests with *Sphagnum* ssp. and typical bog plants.

On the upper terraces of the Komarovsky Coast (Bereg) along the Gulf there are also remnants of *Pinus* forests with a dense herb layer of *Convallaria majalis* on mezopodzolic iron pan sandy soils. *Pinus, Populus tremula, Betula* (with occasional *Tilia, Acer* and *Quercus)* occur in some of the *Picea* forests also growing on mezopodsol iron pan soils with *Oxalis acetosella* dominant in the herb layer. In these forests *Prunus avium, Corylus avellana* and *Sorbus aucuparia* form the shrub layer while typical woodland herbs include *Stellaria nemorum, Aegopodium podograria, Pulmonaria obscura* and *Asarum europaeum.* Some of the *Picea* trees in these forests can be up to 30 m tall.

About 11,000 ha are occupied by different types of native *Betula* forests, which established in "open" ground resulting from the clear felling of coniferous forest, the destruction of the forests by fire or abandonment of agricultural land. The more fertile soils support *Betula* forest with a herb layer dominated by *Convallaria majalis, Rubus saxatilis* and *Stellaria holostea.* Poor soils encouraged the development of *Betula* forests with *Deschampsia caespitosa, Melampyrum pratense* and *Dryopteris carthusiana.* Wet habitats support *Betula* forests with a hygrophytic herbaceous layer including *Filipendula ulmaria, Geum rivale, Athyrium filix-femina, Juncus effusus, Scirpus sylvaticus* and *Solanum dulcamara. Athyrium filix-femina* and *Dryopteris expansa* are quite typical of birch-fern forests on floodplains.

Small remnants of mature *Alnus glutinosa* forests have survived along the coast of the Gulf (for example, in the Nature Monument "Strelninsky Bereg"). These forests also contain small amounts of *Fraxinus excelsior, Ulmus glabra, Tilia cordata, Acer platanoides, Prunus avium, Ribes nigrum* and *Rhamnus cathartica.* The margins of some rivers are lined by *Alnus incana* with some *Prunus avium;* the herb layer comprises an abundance of *Aegopodium podograria, Athyrium filix-femina, Equisetum pratense* and *Geum rivale.*

Swamps (mostly upland *Sphagnum* and transition moors) occupy over 3% of St. Petersburg. There is a unique "Lachtinskoye" swamp in the Yuntolovsky Reserve (700 ha) with *Carex* spp., *Calamagrostis canescens,* low growing *Salix* and a relict population of the rare *Myrica gale*. This reserve also has a relatively undisturbed *Carex-Sphagnum* bog that is virtually treeless. In the Kurortny District along the Gulf, there are still picturesque sandy dunes with single trees of *Pinus* spp.; the typical grassland vegetation includes *Leymus arenarius, Festuca arenaria* and *Carex arenaria.* Scrub habitats dominated by *Salix* including *Salix caprea, S. cinerea, S. phylicifolia* and *S. myrsinifolia* commonly occur in wetlands. In many places, the shallow waters of the Gulf are still dominated by stands of *Phragmites australis* and *Scirpus lacustris.*

A particularly important remnant habitat is the broadleaved forest of the Duderhofskye Hills. The forests, which are growing on sod-podsol light loamy soils and sod calcareous soils with good aeration, have a canopy that mainly comprises *Acer, Fraxinus, Ulmus* and *Quercus* with a shrub layer of *Corylus avellana, Ribes alpinum, Lonicera xylosteum* and *Daphne mezereum.* The herb layer includes *Hepatica nobilis, Lathyrus vernus, Viola mirabilis, Asarum europaeum, Pulmonaria obscura, Polygonatum multiflorum,* many orchids (*Cypripedium calceolus, Epipactis atrorubens, E. helleborine, Listera ovata, Neottia nidus-avis, Dactylorhiza maculata*) and rare *Hieracium* species *(H. caesium, H. oistophyllum, H. prolatatum, H. pellucidum, H. subholophyllum*). In the moss layer, there are a number of typical woodland species such as *Oxyrrhynchium hians, Thuidium assimile, Cirriphyllum piliferum* and *Fissidens taxifolius.* On the tree trunks there is also *Orthotrichum pumilum,* a protected species of the Leningrad region.

Semi-natural Habitats

During Soviet time, the "forest park" (based on the native forests) was introduced as a special type of green area within the city and its suburbs. One of the goals of the Soviet urban planning strategy was to provide short-term rest areas and a healthy environment for people working in the cities and nearby suburbs. Leningrad pioneered the planning of forest parks, the creation of the forest park zone (greenbelt) surrounding St. Petersburg dates back to 1932. The creation of the "Nevsky" forest park in 1936 was the first such project in the former USSR.

Fig. 5 A forest park "Toxovo" in the greenbelt zone with *Picea abies* and *Betula pendula* plant communities

By 2003 the city was surrounded by a vast forest, with a protected forest park zone of 142,500 ha close to the urban area. At present five forest parks are located within intensively urbanised landscapes (Fig. 5). The green zone of St. Petersburg includes forests within a 60 km radius of the city centre, it includes forest parks and parks in urbanised as well as in sub-urban zones. Forest parks are native forests, which have been modified by planning and landscape design to provide opportunities for mass public recreation (Kuznetsov and Ignatieva 2003).

The planting and maintenance works undertaken in the forest parks include the removal of some trees and shrubs, land drainage and the creation of lawns for recreation, maintenance of the trails and roads, sanitary felling, felling for edge formation, decorative planting of trees and shrubs at critical locations (entrances, central points of crossing main roads, viewing points, areas around benches and shelters), plant protection (fencing), creating new meadows by enrichment of the grass cover and sowing of lawn grasses and support for recreational infrastructure construction (Kuznetsov and Ignatieva 2003).

The influences of recreation on natural ecosystems of forest parks include changing of the structure, composition, fauna and soil moisture regime (as a result of soil compaction). The most sensitive in this sense are lichen and moss conifer forests. In forest parks based on *Picea* and *Pinus* communities ("Sosnovka" and "Piskarevka"), the transformation of conifer forests to pioneer deciduous woodlands (mostly *Betula* forests) is striking. Air pollution is an additional factor that influences the anthropogenic succession towards deciduous plant communities in urban areas. In "Sosnovka" forest park, for example, in 1948 *Pinus sylvestris* covered 99% of the forest area but by 1980 it only occupied 74% of the area. The lack of natural regeneration (because of trampling) is another concern in forest parks.

Floristic investigation in Piskarevka forest park shows that *Betula, Pinus* and mixed *Pinus-Betula* communities with small amounts of *Populus tremula* are the most common. In the shrub layer *Prunus avium* and *Sorbus aucuparia* are dominant and occur frequently. Among other plants *Rubus idaeus* and *Geum urbanum* are abundant and occupy up to 40% of the park area. These species are typical indicators of woody disturbed areas. In some remote places forest and weedy forest species such as *Vaccinium myrtillus, V. vitis-idaea, Athyrium filix-femina, Dryopteris carthusiana, Mycelis muralis, Solidago virgaurea, Hieracium umbellatum* and *Epilobium montanum* are still very common.

In *Betula* and mixed *Pinus-Betula* forests (especially on their margins) grasses, pioneer and weedy species are typical; *Elytrigia repens, Chamerion angustifolium, Urtica dioica, Agrostis capillaris, Calamagrostis epigeios* and *Dactylis glomerata.* Recreational activity has caused changes to the herb layer from the disappearance of typical forest species (*Vaccinium*) towards more marginal and anthropogenic species (*Rubus* and *Geum*) and sparse vegetation.

In moist areas (wet meadows and ditches), *Epilobim adenocaulon, Deschampsia cespitosa, Persicaria hydropiper, Geum rivale* and *Glechoma hederacea* are very common. Artificially created open areas (meadows and forest fringes) are dominated by native meadow and ruderal species such as *Achillea millefolium, Artemisia vulgaris, Plantago major, Ptarmica vulgaris, Cirsium arvense, Atriplex prostrata* and *Rumex acetosella.*

The most common non-native species planted in the forest parks are *Larix sibirica, Abies sibirica, Aesculus hippocastanum, Tilia* x *europaea, Fraxinus pennsylvanica, Acer ginnala, A. tataricum, Spiraea chamaedrifolia, Rosa rugosa, R. glabrifolia, Crataegus* spp., *Cotoneaster lucidus, Syringa vulgaris, S. josikaea* and *Chaenomeles japonica.* Broad-leaved native species, such as *Tilia, Acer, Ulmus* and *Quercus* species are planted occasionally. Special forest park nurseries have 70 species of trees and shrubs on their lists (Ignatenko et al. 1980).

Flora of Green Areas

The current green areas of the city include parks, gardens, squares, boulevards, street plantings and plantings in residential and industrial areas. By 2000 St. Petersburg had about 19,000 ha of public green areas, close to 32% of the total city area. There is a spatially uneven distribution of green areas over the city, which reflects its history. Central districts have only 6.3% with 93.7% occurring in other districts (Nilsson et al. 2007).

Tsar Peter dreamed not only about a European city but also a green city with a healthy environment; his formal parks and gardens were admired and imitated by the nobility. Numerous gardens were located next to palaces and mansions along the Neva, Fontanka and Moika rivers. Even at the end of the eighteenth century St. Petersburg had quite clear greenbelts because of its gardens in the central part of the city. Development in the nineteenth and early twentieth centuries destroyed most of these private green sites in the central part of the city transforming the image of the city into a "stone jungle". In the nineteenth century some new public parks, gardens and boulevards were built in the central part of the city (Alexandrovsky Garden and Konnogvardeisky Boulevard). Among the most significant twentieth century additions to the green areas were parks of culture and recreation and new plantings in residential neighbourhoods.

Between 1988 and 1992 the flora of two central and two new districts (built in the eighteenth and nineteenth centuries and post-1945, respectively) was investigated by Ignatieva (1994). All existing habitats, which included lawns, hedges, flower-beds, shrubberies, palisadnik (small planted areas 3–5 m wide in front of an apartment building), factory gardens, historical parks and gardens, roadside plantings, boulevards and forest parks, wastelands, cracks in pavements and granite embankments, were investigated. Floristic analyses and distribution maps of the rare decorative plants were also produced. Six hundred and fourty-five species of vascular plants were recorded, comprising 447 indigenous species, 55 naturalised introduced species (at different stages of naturalisation), 21 non-native, adventive species and 122 non-naturalised introduced species (71 woody and 48 herbaceous).

The taxonomic analysis of spontaneously growing urban plants (which includes indigenous, adventive and naturalised introduced plants) has shown the dominance of ten Families: the Asteraceae (52 species), Poaceae (44), Brassicaceae (33), Rosaceae (23), Cyperaceae (22), Fabaceae (18), Scrophulariaceae (17), Ranunculaceae (15), Caryophyllaceae (14) and Apiaceae (14). In the natural floristic boreal zone that existed prior to the establishment of St. Petersburg Cyperaceae was ranked third in species abundance, it is now ranked sixth having been replaced by the Brassicaceae; clearly indicating the synanthropic character of the urban flora. Among non-naturalised introduced plants, representatives of the Rosaceae family are the most dominant (mostly shrubs and trees) followed by Liliaceae and Salicaceae.

Adventive species were divided into four categories of naturalisation devised by Chichev (1985):

1. Ephemerophytes – 25 species, the majority (15) being introduced plants, such as *Calendula officinalis, Solanum tuberosum* varieties and *Helianthus annuus*.
2. Colonophytes – 31 species, 23 being "garden escapees" (for example, *Veronica filiformis* and *Cicerbita macrophylla,* the latter is a noxious pest in the Komarov Botanical Garden).
3. Epecophytes – 12 species, which are only common in a particular habitat or in different urban habitats, for example, *Hesperis matronalis* and *Fallopia sachalinensis*.
4. Agriophytes – non-native species, which have completely naturalised not only in urban biotopes but also in indigenous plant communities, they include *Impatiens glandulifera* and *Elodea canadensis*.

Historical City Centre

The historical city centre comprises "terraces" of buildings (mostly 3–6 storeys) of different architectural styles, which have created a very dense built environment. Most of the city centre is protected as part of a UNESCO World Heritage Site. The buildings are constructed of many different materials including stone, sandstone, brick, local limestone, granite and marble with iron roofs. They are bordered by concrete and granite embankments and asphalt roads.

Of all the habitats in the historic centre, the parks and gardens are the most extensive. Most of them have the status of historical gardens and parks and are protected by the Committee for the Control, Exploitation and Protection of Cultural and Historic Monuments. The largest complex is the garden "Heart" of St. Petersburg, which includes the Summer Garden, Mikhailovsky Garden, the Mars Field and the garden of the Art Square (Ploschad Iskusstv).

There are several plant habitats related to "hard surfaces" in St. Petersburg, they include joints in walls, basements and roofs of buildings, pavements and the granite blocks of the embankments and similar structures (for example, the Peter and Paul Fortress). Among other types of habitats in the historical centre are the boulevards and rows of street trees; verges adjacent to the granite embankments and roads, courtyard mini gardens and playgrounds; temporary wastelands where buildings have been demolished and construction works are in progress or have yet to start; railways, canals and rivers.

Granite Embankments, Walls and Cracks in Pavements

To stabilise the river banks and provide flood protection, embankments were built along both sides of the Neva and other rivers. Wooden embankments were used initially but between 1763 and 1788 (during the reign of Catherine the Great) the banks of the Neva were re-enforced with granite from large quarries on islands in the Gulf. The re-enforcement of the banks of the rivers continued into the nineteenth and twentieth centuries; resulting in 130 km of re-enforced embankments, made of a variety of different materials, including granite, diorite, concrete and wood. Granite embankments are not only important in flood control they also have a decorative function that has transformed the appearance of St. Petersburg and helped to create the city's identity (Ignatieva et al. 2003). Vascular plants mainly occur in the joints between the granite blocks, for example, there are at least 20 species growing in the granite embankments of the Fontanka and Moika rivers (Goryshina and Ignatieva 2000), the most common being ferns of calcareous substrates, for example, *Cystopteris fragilis* (Fig. 6), *Dryopteris carthusiana* and *Athyrium filix-femina*. *Sagina procumbens, Gnaphalium uliginosum, Potentilla anserina, Epilobium montanum, Poa compressa* as well as woody plants (*Betula pendula, Salix caprea, Sorbus aucuparia* and *Sambucus racemosa*) are very common. *Cystopteris fragilis* appears regularly on old building foundations made

Fig. 6 *Cystopteris fragilis* is one of the most common fern species in granite embankments in St. Petersburg

from local limestone (pudostski isvestnyak). Joints in roofs support typical pioneer plants such as *Betula* spp., *Salix caprea, S. myrsinifolia* and *Chamerion angustifolium.* Cracks in walls, concrete, stone and asphalt support *Betula pendula, Salix caprea, Salix myrsinifolia, Poa annua, Cerastium fontanum* ssp. *holosteoides, Campanula rotundifolia, Campanula rapunculoides, Achillea millefolium, Erysimum cheiranthoides, Stellaria media* and *Sagina procumbens.*

The most common bryophytes in such biotopes include *Bryum argenteum* and *Ceratodon purpureus* while *Barbula convoluta, Bryum caespiticum, Leptobryum pyriforme, Schistidium apocarpum* are among the less common species. Because of high moisture, desiccation by wind and solar radiation, and extreme fluctuations in daily temperature, lichens are confined to the joints or cavities. The lichens grow mostly on horizontal surfaces where water stays for only a short time. *Candelariella aurella, Lecanora crenulata, L. dispersa* and *Verrucaria muralis* are commonly found in the joints.

Other Habitats in the Historical Centre

Most of the boulevards were created in the nineteenth century and re-constructed several times after 1945 and are mainly planted with *Tilia x europaea* and *Ulmus laevis*. The lawns below the trees were originally sown with seed mixtures containing common "lawn species" such as *Poa pratensis, Lolium perenne* and *Festuca pratensis.* Trampling is one of the major disturbances in such habitats consequently the grassland species have been and are being replaced by weedy and trampling-resistant

Fig. 7 A courtyard with very few plants is a typical St. Petersburg centre habitat

species including *Poa annua, Plantago major, Capsella bursa-pastoris, Polygonum aviculare* agg., *Potentilla anserina* and *Trifolium repens*.

Populus hybrids are very common in the shaded courtyards. In the boulevards, courtyards lawn habitats and road verges *Sisymbrium officinale, Stellaria media, Aegopodium podagraria, Poa annua* and *P. pratensis* are among the most common herbaceous species (Fig. 7). On the bark of trees that have been planted in rows next to the rivers and canals, nitrophilous lichens such as *Scoliciosporum chlorococcum, Lecanora hagenii, Hypogymnia physodes* and *Phaeophyscia orbicularis* are very common.

Parks and Gardens

History of Planting Design

From the beginning broad-leaved trees such as *Tilia cordata, Acer platanoides, Quercus robur, Fraxinus excelsior, Ulmus laevis* and *U. glabra* were favoured in preference to other species. St. Petersburg is within the natural distribution of all these species, although some are very close to their northern limit. Before the construction of the city these species were never abundant in the area and only occurred where the environment was favourable, such as floodplains, warm slopes, gullies and fertile calcareous soils where they occurred in small numbers and never in good mature stands. Therefore most of the trees were brought to St. Petersburg from different regions of Russia (including Novgorod and Moscow) where the species are more common in the natural environment. For example, 8,000 trees were planted in the Summer Garden.

The preference for these tree species is often explained by the formal garden fashion of the eighteenth century but it probably also reflected a kind of "nostalgia" of Tsar Peter for his childhood in the landscape of Moscow with its typical broad-leaved forests. *Tilia* spp. were always the dominant species in all formal parks where it was used in hedges, groves and bosquets; maybe the main reason was its high decorative qualities, hardiness and excellent response to clipping. Species of *Acer, Fraxinus, Quercus* and *Ulmus* were very common in formal groves. *Picea* was also used in small groves. Two of the first introduced species to arrive *en masse* in St. Petersburg were *Buxus sempervirens* and *Taxus baccata* for use in formal parterres. However, the difficulty in growing both species for topiary forced St. Petersburg's gardeners to experiment with native species such *Vaccinium vitis-idaea* and *Juniperus communis* (for parterre decoration), and *Betula, Corylus avellana* and *Alnus incana* for hedges. During the time of Tsar Peter some Siberian conifers (*Abies sibirica* and *Pinus sibirica*) were planted for the first time in some private gardens. *Caragana arborescens* (a native of Siberia) and the European *Berberis vulgaris* were very popular non-native plants for hedges.

Lilac (*Syringa vulgaris*) appeared in St. Petersburg in the early eighteenth century, when it was also commonly used for hedges. Many fruit trees, cultivars of *Malus domestica* ssp. *domestica, Pyrus communis* and *Prunus domestica* were grown in garden bosquets. The presence of *Aesculus hippocastanum, Spiraea salicifolia* and *Lonicera tatarica* also reflects the Tsar's gardening interests.

In the eighteenth century, varities of *Tulipa* (Fig. 8) and *Narcissus* and *Hyacinthus* (in spring) and *Lilium* spp. (in summer) were among the most popular decorative herbaceous species.

With the development of the Picturesque style at the end of the eighteenth and beginning of the nineteenth centuries and experiments with newly introduced plants, other species appeared in private and public gardens. *Salix alba* and *S. fragilis* were popular trees, which were used for decorating landscapes around ponds and lakes.

Most public central city parks of the nineteenth century were based on the traditional broad-leaved species but later *Larix sibirica* became one of the most popular coniferous species. Because of the successful activities of the Imperial Botanical Garden, the Imperial Forest Institute and the Imperial nurseries, city parks and gardens were planted with many non-native species. For example, the published list of woody plants to be found in St. Petersburg parks and gardens consisted of 335 species; among them were *Populus balsamifera, Ribes odoratum, Thuja occidentalis* and *Elaeagnus argentea* from North America.

The Gardenesque style influenced the landscape and floristic composition of the city by creating fashionable and complex annual and perennial plants, which included species such as *Senecio bicolor, Lobelia erinus, Iresine* spp., *Ageratum houstonianum, Viola x wittrockiana* and *Phlox drummondii* with *Dracaena* and

Fig. 8 *Tulipa* varieties have always been one of the most popular plants for spring displays in flowerbeds

Agave being used as features in flowerbeds. By the end of the nineteenth century herbaceous borders of *Monarda didyma, Stachys byzantina, Dicentra spectabilis, Phlox subulata, Bellis perennis, Cerastium tomentosum* and *Sedum spurium* became very popular in public and private parks. The choice of introduced herbaceous decorative species available for planting in flowerbeds was enormous (more than 300 species). The beginning of the twentieth century was the most prosperous time for the introduction of non-native plants into St. Petersburg.

The 1917 revolution and the Civil War were detrimental to the development of gardens and the testing of new non-native species. The dominant woody plants in post-revolution planting were fast growing species, for example, *Populus balsamifera, P.* x *berolinensis, Acer negundo* and *A. tataricum*. Among shrubs the most common species were *Lonicera tatarica* and *Caragana arborescens*. *Cotoneaster lucidus,* a species native to the Baikal Lake Region, was successfully introduced to St. Petersburg and was soon the most preferred shrub in most new plantings of hedges and shrubberies. By 1936, the list of plants in government nurseries contained only 30 tree and 37 shrub species. The number of broad-leaved trees and the nineteenth century shrub "favourites" (*Syringa* spp., *Spiraea* spp., and *Philadelphus coronarius*) declined dramatically. The variety of perennial and annual species at that time was very small. Only about 20 species of decorative plants were used in the landscape design of pre-1939 Leningrad. The most popular bedding plants of the 1930s and 1940s were *Begonia semperflorens, B. tuberosa, Lobelia erinus, Pelargonium zonale, Ageratum houstonianum, Heliotropium arborescens* and *Petunia x hybrida,* all typical "global" Gardenesque bedding plants. Floral displays such as "clocks" and portraits of communist leaders were very popular and made from planting *Senecio bicolor, Sedum carneum* and species of *Iresine, Alternanthera* sp. and *Santolina* sp. (Ignatieva and Khodakov 1991). For central features of such floral displays, *Canna indica, Jucca gloriosa, Zea mays, Livistonia australis* and *Phoenix canariensis* were used. This "carpet bedding" fashion was to last for 100 years, from the middle of the nineteenth century to the 1950/60s.

By 2000 St. Petersburg had 71 parks, 176 gardens, 738 squares and 234 boulevards. Today there are 217 species of woody plants used in landscape design, 35 of them are native and 182 are introduced. The 63 most used species include *Tilia cordata, T. platyphyllos, T.* x *vulgaris, Acer platanoides, Quercus robur, Fraxinus pennsylvanica, Acer tataricum, A. negundo, Betula pendula, Ulmus laevis* and *U. glabra, Populus* hybrids, species of *Crataegus, Lonicera tatarica, Caragana arborescens, Berberis vulgaris, B. thunbergii, Syringa vulgaris* and *Spiraea chamaedryfolia.*

Flora of Historical Parks and Gardens

From 1989 to 1998 comprehensive floristic and phytocoenological investigations of 18 (2,378 ha) of the city's most famous historic gardens were undertaken (Ignatieva and Konechnaya 2004). The habitat types studied included lawns, hedges, woodlands (in landscape parks), formal park bosquets and parterres, flowerbeds, aquatic habitats (canals, ponds and lakes), roads and cracks in hard surfaces.

A total of 646 species of vascular plants were recorded, belonging to 307 genera and 98 families. This comprised 576 species of wild-growing plants (515 native, 25 non-native and 36 "garden escapees") and 70 species of non-naturalised non-native woody plants. The genus with the most species was *Carex* (33 species). Among the non-native woody plants, North American species were the most represented (20 species), including *Thuja occidentalis, Picea pungens, Pinus strobus, Populus balsamifera, Quercus rubra* and *Ribes odoratum*. The next group (19 species) comprised European woody species (for example, *Larix decidua, Salix alba, S. fragilis* and *Philadelphus coronarius*) followed by ten Siberian and Far Eastern species (including *Larix sibirica, Pinus sibirica, Caragana arborescens, Berberis thunbergii, Cotoneaster lucidus* and *Acer ginnala*).

There is a clear trend in the declining numbers of species in parks from the outskirts to the city centre. The most species-rich parks were the sub-urban historic parks such as Oranienbaum (400 species) and Pavlovsky Park (398). The city centre gardens (Summer Garden (163), Tavrichesky Garden (149) and Mikhailovsky Garden (147)) had the lowest number of species. The flora of gardens in the centre of St. Petersburg consisted mainly of urbanophile and urbanoneutral species belonging to the weedy and meadow ecological groups.

Rare herbaceous species were recorded in almost all the sub-urban historic parks. For example, *Poa chaixii* was found in Gatchina, Pavlovsk, Peterhof and

Oranienbaum parks. *Luzula luzuloides* grew in almost all parks except the central ones (the Summer Garden, Tavrichesky and Mikhailovsky gardens), Shuvalovsky Park and the park of the Botanical Institute. German and Scandinavian botanists believe that *Poa chaixii* and *Luzula luzuloides* appeared in European parks via grass-seed mixtures that were imported during the late eighteenth and early nineteenth centuries; a period of the Picturesque style in Europe. European botanists considered that grasses such as *Trisetum flavescens* and *Arrhenatherum elatius* (also found in almost all historical parks of St. Petersburg) were introduced to the parks via the same means. There is also a theory that all these species were brought in from central and southern Europe.

The presence of other rare herbaceous species that only occur in a particular park can be explained by historical peculiarities and environmental conditions. These species include *Colchicum autumnale, Phyteuma orbiculare, Valeriana dioica* and *Carex paniculata, C. flacca* and *C. hartmanii*, which were found only in Zverinets (Gatchina), *Phyteuma nigrum* only in Oranienbaum and *Phyteuma spicatum* only in Zverinets and Oranienbaum. The St. Petersburg botanist A. Haare (1978) speculated that some of the rare park species such as *Primula elatior, Phyteuma spicatum, P. orbiculare* and *Colchicum autumnale* and some other species are natural relicts of the original meadows that somehow survived in remote places within the parks.

In all sub-urban historical parks, the spring flora is very rich and creates a beautiful carpet of vernal native species such as *Ficaria verna, Gagea lutea, G. minima, Anemone nemorosa, A. ranunculoides* and *Corydalis solida.* In the gardens in the centre of St. Petersburg, the profuse blooming of *Gagea lutea, G. minima* and *Ficaria verna* (>70% groundcover) has only been seen in the Summer Garden. *Hepatica nobilis* has been found in abundance only in Gatchina and in Pavlovsky Park and *Viola odorata* with *Primula elatior* have been found only in Dvortsovy Park and Zverinets in Gatchina.

Taking into consideration St. Petersburg's wetland "past", the large percentages of wetland and aquatic plants present in most suburban parks is not surprising. The presence of these plants also indicates that disturbances, such as the effects of heavy machinery during construction work or poor management practices have destroyed the parks' drainage systems. In the city centre parks, typical urbanophile species such as *Plantago major, Trifolium repens* and *Poa annua* were dominant reflecting the influence of disturbances, such as trampling, mowing and construction.

Ignatieva and Konechnaya (2004) identified ten indicator groups, which were recommended for monitoring the environmental conditions in the historic parks. These groups reflect the ecological origin of the plants, their migration history and the management history of the parks:

1. Woodland groundcover species typical of natural European broad-leaved forests: *Convallaria majalis, Fragaria moschata, Anemone nemorosa, A. ranunculoides, Corydalis solida, Gagea lutea* and *G. minima.*
2. Boreal plants typical of the southern Taiga forests: *Trientalis europaea.*

3. Natural meadow plants: *Agrostis capillaris, Anthoxanthum odoratum, Alopercurus pratensis, Alchemilla monticola, Achillea millefolium, Campanula patula* and *Vicia cracca.*
4. Non-native herbaceous plants that arrived in grass-seed mixtures: *Trisetum flavescens, Arrhenatherum elatius, Luzula luzuloides, Poa chaixii, Phyteuma nigrum, P. spicatum* and *Pimpinella major.*
5. Garden escapees: *Scilla siberica* and *Gagea granulosa.*
6. Plants typical of anthropogenic disturbance: *Plantago major, Trifolium repens, Poa annua, Potentilla anserina* and *Ranunculus repens.*
7. Plants of fertile and well-drained soils: *Aegopodium podagraria, Anthriscus sylvestris* and *Dactylis glomerata.*
8. Plants typical of wet and poorly drained soils in woodlands, edges and lawns: *Filipendula ulmaria, Lysimachia vulgaris, Calamagrostis phragmitoides, Carex vesicaria, C. nigra, Juncus conglomeratus, Viola palustris* and *Deschampsia caespitosa.*
9. Weed species: *Capsella bursa-pastoris, Chenopodium album, Artemisia vulgaris* and *Arctium tomentosum.*
10. Wetland and aquatic plants: *Glyceria maxima, Carex acuta, Potamogeton natans* and *Alisma plantago-aquatica.*

The results of this study confirmed, once again, the broad-leaved origin of almost all of the planted historical green areas in the city. The present-day plant communities for 10 of the 18 historical parks sampled were dominated by European park species (for example, *Acer platanoides. Tilia cordata. Quercus robur. Ulmus laevis. Ulmus glabra* and *Fraxinus excelsior*). The plant associations found in the ten parks can be summarised as follows:

1. Alexandrovsky Park in Tsarskoye Selo: (a) *Ulmus laevis – Filipendula ulmaria, Aegopodium podagraria* (b) *Acer platanoides – Dactylis glomerata – Aegopodium podagraria* and (c) *Quercus robur – Tilia cordata – Aegopodium podagraria* and *Dactylis glomerata.*
2. Central city gardens; dominant communities in the Mikhailovsky Garden, *Tilia cordata – Acer platanoides – Ulmus glabra – Poa annua – Plantago major* and *Taraxacum officinale.* In the Tavrichesky Garden the dominant plant communities comprise *Ulmus glabra – Quercus robur – Tilia cordata – Acer platanoides – Poa annua – Plantago major – Polygonum avicular* and *Stellaria media.*
3. Nizhny Park in Peterhof; represented by the "wet versions" of broad-leaved plant communities (a) *Acer platanoides – Deschampsia caespitosa,* (b) *Tilia cordata – Acer platanoides – Aegopodium podagraria,* (c) *Betula pubescens – Anthriscus sylvestris – Aegopodium podagraria; Quercus robur – Ranunculus cassubicus – Filipendula ulmaria* and *Tilia cordata – Alnus glutinosa – Equisetum palustre.*

The artificially created meadow plant communities that have been established in some historical parks need relatively frequent and regular management, without it they would be replaced in just a few years by woody pioneer species such as *Alnus incana, Betula pendula, Salix phylicifolia, S. caprea* and *S. myrsinifolia.*

Ephemeral plants such as *Gagea lutea, G. minima* and *Ficaria verna,* along with *Aegopodium podagraria* and a group of weed and meadow-forest species (*Taraxacum officinale, Poa annua* and *Plantago major*) dominated the ground cover of the Summer Garden, the oldest garden in St. Petersburg (founded in 1704). The mesophyte grasses traditionally planted in this park, such as *Poa pratensis, Festuca pratensis* and *Lolium perenne,* have never managed to persist due to the shady conditions. The success of the ephemeroides and *Aegopodium podagraria* can be seen as a stabilising phase of the park's ecosystem, which is very important for extending the life of the old trees (some of the trees in the Summer Garden are 250–300 years old) and should be nurtured.

In some parks (for example, Nizhny Park and Alexandria Park in Peterhof and Dvortsovy Park in Gatchina) the plant communities are dominated by *Quercus* spp. and other broad-leaved trees in the canopy with *Filipendula ulmaria* on the ground. This combination is typical of artificially created communities in damp conditions; it has no equivalent in the native vegetation. The abundance of *Filipendula ulmaria* in many suburban parks indicates a high watertable and impeded drainage, probably caused by a dysfunctional drainage system.

Distribution maps for rare herbaceous species (and spring ephemerals) found in St. Petersburg's historical parks are a compulsory requirement for the restoration and management organisations as an important tool for biodiversity protection in the historic parks.

Residential Neighbourhoods

From 1913 to 1939 the population of St. Petersburg increased from 2.1 million to more than 3 million inhabitants. A new city plan (which was inspired by the Moscow Plan) was produced in 1935; it included the reconstruction and development of new residential areas with the provision of public services and recreational areas. The Plan and the construction of new apartment blocks, which had begun in the southern suburbs of the city, were postponed because of World War II. The concept of "Microrayon" or "Residential Neighbourhoods" changed the appearance of the post-war Soviet cities and provided the urban-spatial framework for harmonious "brotherly" living of equal opportunity which was the core of Socialist ideology. The residential neighbourhoods occupy 30–50 ha and contain 10,000–20,000 people. They were seen as self-contained communities comprising apartment blocks, kindergartens, schools, local medical facilities, shopping and other services. The physical form of the residential neighbourhoods became one of the basic units of new housing construction in the former USSR and other socialist countries. The new plan for the development of Leningrad was approved in 1966, which resulted in the construction (between 1965 and 1973) of many residential neighbourhoods (mainly in the suburbs) with a total of 393,000 individual apartments.

Large-scale planning of the residential neighbourhoods provided the opportunities for the provision of open green spaces. In principle, governmental organisations should have provided the landscape design plans but in reality, the centralised design and planting was done mainly in relation to streets, schools and kindergartens and public parks. In the 1950s and 1960s, the general tendency in the landscaping of all new district developments was to use *Acer negundo* and the fast-growing hybrids of *Populus*. It was a real "poplar boom" – during weekends thousands of people planting species and hybrids of *Populus* next to their houses. The widespread use of *Populus* together with the standardisation of the "architecture" resulted in the uniformity of the new residential environments not only in Leningrad but throughout the former USSR. In many cases female *Populus* were planted, which resulted in the windblown seeds creating fire risks and causing strong allergenic problems.

In the 1970s, the species used in the landscaping of new areas was more diverse compared to the immediate post-war period. *Populus* and *Betula* were still the most common; the other 63 species included *Ulmus, Tilia, Quercus, Acer* (including *A. ginnata* and *A. tataricum*) and *Salix viminalis.* Among the shrubs, *Cotoneaster lucidus, Symphoricarpos albus, Spiraea chamaedryfolia, Crataegus almaatensis* and *Caragana arborescens* were the most common; *Viburnum lantana, Spiraea salicifolia, Syringa vulgaris* and *Ribes odoratum* were used less frequently.

The flowerbeds along the side of some of the roads were planted with decorative plants, the most popular being *Calendula officinalis, Dahlia* spp., *Begonia semperflorens, Petunia* x *hybrida, Salvia splendens, Iberis* spp., *Viola* x *wittrockiana, Bellis perennis* and *Tulipa* varieties. A few climbing species were used for vertical greening, the most common being the North American species, *Parthenocissus quinquefolia.*

The residents of the apartment blocks planted the palisadniks with different decorative tree, shrub and herbaceous species, which is why these areas are much more diverse compared with other residential neighbourhood biotopes. The most common tree species included *Betula, Sorbus aucuparia, Prunus avium, Prunus* spp. and *Malus domestica* ssp. *domestica, Philadelphus coronarius, Rosa rugosa, Symphoricarpos albus, Syringa vulgaris* and *S. josikaea.* With time the "palisadnik" plantings matured and became shady woodlands. Among herbaceous species *Aegopodium podograria* was the dominant plant, which indicated quite good soil conditions. Other plants of the shady palisadniks are *Anthriscus sylvestris, Taraxacum officinale, Urtica dioica, Dactylis glomerata, Ranunculus repens* and *Artemisia vulgaris* (Fig. 9).

The residents also planted a lot of spring geophytes such as *Galanthus nivalis, Scilla siberica, Chionodoxa luciliae* and *Narcissus* species and varieties and summer-flowering decorative plants, for example, *Sedum* spp., *Dicentra* spp., *Cosmos bipinnatus, Saponaria officinalis, Iris* spp., *Lilium tigrinum, Consolida* spp., *Aconitum* spp. and *Symphytum* spp. Non-native plants comprised up to 30–40% of the total flora of the "palisadnik"; however, sometimes orchids that are rare in urban St. Petersburg can be found in this biotope, for example, *Listera ovata*

Fig. 9 "Palisadnik" vegetation dominated by *Aegopodium podagraria* is a very typical habitat in new microdistricts

(Goryshina and Ignatieva 2000). The residential neighbourhoods provide a "refuge" for other quite rare plants such as the protected species *Dactylorhiza baltica,* which was found in damp wastelands in the Kalininsky District.

Another important biotope in the residential neighbourhoods is grassland (lawn) which can be found in open areas between apartment buildings. Some open areas have been colonised by spontaneous ruderal vegetation while others were sown with grass-seed mixtures, including species such as *Poa pratensis, Festuca pratensis* and *F. rubra.* In areas of high recreational pressure, for example, in playgrounds such mixtures were replaced very quickly by species tolerant of trampling such as *Polygonum aviculare* agg., *Potentilla anserina, Trifolium repens, Plantago major, Poa annua* and *Tripleurospermum inodorum.* In areas of moderate and light anthropogenic pressure, more meadow species and typical ruderals appeared, such as *Dactylis glomerata, Phleum pratense, Leontodon autumnalis, Taraxacum officinale, Achillea millefolium, Vicia cracca, Lathyrus pratensis, Ranunculus acris, Leucanthemum vulgare, Artemisia vulgaris, Lamium album* and *Capsella bursa-pastoris.*

Compared with many other countries, the landscape maintenance policy in residential neighbourhoods was limited and mainly restricted to the pruning of woody vegetation. The grassland areas were only mown occasionally. Because of the lack of maintenance, vegetation at different stages of succession has become typical of the residential neighbourhood wastelands, some of which support characterisitic meadow, weed and ruderal species and quite a few garden escapees, for example, *Aster salignus, Helianthus annuus, Pyrethrum parthenium, Impatiens glandulifera,*

Symphytum asperum and *Solidago canadensis*. There are also some small fragments of native *Salix* scrub, which were left during the construction works. Because of constant disturbance caused by the repairing of Roads, and other works, the vegetation contains numerous weedy plants such as *Barbarea vulgaris, Urtica urens, Stellaria media, Arctium tomentosum, Cirsium arvense* and *Chenopodium album.*

In post-perestroika St. Petersburg, the deterioration in landscape maintenance has had botanical benefits. Most of the habitats have been left to "go wild" consequently the biodiversity of neighbourhood floras is quite high compared to that of central districts. The flora of several residential neighbourhoods was studied by Eremeeva (2000), who recorded 551 species of vascular plants, including 102 species of cultivated plants that have naturalised.

The bryophytes that occur in the neighbourhoods include the mosses *Orthotrichum obtusifolium, O. speciosum* and *Leskea polycarpa* (very common on tree bark). On the ground (especially where there have been fires), *Polytrichum piliferum* and *Marchantia polymorpha* can be found.

Dachas

The typical Russian home estate called "usadba" was a dwelling with a fenced area comprising an orchard, vegetable garden and utilitarian structures to support family life. During the eighteenth century the usadbas of the St. Petersburg nobility and merchants were used for growing vegetables (on open ground and in hothouses), fruit trees (cultivars of *Malus domestica, Cerasus vulgaris, Prunus domestica* and *Pyrus commuis*), soft fruits (*Ribes nigrum, R. rubrum, Grossularia reclinata, Fragaria* x *ananassa* and *Rubus idaeus*) and some decorative flowers. Vegetable gardens (*ogorodi*) were another very visible feature of the eighteenth century urban fabric; they could be seen even next to the Tsar's palace in the middle of the city (for the court food supply). Vegetable gardens were also a common feature of the city's outskirts during the nineteenth and twentieth centuries (Goryshina 2003). The most common vegetables were *Brassica oleracea* var. *capitata, Solanum tuberosum* varieties, *Beta vulgaris* ssp. *vulgaris, Daucus carota* ssp. *sativus* and *Pisum sativum.* By the end of the nineteenth century, the species list available from St. Petersburg's hothouses was very large and impressed even western visitors.

As the urbanisation process continued, most usadbas within the city boundaries disappeared. Very similar to the "usadba" was another traditional Russian dwelling often located on the urban fringe and near suburbs – the "dacha". Dacha is the Slavic equivalent of a holiday house or second home with a small garden. The main purpose of dachas was summer recreation, but they also had a practical function and provided a "close to nature" life-style. The first dachas were probably built in the early eighteenth century, along the Peterhof Road, as out-of-town residences for the aristocracy. Although they were located beyond the city limits they did not merge with the surrounding rural landscapes – "*halfway house between the metropolis and the countryside*" (Lovell 2003). By the second part of the nineteenth century, dachas started to serve a broader category of St. Petersburg society such as lawyers, journalists, doctors and merchants. Interestingly dachas survived the 1917 Revolution and became more common, especially in the later years of the Soviet era; before 1939 their ownership was more or less restricted to Communist Party leaders.

Between 1955 and 1958 the Soviet Government was under pressure from urban citizens to have the opportunity of spending time away from their city apartments. As a consequence of the permanent shortage of fresh vegetables and fruits; the government approved legislation that allowed the construction of dachas, which were simple wooden houses. The plots were not more than 600 m^{-2} for Soviet collective farm labourers; they rarely exceeded 1,200 or 1,500 m^{-2} and virtually never exceeded 9,600 m^{-2}. By the early 1990s about 60% of the inhabitants of all major Russian cities had access to dachas. The size of the plot and the house were restricted.

Almost every square centimetre of the land is used for food production with only small areas being used (in some cases) for growing decorative plants. The most popular vegetables and herbs in St. Petersburg dachas are *Brassica oleracea* var. *capitata, Brassica oleracea* var. *botryis, Solanum tuberosum* varieties, *Beta vulgaris* ssp. *vulgaris, Daucus carota* ssp. *sativus, Pisum sativum* and *Brassica rapa, Cucumis sativus, Cucurbita pepo, C. maxima, Lycopersicon esculentum, Raphanus sativus, Allium cepa, A. sativum, Anethum graveolens, Petroselinum crispum, Apium graveolens, Rumex acetosa, Lactuca sativa* and *Rheum* x *hybridum.* The most common fruits are cultivars of *Malus domestica, Prunus domestica, Cerasus vulgaris* and *Pyrus communis* and soft fruit, *Ribes nigrum, R. rubrum, Grossularia reclinata, Amelanchier spicata, Aronia melanocarpa, Rubus idaeus* and *Fragaria ananassa,* see Fig. 10. The most common decorative plants are varities of *Narcissus* spp., *Tulipa* spp., *Iris* spp., *Lilium* spp., *Dahlia* spp., *Aconitum* spp., *Hemerocallis* spp., *Rudbeckia laciniata, Gladiolus* spp., *Phlox paniculata, Syringa vulgaris, Rosa* spp., *Spiraea* spp., *Philadelphus* spp., *Berberis vulgaris, Viburnum opulus, Prunus avium* and *P. virginiana.* The favourite climbers are *Parthenocissus, Clematis* and *Lonicera caprifolium.* Annuals such as *Calendula officinalis, Lathyrus odoratus, Tagetes* spp. and *Centaurea cyanus* are always very much loved among Russian "dachniki" (summerfolks).

Some native herbaceous plants are well established, as weeds, for example, *Aegopodium podagraria* (a highly aggressive competitor of vegetables and decorative flowers), *Anthriscus sylvestris, Rumex obtusifolius, Equisetum arvense, Achillea millefolium, Taraxacum officinale, Tussilago farfara, Geum urbanum* and *Vicia sepium.* There is also a group of typical weeds that are imported in soil mixtures and manure; *Tripleurospermum inodorum, Chenopodium album, Myosotis arvensis, Barbarea vulgaris, Capsella bursa-pastoris, Erysimum cheiranthoides, Rorippa palustris, Cerastium fontanum* ssp. *holosteoides, Stellaria media, Spergula arvensis, Galeopsis tetrahit, Polygonum aviculare* agg., *P. scabrum, Rumex acetosella, Plantago major, Ranunculus repens, Agrostis capillaris, A. stolonifera, Elytrigia repens, Poa annua* and *Viola arvensis.*

Fig. 10 Traditional dacha: every piece of land is used for growing vegetables and fruits

A small number of native plants from previous vegetation types can also be found in dacha gardens; *Leucanthemum vulgare, Campanula patula, Viola canina, Veronica chamaedrys, Scrophularia nodosa* and *Stellaria nemorum* from meadows and previous forests of *Alnus incana.*

Dachas in Perestroika and Post-perestroika times

As the result of rapid urbanisation, many old dachas built in Soviet times were incorporated in the city's boundaries, eventually destroyed and subsequently re-developed by the new affluent Russian "elite".

The twenty-first century villa garden differs dramatically from the traditional Russian "dacha", which was mainly based around a productive landscape. The large gardens of the new houses are landscaped purely for recreation and "western" aesthetics. These new villas are huge, occupying a one ha plot or more. The cost of these "palaces", which are built of brick or concrete (not wood as in traditional dachas), often reaches several million US dollars. The garden design is entirely based on the modern Anglo-American "model" of large lawns, rock gardens and numerous non-native conifers (*Platycladus orientalis, Chamaecyparis lawsoniana, C. obtusa, C. pisifera, Juniperus chinensis, J. squamata, Pinus mugo, P. parviflora*) and fashionable varieties of *Rhododendron* (including "azaleas"), decorative flowering cherries (*Prunus cerasifera*), *Erica carnea* and *Prunus laurocerasus.* The lawns are dominated by *Lolium perenne* and *Festuca rubra.* Such large and

Fig. 11 Villa of the New Russians – dominance of lawn and non-native "*global*" conifer plants

well-maintained lawns "*bespeaks a rejection of Soviet agricultural imperative and a turn towards Anglo-American civilisation*" (Fig. 11).

All the non-native species that are planted in the gardens are from nurseries in Germany, Poland and The Netherlands. Most of this "global" non-native plant material is not suitable for and does not survive in harsh St. Petersburg winters.

Cemeteries

In 2008 there were 85 cemeteries in the city, 13 categorised as historical cemeteries and 3 as memorial cemeteries. The largest and the most impressive among them is the Piskarevskoye Memorial Cemetery where 500,000 people who died during the Siege of Leningrad were buried in 186 mass graves. A Memorial to the victims of the Siege was opened in 1960. The main woody plants used in the landscape design were *Ulmus glabra, Betula pendula, Salix alba, S. fragilis, Tilia cordata, Acer platanoides, Fraxinus* spp., *Aesculus hippocastanum, Malus domestica, Larix* spp. and *Picea pungens*.

Most of the cemeteries are of the Russian Orthodox Church and have a typical general design (rows of burial sites) and detailed design of individual graves (Fig. 12).

Compared with most Catholic and Anglican cemeteries with a very simple design of graves in lawns, the Russian Orthodox cemeteries have a small place in

Fig. 12 A typical cemetery with a small place in the middle of the grave and places around the graves, which are specially designed for the planting of decorative spring and summer flowering herbaceous species

the middle of the grave and places around the graves, which are specially designed for the planting of decorative spring and summer flowering herbaceous species. The most common of the non-native spring flowering species are *Primula* spp., *Scilla siberica, Muscari botryoides, Tulipa* spp. and *Narcissus* spp., *Bellis perennis, Tagetes patula, T. erecta* and *Viola x wittrockiana.* For summer display, *Aquilegia hybrida, Saxifraga caespitosa, Sedum acre, S. spurium, Bergenia crassifolia, Phlox subulata* and *Begonia semperflorens* are very popular. Some of these introduced plants escape from the graves and spread into other parts of the cemeteries. *Convallaria majalis* is the most popular native species to be planted. Shrubs such as *Symphoricarpos albus, Syringa vulgaris, Berberis vulgaris, Crataegus* spp., *Spiraea* spp. and *Prunus avium* can be found in almost all cemeteries in which *Thuja occidentalis* is a rare "guest".

Most modern civic cemeteries are located in native forests, for example, Severnoe (The Northern) Cemetery was established in a *Picea-Pinus* forest (mixed with *Betula pendula* and *Populus tremula*). In this cemetery, typical forest and some weed species such as *Vaccinium myrtillus, Rubus idaeus, Calamagrostis canescens, Oxalis acetocella, Solidago virgaurea, Anthriscus sylvestris, Taraxacum officinale* and *Equisetum arvense* are very common. Species such as *Aegopodium podagraria, Taraxacum officinale, Calystegia sepium, Glechoma hederacea, Agrostis capillaris, Bromopsis inermis, Dactylis glomerata, Festuca pratensis, Poa annua, P. pratensis, Achillea millefolium, Artemisia vulgaris, Hieracium umbellatum, Leucanthemum vulgare, Solidago canadensis, Lathyrus pratensis, Vicia cracca* and *Elytrigia repens* are common along the paths and between the graves. *Phragmites australis, Epilobium adenocaulon, Lysimachia vulgaris* and *Odontites vulgaris* are more typically found in the damp habitats while *Sisymbrium loeselii, Cirsium arvense, Silene latifolia, Tripleurospermum inodorum, Chenopodium album* and *Melilotus albus* are the most characteristic plants of the recently disturbed areas. In old cemeteries, the presence of pioneer willows (*Salix caprea, S. myrsinifolia* and *S. phylicifolia*) indicates the successional processes that are taking place. In Pargolovo Cemetery *Solidago canadensis* and *Lupinus polyphyllus* (all naturalised introduced species) occur *en masse* along the ditches and forest fringes.

Thirty-eight species of lichens were found in the historical cemeteries, which contain monuments made from many rock types. The most representative ones are in the genera *Lecanora* (seven species) and *Physcia* (four). The most common are nitrophilous species: *Lecanora hagenii, Phaeophyscia orbicularis, Physcia dubia, Scoliciosporum chlorococcum* and *Xanthoria polycarpa*. Fifteen species of lichen grow on the trees, two species on stone substrates and one which was found on steel and concrete. *Melanelia exasperata, Physcia dubia* and *P. stellaris* were seen growing on granite.

Transport Routes

Railways

The first railway line was constructed in St. Petersburg in 1837 (St. Petersburg-Tsarskoye Selo), followed by the creation of the line between St. Petersburg and Moscow (650 km). At present there are five main railway stations and numerous other local stations within the administrative boundary of the city. From the beginning, railways were one of the most important vectors for the introduction of new adventive species. By the 1970s, approximately 186 adventive plants were observed along railway lines in the Lenningrad region.

The most common railway habitat is the track of which the following are typical, *Amaranthus albus, A. retroflexus, Artemisia campestris, Crepis tectorum, Conyza canadensis, Senecio viscosus, S. vulgaris, Sisymbrium loeselii, S. wolgense, Oenothera rubricaulis, Anisantha tectorum, Poa compressa, Setaria viridis, Galium aparine, Chaenorhinum minus* and *Linaria vulgaris.*

The halophyte *Tripolium vulgare* appears along the railway ditches of the Baltic and Warsaw railway tracks. Another halophyte, *Salsola tragus,* is also occasionally found along the tracks.

Harbour

According to Tsar Peter's plan, the primary purposes of St. Petersburg were to be a major port and "The Window to Europe" with access to the Baltic Sea.

The location of the port has changed several times since the eighteenth century; by the end of the nineteenth century it was clear that the city needed a new seaport, so it was constructed on Gutuev Island and opened in 1885.

Because of the nature of imported goods during the eighteenth century and probably the nineteenth century, the port was not a significant source of adventive plants. However, by the late twentieth century 98 species of adventive plants had been identified in the modern port (Popov 1995). The ephemerophytes were the dominant group (47.5%) and 85% of the adventive species were herbaceous annual or short-lived species. The most common families represented are the Poaceae, Brassicaceae and Asteraceae. The geographical "core" of the species found in the port are ancient Mediterranean species (46%), which are directly related to the importation of crops.

It is very interesting that only a small portion of the adventive plants are capable of naturalising and competing with the native flora. Most species are only temporary occupants of disturbed habitats. The most "likely" species in the port to naturalize are:

1. Plants from the temperate Northern Hemisphere:
   (a) North American species: *Epilobium adenocaulon, Oenothera biennis* and *O. rubricaulis.*
   (b) Western European species such as *Polygonum bellardii* and *Erucastrum gallicum.*
   (c) Steppe species: *Berteroa incana, Sisymbrium loeselii* and *Salsola tragus.*
   (d) Southern Asian Species: *Impatiens parviflora.*
2. Species from the ancient Mediterranean basin: *Lolium perenne, L. multiflorum, Sisymbrium altissimum* and *Bunias orientalis.*
3. Species that can be considered as almost cosmopolitan: *Setaria pumila, S. viridis* and *Amaranthus retroflexus.*

Two noxious weeds *Ambrosia artemisiifolia* and *Solanum triflorum* have been found within the St. Petersburg Port area.

Aquatic

Aquatic habitats are especially sensitive to pollution. During the eighteenth century, some rare species that survived only in clean water were still abundant in the city's rivers and canals. Rapid industrialisation and the resultant water pollution in the nineteenth and twentieth centuries dramatically changed the composition of the aquatic flora (Goryshina and Ignatieva 2000). Despite pollution some submerged species (*Potamogeton perfoliatus, P. pectinatus* and *Sagittaria sagittifolia*) are still growing in the Neva river even in the downtown area next to the Peter and Paul Fortress and the Hermitage. However, the submerged species that require clean water have now disappeared from the rivers and lakes, for example, *Isoëtes lacustris*. Nevertheless, there are also some rare "Red Book" species which appear in the shallow waters of the northern part of the Gulf (the Neva Mouth), namely, *Isoëtes echinospora, I. lacustris, Alisma gramineum* and *Caulinia tenuissima*.

As previously described, in addition to the Gulf, the city has an extensive network of rivers and canals, which form an important part of the urban landscape. There are several small natural lakes in the northern part of the city (Susdalskie Lakes) with some small lakes also located in other city suburbs (Dudergof, Krasnoe Selo and in Komarovo). Most urban and sub-urban parks have ponds and streams.

The aquatic flora of St. Petersburg is diverse. The more common species of the submerged zone include *Ceratophyllum demersum, Elodea canadensis, Myriophyllum verticillatum, M. spicatum, Potamogeton perfoliatus, P. praelongus, P. lucens, P. berchtoldii, P. compressus, Utricularia* spp., *Lemna trisulca* and *Ranunculus aquatilis* agg. Small plants that can create an underwater "carpet" (and are capable of flowering when the water recedes) have been recorded in the city, they are *Eleocharis acicularis, Ranunculus reptans, Limosella aquatica* and *Elatine hydropiper.* The dominant species of floating-leafed plants are *Nuphar lutea, Nymphaea candida, Stratiotes aloides, Hydrocharis morsus-ranae, Lemna minor, Spirodela polyrhiza, Persicaria amphibia* and *Potamogeton natans.* The emergent/marginal species that occur along the Gulf coast and on the margins of rivers and lakes include *Phragmites australis, Scirpus lacustris, S. sylvaticus, Glyceria maxima, Alisma plantago-aquatica, Sparganium emersum, Typha angustifolia, T. latifolia, Eleocharis* spp., *Carex vesicaria, Iris pseudacorus, Butomus umbellatus, Scutellaria galericulata, Lysimachia vulgaris, Lycopus europaeus, Cicuta virosa, Sium latifolium* and *Bidens cernua.*

Nature Conservation, Environmental Planning and Education

To improve nature protection in the city, the St. Petersburg Administration in 2005 accepted the "Master Plan of St. Petersburg and the boundaries of protected subjects of cultural heritage within St. Petersburg area" and the resolution "Certification of The Master Plan of St. Petersburg". These documents provided the framework for increasing the network of "Special Protected Areas" in St. Petersburg. For example, Elagin Island was recommended as a "Special Protected Area" by 2010.

Several international projects (Russian-Danish and Russian-Finish) were carried out from 1999 to 2004. These projects analysed urban green areas and proposed different options for the development of the green belt and for sustainable management of forestry and biodiversity protection. One of the most important goals of these projects is the provision of ecological education for children and adults. New ecological educational displays, "Demonstration

Sustainable Forest" and ecological trails are organised in forest parks (for example, in Sosnovka).

The project "Representation of valuable biotopes and Red Book Species in the ecological framework of St. Petersburg" was undertaken in 2007 and 2008 using resources of different research institutes of the Russian Academy of Sciences and St. Petersburg State University. Several nature protection projects helped establish a GIS database, aided publication of the Red Data Book of Nature of St. Petersburg (Noskov 2004) and provided suggestions for new protected areas.

Due to urbanisation, many plants which are common in the Leningrad region have disappeared or become rare in St. Petersburg. The large number of water bodies provides a wide diversity of aquatic habitats, some of which are important for rare plant species. Because of anthropogenic pressure and dramatic changes to the native landscapes within the city, Red Data Book of Nature of St. Petersburg was prepared and published in 2004. The threat categories used are in accordance with those published by the IUCN in 2001 (IUCN Red List Categories: Version 3.1), see Appendix VIII of this book.

Criteria for inclusion of a species in the list are: (a) rarity in the city and the Leningrad Region and (b) an indicator of the natural conditions of the area before St. Petersburg was built. The rare plant species within the city limits generally occur in the protected areas. One hundred and twenty-four species of plants and fungi found in the city boundaries are protected, comprising 48 species of vascular plants, 18 bryophytes, 13 algae, 16 lichens and 29 fungi. For example, *Isoëtes echinospora,* which belongs to the "Vulnerable" category, requires clean water. *Lycopodiella inundata* (Near Threatened) is rapidly disappearing as a result of wetland drainage. *Euphorbia palustris* (Critically Endangered) is threatened by the development of the seashores and the drainage of coastal swamps. *Dactylorhiza baltica* (of Least Concern) is disappearing as a result of increasing development. "Endangered" species include *Lathyrus linifolius* while critical endangered/ endangered species include quite a few native orchids, for example, *Neottia nidus-avis*. *Galium hercynicum* is an example of a "Data Deficient" species. Twenty-three species of vascular plants have become extinct in the St. Petersburg area, including *Calypso bulbosa, Lobelia dortmanna* and *Lathyrus pisiformis* (Noskov 2004).

Compared to other Russian cities the protected areas of the city cover a relatively small area. By 2008 six protected areas of unique native habitats covered only 1.5% (2,150 ha) of the city (Fig. 13). Most rare species and "pure" undisturbed examples of natural habitats only occur in the protected areas. However, it has recently been suggested that the network of nature reserves should be expanded and 16 new protected areas established.

Ecological education in St. Petersburg is quite strong; all school programmes have a "Regional Part" where students study local nature and the Red Data Book. There are quite a few children's ecological clubs and schools who even help in gathering and analysing field information for several international research nature protection and biodiversity projects.

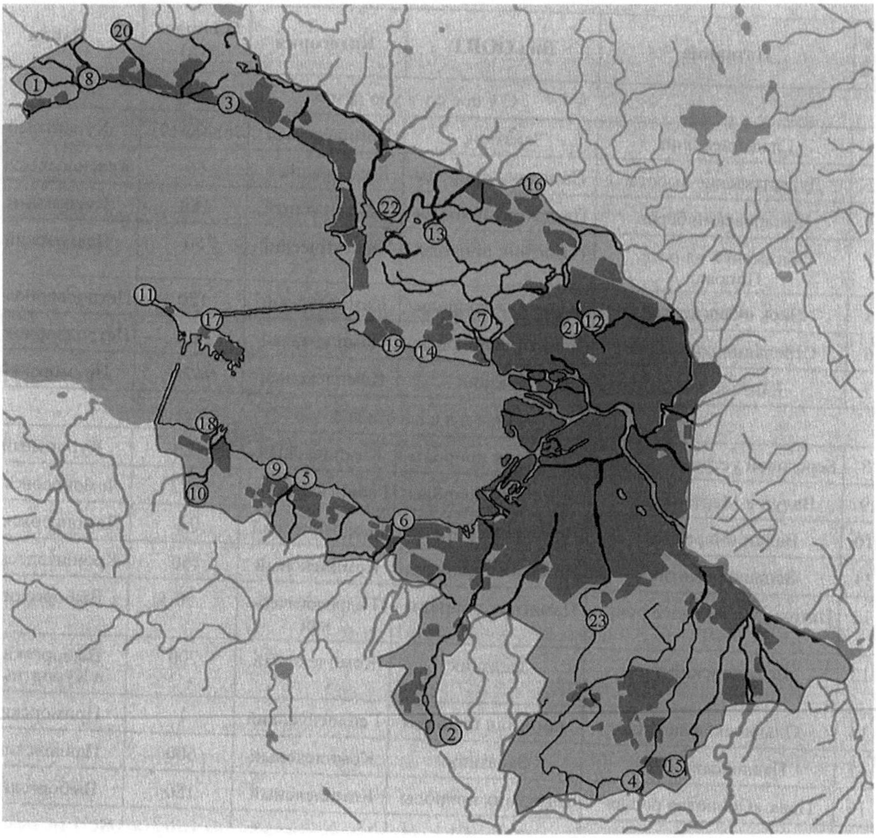

Fig. 13 Existing (1–6) and proposed (8–23) reserve areas in St. Petersburg (Noskov 2004)

Closing Comments

Peter the Great initiated a gigantic experiment to change an inhospitable natural wetland landscape that was subject to frequent flooding into a major port and the most European of the existing Russian cities. As a result of drainage, the construction of canals and buildings, the spreading of fertile soil and the planting of millions of broadleaved trees, the city landscapes were dramatically transformed. During Soviet times, the city was surrounded by self-contained residential neighbourhoods of high-rise apartment blocks each covering 30–50 ha and with a population of 10,000–12,000 people.

Because of its numerous historic buildings, parks and gardens and cultural significance, the city has been designated a UNESCO World Heritage Site, which makes it unique and different from all other Russian cities. The historic parks and gardens have the highest plant biodiversity of all the urban biotopes of the historic city centre. Numerous waterways and their architectural decoration resulted in the creation of a unique St. Petersburg type of biotope – granite embankments. Biotopes connected with different hard surfaces (walls, buildings, pavements, embankments, roads) are the most common in the centre.

The wetland origin of St. Petersburg has influenced the floristic composition of historical parks as well as the vegetation in residential neighbourhoods. The proportion of species that have their origins in different wetland plant communities is always quite high. Within the administrative boundaries there are also remnants of the native forests (forest parks).

The Red Data Book of Nature of St. Petersburg was completed in 2004 (Noskov 2004). By 2008 there were six protected areas with unique native habitats covering 1.5% (2,150 ha) of the city. Most rare species and "pure" undisturbed examples of natural habitats are located in these protected areas.

The recent shift to a market economy and the consequential increase in air pollution (mostly from private vehicles) has seen a decrease in lichen biodiversity and degradation of urban soils. The change in emphasis from designing public green space (as in the Soviet era) predominantly to the gardens of villas of very rich people has resulted in the sub-urbanisation of St. Petersburg. Today the "Venice of the North" is following international trends in landscape design. All plant material for new public and private sectors is sourced from "western" nurseries and based on a mostly non-native, fashionable "global" pool of plants as a consequence of which the urban flora is also becoming standardised.

Because of recent changes in the centre of St. Petersburg (removal of green areas and replacement with new buildings) many new businesses and shops have started to use containerised conifers on the pavements outside. One of the most recent practices has been the removal of the old *Populus,* which are no longer used in new landscape schemes. The latest fashion in the microdistrict "palisadniks" is the removal of the old shrubberies, which include species such as *Syringa* and *Berberis* and replacing them with more globally fashionable rock (alpine) gardens and even small ponds.

Plant names of higher vascular plants were used according to the most important and complete sources in the north-west of the Russian Federation (Tsvelev 2000; Minyaev et al. 1981).

Literature Cited

Ageeva O (1999) Petersburg in Russian social consciousness in early 18th century. Russko-Baltiiski informatsionni Tsentr, Sankt Petersburg (in Russian)

Babikov B, Melnichuk I (2003) Main direction of the Department of soil science and hydro melioration in 21st century. Bulletin of Saint Petersburg Forest Technical Academy 170: 58–66 (in Russian)

Balashova N, Zavarsina A (eds) (1999) Biodiversity of the Leningrad Region (Algae, Fungi, Lichens, Bryophytes, Invertebrates, Fishes, Fish-like vertebrates. Trudi Sankt-Petersburgskogo Obschestva estestvoispitatelei. Ser. 6.T.2. St. Petersburg State University, St. Petersburg. (in Russian)

Bogovaya I, Fursova L (1988) Landscape Art. Agropromisdat, Moscow (in Russian)
Bulygin N, Firsov G (1983) Introduction of maples in the North West part of Russian Federation. Report. Leningrad (in Russian)
Bulygin N, Firsov G (1995) Woody plants of indigenous flora in urbanophytocenoces of St. Petersburg. Bulletin of Main Botanical Garden 172: 3–7 (in Russian)
Chichev A (1985) Adventive flora of railways in Moscow region. Dissertation, Moscow State University (in Russian)
Dubyago T (1963) Russian formal gardens and parks. Stroiisdat, Leningrad (in Russian)
Eremeeva E (2000) Results of flora research of microdistricts in St. Petersburg. Formation of vegetation cover of urbanized areas. Materials of international conference. Veliky Novgorod. 9–10 June 2000: 69–72 (in Russian)
Goryshina T (2003) Green world of old Petersburg. Iskusstvo SPB, St. Petersburg (in Russian)
Goryshina T, Ignatieva M (2000) Botanical excursions around the city. Chimisdat, St. Petersburg (in Russian)
Gusev Yu (1968) Changes of ruderal flora of Leningrad region during 200 years. Botanical Journal 53 (11): 1569–1579 (in Russian)
Haare A (1978) Specierum relictarum locus novus in provincia Leningradensi. Novitates systematicae plantarum vascularium, 15: 240–247. (in Russian)
Ignatenko M, Gavrilov G, Karpov L (1980) Forest-parks of Leningrad. Stroiisdat, Leningrad (in Russian)
Ignatieva M (1982) Whortleberry (*Vaccinium vitis-idaea* L.) in parks. Leningrad Panorama Journal 9:36–38 (in Russian)
Ignatieva M (1994) Flora of green areas of St. Petersburg. Bulletin of Main Botanical Garden 169: 31–35 (in Russian)
Ignatieva M (2005) Case Study: Heritage Landscapes in St. Petersburg, Russia: Past and Present. Proceedings of NZILA "Looking forward to HERITAGE LANDSCAPES" Conference. New Zealand Institute of Landscape Architects Conference Dunedin, 28–30th April 2005: 337–347
Ignatieva M and Khodakov Yu (1991) Decorative plants of St. Petersburg: History of the Development and modern conditions. Unpublished (in Russian)
Ignatieva M, Konechnaya G (1996) Wild herbaceous plants of Komarov Botanical Institute Park. Botanical Journal Russia 81 (3): 96–105 (in Russian)
Ignatieva M, Konechnaya G (2004) Floristic Investigations of Historical Parks in St. Petersburg, Russia. Urban Habitats 2 (1): History, Ecology, and Restoration of a Degraded Urban Wetland
Ignatieva M, May R, Rolle N (2003) The Neva Project: River and the City. Peter the Great and the Neva River Delta. http://enspire.syr.edu/nevaproject/river&city/Peterdelta.html. Accessed 22 January 2009
Kurbatov Yu (2008) Petrograd Leningrad Sankt-Peterburg. Architectural and Urban Planning Lessons. St. Petersburg, Iskusstvo-SPB (in Russian)
Kuznetsov E, Ignatieva M (2003) St.Petersburg Forest Greenbelt Status Report completed for the Danish Forest and Landscape Research Institute
Lovell S (2003) Summerfolk: A History of the Dacha, 1710–2000. Cornell University Press, Ithaca/London
Malysheva N (2003) Lichens of St. Petersburg. St.Petersburg State University. Trudi Sankt-Peterbrgskogo Obschestva ispitatelei prirody 79(3) (in Russian)
Minyaev N, Orlova N, Schmidt V (1981) Key to higher vascular plants of north-west of RSPHSR (Leningrad, Pskov and Novorod regions). LGU, Leningrad (in Russian)
Nekrasova V (1959) Flora of St. Petersburg and its nearest suburbs in 18th century. Botanical Journal 44 (2) 249–261 (in Russian)
Nilsson K, Åkerlund U, Konijnendijk C, Alekseev A, Caspersen O, Guldag S, Kuznetsov E, Mezenko A, Selikhovkin A (2007) Implementing urban greening aid projects – the case of St. Petersburg, Russia. Urban Forestry and Urban Greening (6) (2): 93–101
Pokrovskaya T, Bychkova A (1967) Climate of St. Petersburg and its suburbs. Gidrometeoisdat, Leningrad (in Russian)

Popov V (1995) Analysis of adventive element of St. Petersburg sea port. Botanical Journal 80 (12): 104–106 (in Russian)
Noskov G (ed) (2004) Red Data Book of Nature of Saint-Petersburg. Professional, St. Petersburg (in Russian)
Tsvelev N (2000) Key to higher vascular plants in north-west part of Russia (Leningrad, Pskov and Novorod regions). SPKXPHA, St. Petersburg (in Russian)
Volkova E, Isachenko G, Khramtsov V (eds). (2007) Nature of Elagin Island, St. Petersburg. Elagin Island, St. Petersburg (in Russian)

Sofia

Dimitar Dimitrov, Maya Stoyneva, and Dobri Ivanov

Fig. 1 The St. Alexander Nevski Cathedral

Abstract The flora of Sofia, which is formed by Centro-European species, is highly influenced by the flora of the surrounding mountains and the Sofia Plain. The city contains 920 vascular species (21.26% of the Bulgarian flora). In terms of its history, the flora comprises 552 native species, 275 archaeophytes and 83 neophytes. The city's flora includes 12 statutorily protected species, 14 Bulgarian Red List species and 11 Bulgarian/Balkan endemics. *Fallopia japonica* is the most wide-spread non-native taxon, other non-native species include *Eleusine indica,* which is

Dimitar Dimitrov (✉)
National Natural History Museum, 1 Tsar Osvoboditel Blvd., Sofia, 1000, Bulgaria
e-mail: dimitroff@mail.bg

J.G. Kelcey and N. Müller (eds.), *Plants and Habitats of European Cities*,
DOI 10.1007/978-0-387-89684-7_13, © Springer Science+Business Media, LLC 2011

commonly found around the markets, and *Ambrosia artemisiifolia,* which is found along the railway tracks and on refuse dumps. The algal and lichen floras of Sofia have been poorly studied with little attention paid to them over the last few decades. This particularly applies to the highly urbanised city centre, the walls of building, monuments, tree bark and soils, which are now practically ignored in respect of algological and lichenological studies; new, in-depth studies are needed urgently. Virtually nothing is known about the fungi of the city, except those associated with lichens and those that cause 'diseases' in humans.

Natural Environment of the City

Sofia, the capital of Bulgaria, occupies 1,311 km². It is located at 42°48′ north: 23°20′ east on the western side of Bulgaria and in the southern part of the Sofia Plain and lies at or about 500 m a.s.l.; it is surrounded by mountains – Vitosha (2,290 m) overlooks the city with the lower slopes extending into the southern part. A schematic plan of the city is shown in Fig. 2.

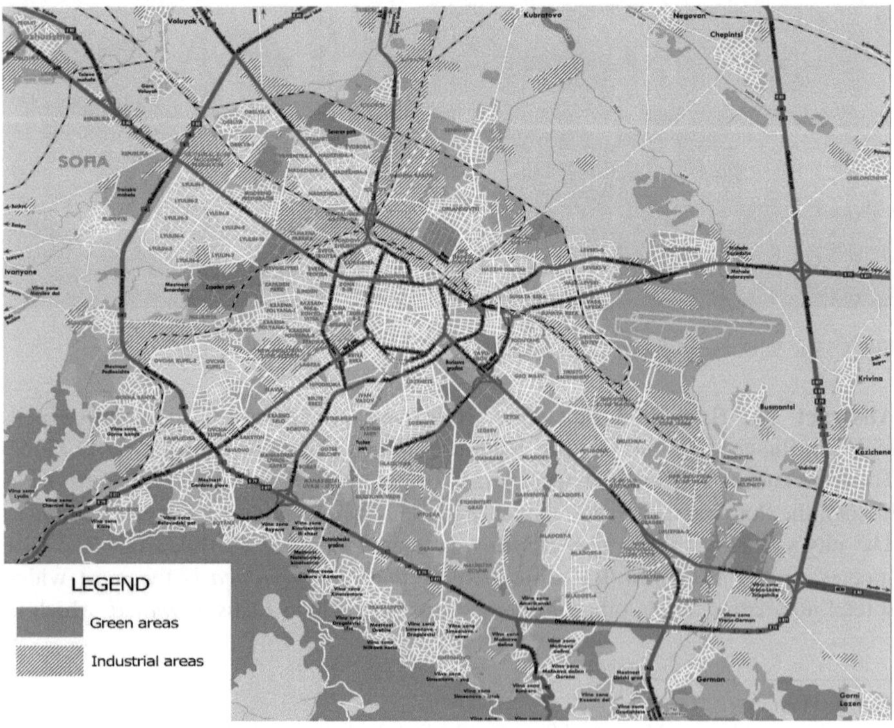

Fig. 2 Schematic plan of Sofia

Geologically, the Sofia Plain is a depression (graben) underlain by Pliocene lake sediments (Pontic gravels, sandy clays and sands). The city is crossed by a number of small tributaries of the Iskar River – Vladayska, Perlovska, Dragalevska, Slatinska, Blato and Kakach. The surface of the city undulates gently with some 300 m between the highest point (Knyazhevo, ca. 700 m) and the lowest point (Iliyantsi, ca. 400 m).

The metropolitan area is mostly covered by alluvial and deluvial meadow soils, and the surrounding area is covered by haplic-chernozems, partially influenced by the limestone of the Balkan Range to the north of the city. A small portion of the eastern part of Sofia (along the Iskar River) is covered by sandy soils and gleysols. Their soil-moisture is extremely high almost all year round, hence their heavy mechanical composition and low natural fertility. In order to be used for agricultural purposes, they have to undergo some chemical reclamation and groundwater re-adjustment.

The typical continental climate of the city is influenced by the climate of the Sofia Plain. The average annual temperature is 10.5°C. January is the coldest month with an average temperature of −1.5°C; July is the warmest month, with an annual average temperature of 20.5°C. The average annual temperature amplitude is 22.9°C, the average annual precipitation is 640 mm; the average annual humidity is 72% and the average number of days with snow cover is 55. The prevailing wind is from the west.

Historical Development

First Settlement to AD 1200

Sofia has a long history of nearly 7,000 years and is therefore one of the oldest European cities. Remnants from the Stone and the Bronze Ages and Neolithic artifacts can still be found in what is today the city centre. It is thought that the first settlers were attracted by the numerous mineral springs. The first settlement found in the city was a Neolithic village, dating to the 6th millennium BC.

Thracians were the first to establish the city, which was then known as Serdica. In the fourth century BC the city was conquered and ruled for a short time by Philip II of Macedonia and his son, Alexander the Great. About 500 BC the area was invaded by the Odrisses who established their own kingdom. In AD 29 under the Roman occupation, the city (then called Ulpia Serdica) became the administrative centre of the area. New structures such as towers, fortress walls, baths, administrative structures, a basilica and a large amphitheatre were built during that period. It is said that Constantine the Great used to call Serdica "my little Rome". A period of rapid growth occurred under the rule of Justinian; during this time enormous walls (parts of which still remain) were built around the city. A well-preserved Roman structure, which was turned into a Christian church, is the St. George Rotunda next to what is now the Sheraton Hotel.

In the fifth century AD the city was conquered by Attila; shortly after his death it became part of the Byzantine Empire. When Danubian Bulgaria was established in AD 681, many Bulgarian kings became interested in the city because of its strategic position. In AD 809 Khan Krum occupied the city and named it Sredets (the Slavic for "centre" or "hub"). In 1018, when Bulgaria was again part of the Byzantine Empire, the city was called Triaditsa (which means "in between the mountains").

Sofia has been besieged several times by the Magyars, Serbians and Crusaders. When the Bulgarian Kingdom was restored in AD 1194, the city was named after St. Sofia temple's, which still exists close to the St. Alexander Nevsky Cathedral. Arts and trade began to flourish and many new buildings and churches were built in the city and its suburbs. The most famous of those is the Boyana church, which has been designated by UNESCO as a World Heritage Site.

1200–AD 1900

In 1382, Sofia was conquered by the Ottomans, resulting in the Turkish authorities rapidly changing the city's architecture. Most Christian temples were abandoned or destroyed and Turkish administrative structures, mosques, baths and markets were built. The Turkish authorities were aware of the important geographical location of Sofia as a major crossroads in the Balkan Peninsula. In the seventeenth century, Sofia was the biggest market place in the region; it was so important that in the following century a stone road was built to Asia Minor. In 1879 Sofia, which then had about 12,000 inhabitants became the capital of Bulgaria. By 1900 the population of the city had grown to 67,789. The 400 ha central park, also known as the Boris' Park, was created by Daniel Neff and Joseph Fry between 1908 and 1934.

1900–AD 1945

During the period 1900–1945, the city's population increased dramatically, resulting in major changes to the architecture and landscape. The narrow "Turkish" streets were replaced with paved avenues; a new sewerage system and telegraph and telephone lines were provided by 1938. As a result the city began to look more like a European city of the 1920s. Under the rule of King Boris III Sofia became a modern city with a unique appearance. The city centre and the area between what is now the Lions Bridge and the Sheraton Hotel are still dotted with buildings from the first half of the twentieth century. Although in the 1930s and 1940s there were numerous strikes, political rumblings and demonstrations, Sofia was the scientific, art and cultural centre of Bulgaria. Between late 1943 and early 1944, the central part of the city was destroyed by American bombs. By 1945 the city had a population of 366,801.

1945–2000

After the communists took power in 1944, the city architecture, especially the central part, was greatly influenced by the urban "Stalinist style", for example, the former Party House, the former Balkan Hotel and the former Central Department Store in the downtown. The President's House is located in the Hotel Atrium; adjacent to the hotel is the Council of Ministers building while the former King's Palace is the National Art Gallery. Typically for all cities and towns under communist control, Sofia contains a large number of high-rise flats. The National Palace of Culture and its landscape setting were built in 1980. Sofia is now the industrial centre of Bulgaria producing about a sixth of the total national output and supporting about an eight of the total population. This is resulting in continual changes in the environment, including the construction of new buildings and the restoration of old ones. By 2000 the city occupied 1,310.8 m^{-2} and supported almost 2 million people. Since that time the population and the area of the city have continued to increase.

Changes of the Environment Due to City Growth

The climate of Sofia has been more affected by human impact than any other city in Bulgaria mainly as a consequence of its economic and population growth. In addition, the structure of the city has a strong influence on wind speed and direction. Over the last 30 years, the amount of sunshine has decreased. This tendency, which is more obvious in mid-summer, is caused by increased cloudiness and a high level of air pollution. From 1951 to 1995, the ambient air temperature increased in the winter and spring, whereas in the autumn there was a slight decrease, especially in November, resulting from human impact, activity and air circulation.

Fog and smoke are typical of the Sofia Plain; the average annual number of foggy days is about 30 but varying from a minimum of 5–10 to a maximum of 79 (recorded in 1914). Normally, fogs are characteristic of the low north-east industrial section of the city, mostly in the winter. The "heat island" that has formed in the city centre prevents foggy days to a certain extent, compared to the surroundings. The average annual number of snow days is about 50–60, which occur between the end of November and mid-March.

Natural landscapes have been changed by intensive human activity brought about by the significant natural resources and favourable working environment. Various types of land use have turned most of the natural landscapes into human landscapes. The main types being agricultural (which still occurs in the suburbs), urban, industrial and recreational areas.

The industrial growth and the existing infrastructure do not comply with the environmental and natural characteristics of the city area, which leads to environmental issues and a need for alternative solutions. The abiotic environmental issues in Sofia include:

1. Air and soil pollution associated with construction, industrial and solid waste.
2. Sodium and chlorine ions released by the use of salt on the roads and pavements in winter inevitably leads to the pollution of the soil and run-off.
3. Natural water basins (rivers and ponds) are drained.
4. High emissions of CO_2, NO_x and SO_2 due to the excess traffic and the industrial plants such as Kremikovtsi. The reduced particulate pollution in the ambient air in the areas of high traffic density creates favourable conditions for ozone formation and subsequent interaction with ambient SO_2 (generated both by Kremikovtsi and the city traffic). Consequently, increased levels of photochemical pollutants with already high SO_2 levels or increases in SO_2 concentrations in the presence of oxidants result in a potential health hazard.
5. Higher average annual temperature in the administrative area.
6. Aridification of the city climate due to the destruction of the vegetation.

Flora

Angiosperms, Gymnosperms and Pteridophytes

The vascular flora of Sofia comprises a total of 900 taxa from 103 families and 450 genera. The statutorily protected plants and those of nature conservation concern are listed in Table 1.

Two protected species (*Hottonia palustris* and *Acorus calamus*) have become extinct. *Hottonia palustris* became extinct in the 1930s when the Kazichene marsh was drained. *Acorus calamus* disappeared when the marshlands along the Iskar River were drained in the 1920s to prevent malaria. The ten Balkan and the one Bulgarian endemic species found within the city are listed in Table 2. The fifty most common plants species of Sofia are listed in Table 3.

Planted Trees and Shrubs

The total number of anthropophytes occurring in the city is 275, including 93 neophytes. The 158 tree and shrub species that are known to have been planted in Sofia are listed in Table 4.

Algae, Mosses and Liverworts

Twenty-six algal taxa have been found in the ponds of the Central (or Boris') Park, including species in the Chlorophyceae, Chrysophyceae (for example, *Mallomonas*

Table 1 Statutorily protected plants and those of nature conservation concern in Sofia

Species	Statutorily protected	Bulgarian Red List	CITES
Aesculus hippocastanum		Endangered [B1a; C2a (iii)]	
Castanea sativa	+	Endangered [B1ab (iii)]	
Cephalanthera damassonium			+
Cephalanthera longifolia			+
Cercis siliquastrum	+		
Epipactis palustris			+
Ficus carica	+	Slightly affected	
Galanthus elwesii	+	Endangered [B1ab (ii, iii, iv, v); C2a (i)]	
Ilex aquifolium	+	Endangered [A1c; B1ab (iii) + 2ab (ii)]	
Juniperus sabina		Endangered [D1+2]	
Laurocerasus officinalis	+		
Leucojum aestivum	+	Vulnerable [A2abcd; B1ab (ii, iv); D2]	
Merendera sobolifera	+	Vulnerable [Bib (iii) + 2b (iii)]	
Nymphaea alba	+	Endangered [B2ab (i, ii, iii, iv)] Annex 2, Annex 3	
Platanthera chlorantha			+
Pyracantha coccinea	+	Nearly EN	
Spiranthes spyralis		Vulnerable [B2ab (iv)]	+
Stachys milanii	+	Endangered [A3C; C2a]	
Taxus baccata	+	Endangered [B1ab (ii, iii, iv, v); C2, a(i)]	
Trifolium rubens		Critically Endangered [A3c; B1a 2(ii)]	+

The threat category and sub-category definitions are those of the IUCN (2001), which are given in Appendix VIII of this book *CITES* = Convention on International Trade in Endangered Species

Table 2 Balkan and Bulgarian endemics that occur in Sofia

Bulgarian
- *Carex otrubae*

Balkan
- *Aesculus hippocastanum*
- *Angelica pancicii*
- *Campanula sparsa*
- *Centaurea affinis*
- *Centaurea uniflora*
- *Cerastium petricola*
- *Cirsium candelabrum*
- *Peucedanum aegopodioides*
- *Scabiosa triniaefolia*
- *Trifolium trichopterum*

pyriformis and *Chromulina* sp.) and the Dinflagellate *Peridinium* sp. Conjugate algae and the euglenoid, *Phacus* sp, have been recorded in some of the ponds in the South Park. Studies of the phytoplankton of two ponds in the residential district of Boyana have shown that one (now dry) contained 56 taxa, while the other (which is in poor condition) contained 367 taxa.

Table 3 The 50 most common plant species in Sofia

Acer negundo	*Juncus compressus*
Acer platanoides	*Knautia arvensis*
Aesculus hippocastanum	*Kochia scoparia*
Amaranthus hybridus	*Lotus corniculatus*
Arenaria serpillifolia	*Malva sylvestris*
Arrhenatherum elatius	***Morus alba***
Atriplex hortensis	*Plantago lanceolata*
Ballota nigra	*Poa annua*
Berteroa incana	*Polygonum aviculare* agg.
Cephalaria transsylvanica	*Populus nigra*
Chelidonium majus	*Prunella vulgaris*
Chenopodium glaucum	*Quercus robur*
Cichorium inthybus	***Robinia pseudoacacia***
Convolvulus arvensis	*Sambucus ebulus*
Conyza canadensis	*Saponaria officinalis*
Cuscuta campestris	*Senecio vernalis*
Cynodon dactylon	*Sinapis arvensis*
Dactylis glomerata	***Sisymbrium altissimum***
Daucus carota	*Sonchus oleraceus*
Erodium cicutarium	*Stellaria media*
Fraxinus excelsior	*Syringa vulgaris*
Galinsoga ciliata	*Tilia cordata*
Geum urbanum	*Torilis arvensis*
Hordeum murinum	*Trifolium repens*
Humulus lupulus	*Vicia grandiflora*

Bold = neophytes

Table 4 List of the trees and shrubs that have been planted in Sofia

Abelia grandiflora	*Caryopteris incana*
Abies cephalonica	*Castanea sativa*
Abies concolor	*Catalpa bignonioides*
Acer ginnala	*Cedrus libani*
Acer negundo	*Celastrus orbiculatus*
Acer palmatum	*Celtis australis*
Acer saccharatum	*Celtis occidentalis*
Aesculus hippocastanum	*Cercis siliquastrum*
Ailanthus altissima	*Chamaecyparis lawsoniana,*
Albizia julibrissin	*Clematis integrifolia*
Amorpha fruticosa	*Corylus avellana*
Ampelopsis veitchii	*Cotoneaster horizontalis*
Berberis vulgaris	*Cotoneaster monopyrenus*
Betula pendula	*Cotoneaster salicifolius*
Buddleja davidii	*Crataegus mollis*
Buddleia variabilis	*Crataegus monogyna*
Buxus sempervirens	*Cupressus sempervirens*
Caragana frutex	*Cydonia oblonga*

(continued)

Table 4 (continued)

Deutzia crenata	*Malus niedzwetzkyana*
Elaeagnus angustifolius	*Malus sylvestris*
Elaeagnus multiflora	*Morus alba*
Erica carnea	*Parthenocissus quinquefolia*
Euonymus europaeus	*Paulownia tomentosa*
Euonymus japonicus	*Phellodendron amurense*
Euonymus radicans	*Philadelphus coronarius*
Exochorda aubertii	*Philadelphus grandiflorus*
Exochorda grandiflora	*Philadelphus* x *virginalis*
Fagus sylvatica	*Physocarpus amurensis*
Ficus carica	*Picea pungens*
Fontanesia phyllireoides	*Pinus excelsa*
Forsythia suspensa	*Pinus strobus*
Fraxinus excelsior	*Platanus* x *hybridus*
Gingko biloba	*Populus alba*
Gleditschia triacanthos	*Populus nigra*
Clarkia grandiflora	*Populus tremula*
Gymnocladus canadensis	*Potentilla fruticosa*
Haenomeles japonica	*Prunus avium*
Hibiscus syriacus	*Prunus mahaleb*
Hippophae rhamnoides	*Prunus padus*
Hydrangea arborescens	*Prunus persica*
Hydrangea breitscheinderi	*Prunus virginiana*
Hydrangea hortensis	*Pseudotsuga glauca*
Jasminum revolutum	*Ptelea trifoliata*
Juglans cinerea	*Pyracantha coccinea*
Juglans nigra	*Pyrus communis*
Juglans regia	*Quercus cerris*
Kerria japonica	*Quercus longipes*
Koelreuteria paniculata	*Quercus robur*
Kolkwitzia amabilis	*Quercus rubra*
Laburnum anagyroides	*Quercus schumardii*
Larix sibirica	*Quercus petraea*
Laurocerrasus officinalis	*Rhodotypus kerrioides*
Libocedrus deccurens	*Rhus coriaria*
Ligustrum vulgare	*Ribes multiflorum*
Liquidamber stiraciflua	*Robinia pseudoacacia*
Liriodendron tulipifera	*Rosa jundzilii*
Lonicera maackii	*Rosa multiflora*
Lonicera tatarica	*Rosa turcica*
Maclura aurantiaca	*Rosa caesia spp. glauca*
Magnolia cobus	*Rubus discolor*
Magnolia grandiflora	*Schizandra sinensis*
Magnolia liliiflora	*Securinega suffruticosa*
Mahonia japonica	*Sequoia sempervirens*
Malus floribunda	*Sophora japonica*

(continued)

Table 4 (continued)

Sorbaria sorbifolia	*Thuja gigantea*
Sorbus aucuparia	*Thuja occidentalis*
Sorbus torminalis	*Tilia argentea*
Spartium junceum	*Tilia cordata*
Spiraea japonica	*Tilia rubra*
Spiraea salicifolia	*Ulmus laevis*
Spiraea thunbergii	*Ulmus minor*
Spiraea vanchouttei	*Viburnum lantana*
Stephanandra tanakae	*Viburnum opulus*
Symphoricarpos albus	*Viburnum rhytidophyllum*
Symphoricarpus henaultii	*Vinca herbacea*
Symphoricarpus orbiculatus	*Vitex agnus-castus*
Syringa vulgaris	*Vitis vinifera*
Tamarix tetrandra	*Weigela japonica*
Taxodium distichum	*Wisteria sinensis*

The algal flora of the Kazichansko Blato pond (in the south-east part of the city) includes two new interesting parasitic euglenoid species, *Astasia bulgarica* and *A. sophiensis.* Studies of the algae of some temporary ponds and swampy areas (including those formed by river floods) and several residential districts have been carried out. In total, more than 100 taxa were found from the *Euglenophyta, Ochrophyta, Chlorophyta* and *Streptophyta,* including a new species *Oedogonium parvulum.*

Data on the phycoflora from other aquatic habitats in Sofia were obtained by a combination of field studies and algal cultures. The canalised watercourses within the city contain sewer grills that are mainly occupied by mud, algae and bryophytes (for example, Fontinalis antipyretica). A total of 64 taxa were recorded mainly distributed in the Cyanoprokaryota (=Cyanophyta), Rhodophyta, Ochrophyta, Chlorophyta and Streptophyta. The cultures made from the water supply pipes were found to contain a small number of diatoms (Bacillariophyceae).

The first studies of the algae of the hot (41–47.5°C), warm (37°C) and tepid (29.3–31°C) thermal springs were reported by Petkoff in 1922. He found a total of 109 taxa (including three taxa new to science) from the Cyanoprokaryota, Rhodophyta, Ochrophyta, Chlorophyta and Streptophyta. The new taxa were *Oedogonium cardiacum* f. *thermalis, Chara foetida* f. *thermalis* and *C. foetidaa* f. *variabilis.*

Chara spp were abundant in the lakes and aquaria in the botanical garden of the Sofia University. *Chara foetida* var. *subinermis* f. *normalis* colonised the pond in the central part of the garden, which also contained nine flagellate species, while the aquaria contained an abundance of *Chara coronata* var. *maxima.*

The earliest reports of the aero-terrestrial algae in the Botanical Garden show that these organisms occur on its walls of the garden, some wooden walls of different buildings, as well as of the moist roofs tiles and tree bark. A study of the soils of different plant pots in the Botanical Garden reported 46 Cyanoprokaryota species and one form from 14 genera.

The soil algae of the many residential districts, Boris' Park, alluvial meadows and in the vicinity of the Boyana ponds have been investigated as have the maroon forest soils

(=leached brown Mediterranean soils) and leached chernozems. About 90 taxa were recorded from the Cyanoprokaryota, Euglenophyta, Ochrophyta and Chlorophyta.

Recent data on the aerophytic algae of Sofia monuments are reported by Gärtner and Stoyneva (2003) and Uzunov et al. (2007, 2008). Gärtner and Stoyneva studied the two copies of the "Debelyanov's Mother" sculpture in the centre of Sofia and compared them with the original and a third copy in Koprivshtitsa (a town in southern Bulgaria). The marble copy is at the rear of the National Gallery in Sofia, where there are "peculiar" micro-climatic conditions. The sculpture contained layers of the green alga *Apatococcus lobatus*. The absence of aerophytes on the bronze copy in the city centre was attributed to the material, the generally drier climate and severe air pollution. Three green algae and two lichen species were found on the original granite sculpture (which is situated in a shaded location) in Koprivshtitsa. The three algae were *Apatococcus lobatus* (in a large quantity), *Trebouxia arboricola* and *Coccomyxa* sp. The lichens were *Candelariella vitellina* and *Lepraria* cf. *neglecta.* The second Koprivshtitsa sculpture, which is in a more isolated location than the original supported only *Apatococcus lobatus,* with lichens on the surrounding stonework.

Fungi (including Lichenised Fungi)

Lichenised Fungi

Information about the lichens of Sofia was first reported by Nikoloff in 1931. He discovered the almost total absence of lichens from the city's flora (especially in the central area), which he attributed to very poor air quality. The few species he recorded included, *Caloplaca saxicola* on the fences in the city centre; *Protoparmeliopsis muralis* on the concrete walls of the Perlovska river and 11 species (including *Xanthoria parietina*) on tree bark in the Boris' Park, subsequently forms of the genera *Parmelia* and *Physcia* and two more species were found on the trees. Years later and except for *Physcia adsendens* (on tree bark on the edge of the park) and several small "spots" of *Caloplaca* sp. (on the concrete walls of the building that is now a seminary), there was almost a total lack of lichens in the park. Except for *Rinodina bishoffii* (which is difficult to see), some of the calcareous lichens of the Alpine stone have disappeared. Likewise, the lichen *Caloplaca saxicola* disappeared from the fences of downtown Sofia.

Nikoloff also recorded the lichens along the roads to and in the area of three settlements situated in the foothills of the Vitosha mountain and which are now residential districts of Sofia. The species recorded include *Xanthoria candelaria* on stones and fences in the residential area of Bistritsa, *Candelariella aurella* on fences and *Caloplaca decipiens* on the moist tiles in Knyazhevo and two species (*Xanthoria parietina* and *Phaeophyscia orbicularis*) on old *Salix* trees along the roads.

The epilithic lichens from the calcareous stones in the University Botanical Garden in the city centre disappeared 2–3 years after the stones were brought from the region of Dragoman. The lichens on the stone monument in front of the Faculty of Biology of the Sofia University have also disappeared. Several small spots of badly developed thalli of *Physcia* sp. have been found on the bark of a tree in the South Park.

The first ecological study of the human impact on the distribution and development of lichens was carried by Philipova (1963). She studied a large region including the central parts and most densely developed part of the city, the southern and sparsely developed quarters and the fields in the foothills of the Vitosha mountain. The results revealed four different zones, clearly defined by the species composition and distribution of lichens:

> Zone 1 – located in the foothills of the Vitosha mountain, which was characterised by the highest lichen diversity (more than 15 epilithic and corticolous lichen taxa). A few lichens have been reported from the residential areas at the foothills of the Vitosha mountain (for example, *Xanthoria fallax* on tree barks in Dragalevtsi (Popnikolov and Zhelezova 1964).
> Zone 2 – between Zone 1 and the city area.
> Zone 3 – the suburbs, where there are very few houses or small houses with big courtyards.
> Zone 4 – the central area of Sofia, which was characterised by an extremely poor lichen flora (only three species were found – in poor condition). The only species that occurs "deeply" into the city is *Parmelia sulcata.*

The diversity of the lichen flora gradually decreased from Zone 3 to 2; the greatest decrease in the number of species and cover occurred in the city centre. Lichen diversity on tree bark decreased faster than that on stone walls and rocks. In addition, Philipova (1963) studied the change in the position of lichens on tree trunks and found that close to the city centre lichens developed mostly at the base of trees and were almost sterile.

In total, about 70 lichen and 600 algal taxa have been reported in various habitats and regions of Sofia during the last century. The distribution and abundance of lichens were generally in accordance with the well-known situation of city centres being "lichen deserts" with species-richness and abundance increasing towards the outskirts.

Most of the algae occur in various aquatic habitats (ponds, lakes, fens, river floods, aquaria, thermal springs, etc.). The species-richness of the algal flora is attributed to the different types of aquatic habitats, although some have disappeared due to the growth of the city.

Habitats

The following habitats (using the Classification of Palaearctic Habitats, Interpretation Manual of European Union Habitats, EUH 15/2, October 1999, European Commission DG Environment) (1999) occur in the administrative area of Sofia:

1. Ponto-Pannonic river bank dwarf sedge communities (22.35 ha) (EUH Code 3130, Corine Code 4.1.1).

Found in the north-east section of the city along the Iskar River, its western tributaries and the lakes formed in the nearby gravel pits.

2. Continental humid meadows (37.26 ha) (EUH Code 6420, Corine Code 38.2521).
   Occurs in the south and the south-east parts of the city.
3. Moeso-Thracian mesophile foothill meadows (38.25 ha) (EUH Code 6510, Corine Code 38.2523).
   This community occurs in the south-west, west and north sections of Sofia.
4. Riparian *Salix* formations (44.10 ha) (EUH Code 91EO, Corine Code 44.1).
   Found along river banks throughout the city.
5. Balkan Range *Alnus glutinosa* galleries (44.22 ha) (EUH Code 91EO, Corine Code 42.216).
   This community type is found in parts of the southern area of Sofia.
6. Ponto-Sarmatic mixed *Populus* riverside forests (44.66 ha) (EUH Code 92AO, Corine Code 44.16).
   Occurs along the Iskar River and its tributaries throughout the city.
7. Water-fringe vegetation (53.00 ha) (EUH Code 3150, Corine Code 53).
   Found around the artificial lakes in the city, mostly in the south part.
8. Lowland hay meadows (*Alopercurus pratensis, Sanguisorba officinalis*) (65.10 ha) (EUH Code 38.2, Corine Code 38.2).

   This community occupies areas in the east section of the city, mostly around the ten Thracian burial mounds. The vegetation around the mounds in Krivina is in the EUH code 6210 (Semi-natural dry grasslands and scrubland facies: on calcareous substrates (*Festuco-Brometalia*)) and the Corine code 34.31 comprising grass species such as *Chrysopogon gryllus, Stipa capillata, Anisantha tectorum, Dichanthium ischaemum, Anthoxanthum odoratum, Elytrigia repens, Festuca valesiaca, Dantonia alpina, Poa pratensis* and *Phleum phleoides*. The herbaceous species include *Trifolium alpestre, Lotus corniculatus, Potentilla argentea, Centaurea salonitana, Eryngium campestre, Verbascum phoeniceum, Asperula cynanchica, Artemesia arvensis, Sideritis montana, Scabiosa, triniaefolia, Hypericum perforatum, Achillea millefolium, Consolida regalis, Rhinanthus rumelicus, Filipendula vulgaris, Falcaria vulgaris* and *Chenopodium glaucum*.

   The vegetation around the mounds by the residential area of Levski-G includes species such as *Arrhenatherum elatius, Koeleria splendens, Agropyron intermedium, Stipa capillata, Festuca pratensis, Chrysopogon gryllus, Poa pratensis, Cynodon dactylon, Hordeum murinum* and *Festuca valesiaca* and species within the Fabaceae like *Astragalus onobrychis, Coronilla varia, Trifolium campestre, T. repens, T. pratense* and *T. incarnatum*.

City Centre

As with all towns and cities the centre is the most intensively developed part of the city with few areas of green space. Although the large Boris' Park extends from the city centre to the south-east, and the centre is fringed by some relatively large parks, most of the central green space comprises small squares and other areas of incidental green space, which mainly comprise mature trees (with some shrubs) in mown grass. The most common tree and shrub species in these small "parks" include *Fraxinus excelsior, Quercus rubra, Acer pseudoplatanus, Platanus* x *hispanica, Ailanthus altissima, Sophora japonica, Syringa vulgaris* and *Gingko biloba.* The herbaceous vegetation is dominated by grasses such as *Poa pratensis, Arrhenatherum elatius, Festuca rubra* ssp. *commutata, F. pratensis, Elymus hispidus* and *Dactylis glomerata* with herbs including *Geum urbanum, Fragaria vesca, Matricaria chamomilla, Hypericum perforatum* and *Plantago major.*

Residential Areas

High Density

High-rise apartment blocks occur around most of the periphery of the city. Typically the blocks are separated by small squares of mature trees in grass and some are "edged" by trees, shrubs and climbers. The more common tree species found in these areas include: *Populus nigra, P. tremula, Juglans regia, Quercus cerris, Cydonia oblonga, Betula pendula, Aesculus hippocastanum, Ligustrum vulgare, Morus alba, Prunus cerasifera, Spiraea thunbergii, Tilia argentea, T. cordata, Ulmus laevis* and *Celtis australis.* The predominant herbs and grasses to be found in these areas include *Poa pratensis, Poa annua, Arrhenatherum elatius, Festuca rubra* ssp. *commutata, Dactylis glomerata, Alopercurus myosuroides, Bromus arvensis* and *Hordeum murinum* with *Vicia grandiflora, Trifolium hybridum, Medicago lupulina, Senecio vernalis, Taraxacum officinale, Geum urbanum* and *Vicia lathyroides* occurring occasionally or rarely.

Low Density

The low-density areas fall into two general age classes, first houses built before ca. 1939 and the larger houses built after ca. 1990. Both age groups generally have relatively large gardens. The older age class contains species such as *Vitis vinifera, Prunus avium, Pyrus communis, Cydonia oblonga, Juglans regia* and *Prunus cerasifera.* The gardens of the newer houses mainly comprise lawns with a variety of non-native trees and shrubs, including *Juglans cinerea, Wisteria sinensis, Lonicera maackii, Symphoricarpos albus, Viburnum rhytidophyllum* and *Cornus alba.* Other plant species found in these areas include *Alnus glutinosa, Salix alba, Populus nigra, Clematis vitalba, Pinus sylvestris, Picea abies, Forsythia* x *intermedia, Salix babylonica* and *Pseudotsuga glauca.*

Industrial and Commercial Areas

The industrial and commercial areas also occur in three general age classes; pre 1939, "communist period" and post-1990. Much of the pre-1990 industrial and commercial areas are now derelict or in need of extensive renovation; as a consequence they are being colonised by ruderal/ephemeral taxa such as *Aesculus hippocastanum, Elaeagnus angustifolius, Thuja occidentalis, Robinia pseudoacacia* and *Gleditschia triacanthos*. The new commercial developments include many large international hotels surrounded by landscaped areas generally dominated by non-native species such as *Magnolia cobus, Cercis siliquastrum, Laburnum anagyroides, Kerria japonica, Malus floribundus, Koelreuteria paniculata* and *Sorbaria sorbifolia*.

Transport Routes

Road Verges

Some of the roads are bordered by verges or are in cuttings. The vegetation mosaic comprises individual or groups of trees in grassland, scrub and mown grassland. Some of the trees and shrubs have been planted while others (both alien and native species) have colonised naturally. The trees and shrubs that occur include *Populus nigra, Acer pseudoplatanus, Juglans regia, J. nigra, Ailanthus altissima, Elaeagnus angustifolius, Platanus* x *hispanica, Betula pendula, Tilia cordata* and *Aesculus hippocastanum***.** The grasses and herbs include *Bromopsis inermis, Dactylis glomerata, Hordeum murinum, Arrhenatherum elatius, Poa compressa, Festuca pratensis, Cynodon dactylon, Digitaria sanguinalis, Atriplex patula, Convolvulus arvensis, Sinapis arvensis, Cephalaria transsylvanica, Kochia scoparia, Knautia arvensis* and *Euphorbia cyparissias*.

Railway Land

The main railway network runs across the north-east side and close to the boundary of the city. Smaller networks occur elsewhere (for example, in the south-west). The vegetation is dominated by species of open conditions, especially ruderal and ephemeral species such as *Euphorbia taurinensis* and *Ambrosia artemisiifolia*. Railway land is the main habitat in Sofia for species such as *Senecio vernalis, Tribulus terrestris, Lithospermum arvense, Silene latifolia* and *Torilis arvensis*.

Airport

The vegetation associated with the airport, which is within the city limits, is similar to that of the road verges.

Recreation Areas

Parks

The city contains 21 parks with a total area of 1,372.35 ha (see Table 5). The vegetation in the city parks comprises mostly *Fraxinus excelsior, Sambucus nigra, Quercus robur, Ulmus laevis, Robinia pseudoacacia, Acer pseudoplatanus, Aesculus hippocastanum, Laburnum anagyroides, Picea pungens, Symphoricarpos albus, Gleditschia triacanthos, Biota orientalis* and *Quercus cerris*. Among the rarer species are *Sophora japonica, Betula pendula, Magnolia cobus, Picea abies, Taxodium distichum, Platanus* x *hispanica* and *Acer platanoides.*

Allotments and Cemeteries

Allotments and cemeteries cover a total of 120.4 ha and are located in the western and the northern parts of the city. The more common trees and shrubs include *Biota orientalis, Cupressus sempervirens, Populus nigra, Quercus frainetto, Ulmus laevis, Juglans regia, Acer pseudoplatanus, A. negundo, Gleditsia triacanthos, Salix alba, S. babylonica, Tilia argentea, Robinia pseudoacacia* and *Syringa vulgaris.*

Sports Centres

The larger sports centres of Sofia are listed in Table 6. The typical plant species found in the larger recreational areas of the city include *Populus alba, P. nigra,*

Table 5 The parks of Sofia

	ha		ha
Bakarena Fabrica Cemetery	17.5	Loven Park	243.04
Boris' Park	400	Malashevtsi Cemetery	49
BAS Garden	18.95	North Park	2.14
Boyana Residence	70.07	Oborishte (Zaimov) Park	11.76
Central Cemetery	53.9	Queen Joanna Hospital Garden	53.9
Doctor's Garden	4.54	South Park	142.1
Garden by Opalchenska Street	1.96	St. Troitsa Park	3.92
Garden by the St. Cyril and Methodius National Library	2.16	Student Park	14.7
		University Botanical Garden	0.72
Garden by the Nova Denitsa Store	24.5	West Park	167.09
Garden by the Universiada Hall	5.88		
Garden by the University of Sofia	5.63		
Geo Milev Park	27.44		
Green areas by the Lomsko Shosse Street	51.45		

Table 6 The larger sports centres of Sofia

	ha
CSCA sport complex	4.22
Garden by the Slavia Stadium	34.30
Levski Sport Centre	14.70
Rakovski Sport Centre	7.84
Septemvri Stadium	9.80

Carpinus betulus, Fagus sylvatica, Salix babylonica, Pyracantha coccinea, Taxus baccata, Pinus sylvestris, P. nigra, Acer negundo, A. saccharatum, Quercus petraea, Q. rubra, Q. cerris, Taxodium distichum, Chamaecyparis lawsoniana, Cedrus libani, Forsythia x *intermedia, Kerria japonica, Laburnum anagyroides, Magnolia cobus, M. liliiflora, Cotoneaster monopyrenus, Laurocerasus officinalis, Sorbus aucuparia, Spiraea salicifolia, Paulownia tomentosa* and *Catalpa bignonioides.*

Open Land

Unused Land

This category of open land is generally called "wasteland" but this is not so; much of it is previously developed land (often called "brownfield") that is awaiting re-development. The plant species that occur in these areas depends on the type of hard surface (for example, concrete), the presence or absence of buildings, the degree of contamination and the length of time the land has been unused.
The older land has been colonised by trees and shrubs (mainly non-native), which have become dominant in terms of cover and frequency. These species include *Ailanthus altissima, Robinia pseudoacacia, Amorpha fruticosa, Salix alba, Acer negundo, Prunus cerasifera, Clematis vitalba, Rosa canina, Ambrosia artemisiifolia, Datura stramonium* and *Amaranthus albus.* On the other hand, the newer sites are dominated by ruderal/ephemeral taxa such as *Conyza canadensis, Centaurea diffusa, Cichorium inthybus, Matricaria trichophylla, Amaranthus albus, Crepis setosa, Sisymbrium altissimum, Anthemis arvensis, Xanthium strumarium, Daucus carota, Melilotus officinalis, Arabidopsis thaliana, Lamium purpureum* and *Malva neglecta,* which although frequent provide little plant cover.

Open habitats supporting similar plant communities occur in association with other areas such as railway land. As described earlier, there is now very little agricultural land in the administrative area of the city.

Most of the former agricultural management has been abandoned and the land left fallow. This fallow land is now being developed for industrial, commercial and residential purposes. The following provides an indication of the process of change in the plant communities from agricultural use to the construction phase of development.

The crop plants grown in the agricultural areas include *Triticum aestivum, Zea mays, Beta vulgaris, Brassica oleracea, Daucus carota, Helianthus annuus, Allium porrum, Solanum tuberosum* varieties, *Lycopersicon esculentum, Capsicum annuum, Prunus avium, Pyrus communis* spp. *communis* var. *sativa, Cucurbita pepo* and *Cucumis sativus.* The former pastures, which were grazed by sheep, cattle and goats, supported plant species such as *Verbascum longifolium, Salvia virgata* and *Stachys germanica.* The meadows, which were cut for hay, contained species including *Poa pratensis, P. bulbosa, Phleum pratense, Alopercurus pratensis, Agrostis capillaris* and *Festuca valesiaca.*

When the arable land, pastures and meadows were abandoned, the "release" of management resulted in a change in the plant species to be found. Those species that occurred in the former arable land included *Cirsium arvense, Knautia arvensis, Amaranthus hybridus* and *Gypsophila muralis;* the species that colonised the pastures and meadows included *Setaria viridis, Abutilon theophrasti* and *Eryngium campestre.* Construction works result in the destruction of the existing vegetation and its replacement by species that include *Chenopodium glaucum, Convolvulus arvensis, Datura stramonium* and *Saponaria officinalis,* which are themselves "lost" as a result of the buildings and the associated landscaping works.

Walls and Pavements

The species that occur on walls are mostly *Cymbalaria muralis, Fallopia baldschuanica* and *Ampelopsis veitchii*. Among those species that occur in the cracks in pavements and concrete blocks are *Polygonum aviculare* agg., *Eragrostis pilosa, Setaria viridis* and *Puccinellia distans*.

Aquatic

Still Water (Lakes and Ponds)

In the artificial lakes and the abandoned gravel pits in the north-east area, there are water plants such as *Potamogeton natans, P. crispus, Elodea canadensis, Lemna minor, L. trisulca, Myriophyllum spicatum, Ceratophyllum demersum, Persicaria amphibia* and *Lycopus europaeus*. For most of the 1990s, many of the ponds in the parks were dry. In the few ponds left today *Nymphaea alba* and *Ceratophyllum demersum* can be found.

Moving Water (Rivers and Smaller Watercourses)

The common species that occur along the riversides' are *Veronica beccabunga, V. anagalis-aquatica, Sparganium erectum, Alisma plantago-aquatica, Typha latifolia,*

Scirpus lacustris, Phragmites australis and *Glyceria fluitans.* The floating-leafed species found include *Potamogeton natans, Myriophyllum spicatum, Ceratophyllum demersum* and *Spirodela polyrhiza.* Many of the watercourses in the city have been canalised and flow in relatively deep channels with steeply sloping or vertical concrete sides. The silt that has accumulated along both sides of the channel has been colonized by trees, shrubs, climbers and other plants, including *Salix alba, Alnus glutinosa, Populus nigra* and *Clematis vitalba.* The marginal vegetation comprises such species as *Phragmites australis, Typha latifolia, Scirpus sylvaticus, Lycopus europaeus, Juncus articulatus* and *J. compressus.* The floating-leafed and submerged species include *Ceratophyllum demersum, Myriophyllum spicatum, Lemna minor* and *Spirodela polyrhiza.*

Nature Conservation, Environmental Planning and Education

According to the Master Plan, which was proposed by the City Council in 2002, the green areas in the city should be increased. It is therefore hoped that new green areas will appear in the near future.

Pollution is one of the most serious environmental problems of the city. In order to improve the situation, it is necessary (*inter alia*) to:

1. Limit the pollution within the maximum permitted values and reduce the effects of inherited pollution.
2. Increase the use of renewable resources.
3. Reclaim the degraded landscapes in the urban and industrial zones.
4. Work out new regional development solutions that meet the objectives of sustainability.
5. Remove the sediment from the riverbeds frequently because it is still polluted with construction and solid waste.

Several environmental improvement projects (which are likely to benefit the flora and fauna of the city) are being considered or have been approved by the authorities. The projects include:

1. The possible demolition of the Kremikovtsi industrial plant and its replacement with new residential and green areas.
2. A new waste treatment plant should be built by 2011 and the efficiency of Kubratovo Wastewater Treatment Plant will be improved to contribute to the sustainable development and facilitate waste collection, which is currently one of the major issues in the city.
3. By the end of 2012 the construction of the metro that will connect the north-west and the south-east areas of the city will be completed. It is expected that the use of the metro will significantly reduce the amount of traffic that uses the main Ring Road and therefore improve air quality. In addition there will be more green areas in the city, for example, new underground parking areas are to be built and landscaped along the Tsar Boris III Boulevard.

4. Some of the lakes in the north-east section that have dried out will be restored.
5. The felling of some of the Iskar River woodlands should be stopped with new laws to be passed by the Ministry of Environment and Water.

Flood protection measures are desirable, for example, one of the heaviest floods was in 2005, when part of Novi Iskar remained under water for nearly 6 weeks.

The objective of the NATURA 2000 policies and projects in the city area is to preserve the natural habitats along the rivers, ponds and meadows as well as the *Leucojum aestivum* community in the wetlands around the residential area of Mladost-2, along the Vartopa River and the *Trifolium rubens* community in the residential area of Simeonovo. In the east section of Sofia, the 10 Thracian mounds in the residential areas of Levski-G and Krivina have been designated by the Government as "Monuments of Culture". As a result the natural vegetation around them must be protected – no digging, ploughing or cultivation is allowed.

The gravel pits that have been created adjacent to the west side of the Iskar River support an abundance of hydrophilic vegetation. The lakes are used for fishing, as nesting places by some water birds and as a recreation area by the people from the Levski district. Some ponds like those around Dolni Bogrov are protected areas. The lakes in the gravel pits near Chepintsi and Negovan in the north-east part of Sofia will also be designated protected areas in the near future.

The parks, the green areas and the gardens will be maintained, fallen trees and shrubs will be removed and new ones will be planted. A number of venerable trees in the city parks are protected by the Nature Protection Act. The oldest tree (more than 500 years old) in Sofia is an *Acer pseudoplatanus* in the residential area of Ovcha Kupel.

Closing Comments

In general, the flora of Sofia, which is formed by Centro-European geo-elements, that is by typical Centro-European species, is highly influenced by the flora of the surrounding mountains and the Sofia Plain. Excluding the lower plants, Sofia contains 21.26% of the Bulgarian flora, which can be divided as shown in Table 7.

Table 7 Number of vascular plants in Sofia compared with Bulgaria

Angiosperms	Sofia	Bulgaria
Tree	98	130
Shrubs	92	164
Herbs	64	254
Grasses	617	3,474
Sedges, rushes, etc.	20	132
Gymnosperms	25	27
Pteridophytes	4	51
Total	920	4,232

Table 8 Origin of the vascular plants of Sofia

Native	552
Archaeophytes	275
Neophytes	83
Total	920

In terms of its history, the flora can be divided into the categories shown in Table 8.

The city flora includes 12 statutorily protected and 14 Bulgarian Red List species. 11 species found in the city limits are Bulgarian/Balkan endemics.

Fallopia japonica is the most widespread non-native taxon; others non-native species include *Eleusine indica,* which is commonly found around the markets and *Ambrosia artemisiifolia,* which is found along the railway tracks and on refuse dumps.

Local trees and shrubs should be used when expanding the green areas of the city, for example, *Ulmus minor, Quercus cerris, Q. frainetto, Tilia argentea, Acer campestre, A. tataricum, Fraxinus ornus, Prunus mahaleb, Castanea sativa, Syringa vulgaris, Corylus avellana, Tamarix tetrandra, T. ramosissima, Sambucus racemosa* and *Euonymus europaeus*. Invasive alien species such as *Amorpha fruticosa, Ailanthus altissima, Quercus rubra* and *Fallopia japonica* should not be planted. Protected plants should be planted in the rock-gardens, so they can be preserved within the green areas.

It is evident that the algae and lichen floras of Sofia have been poorly studied with little attention being paid to them over the last few decades. This particularly applies to the highly urbanised city centre, the walls of building, monuments, tree bark and soils, which are now practically ignored in respect of algological and lichenological studies; new, in-depth studies of both will become urgently necessary in the not too distant future. Virtually nothing is known about the fungi of the city, except those associated with lichens and those that cause "diseases" in humans.

Literature Cited

European Commission (DG Environment) (1999) Interpretation Manual of European Union Habitats, EUR 15/2

Gärtner G, Stoyneva MP (2003) First study of Aerophytic Cryptogams on Monuments in Bulgaria. Ber. nat.-med. Verein Innsbruck 90:73–82

IUCN (The World Conservation Union) (2001) Red List Categories and Criteria. Version 3.1

Nikoloff A (1931) Die Flechtenflora von Witoschagebirge. Godishnik na Sofiyskiya Ouniversitet, FMF, Kniga 3 – Estestveni Naouki 27:29–74 (In Bulgarian, with German summary)

Philipova L (1963) L'influence des conditions écologiques de la ville de Sofia sur le dévelopment et la distribution des lichens. Godishnik na Sofiyskiya Ouniversitet, BGGF, Kniga 1 – Biologiya (Botanika) 56:83–96 (In Bulgarian, with Russian and French summaries)

Uzunov B, Stoyneva MP, Gärtner G (2007) Review of the studies on aero-terrestrial cyanoprokaryotes and algae in Bulgaria with a Checklist of the recorded species. I. Phytologia Balcanica 13 (1):65–73

Uzunov B, Stoyneva M P, Gärtner G (2008) Review of the studies on aero-terrestrial cyanoprokaryotes and algae in Bulgaria with a Checklist of the recorded species. II. Phytologia Balcanica 14 (1):11–18

Further Reading

Decheva R (1965) On the Study of the Soil Protozoa in Bulgaria. Izv. Zool. Inst 18: 199–223 (in Bulgarian,Russian and English summary)

Dimitrov D (in press) Conservation important vascular plants and endemics from the flora of Sofia City. Symposium on the flora of Southeastern Serbia and neighboring regions. Nis, 2005

Draganov S, Dimova E (1971) Investigation on composition and quantity of algae in soil samples, kept long time in air-dry conditions. Ann. Univ. Sof., BF 63:55–64 (Russian summary)

Draganov S, Genova E (1966) Untersuchungen über die Algenflora der Böden Bulgariens. III. Vertikale Verbreitung der Blaualgen an den Nord abhängen des Vitoscha – Gebirges. Ann. Univ. Sof., BF 59:153–163 (in Bulgarian, with Russian and German summary)

Draganov S, Witschewa A (1966) Boden Blaualgen aus dem Botanischen Garden der Sofioter Universität. Ann. Univ. Sof., BF 59:27–34 (in Bulgarian, with Russian and German summary)

Georgieff S (1906) Contribution a l'etude des Diatomees, des Champignons, des Filicinees et des Phanerogames de Bulgarie. Ann. Univ. Sof., BF 2:83–124 (in Bulgarian)

Ivanoff H (1912) Essai d'etude des Flagellates de Bulgarie. Troudove na Bulgarskoto Prirodoizpitatelno Drouzhestvo 5:114–141 (in Bulgarian, with French summary)

Kristiansen J, Stoyneva M (1998) Silica-scaled chrysophytes in Bulgaria. Cryptogamie, Algol 19:19–28

Kristiansen J, Stoyneva MP & Valchanova M (2002) Review of Bulgarian chrysophyte studies with appendix of recorded species. Ann. Univ. Sof., BF 90:43–57

Kuzmanov B, Kozuharov S (1971) Aliens in the Bulgarian flora. Boissiera 19:319–327

Michajlow W (1965) *Astasia bulgarica* sp. n. and *Astasia sophiensis* sp. n., Euglenoidina parasitica found in Bulgaria. Acta Parasitol. Pol. 13:7–17

Michev TM, Stoyneva MP (eds) (2007) Inventory of Bulgarian Wetlands and their Biodiversity. Part 1: Non-Lotic Wetlands. Publ. House Elsi-M, Sofia

Petkoff S (1898–1899) Contribution to the investigation of Bulgarian single-celled green freshwater algae. Period. Spis. Bulg. Knizh. Druzh 57:111–135 (in Bulgarian)

Petkoff S (1900) Deusième contribution à l'étude des algues d'eau douce en Bulgarie. Troud. Bulg. Prirodoizp. D-vo 1:1–21 (in Bulgarian, with French summary)

Petkoff S (1904) Troisième contribution à l'étude des algues d'eau douce de Bulgarie. Period. Spis. Bulg. Knizh. D-vo 16 (1–2):385–416 (in Bulgarian, with French summary)

Petkoff S (1907) Quatrième contribution à l'étude des algues d'eau douce de Bulgarie. Sbornik Narodni Oumotvoreniya 22–23(3):1–23 (in Bulgarian, with French summary)

Petkoff S (1908) Cinquème contribution à l'étude des algues d'eau douce de Bulgarie. Period. Spis. Bulg. Knizh. D-vo 68:603–624 (in Bulgarian, with French summary)

Petkoff S (1908–1909) Les algues de la Bulgarie du SO et leur dispersion. Ann. Univ. Sof., FMF 2:1–88 (in Bulgarian, with French summary)

Petkoff S (1911) Contribution supplémentaire à l' étude des algues du sommet Kom et ses environs. Ann. Univ. Sof., FMF 6:1–15 (in Bulgarian, with French summary)

Petkoff S (1913) Les Characees de Bulgarie. Spisanie BAN 7:1–44 (in Bulgarian, with French summary)

Petkoff S (1914) Les Characees de Bulgarie. Nuova Notar. 25:35–56

Petkoff S (1922) La vegetation des eaux de Vitocha. Contribution a l' hydrologie et l' hygiene de la capital. Ann. Univ. Sof., FMF 18: 1–272 (in Bulgarian with French summary)

Petkoff S (1934a) Contribution Supplementaire aux Characees de Bulgarie. Spis. BAN 51:1–67 (in Bulgarian, French summary)

Petkoff S (1934b) Deux genres nouveaux pour la flore algologique de Bulgarie. Izv. Bulg. Bot. D-vo 6:117–120

Petkoff S (1934–1935) Les Zygnemales de la Bulgarie et leur Dispersion. Ann. Univ. Sof., FMF 31:1–13

Petkoff S (1942) Note sur une forme partikuliere d' *Ulothrix zonata* Kütz. var. *valida* (Näg.) Rabenh. ne formant que des hypnospores. Spis. BAN 65:131–137 (in Bulgarian, with French summary)

Popnikolov A, Zhelezova B (1964) Flora of Bulgaria – Lichens. Narodna Prosveta, Sofia (in Bulgarian)

Stefanov B, Kitanov B (1962) Cultigene plants and cultigene vegetation in Bulgaria, BAS, Sofia, 274 pp (in Bulgarian)

Stoyneva MP (2003a) Steady-State Phytoplankton Assemblages in Shallow Bulgarian Wetlands. Hydrobiologia 502:169–176

Stoyneva M (2003b) Survey on green algae of Bulgarian thermal springs. Biologia (Bratislava) 58(4):563–574

Stoyneva MP, Gärtner G (2004) Taxonomic and ecological notes to the list of green algal species from Bulgarian thermomineral waters. Ber. nat.-med. Verein Innsbruck 91:67–89

Stoyneva M, Gärtner G (2009) Remarkable and newly recorded aeroterrestric cyanoprokaryotes and algae in Bulgaria. In: Ivanova D (ed) Proceedings of IVth Balkan Botanical Congress, Sofia

Stoyneva M, Temniskova-Topalova D, Uzunova K & Nedelkova M (2001) Long-term compositional response of the algal flora of the Boyana Swamp (Western Bulgaria) to the environmental changes. Ann. Univ. Sof., BF 93:43–71

Temniskova-Topalova D (1965) Über die Flora von Euglenophyta und Volvophyceae in Bulgarien. II. Ann. Univ. Sof, BF 58:49–58 (in Bulgarian, with Russian and German summary)

Temniskova-Topalova D (1966) Über die Flora von Euglenophyta und Volvophyceae in Bulgarien. III. Ann. Univ. Sof., BF 59: 43–57 (in Bulgarian, with Russian and German summary)

Valkanov A (1926) Beitrag zur Kenntniss der Flagellaten Bulgariens. Izv.Bulg. Bot. D-vo 1:105–120 (in Bulgarian, with German summary)

Valkanov A (1928) Protistenstudien ii. Notizen über die Flagellaten Bulgariens. Arch. Protistenk 63:419–450

Valkanov A (1968) Notizen über *Volvulina steinii* Playfair 1915, Gefunden in Bulgarien. Izv. Zool. Inst. 27:5–11 (in Bulgarian, with Russian and German summary)

Vatev S (1904) Natural Curative Resources of Bulgaria. Vidin (in Bulgarian)

Vodeničarov D (1958) Beitrag zur Algenflora Bulgariens. Izv. Bot. Inst 6:431–438 (in Bulgarian, with Russian and German summary)

Vodeničarov D (1962) Beitrag zur Algenflora Bulgariens. IV. Izv. Bot. Inst 10:145–159 (in Bulgarian, with Russian and German summary)

Vodenicharov D, Draganov S & Temniskova D (1971) Flora of Bulgaria. Vol. 1, Algae. Narodna Prosveta, Sofia (in Bulgarian)

Wodeničaroff D (1959–1960) Beitrag zur Erforschung der Wasser-und Algenflora und Vegetation auf dem Ljulingebirge. Ann. Univ. Sof., BGGF 51:61–82 (in Bulgarian, with Russian and German summary)

Wodenitscharov D (1960) Beitrag zur Algenflora Bulgariens. iii. Ann. Univ. Sof., BGGF 52:137–151 (in Bulgarian, with Russian and German summary)

Wodenitscharov D (1961) *Oedogonium parvulum* sp. nov. Feddes Repertorium (Berlin) 64:82–85

Yankova R (1992) Aeropalinological investigation in Sofia with reference to the hay fever. Ann. Univ. Sof. FB, 82(2): 89–98 (in Bulgarian)

Vienna

Alexander Mrkvicka

Fig. 1 Stephansdom – the town's landmark

Abstract The flora of Vienna is very rich in species both nationally and in the European context. About two-thirds of the species occur in natural and semi-natural areas, including forests, formerly extensively used steppe pastures as well as arable land, meadows and wetlands. Many of these biotopes are protected by different regulations and are managed and used in sustainable ways. The situation of man-made habitats is quite different. Large areas have been altered or lost, especially

Alexander Mrkvicka (✉)
Gruppe Stadtwald, Erholungsgebiete und Nationalpark, MA 49 – Municipial Forestry Office
Vienna, A – 1082 Wien, Volksgartenstraße 3, Austria
e-mail: alexander.mrkvicka@wien.gv.at

J.G. Kelcey and N. Müller (eds.), *Plants and Habitats of European Cities*,
DOI 10.1007/978-0-387-89684-7_14,

in the economic boom of the 1990s. There is also a trend to tidy up places with ruderal vegetation. Because these habitats play an important role in contributing at least a third to the Viennese flora, it will be necessary to focus intensive efforts on them in order to preserve their vegetation and component flora and therefore the biodiversity of the city.

Natural Environment

Vienna is situated at the geographical co-ordinates 16° 22′ 23″ east and 48° 12′ 31″ north (WGS 84) in a depression between the eastern Alps and the Carpathian Mountains, the highest point of the city is 542 m a.s.l., the lowest is 152 m a.s.l. and the human population is 1.8 million, see Fig. 2. The area of 415 km^2 within the borders of Vienna is roughly divided by the Danube River into three quite different parts. The western and south-western parts are characterised by the hills of the Wienerwald, in the south the landscape is dominated by river terraces that are relicts of the Ice Ages, while in the north-east and east the landscape comprises the relatively flat land of the floodplains of the Danube and its tributaries.

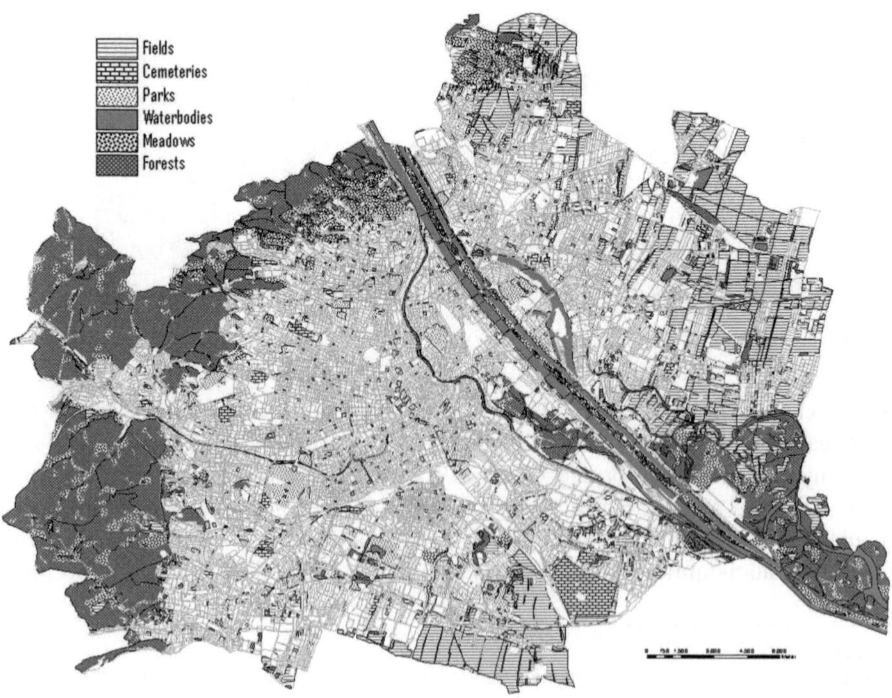

Fig. 2 Vegetation/land use in Vienna today

Soils

The north-eastern and eastern parts of the city are dominated by chernozems ("*Schwarzerde*"), in the Wienerwald hills cambisols ("*Braunerde*") are widespread on flysch. Rendzina soils, which are typically found on dolomite and limestone, are limited to a small area in the south-western part of Vienna. In addition to these natural soils, many different "artificial" soils of various provenances can be found. Railway areas are frequently dominated by serpentinite which comes from the Styrian Alps. Wastelands are often covered by bricks and calcareous grit – the remains of concrete. Up to the 1970s, the soil for parks and gardens was mostly imported from local sources but nowadays it is imported from up to 50 km and is the source of newcomers to the flora.

Climate

Owing to its geographical situation, the influences of oceanic (from western Europe) and continental climate (from eastern Europe) are responsible for regional climatic differences within the city. For example, the mean minimum temperature in the centre and the south is –12°C, in the east it is –15°C and in the west it is –18°C. The mean minimum temperature is –0.7°C and occurs in January; the mean maximum temperature is 19.9°C and occurs in July. The mean annual precipitation is 800 mm in the western parts but only 600 mm in the eastern parts.

The air quality in Vienna depends mainly on the wind direction, for two-thirds of the year the westerly winds supply the city with fresh, clean air from the Wienerwald. Temperature inversions during the winter result in a reduction in air quality but as a consequence of considerable effort and expense of local air pollution, control measures and actions by the federal government the air quality have much improved since the 1970s.

Since the early 1990s, some species of southern or south-eastern Europe such as *Senecio inaequidens* and *Geranium purpureum* have started to colonise central Europe. During the last 20 years, it has been possible to successfully cultivate Mediterranean species, for example, *Arbutus unedo* and *Quercus ilex* outside in parks and gardens. However, much more significant than plants is the increasing spread of animal species, especially insects, snails and spiders, of southern Europe.

Historical Development

Vienna is situated at a junction of prehistoric trade routes along the Danube from the Black Sea to western Europe, the River March to the Baltic Sea ("Bernsteinstraße") and an ancient Roman road to the Adriatic Sea. Human influence on the flora and fauna started in the Palaeolithic. By the time of the Neolithic (about 7,600 years ago), the first agricultural settlements in the region were founded. From this time on, the flora and fauna were modified by human influence, at least in parts of the area. After the Roman invasion (10 BC), a military camp named Vindobona and some Roman villae were built, each of which supported about 100 inhabitants together with many agricultural plants. Between the end of the Roman Empire and 1150, Vienna was only a small village. From 1196 onwards, fortifications were built. The former suburbs, most of which were destroyed during the Osmanic invasion of 1683 formed a circle around the outside of the fortifications.

After 1700, Vienna expanded rapidly; between 1860 and 1918, the human population reached 2.5 million. Between 1918 and 1923, huge areas of arable land, meadows and even forests were divided into allotment gardens ("Schrebergärten") to provide land for new buildings and to improve the food supply for the inhabitants. Since 1945, the population of Vienna has been more or less constant at or about 1.7 million people; however, the area of developed land has doubled in the intervening 65 years. Today about 50% of Vienna is occupied by "green areas", for example, arable land (5,646 ha), vineyards (690 ha), market gardens (569 ha), meadows (2,512 ha), forests (6,917 ha), parks (892 ha) and allotment gardens (1,230 ha).

Flora

The flora of the city (and State) of Vienna comprises 2,399 wild plant species and subspecies – it is therefore considered to be very rich in species both nationally and in the European context. The species-richness is mainly due to the location of the city within the boundaries of four large floristic regions, central European, Alpic, Pontic-Pannonian and sub-Mediterranean, which comprise substantial proportions of natural and semi-natural areas, including natural forests, formerly extensively used steppe pastures as well as arable land, meadows and water. In addition, man-made habitats that are typical of large cities, for example, waste places, housing areas, transport routes (railways, docks, streets and canals) and fallow land (including industrial and commercial areas) also play an important role in determining the flora.

Out of a total of 2,399 species, 1,464 (61.0%) are indigenous flora while 140 (5.9%) are neophytes and 795 (33.1%) are immigrants introduced with purpose or unintentionally but have not (yet) naturalised in Vienna. The 50 most common taxa found in Vienna are listed in Table 1.

Planted Trees and Shrubs

The first scientific record of an introduced tree species goes back to the time of Clusius who introduced *Aesculus hippocastanum* into Vienna at about 1576. The Habsburg Emperors were very interested in flora and fauna, especially in the eighteenth and nineteenth centuries, and imported a large variety of tree and shrub species from four

Table 1 Most common taxa found in Vienna

Achillea collina	*Lapsana communis*
Ailanthus altissima*	*Lathyrus pratensis*
*Alliaria petiolata**	*Leontodon hispidus**
Amaranthus powellii	*Ligustrum vulgare**
*Anthriscus sylvestris**	*Lolium perenne**
*Arctium lappa**	***Lycium barbarum***
*Arenaria serpyllifolia**	*Malva neglecta**
*Arrhenatherum elatius**	*Medicago lupulina**
*Artemisia vulgaris**	*Melilotus officinalis**
*Ballota nigra**	*Mercurialis annua*
*Bellis perennis**	*Mycelis muralis*
Berteroa incana	*Parietaria officinalis*
*Anisantha sterilis**	*Plantago lanceolata**
Calamagrostis epigejos	*Plantago major**
*Capsella bursa-pastoris**	*Plantago media*
Lepidium draba	*Poa angustifolia**
Carduus acanthoides	*Poa annua**
*Carex pilosa**	*Polygonum arenastrum**
Cerastium fontanum ssp. *holosteoides*	*Ranunculus ficaria*
Chenopodium album	*Ranunculus repens*
*Cirsium arvense**	*Rosa canina**
*Convolvulus arvensis**	*Rubus caesius**
Conyza canadensis*	*Rumex crispus**
*Cornus sanguinea**	*Rumex sanguineus*
*Corydalis cava**	*Sambucus nigra*
*Dactylis glomerata**	*Saponaria officinalis*
Eragrostis minor	*Scabiosa ochroleuca**
Euphorbia peplus	*Senecio vulgaris*
Fallopia baldschuanica*	*Sisymbrium loeselii**
*Festuca pratensis**	***Solidago gigantea***
*Fumaria officinalis**	*Sonchus oleraceus*
Galinsoga ciliata*	*Stellaria media (sensu stricto)**
*Galium aparine**	*Taraxacum officinale* agg.*
Galium sylvaticum	*Trifolium pratense*
*Geranium pyrenaicum**	*Trifolium repens**
*Geranium robertianum**	*Tripleurospermum inodorum**
Glechoma hederacea	*Tussilago farfara*
Hieracium murorum	*Urtica dioica**
Impatiens parviflora*	*Veronica persica**
*Lactuca serriola**	*Veronica hederifolia* ssp. *lucorum**
Lamium maculatum	*Viola arvensis*

*Fifty most common species
Bold = neophytes

continents for planting in the parks and gardens of the city. Owing to limiting frosts and dry summers, only some of them survived and were cultivated for further use. Today, about 50 species of trees and shrubs are frequently planted in alleys, parks and gardens; see Table 2; the less common taxa are listed in Table 3.

Table 2 Fifty species of trees and shrubs that are most frequently planted in alleys, parks and gardens

Acer monspessulanum	*Parthenocissus quinquefolia*
Aesculus hippocastanum	*Parthenocissus tricuspidata*
Amelanchier spicata	*Philadelphus pubescens*
Berberis thunbergii	*Platanus* x *hispanica*
Buxus sempervirens	*Populus nigra cv.* 'Italica'
Catalpa bignonioides	*Potentilla fruticosa*
Catalpa ovata	*Prunus laurocerasus*
Celtis australis	*Pyracantha coccinea*
Celtis occidentalis	*Ribes odoratum*
Chamaecyparis pisifera	*Ribes sanguineum*
Cornus sanguinea	*Rosa spp.*
Corylus colurna	*Salix alba* x *babylonica,* *Salix fragilis* x *babylonica*
Corylus maxima	*Sophora japonica*
Cotinus coggygria	*Sorbus intermedia*
Deutzia crenata	*Spiraea chamaedryfolia*
Forsythia x *intermedia*	*Spiraea* x *arguta*
Fraxinus ornus	*Spiraea* x *cinerea*
Ginkgo biloba	*Spiraea* x *vanhouttei*
Gleditsia triacanthos	*Symphoricarpos orbicularis*
Hibiscus syriacus	*Thuja occidentalis*
Viburnum opulus	*Tilia tomentosa*
Kolkwitzia amabilis	*Tilia* x *euchlora*
Laburnum x *watereri*	*Viburnum farreri*
Ligustrum ovalifolium	*Viburnum rhytidophyllum*
Malus baccata	*Weigelia florida*

Table 3 The less common taxa that have been planted along roads and in parks and gardens

Acer ginnala	*Jasminum nudiflorum*
Aesculus x *carnea*	*Juglans nigra*
Campsis radicans	*Ligustrum obtusifolium*
Caragana arborescens	*Liquidambar styraciflua*
Caragana frutex	*Liriodendron tulipifera*
Cercis siliquastrum	*Lonicera nitida*
Chaenomeles speciosa	*Miscanthus* x *giganteus*
Cotoneaster bullatus	*Populus deltoides*
Cotoneaster franchetti	*Populus simonii*
Cotoneaster lucidus	*Populus* x *canadensis*
Cotoneaster multiflorus	*Prunus serotina*
Cotoneaster niger	*Rubus occidentalis*
Cytisus scoparius	*Rubus phoeniculasius*
Forsythia suspensa	*Sorbaria sorbifolia*
Forsythia viridissima	*Spiraea japonica*
Fraxinus americana	*Spiraea salicifolia*
Fraxinus excelsior 'Diversifolia'	*Ulmus pumila*
Fraxinus excelsior 'Pendula'	*Viburnum* x *burkwoodii*
Iberis sempervirens	*Wisteria sinensis*

Algae, Mosses and Liverworts

The algal flora comprises a high diversity of species because of the many different habitats. Recent research has been carried out in the floodplain of the National Park "Donauauen", while some species that occur in the bathing ponds are monitored in accordance with the EC Bathing Water Directive (76/160/EEC).

Historic data for the bryophyte flora of Vienna contain records of 56 species of liverworts (Hepaticae), two species of hornworts (Anthocerotae) and 250 species of mosses (Musci) in Vienna, Zechmeister et al. (1998a, b, 2001). Most of the species occur in natural and semi-natural habitats such as the Wienerwald and Donauauen. In the urban areas, mosses are decreasing as the result of the increasing use of concrete, steel and glass instead of "moss-friendly" materials like brick and stones.

Fungi (including Lichenised Fungi)

The lichen flora of the city has changed considerably since 1853. Terricolous lichens have disappeared almost completely, whereas the diversity and abundance of saxicolous and epiphytic species has developed because of climatic conditions, air pollution and the presence of suitable substrata. Ninety-seven taxa have been recorded in the city (Türk 2006). Between 1981 and 1987, 1,241 species of fungi were found in Vienna by Krisai-Greilhuber (1992); of these 11 were taxa new to science, 4 were recorded in Europe for the first time and 70 were new records for Austria and over 300 species were new records for Vienna. Mycorrhiza form 25.6% of the 1,241 species, 34.4% occur on wood and 40% in leaf litter.

Habitats

Querco-Carpinetum (EUH code 9170; Corine code 3.1.1.) and *Asperulo-Fagetum* (EUH code 9130 and Corine code 3.1.1.) forests are characteristic of the Wienerwald hills in the western parts of Vienna.

Mixed thermophilous stands of *Quercus petraea* and *Carpinus betulus* cover about 1,600 ha of the city. Tree and shrub species such as *Prunus avium, Sorbus torminalis, S. domestica, Acer campestre, Corylus avellana, Cornus mas, Crataegus laevigata, Ligustrum vulgare* and *Lonicera xylosteum* are frequently found in this forest type. The herb layer is characterised by *Galium sylvaticum, Convallaria majalis, Stellaria holostea* and *Hepatica nobilis*. As a consequence of light conditions, *Quercus-Carpinus* forests have a two- or three-layered vertical structure and provide an important habitat for numerous animal species. In the forest reserve "Johannser Kogel", *Quercus* stands that are 400–500 years old have been allowed to develop without any direct human influence since 1972, see Fig. 3.

Forests dominated by *Quercus cerris* occurred on the slopes of the Viennese forest in ancient time, most being replaced by vineyards in Medieval times.

Fig. 3 Four hundred-year-old oak in the forest reserve "Johannser Kogel"

Thermophilous species of calcareous soils including *Cornus mas, Lithospermum purpureocaeruleum, Laser trilobum* and species of slightly acidic soils, for example, *Cytisus nigricans* and *Lathyrus niger* can be found together. In former times, *Quercus cerris* was also planted in the beech forests for the purpose of feeding *Sus scrofa* with the large acorns.

Acidophilous forests of *Quercus petraea* and *Quercus cerris* occur on acid sandstones, mainly on slopes exposed to the west or south-west. Other tree species frequently found in these forests include *Sorbus aria, Pinus sylvestris* and *P. nigra.* The shrub and herb layers are poorly developed; the characteristic low shrub and herb species are *Calluna vulgaris, Vaccinium myrtillus, Luzula pilosa* and *L. albida, Hieracium* spp. and mosses such as *Lecobryum glaucum* and *Polytrichum* spp.

Beech forests are widespread on cambisols in the Wienerwald region, which is the largest area with sub-montane *Fagus* forests in middle Europe. Within the borders of Vienna, *Fagus* forests (*Asperulo-Fagetum*) cover about 1,500 ha. They prefer more humid conditions and occur above 400 m on sunny slopes and down to 200 m on shaded slopes and in valleys. The herb layer is often sparse and characterised by *Galium odoratum, Mercuralis perennis, Lathyrus vernus, Hordelymus europaeus, Viola reichenbachiana, Neottia nidus-avis* and *Sanicula europaea.* On humid soils with a large component of clay, *Allium ursinum* can also be found.

Fig. 4 *Quercus pubescens* on the southern slopes of Mount "Léopoldsberg"

The *Luzulo-Fagetum forests* (EUH code 9110; Corine code 3.1.1.) mainly occur on ridges and summits with acid soils. The herb layer resembles those of the acidophilous *Quercus* forest. The *Cephalanthero-Fagetum* forests are restricted to a small area in the south-western part of Vienna. The herb layer is characterised by *Carex digitata, C. alba, Sesleria albicans, Cephalanthera damasonium* and *C. longifolia.*

Forests of *Quercus pubescens* (EUH code 91H0; Corine code 3.1.1.) (Fig. 4) are limited to a few locations with dry soils on limestone. Other tree species found in this forest type include *Sorbus danubialis, S. graeca, Acer campestre, Pyrus pyraster. P. nivalis, Viburnum lantana* and *Euonymus verrucosa*. In clearings within the forests and at the ecotone with adjacent dry grasslands, thermophilous forest fringes occur containing the characteristic species *Dictamnus albus, Geranium sanguineum, Vincetoxicum hirundinaria, Limodorum abortivum, Lithospermum purpureocaeruleum, Clematic recta, Veronica austriaca* and *Rosa pimpinellifolia.*

In Vienna, Black Pine Forests (*Pinetum nigrae*) are limited to two ridges in the south-west, where the limestone Alps extend into the city area. *Pinus nigra* (Fig. 5) is frequently accompanied by *Sorbus aria, Amelanchier ovalis* and *Cotoneaster* species. The herb layer is dominated by *Sesleria albicans.*

About 3,000 ha of floodplain forests occur on the banks of the Danube, mainly in the north-eastern part of the city. The forests are protected as the "Donauauen National Park". Before the 1870s, the riverbed of the Danube, which in Vienna has the hydrological characteristic of an alpine river, was braided into dozens of small channels separated by sand and gravel banks and gallery forests (forests along the river margins) of *Salix alba, Populus nigra* and *P. alba* (*Salicetum albae*) (EUH

Fig. 5 *Pinus nigra* forest on dolomite in the south of Vienna

code 92A0; Corine code 3.1.1.). In the 1870s, a new, straightened channel was excavated and embankments created on both sides of the river. The consequential changes in the hydrology has resulted in hardwood forests (*Pruno-Fraxinetum, Ulmo-Fraxinetum*) with *Fraxinus excelsior, Prunus padus, Acer campestre* and *A. pseudoplatanus, Tilia platyphyllos* and also *Quercus robur* slowly replacing the former gallery forests. Two remarkable species of the riparian forests in the "Donauauen National Park" are *Ulmus laevis* (which has its largest Austrian population in the park) and *Vitis vinifera* ssp. *sylvestris,* which is very rare in central Europe. The herb layer of the hardwood floodplain forests is characterised by large populations of geophytes, for example, *Allium ursinum, Galanthus nivalis* (Fig. 6) and *Gagea* spp.

As a result of the regulation of the Danube, the dynamic habitats formed by the gravel or sand banks are now restricted to a small area between the river and the embankments, see Fig. 7.

Behind the embankments, all stages of the plagiosere succession can be found including scrub and woodland developed on stream gravels. The few remaining "Heißländs" (for example, gravel and sand banks with steppe (EUH code 6240 and 6260; Corine code 3.3.1.)) are managed to preserve their existing vegetation, which is a unique mixture of Alpine, Pannonian and sub-Mediterranean species, for example, *Selaginella helvetica, Teucrium botrys, Orchis coriophora, Euphorbia seguieriana, Himantoglossum adriaticum, Stipa joannis* and *Thymelea passerina.*

Fig. 6 Early spring in the "Donauauen national park": *Galanthus nivalis*

Fig. 7 Gravel banks with *Salix purpurea* at the river Danube

As the result of the lack of inundation, the habitats of the former watercourses behind the embankments have also changed since the river was regulated. Nowadays, most of them are characterised by the plant communities associated with ponds and lakes including free and rooted floating vegetation (*Lemna minor* and *Nymphaea alba*), submerged vegetation (for example, *Ranuncululus fluitans* and *Sagittaria* communities (*Sagittario-Sparganietum ermersi*)) and emergent vegetation such as reed beds (*Phragmitetum* and *Carex elata*). *Chara* spp. and *Nitella* spp. (for example, *Charetea fragilis*) meadows frequently develop where cold groundwater emerges from the gravel bed.

Vegetation of the Cultural Landscape

There are about 9,000 ha of cultural landscapes in Vienna including meadows, arable fields and vineyards. Up to the 1960s, cattle were raised in the rural parts of the city to provide milk for the markets and meadows were mown to produce hay to meet local demands. Since then, the use and purpose of the meadows have changed from agricultural production to recreation and nature conservation as a consequence of the cost of mowing has to be met by government aid.

Very dry calcareous grasslands (*Xerobromion* – EUH code 6210; Corine code 2.3.1.) only occur on a few hills in the western parts of the city. They are characterised by sub-Mediterranean and Pannonian species, including *Stipa* spp., *Koeleria pyramidata, Pulsatilla grandis, P. pratensis* ssp. *nigricans, Adonis vernalis, Potentilla arenaria, Iris pumila, I. variegata* and geophytes, for example, *Gagea* spp., *Ornithogalum* spp. and *Muscari* spp. Semi-dry grasslands with *Brachypodium pinnatum* and *Bromus erectus* are more widespread, especially in the eastern parts of Vienna. The important species in this habitat include *Polygala major, Euphorbia verrucosa, Dianthus pontederae, Inula ensifolia, I. hirta, Orchis militaris, O. ustulata, Ophrys sphegodes, O. fuciflora* and *Himantoglossum adraticum.*

Lowland hay meadows – *Arrhenatherion* (EUH code 61XX; Corine code 2.3.1.) – are widespread throughout the city. The dry variations with low nutrient levels have many species in common with semi-dry grasslands, for example, *Salvia pratensis, Leucanthemum vulgare, Knautia arvensis* and *Centaurea scabiosa*. Traditionally, these meadows were mown once in June for hay and then grazed by cattle from August until the winter. Lowland hay meadows are mown two or three times per year. The characteristic species are *Arrhenatherum elatius, Trisetum flavescens, Tragopogon orientalis, Dactylis glomerata, Galium mollugo* ssp. *erectum, Campanula patula, Ranunculus acris, Centaurea jacea* and *Trifolium pratense.*

Humid grasslands with nutrient-poor soils (*Molinia caerulea* ssp. *caerulea*, EUH code 6410; Corine code 2.3.1.) were common in the Wienerwald area until the 1970s. Today, only a few meadows of this type, most of them nature reserves, are left. The characteristic species are *Molinia caerulea* ssp. *arundinacea, Gentiana pneumonanthe, Iris sibirica, Inula salicina* (Fig. 8), *Dianthus superbus, Gratiola officinalis* and *Dactylorhiza* spp. These meadows are mown once a year between August and October.

Since the 1980s, an increasing number of small ponds in private gardens and other artificial waters contribute to the plant and habitat biodiversity of Vienna, especially in the western parts where no natural ponds and lakes occur. The species frequently found in this habitat are *Phragmites australis, Iris pseudacorus, Typha*

Fig. 8 Humid grassland with *Inula salicina* in the south-west of Vienna

latifolia, T. angustifolia, Nymphaea spp. *Potamogeton lucens, Myriophyllum spicatum* and *M. verticillatum*. The less common species include *Nymphoides peltata, Utricularia vulgaris, Hottonia palustris, Stratiotes aloides* and *Riccia fluitans*.

The aquatic vegetation of ponds in parks depends mainly on their age and construction. The ornamental ponds, for example, in the Stadtpark and Schönbrunn Park are made of concrete and heavily polluted by the faeces of water birds. In these conditions, submerged macrophytes are unable to survive. Ornamental plants, such as *Nymphaea* spp., have been planted frequently in these ponds. The watercourses "Wienfluss" and "Donaukanal" are virtually free of aquatic vegetation because of the lack of structures and suitable habitats, see Fig. 9.

The water of the Neue and Alte Donau (Fig. 10) is still for most of the year, consequently it contains dense populations of submerged macrophytes, such as *Potamogeton* spp., *Myriophyllum* spp., *Elodea nutallii, Najas marina* and *N. minor*. Extensive *Chara* meadows are also present.

The south-eastern and north-eastern parts of the city include open landscapes of arable land with small areas of field margins and wasteland, see Fig. 11. The main crops are cereals *Pisum sativum, Beta vulgaris* ssp. *vulgaris* and *Helianthus annuus*. Since the 1990s, arable weeds such as *Agrostemma githago, Centaurea cyanus, Legousia speculum-veneris, Androsace maxima, Nigella arvensis, Consolida regalis* and *Papaver rhoeas* have increased again as a result of the increase in organic farming.

The flora associated with traditional vineyards (Fig. 12) bounded by margins, walls, hedges and thermophile fringes comprises mainly thermophilous species, for example, *Gagea villosa, Muscari neglectum, M. comosum, Stachys recta, Lathyrus latifolius, Rosa pimpinellifolia* and *Prunus fruticosus*. Since the 1980s, the biodiversity of the vineyards has steadily declined as the result of increasing intensive management, especially the use of herbicides.

Fig. 9 River "Wienfluss" in the city

Fig. 10 Beach at the "Neue Donau" near the centre of the city

Settlement Areas

In the city centre and high density housing areas, only few remains of spontaneous vegetation can be found. Most of the plants such as *Polygonum avenastrum, Chenopodium vulvaria, C. murale, Sisymbrium loeselii, Herniaria glabra* and

Fig. 11 Agricultural landscape in the south-east of Vienna

Fig. 12 View from the vineyards on the foothills of the "Wienerwald"

Asplenium ruta-muraria are limited to cracks in pavements and walls. Since the 1970s, the dog population has been increasing (including in the high density housing areas), this has changed the vegetation that occurs on the soil or gravel at the base of street trees ("Baumscheiben") and small lawns dramatically. These habitats, where Forstner and Hubl (1971) found adventive species in the 1960s, are nowadays dominated by nitrophilic herbs such as *Malva neglecta* and species of *Chenopodium.*

Because of their age, the low density housing areas contain a very different vegetation and biodiversity than the high density housing areas. Low density housing mainly occurs in the western part of Vienna where settlement took place in the late nineteenth century. The parks and gardens contain old trees and mixed hedges, which contribute to the biodiversity. In the new settlements, which are located in the north-east and south-east, many of the gardens are dominated by *Thuja occidentalis* and frequently mown lawns

Until the 1990s, many of the industrial areas rich in plant species comprised unused or poorly maintained land. They supported and were the only locations in Vienna with species such as *Allium carinatum, Gypsophila scorzonerifolia* and *Anchusa ochroleuca* (see Fig. 13).

Since then, the demand for industrial properties has increased and many new buildings have been constructed. Consequently, the flora of industrial areas is decreasing quickly and has become totally absent in some parts of the city.

Transport routes (railways, roads and the Danube) directly connect Vienna to western, northern and eastern Europe and the Balkans. The first documented plant immigrations were along the railway tracks, the species included *Geranium purpureum, Senecio inaequidens, Tribulus terrestris, Galeopsis angustifolia* and *Chaenorrhinum littorale.* The common species of the "railway flora" are *Senecio vernalis, Echium vulgare* (see Fig. 14), *Geranium robertianum, Galeopsis angustifolia* and species which originated from spilled cargoes, for example, *Triticum* and

Fig. 13 Industrial area with spontaneous trees of *Populus nigra*

Fig. 14 Railway area with *Echium vulgare*

Brassica species. Less common are *Saxifraga tridactylites, Petrorhagia prolifera* and *Vulpia myosuruides*.

Increasing road transport has resulted in the introduction of several species such as *Dittrichia graveolens, Atriplex heterosperma* and *Polycarpon tetraphyllum*. There are two commercial harbours on the Danube. Most ships and their cargo come from eastern Europe, and this has resulted in the occurrence of the eastern European species *Dianthus giganteus* and *Pimpinella peregrina* in the harbour areas.

Recreational areas contain a variety of environmental conditions, the dominant vegetation being influenced by their age and location. Old Parks and cemeteries like Schönbrunn, Türkenschanzpark, Prater (see Fig. 15) and the "Zentralfriedhof" are "hot spots" of biodiversity.

At the time of their creation, they were situated within the areas of natural vegetation, fragments of which can still be found in them. The largest populations of geophytes, such as *Tulipa sylvestris, Ornithogalum nutans* and *O. umbellatum,* occur in these areas. The very endangered *Apium repens,* which normally grows along rivers, also frequently occurs in cemeteries. The more recently established parks, which are often situated in high density housing areas and are heavily used, comprise lawns (created by laying turf) and a small number of tree, shrub and perennial herbaceous species, such as *Thuja occidentalis, Chamaecyparis* spp. and *Juniperus* spp.

As the result of the application of fertilisers, irrigation and frequent close mowing, sports fields contain few plant species. Since the 1990s, in order to reduce the maintenance costs, many of these areas have been reconstructed with artificial surfaces.

Fig. 15 "Hauptalle" of *Aesculus hippocastanum* in the Prater Park

Fig. 16 The former clay pits at "Wienerberg" were designed as a recreational area in the 1980s

Waste ground and refuse disposal sites in Vienna like Laaerberg and Wienerberg, where Forstner and Hübl (1971) found many adventive species no longer exist. Many of these sites have been closed and reconstructed to prevent groundwater contamination. Most of them were restored and designed to create recreational areas, see Fig. 16. The city government followed two strategies:

1. Two former disposal sites were designed as ornamental landscapes (WIG-Wiener Internationale Gartenschau 1964 and 1974).
2. The others were designed as “Second Hand Nature” using local species. Probably, the best example of these is the recreational area Wienerberg, where endangered species including *Ononis arvensis, Veronica austriaca, Isatis tinctoria, Orobanche arenaria, Lotus glaber, Cyperus fuscus, Verbascum speciosum* and *Lavatera thuringiaca* were successfully established.

Nature Conservation, Environmental Planning and Education

Environmental planning has a long tradition in Vienna. In the late nineteenth century, the city expanded rapidly to accommodate the immigration of about two million people from the eastern parts of the Austro-Hungarian Empire. In 1905, a legally protected 600 m wide green belt was established around the whole of the city. In the same declaration, the Wienerwald area and the floodplains along the Danube were protected. Since 1945, two parts of the green belt have been used for the allotment for gardens and the construction of new buildings.

In the 1980s, the Environment Office of the City Council started a detailed mapping programme of the city. Since 1988, data on phytotopes of the whole city have been available digitally in a database and GIS. Parallel to the scientific publication of the data an exhibition called “BLUBB – Biotope, Landschaften, Utopien, bewusst beleben” was held in the city centre to inform people interested in environmental matters about the programme and to inspect the data. In 2008, a revision of the biotope mapping was started with the main focus on NATURA 2000 (Flora-Fauna Habitat) Biotopes. Since 1998, the rare and endangered biotopes and species identified during the mapping exercise have been protected by law.

Every 10 years, starting in 1984, the City Council publishes a general development plan for the city including proposals for the conservation, improvement and protection of the green areas. In 1995, the Council produced a special plan for the protection and management of the green belt. The aim of the Council is to buy strategically important parts of the green belt and to develop a network of multifunctional recreational and ecological areas in the outer parts of the city.

About 50% of Vienna is subject to nature conservation/landscape protection, see Table 4 and Fig. 17.

Table 4 Area of protected land in Vienna

National parks	5.4%
Nature reserves	10.5%
Forest reserves	0.4%
Protected landscapes (including four Nature 2000 sites)	13.8%
Biosphere reserve (Wienerwald)	25.0%

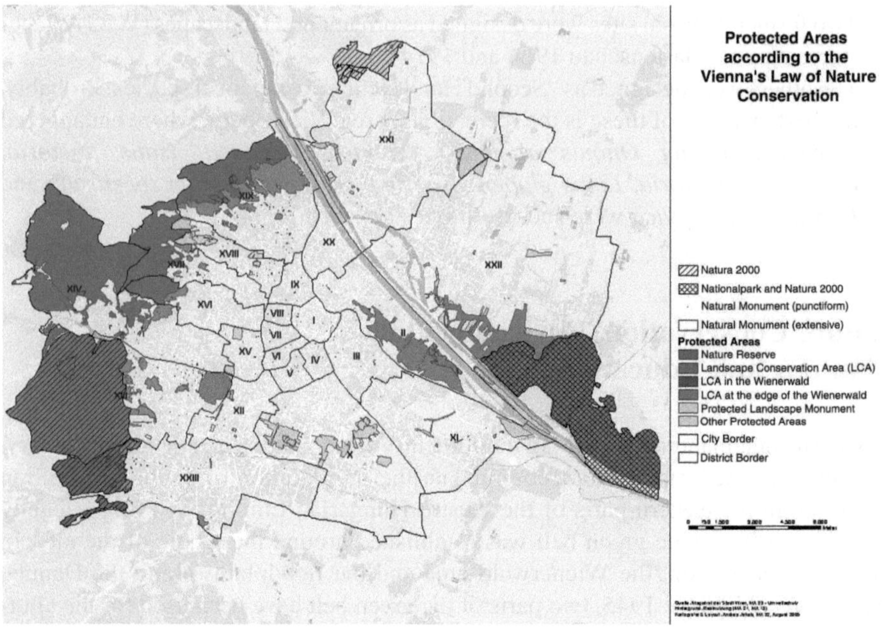

Fig. 17 Protected Areas in Vienna

In addition to the Natura 2000 programme, initiatives including "Netzwerk Natur" ("Network of Nature") and projects for sustainable development are implemented by the City Council to preserve and increase the nature conservation value of habitats in the city. About 7,600 ha of forests and meadows in the Wienerwald, 2,980 ha of floodplain meadows and forests and 373 ha of the green belt are owned by the City Council and managed by the municipal Department of Forestry according to management plans developed in co-operation with the municipal Departments of Nature Protection and Forestry and experts from universities and non-governmental organisations. The municipal Department of Agriculture cultivates 1,900 ha of arable land; since the 1990s, over 800 ha of this land has been strictly reserved for organic farming.

Vienna contains 765 (ca. 32%) of the plant species listed in the Austrian Red Data List and which are therefore endangered in some way. A comparison of the endangered species with their typical habitats shows that the majority of the more severely endangered species are not part of the natural ecosystems (like forests, thus maintaining some equilibrium by self-regulation) but occur in habitats extensively altered and influenced by human activities.

Since the 1970s, the number of nature trails and environmental education projects in the city has increased. The first trails were relatively simple and usually restricted to providing names of the trees and shrubs; nowadays, the focus lies on the provision of panels with more detailed information and pictures of plants, animals and their habitats and interactive trails, especially for children. In addition, many books, booklets and folders on environmental themes especially designed for Vienna have been published and are currently available, for example, "Wildwuchs", "Wiens Pflanzenwelt", "Natur – Wanderführer Wien", and "Abenteuer Natur Wien".

In the 1990s, the municipal Department of Forestry established the "Wiener Waldschule", where pupils spend a whole day guided by pedagogically skilled foresters who provide them with adventures in the forest and information about nature (including animals and plants and their habitats) of Vienna. The programme has been a great success; as a result, since 2000, more than ten environmental educational facilities have been established by different municipal departments in all parts of Vienna.

The city also includes several natural and semi-natural habitats and a large number of plant species, many of which are endangered. As a result of this combination, Vienna has good conditions for making an important contribution to the protection of various plant species and communities. For that reason, the city has considerable responsibility for plant and habitat protection beyond its borders.

Closing Comments

The flora of Vienna is very rich in species both nationally and in the European context. About two-thirds of the species occur in natural and semi-natural areas, including forests, formerly extensively used steppe pastures as well as arable land, meadows and wetlands. Many of these biotopes are protected by different regulations and are managed and used in sustainable ways.

The situation of man-made habitats is quite different. Large areas have been altered or lost, especially in the economic boom of the 1990s. There is also a trend to tidy up places with ruderal vegetation. Because these habitats play an important role in contributing at least a third of the Viennese flora, it will be necessary to focus intensive efforts on them in order to preserve their vegetation and component flora and therefore the biodiversity of the city.

Literature Cited

Forstner W, Hübl E (1971) Ruderal-, Segetal- und Adventivflora von Wien. Notring Verlag, Wien.

Krisai-Greilhuber I (1992) Die Makromyceten im Raum von Wien: Ökologie und Floristik. IHW, Eching.

Türk R (2006) Die Flechtenflora in Wien-Veränderungen im Zeitraum 1853 bis 2004 – The Lichen Flora in Vienna – Changes in the time-period 1853 to 2004. Verhandlungen der Zoologisch-Botanischen Gesellschaft in Österreich 143. Wien.

Zechmeister HG, Humer K, Hohenwallner D (1998) Historische Moosflora von Wien. Teil 1: Leber- und Hornmoose (Hepaticae, Anthocerotae). Verh. Zool.-Bot. Ges. Wien 135: 343–351.

Zechmeister HG, Hohenwallner D, Humer K (1998) Historische Moosflora von Wien. Teil 2: Laubmoose (Musci). Verh. Zool.-Bot. Ges. Wien 135: 353–379.

Zechmeister, HG, Hohenwallner D, Humer-Hochwimmer K. (2001) Die Erforschung der Moosflora von Wien. Berichte der Reinhold-Tüxen-Gesellschaft (RTG) 13: 291–295.

Further Reading

Adler W, Mrkvicka AC, Zuna-Kratky T (1997) Natur-Wanderführer Wien. Verlag Bohmann, Wien.
Adler W, Mrkvicka AC (2003) Die Flora Wiens gestern und heute: Die wildwachsenden Farn- und Blütenpflanzen in der Stadt Wien von der Mitte des 19. Jahrhunderts bis zur Jahrtausendwende. Naturhistorisches Museum, Wien (Austria).
ARGE Biotopkartierung Wien (Ed.) (1998) Erhebung schutzwürdiger und entwicklungsfähiger Landschaftsteile Wiens (Biotopkartierung Wien). Im Auftrag der Magistratsabteilung 22, Wien.
Brunner K, Schneider P (Eds.) (2005) Umwelt Stadt. Geschichte des Natur- und Lebensraumes Wien. Böhlau Verlag, Wien.
Berger R, Ehrendorfer F (Eds.) (2010) Ökosystem Stadt. Die Naturgeschichte Wiens. Böhlau Verlag, Wien.
Lipka D (2005) Abenteuer. Natur. Wien. Unterwegs zu Biber, Zauberpflanze & Co. MA 22, Wien.
Mikocki J et al. (2003) Wildwuchs. Amt der Wiener Landesregierung.
Steiner GM, Punz W et al. (Eds) (1990) BLUBB – Biotope, Landschaften, Utopien, Bewusst beleben, Wien.
Vitek E et al. (2004) Wiens Pflanzenwelt. Verlag des Naturhistorischen Museums, Wien.

Warsaw

Barbara Sudnik-Wójcikowska and Halina Galera

Fig. 1 The Old Town: the Castle Square, St. John's Cathedral, remains of the city wall (*left side*) and the Royal Castle (*right side*) (Photo I. I. Moysiyenko)

Abstract As in the case of other metropolitan areas, the size and the structure of land use in Warsaw have changed substantially both spatially and temporally. Areas of natural and semi-natural habitats and agricultural land have declined, especially in

Barbara Sudnik-Wójcikowska (✉) and Halina Galera
Department of Plant Ecology and Environmental Conservation, University of Warsaw, Al. Ujazdowskie 4, 00-478, Warsaw, Poland
e-mail: barbara.sudnik@uw.edu.pl; h.galera@uw.edu.pl

J.G. Kelcey and N. Müller (eds.), *Plants and Habitats of European Cities*,
DOI 10.1007/978-0-387-89684-7_15, © Springer Science+Business Media, LLC 2011

the peripheral parts of the city. At the same time, new habitats have appeared, such as industrial areas, tramlines and railways. After 1945 a considerable part of the central area of the city was occupied by rubble that was initially devoid of vegetation but which was soon to be colonised by many plant species. The area of rubble decreased as the city was re-constructed. As a result of spatial planning policies gradual changes occurred in the structure of the spontaneous and cultivated flora. Future trends in the flora of the city can be predicted based on the observations made so far. The establishment of protected areas within the city has only limited success in preventing the disappearance of native species. The conversion of agricultural land to urban uses has led to the elimination of some native weed species and archaeophytes (especially segetal species). It appears, therefore, that the importance of alien species in the flora of Warsaw will continue to increase as will the number, frequency and abundance of thermophilous and xerothermic species with short life cycles.

Natural Environment

Location

Warsaw, which is the capital of Poland, lies between coordinates 52°05′48″ to 52°22′11″ north and 020°51′04″ to 021°16′26″ east, in the central part of Warsaw Basin on the River Vistula at elevation 77–116 m a.s.l. The city is located in the sub-province "Central Poland Lowland" and in the macro-region Central Mazovian Lowland. The River Vistula is a characteristic feature of the city dividing it into two environmentally and historically different parts located on opposite banks: Warsaw on the left bank of the Vistula – and – Praga on the right bank.

Geomorphology

There are three distinctive units in the relief of Warsaw: the valley of the Vistula, Warszawska Plain and a small part of the Wołomińska Plain (Fig. 2).

The Vistula valley near Warsaw was created by fluvial processes that occurred during the Quaternary period, in the final stages of the last glaciation and the Holocene (Encyklopedia Warszawy 1994; Pawlak and Teisseyre-Sierpińska 2006). In the river valley, the following terraces can be distinguished:

1. The upper terrace is covered by sand dunes (called Kampinoski Terrace on the left bank and Otwocki Terrace on the right bank). The altitude of the terraces is 7–10 m higher than the mean river level.
2. The lower terrace (Praski Terrace) on both banks of the River Vistula; it has a uniform topography and an elevation 4–7 m higher than the mean river level.
3. The topography of the floodplain, which is 0.5–3 m higher than the mean river level, is flat. The recent floodplain terrace is limited by flood protective bunds.
4. The river channel, which has changed its alignment during historical times (Fig. 2).

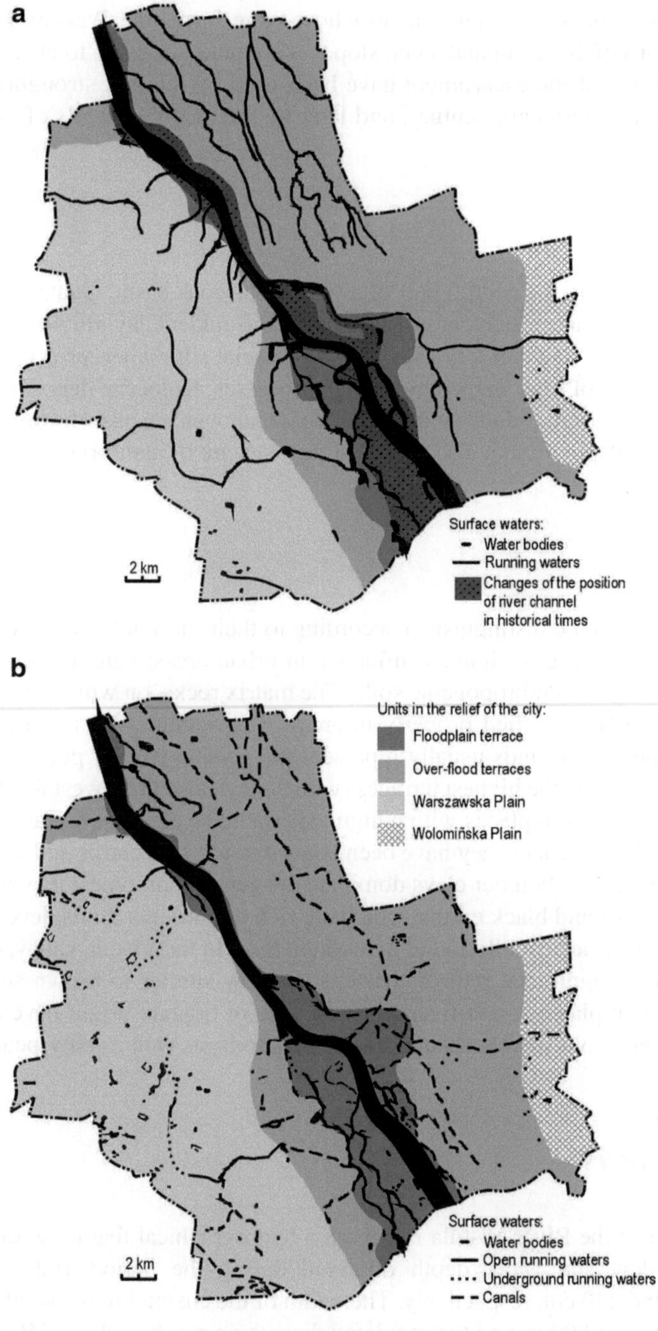

Fig. 2 Geomorphology and hydrographical layout of Warsaw: (**a**) in historical times (natural surface waters and changes of the position of river channel); (**b**) in the twenty-first century

Warszawska Plain is a glacial plateau whose edge forms the Warsaw Escarpment with elevations of 10–25 m and steep slopes with inclinations up to 60°. Favourable defence features of the escarpment have been used by the first strongholds, which were built in the thirteenth century, and later by the medieval walls of the city.

Geology

Warsaw is located in the central part of a vast Cretaceous basin. Quaternary deposits form a glacial plateau, which comprises disturbed Boulder Clay, fluvio-glacial sands, interglacial gravel, stagnant lake clay and peri-glacial silt. Older terraces in the river valley are formed of sand shaped by aeolian processes. Holocene deposits are mainly soils of fluvial origin, sands and peat with organogenic soils (formed in poorly drained terrain depressions). The youngest deposits are of anthropogenic origin.

Soils

Two soil types can be distinguished according to their morphological properties and depending on the level of human influence in urban areas: natural soils (with preserved horizons) and anthropogenic soils. The matrix rocks, on which the soils developed, can only be classified properly in areas of agricultural land and forests. The soils developed from sands usually represent the genetic type of podzolic soils and podzols. The soils on the highest terraces (with dunes) have the lowest fertility. Boggy soils have developed in places with a high watertable level; however; as the result of land reclamation practices they have been converted to black earths and marshy soils. Soils developed from boulder clays dominate the genetic soil types; they are complemented by brown and black earths, which are rich in nutrients. Soils developed from alluvial sediments are usually found in medium fens. In the Vistula valley, outside the flood terrace, a significant part of the fens are now similar to brown soils. In wet places of glacial plateaux and river terraces, soils of organic origin have developed, including boggy soils, peat soils and boggy peat soils, as well as silty-peat soils.

Surface Waters

The 31.6 km of the River Vistula is the main hydrographical feature within the city. The mean flow and water depth observed during the period 1921–2000 were 568 $m^{-3} s^{-1}$ and 240 cm, respectively. The width of the channel in the southern part of the city is about 1,000 m and the regulated downtown reach is about 350 m wide.

The major tributary of the River Vistula is Wilanówka, which has a total length of 16.5 km (9 km within the city area). It rises in the south of the city and flows through old oxbows located on the left bank of the Vistula floodplain. Other watercourses that drain the Warszawska Plain (for example, Bielański stream) have been partly culverted (Fig. 2). In the south of the city, a relatively well-preserved natural river network and oxbow lakes occur on the floodplain terraces on both sides of the river. There is also a dense river network in the north-east part of the city. Throughout the city, there are a considerable number of small water bodies originating from the man-made excavation pits and the nineteenth century fortification moats.

Climate

Because of its geographical position, the weather in Warsaw is changeable. The climate is characteristic of central Poland and belongs to the moderate temperate climatic zone with transitional features between maritime (Atlantic) and continental conditions. It shows considerable variability with a predominant western circulation bringing polar-maritime air masses from the Atlantic Ocean, which produces rainy and cold weather in summer and mild weather in winter. The advection of continental air masses from Asia and eastern Europe brings sunny and dry weather in summer and frosty weather in winter. A sudden drop in temperature associated with the advection of arctic air masses (often observed in April and May) occur as may heat waves caused by inflow of tropical air masses from the south in summer (Gutry-Korycka 2005).

The average annual total solar radiation between 1961 and 2000 amounted to 3,538 MJ m^{-2}. The coldest month of the year is January with an average air temperature −3°C, while the warmest is July when the temperature exceeds 18°C. The average annual temperature is roughly 8°C. Extreme temperatures recorded in Warsaw are 36.8°C (August 22, 1943) and −32.6°C (February 10, 1929). Owing to the heat-island effect, there is a change in the length of the thermal seasons of the year (winters shorter by up to 2 weeks) and the growing season is longer by a few days.

Between 1951 and 2000, the average annual precipitation ranged from 510 to 570 mm. Most of the precipitation took place in June and July with the lowest amounts occurring in February and March. Snow cover lasts about 70 days and is about 5 cm thick. The pattern of urban structures and the orientation of the River Vistula valley are also important factors that influence the local climate of the city.

Historical Development of the City

Natural Factors Determining the Rise of the City

In geographical terms, Warsaw lies almost in the centre of Poland's historical territory; its location determined the city's political role and the decision to establish the capital there. It is situated at the intersection of trade routes running from the east to west and south to north of Europe, thus forming an important communication crossroads. The city's location was not accidental but arose as a result of the natural conditions prevailing in the region. The River Vistula, as well as the

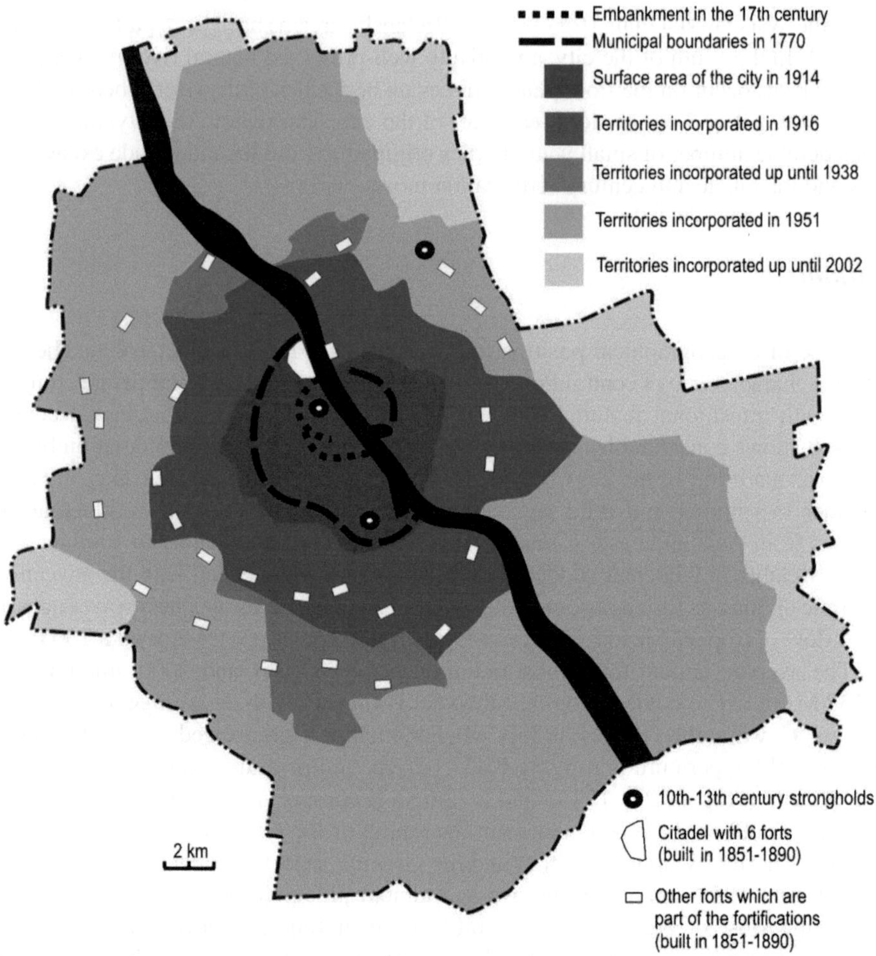

Fig. 3 Territorial development of Warsaw

steep banks of the escarpment, with its deep ravines, provided good precipices. The beginnings of human settlement on the site of the modern-day Warsaw go back to around 10,000 years BC. The three strongholds that were built on the Vistula in the early Middle Ages (tenth to thirteenth centuries) can be considered to be the earliest beginnings of urban development (Fig. 3).

Fourteenth to Sixteenth Century

The city was founded in the first half of the fourteenth century (Tables 1 and 2); about that time two rows of defensive walls were built around the city and its stronghold. In 1408, the earliest suburbs were built, the so-called New Town, situated north of the walls of the Old Town (Fig. 1). In the sixteenth century, the city was expanded

Table 1 The size of Warsaw from the thirteenth to twenty-first century

Date	Surface area (excluding the Vistula corridor)(km^2)
Thirteenth to fourteenth century	0.1
Fifteenth century	0.2
1564	0.3
1655	3.4
1770	16.3
1826	21.5
1913	32.7
1917	114.8
1930	118.1
1938	134.7
1951	411.7
1957	430.3
1977	469.6
1992	479.1
2002	502.3

Table 2 The number of permanent residents in Warsaw from the fourteenth to the twenty-first century

Date	No. of inhabitants
Fourteenth century	1,500–2,000
Fifteenth to sixteenth century	5,000
1564–1565	10,000
1655	18,000
1792	120,000
1795	70,000
1830	145,000
1861	200,000
1897	594,000
1916	758,000
1921	937,000
1939	1 310,000
September 1945	378,000
1949	605,000
1978	1,552,000
1993	1,644,000
September 2007	1,707,000

beyond the boundaries of Old and New Warsaw. The sanitary conditions were vastly improved, mains water supplies were constructed and the main roads were cobbled. The first permanent wooden bridge spanning the Vistula was built between 1568 and 1573; it can be seen on the first image of Warsaw that has been preserved. The woodcut (Fig. 4), which dates from 1581, also depicts the first known view of private gardens laid out in Poland (that is the Renaissance parterre garden, along the escarpment below the Royal Castle) and the first rubbish tip in Warsaw.

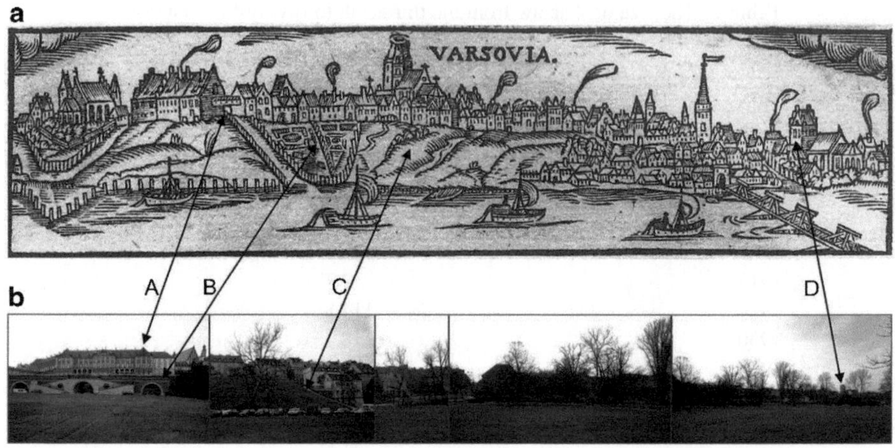

Fig. 4 Warsaw Escarpment in the vicinity of the Royal Castle – viewed from the River Vistula: (**a**) state in the sixteenth century (woodcut published in 1581); (**b**) the same area in the twenty-first century (Photo M. Skrzecz). (A) Royal Castle, (B) Royal Garden, (C) first rubbish tip in Warsaw, (D) the oldest church in the city

Seventeenth and Eighteenth Centuries

At the turn of the seventeenth century, the city became the formal residence of the Royal Family and the capital of the Polish state. During the seventeenth and eighteenth centuries, magnates (as well as the nobility and clergy) began to build their residences to the north and south of the Old and New Towns. Particularly, magnificent palaces were erected along the Warsaw Escarpment. Administrative units, privately owned and independent of the urban authorities, helped shape modern Warsaw. Although the city was affected by natural disasters (fires, hurricanes and plagues), it became wealthy and expanded. Two settlements on the right bank of the Vistula gained urban rights: Skaryszew (1641) and Praga (1648). As a result of economic development and religious tolerance in Poland, the influx of foreigners increased (Table 2).

Warsaw's dynamic development was halted by the Swedish invasions in the mid-seventeenth century. Military action led to the capital's devastation. Although several decades were spent rebuilding the city, the Northern Wars (1702–1709) led to a similar decline, during which the Swedes occupied and destroyed the city several times. The long period of peace after 1716 made it possible to rebuild and organise the capital.

The eighteenth century witnessed a pronounced growth in the political and cultural significance of the capital. The partitions that divided the country in 1772, 1793 and 1795 put a stop to the reforms and the State was divided among three powers, Prussia, Russia and Austria.

1795–1918

In 1795, Warsaw became part of the Prussian Empire. Then, following the Napoleonic wars in 1815, it became the capital of the Congress Kingdom, which was totally dependent on Russia. In retaliation for the armed opposition by the Polish nation, the Tsarist authorities ordered the building of a fortress in Warsaw – a system of fortifications comprising a mighty citadel with six forts and one on the other side of the river and a double ring with 27 forts surrounding the city (Fig. 3). The system of fortifications built between 1832 and 1890 prevented Warsaw's spatial development (contrast the city boundaries with those of 1914 and 1916). Whole settlements were destroyed by the construction of the fortifications, and it was forbidden to construct brick houses within a wide area surrounding them. The outbreak of the First World War had serious economic repercussions. After the Germans invaded Warsaw in 1915, the city regained a certain amount of autonomy, which made it possible to modernise the city. In 1916, the first modern urban plan was prepared resulting in the city being restructured and enlarged threefold (Table 1 and Fig. 3).

1918–1945

Following independence in 1918 substantial reconstruction of the city's infrastructure was carried out. At this time, many political, cultural, scientific and academic institutions were either created or opened. New residential housing estates were built, as well as water mains and gas networks. Industrial plants and factories and communications were developed.

The process of reviving Warsaw as a capital city was accompanied by the threat of a Russian invasion, which was stopped on the right bank of the city in 1920. However, there were soon new threats from Germany and the Soviet Union. Between 1939 and 1945, 84% of Warsaw was destroyed including the whole of the left bank. Of the 25,500 buildings (not counting industrial premises), more than 11,000 were totally destroyed and another 10,000 were badly damaged. The urban infrastructure was almost entirely destroyed or damaged, including all the railways stations and bridges over the Vistula.

1945–1989

The rebuilding of Warsaw began immediately after 1945 when work was undertaken to clear the rubble (about 20 million m^{-3} were eventually removed). The reconstruction of the city's monuments lasted until the 1980s; the Royal Castle (Fig. 1) was reconstructed between 1971 and 1984. At the same time, extensive rebuilding work was undertaken in some of the districts.

The resulting changes contradicted the city's pre-war character, imparting it with "the new socialist realism". The greatest changes were carried out in the city centre, where many homes, which had been salvaged after the war, were destroyed in order to build the Palace of Culture and Science (1952–1955) – a "gift" from the Soviet Union to Poland. The "10th Anniversary Stadium", which is on the right bank of the Vistula, was built on rubble; however, after 20 years it became disused and derelict. The city's development was interspersed with periods of increased investment and periods of stagnation caused by political repressions and economic crises. Despite the laws (which were binding) forbidding immigrants to obtain permanent residence in Poland between 1954 and 1984, and also the wave of emigration after the political repressions, the population of Warsaw continued to grow (Table 2).

Warsaw on the Eve of the Twenty-first Century

The city's intensive development began in the 1990s following political changes thanks to which Poland regained its full sovereignty. Modern investments have been carried out somewhat chaotically. Coherent city planning can only be achieved by preparing a detailed Urban Plan and strictly enforcing it (and the relevant regulations), including those for the provision of public open space and the allocation of land for large-area commercial developments. It is therefore unfortunate that the Spatial Development Conditions and Directions Study were only presented to the public for consultation in 2007. Currently, the city has an area of 517 km^2 (together with the river corridor), with 33,000 inhabitants per 1 km^2 (compare Tables 1 and 2).

Changes of the Environment Due to City Growth

Forests, rivers and wetlands constituted the predominant elements of the landscape in Warsaw until the thirteenth and fourteenth centuries. Forest clearance to provide land for buildings and agriculture was one of the earliest forms of human interference. By the middle of the nineteenth century, changes in the landscape had mainly resulted in the reduction of agricultural and wooded areas. In addition, improvements in living conditions resulted in the reduction of pollution caused by organic waste but as the city developed new hazards, typical of urban areas, appeared.

Changes in Layout of the Land

Several "rubble hills" (up to 30 m high) were created during the rebuilding of the city after 1945. Parts of the Warsaw Escarpment were covered by a thick layer of rubble, while some rubble was transported to the right side of the Vistula to form the foundations of the 10th Anniversary Stadium. Other structures that are typical components of the anthropogenic topography of urban areas also occur, for example, banks and moats within forts, railway embankments, tunnels and pits formed by mineral extraction.

Changes in Mechanical and Chemical Composition of Soil

As the result of war damage, the disposal of rubble and other major disturbances, the physical structure and chemical composition of the soils have been substantially altered from those that existed previously, which were considerably different from the original, natural soils. These substrates have a large carbonate content, which when coupled with the carbonates and other alkalis produced by power stations result in soils with a pH higher that 7.2 (Biernacki et al. 1990). The application of road de-icing salts, for example, NaCl and $CaCl_2$, has caused the roadside soil to become saline. In addition, the chemistry of some soils has been influenced by heavy metals and other pollutants from motor vehicle exhaust emissions and industrial processes.

Changes of Hydrological Network

The effects of the development of Warsaw on the city's hydrology can be seen by examining the older parts of the city (Fig. 2a). The springs in this area have dried up and in the fifteenth century the Dunaj stream, which flowed between the Old and New Towns, was included in the city's moat. The Bełcząca (a small river) was culverted in the eighteenth century; in the following century, various streams including the Drna disappeared. Natural watercourses in the area occupied by buildings on the right side of the Vistula disappeared or were included in the city sewerage in the nineteenth and twentieth centuries. All the springs in the Warszawska Plain dried up and only a few springs have remained on the Warsaw Escarpment (they are only active periodically). Extensive wetlands and some bodies of water have disappeared.

The watercourses that are used to irrigate the right side of the city are supplied from new artificial canals and drainage ditches (Fig. 2b). The moats around the forts and many ponds in the city parks are anthropogenic water bodies as are those resulting from the flooding of mineral workings such as clay, sand and gravel pits as well as those resulting from peat workings. Radical changes can also be observed in the Vistula River, whose present course was created in the nineteenth century (Encyklopedia Warszawy 1994).

Water Contamination

In the fourteenth and fifteenth centuries, the inhabitants of Warsaw obtained water from wells or directly from the Vistula. The pollution of shallow groundwaters increased rapidly in the nineteenth century; consequently, it was necessary to provide the inhabitants with clean water from the deep water-bearing strata that discharged below the Warsaw Escarpment. The major problems relating to water quality were resolved at the turn of the nineteenth century as a result of the installation of sewage disposal and water supply systems. At present, 95% of the inhabitants of Warsaw are connected to a potable water supply, which is extracted from alluvial sediments below the channel of the Vistula. Today, the river water is contaminated by physico-chemical and bacteriological agents, which also contaminate other surface waters (Encyklopedia Warszawy 1994).

Changes in Climate Conditions

The urban heat-island phenomenon has been found to have a considerable effect on temperature trends in the city. The average annual temperature in the centre is higher than the surrounding area by ca. 0.8°C. Depending on the meteorological conditions, the difference in temperature between the city centre and the urban fringe is 7–8°C and during anticyclone, windless and cloudless weather, it may reach 10°C.

The increase in precipitation and the decreasing penetration of sunlight result from the impact of the so-called urban aerosol. The average annual precipitation in the centre of the left bank of Warsaw is 40–70 mm higher than that in the sub-urban areas. In the centre of the city, the total annual insolation is lower by 150 hours than that in the suburbs (Gutry-Korycka 2005).

Air Pollution

Air pollution in Warsaw increased in the nineteenth century as a result of the use of brown coal for residential heating and industrial furnaces. Industry and motor vehicles have been the main source of air pollution in recent times. Carbon monoxide, oxides of nitrogen and hydrocarbons sometimes exceed permitted levels. However, pollution caused by particulate matter has decreased to an acceptable level, as a result (among other reasons) of the installation of filters at the "Warsaw" Steelworks (Gutry-Korycka 2005). The level of air pollution can change not only through decreasing emission of pollutants but also as a result of efficient air exchange over the city.

Flora

Floristic studies conducted within the vicinity of Warsaw have a long history. Martin Bernhardi, the doctor of the Polish king Jan Kazimierz, was the author of the first work on the flora of Warsaw (1652). In the work: *"Catalogus plantarum tum exoticarum quam indigenarum que anno MDCLI in hortis Regiis Varsaviae et circa eandem in locis sylvaticis, pratensibus, arenosis et paludosis nascuntur collectarum* [...] *a chirurgo Martino Bernhardo"* he described 477 species of plants. Eighty years later, Erndtel published the flora of Warsaw ("*Vividiarum Varsaviense*" 1730).

Intensive floristic studies within the municipal boundaries and vicinity of Warsaw were carried out in the nineteenth century, especially during the second half. Investigations were conducted within the valuable natural areas and anthropogenic habitats, such as railway areas. During the interwar period, some areas within Warsaw city limits were proposed for protection. Therefore, there is a large archive of material including books, herbarium specimens with which the present-day data can be compared.

Studies involving bryophytes, lichens and fungi are only fragmentary, having only been carried out in parts of the city or in specific habitats, for example, the bryophyte flora of some cemeteries and parks and the algal flora of some water bodies. Unpublished material exists about the lichens and fungi in the Vistula valley and the forests within the borders of the city.

Vascular Plants

Detailed floristic investigations were carried out within the boundaries of Warsaw in the 1980s and 1990s (Sudnik-Wójcikowska 1987–1998). The plant distribution was recorded using a grid comprising of 225 squares of 1.5 km^2 (which cover an area of 2.25 km^2).

Figure 5 compares the species-richness of Warsaw (1,279 taxa) with the flora of Poland (3,555 taxa). As expected, alien species play a much more important role in

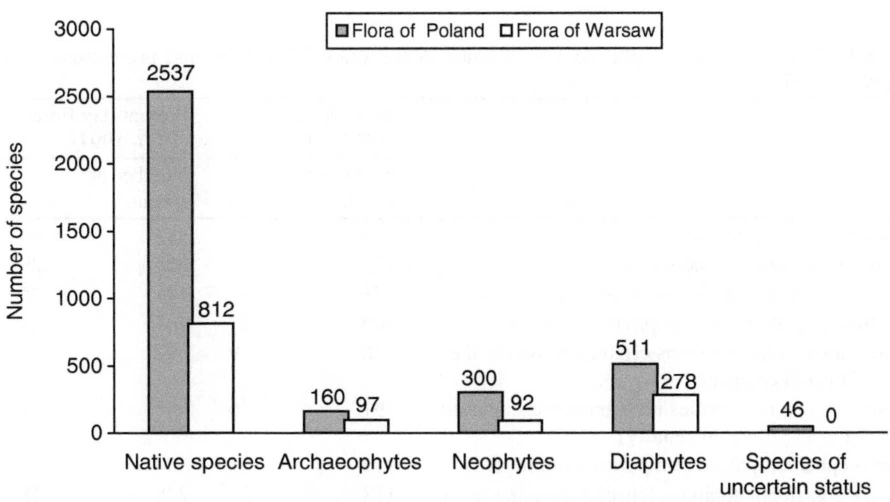

Fig. 5 The species-richness of the native and alien in the flora of Warsaw compared with the flora of Poland

the flora of Warsaw, constituting about 40% of the flora of the city and 30% of the flora of Poland. The number of archaeophytes recorded in the city is similar to the number of neophytes while neophytes are more numerous in the flora of Poland. The diversity of the present-day flora of Warsaw is comparable with that of two other large Polish cities whose flora was investigated by other authors using similar methods; Poznań (229 km^2) comprises 1,223 species, Łódź (294 km^2) comprises 1,072 species.

Changes in the Flora of Warsaw over the last 150–200 years

Warsaw contains a large collection of archival materials (about 15,000 herbarium specimens – the oldest being 1824) and historical publications dealing with the flora of some habitats or parts of the city. Table 3 and Fig. 6 present quantitative data of the total flora of Warsaw between 1824 and 1997 and the present-day flora. A total of 1,462 species have been identified within Warsaw over the last 150–200 years. A comparative analysis of the historical and present-day data has enabled the changes in the flora of the city to be determined, for example, the rate of introductions of alien species and loss rate of native species.

During the last two centuries, two species groups have shown a decline. It is estimated that the number of native species has been reduced by 25–30%. Species that were recorded from a few localities 100–200 years ago but have not been recorded recently include *Anemone patens, Dracocephalum ruyschiana, Scorzonera purpurea* and *Hippuris vulgaris.* The approximate size and location of areas in Warsaw where the greatest losses of native species had been recorded were presented in an earlier work (Sudnik-Wójcikowska 1987–1998).

Table 3 The structure of the flora of Warsaw in the years 1824–1997 and present-day flora (1977–1997)

	Total flora (1824–1997)		Present-day flora (1977–1997)	
	Number of species	(%)	Number of species	(%)
Native species, including:	953	65	812	63
Ns – non-synanthropic species	475	32	368	29
Ap – apophytes (native synanthropic species)	478	33	444	35
Alien species or anthropophytes, including:	509	35	467	37
Ar – archeophytes (aliens established before the fifteenth century)	128	9	97	8
Ne – neophytes (species permanently established after the fifteenth century)	91	6	92	7
Ef – ephemerophytes (temporarily introduced)	112	8	50	4
Eg – ergasiophygophytes (species escaping from cultivation, not established)	178	12	228	18
The number of species recorded in the city	1,462	100	1,279	100

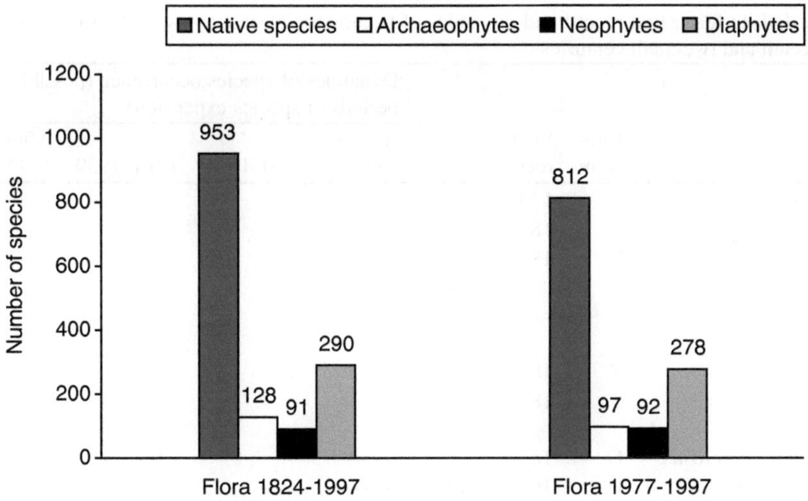

Fig. 6 Structure of the flora of Warsaw between 1824 and 1997 and between 1977 and 1997

The number of archeophytes has decreased by 23%; however, they do not form a homogenous group. Some of the ruderal archeophytes and most of the segetal species associated with traditional agricultural practices have disappeared altogether, for example, *Chenopodium opulifolium, Camelina alyssum* and *Aphanes australis*. However, there is a group among archeophytes, mainly ruderal species, which are able to persist and are quite common in Warsaw, even in areas under strong human pressure, for example, *Ballota nigra, Descurainia sophia, Atriplex sagittata, Lepidium ruderale, Solanum nigrum, Sinapis arvensis, Sonchus oleraceus* and *Setaria viridis*.

Another noteworthy group is the neophytes (Table 4); all the species noted in the nineteenth century are currently found within the Warsaw area.

The degree of invasiveness of some of the species can be determined by comparing their current localities with those of the nineteenth and twentieth centuries, for example, *Impatiens parviflora* and *Eragrostis minor* (Sudnik-Wójcikowska 1998). Only tentative conclusions can be drawn from such observations. Much longer-term observation is required to determine whether or not any of these species become invasive. Other species, such as *Cyclachaena xanthiifolia, Lepidium virginicum, Bidens frondosa, Echinocystis lobata* and *Fallopia japonica*, have been introduced in the last 50 years and spread rapidly throughout the city. *Cyclachaena xanthiifolia* (Fig. 7), a prairie species, was noted for the first time in 1959.

The plant had probably been introduced with American grain and later spread very quickly, especially throughout the more central parts of the city. The number of current localities of the species appears to have stabilised or slightly decreased as the result of the modernisation of the city. *Bassia scoparia*, an Irano-Turanian species, is another taxon whose population in Warsaw has increased in the last 10–15 years. *Eragrostis albensis* was noted in Warsaw for the first time in the middle of the twentieth century on the sandy banks of the Vistula. At the beginning

Table 4 Life form, first record and dynamics of occurrence of neophytes in Warsaw during the nineteenth and twentieth centuries

			Dynamics of species occurrence (possible periods of species expansion)			
Species	Life-form	First record	Before 1850	1850–1914	1914–1939	After 1939
Conyza canadensis	T	1730	--□--	-------	-------	-------
Senecio vernalis	T	1824	-------	--□--	-------	-------
Sisymbrium loeselii	T	1824	-------	-------	--•--	--□--
Atriplex tatarica	T	First half of nineteenth century	-------	----	- - - -	--□--
Galinsoga parviflora	T	1870	- - - -	--□--	-------	-------
Amaranthus retroflexus	T	1873	- - - -	--□--	-------	-------
Helianthus tuberosus	G	1873	-------	-------	--•--	--□--
Reseda lutea	H	1652?,1873	-------	-------	--•--	--□--
Acer negundo	F	1880	- - - -	- - - -	-------	--□--
Erigeron annuus	T	1867	- - - -	--□--	-------	-------
Parthenocissus quinquefolia	F	1880	- - - -	-------	-------	--□--
Robinia pseudoacacia	F	1873	- - - -	--•--	-------	--□--
Bunias orientalis	H	1883		-------	-------	--□--
Oxalis stricta	G,T	1873		-------	--□--	-------
Matricaria discoidea	T	1884		--□--	-------	-------
Diplotaxis muralis	T	1895		-------	--•--	--□--
Eragrostis minor	T	1894		-------	-------	--•--
Oenothera depressa	H	1894		-------	-------	--□--
Sisymbrium altissimum	T	1894		-------	--•--	--□--
Impatiens parviflora	T	1884		-------	-------	--□--
Solidago gigantea	H,G	1894		-------	--•--	--□--
Solidago canadensis	H,G	1922		-------	--•--	--□--
Xanthium albinum	T	1869		--□--	-------	-------
Galinsoga quadriradiata	T	1917		-------	--•--	--□--
Medicago sativa ssp. *varia*	H	1918		-------	-------	--□--
Cyclachaena xanthiifolia	T	1959				--□--
Lepidium virginicum	T	1946				--□--
Bidens frondosa	T	1960				--□--
Echinocystis lobata	T	1964				--□--
Fallopia japonica	G	1964				--□--

T therophyte, *G* geophyte, *H* hemicryptophyte, *F* phanerophyte; (•) more frequent occurrence of the species; (□) expansion of the species within the city limits: species occurrence: - - - very likely, -------- confirmed by historical data

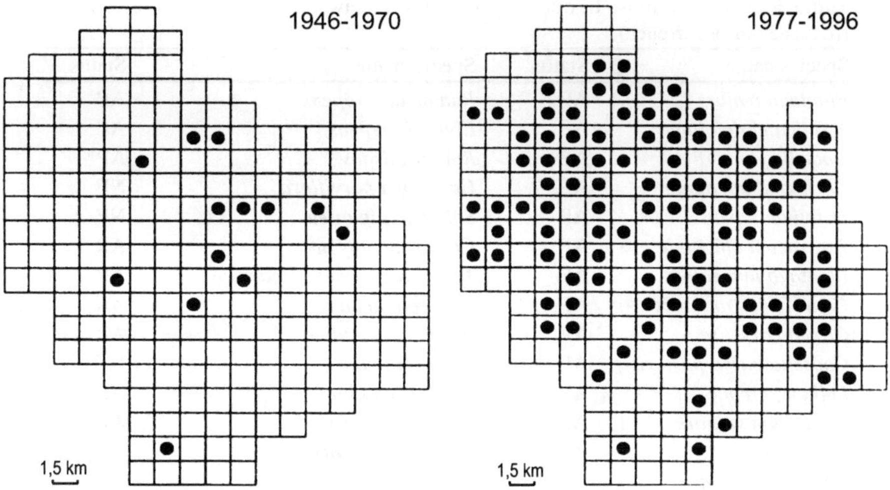

Fig. 7 Changes in the distribution of *Cyclachaena xanthiifolia* – alien species recorded for the first time in Warsaw after 1945

of the 1980s, the species began to spread rapidly in anthropogenic habitats. At this time, it was recorded as *E. pilosa* and in 1995 described by Scholz as *E. albensis* – a neo-endemic species in Europe (however, its status is unclear).

The expansion of some native species is also worthy of attention, for example, *Viscum album* ssp. *album,* a hemi-parasite which colonises more and more trees in cities. Detailed studies conducted in Warsaw since 2003 indicate that the species has been found abundantly in green spaces within housing developments built in the 1960s and 1970s, and along roadsides. The most frequent hosts of the *Viscum* in the city are three alien species, *Acer saccharinum, Populus* x *canadensis* and *Robinia pseudoacacia.*

The Most Frequent Species

The 50 most frequent species that occur within the administrative boundaries of Warsaw are listed in Table 5. Native species are dominant (72%), with archeophytes and neophytes occurring in equal proportions – 14% each.

Planted Trees and Shrubs

Urban green space (excluding natural and semi-natural green spaces) comprises about 9% of the total area occupied by vegetation in Warsaw. It includes, among others, the network of parks, green squares, street-side greenery, children's play

Table 5 The fifty most frequent species in Warsaw (species list in the order of frequency of occurrence)

Species name	Status	Species name	Status
Plantago major	AP	*Ranunculus repens*	AP
Achillea millefolium	AP	*Elytrigia repens*	AP
Dactylis glomerata	AP	*Rumex crispus*	AP
Chenopodium album	AP	***Galinsoga parviflora***	NE
Trifolium repens	AP	***Solidago gigantea***	NE
Taraxacum officinale	AP	*Cichorium intybus*	AR
Plantago lanceolata	AP	*Tripleurospermum inodorum*	AR
Convolvulus arvensis	AP	*Berteroa incana*	AR?
Cirsium arvense	AP	*Medicago lupulina*	AP
Capsella bursa-pastoris	AR	***Acer negundo***	NE
Lolium perenne	AP	*Agrostis capillaris*	AP
Tanacetum vulgare	AP	*Linaria vulgaris*	AP
Poa annua	AP	*Armoracia rusticana*	AP?
Trifolium pratense	AP	*Stellaria media*	AP
Silene latifolia	AP	*Vicia cracca*	AP
Urtica dioica	AP	*Bromopsis inermis*	AP
Conyza canadensis	NE	*Sambucus nigra*	AP
Sisymbrium loeselii	NE	*Atriplex patula*	AP
Potentilla anserina	AP	*Senecio vulgaris*	AR
Artemisia vulgaris	AP	***Amaranthus retroflexus***	NE
Poa pratensis	AP	*Ballota nigra*	AR
Bromus hordeaceus	AP	*Cerastium fontanum* ssp. *vulgare*	AP
Daucus carota	AP	*Sisymbrium officinale*	AR
Leontodon autumnalis	AP	*Carex hirta*	AP
Matricaria discoidea	NE	*Rumex acetosella*	AP

Key: *AP* apophytes, *NE* neophytes, *AR* archaeophytes, ? = doubtful

spaces, allotments, cemeteries, outdoor sports facilities, botanic and zoological gardens. It is estimated that the number of cultivated species growing in the city's green areas is about 200 (sub-species and varieties are not included).

The most frequently planted trees are *Acer saccharinum, A. negundo, A. platanoides, A. pseudoplatanus, Populus x canadensis, Tilia x euchlora, T. cordata, Aesculus hippocastanum, Robinia pseudoacacia* and *Sorbus aucuparia*. Many of the trees are in poor or very poor condition, especially those in the city centre. In 1986, an inventory of trees was made in the district of Śródmieście (downtown area). The total length of the streets surveyed was approximately 240 km. The study indicated that 54% of the trees were healthy, 37% were in poor condition and 9% of the trees were dying. The survey also found that 43% of the trees growing along the main streets had disappeared over the last 23 years. Newly planted trees and shrubs along busy streets do not have a good chance of survival, although the different species can vary in their tolerance.

It should be noted that some parts of the tree-lined avenues of the city have been designated as nature monuments. In many places, *Acer saccharinum, A. negundo* and *Populus* spp. are no longer planted. It is therefore necessary to plant more highly resistant taxa, such as *Platanus x acerifolia,* which used to be a rare species in Warsaw, and has been planted extensively in the last 10 years. The following species are planted because they appear to tolerate increased salinity levels: *Elaeagnus angustifolius, Amorpha fruticosa* and *Tamarix* sp.; preliminary studies indicate that they have not yet posed a threat to the native flora of Warsaw.

Habitats

The land use in the city (based on data from 2006) is presented in Table 6. As stated earlier, complex studies on the plant cover of Warsaw, including the charting of the flora and vegetation within the city boundaries, were conducted in the 1980s and 1990s (for example, Sudnik-Wójcikowska 1987–1998; Chojnacki 1991). At the end of the 1990s and at the beginning of the twenty-first century, the flora of some parks and habitats of the city was analysed, as well as the dynamics of some species and syntaxa (for example, Sudnik-Wójcikowska 1998; Sudnik-Wójcikowska and Galera 2005).

Surveys have been carried out of some of the city's existing and planned nature reserves (Wojtatowicz 2006). The Bielański Forest, which is located about 6 km from the city centre, has been surveyed in the greatest detail. Floristic studies were also conducted within the Kabacki and Bemowski forests, the floodplain forests along the Vistula and the meadows below the Warsaw Escarpment. In the 1960s, investigations were conducted on the vegetation and flora of all the bodies of water in Warsaw. Owing to drastic man-made changes to the environment, these studies need to be repeated and extended.

With the exception of the surveys of the dendroflora, which were carried out before 1939, only a few studies of the parks have been undertaken. However, the

Table 6 The structure of land use in 2006

Land-use type		Percentage in the city area (517 km^2)	
Natural and semi-natural habitats – forests		14	
Natural and semi-natural habitats – non-forest vegetation		9	
Crops and abandoned fields		12	
Urban green spaces	Parks	9	2.36
	Botanic gardens and zoo		0.16
	Green areas and squares		0.09
	Allotments		3.61
	Cemeteries		0.90
	Roadside and streetside strips of greenery		1.80
Biologically inactive areas and areas occupied by ruderal vegetation		47	
Water		3	
Other land-use types		6	
Total		100	

spontaneous flora of cemeteries, allotments and botanic gardens have been recorded (Sudnik-Wójcikowska and Galera 2005). A floristic inventory of military facilities (old forts and abandoned military polygons) has been initiated recently.

An inventory of the vegetation and flora of the railway areas was carried out in the 1960s and repeated recently (but only of a very small area). At the end of the 1990s, the flora and vegetation of the tramways were investigated in greater detail. Because of the recent modernisation of the tramways, this information is now only of historical value. During the last 15 years, floristic investigations have been carried out on such interesting structures as the flat roofs of the Palace of Culture and Science, the horizontal surfaces of disused factories, bridges and sports facilities. In addition, surveys of the flora of domestic refuse sites have been carried out.

Natural and Semi-natural Habitats

The map of the present-day potential natural vegetation of Warsaw (Chojnacki 1991) indicates that the housing and industrial areas in the central, western and southern parts would be dominated by deciduous forest communities (with *Acer* spp. and *Robinia*) related to the species-rich, mesic *Quercus-Carpinus* forests (with elements from the alliance *Carpinion* and from the class Robinietea). Mixed *Pinus-Quercus* forests would occur mainly in the east and north parts of the city and floodplain forests would grow along the Vistula River valley, which is very wide, especially in the southern part of the city.

The city fringe, as well as the Warsaw Escarpment and the Vistula River valley are areas where patches of semi-natural vegetation and vegetation resembling natural ones (forests, scrubs, grasslands and waterside communities) have survived to this day. It is estimated that in the 1990s this type of vegetation occupied about 25% of the city area; however, general observation suggests that this is now lower.

The Warsaw Escarpment, which is a unique feature of the city's landscape, had and has a major influence on the spatial development of Warsaw. Technical problems that have prevented building close to the escarpment and the Vistula valley have enabled many interesting and valuable plant communities to survive. The most interesting semi-natural habitats, especially those listed in the EC Habitat Directive, are described in the following paragraphs.

Forests

Some of the forests within the boundaries of Warsaw, for example, Bielański and Natoliński Forests are the remnants of the former Mazovian virgin forest. The deterioration of these forests accelerated during the first 25 years of the twentieth century. Nowadays, forests, including degraded forests and monocultures, cover an area of

7,260 ha (14%) of the municipality. Forest complexes are mainly found in the peripheral parts of the city and are usually contiguous with forests outside the city, which form a green ring around it. Smaller fragments of forest occur within built-up areas.

About 40% of Warsaw's forests are 20–50-year-old monocultures mainly of *Pinus sylvestris*, which have been planted on former agricultural, and industrial land; only 20% of the total forest area is covered by plant communities resembling natural ones. At present, eleven forest associations and five scrub associations have been identified within the city (Chojnacki 1991). It is thought that the plant associations are those that dominated the prehistoric Mazovian Forest centuries ago. The deciduous forests and scrub are represented by nine associations, whereas *Pinus* and mesic mixed *Pinus-Quercus* forests have four associations. Most of these communities (habitats) are included in the Natura 2000 network. They have been preserved to varying degrees, and therefore require special protection. In many cases, the communities have been severely degraded and should be defined only as substitutive communities from the alliance *Carpinion* (*Carpinus* forest or floodplain forest).

Deciduous Forests and Shrubs

1. *Tilio-Carpinetum* (EUH code: 9170-2; Corine code 41.262)
   This type of forest has persisted in small, fertile and relatively moist areas on Warszawska Plain (for example, in some parts of Kabacki Forest), and on the over-flood terrace on the left bank of the Vistula River (for example, Bielański and Natoliński Forests), where richer and older *Quercus-Carpinus* stands have survived within floristically poor *Quercus-Carpinus* forests. In many places, the structure of the canopy has been altered as the result of cultivation of *Pinus sylvestris, Betula pendula* and the introduction of alien species such as *Acer negundo, Robinia pseudoacacia* and *Prunus serotina.*
2. *Potentillo albae-Quercetum* (EUH code: 9110-1; Corine code 41.11)
   This is a rare forest community in Warsaw. It occurs in nutrient-rich and relatively dry habitats, along the edge of the Warsaw Escarpment (from Powsin to Wolica) and in the eastern part of the city (King John III Sobieski Reserve). Patches of relatively young stands of *Quercus* spp. are also found in the districts of Anin and Wawer. It seems that forest management and grazing have contributed to the survival of patches of the thermophilous *Quercus* forest, which are syntaxonomically diverse.
3. *Ficario-Ulmetum* (EUH code: 91F0; Corine code 44.4)
   This forest type occurs on thick, fine-grained alluvial soils on the floodplain terrace of the Vistula River and in some depressions within the Praski Terrace. The community is relatively well-preserved in nutrient-rich habitats such as Bielański and Młocinski forests and in Morysinek.
4. *Salicetum albae* (EUH code: 91E0-1; Corine code 44.13) and *Populetum albae* (EUH code: 91E0-2; Corine code 44.13)

Both types of floodplain forest occur on the left and right sides of the Vistula in those areas furthest from the channel that are flooded annually. The better developed communities are found on the northern and southern outskirts of the city. Those forests located closer to the centre serve as parks (and therefore subject to trampling and eutrophication). Trees are felled occasionally to facilitate ice flow.

5. *Salicetum triandro-viminalis* (No EUH or Corine codes)
   This vegetation type develops best along most of the Vistula River within the city. It occurs on banks close to the river channel, where the ice flows every winter. In some sections, it replaces degraded *Salix* and *Populus* floodplain forests.

6. *Fraxino-Alnetum* (EUH code: 91E0-3; Corine code 44.3)
   This community occurs on the Praski Terrace in the peripheral parts of the city, in the vicinity of small watercourses and drainage ditches or within larger forest complexes. This community is relatively well developed in Bemowski forest and in "Zakole Wawerskie". Frequently, these types of forests developed as a result of afforestation of wet meadows and wasteland after 1945. The trees are fairly young and all about the same age. The undergrowth beneath the trees has been disturbed.

7. *Ribeso nigri-Alnetum* (No EUH or Corine codes)
   This type of forest occurs in periodically flooded depressions filled with a thin layer of peat and with no inflow or outflow. Such forests are found in the peripheral parts of the city, mainly in the Vistula River valley, on the Praski Terrace (for example, in the north and east of the Bródno and Grochów districts) and at the base of the escarpment in the district of Ursynów but rarely on the floodplain terrace (Zakole Wawerskie).

8. *Salicetum pentandro-cinereae* (No EUH or Corine codes)
   This vegetation type occurs in abandoned meadows; it precedes the previous community type during the succession process or forms as a result of degradation of such forests. It is found in the north-eastern peripheral areas of the city.

Mixed *Pinus-Quercus* and *Pinus* Forests

1. *Querco roboris-Pinetum* (No EUH or Corine codes)
   These forests occur on relatively rich loose sands or sandy clay soils, on dune terraces, and in some places within the Warszawska and Błońska Plain. The plant community varies in structure and species composition depending on the intensity of human interference or other activities (for example, forest management, hiking). Some of the tree stands developed spontaneously from monocultures of *Pinus* spp. They occur, among other places, on sand dunes between the districts of Białołęka and Choszczówka, in Rembertowski, Aniński and Młociński Forests.

2. *Peucedano-Pinetum* (No EUH or Corine codes)
   This type of forest occurs mainly on dune terraces (from Rembertów to Falenica) and at the top of the fixed dunes on the Praski Terrace, in the eastern

part of the city (for example, Białołęka, Choszczówka). In many places, sandy grassland species are over-represented because of excessive exploitation of timber.

3. *Molinio-Pinetum* (No EUH or Corine codes)
   This type of forest is rarely found, it occurs in moist depressions between the dunes on the eastern city fringe (for example, near Wiśniowa Góra). The ground cover, which is dominated by *Molinia caerulea* ssp. *caerulea,* contains a high proportion of *Ericaceous* species.

4. *Vaccinio uliginosi-Pinetum* (EUH code: 91D0-2; Corine code 41 A)
   This forest type has been reported from single localities only. It occurs close to the eastern boundaries of the city of Warsaw, accompanied by patches of poor fen near Zbójna Góra.

Non-forest Habitats

Non-forest habitats within the boundaries of big cities have been particularly subject to alteration and disturbance, especially during the last 20 years. Most of the plant communities described below are declining and occupy only small areas. Therefore, they may gradually disappear.

Aquatic and Reedswamp Communities

1. *Lemnetea* and *Potametea* communities (No EUH or Corine codes)
   The better developed areas of these communities occur in large water bodies such as old riverbeds, mainly in the southern part of the floodplain terrace of the Vistula river, for example, in Lisowskie and Powsińskie lakes, on the edge of existing clay pits, for example, in the district of Jelonki, in lakes near Dawidy and Jeziorki and in channels and ponds within the parks. The diversity of the plant communities found within a given area depends on factors such as: size and depth of the body of water and degree of its contamination. The large, old riverbeds support the greatest number of plant communities.

2. *Phragmition* (No EUH or Corine codes)
   These communities are usually found in littoral habitats in standing bodies of water and in slow-flowing watercourses, more rarely in periodically flooded terrain depressions. They occur abundantly in the floodplain terrace of the Vistula river valley, especially in the southern part of the city. Communities dominated by *Typha latifolia* and *Glyceria aquatica* occupy the largest area. Phytocoenoses of the communities are also found growing along the edge of some park ponds.

3. *Bidentetea tripartiti* (No EUH or Corine codes)
   These communities occur along both sides of the channel of the Vistula river that are exposed during low water levels. They also occur on the exposed margins of water bodies including lakes, ponds and old riverbeds. The communities comprise a high proportion of neophytes such as *Bidens frondosa* and *Xanthium albinum* as well as the commonly occurring *Echinocystis lobata* in the neighbouring riverside herb and "veil" communities.

Heath and Nardus stricta Moorlands

Relict heath and *Nardus stricta* moorland communities and other phytocoenoses from the class *Nardo-Callunetea* occur in the outskirts of the city.

1. Calluno-Arctostaphylion (EUH code: 4030; Corine code 16.24)
   These communities, which are poor in species and highly degraded, occupy very small areas near Wólka Węglowa and in Wawerski Forest.
2. Lowland grasslands with *Nardus stricta* from the alliance *Violion caninae*
   It is difficult to univocally identify the above plant communities, whose structure and composition are highly variable. They are found on shallow, reclaimed peat soils in the depressions within the Praski Terrace and Błońska Plain. In the north-eastern part of the city, these communities probably developed from meadows degraded because of excessive grazing. The extent of these communities has decreased because of urbanisation.

Communities of Sandy Grasslands

These sandy communities have been degraded and are poor in species or have a higher contribution of ruderal species. They include species such as *Calamagrostis epigejos* and *Leymus arenarius,* which have been planted to stabilise dunes.

1. *Corynephorus canescens* grasslands – *Corynephorion canescentis* (No EUH or Corine codes)
   These communities develop optimally under oceanic and sub-oceanic climate conditions. The enclaves of this type of vegetation are found on dunes of the over-flood terraces and occupy small areas in coniferous and mixed forests.
2. *Koelerion glaucae* (EUH code:*6120; Corine code 34.12)

   These communities, with a sub-continental centre of distribution, corresponds (to some degree) to the sandy steppes of south-eastern Europe. In Warsaw, they contain a small proportion of continental species and occur in habitats that are more fertile and most the *Corynephorus canescens* grasslands. The communities occupy small areas of river and fluvioglacial sands, mainly on over-flood terraces.

Meadow Communities

1. *Molinion* (EUH code: 6410; Corine code 37.311)
   The area of these communities, which once occurred on dune terraces and on the Błońska Plain, has been greatly reduced over the last few years. In spite of the efforts of numerous ecologists from non-governmental organisations, the meadows have practically disappeared from the north-western part of the city (for example, Bemowo and Radiowo); however, because of active protection, they have managed to survive in the Kalinowa Łąka reserve.

2. *Filipendulo-Petasition* (No EUH or Corine codes)
   These types of communities occupy very small areas on the city fringe.

3. *Arrhenatherion* (No EUH or Corine codes)
   Communities of this type occur mainly on the rich alluvial soils of the floodplain terrace of the Vistula valley and on the lighter alluvial soils of the Praski Terrace. The species composition may be altered by the sowing of large quantities of pasture grasses or introducing nitrophilous ruderal species.

4. *Cynosurion* (No EUH or Corine codes)
   These pasture communities occur on various organogenic and mineral soils in the sub-urban zone in different parts of the city. They are spatially associated with trampled grasslands (*Plantaginetalia*) or with mesic meadow (*Arrhenatherion*) communities, depending on their form and intensity of human impact.

City Centre and Residential Areas

It has been estimated that in the 1990s, about 39% of the Warsaw area was occupied by synanthropic vegetation (including ruderal, segetal and urban green areas) and about 29% covered by various other forms of hard surface.

The spatial structure of Warsaw is similar to that of other urban agglomerations and is largely the result of historical development as well as changes in its natural environment. The administrative centre (the oldest part of Warsaw) is at the centre of four concentric zones, each representing more intensive levels of human activity. Quantitative and qualitative studies of the flora have shown that the older central part of the city is distinctive. It is characterised by high density housing (high-rise and low-rise buildings), the highest population density and the most developed transportation network. At the same time, the heat-island effect on the plant cover in the central parts of Warsaw are more pronounced, which is indicated on satellite infrared photographs of temperature distribution in the city.

Trends in the development of plant cover were observed in relation to:

(a) Timing of the phenological events in the case of some species.
(b) Distribution patterns of particular species.
(c) Species composition.

It was found that cultivars of *Forsythia* are a good indicator of the temperature conditions in the city. The time when buds open or flowers are in full bloom in early spring was analysed, along a north-to-south transect running through the city centre. It was found that the phenological events described above occurred 5–7 days earlier in the city than they did in the suburbs.

Particular attention was paid to a group of alien species which occurred exclusively or more frequently in the central parts of the city (Sudnik-Wójcikowska 2000). They were mainly thermophilous species (with high temperature indicator value, the so-called Ellenberg's indicators) introduced accidentally from zones warmer than those in Poland, for example, *Ailanthus altissima* (Fig. 8a), *Hordeum murinum, Eragrostis minor* (Fig. 8b) and *Cyclachaena xanthifolia* (Fig. 7), which are indicators of the urban heat island in Warsaw.

Studies have indicated that the structure and species composition of the flora were more indicative of the temperature conditions in the city than the distribution of individual species. The central part of Warsaw is subjected to the strongest human impact (anthropopressure, Fig. 9a). The richness of the flora of particular grid squares decreased towards the centre of the city while the proportion of neophytes increased (Fig. 9b). The Ellenberg's indicators showed that both the mean temperature indicator values for the flora of particular grid squares (Fig. 9c) and the proportion of species with higher temperature requirements distinguished the centre from other parts of the city. It should be noted that alien species with higher temperature demands (but having narrower ecological amplitude) prevail in the flora of the city centre. On the other hand, ubiquitous species occurring commonly both in the centre and peripheral parts of the city, which showed a wide ecological amplitude and tolerated higher temperatures, were highly represented in the native flora of the city centre.

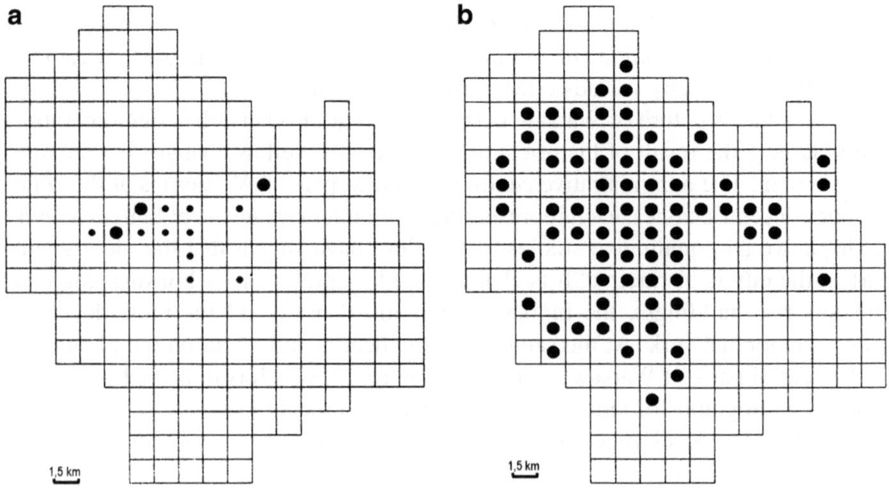

Fig. 8 The distribution of two thermophilous species in Warsaw: (**a**) *Ailanthus altissima*; (**b**) *Eragrostis minor*

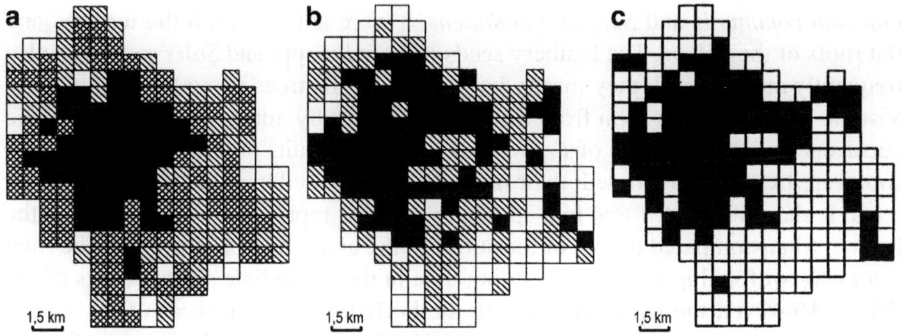

Fig. 9 The effect of anthropopressure and urban heat island on the species composition of the flora of Warsaw: (**a**) four anthropopressure zones in Warsaw; (**b**) an increase in the proportion of neophytes in the flora of squares towards the city centre; (**c**) the division of Warsaw into two zones based on the mean temperature indicator values obtained for the flora of each sample square. The darker the hatch, the higher is the man-made transformation of habitats (**a**), the role of neophytes (**b**) and the mean temperature indicator values (**c**)

Crevices, cracks and joints in walls, paving and other structures are typical city centre habitats.

The species-richness of the plant cover of such habitats depends on the age, exposure, frequency of renovation practices and degree of conservation. Conditions for growth are poor because the plants are exposed to extreme temperature fluctuations, dry conditions and do not have easy access to soil, which is frequently contaminated and has poor gas exchange. Heavily trampled areas, for example, cracks in steps and flagstones, gaps between kerbs and cracks in asphalt contain species with particular growth forms. These species, which occur frequently in the city, include species with rosettes of basal leaves (for example, *Plantago major*), tussocky hemicrypophytes (*Lolium perenne*) and therophytes with procumbent stems that are resistant to mechanical damage for example, *Polygonum aviculare* agg. and *Poa annua*. These habitats also contain annual grasses such as *Eragrostis minor* (since the nineteenth century) and *E. albensis*.

The flora of the Palace of Culture and Science, which is an architectural symbol of the past epoch, is worthy of attention. The characteristic silhouette of this huge building used to dominate the city centre. Although it is slowly being eclipsed by modern high-rise commercial buildings, it is still a spatially isolated structure that is exposed to the impacts of the heat island, alternating temperatures and air pollution. In the 1990s, a botanical survey of the external horizontal surfaces (roofs, terraces, stairs and balconies) was carried out. A total of 111 species of vascular plants were found, a high proportion of which were thermophilous and heliophilous plants.

The surprisingly high contribution of higrophilous species was attributed to the fact that the disposal of run-off from flat roofs was not working properly. The high proportion of anemochorous species (85%) in the flora of the Palace is a result of the spatial isolation of the building. The pappus-bearing achenes of the species from the Asteraceae (*Lactuca serriola, Taraxacum officinale, Conyza canadensis,*

Lapsana communis and *Solidago canadensis*) were able to reach the terraces and flat roofs of the Palace. The feathery seeds of *Populus* spp. and *Salix* spp. were also frequently encountered; they may have come from the trees along the banks of the Vistula, which is about 2 km from the Palace. Especially noteworthy is the sporadic occurrence of the seedlings of *Typha* sp. The winged fruits of *Betula* spp. and *Acer* spp. were frequently recorded; their presence is attributed to strong upward air currents. It should be noted that tree species were over-represented in the flora of the Palace, when compared to that of the surrounding area; their proportion of the total flora was twofold higher on the building than in the immediate surroundings of the Palace. However, the trees recorded on the horizontal roof surfaces of the Palace were represented exclusively by seedlings. Weeding activities, which are performed periodically, do not allow the trees to develop further.

The seeds of other plants, such as *Sambucus nigra* and *Solanum nigrum,* were dispersed onto the terraces or rooftops by birds (endozoochory). The presence of diaspores of cultivars of *Malus, Prunus* and *Lycopersicon esculentum* on the terraces is attributed to human activities.

Transport Routes

Railways

The first railway reached Warsaw in 1848 (the Warsaw-Vienna line), with additional lines being built in the 1860s and 1870s. Freight trains were not used until the 1990s. At the same time, intensive grain import, mostly from America and southern Europe facilitated the dispersal of weed diaspores. In those days, the flora of railway areas was richer in species of alien origin, which were only temporarily introduced, including numerous exotic weeds associated with grain.

Botanical studies of the railway flora were carried out in the 1960s, 1980s and 1990s. The latter was of the railway tracks leading to and from a large grain silo. The results of these studies indicated the presence of many rare species, which were recorded for the first time in Warsaw, for example, *Aegilops cylindrica, Ambrosia artemisiifolia, A. trifida, Ammi majus, Bifora radians, Consolida orientalis, Gypsophila pilosa, Hirschfeldia incana, Myagrum perfoliatum, Rapistrum rugosum, Silene gallica* and *Sorghum halepense*. Present investigations indicate a considerable decline in the number of ephemeral species growing on the railway tracks.

Railway land comprises a mosaic of microhabitats including tracks, inter-rail areas, goods yards, embankments, cuttings, railway stations and loading platforms. Plant growth can be impeded by unfavourable environmental conditions, for example, fluctuations in temperature, moisture and pollution (especially hydrocarbons), as well as continuous mechanical damage, which change over time and short distances. In addition, the movement of air caused by the trains facilitates the spread of some plants, for example, anemochorous or boleachorous species such as

Tragopogon dubius and *Eragrostis minor* (which occur abundantly in railway areas), *Cirsium vulgare, C. arvense, Carduus acanthoides, Senecio vernalis, S. viscosus, S. vulgaris, Solidago canadensis* and *S. gigantea.*

Another factor that affects the distribution and growth of plants is the variable substrate, which may consist of a hard rock, aggregate or sand sometimes covered with a thin layer of humus. These areas are mainly inhabited by species associated with gravel pits and stones, such as *Geranium robertianum* and *Galeopsis angustifolia,* which have only recently been recorded in Warsaw. In addition, the periodic application of herbicides (salts of glyphosate) reduces the number of species, especially after the first round of spraying. However, some species (usually deep-rooted perennials) are resistant to some herbicides; they include *Equisetum arvense, Carex hirta, Convolvulus arvensis, Hypericum perforatum, Poa compressa, Calamagrostis epigejos* and *Cirsium arvense.*

The exposed and very dry substrates result in the plants being subject to extreme temperature fluctuations, excessive light intensity, wind and frost action, as well as erosion. Where these conditions prevail, short-lived species (therophytes) are abundant, especially summer annuals that require high germination temperatures, for example, *Eragrostis minor, Digitaria sanguinalis, Setaria viridis, Amaranthus retroflexus, A. albus* and *Bidens frondosa.* A number of annual and perennial Irano-Turanian, Pontic, Pannonic, Mediterranean, Sub-Mediterranean species were also recorded; they include *Bunias orientalis, Lepidium draba, Salsola kali* ssp. *ruthenica, Sisymbrium altissimum, S. wolgense* and recently *Bassia scoparia.* Other species, particularly worthy of mention, are those that were planted at the turn of the nineteenth century to strengthen railway embankments and which have persisted along some railway tracks, for example, *Sisymbrium wolgense* and *Leymus arenarius.* Old localities of these two species were recently found in the vicinity of eastern Warsaw station.

Tramways

The first horse-drawn tram in Warsaw was introduced in 1866. Twenty years later, the system comprised 22 km of track, 150 tramcars drawn by 390 horses. Electric power eventually replaced horsepower and by 1939 the length of track had increased to 120 km. After 1945, the network was significantly extended. It now comprises 34 lines with a total of 470 km of track. The flora of tramway tracks is far less interesting than that of railways. It is exposed to similar unfavourable conditions for plant growth, such as: mechanical damage, extreme temperatures, dry conditions, contamination by hydrocarbons and the use of herbicides (salts of glyphosate are applied twice during the growing season). The modernisation of the tracks has further limited the space available for plants.

Detailed studies of the flora and vegetation of tracks were conducted in 1995 and 1996, before their modernisation (Sudnik-Wójcikowska 2001). A total of 237 phytosociological relevés and 199 species, including 104 native and 95 aliens, were recorded. *Cerastium semidecandrum* was dominant during the spring season, *Senecio vernalis* and *Capsella bursa-pastoris* were abundant as well. During the summer and after the first spraying of herbicide, *Eragrostis minor* and *Digitaria sanguinalis* were dominant. After the second spraying, there were fewer species, most of which were autumn annuals such as *Atriplex tatarica.*

Species from far away places, which were abundant and are still found in railway areas, spread along tramway lines mainly from ruderal habitats. However, the seedlings of useful plants were found more frequently along tramway lines than railway tracks. *Eragrostis minor* and *Atriplex tatarica* occurred frequently and abundantly along the tracks. Growing previously on the Vistula banks, *Eragrostis albensis* has been spreading into ruderal habitats since the beginning of the 1980s; at present, it is common on the tramway tracks in many districts. Thermophilous tree species, which are rare in Poland, occur as single specimens, for example, seedling of *Ailanthus altissima* and *Gleditsia triacanthos.*

Roads, Expressways, Streets and Trampled Areas

Trampled areas including the sides of roads and paths are specific types of habitats in which the substrate is very compact and usually poor in oxygen. In addition, mechanical factors eliminate some species and favour the development of low-growing vegetation comprising anthropogenic, moderately nitrophilous and relatively species-poor communities. The dominant species are light-loving common, cosmopolitan annuals with poor competitive abilities such as *Poa annua, Plantago major* and *Polygonum aviculare* agg. Other species which occur abundantly include *Eragrostis minor* and recently *E. albensis.*

In addition to mechanical factors, the application of herbicides and chemical factors such as pollution (from vehicle exhaust emissions and other sources) determines the species growing along roadsides. During the winter, salts (mainly NaCl with some $CaCl_2$ mixed with sand) are used to de-ice the roads and footpaths. As a result, salt-rich (*solonchak* and *solonetz*) soils comprise about 15% of the total soil area of the city (Biernacki et al. 1990). The concentration of soluble salts in the roadside soils varies within the seasons and spatially. Seasonally, the concentration is highest in the spring and then declines as the salts are gradually washed away by rain. However, during prolonged dry periods, the salt level may increase as a result of capillary action bringing salt from lower levels in the soil. The greatest concentration occurs 50–70 cm from the carriageway declining with increasing distance from the road. This strip, which receives the greatest concentration of salt and mud, has been colonised by sub-halophilous species that germinate mainly in late spring or early summer, for example, *Atriplex tatarica, Puccinellia distans, Elytrigia repens* and *Polygonum aviculare* agg. They are accompanied by species that occur at a greater distance from the carriageway such as *Lolium perenne, Taraxacum officinale, Convolvulus arvensis, Digitaria sanguinalis* and *Leontodon autumnalis.*

Roadside trees are sensitive to and affected by increasing salinity in the soil. For example, damage to the leaf surface or the timing of some phenological events of *Tilia x euchlora* were observed in Warsaw. As described earlier, the roadside trees in the central districts are in poor condition. The estimated average life span of roadside trees growing under the worst conditions is about 7 years.

Recreation Areas

Managed landscapes used for recreation include 84 parks and historical gardens, two botanic gardens and one zoological garden, 28 nineteenth century forts, 280 allotments, 45 cemeteries, several larger sports facilities, numerous green squares, lawns and playgrounds. These areas occupy a total of about 9% of the city (Table 6).

Parks and the Warsaw Escarpment

The oldest parks and palace gardens in Warsaw were established in the sixteenth and seventeenth centuries. The first garden (Saxon Garden) was opened to the public in 1727. Starting from the end of the seventeenth century, a number of park and palace complexes and manor houses with gardens (for example, Łazienkowski, Wilanowski, Ursynowski and Natoliński Parks) were established in the immediate vicinity of Warsaw. Most of them gave rise to the creation and further development of urban parks in the city. Two botanic gardens (one of which does not exist now) were founded in the nineteenth century. In the second half of the nineteenth century, three new parks were established, namely, Praski, Ujazdowski and Skaryszewski. Żeromski, Traugutt and Dreszer Parks, as well as the Warsaw zoo, were opened after soon after 1918. Most of the parks were almost completely destroyed between 1939 and 1945. Several new parks were created after 1945 including Szczęśliwicki, Moczydło and Citadel. Today, the 84 parks in Warsaw occupy 1,213 ha.

The topographical conditions determine the location of the parks, which can be divided into three main groups: (a) parks located on the Warsaw Escarpment (Kaskada, Arkadia, Królikarnia and Ursynowski Parks); (b) parks within Warszawska Plain (Żeromski Park, Saxon Garden, Krasiński Garden and the Szwajcarska Valley) and (c) parks situated below the escarpment in the Vistula Valley (Kępa Potocka, Łazienkowski, Praski and Wilanowski Parks).

Historic parks located in the Escarpment usually comprise two parts: the upper part with the palace situated on the edge of the Escarpment and a lower garden on the slopes below the Escarpment. The palace is usually centrally placed with respect to the axis of the park. The park areas within the Warszawska Plain have a very flat and quite uniform topography and were usually designed in the baroque style. The old parks established within the Vistula Valley contain channels and bodies of water; two of them are the largest parks in Warsaw: Łazienkowski (76 ha) and Skaryszewski Park (50 ha). Before 1939, dendrological studies were conducted in three large parks, Paderewski Park, Krasiński Park and Saxon Garden. Floristic investigations have been carried out in Wilanowski and Natoliński Parks. However, the spontaneous flora of the parks in Warsaw has not been investigated in detail.

Most of the forts (Fig. 3) are covered by spontaneous vegetation. Others are utilised as green areas (for example, the area surrounding the Citadel – about 17 ha). The flora of some forts is the subject of active research.

Botanic Gardens

There are two botanic gardens in Warsaw: the Botanic Garden of the University of Warsaw, which was established in 1818 in the city centre, covers 3.5 ha, and the Botanic Garden of the Polish Academy of Sciences, which was established in 1974, occupies 40 ha near the southern border of the city. Botanic gardens contain structurally and biologically diverse habitats, which is attributed to the various forms of land use. In addition to areas that are subject to direct anthropogenic influences (for example, intensively cultivated flowerbeds and heavily trampled roadsides), there are areas that have been only slightly transformed by human activities (for example, parkland areas). Botanic gardens are areas of intensive cultivation of plants belonging to many taxa – species, varieties and hybrids. The number of taxa that are cultivated in the gardens changes every year. Moreover, the plants are introduced from various sources. There has been a regular international exchange of plant material (diaspores) between the botanic gardens for many years. Sometimes, the diaspores of weeds are introduced accidentally to the gardens in imported soil or with cultivated plants.

Studies of the spontaneous flora of the two gardens were carried out between 1992 and 1999 by Sudnik-Wójcikowska and Galera (2005), who recorded 675 taxa – 55% were aliens of which 27% occurred locally and temporarily having been introduced accidentally or escaped from cultivation. Particularly noteworthy were wild-growing species, which had persisted only in botanic gardens for years and have not so far been noted in other parts of the city, for example, *Eragrostis multicaulis, Veronica peregrina, Euphorbia maculata* and *E. humifusa*.

Allotments

The first allotment gardens (called "Obrońców Pokoju") were established in the peripheral parts of Warsaw in 1905. During the interwar period and after 1945, the area of the allotments increased considerably. There are currently 280 allotment gardens covering 1,858 ha of the city. They are an important element of the urban landscape and are under threat from urban expansion. From 2004 to 2007, detailed floristic investigations were undertaken in 64 allotments. A total of 160 taxa, which had "escaped" cultivation, were recorded in the immediate vicinity of the allotments. Among the garden escapes, the most frequently occurring taxa (recorded in the vicinity of at least 33 allotments) were *Lunaria annua, Muscari neglectum, Fragaria* x *ananassa, Iris* x *hybrida, Tulipa* 'Darwin Hybrid', *Helianthus tuberosus* and *Fallopia japonica.* The last two have become permanently established invasive species and are widespread in the city. The allotments were probably the main source of diaspores (vegetative diaspores) of both species. Those taxa found more rarely include *Parthenocissus quinquefolia, Echinops sphaerocephalus, Myosotis sylvatica* and *Papaver orientale.*

Cemeteries

Although cemeteries are an integral part of the city, they undergo much slower changes than the surrounding area. The 45 cemeteries cover a total area of 449 ha and vary in age, size, degree of preservation and intensity of use. Very large cemeteries have survived within the city limits. Komunalny Północny (143 ha) and Bródnowski (113 ha) cemeteries are the biggest in Warsaw. The third largest and the oldest cemetery in the city is Powązki Cemetery (43 ha), which was founded in 1790. The flora of 24 urban and sub-urban cemeteries covering 280 ha and located on the left bank of the Vistula River was surveyed between 1989 and 1991 (Sudnik-Wójcikowska and Galera 2005). The number of species recorded in the cemeteries ranged from 79 to 209. The urban and sub-urban cemeteries each contained a similar number of species (503 and 509 species, respectively), the range of synanthropic species groups, which was associated with similar patterns of land use.

A total of 617 species (which constitutes about 50% of the total flora of the city) were recorded within the cemeteries studied. A similar percentage of permanently established species (apophytes, archaeophytes and neophytes) was noted both in the flora of Warsaw and the cemeteries. However, species that "escaped" cultivation and established temporarily, particularly alien ornamentals, were over-represented in the flora of the cemeteries. The plants usually grew around waste dumps, along fences, on roadsides and in cemetery "wastelands". The most frequent species that were growing wild included *Ornithogalum umbellatum, Iris* x *hybrida, Hemerocallis fulva, Tanacetum parthenium, Cerastium tomentosum, Phedimus spurius, Viola* x *wittrockiana, Cosmos bipinnatus, Rudbeckia laciniata* and *R. hirta.* Indigenous cultivated species that "escaped"' from cultivation included: *Convallaria majalis, Hedera helix, Vinca minor, Sedum rupestre* and *Myosotis sylvatica.* Therefore, cemeteries, like gardens, are also centres of dispersal of alien species (a constant and abundant source of diaspores). However, the contribution of ephemeral species to the flora of cemeteries was much lower than in the case of the flora of Warsaw as a whole. This is attributed to the limited influx of diaspores of ephemeral species into cemeteries.

Cemeteries constitute "green islands" in the city landscape. However, they are distinctly different from outdoor recreation areas. The observed human impacts on cemeteries include changes in soil structure, contamination, lowering of the level of groundwater and conservation works. Some forest species were found in the cemeteries located near the city centre, these species have no ornamental value and were probably not planted; they include *Galium odoratum* and *Cystopteris fragilis.* Although the possibility of accidental introductions cannot be disregarded, it is more likely that they were present when the cemeteries were established, providing evidence that old cemeteries can play an important role as refugia for certain plant species.

Sports Facilities

The first large sports facilities were built in Warsaw at the beginning of the twentieth century (Agrykola, the former "J. Sobieski School Park"). In the 1990s, there were 15 stadia within the Warsaw area, including Warsaw's 10th Anniversary Stadium. In the 1980s, the use of the latter for sport was abandoned and it was turned into an outdoor market, which became one of the biggest in Europe, with many traders coming from Asia. The stadium was demolished in 2008.

In 2007 and 2008, the 10th Anniversary Stadium was subjected to detailed floristic research (Ostrowski et al. 2008). However, because it was demolished in 2008, the data are only of historic value. The crumbling tribunes of the stadium were colonised in many places (gaps between flagstones and crevices on stairs) by vegetation that locally resembled a young, open forest. Most of the trees were about 10 years old and comprised species with light seeds from the nearby floodplain forests, for example, *Populus* species and hybrids, *Acer platanoides, A. pseudoplatanus, A. negundo, Morus alba, Prunus avium, P. serotina, Prunus cerasifera* and *Parthenocissus quinquefolia,* whose diaspores were probably dispersed by birds.

A high percentage of heliophilous and thermophilous species was noted among herbaceous plants, for example, *Eragrostis minor, Anisantha tectorum, Conyza canadensis* and *Galeopsis angustifolia.*

Common native species dominated among perennials, for example, *Calamagrostis epigejos, Convolvulus arvensis* and *Artemisia vulgaris,* whereas *Solidago canadensis* prevailed among alien species. Small specimens of exotic species, such as *Punica granatum, Celtis occidentalis* and *Ailanthus altissima,* as well as *Fragaria* x *ananassa, Vitis* sp. and *Lycopersicon esculentum,* were also found. The above plants, as well as nitrophilous species, which occurred abundantly within the upper part of the tribunes, (for example, *Solanum nigrum, S. dulcamara, Atriplex sagittata, A. patula* and *Chelidonium majus*), were probably connected with the "activities" of traders and their customers.

Open Land

Landfill

There are two large municipal domestic refuse tips in and around Warsaw. They are situated in the peripheral parts of the city: the site at Radiowo, which was established in 1978, covers 15 ha; the site at Łubna covers 20 ha. There are also three big industrial waste disposal sites within the city. Industrial waste is generated by two power stations (which occupy 48 and 45 ha, respectively) and the Luccini steelworks (34 ha). In addition, there are about 100 illegal waste disposal sites, each of them covering an area of less than 100 m^{-2}.

Floristic investigations of the two municipal sites were undertaken between 1998 and 2007. A total of 125 species of vascular plants, including 30 species on top and 60 species on the side slopes, were recorded at Radiowo. The two sites have a combined total flora of 245 species. Reclamation work has been carried out at the two sites for many years; this includes increasing the soil thickness and stability and improving air–water relations in the soil. Plant cover development which takes place within the landfill sites must be accompanied by reasonable human actions. The assortment of species used in the plantings on the landfills should change during the reclamation process of these sites.

The top (the "youngest") parts of the site were not very stable and therefore largely devoid of plants. The upper part of the slope was colonised first by nitrophilous annual species, mainly the Chenopodiaceae, including *Bassia scoparia, Atriplex patula* and *A. nitens*. *Bassia scoparia* which is not frequently found in the city was especially abundant, covering several hundreds of square meters. Gradually, the older patches were colonised spontaneously by other Chenopodiaceae (for example, *Atriplex tatarica* and *Chenopodium album*) and Solanaceae (for example, *Solanum nigrum*) species. However, these species do not form a permanent dense cover of vegetation.

The application of a 30 cm layer of humus or top soil on the slopes allowed perennial species to colonise gradually, the most common being perennial native species with a wide ecological amplitude, for example, *Artemisia vulgaris, Calamagrostis epigejos* and *Tanacetum vulgare* as well as alien species (for example, *Solidago canadensis* and *S. gigantea*) and cultivated taxa whose diaspores have or could have been introduced with the waste. Re-cultivation of the top and slopes require species that will form a continuous cover, consequently they are sown with a commercial seed mix containing species such as *Lolium perenne, Calamagrostis epigejos* and *Agrostis capillaris*. This limits the number of sites available for colonisation by *Chenopodium* spp. and *Artemisia* spp. In addition, many Fabaceae species are sown, mainly from the genera *Lupinus, Trifolium* and *Vicia* which fix atmospheric nitrogen and, therefore, facilitate gradual colonisation by shrubs and trees, such as *Sambucus nigra, Populus tremula* and *Acer negundo*.

Subsequently, the slopes are additionally reinforced by the planting of various tree species including *Salix* spp., *Populus* spp. and *Prunus serotina*, which take up considerable amounts of nutrients, increase their biomass and are less vulnerable to fungal disease and pests. Polluted water discharges at the base of the landfill resulting in less sensitive aquatic and marginal species appearing spontaneously (or being artificially introduced).

Crops and Waste Grounds

One of the consequences of political and economical changes in Poland is the increase in land price within the city limits, which resulted in the agricultural land in the peripheral areas being sold for development. In the 1970s, arable areas (including orchards and gardens) occupied 34.5% of Warsaw (which then covered ca. 430 km^2). Today, arable land comprises about 12% of the total land area of the city (which occupies about 502 km^2). The largest remaining parcels of agricultural land are located on the eastern, north-eastern (Wawer, Białołęka) and southern (Wilanów, Ursynów) peripheral parts of the city. However, some of the parcels of arable land are being gradually fragmented or abandoned – in 2002, the latter comprised 57% of the total agricultural land. The abandoned fields are frequently colonised by species from the class *Artemisietea,* mostly by American and Asian neophytes. During the last few decades, *Solidago canadensis* and *S. gigantea* (both prairie species) have played a significant role in ruderal plant communities including post-industrial sites, railway land and the flora of the abandoned fields. In some places, they form strongly compact patches, where native ruderal plants, for example, *Artemisia vulgaris* and *Tanacetum vulgare,* do not receive enough sunlight. Another alien species, *Fallopia japonica,* is dominant and forms "single species stands" in wasteland and railway habitats and on the margins of water bodies where it is highly invasive and displaces native species.

Aquatic – Vistula River

The inter-embankment zone of the middle Vistula River valley, located between the towns of Dęblin (to the south of the city) and Płock (to the north), falls within a Natura 2000 "Special Protection Area". There are 31.6 km of the Vistula River in the city, where it is about (350) 700–1,000 m wide. During periods of low flow, alluvial islands are exposed and sandy beaches are formed on the lower banks. Alluvial soils develop on the flood terrace, which are some of the richest soils in Warsaw. Although the waters of the Vistula below Warsaw cannot be classified into any of the "purity" classes, most of the alluvial areas within Warsaw are covered by vegetation that resembles the natural vegetation cover (Fig. 10). Except for the re-enforced sections and the areas within the vicinity of the outfall sewers, a pronounced vegetation zonation occurs on the river banks throughout the city. Species from the Class *Isoëto-Nanojuncetea* and *Bidentetea tripatiti* are found closest to the river channel. *Salix* scrub and *Salix-Populus* forests, which are better preserved near the northern and southern boundaries of the city, occur further away from the river.

The plant communities occurring in the vicinity of the outfalls between Młociny and Bielany, which are highly polluted and highly eutrophic, comprise several common therophytes form the *Bidentetea tripartiti* class, mostly autumn annuals, such as *Chenopodium album, Rorippa palustris, Rumex palustris* and *Persicaria lapathifolium.* Single specimens of *Rumex maritimus, Chenopodium rubrum* and *C. glaucum* were noted. *Eragrostis albensis,* which is recognised as a European neo-endemic, grows frequently on the sand. *Bidens frondosa* (from North America) occur abundantly, whereas *B. tripartita* (an indigenous plant which occurred frequently in the past) is found only very rarely. Other alien species include mainly *Echinocystis lobata* and *Lycopersicon esculentum.* The seedlings of the latter species occur abundantly and form dense populations in the immediate vicinity of a sewage collector. Sometimes, single specimens of *Cannabis sativa* ssp. *indica* are found.

Fig. 10 View over the oldest part of Warsaw (the Old and the New Town) from the right bank of the River Vistula (Photo B. Sudnik-Wójcikowska)

Comparison of the Flora of some anthropogenic habitats within Warsaw

The various types of habitat that occur in the city are characterised by disturbed soils and moisture content, changes in the micro-climate conditions and increased levels of air pollution all caused by increasing urbanisation. Six types of habitat that were or are subjected to different types and intensities of human activity were studied by Sudnik-Wójcikowska and Galera (2005), also see Fig. 11. They are as follows: landfill, tram tracks, an abandoned stadium (the 10th Anniversary Stadium), the horizontal surfaces of the Palace of Culture and Science, cemeteries and botanic gardens.

The habitats studied covered different surface areas and were accessible to plants to varying degree. Field research was restricted to areas of homogenous land use and aimed to obtain complete floristic lists. Only spontaneous and self-perpetuating species were considered, especially in the botanic gardens and cemeteries. The results of floristic investigations conducted within the six habitats were compared. Particular attention was paid to the structure of the flora, the qualitative and quantitative similarity between floras were of lesser importance. The following parameters were considered (Fig. 12), namely, the richness of species and the percentage of:

(a) Permanently established alien species.
(b) Garden escapes (ergasiophygophytes).
(c) Annuals (therophytes).
(d) Trees and shrubs (phanerophytes).

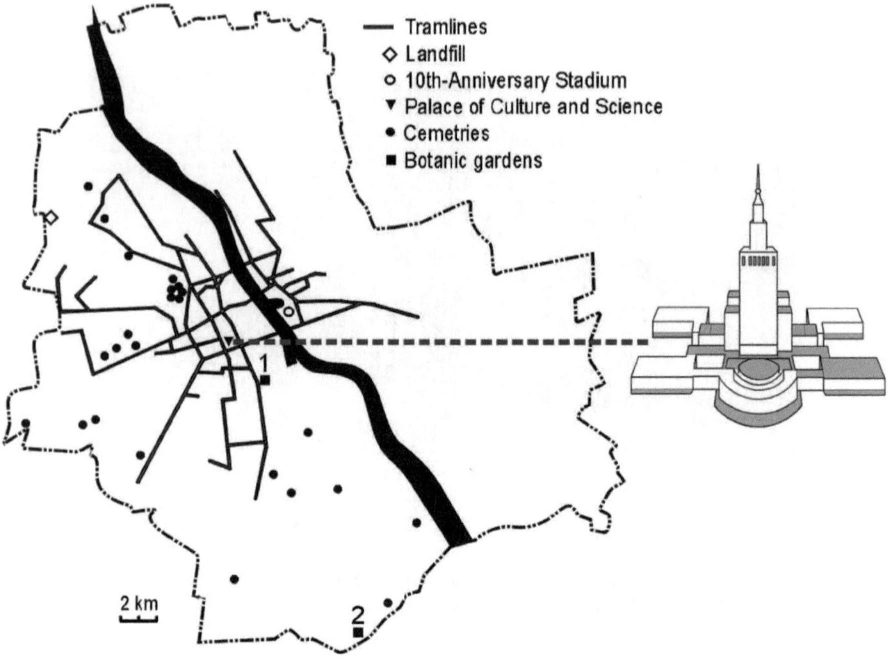

Fig. 11 Habitats studied in Warsaw: tramline network, the landfill sites in Radiowo, the Palace of Culture and Science (in grey: areas sampled in the Palace), the 10th Anniversary Stadium, 24 cemeteries on the left bank of the Vistula River, two botanic gardens: Botanic Garden of University of Warsaw (1) and Botanic Garden of Polish Academy of Sciences (2)

(e) Species producing diaspores dispersed by wind (anemochores).
(f) Species producing diaspores dispersed by animals (endozoochores).

The botanic gardens and cemeteries contained the highest number of species. Alien species constituted about 50% of the flora, virtually half of them being garden escapes, which is twice the proportion found in the other habitats studied. In addition, the flora of these two habitats was much more diverse with respect to dispersal modes, the proportion of anemochores in the flora being much lower than in the other four habitats.

The flora of landfill sites and tram tracks was relatively rich in species. The lower proportion of phanerophytes is attributed to the extreme environmental conditions (mechanical damage, herbicides and contamination). The two habitats also differed from each other with respect to the species-richness and relative abundance of therophytes. The tram lines were more species-rich but their abundance was low, whereas the species-richness of the landfill sites was lower but the plants occurred more abundantly.

The lowest number of species was recorded in the abandoned stadium and the Palace of Culture and Science. It should be noted, however, that native species, which comprised over 60% of the flora, were dominant in all the above habitats.

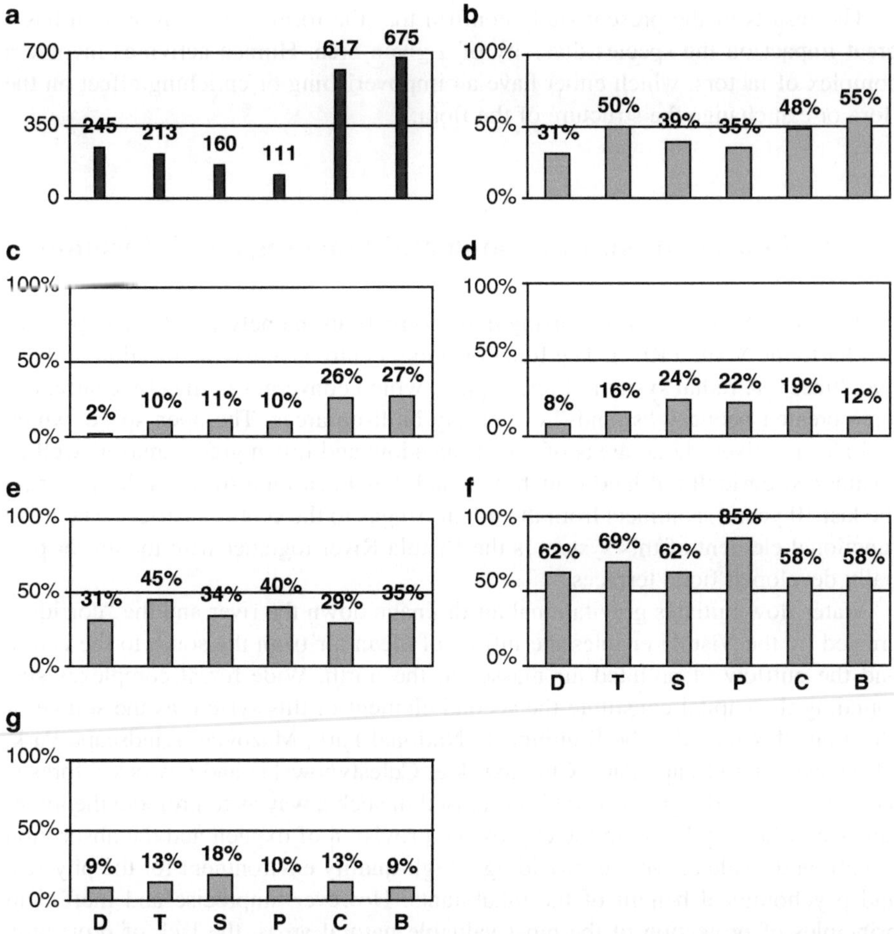

Fig. 12 Species-richness and structure of the flora (percentage of particular species groups) of the six types of habitats studied in Warsaw: D – landfill, T – tramlines, S – 10th Anniversary Stadium, P – Palace of Culture and Science, C – cemeteries, B – botanic gardens: (**a**) the richness of species; (**b**) the percentage of permanently established alien species; (**c**) the percentage of garden escapes; (**d**) the percentage of trees and shrubs species; (**e**) the percentage of annual species; (**f**) the percentage of species dispersed by wind; (**g**) the percentage of species dispersed by animals (endozoochores)

The flora of the stadium and the Palace also contained the highest proportion of trees. The high (85%) contribution of anemochores to the flora of the Palace is striking. It is attributable to the spatial isolation of this large free-standing building, which is subjected to winds and strong upward air currents and is virtually inaccessible to people and other mammals and the diaspores of plants other than anemochores. The proportion of anemochorous species in the flora of the stadium declined in favour of species dispersed by birds from the nearby floodplain forests occurring along the Vistula River and parks.

The results of the present study confirm that the form of land utilisation has a great impact on the species diversity of a given area. Human activities involve a complex of factors, which either have an impoverishing or enriching effect on the flora or can change the structure of the flora.

Nature Conservation, Environmental Planning and Education

In Warsaw, urbanisation is occurring in two directions, namely, parallel and perpendicular to the Vistula River. The functional city centre is at the intersection of these two "lines". A radial system of open space, which converges in the city centre, has been created between "strands" of densely built-up areas. The open space, which includes relatively large areas of forest, meadow and urban green areas as well as abandoned agricultural land (much of which has been intensively built up during the last 10 years), contracts from the urban fringes to the central districts. The main functional element of this system is the Vistula River together with the un- or partially developed flood terraces.

Water flow initiates gravitational air drainage down the river and the "corridor" created by the Vistula enables the inflow of clean air from the south to the centre and the outflow of polluted air masses to the north. Wide forest complexes surrounding the capital constitute the second element of this system as the source of clean air; this includes the Kampinoski National Park, Mazovian Landscape Park, Chojnowski Landscape Park, Chojnowskie, Celestynowskie and Otwockie forests. The structure of the green space is organised in such a way as to promote the maintenance of healthy living in the city by the provision of oxygen and the absorption of carbon dioxide as well as providing a high-quality environment for the physical and psychological benefit of the inhabitants. However, imprecise and inefficient principles of protection of the most valuable natural areas, the lack of protection zones coupled with the strong investment and development pressures constitute important problems for the city authorities.

Red List Species

Eighteen species listed in the Polish Red Data Book occur or have occurred within the present boundaries of Warsaw (Table 7). Five species of orchid have not been seen since the middle of the nineteenth century; they are *Cypripedium calceolus, Herminium monorchis, Liparis loeselii, Orchis coriophora* and *O. ustulata*. Only five species from the Polish Red Data Book were still recorded in Warsaw in the 1980s and 1990s, four of them are within the threat category "Vulnerable" and one in the category "Endangered". Most of the species occurred as single populations, for example, *Succisella inflexa* and *Pycreus flavescens*.

Table 7 The flora of Warsaw: the past and present occurrence of species included in the Appendix II of the Habitat Directive and in the Polish Red Data Book (2001)

		Species included in:	
Species	Year	Appendix II of the Habitat Directive	Polish Red Data Book (2001)
Botrychium matricariifolium	1866	–	+ (CR)
Camelina alyssum	1897	–	+ (EX)
Carex limosa	1885	–	+ (LR)
Pycreus flavescens	+	–	+ (VU)
Cypripedium calceolus	1873	+	+ (VU)
Elatine alsinastrum	+	–	+ (VU)
Herminium monorchis	1873	–	+ (CR)
Liparis loeselii	1873	+	+ (VU)
Lythrum hyssopifolium	1964	–	+ (LR)
Nymphoides peltata	1969	–	+ (VU)
Orchis coriophora	1878	–	+ (EX)
Orchis ustulata	1892	–	+ (EN)
Angelica palustris	+	+	+ (EN)
Polemonium caeruleum	1873	–	+ (VU)
Anemone patens	1933	+	+ (LR)
Saxifraga hirculus	1867	+	+ (EN)
Silene borysthenica	+	–	+ (VU)
Succisella inflexa	+	–	+ (VU)
Thesium ebracteatum	1933	+	–

Red list categories: *EX* extinct, *CR* critically endangered, *EN* endangered, *VU* vulnerable, *LR* lower risk

The species listed in Appendix II of the EC Habitats Directive are poorly represented in the flora of Warsaw (Table 7). Among the six species belonging to this group, three species had become extinct by the end of the nineteenth century. Two other species, *Anemone patens* and *Thesium ebracteatum,* disappeared sometime between 1918 and 1938. Only *Angelica palustris* has persisted up to the present time – in a wet meadow below the Warsaw Escarpment. The presence of the species at this locality was confirmed in 2002. However, the area is now threatened by residential development.

Protected Areas

The most valuable natural areas are found in the peripheral parts of the city. Most of the protected areas are located within the north-western, southern and eastern city boundaries (Fig. 13). Thirty percent (15,500 ha, including the Vistula River bed) of the city area (51,700 ha) is protected, which is a larger area than that of many other European urban agglomerations.

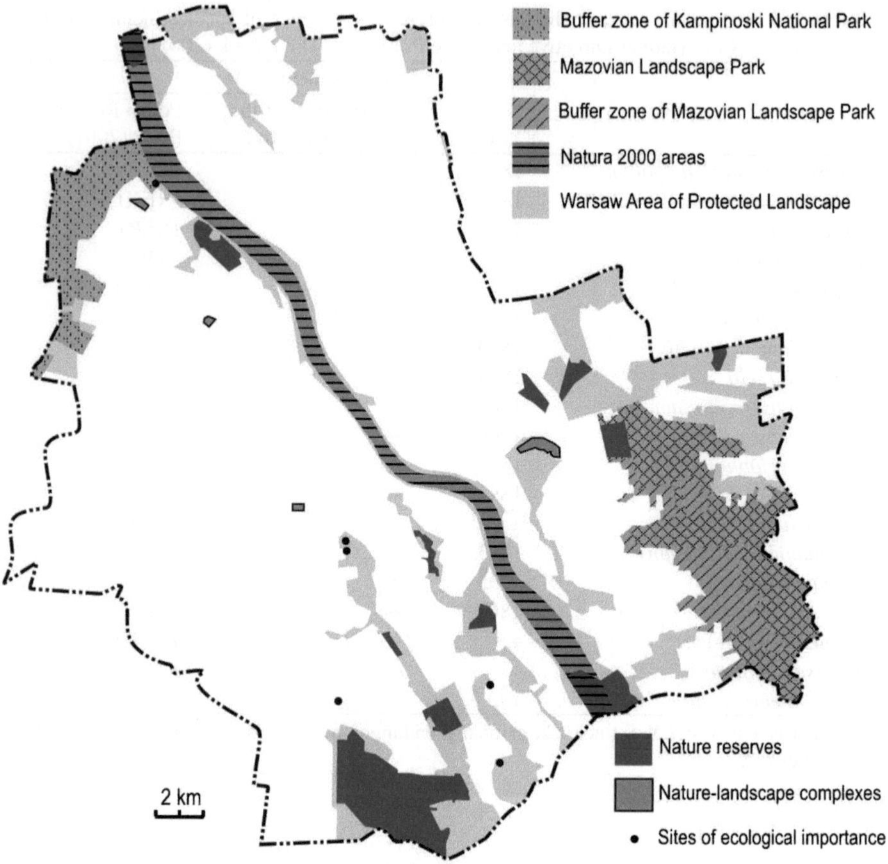

Fig. 13 Various forms of nature conservation protection applied within the municipal boundaries of Warsaw

Kampinoski National Park

Warsaw is the only European capital city that borders the territory of a national park. The Kampinoski National Park is just a few kilometers from the north-eastern boundary of the city. The park covers about 38,500 ha and has a buffer zone of a further 37,800 ha of which 1,073 ha is located within the city. The Park, which was established in 1959, was designated a UNESCO Biosphere Reserve in 2000. In 2004, the Park was identified for protection under European Union Directive, Natura 2000 network (area code: PLC14001 Kampinoski Forest). The Park protects one of the best preserved complexes of inland dunes in Europe. However, the largest part of the park is occupied by forests, mainly sub-continental and mesic *Pinus* forests, sub-continental *Quercus-Carpinus* forests, *Alnus* forests and floodplain forests. The strictly protected areas occupy 4,600 ha; the remaining area is actively protected and has unique landscape values.

Special Protection Area for Birds (Natura 2000): The Middle Vistula River Valley

The inter-embankment of the Middle Vistula River valley situated between the cities of Dęblin (to the south of Warsaw) and Płock (to the north) was included in the Natura 2000 network (area code: PLB 140004) in 2004. The site has a total area of 27,000 ha (including 1,050 ha within the city). The section of the river within the site is braided with numerous islands. The area contains *Salix* scrub, meadows and pastures as well as fragments of well-preserved *Salix* alluvial forests and *Populus* alluvial forests. The Special Protection Area provides a refuge for at least 22 bird species listed in Table 7 of the EC Birds Directive and nine bird species included in the Polish Red Data Book (2001).

Mazovian Landscape Park

The Czesław Łaszek Mazovian Landscape Park was established in 1987. It is situated within the confines of two geomorphological units – the Wołomińska Plain and the Valley of the Vistula. The Park, including the buffer zone, covers an area of 23,702 ha of which about 4,100 ha is within the city. The primary objective of establishment of the Park was to protect a large forest complex with sand dunes (up to 20 m high), lakes and fen. The complex comprises mainly *Pinus* forests, well-preserved *Alnus* woodland and *Quercus-Carpinus* forests as well as small fragments of thermophilous *Quercus* and floodplain forests.

The southern part of the Park encompasses the bog and meadow complex of the "Całowanie" peatland (3,500 ha), which contains, among others, stands of *Betula humilis*. The most important plant species are *Drosera rotundifolia, D. anglica, Iris sibirica, Ophioglossum vulgatum, Pedicularis sceptrum-carolinum, Polemonium caeruleum* and *Andromeda polifolia.* The Park is one of the largest wildlife refuges located in the immediate vicinity of Warsaw; it contains nine nature reserves and over 50 nature monuments including individual trees (*Quercus, Pinus, Tilia* and *Fraxinus*), 40 sites of ecological importance and 12 educational trails with a total length of 90 km.

Nature and Landscape Reserves and other Protected Areas

There are 14 partially protected nature reserves (a total area of 1,830 ha) within the municipal boundaries of Warsaw (within the reserves there are six nature trails). They comprise:

1. Six landscape reserves: Bielański Forest (130 ha; Fig. 14), Kabacki Forest (903 ha), Olszynka Grochowska (57 ha), Czerniakowskie Lake (48 ha), Ursynów Escarpment (23 ha) and Morysin (54 ha).
2. Two forest reserves: King John III Sobieski Forest (114 ha) and Natoliński Forest (105 ha).

Fig. 14 The Bielański Forest reserve – remnant of the former Mazovian virgin forest, well-preserved species-rich *Quercus-Carpinus* forest, which is located only about 6 km from the city centre (Photo B. Sudnik-Wójcikowska)

3. Two peat reserves: Bagno Jacka (19 ha) and Łosiowe Błota (31 ha).
4. Two faunistic reserves: Wyspy Zawadowskie (530 ha, including 185 ha within the city) and Ławice Kiełpińskie (803 ha, including 88 ha within Warsaw).
5. Two floristic reserves:
   (a) Kawęczyn (70 ha), which includes relatively young *Quercus* stands and small fragments of a thermophilous *Quercus* forest with important xerothermic species, as well as fragments of sub-continental *Quercus-Carpinus* and mixed *Pinus-Quercus* forests. The most noteworthy species are *Serratula tinctoria, Campanula bononiensis, Lathyrus niger, L. linifolius, Hypericum montanum, Melittis melisophyllum, Thalictrum aquilegiifolium* and *T. minus*.
   (b) Kalinowa Łąka (3.5 ha) is located in the Bemowski Forest, near the city's western border. It protects a mid-forest wet meadow with *Trollius europaeus* and other rare species, *Gentiana pneumonanthe, Crepis mollis, Dactylorhiza majalis* and *D. incarnata.*

Sites of Ecological Importance and Nature–Landscape Complexes

The six sites of ecological importance within Warsaw were created to protect the remaining remnants of natural ecosystems and their biological diversity, they are Przy

Lesie Młocińskim (5 ha), Imielińskie Lake (4 ha), Powsin (2 ha), Powsinek (3 ha), Jan Kusociński Escarpment (0.6 ha) and The Czesław Łaszek Escarpment (0.4 ha).

Nature–Landscape Complexes

Nature–landscape complexes protect unique and valuable fragments of natural and cultural landscapes in Poland, which have special aesthetic and scenic values. The following four areas are part of them; Dęby Młocińskie (9 ha), Olszyna (2 ha), Zakole Wawerskie (54 ha) and the park of the Warsaw University of Life Sciences (2 ha), which is protected for its cultural landscape as an old park, as well as some old, rare native and non-native trees.

The Warsaw Area of Protected Landscape

The Warsaw Area of Protected Landscape was established in 1997 which includes the zone of special environmental protection and the zone of urban protection. It covers a total area of 149,000 ha (8,500 ha within the city) and encompasses spatially connected areas surrounding the valleys of the Vistula and Narew rivers (including its tributaries).

Important Plant Areas, Statutorily Protected Plant Species and Protected Objects

There are no floristic refuges within the Warsaw area. However, several plant refuges are found in the vicinity of the city: Kampinoski Forest, Bolimowski Forest, Gostynińsko-Włocławska Refuge and Chojnowska Refuge.

Nature Monuments

According to Polish law, a protected nature monument is an animate or inanimate natural object (individual or groups of objects) of scientific, historical, cultural, aesthetic or scenic value. There are about 459 protected nature monuments in Warsaw, including 284 individual trees and 103 groups of trees. About 2,300 trees are recognised as nature monuments within the city. These are mainly representatives of the following species, *Quercus robur* (a 600-year-old oak tree called "Mieszko I" is the oldest tree in Warsaw), *Tilia cordata, Fraxinus excelsior, Ulmus laevis, Fagus sylvatica, Acer platanoides, Populus alba* and *P. nigra*.

Legally Protected Plant Species

Sixty plant species that are found in the city are fully (40) or partially (20) protected by Polish law.

Closing Comments

As in the case of other metropolitan areas, the size and the structure of land use in Warsaw have changed substantially, areas of natural and semi-natural habitats and agricultural land have declined, especially in the peripheral parts of the city. At the same time, new habitats have appeared, such as industrial areas, tram lines and railways. After 1945, a considerable part of the central part of the city was occupied by rubble that was totally devoid of vegetation but which was soon to be colonised by many plant species.

The area of rubble decreased as the city was reconstructed. During the political–economic transition in the 1990s, planning policies relating to the spatial structure of the city changed (for example, restricting the use of space for housing, intensification of investment activity). As a result, gradual changes in the structure of the spontaneous and cultivated flora occurred.

Future trends in the flora of Warsaw can be predicted based on the observations made so far. It seems that the establishment of protected areas within the city has only limited success in preventing the disappearance of native species. The abandonment and conversion of agricultural land to urban uses have led to the elimination of some native weed species and archaeophytes (especially segetal plants). It appears, therefore, that the importance of alien species in the flora of Warsaw will continue to increase as will the number, frequency and abundance of thermophilous and xerothermic species with short life cycles. On the other hand, more compact development, revitalisation of degraded areas and intensive management of urban greenery will limit the rate of influx of alien species. It seems, therefore, that the period of floristic "prosperity" in Warsaw has come to an end.

Literature Cited

Biernacki Z, Kazimierski J, Wróblewski A (eds) (1990) Środowisko przyrodnicze Warszawy. Państwowe Wydawnictwo Naukowe, Warszawa

Chojnacki J (1991) Zróżnicowanie przestrzenne roślinności Warszawy. Wydawnictwa Uniwersytetu Warszawskiego

Encyklopedia Warszawy (1994) Wydawnictwo Naukowe PWN, Warszawa

Gutry-Korycka M (ed) (2005) Urban sprawl. Warsaw Agglomeration. Case study. Warsaw University Press, Warsaw

Ostrowski M, Sudnik-Wójcikowska B, Galera H (2008) To the eyes of the explorers there appeared a green isle. In: Warsza J (ed) Stadium X – a place that never was. Bęc Zmiana Foundation, Warszawa, Kraków

Pawlak J, Teisseyre-Sierpińska M (2006) Opracowanie ekofizjograficzne do Studium uwarunkowań i kierunków zagospodarowania przestrzennego m. st. Warszawy. Urząd Miasta Stołecznego Warszawy, Warszawa

Sudnik-Wójcikowska B (1987–1998) Flora miasta Warszawy i jej przemiany w ciągu XIX i XX wieku. Część 1–3. Wydawnictwa Uniwersytetu Warszawskiego, Warszawa

Sudnik-Wójcikowska B (1998) Czasowe i przestrzenne aspekty procesu synantropizacji flory na przykładzie wybranych miast Europy Środkowej. Wydawnictwa Uniwersytetu Warszawskiego, Warszawa

Sudnik-Wójcikowska B (2000) The role of flora in bioindication of the temperature conditions in urban areas. In: Jackowiak B, Żukowski W (eds) Mechanisms of anthropogenic changes of the plant cover. Publications of the Department of Plant Taxonomy of the Adam Mickiewicz University in Poznań 10: 271–279

Sudnik-Wójcikowska B (2001) The role of flora in bioindication of the temperature conditions in urban areas. In: Jackowiak B, Żukowski W (eds) Mechanisms of anthropogenic changes of the plant cover. Publications of the Department of Plant Taxonomy of the Adam Mickiewicz University in Poznań 10: 271–279

Sudnik-Wójcikowska B, Galera H (2005) Floristic differences in some anthropogenic habitats in Warsaw. Ann. Bot. Fennici 42: 185–193

Wojtatowicz J (ed) (2006) Warszawska przyroda. Obszary i obiekty chronione. Wydział Ochrony Środowiska Urzędu Stołecznego Miasta Warszawy, Warszawa

Zurich

Elias Landolt

Fig. 1 Grossmünster (Photo Walter Lämmler, Zürich)

Elias Landolt (✉)
Institut für Integrative Biologie (Swiss Federal Institute of Technology), ETH Zürich,
Universitätsstr. 16, CH-8092 Zürich, Switzerland
e-mail: landolt@env.ethz.ch

J.G. Kelcey and N. Müller (eds.), *Plants and Habitats of European Cities*,
DOI 10.1007/978-0-387-89684-7_16, © Springer Science+Business Media, LLC 2011

Abstract Zurich, which is the largest city in Switzerland, is situated to the north of the Alps and at the northern end of Lake Zurich in an area with a rather cool sub-oceanic climate. More than 1,400 indigenous or naturalised species of higher plants have been recorded in the city, of which almost 200 (mainly archaeophytes) have become extinct within the last 160 years. The extinctions were caused by habitat loss resulting from the intensification of agricultural production and residential and other forms of development. On the other hand 300 new species have been able to colonize and become established in the city. The reasons for this being general climatic warming and the "heat-island effect" that is a characteristic of the urban environment. The previously relatively cool summers prevented many species from achieving their life cycle, which the recent warmer temperatures now allow them to. Also the milder winters enable many species to survive in the city, over winter, particularly in sheltered places. The absence of really cold winters for more than 20 years has enabled some sub-tropical plants to survive throughout the winter and even set fruit and regenerate from seeds, for example, the palm *Trachycarpus fortunei*.

Natural Environment

Zurich is situated in the north-eastern part of Switzerland at 47°23′ north and 8°12′ east. The administrative city area comprises about 88 km^2 (the Canton is much larger), it is located in the Limmat Valley at the northern end of Lake Zurich (Fig. 2). The valley is the extension of the lake region to the north-west between the mountain ranges of Albis to the south-west and Pfannenstil to the north-east, see Fig. 3. The surroundings of the city are known as the "Swiss Midlands".

Fig. 2 City centre towards the south with the Lake Zurich; the Alps are in the background

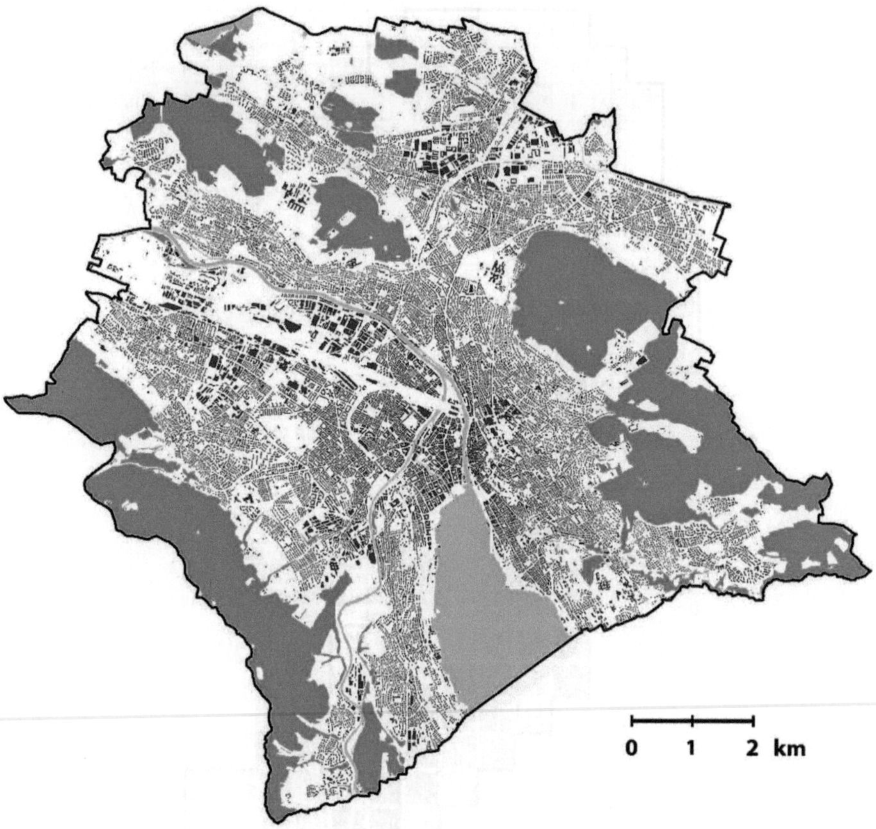

Fig. 3 Map of the city (from "Grün Stadt Zürich" after GEOZ, Zurich). Dark shading is forest; the light shading is Lake Zurich. The area corresponds to the upper part of the distribution maps

The north-eastern part of the city lies in the Glatt valley. The lowest part of Zurich is 390 m a.s.l. (in the Limmat valley), the highest point is 870 m a.s.l. on Mount Uto to the west (Fig. 4). Most of the city has been built between 400 m a.s.l. and 600 m a.s.l. The hills are mainly forested (Fig. 5), while most of the sealed surfaces occur in the flat part of the valley (Fig. 6). The percentage of forest and sealing are negatively correlated; within the administrative city area, the forest occupies 24% and the sealed part 18%. Fifty-two percent of the area is covered by fields, meadows and gardens. The height of the mountains gradually increases southwards until about 60 km from the city; they reach an altitude of about 3,000 m a.s.l.; on clear days, the mountains of the Alps can be seen from the city.

The solid geology north of the Alps comprises sedimentary rocks and deposits from the Tertiary and Quaternary Periods. The mountain ranges comprise mostly horizontal layers of sandstone, marl and conglomerates.

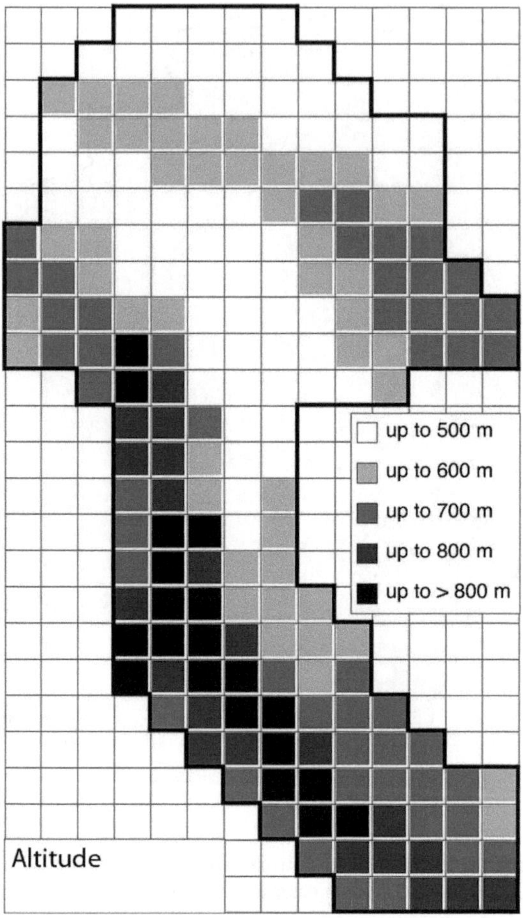

Fig. 4 Altitudinal differences of the map squares within the city limits and in a narrow strip about 10 km outside and to the south of the city (estimated from topographical maps)

During the Ice Ages, glaciers left extensive moraines on which most of the city has been built. In the valleys, sand and gravel deposits of the rivers predominate. The main soils in the surroundings of the city are clayey young brown earths, which are rich in nutrients and slightly acidic in the upper layers. Within the city, virtually all of the soils have been substantially disturbed by millennia of human activity.

The general climate is sub-oceanic, the mean annual temperature is 7–9°C (January −1 to +1°C, July 17–19°C); the mean annual precipitation is 100–130 cm. The high precipitation is characteristic of the mountains on the southern side of the city, whereas the lower parts of the north are slightly drier and the climate is more continental, see Fig. 7.

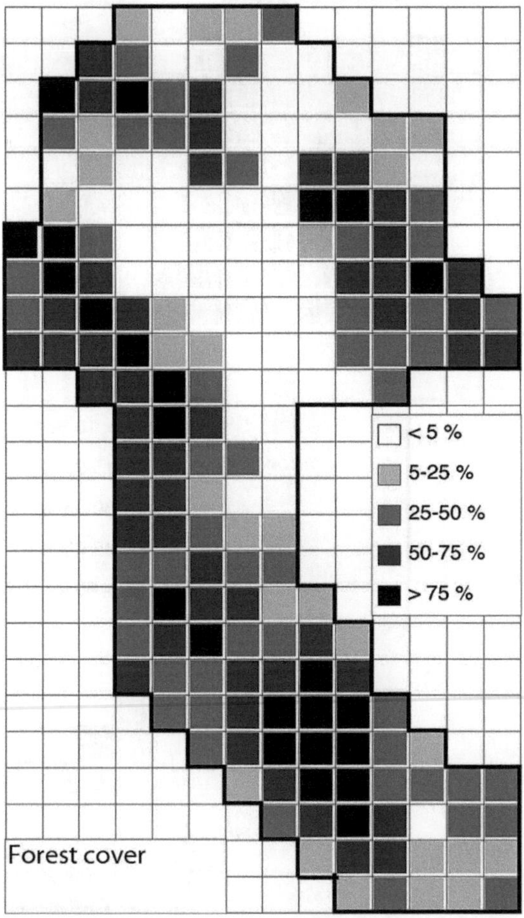

Fig. 5 Forest cover (%) within the city limits and in a narrow strip about 10 km outside and to the south of the city (estimated from topographical maps)

Historical Development

After the retreat of glaciers 15,000 years ago, the area now occupied by the city was colonised by hunter-gatherers and fishermen. About 6,000–7,000 years ago, people established permanent settlements along the lakeside, mainly houses on stilts. They began to clear the forests to provide grazing land for domesticated animals and to cultivate around their houses for the growing of cereals and other edible plants. However, their main food was fish from the rivers and the lake. Gradually, people spread from the lakeside into the surrounding areas and occasionally traded with other people as far south as the Mediterranean. It was during these times that many cultivated plants and the first weeds were introduced into this part of what is now Switzerland. In 1000 BC, the settlement was inhabited by the Helvetians, a Celtic tribe who called it Turus.

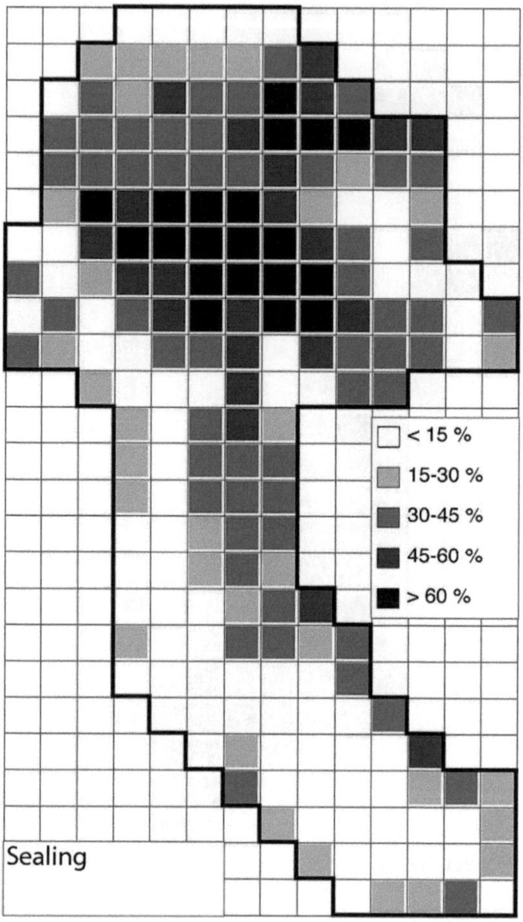

Fig. 6 Area (as %) of soil that has been built over or sealed within the city and in a narrow strip about 10 km outside and to the south of the city (according to "Umweltfachstelle of the City of Zurich" and estimated from topographical maps)

Later, it was renamed Turicum by the Romans who invaded and occupied the area about 15 BC and built a fortified custom station and market place. Following their defeat by the German tribes, the Romans left the region in about AD 400, after which one of the tribes (the Alemans) settled in the region and mixed with the few remaining Romans and the original population of Celts. Zurich was first mentioned as a city in AD 929. It was then an important city of southern Germany. In 1351, it formed an alliance with the "Eidgenossenschaft", a federation of some small regions of the central Alps and surrounding cities. This federation was an early precursor of Switzerland. Zurich was then a fairly rich town, so the city council was able to buy surrounding areas from the aristocracy who were in need of money. This land, which occupied almost 2,000 km^2, now forms the Canton of Zurich. In AD 1200 or thereabouts, the city contained about 4,000 people.

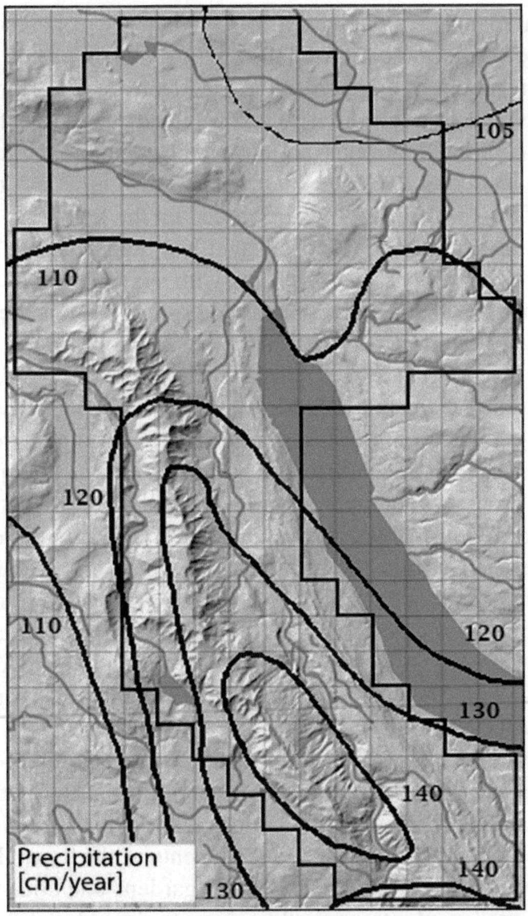

Fig. 7 Precipitation in cm/year within the city and about 10 km outside and to the south of the city

By AD 1820, the number had risen to about 20,000. As the result of industrialisation during the nineteenth century, the human population increased to 100,000. During this period, the most important sources of income came from the textile industries and the manufacture of various types of machinery. In 1893, and again in 1934, some surrounding communities were incorporated into the city, resulting in the number of people in the city rising to about 400,000. Today, the agglomeration of Zurich contains more than a million people.

Changes of the Environment because of City Growth

Up to about 1700 the influence of the city on the environment was relatively small and localised. Outside the town wall, the village structures, arable fields, meadows and pastures formed new niches for colonisation by plants (and animals).

Weeds and ruderal plants occupied the strips between rows of crop plants and along roadsides. Plants of meadows colonised the area from open places within the forest, from the natural meadows at higher altitudes and from eastern European and/or the Mediterranean steppes. In addition, new taxa evolved as the result of hybridisation between the indigenous and introduced plants and between two introduced species.

Generally, the land was overused, nutrients were scarce and the manure from cattle kept at the farms was not stored and spread on the fields. As a result, the soils became poor in nitrogen and phosphorous and lost their fertility. The forests were restricted to the steeper slopes and to places with very poor soils. However, some farmers scraped the soil off some forest areas and spread it on the fields as a fertiliser. Because of over-exploitation for fuel and other purposes parts of some forests remained open and supported a rich herb layer. On the other hand, the use of other forests for grazing domesticated animals, the use of the twigs for fodder and dry leaves for bedding changed many forests into rather unproductive ecosystems. In the forests, growing on more undisturbed grounds, the trees (especially *Quercus* spp.) were left to grow into the upper tree layer and then used as timber.

Although the available habitats for plants were not favourable, nevertheless a unique flora developed within the urban area. Roman and Medieval cities were rather compact, the streets were mostly paved and gardens were rare – and small. Some weed and ruderal species colonised the sides of roads and areas covered by loose material. In general, soils within the city were rich in nitrogen and lime from houses and other rubble. Walls were colonised by plants that originated from the rocks and cliffs of the adjacent mountains.

Since the seventeenth century, many rich people built summerhouses outside the city to avoid the stinking and unpleasant city area during the hot season. Around the summerhouses, they created exotic gardens with many newly introduced plants from other parts of the world. In the nineteenth century, the expanding city was still rather compact, even so most houses had small gardens and courtyards with many opportunities for garden plants as well as for weeds and ruderals. Later, in the twentieth century, the surfaces of most of the courtyards were sealed and used as parking places, which has prevented the growth of many plants. Rich people built villas and often surrounded them with extensive gardens or parks in which could be found plants of the surrounding countryside, including some that were associated with the original forest. Today, many of these gardens have disappeared because they are often too time-consuming and expensive to maintain. In addition, land values in Zurich are high; this combined with the high maintenance costs means that is in the financial interest of owners to sell parts of their gardens for residential development.

Up to the industrial development during the nineteenth century, the influence of the city on the climate was only local. Places along the walls of the houses or between the pavements became warmer, enabling some plant species from southern countries to exist much further north. With the extensive increase in the number of houses and roads since the second half of the nineteenth century, the general climate became warmer and the air humidity lower. The higher level of solar radiation continues to enable some plants from more continental areas to become established in the city.

Flora

General Aspects

Although the distribution and other information about higher plants in Zurich have been well documented by Landolt (2001), little is known about the algae, mosses, liverworts, lichenised fungi and fungi, which have not been recorded systematically.

The administrative area of the city contains 1,211 indigenous or naturalised species of vascular plants – excluding cultivars and ephemerophytes. The figure includes about 300 species that have been introduced into the city within the last 160 years; 188 additional species became extinct during this period including a few newly introduced plants that grew for at least 10–30 years and then disappeared. A further 600 introduced species and garden escapes are found only occasionally introduced or locally escaped from cultivation; included in this number are the cultivated species that very often occur in parks and gardens but rarely escape. Of the native or more or less naturalised species, 58% are indigenous, 19% archaeophytes and 23% neophytes. However in the more urbanised area, more than half of the species are neophytes.

The average number of species per km^2 is 451, not counting the three squares that are mainly (>75%) covered by the lake. The lowest numbers occur within forests squares of rather flat areas (294 and 315, respectively). The highest numbers, ranging from 490 to 607, are found in the mixed stands, the outer urban areas, semi-natural areas of wetland, steep mountain slopes and railway land. The distribution maps of the city's flora include a narrow strip of about 10 km long to the south-west of the lake. This area was included in order to be able to compare the urban flora with the flora adjacent to but outside the city where there are fewer people and the climate is cooler and more humid.

Twenty-five percent of the species are frequent within the administrative area growing in more than 50% of the km^2 squares, while 18% of the species occur in more than 90% of the squares. The most frequent species, those growing in 95% or more of the square, are listed in Table 1. The species are generalists that have to be considered as being the best adapted to the special conditions of the region. The species are divided into two categories – moderately urbanophobic and moderately urbanophilic.

The urbanophobic species have the tendency to avoid the more densely developed areas and mainly grow outside the urban area; however, they can also be found in allotment gardens and parks within the city. Many of the species are indigenous. The strongly urbanophobic species can only be found on natural or semi-natural places (forests, rocks, wet meadows and waters). *Veronica montana* (Fig. 8) is an example of an urbanophobic species that grows in forests on wet and basic soils. The construction of forest roads has enabled the frequency of the species to expand considerably. Water-logged and basic conditions, which are favourable to the establishment and spread of the species, have been created adjacent to the roads – conditions that are completely absent from developed areas.

The urbanophilic species, which prefer urban sites and avoid closed forests, have their centre of distribution within the urban area but are spreading along roads and

Table 1 One hundred of the most frequent species growing in the region of Zurich

**Acer platanoides*[a]	**Hedera helix*[a]
**Acer pseudoplatanus*[a]	*Holcus lanatus*[b]
**Aegopodium podagraria*[a]	***Impatiens parviflora***[b]
Agrostis stolonifera[a]	**Lolium perenne*[b]
**Ajuga reptans*[a]	**Lotus corniculatus*[a]
**Alliaria petiolata*[a]	*Lysimachia nummularia*[a]
**Allium ursinum*[a]	*Medicago lupulina*[b]
Arrhenatherum elatius[b]	***Panicum capillare***[b]
**Bellis perennis*[a]	**Persicaria maculosa*[b]
Betula pendula[a]	*Phleum pratense*[a]
Brachypodium sylvaticum[a]	*Plantago lanceolata*[a]
Bromus hordeaceus[b]	**Plantago major*[a]
**Calystegia sepium*[a]	**Poa annua*[a]
**Capsella bursa-pastoris*[b]	*Poa nemoralis*[a]
Cardamine flexuosa[a]	*Poa pratensis*[a]
**Cardamine hirsuta*[b]	**Poa trivialis*[a]
**Cardamine pratensis*[a]	**Polygonum arenastrum*[b]
Carex sylvatica[a]	*Polygonum aviculare* agg.[b]
Carpinus betulus[a]	*Potentilla reptans*[a]
**Cerastium fontanum*[b]	*Potentilla sterilis*[a]
**Chaenorhinum minus*[b]	*Prunella vulgaris*[a]
**Chelidonium majus*[b]	*Prunus avium*[a]
Chenopodium album[b]	*Quercus robur*[a]
Chenopodium polyspermum[b]	**Ranunculus ficaria*[a]
**Circaea lutetiana*[a]	**Ranunculus frieseanus*[b]
**Cirsium arvense*[b]	**Ranunculus repens*[a]
Clematis vitalba[a]	***Rubus armeniacus***[b]
Convallaria majalis[a,b]	*Rubus caesius*[a]
Convolvulus arvensis[b]	*Rumex acetosa*[b]
Conyza canadensis[b]	*Rumex obtusifolia*[b]
Cornus sanguinea[a]	**Sagina procumbens*[b]
Corylus avellana[a]	**Salix caprea*[a]
**Crepis capillaris*[b]	*Sambucus nigra*[a]
**Dactylis glomerata*[b]	**Senecio vulgaris*[b]
Digitaria sanguinalis[b]	***Setaria viridis***[b]
Dryopteris filix-mas[a]	***Solidago canadensis***[b]
Echinochloa crus-galli[b]	***Solidago gigantea***[b]
Elytrigia repens[b]	*Sonchus asper*[a]
**Epilobium montanum*[a]	**Stellaria media*[a]
Erigeron annuus[b]	**Taraxacum officinale*[a]
**Euphorbia peplus*[b]	*Taxus baccata*[a]
**Fagus sylvatica*[a]	*Trifolium pratense*[a,b]
Festuca gigantea[a]	**Trifolium repens*[a]
**Fragaria vesca*[a]	*Ulmus glabra*[a]
**Fraxinus excelsior*[a]	*Urtica dioeca*[a]
Galinsoga quadriradiata[b]	***Veronica filiformis***[b]
Galium mollugo ssp. *erectum*[a]	*Veronica hederifolia*[b]
**Geranium robertianum*[a]	***Veronica persica***[b]

(continued)

Table 1 (continued)

Geum urbanum[a]	*Vicia sepium*[a]
**Glechoma hederacea*[a]	*Viola reichenbachiana*[a]

The species are found in more than 95% of the km² of the administrative area of Zurich
The 50 most frequent are indicated by an asterisk (*); Bold = neophytes
[a]Moderately urbanophobic
[b]Moderately urbanophilic

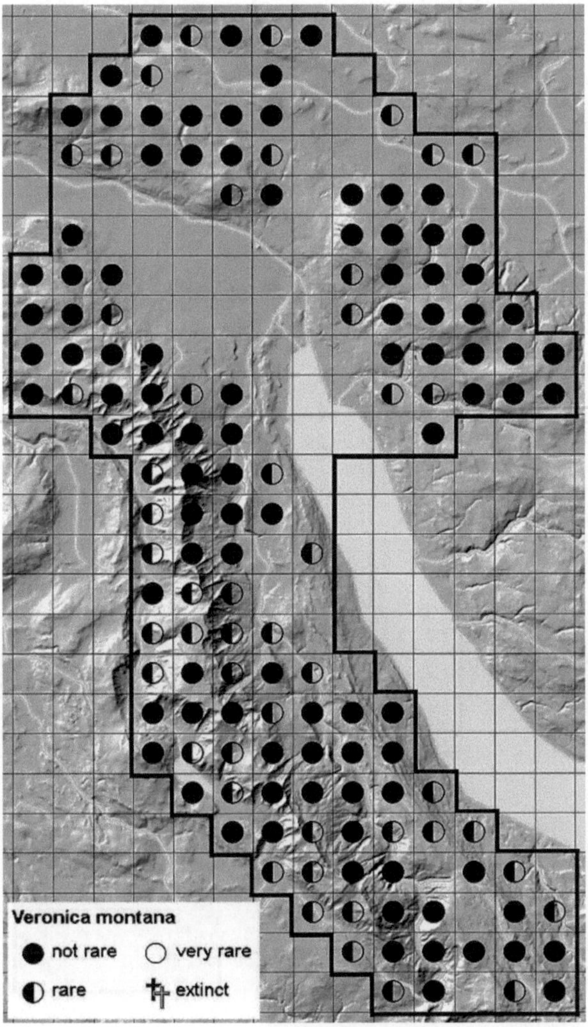

Fig. 8 Distribution map of *Veronica montana*, a strongly urbanophobic species. *Not Rare*: occurs in 1.0 km square with >200 individuals and covers >1.0 ha; *Rare*: occurs in 1.0 km square with >20 individuals and covers 0.01 to 1.0 ha; *Very Rare*: occurs in 1.0 km square with <20 individuals or covers <0.01 ha

colonising open places outside the built areas. Strongly urbanophilic plants are restricted to the sealed surfaces (for example, impermeable tarmac and paved areas) and the warmest parts of the inner city, for example, *Polygonum calcatum* (Fig. 9), which grows mostly in paved areas. In Zurich, *Carex divulsa* is strongly urbanophilic. It is temperature dependent and therefore restricted to the inner city area where it can be 1–2°C warmer than outside the built-up areas. In the warmer regions of the Southern Alps, the species is urbanoneutral; it is also frequent in open forests.

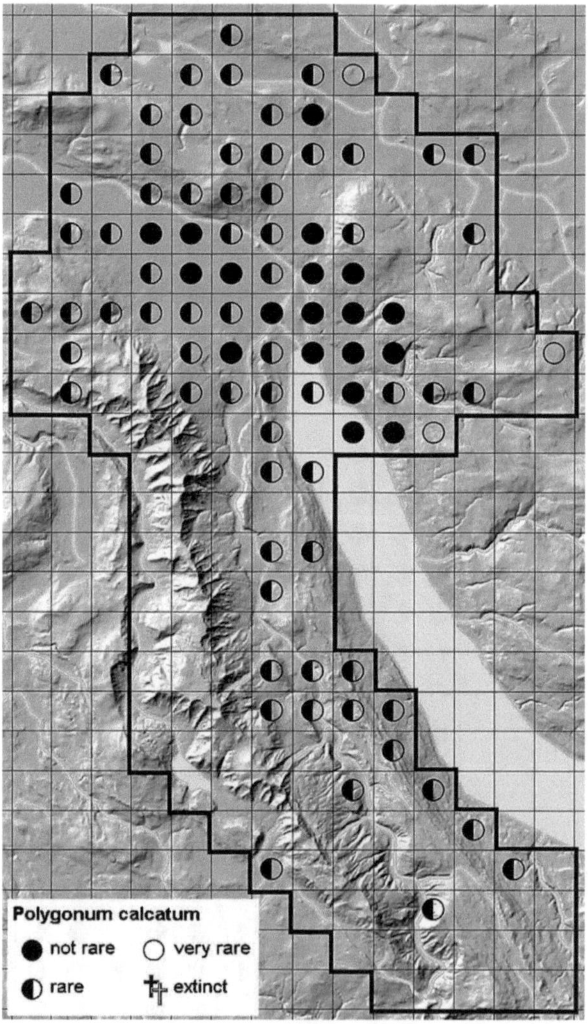

Fig. 9 Distribution map of *Polygonum calcatum*, a strongly urbanophilic species. *Not Rare*: occurs in 1.0 km square with >200 individuals and covers >1.0 ha; *Rare*: occurs in 1.0 km square with >20 individuals and covers 0.01 to 1.0 ha; *Very Rare*: occurs in 1.0 km square with <20 individuals or covers <0.01 ha

Origins of the Indigenous Flora

The indigenous flora of Zurich contains several different elements. First, the more continental species entered the region from the north, which is drier and sunnier in the spring and summer, see Fig. 7. Some of these species (for example, *Ranunculus auricomus* sensu lato, see Fig. 10) just reach the northern part of the administrative

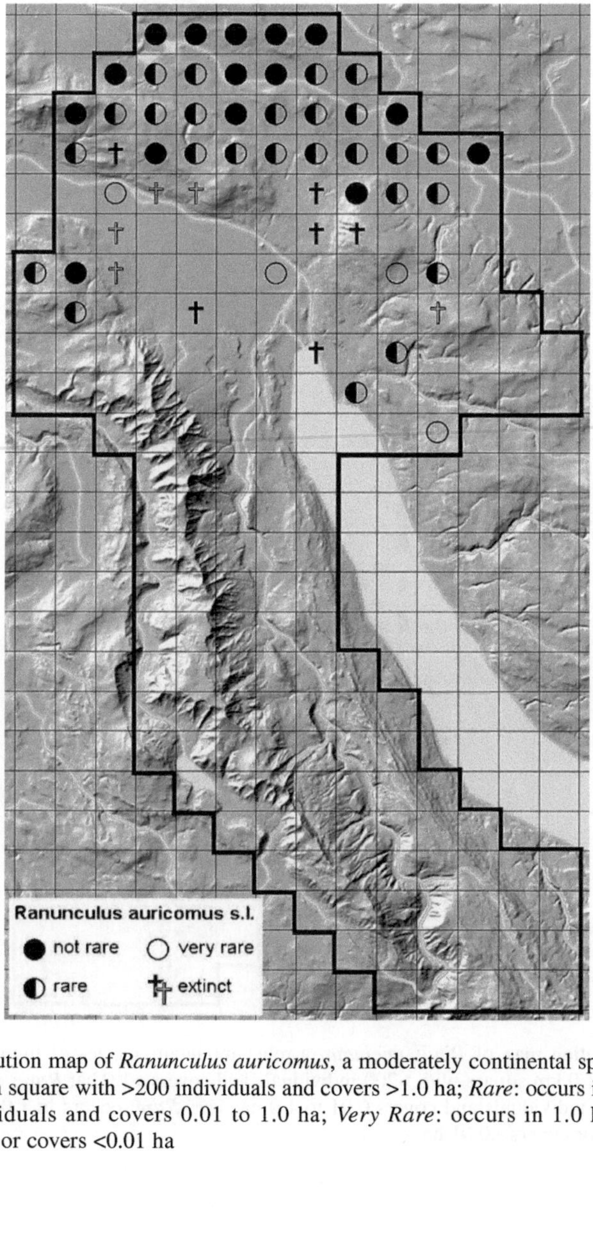

Fig. 10 Distribution map of *Ranunculus auricomus*, a moderately continental species. *Not Rare*: occurs in 1.0 km square with >200 individuals and covers >1.0 ha; *Rare*: occurs in 1.0 km square with >20 individuals and covers 0.01 to 1.0 ha; *Very Rare*: occurs in 1.0 km square with <20 individuals or covers <0.01 ha

area, while other species (*Pulmonaria obscura*) proceeded a little further to the south. Second, many species whose main distribution is in the central European lowlands, such as most forest trees, of which *Fagus sylvatica* is an example. Third, montane species of open habitats that occur mainly in rocky places or in the meadows of the Alps and other European mountains; this group includes *Pinus uncinata* (Fig. 11). Fourth, species of mountain forests, which are mostly widespread and frequent, for example, *Lonicera nigra,* see Fig. 12.

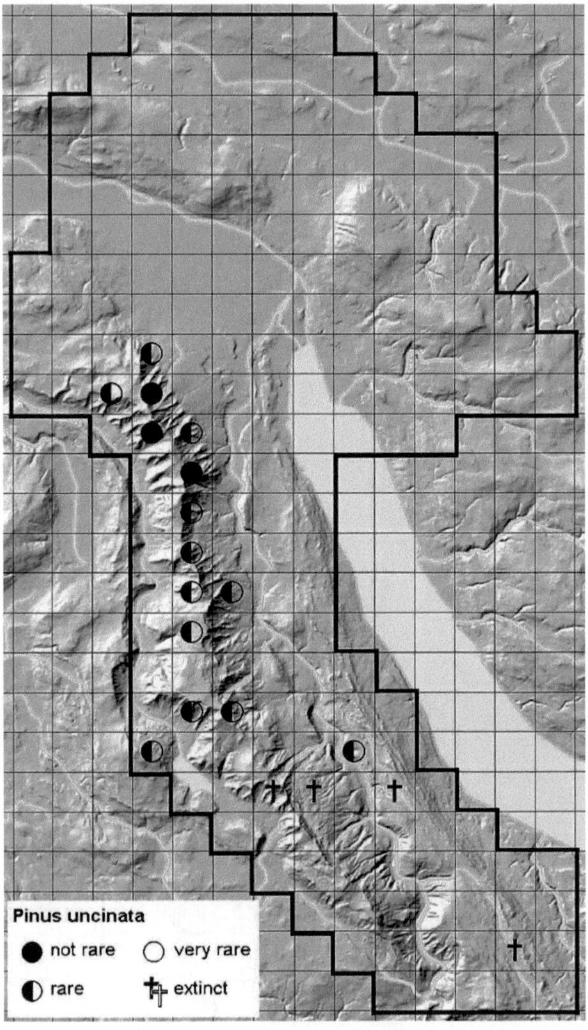

Fig. 11 Distribution map of *Pinus uncinata*, a relict mountain species of higher altitudes. *Not Rare*: occurs in 1.0 km square with >200 individuals and covers >1.0 ha; *Rare*: occurs in 1.0 km square with >20 individuals and covers 0.01 to 1.0 ha; *Very Rare*: occurs in 1.0 km square with <20 individuals or covers <0.01 ha

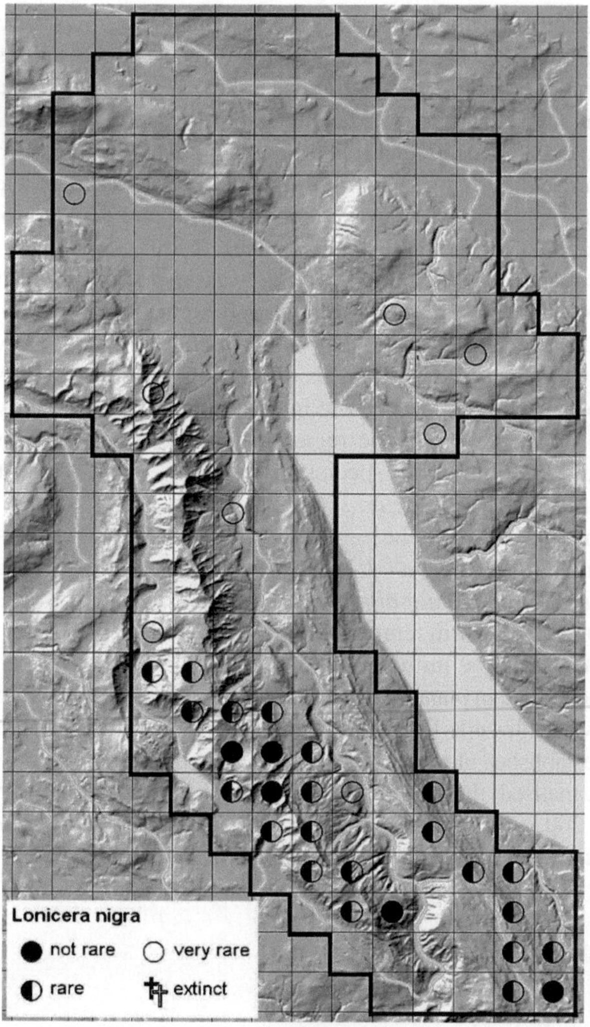

Fig. 12 Distribution map of *Lonicera nigra*, a forest species of higher altitudes. *Not Rare*: occurs in 1.0 km square with >200 individuals and covers >1.0 ha; *Rare*: occurs in 1.0 km square with >20 individuals and covers 0.01 to 1.0 ha; *Very Rare*: occurs in 1.0 km square with <20 individuals or covers <0.01 ha

The first montane species of open habitats colonised the areas surrounding the Alps shortly after the retreat of the glaciers. Most of the species disappeared when the temperature warmed up and the first trees colonised and formed closed forests. Only on the steep slopes of the Albis range west of the city where the forest was not able to form closed stands, have some of the species survived, they include *Carduus defloratus, Carlina bibersteinii, Campanula cochleariifolia, Buphthalmum salicifolium, Leucanthemum adustum, Digitalis grandiflorum, Thesium alpinum,*

Polygala chamaebuxus, Vicia sylvatica, Stachys alpina, Saxifraga mutata, S. aizoides and *Pinus uncinata* (see Fig. 11). Many of these species need some solar radiation during summer, the duration of which is less in the region to the south. Therefore, the typical montane species of open habitats are restricted to the north of the Albis range. They are called "postglacial relicts" and are generally endangered because of human interference in seeking to stabilise the unstable slopes and the eutrophication of the soil by air pollution, which results in the faster and denser growth of forests and the dominance of widely distributed species.

On the hills, especially of the Albis range west and south-west of the city area (see Fig. 4), forest species of cooler regions and higher altitudes can be found, for example, *Lonicera nigra* (see Fig. 12), which is rarely found below 650 m a.s.l. The situation with *Lonicera alpigena* is slightly different, only growing at lower altitudes on north-facing exposures and in cool ravines. Both species are generally missing in the northern part of the region where the warmer climate of the city restricts their distribution. Other species that are rather rare within the city limits are *Sambucus racemosa* and *Luzula sylvatica*.

Some mountain plants have become characteristic wetland species at lower altitudes, they include *Aster bellidiastrum, Gentiana verna, Primula farinosa, Parnassia palustris, Pinguicula alpina* and *Dianthus superbus*. Because of the lack of wetland habitats and warm temperatures, most of these species are now very rare and endangered within the limits of the city.

Finally, some mountain plants reach the city along the River Sihl, which originates in the Alps and flows northwards. Propagules are washed down periodically, some germinating and becoming established along the river banks. Since most of the river within the urban area has been canalised, these plants have become very rare or extinct within the city during the last 200 years; they include *Carduus personata, Ranunculus aconitifolius* and *R. montanus,* see Fig. 13. The latter is frequent in the meadows of the sub-alpine and alpine belt in the Alps but is rarely found below 1,200 m.

Forests

The forests of Switzerland have been statutorily protected since 1874. It is not allowed to clear even small forest patches without planting new trees in the same place. Therefore, about 25% of the administrative area of the city is still occupied by forests. The valley bottom and the more gentle slopes of the valley sides were converted to agricultural use and later developed for housing, industrial and commercial purposes. Kuhn (1967) investigated and mapped the potential natural forest vegetation of Zurich using the methods of Braun-Blanquet. If his results are translated into the nomenclature of the "Interpretation Manual of European Union Habitats" (Natura 2000. EUR 15/2. European Commission DG Environment. 179 pp.), the city lies within the climax vegetation of the "*Asperulo-Fagetum* forests". More than half of the forests belong, potentially, to the mesophilic *Fagus* forests. However, the planting of non-native trees species, especially *Picea abies,* has

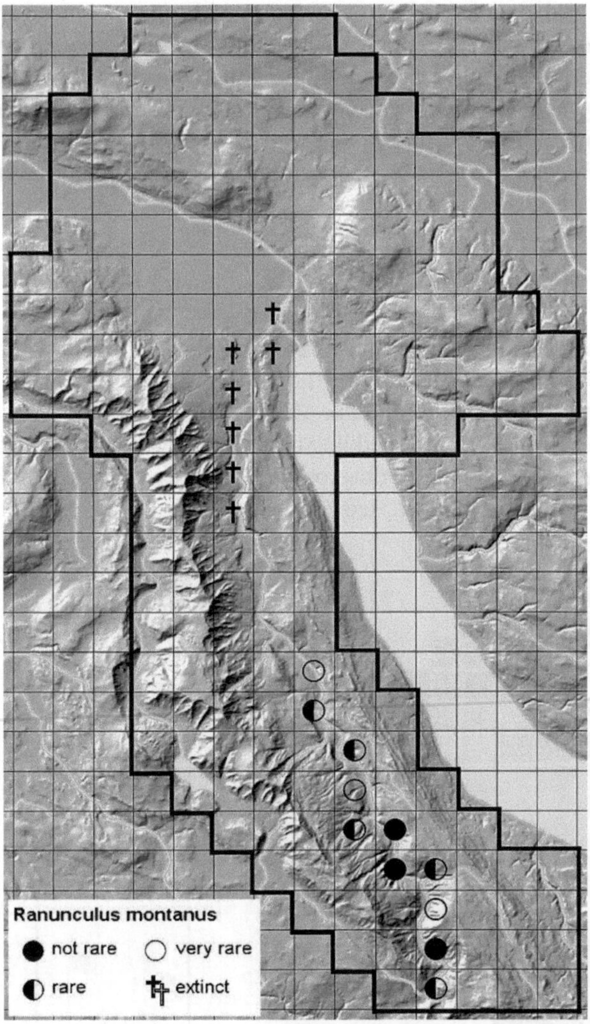

Fig. 13 Distribution map of *Ranunculus montanus*, a mountain species washed down by the river Sihl. *Not Rare*: occurs in 1.0 km square with >200 individuals and covers >1.0 ha; *Rare*: occurs in 1.0 km square with >20 individuals and covers 0.01 to 1.0 ha; *Very Rare*: occurs in 1.0 km square with <20 individuals or covers <0.01 ha

changed the ecology of many areas of this forest type. At higher altitudes and on north-facing slopes *Abies alba* is an important element of these forests.

The sub-Atlantic and medio-European *Quercus* or *Quercus-Carpinus* forests of the "*Carpinion betuli*" occur in some of the damp areas. The forests of the rounded hilltops belong to the "*Luzulo-Fagetum*" forests, while on the steeper slopes medio-European limestone beech forests of the "*Cephalanthero-Fagion*" predominate mixed with stands of *Taxus baccata*. Fragments of *Pinus sylvestris* forest can be found on very steep and exposed situations. Open vegetation with some montane species, such as *Pinus uncinata* (see Fig. 11), occurs on landslips and in clearances. Woodland fragments occur, although only rarely, along the watercourses. Alluvial *Alnus-Fraxinus* woodland grows along the rivers and creeks where the watertable is relatively high, while Riparian mixed forests of *Quercus robur-Ulmus glabra* and *U. minor-Fraxinus* ("*Ulmenion minoris*") occur on the sides of the major rivers.

Wetlands and Unfertilised Meadows

In the nineteenth century semi-natural river- and lake side and wetland vegetation was still widespread and abundant in the bottom of the valleys, on the mountain slopes and along the margins of Lake Zurich. The meadows were mainly cut in early autumn and the hay used as bedding for cattle. More than 95% of these meadows have disappeared as the result of development and the expansion of the city and the intensification of the agriculture, see Fig. 14.

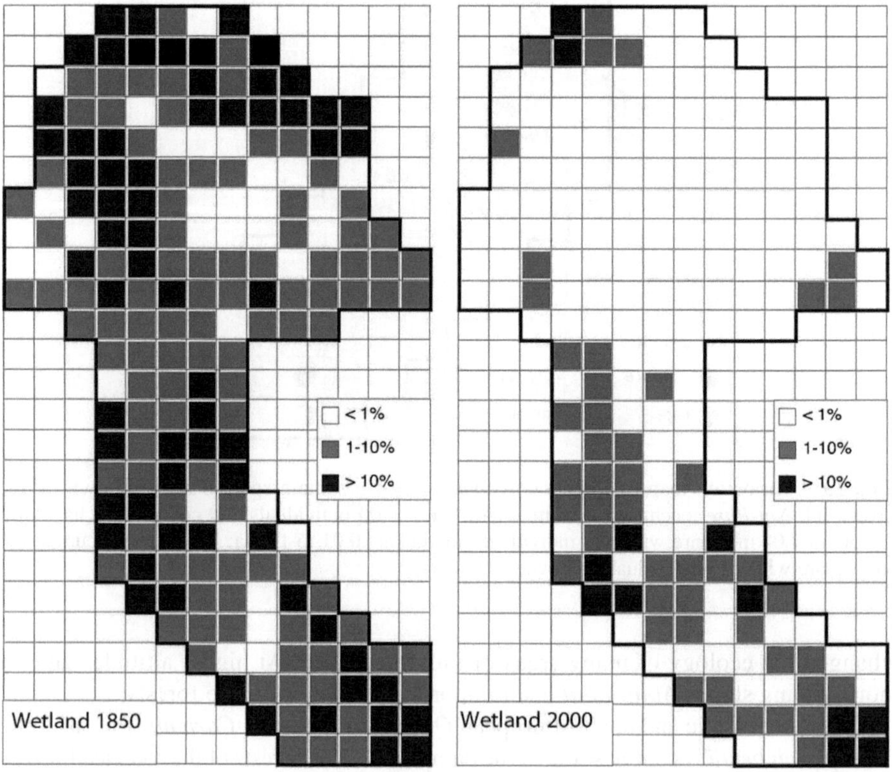

Fig. 14 Area of wetlands in percent of the area within the city limits and in a narrow strip about 10 km outside and to the south of the city. Situation in AD 1850 (*left*) (estimated from old maps). Situation in AD 2000 (*right*) (estimated from recent maps)

The "reed" meadows that formerly existed along the lake have been converted to parks or used for residential and other developments. The shore is now "artificial" and has no natural marginal vegetation. The wet meadows in the valleys have suffered a similar fate – they have been drained and built on or used as pasture for fattening cattle. Originally, these wetlands had a high floristic biodiversity, the very few that remain contain many rare species, such as *Cladium mariscus, Carex limosa, Drosera rotundifolia, D. anglica, Gentiana verna, Iris sibirica, Epilobium palustre, Menyanthes trifoliata, Orchis palustris, Potentilla palustris* and *Pinguicula vulgaris.* Many plants that are characteristic of this habitat have disappeared from the city, for example, *Gentinana pneumonanthe* (Fig. 15), *Carex chordorhiza, Scheuchzeria palustris, Liparis loeselii, Lycopodiella inundata, Senecio paludosus* and *Vaccinium uliginosum.*

Among the very few remnants of wetland within the city is the region of Katzensee in the north-west – a very interesting wetland with a high biodiversity and occupying more than 1 km^2. In the south of the city, much more of the traditional wetland has survived. This region is substantially less populated and because of the high precipitation not suitable for cereal cultivation. Instead of straw from cereal crops, the farmers use the hay of the wetland meadows for cattle bedding. Therefore, they did not convert the meadows to highly productive pastures for the fattening of cattle, as is the case in the drier regions. The predominant wetland vegetation is that of alkaline fens (for example, "*Caricion davallianae*" and "*Molinion*") and to a lesser extent transitional mires and "quaking" bogs. Acidic *Sphagnum* bogs are rare and have been partially destroyed. All areas of wetland vegetation are now subject to strong protection measures and regularly inspected by officials from the appropriate authorities, amongst others.

Semi-natural drier grassland is very rare. The climate is too wet and the few fragments of this vegetation type that remain have been treated with fertiliser or built on, especially on the southern slopes. This has resulted in the disappearance of many interesting plants such as *Carex ericetorum, Gentiana cruciata, Euphorbia verrucosa* and *Helictotrichon pratense.*

Residential Areas

The residential areas comprise buildings of various kinds, courtyards, walls, roofs, unused land (wasteland), paved areas and many other habitats. This matrix provides a high diversity of ecological niches enabling many species to form small isolated populations, for example, many weeds (archaeophytes, otherwise extinct in the region and newly introduced neophytes) including rock plants such as *Pseudofumaria lutea, Asplenium ruta-muraria* and *Cymbalaria muralis.* Typical plants of the pavements of the inner city are *Polygonum calcatum, Sagina procumbens, Plantago major, Herniaria glabra* and *Erophila praecox.*

Ruderal species including *Chenopodium* spp., *Atriplex prostrata* and *Sisymbrium officinale* occur in open, nutrient-rich places around houses. Many of these species, for example, *Chenopodium vulvaria,* have become very rare or extinct in recent times mainly because waste places around houses are much better maintained.

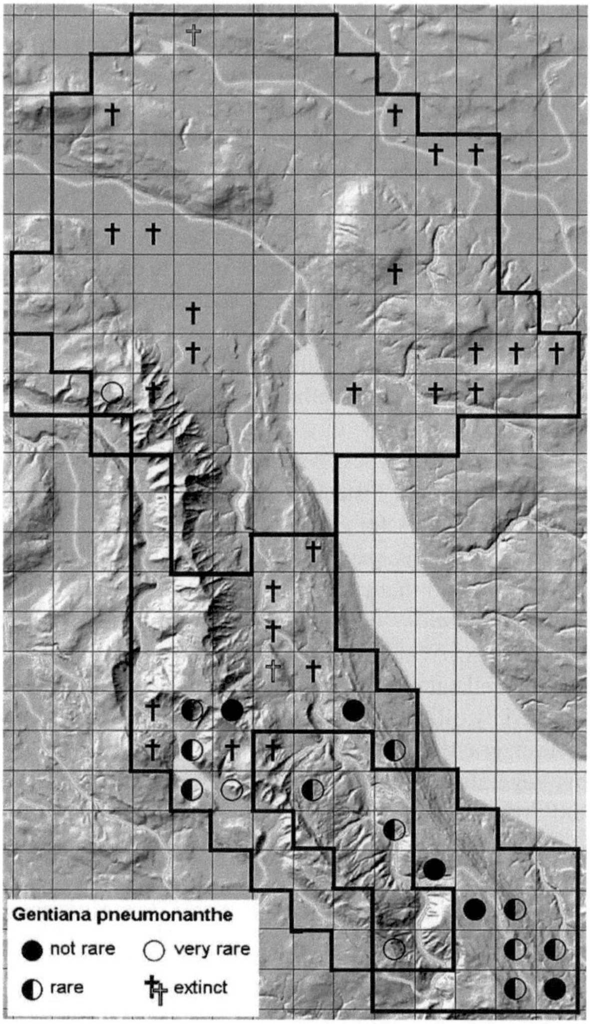

Fig. 15 *Gentinana pneumonanthe*, a species of reed meadows. *Not Rare*: occurs in 1.0 km square with >200 individuals and covers >1.0 ha; *Rare*: occurs in 1.0 km square with >20 individuals and covers 0.01 to 1.0 ha; *Very Rare*: occurs in 1.0 km square with <20 individuals or covers <0.01 ha

Transport Routes

Some of the land associated with railway operations is of considerable botanical interest. Stony habitats, which are naturally rare in the area around Zurich, are widespread on railway land. Consequently, many species that are restricted to "stressed" conditions are also restricted to railways, including plants that are characteristic of scree, which has higher temperatures and lower plant competition than most natural habitats.

The species that are restricted to or mainly occur in association with railways include *Crepis foetida, Tragopogon minor, Papaver rhoeas* sensu lato, *Epilobium dodonaei, Galeopsis angustifolia* and *Scrophularia canina*.

An interesting example of the rapid introduction, colonisation and spread of a species along railway lines is *Geranium purpureum,* a Mediterranean species related to *Geranium robertianum,* which occurs in Zurich only on railway land. The species was first recorded in Zurich by the author in 1985. Since then, it has colonised the gravel along the railway tracks. By 1998, it was distributed throughout the railway network of the city up to 600 m a.s.l., see Fig. 16, which shows the distribution of railways within the city.

Many species growing in fields and waste places in the Mediterranean regions have been introduced to Switzerland and Zurich by motor vehicles. In addition, cuttings and embankments along the highways are sometimes good places to find meadow plants of the *Mesobrometum* (drier meadows with low nutrient content), for example, *Thymus pulegioides, Thlaspi perfoliatum* and *Sanguisorba minor*.

Recreation Areas

Recreation areas outside the surrounding forests are restricted to the city parks (especially along the lake) and cemeteries. They often contain forest plants from the surrounding areas and a variety of meadow species. Some lawns contain *Sherardia arvensis, Aphanes arvensis, Ornithogalum umbellatum, Poa bulbosa, Chionodoxa luciliae, Scilla siberica, Crocus tommasinianus, Oxalis corniculata* and *Leontodon saxatilis*. *Mentha requienii* (a Corsican endemic) occurs in specific localities such as playing fields, where it was probably introduced in commercial grass-seed mixtures. Typical plants of warm, open parks in the inner city are *Carex divulsa* and *Dactylis polygama*.

Aquatic

About 6% of Zurich is occupied by water, comprising Lake Zurich, Katzenseen, and the rivers Limmat, Sihl and Glatt. Except for Katzenseen and some minor watercourses outside the built-up area, most of the lake and river margins are "sealed" with various types of impermeable materials such as paving. As a consequence, there is very little marginal vegetation and what there is contains only a few species. Some submerged macrophytes occur in Lake Zurich and in the Limmat, for example, *Najas marina, Elodea nuttallii, E. canadensis* and nine species of *Potamogeton*.

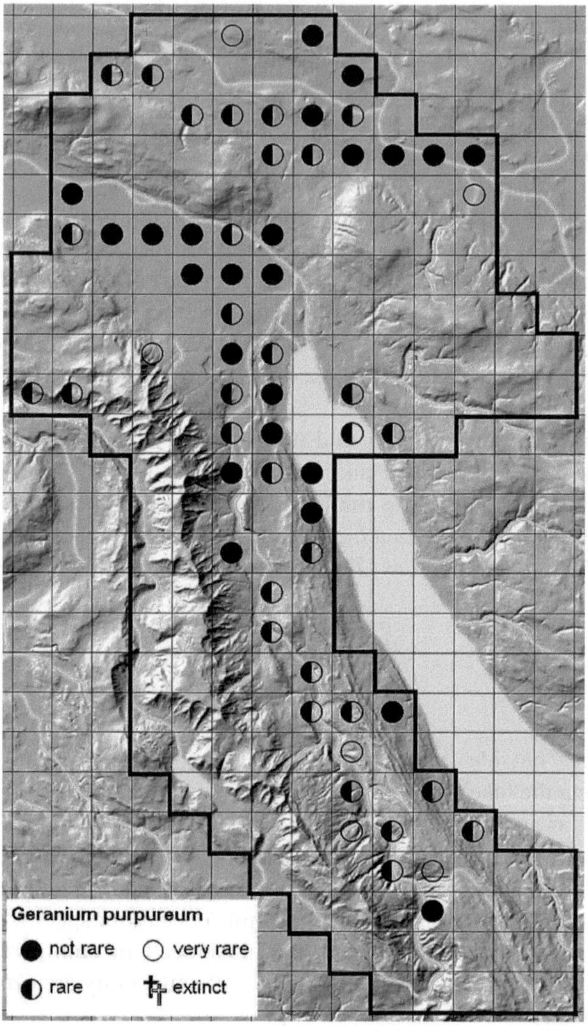

Fig. 16 Distribution map of *Geranium purpureum*, a Mediterranean species spreading rapidly along railway tracks. *Not Rare*: occurs in 1.0 km square with >200 individuals and covers >1.0 ha; *Rare*: occurs in 1.0 km square with >20 individuals and covers 0.01 to 1.0 ha; *Very Rare*: occurs in 1.0 km square with <20 individuals or covers <0.01 ha

Frequently Cultivated Trees Along the Roadsides

More than a hundred tree species have been planted along the roadsides of Zurich. There are at least 100 individual trees of about 30 taxa. Table 2 lists the nine taxa and two hybrids (excluding sub-species and cultivars) of which there are more than 500 individual trees along city roads. The list provides a good indication of the

Table 2 Frequent trees along roads with number of individuals (according to "Grün Stadt Zürich")

Species	No. of individual trees
Acer platanoides (I)	2,757
Robinia pseudoacacia (N)	1,714
Platanus x *hispanica* (N*)	1,660
Aesculus hippocastaneum (N)	1,298
Tilia x *europaea* (I)	1,120
Betula pendula (I)	932
Prunus serrulata (N*)	656
Populus nigra 'Italica' (N*)	651
Acer pseudoplatanus (I)	625
Acer campestris (I)	595
Sophora japonica (N*)	592

I: indigenous plant
N: neophyte occasionally or often naturalised
N*: neophyte rarely or never naturalised

"taxa" that are able to tolerate and survive adjacent to roads and the trendy popularity of the "species" selected by the appropriate city department ("Grün Stadt Zürich"). Species that are indigenous or have naturalised in the region are especially well adapted to the local climatic conditions.

The Frequently Cultivated Neophytic Shrubs and Trees in Gardens and Parks

During the last 200 years, many thousands of exotic trees and shrubs of many taxa have been planted throughout the city. The species, sub-species and cultivars listed in Table 3 can be found in most of the city squares.

Invasive Neophytes

Cities are susceptible to being colonised by invasive plants; that is non-native species that are able to establish successfully and spread "aggressively" in a new environment. The many open and disturbed habitats enable these plants to colonise and spread rapidly from the place or places of their initial introduction to many other places where they establish large populations to the detriment of the indigenous flora. To emphasise these species, they are grouped into Black and Watch Lists.

The Black List comprises foreign species, which are invasive and suppress or damage some established plant species in a certain geographical region or are a human health problem. The Watch List contains foreign invasive species of a region, which might become harmful to other plants or humans and are therefore candidates

Table 3 Selected species, sub-species and cultivars of trees and shrubs that are frequently found in the city squares

Acer saccharinum (N*)	*Lonicera pileata* (N)
Berberis julianae (N)	*Mahonia aquifolium* (N)
Berberis thunbergii (N)	*Parthenocissus inserta* (N)
Buddleja davidii (N)	*Paulownia tomentosa* (N)
Buxus sempervirens (N*)	*Philadelphus coronarius* (N*)
Catalpa bignonioides (N*)	*Picea abies* (I)
Cornus sericea (N)	*Pinus nigra* (N*)
Cotoneaster bullatus (N)	*Populus nigra* 'Italica' (N*)
Cotoneaster dammeri (N*)	*Prunus cerasifera* (N)
Cotoneaster divaricatus (N)	*Prunus laurocerasus* (N)
Cotoneaster horizontalis (N)	*Rosa multiflora* (N)
Cotoneaster salicifolius (N)	*Spiraea japonica* (N*)
Deutzia scabra (N)	*Spiraea ulmifolia* (N)
Forsythia x *intermedia* s.l. (N*)	*Symphoricarpos albus* (N)
Hypericum calycinum (N*)	*Symphoricarpus* x *chenaultii* (N*)
Jasminum nudiflorum (N*)	*Syringa vulgaris* (N*)
Kerria japonica (N)	*Thuja occidentalis* (N)
Ligustrum ovalifolium (N*)	*Viburnum fragrans* (N*)
Lonicera henryi (N)	*Viburnum rhytidifolium* (N)
	Wisteria sinensis (N*)

I: indigenous plant
N: neophyte occasionally or often naturalised
N*: neophyte rarely or never naturalised

for inclusion in the Black List. The following species listed in the Black and Watch Lists of Switzerland are problematic in Zurich.

1. Black list species

The Black List of Switzerland (2008) actually contains 24 species of which the following are important within the area of Zurich

(a) *Rubus armeniacus* is by far the most invasive plant in the city. It is a frequently occurring garden plant that has been spread by birds to almost all parts of the city. Once established and if it is not controlled, it can, within a few years, become dominant in many habitats, including meadows, hedges, waste places and walls. Native brambles may also be invasive in open forests and clearings; however, they rarely spread outside forests and hedges.
(b) *Solidago gigantea* is abundant throughout the city being found in moist places along rivers, creeks and lakes and in wet/damp meadows. It forms closed stands in which no other plant can grow.
(c) *Fallopia japonica* is widespread on moist open soils and in the riparian forests. It is very difficult to get rid of it once it has colonised a place.

Other undesirable, invasive species in the Black List that are detrimental to the local flora or harmful to people include *Heracleum mantegazzianum, Artemisia verlotiorum,*

Robinia pseudoacacia and *Prunus laurocerasus*. Other species in the Black List are either only locally frequent or they have their main distribution area in other regions of Switzerland.

2. Watch list species

   The Watch List currently contains 21 species of which the following behave like Black List plants in the region: *Erigeron annuus, Lonicera henryi* and *Cornus sericea*. The last species escaped from nearby gardens in the north-western part of the city into a 10 ha wetland area and developed within a few years to the most frequent shrub in the area, threatening the survival of all the other plant species. Some years ago it was dug out by the city authorities, this has resulted in there are only a few plants left in the area.

Changes in the Flora

Owing to the loss and alteration of habitats, the introduction of garden plants and the warming up of the climate, the flora of Zurich has changed considerably during the last 160 years. Between 1839 (Kölliker 1839) and 1998, 188 species (15% of the total present flora) have disappeared and 294 (24%) new species have been introduced and become established. It is interesting to compare the different habitat requirements of the extinct species with those of the newly introduced species. Most of the species that have disappeared were indigenous specialists that were probably unable to adapt to changing conditions. On the other hand, the second category comprises non-native generalists that are able to adapt well to change.

If we compare the ecological indicator values of the two categories (see Table 4), the species in the first category show distinctly lower values of temperature and nutrient conditions and higher values for continentality, which means that they mainly occur in cooler and more continental places and on soils with lower nutrient content. On the other hand, it indicates that the newcomers in the second category are better adapted to the higher temperatures and eutrophication, which are a general characteristic of urban conditions. The climate has, apparently, become more oceanic probably because of slightly higher precipitation and more haze within the city area.

As was described earlier in relation to *Geranium purpureum*, the colonisation of the city by neophytes can be very rapid, some other examples are shown in Fig. 17.

Table 4 Ecological indicator values (after Landolt 1977) for extinct and newly introduced species

Ecological indicator values	N	T	C	n
Extinct species	2.7	3.6	3.1	159
Newly introduced species	3.2	4.4	2.8	284

The scale is from 1 to 5, 5 being very rich or high, 1 very poor or low in relation to the factor *N* nutrients, *T* temperature, *C* continentality, *n* number of species

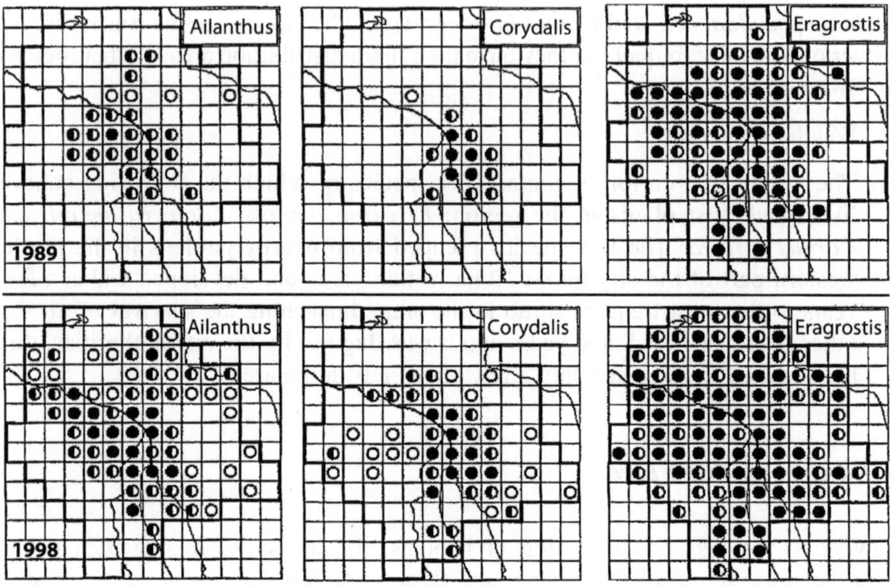

Fig. 17 Colonisation of the city; the increase of distribution of *Ailanthus altissima*, *Corydalis lutea (=Pseudofumaria lutea)* and *Eragrostis minor* between 1989 and 1998 (from Landolt 2000)

The neophytes *Ailanthus altissima, Corydalis lutea* and *Eragrostis minor* have doubled their distribution within 10 years. Originally, *Ailanthus altissima* was a garden tree and remained so for decades until it "escaped" from gardens and became naturalised about 1970. *Corydalis lutea,* a species of limestone scree in the southern Alps, was introduced into gardens but by the eighteenth century it was reported to occur on the city walls. *Eragrostis minor* was introduced around 1870; for some decades, it was a rare plant of pavements in the inner city. The reason for the sudden increase in the expansion and population sizes of the three species in the last few years is attributed to the warming of the city's climate.

The rate of extinction of species within the city was slower than the increase in the number and spread of neophytes. The main reason for the disappearance of species is habitat loss, especially wetlands, dry nutrient-poor meadows, arable land and refuse tips. The eutrophication of meadows, pastures and arable land by the application of fertilisers and the extensive use of herbicides may also have prevented the survival and successful existence of many species that were formerly rather frequent.

Chenopodium vulvaria is an example of the extinction of a species that is characteristic of urban and rural roadsides and places along house walls where it grew under warm, nutrient-rich almost continental conditions. The reasons for its disappearance has not been investigated. The nitrogen content of urban soils is high; therefore, the loss may be one or a combination of less intensive solar radiation or the lack of rubbish tips and open ground along tarmaced roads, which have not allowed the species to form viable populations. *Gentiana pneumonanthe* (Fig. 15) is a typical example of a species that has become extinct entirely because of the loss of suitable habitats. The extensive wetlands, fens and bogs that once occupied the valley bottoms and mountain sides have been developed for housing or subject to different management regimes including improved drainage and the application of fertilisers and herbicides. As a result, the populations of the plant have been totally destroyed or became too small to be viable and died out.

Most of the "Swiss Midlands" shows similar losses of biodiversity. Only in the regions that are not so densely inhabited and intensively managed, for example, in parts of the mountains, the situation is better. The hilly region south of Zurich is a good example of well-conserved wetlands (see Figs. 14 and 15). Most of the existing wetlands in Switzerland are nationally or regionally protected; farmers receive compensation to enable them to manage the vegetation for nature conservation reasons. However, the continual and constant contribution of nitrogen from the air is a major threat to the typical wetland plants because it promotes the growth of some of the common, competitive, nitrophilous species, which are able to outcompete the rare wetland species. So far, no way has been found to prevent or reduce the contribution of nitrogen from aerial sources.

Nature Conservation, Environmental Planning and Education

The legislative framework in Switzerland is rather complicated. There are three levels of Government, the Federation, the Canton and the Community. The last two only have authority within their own administrative areas. The Zurich city council has established a special department responsible for the protection of nature and for management of natural and semi-natural areas, parks, cemeteries, forests and other interesting habitats, including offices for nature conservation and for garden protection.

Some sites containing a high biodiversity, rare organisms, special habitats and characteristic landscapes or some beautiful trees in the city have been mapped and some of the most interesting and important places have become legally protected – others are just noted on a map. At present, 19 sites comprising a total of 29 ha have been legally protected, most of them are outside the urban city in zones where building is not permitted. It is proposed to protect a further 70 ha, again outside the built-up area of the city. The whole map identifies 559 sites and objects (with a total area of 5.78 km^2) of nature conservation value that are worthy of protection. It is not allowed to change any of these features without permission of the city council. If an owner wishes to make any changes to his property, the city authority must decide if it prefers to withdraw the conservation status or if it wishes to maintain it. In the latter case, the Council must compensate the owner financially because he (or she) does not have the freedom to do what he (or she) wishes.

The financial opportunities to protect gardens, meadows and other important areas with high biodiversity within the built city are rather limited. However, some small fragments of protected areas, some beautiful trees and especially some valuable areas owned by the public have been protected and preserved. Generally, the few places with a high biodiversity in the city outside the built-up areas are well preserved and managed.

It is astonishing to note how many nationally very rare or endangered species are able to survive in the city. Seventy-five species that occur in Zurich are rare, vulnerable or endangered in Switzerland (Moser et al. 2002); the species are listed in Table 5.

Table 5 Species in the flora of Zurich, which are protected in Switzerland

Endangered species	CH	Z	H	T
Alisma gramineum	EN	EN	4	I
Allium scorodoprasum	VU	EN	7	A
Alopercurus geniculatus	VU	EN	5	I
Atriplex prostrata	EN	VU	7	N
Blackstonia perfoliata	VU	EN	5	I
Bromus arvensis	VU	EN	7	A
Bromus commutatus	VU	EN	7	A
Butomus umbellatus	VU	VU	4	I
Calamagrostis canescens	VU	EN	5	I
Cardamine dentata	VU	VU	5	I
Carex diandra	VU	EN	5	I
Carex hartmanii	VU	EN	5	I
Carex pseudocyperus	VU	EN	5	I
Carex riparia	VU	EN	5	I
Carex vulpina	EN	EN	5	I
Cerastium glutinosum	VU	VU	7	A
Cirsium tuberosum	VU	VU	5	I
Conium maculatum	VU	EN	7	A
Crepis foetida	VU	VU	7	N
Crepis praemorsa	VU	EN	6	I
Cyperus flavescens	VU	EN	5	I
Cyperus fuscus	EN	EN	5	I
Cypripedium calceolus	VU	VU	1	I
Dipsacus pilosus	VU	VU	7	I
Drosera anglica	VU	EN	5	I
Eleocharis mammilata	VU	EN	5	I
Filipendula hexapetala	VU	EN	6	I
Gentiana pneumonanthe	VU	EN	5	I
Glyceria maxima	VU	VU	4	I
Herniaria hirsuta	VU	EN	3	N
Hieracium bauhinii	EN	EN	6	N
Iris sibirica	VU	EN	5	I/N
Kickxia elatine	VU	EN	7	N
Kickxia spuria	EN	EN	5	I
Lactuca virosa	EN	EN	5	I
Laserpitium prutenicum	VU	EN	7	A
Leersia oryzoides	VU	VU	7	A

(continued)

Table 5 (continued)

Endangered species	CH	Z	H	T
Legousia speculum-veneris	VU	EN	7	N
Linaria repens	VU	VU	7	A
Matteuccia struthiopteris	VU	VU	1	N
Minuartia hybrida	EN	VU	7	N
Najas marina	VU	VU	4	I
Ophrys apifera	VU	EN	6	A
Orchis palustris	VU	EN	5	I
Orchis purpurea	VU	EN	1	I
Osmunda regalis	VU	VU	1	N
Papaver argemone	VU	EN	7	A
Potamogeton coloratus	EN	EN	4	I
Potamogeton friesii	EN	EN	4	I
Potamogeton nodosus	VU	EN	4	I
Potamogeton pusillus	VU	VU	4	I
Ranunculus circinatus	EN	EN	4	I
Ranunculus lingua	VU	EN	4	I
Ranunculus sceleratus	VU	EN	7	I
Reseda luteola	VU	EN	7	A
Schoenoplectus tabernaemontani	VU	EN	4	I
Scleranthus annuus	VU	EN	7	A
Silene noctiflora	VU	EN	7	A
Sisyrinchium montanum	EN	EN	5	N
Sparganium emersum	EN	EN	4	I
Sparganium microcarpum	EN	EN	4	I
Sparganium natans	EN	EN	4	I
Sparganium neglectum	EN	EN	4	I
Spergula arvensis	VU	EN	7	A
Spiranthes aestivalis	VU	EN	5	I
Thalictrum flavum	VU	EN	5	I
Thesium rostratum	VU	EN	1	I
Tragopogon minor	EN	VU	7	N
Typha shuttleworthii	VU	VU	4	I
Utricularia bremii	EN	EN	4	I
Utricularia intermedia	EN	EN	4	I
Valerianella dentata	VU	EN	7	A
Veronica catenata	EN	VU	4	I
Veronica scutellata	VU	EN	5	I
Viola persicifolia	EN	EN	5	I
Zannichellia palustris	VU	EN	4	I

CH threat level in Switzerland (according to Moser et al. 2002), *Z* threat level in the city of Zurich (according to Landolt 2001), H habitat, T time of establishment

Threat categories: EN endangered, VU vulnerable

Habitat: (1) Forests, (2) Mountain slopes, (3) Pioneer sites of lower altitudes, (4) Water areas, (5) Wetlands, (6) Vegetation of dry and poor meadows, (7) Weed and ruderal vegetation.

Time of establishment in the region: A archaeophyte, I indigenous, N neophyte

In addition, there are many other species occurring in Zurich that are endangered in the “Swiss Midlands”. There are three main reasons why this should be so:

1. Some indigenous species (identified as “I” in the Table 5) survive in one of the few protected areas.
2. There are a lot of small sites within the city, for example, railway land and roadsides, which are well suited to the survival of many endangered species and which are not or less intensively treated with herbicides. Many archaeophytes occur in this group.
3. Some species that are rare in Switzerland are cultivated in gardens from which they have spread into the city, for example, *Matteuccia struthiopteris*.

Closing Comments

Zurich, which is the largest city in Switzerland, is situated to the north of the Alps and at the northern end of Lake Zurich in an area with a rather cool sub-oceanic climate. The difference in altitude between the lowest and highest points is nearly 500 m; the annual average precipitation varies from 100 to 130 cm, while the soils vary from basic to moderately acidic and from stable to unstable. The surroundings are called the “Swiss Midlands” and are densely populated – the city agglomeration contains about one million inhabitants. In relation to its size, the city contains a very high number of indigenous or other established vascular plant species. This might be caused by one or a combination of several factors including the high diversity/mosaic of natural or semi-natural habitats, the variation in soil types and slope (from flat to steep) with different orientations and drainage. Fortunately, some interesting wetland with small lakes, fens and fragments of bogs still remain in the north-west part of the city. In contrast, dry, nutrient-poor meadows are rare.

More than 1,200 indigenous or naturalised species of higher plants occur within the city limits. Virtually, 200 additional species that have been recorded in Zurich have become extinct within the last 160 years. Most of the species are of previously well-established archaeophytes, which have become extinct because of the habitat loss caused by the intensification and rationalisation of agricultural production and residential, commercial and other forms of development. However, a few of these species have been able to find alternative opportunities for life in urban habitats, including railway land and roadsides.

On the other hand, many more (300) new species have been able to colonise and become established in the city. The reasons for this being general climatic warming and the “heat-island effect” that is a characteristic of the urban environment. The previous relatively cool summers prevented many species from achieving their life cycle, which the recent warmer temperatures now allow them to. Also, the milder winters enable many species to survive in the city over winter, particularly in sheltered places. There have been no really cold winters in the city for more than 20 years, which has enabled some sub-tropical plants to survive throughout the winter and even set fruit and regenerate from seeds, for example, *Trachycarpus fortunei*. In the southern foothills of the Alps (for example, *Ticino*), this palm species is abundant and gives the feeling of a sub-tropical climate.

Another reason for the high species diversity is the increasing number of non-native garden species. Many of these species have adapted to the local environmental conditions and spread from gardens into the surrounding areas where they have established viable populations. Since most of the garden plants originate from warmer regions, it appears that it is the warming of the climate generally coupled with the "heat-island effect" of the city that has enabled them to naturalise.

It is exciting to observe, nearly every year, the successful expansion of the newcomers. It is to be expected that still more species will become established within the city and that the total number of species will continue to increase despite the continuing extinctions in the surrounding area. The imposition of statutory species protection and the efforts of official authorities and private organisations will help to ensure that a high level of biodiversity can be maintained, at least for the next few decades.

On the whole, the biodiversity of the city is increasing in contrast to the situation in the rural areas, where it is decreasing. However, it is arguable whether the new neophytes, most of which have large ecological and geographical amplitudes, are of the same value to the whole ecosystem than the extinct indigenous species which were restricted to very special habitats and had a closer association with stable environmental conditions. These new plant communities provide exciting opportunities for botanical investigation.

Acknowledgements I am very thankful for the help to the following persons: *Walter Lämmler,* for preparing the figures and for discussions, *Stefan Hose,* for information on the environmental management within the city and for making available the list of street trees of Zurich.

Literature Cited

Kölliker A (1839) Verzeichnis der phanerogamischen Gewächse des Cantons Zürich. Orell/Füssli, Zürich.

Kuhn N (1967) Natürliche Waldgesellschaften und Waldstandorte der Umgebung von Zürich. Veröff Geobot Inst ETH Stiftung Rübel, Zürich, 40.

Landolt E (1977) Ökologische Zeigerwerte zur Schweizer Flora. Veröff. Geobot. Inst. ETH, Stiftung Rübel, Zürich, 64.

Landolt E (2000) Some results of a floristic inventory within the city of Zurich. Preslia 72(2–4): 441–455.

Landolt E (2001) Flora der Stadt Zürich (1984–1998). Birkhäuser Verlag, Basel, Boston, Berlin.

Moser D M, Gygax A, Bäumler B, Wyler N, Palese R (2002) Rote Liste der gefährdeten Arten der Schweiz. Farn- und Blütenpflanzen. BUWAL, Bern.

Conclusions

Norbert Müller

This book covers a wide range of well-known European cities extending from Almería in the south-west to St. Petersburg in the north-east and from Sofia in the south-east to London in the north-west with most cities being concentrated in central Europe, see Fig. 1. There are five major gaps in the geographical representation, Scandinavia, eastern Europe (Moscow to the Urals), Balkan Peninsula, southern Europe and the Iberian Peninsula (including France) – Almería being the single exception to the last two. In terms of population size, London was the world's first mega-city (more than one million people living in it in the nineteenth century), and now Moscow is the largest city in Europe.

While all cities are at least several hundred or even thousand of years old, there is one exception – the newly constructed city of Milton Keynes, which was developed on intensively managed agricultural land starting at the very beginning of the 1970s. For this reason, Milton Keynes is in a unique position in relation to studying the dynamics of urban vegetation and its flora. Because of its recent origin, it is not comparable with the other cities, whose plants and habitats have been influenced by people of many generations.

The Beginning of Urban Development: The Natural Environment of the Cities

The majority of cities described are located in the temperate zone where summer broad-leaved forests predominate. At the extremes are St. Petersburg, which is within the boreal zone of evergreen coniferous forests and Almería, which is in the meridional zone of evergreen broad-leaved forests. London, Milton Keynes,

Norbert Müller (✉)
Department of Landscape Management and Restoration Ecology,
University of Applied Sciences Erfurt, Leipziger Str. 77, 99085 Erfurt, Germany
e-mail: n.mueller@fh-erfurt.de

J.G. Kelcey and N. Müller (eds.), *Plants and Habitats of European Cities*,
DOI 10.1007/978-0-387-89684-7_17,

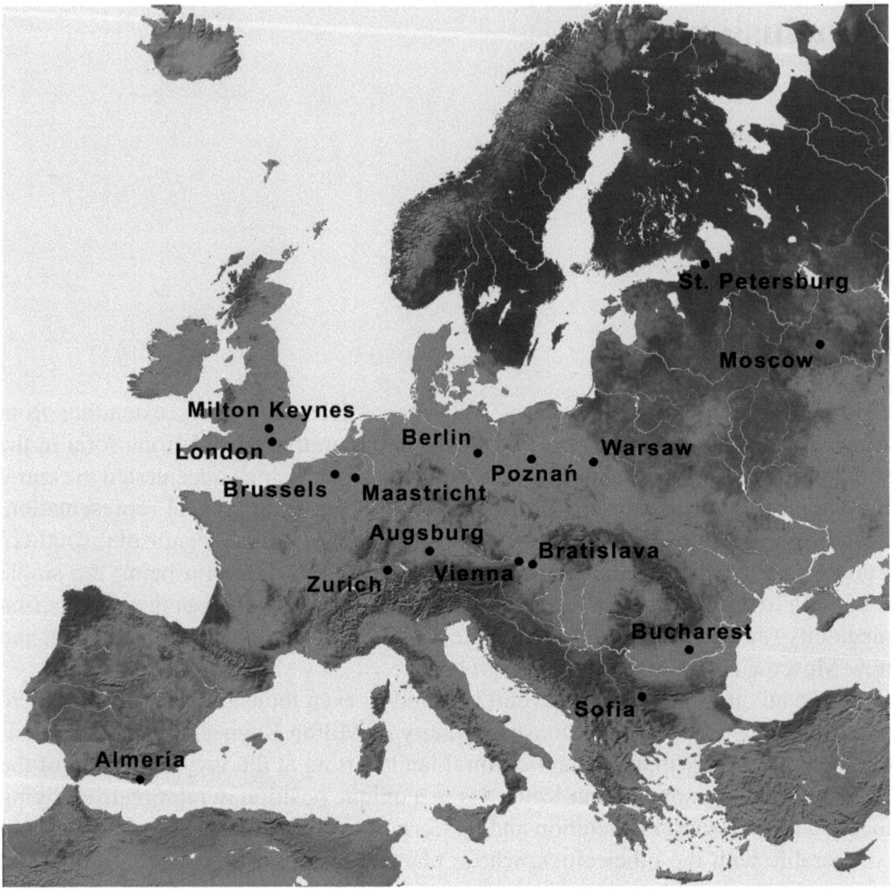

Fig. 1 Location of the cities

Brussels and Maastricht are influenced by an oceanic climate usually moist air and with less difference in the range of summer and winter temperatures and precipitation. The majority of the cities have a sub-oceanic to sub-continental climate. Moscow and Bucharest have the most distinctive continental climate usually drier and with substantial differences between summer and winter temperatures and precipitation. Most cities are situated in lowland areas, namely, Berlin, Brussels, Bucharest, London, Maastricht, Milton Keynes and Poznań. Some areas are adjacent to the sea, namely, Almería and St. Petersburg. Bratislava, Moscow, Vienna and Warsaw occur at higher altitudes, while Augsburg, Sofia and Zurich occur within montane regions. In summary, the altitude of the cities described ranges from sea level (Almería) to 870 m above sea level (Zurich).

The Evolution of Cities in Europe: The History of a Long- Lasting Human Impact on the Environment

Most of the areas occupied by the present cities were settled by people in Neolithic times, when Europe was colonised by agriculturalists. From the point of view of landscape history, it is important to recognise that the re-colonisation of Europe by trees after the last Ice Age was not completed before human influence began to cause local disturbances. An important time for many central European cities was the invasion of the Romans who established many fortifications and towns as the Empire expanded, for example, Augsburg 15 BC, Vienna 10 BC, Bratislava and Maastricht first century AD, Bucharest second century AD, Brussels sixth century AD and London 43 AD. The introduction of new food plants and new agricultural methods can be regarded as one of the main drivers for the further population growth and the development of cities. Augsburg (*Augusta vindelicorum*) and London (*Londinium*) acquired their names during Roman times.

In the Middle Ages (500–1500), when the Roman influence in central Europe was replaced by Christian Feudalism, many cities experienced their first peak of prosperity, many being cited by their present name, Bratislava (903), Zurich (929), Almería (955), Brussels (979), Moscow (1147), Vienna (1150), Sofia (1194), Berlin (1200), Poznań (1253) and Bucharest (1368). The exceptions are St. Petersburg (founded in 1703) and Milton Keynes (founded in 1971). Most cities underwent further rapid development in the late nineteenth and early twentieth centuries during the industrialisation of Europe. An exception was the population growth of London which had 850,000 inhabitants in 1800 and 3.5 million in 1900 – more than 400% growth in 100 years. Today, with 8.8 million inhabitants, Moscow is the largest city in Europe, followed (in relation to the cities in this book) by St. Petersburg, London and Berlin (Fig. 2).

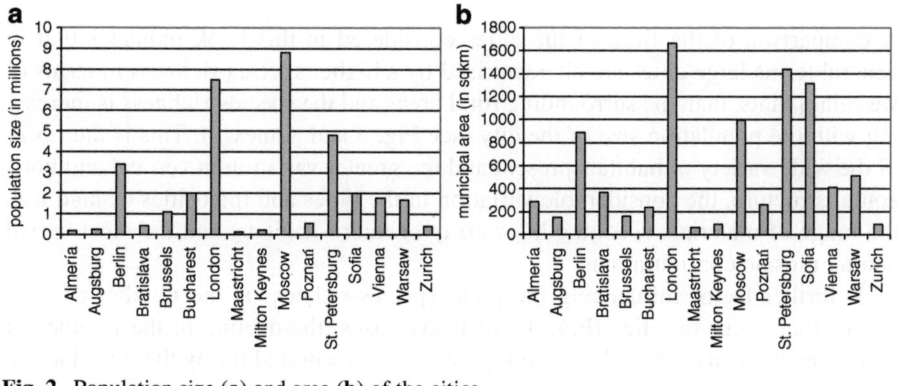

Fig. 2 Population size (**a**) and area (**b**) of the cities

Cities: A New Environment

The establishment and growth of cities result in major environmental changes of which probably the best known and the most extensively and intensively researched are air and water pollution. For example, air pollution in London has been a recurrent problem since the end of the Medieval period when it started to rely on energy from the burning of coal imported from Newcastle and other places. London smog (a combination of smoke and fog), which had severe consequences for the health of its inhabitants and plant life, is legendary. In 1937, emissions from chimneys produced 125 tons of solids per km^2 per year, of which 17 tonnes were sulphates and 2 tonnes tar (Crawley, in this book). Climate change in cities was described for the first time in London in the early nineteenth century and the impact of air pollution during the same period. The adverse effect of air pollution in all large European cities was observed in the nineteenth century as the result of the decline and extinction of lichenised fungi. The use of lichenised fungi in the determination of air quality may be one of the longest traditions of using plants as indicators for environmental change. Today in several cities, for example, Bratislava and London, some lichen species have returned to the urban core following the improvement of air quality, especially the reduction of sulphur dioxide. In some cities, including Bucharest, Poznań and Sofia, air pollution is still a serious problem.

The continuous influence of human activities in cities since ancient times has altered the structure and chemistry of natural soils. Little is known about the evolution of urban soils, which were systematically described in Berlin for the first time in the 1980s. Because similar settlement types are causing similar changes to the natural environment, urban land-use types have been used as ecological units in cities since the 1980s. The distinctiveness of some urban habitats such as waste ground, railway areas and different settlement types were investigated systematically for the first time in the "Biotope mapping of Berlin" and subsequently in many German and other cities of central Europe.

Cities as "Hot Spots" of Plant Diversity

A comparison of the flora of the cities considered in this book indicates that in general terms large cities are characterised by a higher species-richness in terms of vascular plants, than the surrounding rural areas and the species-richness is increasing with the population size of the city (see Fig. 3 and Annex 1). This is the result of the wide variety of habitats present and the greater variation in vertical and horizontal structure, the considerable variation in the types and intensities of land use, the range of materials used, the huge array of micro-habitats, and the most varied habitat mosaic configurations.

A further factor influencing the plant species-richness is the number of neophytes that occur in cities (Fig. 4). In many cases, the decline in the number of native species caused by urban development is compensated for by the introduction and naturalisation of neophytes.

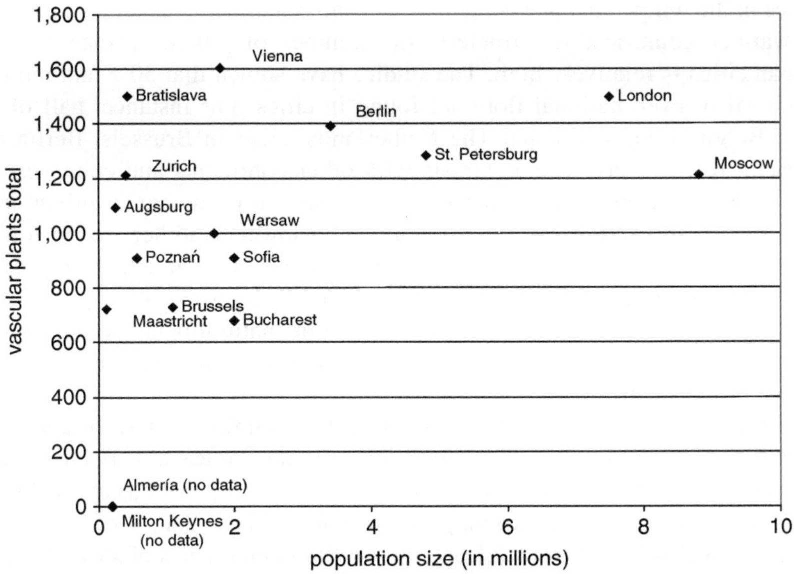

Fig. 3 Species-richness of vascular plants and population size

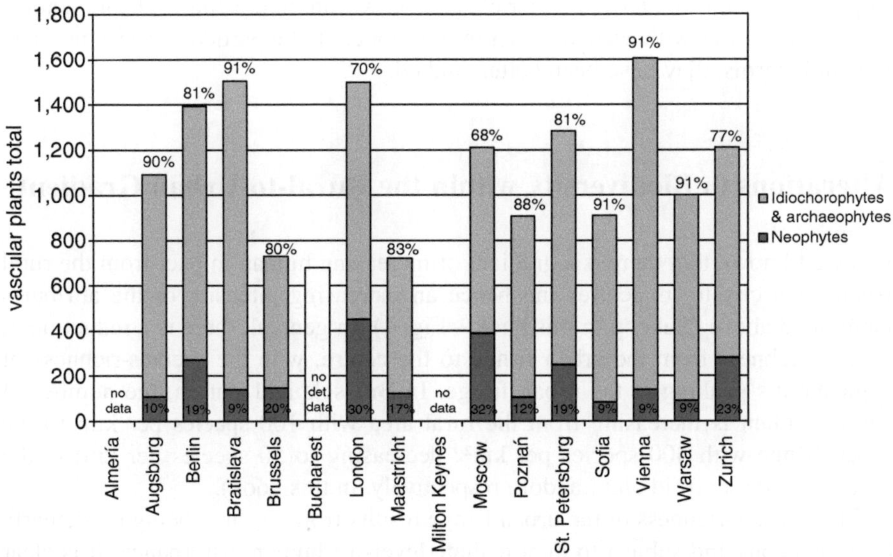

Fig. 4 Species-richness of vascular plants divided into idiochorophytes, archaeophytes and neophytes

Cities in the temperate zone contain a substantial number of naturalised species from warmer countries; nevertheless, the number of native species in large European cities is relatively high. The studies have shown that 50% and more of the regional or even national flora are found in cities. For instance, half of the flora of Belgium, Germany and The Netherlands occur in Brussels, Berlin and Maastricht, respectively. This contrasts with other continents and countries, for example, New Zealand, where non-native species are dominant in urban areas and the number of native species is significantly lower. Another reason for the high biodiversity of European cities is that they have been established along landscape transition zones and rivers in regions that are naturally highly heterogeneous in terms of their landscape or even located in "natural botanical hot spots". Bratislava and Vienna, which have a large number of native species, are good examples of the latter, see Fig. 4.

In addition, the species-richness of the urban flora is influenced by the age of the city and the presence of "special" habitats, for example, the Royal Parks in London and the urban wastelands in Berlin. The long period of time that has elapsed since most of the cities were established may have resulted in the evolution of new "urban species" (see below). Last but not least, most cities contain sites of special importance for nature conservation with respect to the protection of threatened species and habitats. Many contain remnants of pristine vegetation that have survived because of topographical, soil and other features which results in the land being unsuitable for development. Many of these sites contain rare species and species-rich habitats. As discussed below, remarkable examples of pristine habitats occur within virtually all of the cities.

Finally, an analysis of large-scale floristic mapping exercises and the relationship between cities and species richness shows an interesting "phenomenon", namely, that cities with academic institutions appear to be particularly species-rich. In simple terms, they have been better studied.

Alterations to Biodiversity within the Rural-to-Urban Gradient

It is well known that there is a gradient of increasing human impact from the rural fringe of a city to its centre, and hence an increasing intensity of the attributes mentioned above (Sukopp in this book – Fig. 4). In general, there is a reduction in species-richness from the urban fringe to the centre, with the species-richness of angiosperms peaking at the urban fringe. In Brussels and Zurich, the number of vascular plant is increasing from the rural area with 100 species per km^2 to the urban fringe with 400 species per km^2; decreasing to 50 species per km^2 in the urban core (Godefroid and Landolt, respectively, in this book).

The species-richness of the urban fringe results from the area being particularly heterogeneous and subject to intermediate levels of human disturbance. It is clear that there is a strong correlation between the greatest human impact in the central core and the reduction of species-richness.

Ornamental Species and Plant Fashions

It is estimated that about 12,000 plant species have been introduced to central Europe since the Neolithic, mainly for food, medical and ornamental purposes. The latter, especially trees and shrubs, plays an important ecological role in cities. In the urban core, planted trees and shrubs (both native and non-native) constitute the main part of the plant biomass, the number of species is significantly higher as the result of the numerous introduced taxa, which predominate. In Moscow, for example, 370 cultivated woody species have been recorded, while in Milton Keynes at least 1,700 mainly woody non-native taxa were planted between 1971 and 1992. *Aesculus hippocastanum* a tree "endemic" to the Balkans has been recorded as a very common planted tree in most of the cities, including Vienna where it was introduced in 1570. Other frequently planted alien trees include *Acer negundo, Platanus* hybrids and *Robinia pseudoacacia* from North America and *Ailanthus altissima* from China. Several tree species that are native to parts of Europe but non-native in others are frequently planted in cities, they include *Acer pseudoplatanus A. platanoides, Fraxinus excelsior* and *Tilia* species. In general terms, the proportion of native and non-native species in German cities is about equal.

Urban Opportunists: Adaptations of Native Plants to Urban Habitats and Plant Evolution in Cities

Another important group of plants within cities are common "weeds", which occur frequently in virtually all urban habitats. The biological characteristics of these species, which are restricted to open or disturbed conditions, include an annual or biennial life form, the production of large quantities of seed, high genetic variability and phenotypic plasticity. Due to their biomass within the urban core, they have important ecological functions although they are regarded as undesirable species that are frequently controlled by the application of herbicides. An analysis of the list of the 50 most frequent species in all cities given in Annex 2 shows that there is a weak correlation between the size of the city and the frequency of neophytes such as *Acer negundo, Conyza canadensis, Galinsoga ciliata, Amaranthus retroflexus, Erigeron annuus, Robinia pseudoacacia* and *Solidago canadensis.* In larger cities, there seems to be more non-native species than in smaller ones.

It is evident that many more species that are native in Europe are better adapted to urban conditions than the introduced species. Many European species (called apophytes), have found a second important home in urban habitats for example, *Daucus carota, Lolium perenne, Plantago lanceolata* and *Trifolium repens.* Another phenomenon can be observed when analysing the 20 most common urban opportunists listed in Table 1 – in addition to the apophytes and one neophyte from North

Table 1 Status of the most frequent species in all cities (extracted from Annex 2)

Occurrence (total 15 cities)	Species
14	***Poa annua***[a]
14	***Polygonum aviculare* agg.**[a]
13	***Dactylis glomerata***[a]
13	***Stellaria media* agg.**[a]
12	**Conyza canadensis**[b]
12	***Plantago major* agg.**[a]
12	*Trifolium repens*[c]
11	*Artemisia vulgaris*[c]
11	***Capsella bursa-pastoris***[a]
11	*Cirsium arvense*[c]
11	***Convolvolus arvensis***[a]
11	***Urtica dioica***[a]
10	*Plantago lanceolata*[c]
10	***Taraxacum officinale* agg.**[a]
9	*Achillea millefolium* agg.[c]
9	***Chenopodium album***[a]
9	*Lolium perenne*[c]
8	*Aegopodium podagraria*[c]
8	***Bellis perennis***[a]
7	*Ballota nigra*[c]
7	***Elytrigia repens***[a]

[a]***Anecophytes*** (synonym: neogene species), taxa that have evolved in Europe and spread in association with human influence
[b]**Neophytes**, taxa introduced from another country after 1500
[c]*Apophytes*, taxa that migrated into urban areas from natural habitats and are now found frequently in urban areas

America (*Conyza canadensis*), there is a large number of "homeless" species that do not have a habitat in the natural vegetation of Europe. It is suspected that these species evolved somewhere in Europe since the Neolithic, when people started to establish permanent settlements and develop agricultural practices. These neogene species (or anecophytes) include, *Poa annua, Polygonum aviculare* agg., *Dactylis glomerata* and *Stellaria media* agg. (see Table 1).

Earlier, botanical investigations in some of the larger urban agglomerations in the northern hemisphere have shown that these "European anecophytes" are very successful on a world-wide scale and comprise up to 80% of the most frequent plants in large North American cities such as New York, Los Angeles and San Francisco. It is considered that the success of these European urban plants in cities of the United States results from evolutionary processes in Europe over several millennia during which time some taxa (apophytes) have adapted, and other taxa (anecophytes) have evolved to grow and thrive in areas subjected to human disturbance. Consequently, they are "better equipped" than native American species to exploit the conditions created by the continual expansion of urban/industrial development and other human activities.

Evolutionary processes as the consequence of the introduction of non-native species have been observed in cities with increasing frequency during recent times, a well-studied example is *Oenothera* spp. in Europe. During the 1980s, more than 15 taxa have been identified in Europe with two exceptions they are not identical to the introduced North American taxa from which they are descended. These new European taxa have evolved since their American parent species were introduced into Europe about 350 years ago. Accelerated speciation as a result of the dispersal of a number of individuals to a new location, the "founder effect" is in this case expected to be the reason for the fast evolution. Similarly, *Aster novi-angliae, A. novi-belgii, A. lanceolatus, A. laevis* and hybrids, which were introduced from North America, are found in many European cities where they are becoming increasingly variable both morphologically and in their ecological amplitude, which suggests that new taxa are evolving. Another reason for the fast evolution of new species in cities is hybridisation between native and non-native species from the same genus, for example, *Populus div. spec.* As the number of people living in a city increases, so does the number of introduced species; consequently there is an increase in the probability of hybridisation, especially of closely related native and non-native species growing in close proximity to each other.

Losers and Winners: Endangered Species and Habitats

In general, Red Data Lists of city areas have a higher percentage of extinct and endangered species than Red Lists of the whole country. There are two groups of species in cities that have become extinct:

- taxa of several habitats that have decreased throughout Europe as the result of changes in land use, for example, "weeds" of arable land and species of semi-dry grasslands.
- taxa of wetlands.

The decrease in the last group is disproportionately higher in urban areas because of urban impacts such as the general lowering of the watertable, positive drainage works and pollution. Such decreases are reported from several cities, including Augsburg, Maastricht, Poznań and Zurich, and the creation of St. Petersburg was only possible after extensive drainage works.

On the other hand, some urban habitats, for example, walls, railway embankments and waste ground, can provide new habitats for endangered native species:

In London, two ferns that were thought to be extinct in Britain were rediscovered on walls. The best conserved mesophile grasslands in southern Germany have been recorded in the old parks of Augsburg. The old walls and fortifications in the urban core of Maastricht are the most valuable habitats for endangered species.

Threats to Global Biodiversity: Invasive Species

Worldwide, cities are regarded as centres for the importation, naturalisation and spread of non-native species. Deliberate introductions for horticulture, forestry and landscaping purposes play the major role in cities, while unintended introductions in goods are of less importance. A wide range of naturalised non-native species have been recorded in many cities ranging from 83 species in Sofia to 450 in London (compare Fig. 4 and Annex 1). Only a few of the naturalised species in European cities have become invasive and a threat to biodiversity. *Solidago canadensis* and *S. gigantea* from North America and *Fallopia japonica* and *F. sacchalinensis* from Asia are reported to be invasive in most cities. Since these species were introduced into Europe several hundred years ago, it appears that there is a long time lapse between the time of introduction, naturalisation and invasive spreading. The morphology of *Solidago canadensis* and *S. gigantea* in their native North American is different from that in Europe (where both species are aliens). It is suspected that North American species have evolved into new taxa or physiological ecotypes since they were introduced into Europe. *Fallopia japonica* and *F. sacchalinensis,* which were introduced into Europe, have hybridised; the hybrid (*Fallopia* x *bohemica)* has become a major invasive taxon in England and Germany. The mechanisms as to why a non-native or native taxon becomes invasive or expands its distribution and abundance are often unknown and is therefore of considerable botanical interest. In his studies of Sheffield (England), Oliver Gilbert considered that *Fallopia japonica* was making a positive contribution to the city's flora by providing a canopy for woodland species such as *Hyacinthoides non-scripta.* Other non-native trees and shrubs are considered to be of value for birds in providing berries during the winter.

Contributions to Global Biodiversity

Many cities have a long tradition of designating areas for nature conservation protection, for example, the cities in this book contain protected areas (often extensive) with relicts of pristine natural and semi-natural habitats within their borders (Fig. 5). Within the European Commission's Habitat Directive the Member States

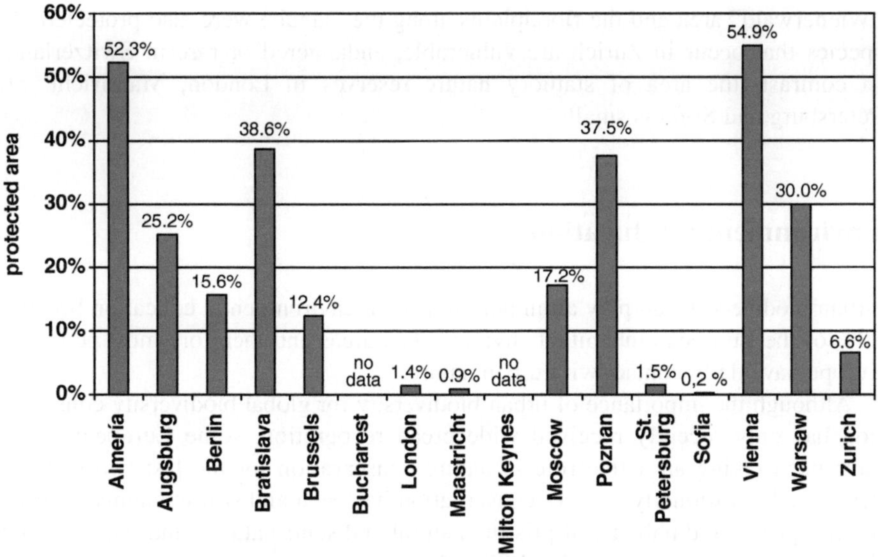

Fig. 5 Percentage of protected areas

of the European Union are required to allocate at least 10% of the area of their country for nature conservation protection. Most of the cities described in the book have a high percentage of nature reserves, which contain species and habitats of national importance.

As an example, Augsburg contains the most extensive protected areas for nature conservation in Bavaria outside the Alps. The areas were protected in 1910 because of the presence of representative habitats of an alpine river; today, the city contains the largest population of *Gladiolus palustris* within the European Union. In Berlin, 25 different "European Union habitats" occur, the highest number of any of the 16 cities described in this book. The endemic species *Dianthus lumnitzeri* and the West Carpathian-Pannonian sub-endemics *Hieracium echioides, Sempervivum hirtum* f. *glabrescens* and *Taraxacum danubium* have their *locus classicus* in Bratislava. A substantial part of the largest forest in north Belgium (1,660 ha of the 4,400 ha Sonian Forest) is located within the municipal area of Brussels. Sofia includes 12 statutorily protected species, 14 Bulgarian Red List species and several Balkan endemics, including *Aesculus hippocastanum, Angelica pancicii, Campanula sparsa, Centaurea affinis, Centaurea uniflora, Cerastium petricola, Cirsium candelabru, Peucedanum aegopodioides, Scabiosa triniaefolia* and *Trifolium trichopterum.*

In Vienna where 56% of the city area has been subject to nature conservation protection since 1905, a statutory green belt 600 m wide was established around the city. In the same declaration, the most important pristine landscapes the

"Wienerwald" area and the floodplains along the Danube were also protected. 75 species that occur in Zurich are vulnerable, endangered or rare in Switzerland. In contrast, the area of statuory nature reserves in London, Maastricht, St. Petersburg and Sofia is small.

Environmental Education

Urban biodiversity can play an important role in environmental education because 80% of the European inhabitants live in urban areas and therefore most people in Europe have daily contact with urban nature.

Although the importance of urban biodiversity for global biodiversity conservation has only recently received widespread recognition, some European cities have been taking an active role in nature conservation for the last two to three decades. "Traditionally" nature conservation has been and still is mainly focused on the species and habitats of pristine natural and semi-natural landscapes within the cities. However, since the 1980s there has been an increasing awareness (in some circles) of the plants and habitats and ecological services in urban areas. For example, during this period the frequency of mowing the lawns in parks in Augsburg was reduced in order to establish species-rich meadows to increase biodiversity and allow people to become more aware of urban nature. Numerous investigations of the value of urban habitats have been carried out in Berlin since the 1970s. They resulted in the production of the first "species and habitat programme" for an entire city; in addition, Berlin was also the first city in the world to protect urban waste ground as a nature reserve. In the 1990s, species found on the walls and fortifications of Maastricht were identified as being threatened nationally, and therefore added to the Red Data List of The Netherlands. Special programmes for urban plants and habitats have been developed in Vienna since the 1990s.

Closing Comments

In global terms, some European cities are the most extensive and intensively researched in relation to urban plants and habitats. However, this does not mean that the research is comprehensive and thorough, only that it is better botanically researched than most cities in Europe and other continents. Although the plants and habitats in cities are appreciated, the importance and possibilities of their contributing to the reduction of the global loss of biodiversity continues to receive little consideration. For example, the European Commission's Habitats Directive (the cornerstone of European Union's nature conservation botanical and habitats policies) disregards urban habitats, as do many (if not most) national nature conservation policies and protection measures. The proper understanding of the urban flora and its habitats of European cities requires the European Commission in collaboration with the other European Governments to develop their research programme in order to investigate the:

(a) comparative biodiversity of all plant groups, especially the very serious lack of knowledge about non-vascular plants;
(b) changes in plant and habitat biodiversity by careful monitoring;
(c) impact that urban design has on urban biodiversity;
(d) influence that urban biodiversity has on the surrounding landscape and flora;
(e) influence that the surrounding landscape and flora has on urban biodiversity;
(f) link between climate change and urban plant diversity. The evolutionary processes operating on plants and their habitats in cities are a neglected area of scientific investigation, although preliminary studies indicate that plants and habitats are changing in response to increasing temperature and human impact.

Annex 1 Summarised data of the 16 European cities (primarily according the information in the single city chapters)

City	1 Almería	2 Augsburg	3 Berlin	4 Bratislava	5 Brussels	6 Bucharest	7 London
First Author	Elias Dana	Norbert Müller	H. Sukopp	Viera Feráková	S. Godefroid	Marilena Onete	Mick Crawley
Date of Investigation	1999–2000	1985–1990	1985–1995	1985–1990 (2009)	1992–1994	2006–2008	2009
Average Altitude of the City (in m a.s.l.)	23.00	494.00	35.00	140.00	398.00	82.00	5.00
Municipal Area (in km²)	291.00	147.00	889.00	367.90	161.00	238.00	1662.00
Population Size	190,000	250,000	3,400,000	425,540	1,100,000	2,000,000	7,500,000
Vascular Plants Total (All Spontaneous and Naturalized, Without Casuals)	no info.	1,092	1,393	1,502	730	680	1,498
Idiochorophytes and archaeophytes	no info.	984	1,122	1,362	584	no info.	1,048
Neophytes (Without Casuals)	no info.	108	271	140	146	no info.	450
Extinct Vascular Plants	no info.	67	203	118	148	no info.	129
Protected Area in km²	152.06	37.00	139.00	142.08	19.99	0.00	23.00
Habitats of Annex I of the European Union Habitat Directive	10	14	25	13	8	12	no detailed information
* Asterisk means **priority habitat**	1210	3130	2310	6110	3150	R4147	oak woodland
	1420	3140	2330	6210	6430	R3122	lowland heaths
	1430	3150	3140	6240	6510	R3420	
	1520*	3240	3150	6440	9120	R3715	
	2210	3260	3160	6510	9130	R3716	
	2230	5130	3260	9110	9150	R5305	
	5220*	6210	4030	9130	9160	R5309	
	6220*	6410	**6120***	9150	**91E0***	R5312	
	6310	6510	6214	**9180***		R8710	
	92D0	**7220***	**6220***	**91E0***		R8704	
		7230	6410	91F0		R2202	
		9110	6430	**91G0***		R2207	
		91E0*	6510	**91H0***			
		91F0	7140				
			7150				
			7220*				
			7230				
			9110				
			9160				
			9170				
			9190				
			91D0*				
			91D1*				
			91D2*				
			91E0*				
Plants of Annex II of the European Union Habitat Directive		Cypripedium calceolus					
		Gladiolus palustris					

8 Maastricht	9 Milton Keynes	10 Moscow	11 Poznan	12 St. Petersburg	13 Sofia	14 Vienna	15 Warsaw	16 Zurich
Eddy Weeda	John Kelcey	A. Shvetsov	B. Jackowiak	Maria Ignatieva	D. Dimitrov	Alex Mrkvicka	B. Sudnik-W.	Elias Landolt
1990–2006	1973–2001	1981–2000	1828–1990	1990–2008	no info.	no info.	1977–1997	1984–1998
60.00	11.00	156.00	96.00	4.00	500.00	203.00	107.00	569.00
60.00	90.00	994.00	262.00	1439.00	1311.00	415.00	517.00	88.00
120,000	207,000	8,800,000	565,000	4,800,000	2,000,000	1,800,000	1,707,000	400,000
721	no info.	1,211	908	1,282	910	1,604	1,001	1,211
599	no info.	823	803	1,033	827	1,464	909	932
122	no info.	388	105	249	83	140	92	279
76	no info.	104	124	23	4	170	149	188
0.51	no info.	171.00	98,37	21.50	3.19	228.00	155.00	5.78
8	no info.	no info.	11	no info.	7	10	8	no info.
3260 3270 6110 6210 6230 6510 6160 **91E0***			2330 3150 3260 3270 **6120*** 6210 6410 9170 9190 **91E0*** 91I0		3130 3150 6210 6420 6510 **91E0*** 92A0	61XX 6210 **6240*** **6260*** 6410 9110 9130 9170 **91H0*** 92A0	4030 **6120*** 6410 9110 9170 **91D0*** **91E0*** 91F0	
			Angelica palustris Liparis loeselii Pulsatilla patens				Angelica palustris	

Annex 2 Most frequent plant species in the urbanised area of 15 cities and their origin (taxa with abundance of 5 and higher; i = idiochorophytes or archaeophyte, n = neophyte)

City		1 Almería	2 Augsburg	3 Berlin	4 Bratislava	5 Brussels	6 Bucharest	7 London	8 Maastricht	9 Milton Keynes	10 Moscow	11 Poznan	12 St. Petersburg	13 Sofia	14 Vienna	15 Warsaw	16 Zurich
Urbanised area (in km²)		9.95	65.00	889.00	67.10	161.00	31.00	300.00	36.00	no info.	994.00	228.60	660.00	47.50	155.00	430.00	80.00
Abundance Related to 15 Cities	Taxa																
14	*Poa annua*	*i*	*i*	*i*	*i*	*i*		*i*	*i*		*i*	*i*	*i*	*i*	*i*	*i*	*i*
14	*Polygonum aviculare agg.*	*i*	*i*	*i*	*i*	*i*	*i*	*i*	*i*		*i*	*i*	*i*	*i*	*i*		*i*
13	*Dactylis glomerata*		*i*	*i*	*i*	*i*		*i*	*i*		*i*	*i*	*i*	*i*	*i*	*i*	*i*
13	*Stellaria media agg.*		*i*	*i*	*i*	*i*		*i*	*i*		*i*	*i*	*i*	*i*	*i*	*i*	*i*
12	*Conyza canadensis*		*n*	*n*	*n*	*n*	*n*		*n*		*n*	*n*		*n*	*n*	*n*	*n*
12	*Plantago major agg.*		*i*	*i*	*i*	*i*	*i*		*i*		*i*	*i*	*i*		*i*	*i*	*i*
12	*Trifolium repens*		*i*	*i*	*i*	*i*			*i*		*i*	*i*	*i*	*i*	*i*	*i*	*i*
11	*Artemisia vulgaris*		*i*	*i*	*i*	*i*		*i*	*i*		*i*	*i*	*i*		*i*	*i*	
11	*Capsella bursa-pastoris*		*i*	*i*	*i*	*i*	*i*				*i*	*i*	*i*		*i*	*i*	*i*
11	*Cirsium arvense*		*i*	*i*	*i*	*i*	*i*	*i*	*i*		*i*	*i*			*i*	*i*	
11	*Convolvulus arvensis*	*i*	*i*	*i*	*i*		*i*		*i*		*i*	*i*		*i*	*i*	*i*	
11	*Urtica dioica*		*i*	*i*	*i*	*i*	*i*		*i*		*i*	*i*	*i*		*i*	*i*	
10	*Plantago lanceolata*		*i*	*i*		*i*	*i*	*i*	*i*			*i*		*i*	*i*	*i*	
10	*Taraxacum officinale agg.*		*i*		*i*		*i*		*i*		*i*	*i*	*i*		*i*	*i*	*i*
9	*Achillea millefolium agg.*		*i*	*i*		*i*	*i*		*i*		*i*	*i*	*i*			*i*	
9	*Chenopodium album*	*i*	*i*	*i*	*i*				*i*		*i*	*i*	*i*			*i*	
9	*Lolium perenne*		*i*	*i*	*i*	*i*			*i*		*n*	*i*			*i*	*i*	
8	*Aegopodium podagraria*		*i*	*i*	*i*	*i*		*i*			*i*		*i*				*i*
8	*Bellis perennis*		*i*	*i*	*i*	*i*	*i*		*i*						*i*		*i*
7	*Ballota nigra*			*i*	*i*			*i*				*i*		*i*	*i*	*i*	

7	*Elytrigia repens*		*i*		*i*				*i*	*i*	*i*	*i*			*i*	
7	*Galium spurium agg.*		*i*		*i*	*i*		*i*	*i*		*i*			*i*		
7	*Geum urbanum*		*i*	*i*		*i*	*i*	*i*		*i*			*i*			
7	*Glechoma hederacea*		*i*	*i*	*i*			*i*	*i*	*i*						*i*
7	*Ranunculus repens*			*i*		*i*			*i*	*i*		*i*			*i*	*i*
7	*Sambucus nigra*		*i*	*i*		*i*		*i*	*i*		*i*				*i*	
7	*Sonchus oleraceus*	*i*	*i*			*i*		*i*	*i*	*i*			*i*			
6	*Acer negundo*			*n*			*n*			*n*		*n*	*n*		*n*	
6	*Acer platanoides*		*i*	*i*		*n*						*i*	*n*			*i*
6	*Acer pseudoplatanus*		*i*	*i*		*i*	*i*	*i*								*i*
6	*Arrhenatherum elatius*			*i*		*i*			*i*		*i*		*i*	*i*		
6	*Daucus carota*				*i*		*i*		*i*		*i*		*i*		*i*	
6	*Fraxinus excelsior*		*i*			*i*	*i*	*i*					*i*			*i*
6	*Medicago lupulina*							*i*		*i*	*i*			*i*	*i*	*i*
6	*Poa pratensis agg.*		*i*	*i*						*i*		*i*		*i*	*i*	
6	*Senecio vulgaris*		*i*			*i*		*i*	*i*						*i*	*i*
6	*Trifolium pratense*		*i*						*i*	*i*	*i*				*i*	*i*
5	*Betula pendula*		*i*	*i*		*i*				*i*		*i*				
5	*Cerastium fontanum agg.*		*i*			*i*						*i*			*i*	*i*
5	*Chelidonium majus*			*i*	*i*		*i*						*i*			*i*
5	*Cichorium intybus*				*i*		*i*				*i*		*i*		*i*	
5	*Galinsoga ciliata*		*n*							*n*			*n*	*i*		*n*
5	*Geranium robertianum*		*i*			*i*		*i*						*i*		*i*
5	*Lactuca serriola*				*i*			*i*	*i*		*i*			*i*		
5	*Matricaria discoidea*		*i*			*n*				*n*	*n*				*n*	
5	*Rumex crispus*		*i*				*i*				*i*			*i*	*i*	
5	*Rumex obtusifolius*		*i*			*i*	*i*		*i*	*i*						
5	*Tanacetum vulgare*			*i*	*i*				*i*		*i*				*i*	
5	*Tripleurospermum inodorum*				*i*					*i*	*i*			*i*	*i*	

Further Reading

Dunn R R, Gavin M C, Sanchez M C and Solomon J N (2006) The pigeon paradox: Dependence of global conservation on urban nature. Conservation Biology 20: 1814–1816

Müller N (2010) On the most frequently occurring vascular plants and the role of non-native species in urban areas – a comparison of selected cities in the old and the new worlds. In Müller N, Werner P and Kelcey JG (eds), Urban Biodiversity and Design Conservation Science and Practice 7, Wiley-Blackwell, Oxford: 227–242

Kunick W (1987) Woody vegetation in settlements. Landscape and Urban Planning 14: 57–78

Stace C A (2010) New Flora of the British Isles. Cambridge University Press, 1266p

Sukopp H (2002) On the early history of urban ecology in Europe. Preslia 74: 373–393

Appendix I

VASCULAR TAXA REFERRED TO IN THE CHAPTERS

Because there is no standard European nomenclature for the vascular plants; the names of the taxa are those supplied by the authors. This has resulted in three difficulties – a taxon having more than one name, a name that has been superceded or the use of different spellings. This appendix has been compiled in an attempt to overcome these difficulties mainly by reference to Stace, C.A. (2010) New Flora of the British Isles. 3rd Ed. Cambridge University Press, Cambridge, which contains most of the vascular plant taxa referred to in the chapters. We are indebted to Clive Stace for his considerable help in updating the nomeclature and in identifying taxa that appear under different names but any mistakes are ours. Where a name provided by an author has been given an more recent name it is indicated by =. Best endeavours have been made to list all the taxa mentioned in the chapters; however, the aspiration may not have been achieved. The spellings of some of the taxa in some of the chapters may vary with this appendix, the spellings in the appendix should be the correct ones.

Abies alba
Abies cephalonica
Abies concolor
Abies nordmanniana
Abies sibirica
Abutilon theophrasti
Acacia dealbata
Acacia longifolia
Acacia retinoides
Acacia saligna
Acacia spp.
Acer campestre
Acer ginnala
Acer negundo
Acer palmatum
Acer platanoides
Acer platanoides 'Fassens Black'
Acer platanoides 'Globosa'
Acer platanoides 'Schwedleri'
Acer pseudoplatanus
Acer pseudoplatanus 'Atropurpureum'
Acer saccharinum
Acer spp.
Acer tataricum
Achillea millefolium
Achillea ptarmica
Achillea ptarmica 'Flore Pleno'
Achillea ptarmica cultivars
Aconitum napellus
Aconitum spp.
Acorus calamus
Actaea spicata

DOI 10.1007/978-0-387-89684-7, © Springer Science+Business Media, LLC 2011

Adenocaulon adhaerescens
Adiantum capillus-veneris
Adiantum raddianum
Adonis flammea
Adonis vernalis
Adoxa moschatellina
Aegilops cylindrica
Aegopodium podograria
Aesculus carnea
Aesculus hippocastanum
Aethusa cynapium
Agave americana
Agave sisalana
Agave sp.
Ageratum houstonianum
Agrimonia eupatoria
Agropyron intermedium=*Elytrigia intermedia*
Agropyron pectinatum
Agrostemma githago
Agrostis canina
Agrostis capillaris
Agrostis gigantea
Agrostis spp.
Agrostis stolonifera
Ailanthus altissima
Aira caryophyllea
Aira praecox
Ajuga reptans
Albizia julibrissin
Alcea rosea
Alchemilla mollis
Alchemilla monticola
Alchemilla spp.
Alisma gramineum ssp. *gramineum*
Alisma gramineum ssp. *wahlenbergii*
Alisma plantago-aquatica
Alliaria petiolata
Allium angulosum
Allium ascalonicum=*Allium cepa*
Allium cepa
Allium oleraceum
Allium porrum
Allium sativum
Allium schoenoprasum
Allium senescens
Allium spp.
Allium ursinum
Allium vineale
Alnus cordata
Alnus glutinosa
Alnus incana
Alnus spp.
Alopercurus myosuroides
Alopercurus pratensis
Alopercurus geniculatus
Alopercurus myosuroides
Alopercurus pratensis
Alternanthera sp.
Alyssum montanum ssp. *gmelinii*
Alyssum montanum ssp. *montanum*
Alyssum saxatile=*Aurinia saxatilis*
Amaranthus albus
Amaranthus blitoides
Amaranthus blitum
Amaranthus blitum var. *ascendens*
Amaranthus caudatus
Amaranthus chlorostachys
Amaranthus crispus
Amaranthus deflexus
Amaranthus hybridus
Amaranthus muricatus
Amaranthus powellii
Amaranthus retroflexus
Amaranthus spp.
Amaranthus viridis
Ambrosia artemisiifolia
Ambrosia trifida
Amelanchier lamarkii
Amelanchier ovalis
Amelanchier spicata
Ammi majus
Ammochloa palestina
Ammophila arenaria
Amomum cardamom
Amorpha fruticosa
Ampelopsis veitchii
Amygdalus communis
Anacamptis pyramidalis=*Prunus dulcis*
Anacyclus clavatus
Anagallis arvensis
Anagallis arvensis ssp. *arvensis*
Anagallis arvensis ssp. *foemina*
Anchusa arvensis
Anchusa officinalis
Androcymbium gramineum
Andromeda polifolia
Androsace maxima

Anemone blanda
Anemone nemorosa
Anemone patens
Anemone ranunculoides
Anethum graveolens
Angelica archangelica
Angelica palustris
Angelica pancicii
Angelica sylvestris
Anisantha diandra
Anisantha rubens
Anisantha sterilis
Anisantha tectorum
Anthemis arvensis
Anthemis cotula
Anthemis tinctoria
Anthoxanthum odoratum
Anthriscus cerefolium ssp. *trichosperma*
Anthriscus sylvestris
Anthyllis vulneraria
Antirrhinum mollisimum
Antirrihinum sp.
Apera interrupta
Apera spica-venti
Aphanes arvensis
Aphanes australis
Apium graveolens
Apium graveolens var. *dulce*
Apium nodiflorum
Apium repens
Aquilegia hybrida
Aquilegia vulgaris
Arabidopsis thaliana
Arabis arenosa
Arabis glabra
Arabis hirsuta
Arabis sagittaria
Aralia chinensis
Araucaria araucana
Arbutus unedo
Archangelica officinalis=*Angelica officinalis*
Arctium lappa
Arctium minus
Arctium nemorosum
Arctium tomentosum
Arctostaphylos uva-ursi
Arenaria serpyllifolia
Arenaria serpyllifolia ssp. *leptoclados*= *A. leptoclados*
Aristolochia clematitis
Armeniaca vulgaris=*Prunus armeniaca*
Armeria elongata
Armeria maritime
Armeria vulgaris=*A. maritima* ssp. *maritima*
Armoracia rusticana
Arnoseris minima
Aronia melanocarpa
Arrhenatherum elatius
Artemesia abrotanum
Artemesia arvensis
Artemisia absynthium
Artemisia annua
Artemisia austriaca
Artemisia barrelieri
Artemisia campestris
Artemisia herba-alba
Artemisia pontica
Artemisia repens
Artemisia scoparia
Artemisia sieversiana
Artemisia spp.
Artemisia vulgaris
Arthroemum macostachym
Arum maculatum
Arum orientale
Aruncus vulgaris
Arundo donax
Asarina procumbens
Asarum europaeum
Asclepias syriaca
Asparagus albus
Asparagus officinalis
Asparagus officinalis ssp. officinalis
Asparagus stipularis
Asperugo procumbens
Asperula cynanchica
Asphodelus spp.
Asplenium adiantum-nigrum
Asplenium adiantum-nigrum ssp. *adiantum-nigrum*
Asplenium adiantum-nigrum ssp. *adiantum-nigrum* var. *silesiacum*
Asplenium fontanum
Asplenium marinum
Asplenium ruta-muraria
Asplenium scolopendrium
Asplenium scolopendrium ssp. *scolopendrium*
Asplenium trichomanes
Asplenium trichomanes ssp. *quadrivalens*

Asplenium viride
Aster amellus
Aster laevis
Aster lanceolatus
Aster novae-angliae
Aster novi-belgii
Aster salignus
Aster x *salignus*
Aster spp.
Aster squamatus
Aster tripolium
Aster x *versicolor*
Astilbe x *arendsii*
Astragalus alopecuroides
Astragalus arenarius
Astragalus asper
Astragalus cicer
Astragalus glycyphyllos
Astragalus onobrychis
Astrantia major
Athyrum felix-femina
Atriplex glauca
Atriplex halimus
Atriplex hortensis
Atriplex littoralis
Atriplex oblongifolia
Atriplex patula
Atriplex portulacoides
Atriplex prostrata
Atriplex sagittata
Atriplex semibaccata
Atriplex spp.
Atriplex tatarica
Atropa belladonna
Aubretia deltoidea
Aucuba japonica
Avena barbata
Avena fatua
Avena sativa
Avena spp.
Avena strigosa
Avenella flexuosa
Azolla filiculoides

Ballota nigra
Ballota nigra ssp. *meridionalis*
Barbarea vulgaris
Bassia hyssopifolia
Bassia laniflora
Bassia scoparia
Bassia scoparia ssp. *densiflora*
Beckmannia eruciformis
Begonia semperflorens
Begonia spp.
Begonia tuberosa
Bellis perennis
Berberis darwinii
Berberis julianae
Berberis spp.
Berberis x *stenophylla*
Berberis thunbergii
Berberis verruculosa
Berberis vulgaris
Bergenia crassifolia
Berteroa incana
Berula erecta
Beta vulgaris
Beta vulgaris ssp.*vulgaris*
Betula alba=*B. pubescens*
Betula humilis
Betula nana
Betula pendula
Betula pubescens
Betula spp.
Bidens cernua
Bidens frondosa
Bidens tripartita
Bifora radians
Biota orientalis
Blackstonia acuminata
Blackstonia perfoliata
Blechnum spicant
Blysmus compressus
Bolboschoenus maritimus
Bombycilaena erecta
Bothriochloa ischaemum
Botrychium lunaria
Botrychium matricariifolium
Botrychium virginianum
Bougainvillea spp.
Brachyglottis 'Sunshine'=*B.* x *jubar*
Brachypodium x *cugnacci*
Brachypodium distachyon
Brachypodium pinnatum
Brachypodium pinnatum sensu stricto
Brachypodium retusum
Brachypodium rupestre
Brachypodium sylvaticum

Brassica elongata
Brassica juncea
Brassica napus
Brassica napus ssp. *campestris*
Brassica napus ssp. *oleifera*
Brassica nigra
Brassica oleracea
Brassica oleracea var. *botrytis*
Brassica oleracea var. *capitata*
Brassica oleracea var. *gemmifera*
Brassica oleracea var. *oleracea*
Brassica rapa
Brassica rapa ssp. *oleifera*
Briza maxima
Briza media
Bromopsis erecta
Bromopsis inermis
Bromopsis inermis ssp. *inermis*
Bromus arvensis
Bromus commutatus
Bromus hordeaceus
Bromus secalinus
Brousonettia papyrifera
Brunnera macrophylla
Brunnera sibirica
Bryonia dioica
Buddleia variabilis
Buddleja davidii
Buddleja spp.
Bunias orientalis
Bupleurum affine
Bupleurum rotundifolium
Butomus umbellatus
Buxus sempervirens

Cabomba caroliniana
Cakile maritima
Calamagrostis canescens
Calamagrostis epigeijos
Calamagrostis phragmitoides
Calamagrostis varia
Calendula arvensis
Calendula officinalis
Calendula spp.
Calicotome intermedia
Calla palustris
Callitriche cophocarpa
Callitriche palustris
Callitriche sp.
Calluna vulgaris
Caltha plaustris
Calypso bulbosa
Calystegia sepium
Calystegia sylvatica
Camelina alyssum
Camelina sativa
Campanula bononiensis
Campanula glomerata
Campanula latifolia
Campanula patula
Campanula persicifolia
Campanula poscharskyana
Campanula rapunculoides
Campanula rapunculus
Campanula rotundifolia
Campanula sibirica
Campanula sparsa
Campanula trachelium
Campanula xylorrhiza
Camphorosma monspeliaca
Campsis radicans
Canna indica
Cannabis sativa
Cannabis sativa ssp. *indica*
Capparis ovata
Capsella bursa-pastoris
Capsicum annuum
Capsicum spp.
Caragana arborescens
Caragana frutex
Caralluma europaea ssp. *europaea*
Cardamine amara
Cardamine bulbifera
Cardamine flexuosa
Cardamine hirsuta
Cardamine impatiens
Cardamine leucantha
Cardamine pratensis
Carduus acanthoides
Carduus crispus
Carduus nutans
Carduus tenuiflorus
Carex acuta
Carex acutiformis
Carex alba
Carex appropinquata
Carex arenaria
Carex brizoides

Carex caryophyllea
Carex cespitosa
Carex chordorrhiza
Carex curta
Carex davalliana
Carex diandra
Carex digitata
Carex dioica
Carex disticha
Carex divulsa ssp. *divulsa*
Carex divulsa ssp. *leersii*
Carex echinata
Carex elata
Carex flacca
Carex hartmanii
Carex hirta
Carex humilis
Carex lasiocarpa
Carex ligerica
Carex limosa
Carex melanostachya
Carex michelii
Carex muricata sensu lato
Carex nigra
Carex oederi sensu lato
Carex otrubae
Carex ovalis = *C. leporina*
Carex pallescens
Carex panicea
Carex paniculata
Carex pendula
Carex pilosa
Carex pilulifera
Carex praecox
Carex pseudocyperus
Carex remota
Carex riparia
Carex rostrata
Carex spicata
Carex spp.
Carex strigosa
Carex sylvatica
Carex vaginata
Carex vesicaria
Carex viridula ssp. *oedocarpa* = *C. demissa*
Carex vulpina
Carlina vulgaris
Carlyna corymbosa
Carpinus betulus
Carrichtera annua
Carthamus lanatus
Carum carvi
Caryopteris × *cladonensis*
Caryopteris incana
Castanea sativa
Catabrosa aquatica
Catalpa bignonioides
Catapodium rigidum
Caucalis platycarpos
Caulinia tenuissima
Cedrus libani
Celastrus orbiculatus
Celtis australis
Celtis occidentalis
Centaurea affinis
Centaurea arenaria
Centaurea calcitrapa
Centaurea cyanus
Centaurea diffusa
Centaurea diluta
Centaurea jacea
Centaurea montana
Centaurea nigra
Centaurea salonitana
Centaurea scabiosa
Centaurea solstitialis spp. *solstitialis*
Centaurea spp.
Centaurea stoebe
Centaurea trichocephala
Centaurea uniflora
Centaurium erythraea
Centaurium littorale
Centaurium pulchellum
Centranthus ruber
Cephalanthera damasonium
Cephalanthera longifolia
Cephalanthera rubra
Cephalaria transsylvanica
Cerastium fontanum
Cerastium fontanum ssp. *holosteoides*
Cerastium fontanum ssp. *vulgare*
Cerastium glomeratum
Cerastium petricola
Cerastium pumilum
Cerastium semidecandrum
Cerastium tomentosum
Cerasus avium = *Prunus avium*
Cerasus mahaleb = *Prunus mahaleb*

Cerasus serrulata = *Prunus serrulata*
Cerasus vulgaris = *Prunus cerasus*
Ceratocephala falcate
Ceratocephala orthoceras
Ceratocephala testiculata
Ceratochloa carinata
Ceratonia siliqua
Ceratophyllum demersum
Ceratophyllum submersum
Cercis siliquastrum
Ceterach officinarum
Chaenomeles japonica
Chaenomeles speciosa
Chaenorhinum minus
Chaerophyllum hirsutum
Chaerophyllum temulum
Chamaecyparis lawsoniana
Chamaecyparis obtusa
Chamaecyparis pisifera
Chamaecyparis spp.
Chamaecytisus austriacus
Chamaedaphne calyculata
Chamaerops humilis
Chamaesyce serpens
Chamerion angustifolium
Chaste sp.
Chelidonium majus
Chenopodium album
Chenopodium bonus-henricus
Chenopodium botrys
Chenopodium ficifolium
Chenopodium foliosum
Chenopodium glaucum
Chenopodium hybridum
Chenopodium murale
Chenopodium opulifolium
Chenopodium polyspermum
Chenopodium pumilio
Chenopodium rubrum
Chenopodium strictum
Chenopodium urbicum
Chenopodium vulvaria
Chimaphila umbellata
Chionodoxa luciliae = *Scilla luciliae*
Chlorophytum comosum
Chondrilla juncea
Chorispora tenella
Chrysanthemum coronarium = *Glebionis coronaria*
Chrysopogon gryllus
Chrysosplenium alternifolium
Chrysosplenium oppositifolium
Cicerbita macrophylla
Cichorium intybus
Cicuta virosa
Circaea lutetiana
Cirsium acaule
Cirsium arvense
Cirsium brachycephalum
Cirsium candelabrum
Cirsium eriophorum
Cirsium oleraceum
Cirsium palustre
Cirsium serrulatum
Cirsium spp.
Cirsium tuberosum
Cirsium vulgare
Cistanche phelypaea
Citrullus lanatus
Cladium marsicus
Clarkia grandiflora
Claytonia perfoliata
Clematic recta
Clematis integrifolia
Clematis spp.
Clematis vitalba
Clematis viticella
Clinopodium acinos
Clinopodium arvensis
Clinopodium menthifolium
Clinopodium vulgare
Cochlearia danica
Coeloglossum viride
Colchicum autumnale
Coleus blumei
Colutea arborescens
Commelina communis
Conioselinum tataricum
Conium maculatum
Conopodium majus
Conringia austriaca
Consolida orientalis
Consolida regalis
Consolida regalis ssp. *paniculata*
Consolida spp.
Convallaria majalis
Convolvulus arvensis
Convolvulus tricolour

Conyza bonariensis
Conyza canadensis
Conyza floribunda
Conyza spp.
Conyza sumatrensis
Corallorhiza trifida
Coriandrum sativum
Corispennum sp.
Cornus alba
Cornus mas
Cornus sanguinea
Cornus sericea
Coronopus didymus = *Lepidium didymus*
Coronopus squamatus = *Lepidium coronopus*
Corrigiola litoralis
Cortaderia selloana
Corydalis cava
Corydalis intermedia
Corydalis marschalliana
Corydalis ochotensis
Corydalis solida
Corylus avellana
Corylus colurna
Corylus maxima
Corynephorus canescens
Cosentinia vellea ssp. *bivalens*
Cosmos bipinnatus
Cosmos sp.
Cotinus coggygria
Cotoneaster conspicuus 'Decora'
Cotoneaster dammeri
Cotoneaster dielsianus
Cotoneaster franchetii
Cotoneaster horizontalis
Cotoneaster integrifolius
Cotoneaster lacteus
Cotoneaster lucidus
Cotoneaster monopyrenus
Cotoneaster salicifolius
Cotoneaster salicifolius 'Afterglow'
Cotoneaster simonsii
Cotoneaster x *intermedia*
Cotoneaster x *watereri*
Cotula coronopifolia
Coyncia tournefourtii
Crassula helmsii
Crataegus almaatensis
Crataegus laevigata
Crataegus mollis
Crataegus monogyna
Crataegus persimilis 'Prunifolia'
Crataegus spp.
Crataegus x *media*
Crepis biennis
Crepis capillaris
Crepis foetida
Crepis mollis
Crepis paludosa
Crepis setosa
Crepis sp.
Crepis tectorum
Crepis vesicaria ssp. *taraxacifolia*
Crocus chrysanthus x *biflorus*
Crocus flavus
Crocus napolitanus
Crocus spp.
Cruciata laevipes
Cucumis sativus
Cucurbita maxima
Cucurbita pepo
x *Cupressocyparis leylandii*
Cupressus sempervirens
Cuscuta campestris
Cuscuta epilinum
Cuscuta epithymum
Cuscuta europaea
Cyclachaena xanthiifolia
Cyclamen purpurascens
Cydonia oblonga
Cymbalaria muralis
Cynodon dactylon
Cynoglossum officinale
Cynomorium coccineum
Cynosurus cristatus
Cyperus fuscus
Cyperus rotundus
Cypripedium calceolus
Cypripedium guttatum
Cystopteris fragilis
Cytisus battanieri
Cytisus nigricans
Cytisus x *praecox*
Cytisus scoparius
Cytrus aurantium

Dactylis glomerata
Dactylis polygama
Dactylorhiza baltica

Dactylorhiza fuchsii
Dactylorhiza incarnata
Dactylorhiza longifolia
Dactylorhiza maculata
Dactylorhiza majalis
Dactylorhiza spp.
Dahlia spp.
Damasomium alisma
Danthonia alpina
Daphne mezereum
Datura inoxia
Datura stramonium
Daucus carota
Daucus carota ssp. *sativus*
Deschampsia cespitosa
Deschampsia flexuosa = *Avenella flexuosa*
Descurainia sophia
Deutzia crenata
Deutzia scabra
Dianthus armeria
Dianthus barbatus
Dianthus carthusianorum
Dianthus collinus ssp. *collinus*
Dianthus deltoides
Dianthus fischeri
Dianthus lumnitzeri
Dianthus pontederae
Dianthus praecox ssp. *lumnitzeri*
Dianthus superbus
Dicentra formosa
Dicentra spectabilis = *Lamprocapnos spectabilis*
Dicentra spp.
Dichanthium ischaemum
Dichostylis micheliana
Dictamnus albus
Digitalis purpurea
Digitaria ischaemum
Digitaria sanguinalis
Digitaria schaemum
Dinebra retroflexa
Diospyros lotus
Diphasiastrum alpinum
Diplazium sibiricum
Diplotaxis muralis
Diplotaxis tenuifolia
Dipsacus fullonum
Dipsacus laciniatus
Dipsacus sp.
Dittrichia viscosa
Dorycnium germanicum
Downingia elegans
Dracaena drago
Dracaena sp.
Dracocephalum ruyschiana
Drosera anglica
Drosera rotundifolia
Dryopteris carthusiana
Dryopteris dilatata
Dryopteris expansa
Dryopteris filix-mas
Duchesnea indica = *Potentilla indica*

Echinaria capitata
Echinochloa crus-galli
Echinocystis lobata
Echinops ritro ssp. *ruthenicus*
Echinops sphaerocephalus
Echium creticum
Echium russicum
Echium vulgare
Eichhornia crassipes
Elaeagnus angustifolius
Elaeagnus argentea
Elaeagnus x *ebbingei* = *E* x *submacrophylla*
Elaeagnus multiflora
Elaeagnus pungens 'Maculata'
Elaeagnus umbellata
Elatine alsinastrum
Elatine hexandra
Elatine hydropiper
Elatine triandra
Eleocharis acicularis
Eleocharis palustris
Eleocharis quinqueflora
Eleocharis spp.
Eleusine indica
Elodea canadensis
Elodea nuttallii
Elsholtzia ciliata
Elymus hispidus = *Elytrigia intermedia*
Elytrigia repens
Epilobium ciliatum
Epilobium hirsutum
Epilobium komarovianum
Epilobium montanum
Epilobium palustre
Epilobium parviflorum
Epilobium pseudorubescens

Epilobium rubescens
Epimedium alpinum
Epipactis atrorubens
Epipactis helleborine
Epipactis palustris
Epipactis spp.
Equisetum arvense
Equisetum fluviatile
Equisetum palustre
Equisetum scirpoides
Equisetum sylvaticum
Equisetum telmateia
Equisetum variegatum
Eragrostis albensis
Eragrostis minor
Eragrostis pilosa
Eranthis hyemalis
Erechtites hieraciifolius
Erica carnea
Erica erigena
Erigeron acer = E. acris
Erigeron annuus
Erigeron annuus ssp. *annuus*
Erigeron annuus ssp. *septentrionalis*
Eriophorum angustifolium
Eriophorum gracile
Eriophorum latifolium
Eriophorum vaginatum
Erodium chium
Erodium cicutarium
Erophila verna
Eruca vesicaria
Erucastrum gallicum
Eryngium campestre
Eryngium maritimum
Eryngium planum
Erysimum cheiranthoides
Erysimum cheiri
Erysimum diffusum
Erysimum x *marshallii*
Escallonia macrantha
Eucalyptus gunnii
Euclidium syriacum
Euonymus europaeus
Euonymus fortunei
Euonymus japonicus
Euonymus radicans
Euonymus verrucosa
Eupatorium cannabinum
Euphorbia amygdaloides
Euphorbia characias
Euphorbia cyparissias
Euphorbia helioscopia
Euphorbia humifusa
Euphorbia lathyris
Euphorbia maculata
Euphorbia palustris
Euphorbia paralias
Euphorbia peplis
Euphorbia peplus
Euphorbia pulcherrima
Euphorbia taurinensis
Euphorbia terracina
Euphorbia verrucosa
Euphrasia agg.
Euzomodendron bourgeanum
Exochorda aubertii
Exochorda grandiflora

Fagonia cretica
Fagopyrum esculentum
Fagus sylavtica
Fagus syvlatica 'Purpurea'
Falcaria vulgaris
Fallopia baldschuanica
Fallopia × bohemica
Fallopia convolvulus
Fallopia dumetorum
Fallopia japonica
Fallopia sachalinensis
Fallopia sp.
Fatsia japonica
Festuca arenaria
Festuca arundinacea
Festuca brevipila
Festuca gigantea
Festuca ovina
Festuca pallens
Festuca pratensis
Festuca rubra
Festuca rubra ssp. *commutata*
Festuca spp.
Festuca valesiaca
Ficaria verna
Ficus carica
Ficus elastica
Ficus macrophylla
Ficus retusa

Ficus retusa var. *nitida*
Ficus rubiginosa
Ficus sp.
Filago arvensis
Filago minima
Filipendula ulmaria
Filipendula vulgaris
Foeniculum vulgare
Fontanesia phyllireoides
Forsythia europaea
Forsythia spp.
Forsythia suspensa
Forsythia x *intermedia*
Fragaria x *ananassa*
Fragaria moschata
Fragaria vesca
Fragaria viridis
Frangula alnus
Fraxinus americana
Fraxinus angustifolia
Fraxinus excelsior
Fraxinus excelsior 'Globosa'
Fraxinus excelsior 'Pendula'
Fraxinus ornus
Fraxinus pennsylvanica
Fraxinus sp.
Fumana procumbens
Fumaria capreolata
Fumaria officinalis
Fumaria spp.
Fumaria vaillantii

Gagea granulosa
Gagea lutea
Gagea minima
Gagea pratensis
Gagea spp.
Gaillardia sp.
Galanthus elwesii
Galanthus nivalis
Galanthus nivalis 'Flore Pleno'
Galanthus plicatus
Galanthus sp.
Galanthus woronowii
Galega officinalis
Galeopsis angustifolia
Galeopsis pubescens
Galeopsis speciosa
Galeopsis tetrahit
Galinsoga parviflora
Galinsoga quadriradiata
Galium aparine
Galium ephedroides
Galium glaucum
Galium hercynicum
Galium mollugo
Galium mollugo ssp. *erectum*
Galium odoratum
Galium palustre
Galium parisiense ssp. *anglicum*
Galium saxatile
Galium schultesii
Galium spurium
Galium sylvaticum
Galium uliginosum
Galium verum
Gaultheria shallon
Genista hispanica
Genista monspessulana
Genista tinctoria
Gentiana clusii
Gentiana cruciata
Gentiana pneumonanthe
Gentianella campestris
Gentianella uliginosa
Geranium dissectum
Geranium endressii
Geranium molle
Geranium pratense
Geranium pusillum
Geranium pyrenaicum
Geranium robertianum
Geranium rotundifolium
Geranium sanguineum
Geum macrophyllum
Geum quellyon
Geum rivale
Geum urbanum
Gingko biloba
Gladiolus imbricatus
Gladiolus palustris
Gladiolus spp.
Glaucium corniculatum
Glaucium flavum
Glechoma hederacea
Glechoma hirsutea
Gleditsia triacanthos
Glyceria aquatica

Glyceria declinata
Glyceria fluitans
Glyceria fluitans x *declinata*
Glyceria maxima
Glyceria notata
Gnaphalium sylvaticum
Gnaphalium uliginosum
Goodyera repens
Gossypium spp.
Gratiola officinalis
Groenlandia densa
Grossularia reclinata = Ribes reclinatum
Guizotia abyssinica
Gymnadenia conopsea
Gymnocarpium dryopteris
Gymnocladus canadensis
Gypsophila fastigiata
Gypsophila hispanica
Gypsophila muralis
Gypsophila paniculata
Gypsophila pilosa
Gypsophila repens
Gypsophila struthium

Halogeton sativus
Haloxylum articulatum/Hammada articulata
Haplophyllum rosmarinifolium
Hedera helix
Hedypnois rhagadioloides
Heleochloa alopecuroides
Heleochloa schoenoides
Helianthemum lavandulifolium
Helianthemum nummularium
Helianthemum squamatum
Helianthus annuus
Helianthus sp.
Helianthus subcanescens
Helianthus tuberosus
Helichrysum arenarium
Helictotrichon pubescens = Avenula pubescens
Heliotropium arborescens
Heliotropium curassavicum
Heliotropium europaeum
Helleborus niger
Hemerocallis fulva
Hemerocallis spp.
Hepatica nobilis
Heracleum mantegazzianum
Heracleum sosnowskyi
Hercaleum sphondylium
Herminium monorchis
Herniaria fontanessii ssp. *almeriana*
Herniaria glabra
Herniaria hirsuta
Hesperis matronalis
Hibiscus spp.
Hibiscus syriacus
Hieracium amplexicaule
Hieracium caesium
Hieracium echioides
Hieracium macranthum
Hieracium murorum sensu lato
Hieracium oistophyllum
Hieracium prolatatum
Hieracium speluncarum
Hieracium spp.
Hieracium subholophyllum
Hieracium umbellatum
Himantoglossum adriaticum
Hippophae rhamnoides
Hippuris vulgaris
Hirschfeldia incana
Holcus lanatus
Holcus mollis
Honckenya peploides
Hordelymus europaeus
Hordeum distichon
Hordeum jubatum
Hordeum murinum
Hordeum murinum ssp. *leporinum*
Hordeum secalinum
Hordeum sp.
Hordeum vulgare
Hortensia petiolaris = Hydrangea petiolaris
Hottonia palustris
Humulus lupulus
Hyacinthoides cultivars
Hyacinthoides x *massartiana*
Hyacinthoides non-scripta
Hydrangea arborescens
Hydrangea breitscheinderi
Hydrangea hortensis = H. macrophylla
Hydrocharis morsus-ranae
Hylomecon vernalis
Hymenolobus procumbens
Hyoscyamus albus
Hyoscyamus niger
Hyparrhenia hirta

Hypericum androsaemum x *hircinum*
Hypericum androsaemum x *inodorum*
Hypericum calycinum
Hypericum hirsutum
Hypericum montanum
Hypericum perforatum
Hypericum pulchrum
Hypericum robertii
Hypericum spp.
Hypericum x *inodorum* 'Elstead'
Hypochaeris radicata
Hypopitys monotropa
Hyssopus officinalis

Iberis spp.
Iberis umbellata
Ilex aquifolium
Impatiens balfourii
Impatiens capensis
Impatiens glandulifera
Impatiens noli-tangere
Impatiens parviflora
Inula britannica
Inula conyzae
Inula crithmoides
Inula ensifolia
Inula helenium
Inula hirta
Inula salicina
Inula salicina ssp. *sabuletorum*
Iondraba laevigata ssp. *kerneri*
Iresine spp.
Iris foetidissima
Iris germanica
Iris x *hybrida*
Iris pseudacorus
Iris pumila
Iris sibirica
Iris spp.
Iris variegata
Isatis tinctoria
Isatis tinctoria ssp. *tinctoria*
Isoëtes echinospora
Isoëtes lacustris
Isolepis setacea
Iva xanthiifolia

Jasione montana
Jasminum nudiflorum
Jasminum officinale
Jasminum revolutum
Juglans cinerea
Juglans nigra
Juglans regia
Juncus alpinoarticulatus
Juncus articulatus
Juncus bufonius
Juncus compressus
Juncus conglomeratus
Juncus effusus
Juncus gerardii
Juncus inflexus
Juncus subnodulosus
Juncus tenuis
Juniperus chinensis
Juniperus communis
Juniperus horizontalis
Juniperus sabina
Juniperus spp.
Juniperus squamata
Juniperus virginiana

Kernera saxatilis
Kerria japonica
Kickxia elatine
Kickxia spp.
Kickxia spuria
Knautia arvensis
Kochia scoparia
Kochia scoparia ssp. *scoparia*
Koeleria cristata = *K. macrantha*
Koeleria delavignei
Koeleria glauca
Koeleria macrantha
Koeleria pyramidata
Koeleria splendens
Koelreuteria paniculata
Kolkwitzia amabilis

Laburnum alpinum
Laburnum anagyroides
Laburnum sp.
Lactuca sativa
Lactuca sativa varieties
Lactuca serriola
Lactuca tatarica
Lactuca virosa
Lagarosiphon major

Lamarckia aurea
Lamiastrum galeobdolon
Lamiastrum galeobdolon ssp. *argentatum*
Lamium album
Lamium amplexicaule
Lamium maculatum
Lamium purpureum
Lappula squarrosa
Lapsana communis
Larix decidua
Larix sibirica
Larix spp.
Laser trilobum
Lathraea squamaria
Lathyrus aphaca
Lathyrus japonicus
Lathyrus latifolius
Lathyrus linifolius
Lathyrus niger
Lathyrus odoratus
Lathyrus palustris
Lathyrus pisiformis
Lathyrus pratensis
Lathyrus sylvestris
Lathyrus tuberosus
Lathyrus vernus
Launaea arborescens
Laurocerasus officinalis = *Prunus laurocerasus*
Lavandula dentata
Lavandula x *intermedia*
Lavandula varieties
Lavatera cretica
Lavatera oblongifolia
Ledum palustre
Leersia oryzoides
Legousia speculum-veneris
Lemna minor
Lemna minuta
Lemna triscula
Lens culinaris
Leontodon autumnalis = *Scorzoneroides autumnalis*
Leontodon hispidus
Leontodon saxatilis
Leonurus cardiaca
Leonurus marrubiastrum
Leonurus quinquelobatus
Lepidium draba
Lepidium heterophyllum
Lepidium latifolium
Lepidium ruderale
Lepidium sativum
L. virginatum = *L. densiflorum*
Lerchenfeldia flexuosa
Leucanthemum adustum
Leucanthemum spp.
Leucanthemum x *superbum*
Leucanthemum vulgare
Leucojum aestivum
Leucojum vernum
Leycesteria formosa
Leymus arenarius
Libocedrus deccurens
Ligustrum ovalifolium
Ligustrum vulgare
Lilium martagon
Lilium spp.
Lilium tigrinum
Limnanthes douglasii
Limodorum abortivum
Limoniastrum monopetalum
Limonium spp.
Limonium vulgare
Limosella aquatica
Linaria alpina
Linaria nigricans
Linaria oligantha
Linaria pedunculata
Linaria repens
Linaria vulgaris
Lindernia procumbens
Linnaea borealis
Linum catharticum
Linum usitatissimum
Linum viscosum
Liparis loeselii
Liquidamber stiraciflua
Liriodendron tulipifera
Listera ovata
Lithospermum arvense
Lithospermum officinale
Lithospermum purpureocaeruleum
Livistonia australis
Lobelia dortmanna
Lobelia erinus

Lobularia maritima
Loeflingia baetica
Lolium multiflorum
Lolium perenne
Lolium perenne cultivars
Lolium remotum
Lolium spp.
Lolium temulentum
Lonicera caprifolium
Lonicera henryi
Lonicera japonica
Lonicera maackii
Lonicera nitida
Lonicera periclymenum
Lonicera pileata
Lonicera spp.
Lonicera tatarica
Lonicera xylosteum
Lotus corniculatus
Lotus glaber
Lotus pedunculatus
Ludwigia grandiflora
Lunaria annua
Lunaria rediviva
Lupinus polyphyllus
Lupinus spp.
Luzula campestris
Luzula luzuloides
Luzula multiflora
Luzula pilosa
Luzula sylvatica
Lychnis flos-cuculi
Lycium barbarum
Lycium intricatum
Lycopersicon esculentum = *Solanum lycopersicum*
Lycopodiella inundata
Lycopodioides helveticum = *Selaginella helvetica*
Lycopodium annotinum
Lycopodium clavatum
Lycopodium spp.
Lycopus europaeus
Lygeum spartum
Lygos monosperma
Lysimachia nemorum
Lysimachia nummularia
Lysimachia punctata
Lysimachia vulgaris
Lythrum hyssopifolium
Lythrum salicaria

Maclura aurantiaca
Maclura pomifera
Magnolia cobus
Magnolia grandiflora
Magnolia liliiflora
Magnolia × *soulangiana*
Magnolia spp.
Mahonia aquifolium
Mahonia japonica
Mahonia spp.
Maianthemum bifolium
Malaxis monophyllos
Malcolmia spp.
Malus cultivars
Malus domestica
Malus floribunda
Malus niedzwetzkyana
Malus sylvestris
Malva mauritiana
Malva neglecta
Malva parviflora
Malva pusilla
Malva sylvestris
Marrubium peregrinum
Marrubium vulgare
Marsilea quadrifolia
Matricaria chamomilla
Matricaria discoidea
Matricaria recutita = *M. chamomilla*
Matricaria trichophylla
Matteuccia struthiopteris
Matthiola sinuata
Matthiola tricuspidata
Maytenus senegalensis ssp. *europaeus*
Medicago lupulina
Medicago minima
Medicago monspeliaca
Medicago sativa
Medicago sativa spp. *sativa*
Medicago sativa ssp. *falcata*
Medicago sativa ssp. *varia*
Medicago spp.
Meehania urticifolia
Melampyrum cristatum

Melampyrum pratense
Melia azedarach
Melica nutans
Melica uniflora
Melilotus albus
Melilotus officinalis
Melilotus sp.
Melissa officinalis
Melittis melisophyllum
Melo sativus = Cucumis sativus
Meniocus linifolius = Alyssum linifolium
Mentha aquatica
Mentha x *piperita*
Mentha pulegium
Mentha suaveolens
Mentha x *villosa*
Menyanthes trifoliata
Mercurialis annua
Mercurialis perennis
Mercurialis sp.
Merendera sobolifera
Mesembryanthemum crystallinum
Mesembryanthemum nodiflorum
Mespilus germanica
Metasequoia glyptostroboides
Milium effusum
Mimulus guttatus
Minuartia glaucina
Minuartia hybrida
Minuartia stricta
Misopates orontium
Moehringia trinerva
Moenchia mantica
Molinia caerulea ssp. *arundinacea*
Molinia caerulea ssp. *caerulea*
Monarda didyma
Monotropa hypopitys = Hypopitys monotropa
Montia fontana
Montia fontana ssp. *chondrosperma*
Moricandia arvensis
Morus alba
Morus nigra
Morus sp.
Muscari armeniacum
Muscari botryoides
Muscari comosum
Muscari neglectum
Muscari spp.
Myagrum perfoliatum
Mycelis muralis
Myosotis arvensis
Myosotis discolor
Myosotis scorpioides
Myosotis sylvatica
Myosoton aquaticum
Myrica gale
Myricaria germanica
Myriophyllum alterniflorum
Myriophyllum aquaticum
Myriophyllum sp.
Myriophyllum spicatum
Myriophyllum verticillatum
Myrrhis odorata

Najas marina
Najas minor
Narcissus 'Golden Harvest'
Narcissus 'Ice Follies'
Narcissus poeticus
Narcissus pseudonarcissus
Narcissus varieties
Nardus stricta
Neottia nidus-avis
Nepeta cataria
Nerium oleander
Nicotiana glauca
Nigella arvensis
Nigella damascena
Nothofagus spp.
Nuphar lutea
Nuphar pumila
Nymphaea alba
Nymphaea candida
Nymphoides peltata

Odontites vernus sensu lato
Odontites vernus ssp. *serotinus*
Odontites vulgaris = Odontites vernus ssp. *serotinus*
Oenanthe aquatica
Oenanthe crocata
Oenanthe fistulosa
Oenanthe silaifolia ssp. *silaifolia*
Oenothera biennis
Oenothera cambrica
Oenothera depressa
Oenothera glazioviana
Oenothera parviflora

Oenothera rubricaulis
Oenothera sp.
Oenothera spp.
Olea europaea
Olea europaea ssp. *sylvestris*
Omphalodes scorpioides
Onobrychis arenaria
Ononis x *pseudohircina*
Ononis pusilla
Ononis repens
Ononis spinosa
Ononis talaverae
Onopordum acanthium
Onosma tinctoria
Ophioglossum vulgatum
Ophrys apifera
Ophrys fuciflora
Ophrys insectifera
Ophrys sphegodes
Opuntia ficus-indica
Orchis coriophora = *Anacamptis coriophora*
Orchis coriophora ssp. *coriophora* = *Anacamptis coriophora*
Orchis mascula
Orchis militaris
Orchis morio
Orchis tridentata ssp. *tridentata*
Orchis ustulata ssp. *ustulata* = *Neotinea ustulata*
Orchis ustulata = *Neotinea ustulata*
Oreopteris limbosperma
Origanum vulgare
Orlaya grandiflora
Ornithogalum angustifolium
Ornithogalum kochii
Ornithogalum nutans
Ornithogalum spp.
Ornithogalum umbellatum
Ornithogalum umbellatum ssp. *umbellatum*
Ornithopus perpusillus
Orobanche artemisiae-campestris
Orobanche coerulescens
Orobanche gracilis
Orobanche hederae
Orobanche minor
Orobanche teucrii
Oryza sativa
Oryzopsis miliacea
Osmanthus heterophyllus
Ostericum palustre
Oxalis acetosella
Oxalis articulata
Oxalis corniculata
Oxalis debilis
Oxalis dillenii
Oxalis exilis
Oxalis incarnata
Oxalis latifolia
Oxalis pes-caprae
Oxalis repens
Oxalis stricta
Oxybaphus nyctagineus
Oxyria digyna

Pachysandra terminalis
Paeonia officinalis
Pancratium maritimum
Panicum capillare
Panicum dichotomiflorum
Panicum lindheimeri
Panicum miliaceum
Panicum miliaceum ssp. *agricola*
Panicum miliaceum ssp. *ruderale*
Papaver argemone
Papaver dubium
Papaver hybridum
Papaver orientale
Papaver rhoeas
Papaver somniferum
Papaver spp.
Parapholis strigosa
Parietaria judaica
Parietaria officinalis
Parietaria pensylvanica
Paris quadrifolia
Parrotia persica
Parthenocissus inserta = *P. vitacea*
Parthenocissus quinquefolia
Parthenocissus sp.
Parthenocissus tricuspidata
Passiflora caerulea
Pastinaca sativa
Paulownia tomentosa
Pedicularis sceptrum-carolinum
Peganum harmala
Pelargonium zonale
Pellaea falcate
Pellaea rotundifolia

Pennisetum setaceum
Periploca laevigata
Periploca laevigata ssp. *angustifolia*
Persea americana
Persica vulgaris = *Prunus persica*
Persicaria amphibia
Persicaria amplexicaulis
Persicaria hydropiper
Persicaria lapathifolia
Persicaria maculosa
Persicaria minor
Persicaria mitis
Persicaria orientalis
Petasites hybridus
Petasites officinalis = *Petasites hybridus*
Petrorhagia prolifera
Petroselinum crispum
Petunia x *hybrida*
Peucedanum aegopodioides
Peucedanum arenarium
Peucedanum arenarium ssp. *arenarium*
Peucedanum oreoselinum
Phacelia tanacetifolia
Phaeophyscia orbicularis
Phalaris arundinacea
Phalaris arundinacea 'Picta'
Phalaris canariensis
Phalaris minor
Phalaris paradoxa
Phaseolus coccineus
Phelipanche arenaria
Phellodendron amurense
Philadelphus coronarius
Philadelphus grandiflorus
Philadelphus spp.
Philadelphus x *virginalis*
Phleum bertolonii
Phleum phleoides
Phleum pratense
Phlomis purpurea ssp. *almeriensis*
Phlomis tuberosa
Phlox drummondii
Phlox paniculata
Phlox subulata
Phoenix canariensis
Phoenix dactylifera
Phoenix spp.
Pholiurus pannonicus
Photinia fraseri 'Red Robin'
Phragmites altissimus
Phragmites australis
Phyllitis scolopendrium = *Asplenium scolopendrium*
Physalis alkekengi var. *franchetii*
Physalis spp.
Physcia caesia
Physocarpus amurensis
Physocarpus opulifolius
Phyteuma nigrum
Phyteuma orbiculare
Phyteuma spicatum
Phytolacca americana
Phytolacca esculenta
Picea abies
Picea glauca
Picea omorika
Picea pungens
Picea spp.
Picris echioides
Picris hieracioides
Pilosella praealta sensu lato
Pilosella spp.
Pilularia globulifera
Pimpinella major
Pimpinella saxifraga
Pinguicula vulgaris
Pinus canariensis
Pinus excelsa
Pinus mugo
Pinus nigra
Pinus nigra ssp. *nigra*
Pinus parviflora
Pinus pungens
Pinus radiata
Pinus sibirica
Pinus spp.
Pinus strobus
Pinus sylvestris
Pistacia lentiscus
Pistia stratiotes
Pisum sativum
Plantago altissima
Plantago coronopus
Plantago lagopus
Plantago lanceolata
Plantago major
Plantago major ssp. *major*
Plantago media

Plantago ovata
Platanthera chlorantha
Platanus x *hispanica*
Platanus occidentalis
Platanus orientalis
Platanus sp.
Platycladus orientalis
Poa alpina
Poa angustifolia
Poa annua
Poa bulbosa
Poa chaixii
Poa compressa
Poa nemoralis
Poa pratensis
Poa spp.
Poa trivialis
Polemonium caeruleum
Polycarpon tetraphyllum
Polygala comosa
Polygala major
Polygala mospelliaca
Polygala vulgaris
Polygonatum latifolium
Polygonatum multiflorum
Polygonatum verticillatum
Polygonum arenastrum
Polygonum aviculare agg.
Polygonum bellardii
Polygonum scabrum
Polygonum spp.
Polypodium interjectum
Polypodium vulgare
Polypogon monspeliensis
Polystichum aculeatum
Polystichum braunii
Populus alba
Populus balsamifera
Populus x *berolinensis*
Populus x *canadensis*
Populus x *canadensis* 'Robusta'
Populus nigra
Populus nigra 'Italica'
Populus nigra ssp. *betulifolia*
Populus spp.
Populus tremula
Populus trichocarpa
Portulaca oleracea
Portulaca oleracea sensu lato
Potamogeton berchtoldii
Potamogeton compressus
Potamogeton crispus
Potamogeton lucens
Potamogeton natans
Potamogeton nodosus
Potamogeton obtusifolius
Potamogeton pectinatus
Potamogeton perfoliatus
Potamogeton praelongus
Potamogeton pusillus
Potamogeton spp.
Potamogeton trichoides
Potamogeton zizii = *P.* x *zizii* = *P. angustifolium*
Potentilla anserina
Potentilla arenaria
Potentilla argentea
Potentilla erecta
Potentilla fruticosa
Potentilla inclinata
Potentilla indica
Potentilla neumanniana
Potentilla norvegica
Potentilla pedata
Potentilla reptans
Potentilla rupestris
Potentilla sterilis
Potentilla supina
Primula elatior
Primula spp.
Primula veris
Primula vulgaris
Pritzelago alpina = *Hornungia alpina*
Prunella vulgaris
Prunus armeniaca
Prunus avium
Prunus cerasifera
Prunus cerasifera 'Nigra'
Prunus cerasifera 'Pissardi'
Prunus cerasifera sensu lato
Prunus domestica
Prunus domestica ssp. *domestica*
Prunus dulcis
Prunus fruticans
Prunus laurocerasus
Prunus laurocerasus 'Zabeliana'
Prunus lusitanica
Prunus mahaleb
Prunus padus

Prunus persica
Prunus serotina
Prunus serrulata
Prunus spinosa
Prunus spp.
Prunus subhirtella 'Autumnalis'
Prunus virginiana
Pseudofumaria alba
Pseudofumaria lutea
Pseudotsuga glauca
Pseudotsuga menziesii
Psoralea bituminosa
Ptelea trifoliata
Pteridium aquilinum
Pteris cretica
Pteris multifida
Pteris nipponica
Pteris tremula
Pteris vittata
Pterocarya fraxinifolia
Puccinellia distans
Puccinellia distans ssp. *distans*
Puccinellia fasciculata
Puccinellia rupestris
Pulicaria dysenterica
Pulicaria vulgaris
Pulmonaria obscura
Pulmonaria officinalis
Pulsatilla grandis
Pulsatilla patens
Pulsatilla pratensis ssp. *nigricans*
Punica granatum
Pycreus flavescens = *Cyperus flavescens*
Pyracantha coccinea
Pyracantha sp.
Pyrethrum parthenium = *Tanacetum parthenium*
Pyrola media
Pyrola rotundifolia
Pyrus communis
Pyrus communis ssp. *communis* var. *sativa*
Pyrus malus
Pyrus nivalis
Pyrus pyraster

Quercus cerris
Quercus coccifera
Quercus dalechampii
Quercus frainetto
Quercus ilex
Quercus longipes
Quercus petraea
Quercus polycarpa
Quercus x *pseudosuber*
Quercus pubescens
Quercus robur
Quercus rotundifolia
Quercus rubra
Quercus schumardii
Quercus spp.
Quercus suber
Quercus virgiliana

Ranunculus acris
Ranunculus aquatilis
Ranunculus aquatilis agg.
Ranunculus arvensis
Ranunculus auricomus
Ranunculus bulbosus
Ranunculus cassubicus
Ranunculus circinatus
Ranunculus ficaria = *Ficaria verna*
Ranunculus ficaria ssp. *bulbifera* = *Ficaria verna* ssp. *verna*
Ranunculus ficaria ssp. *ficaria* = *Ficaria verna* ssp. *fertilis*
Ranunculus flammula
Ranunculus fluitans
Ranunculus hederaceus
Ranunculus lateriflorus
Ranunculus marginatus
Ranunculus polyanthemos
Ranunculus polyphyllus
Ranunculus repens
Ranunculus reptans
Ranunculus sardous
Ranunculus sceleratus
Ranunculus trichophyllus
Raphanus raphanistrum
Raphanus sativus
Rapistrum rugosum
Reichardia tingitana
Reseda lutea
Reseda luteola
Reseda phyteuma
Rhamnus cathartica
Rhamnus oleoides ssp. *angustifolia*
Rhamnus saxatilis
Rhamnus saxatilis ssp. *saxatilis*
Rheum x *hybridum* = *R.* x *rhabarbarum*

Rhinanthus alectorolophus
Rhinanthus minor
Rhinanthus rumelicus
Rhododendron scandens
Rhododendron sinogrande
Rhododendron spp.
Rhodotypos scandens
Rhodotypus kerrioides
Rhus coriaria
Rhus typhina
Rhynchophora fusia
Ribes alpinum
Ribes multiflorum
Ribes nigrum
Ribes odoratum
Ribes rubrum
Ribes sanguineum
Ribes spicatum
Ribes spp.
Ribes uva-crispa
Ricinus communis
Robinia pseudacacia
Robinia pseudoacacia 'Bessonioana'
Robinia pseudoacacia 'Monophylla'
Robinia sp.
Rorippa amphibium
Rorippa austriaca
Rorippa nasturtium-aquaticum = *Nasturtium officinale*
Rorippa palustris
Rorippa pyrenaica
Rorippa sylvestris
Rosa agg.
Rosa arvensis
Rosa caesia ssp. *glauca*
Rosa canina
Rosa glabrifolia
Rosa jundzilii
Rosa mollis
Rosa multiflora
Rosa pimpinellifolia = *R. spinosissima*
Rosa rugosa
Rosa turcica
Rosmarinus eriocalix
Rosmarinus officinalis
Rubus armeniacus
Rubus caesius
Rubus discolor
Rubus fruticosus agg.
Rubus hirtus
Rubus idaeus
Rubus laciniatus
Rubus loganobaccus
Rubus saxatilis
Rubus spp.
Rudbeckia hirta
Rudbeckia laciniata
Rudbeckia sp.
Rumex acetosa
Rumex acetosella
Rumex aquaticus
Rumex conglomeratus
Rumex crispus
Rumex cristatus
Rumex hydrolapathum
Rumex maritimus
Rumex obtusifolius
Rumex palustris
Rumex patientia
Rumex pulcher
Rumex sanguineus
Rumex scutatus
Rumex spp.
Rumex stenophyllus
Rumex thyrsiflorus
Ruscus hypoglossum
Ruta graveolens

Sagina apetala
Sagina maritima
Sagina procumbens
Sagittaria sagittifolia
Salicornia sp.
Salix alba
Salix alba var. *caerulea*
Salix babylonica
Salix caprea
Salix cinerea
Salix cinerea ssp.*oleifolia*
Salix elaeagnos
Salix fragilis
Salix myrsinifolia
Salix x *pendulina*
Salix phylicifolia
Salix purpurea
Salix racemosa
Salix repens
Salix x *sepulcralis*
Salix spp.
Salix triandra

Salix viminalis
Salsola collina
Salsola genistoides
Salsola kali ssp. *ruthenica* = *S. kali* ssp. *tragus*
Salsola kali = *S. trag*us
Salsola oppositifolia
Salsola papillosa
Salsola tragus
Salsola vermiculata
Salvia glutinosa
Salvia nemorosa
Salvia pratensis
Salvia splendens
Salvia spp.
Salvia verticillata
Salvia virgata
Salvinia natans
Sambucus ebulus
Sambucus nigra
Sambucus racemosa
Sanguisorba minor = *Poterium sanguisorba*
Sanguisorba officinalis
Sanicula europaea
Santolina sp.
Saponaria officinalis
Sarcocapnos eneaphylla
Sarcocornia fruticosa
Sarcocornia perennis
Sarcocornia perennis ssp. *alpini*
Satureja hortensis
Saxifraga aizoides
Saxifraga cespitosa
Saxifraga granulata
Saxifraga hirculus
Saxifraga tridactylites
Scabiosa canescens
Scabiosa columbaria
Scabiosa stellata
Scabiosa triniaefolia
Scandix pecten-veneris
Scheuchzeria palustris
Schinus spp.
Schismus barbatus
Schizandra sinensis
Schizopepon bryoniifolius
Schlerochloa dura
Schoenoplectus lacustris
Schoenoplectus triqueter
Schoenus ferrugineus
Scilla bifolia
Scilla luciliae
Scilla siberica
Scilla verna
Scilla vindobonensis
Scirpus sylvaticus
Scleranthus annuus
Sclerochloa dura
Scolochloa festucacea
Scorzonera austriaca
Scorzonera hispanica
Scorzonera humilis
Scorzonera parviflora
Scorzonera purpurea
Scrophularia auriculata
Scrophularia nodosa
Scrophularia umbrosa
Scrophularia vernalis
Scutellaria altissima
Scutellaria galericulata
Secale cereale
Securigera varia
Securinega suffruticosa
Securinega tinctoria
Sedum acre
Sedum album
Sedum carneum
Sedum hispanicum
Sedum rupestre
Sedum sarmentosum
Sedum sexangulare
Sedum spectabile
Sedum spp.
Sedum spurium
Selinum carvifolia
Sempervivum hirtum forma *glabrescens*
Senecio aquaticus
Senecio cineraria
Senecio doria
Senecio erucifolius
Senecio inaequidens
Senecio jacobaea
Senecio ovatus
Senecio paludosus
Senecio squalidu
Senecio vernalis
Senecio viscosus
Senecio vulgaris
Sequoia sempervirens

Sequoiadendron gigantum
Serratula tinctoria
Seseli annuum
Seseli libanotis
Sesleria caerulea
Setaria faberi
Setaria italica
Setaria pumila
Setaria verticillata
Setaria viridis
Sherardia arvensis
Sideritis lasiantha
Sideritis montana
Silaum silaus
Silene borysthenica
Silene chlorantha
Silene conica
Silene dichotoma
Silene dioica
Silene gallica
Silene latifolia
Silene noctiflora
Silene nutans
Silene sp.
Silene vulgaris
Silybum marianum
Sinapis alba
Sinapis arvensis
Sisymbrium altissimum
Sisymbrium austriacum ssp. *chrysanthum*
Sisymbrium irio
Sisymbrium loeselii
Sisymbrium officinale
Sisymbrium orientale
Sisymbrium supinum
Sisymbrium volgense
Sisyrinchium striatum
Sium latifolium
Smyrnium olusatrum
Smyrnium perfoliatum
Solanum alatum
Solanum chenopodioides
Solanum diflorum
Solanum dulcamara
Solanum laciniatum
Solanum nigrum
Solanum nigrum ssp. *schultesii*
Solanum physalifolium
Solanum rostratum
Solanum sarachoides
Solanum scabrum
Solanum triflorum
Solanum tuberosum varieties
Solidago canadensis
Solidago gigantea
Solidago spp.
Solidago virgaurea
Sonchus arvensis
Sonchus asper
Sonchus oleraceus
Sonchus tenerrimus
Sophora japonica
Sorbaria sorbifolia
Sorbus aria 'Magnifica'
Sorbus aria agg.
Sorbus aucuparia
Sorbus danubialis
Sorbus domestica
Sorbus graeca
Sorbus intermedia
Sorbus spp.
Sorbus torminalis
Sorghum halepense
Sparganium emersum
Sparganium erectum
Sparganium natans
Sparmannia africana
Spartium junceum
Spergula arvensis
Spergula morisonii
Spergula pentandra
Spergularia bocconnii
Spergularia diandra
Spergularia marina
Spergularia rubra
Spiraea alba
Spiraea chamaedryfolia
Spiraea japonica
Spiraea salicifolia
Spiraea spp.
Spiraea thunbergii
Spiraea vanchouttei
Spiranthes aestivalis
Spiranthes spiralis
Spirodela polyrhiza
Sporobolus cryptandrus
Stachys annua
Stachys byzantina

Stachys germanica
Stachys milanii
Stachys officinalis
Stachys palustris
Stachys recta
Stachys sylvatica
Stellaria graminea
Stellaria holostea
Stellaria media
Stellaria media agg.
Stellaria nemorum
Stellaria palustris
Stellaria uglinosa = *S. alsine*
Stephanandra tanakae
Stipa capensis
Stipa capillata
Stipa joannis
Stipa pulcherrima
Stipa spp.
Stipa tenacissima
Stipagrostis plumose
Stratiotes aloides
Suaeda pruinosa
Suaeda sp.
Suaeda vera
Succisa pratensis
Succisella inflexa
Symphoricarpos albus
Symphoricarpos rivularis
Symphoricarpos x *chenaultii* 'Hancock'
Symphoricarpus orbiculatus
Symphoricarpus spp.
Symphoricarpus x *chenaultii*
Symphytum asperum
Symphytum caucasicum
Symphytum officinale
Symphytum spp.
Symphytum x *uplandicum*
Syringa josikaea
Syringa spp.
Syringa vulgaris

Tagetes erecta
Tagetes patula
Tagetes spp.
Tamarix ramosissima
Tamarix spp.
Tamarix tetrandra
Tamus communis
Tanacetum balsamita
Tanacetum parthenium
Tanacetum vulgare
Taraxacum agg.
Taraxacum danubium
Taraxacum officinale agg.
Taraxacum sect. *Erythrosperma*
Taraxacum sect. *Ruderalia*
Taraxacum serotinum
Taxodium distichum
Taxus baccata
Tecomaria capensis
Teesdalia nudicaulis
Telekia speciosa
Tetradium danielli
Tetragonia tetragonoides
Tetragonolobus maritimus
Teucrium balthazaris
Teucrium carthaginense
Teucrium chamaedrys
Teucrium compactum
Teucrium intrincatum
Teucrium libanitis
Teucrium polium
Teucrium pumilum
Teucrium scordium
Teucrium scorodonia
Thalictrum alpinum
Thalictrum aquilegiifolium
Thalictrum flavum
Thalictrum minus
Thesium ebracteatum
Thlapsi arvense
Thlaspi perfoliatum = *Microthlaspi perfoliatum*
Thuja gigantea
Thuja occidentalis
Thuja orientalis
Thuja plicata
Thuja spp.
Thymus antoninae
Thymus x *loevyanus*
Thymus longiflorus
Thymus marschallianus
Thymus ovatus
Thymus pulegioides
Thymus serpyllum
Tilia argentea
Tilia cordata
Tilia x *euchlora*
Tilia x *europaea*

Tilia intermedia
Tilia platyphyllos
Tilia rubra
Tilia spp.
Tilia tomentosa
Tipuana tipu
Tithymalus amygdaloides
=*Euphorbia amygdaloides*
Torilis arvensis
Torilis japonica
Trachycarpus fortunei
Tragopogon dubius
Tragopogon orientalis
Tragopogon pratensis
Tragopogon pratensis ssp. *minor*
Tragus racemosus
Trapa natans
Tribulus terrestris
Trientalis europaea
Trifolium alpestre
Trifolium campestre
Trifolium dubium
Trifolium hybridum
Trifolium incarnatum
Trifolium medium
Trifolium micranthum
Trifolium montanum
Trifolium pratense
Trifolium repens
Trifolium rubens
Trifolium spp.
Trifolium striatum
Trifolium subterraneum
Trifolium trichopterum
Triglochin palustris
Tripleurospermum inodorum
Tripleurospermum maritimum
Tripolium vulgare = *Aster tripolium*
Trisetum flavescens
Triticum aestivum
Triticum dicoccon
Triticum monococcum
Triticum spelta
Triticum spp.
Trollius europaeus
Tsuga heterophylla
Tulipa 'Darwin Hybrid'
Tulipa gesnerana
Tulipa spp.
Tussilago farfara
Typha angustifolia
Typha latifolia
Typha laxmannii
Typha minima
Typha shuttleworthii
Typha sp.

Ulex europaeus
Ulmus glabra
Ulmus laevis
Ulmus minor
Ulmus plotii
Ulmus procera
Ulmus sp.
Umbilicus rupestris
Urospermum picroides
Urtica dioica
Urtica membranacea
Urtica urens
Utricularia australis
Utricularia intermedia
Utricularia minor
Utricularia spp.
Utricularia vulgaris

Vaccinium myrtillus
Vaccinium oxycoccus
Vaccinium uliginosum
Vaccinium vitis-idaea
Valeriana dioica
Valeriana repens
Valerianella carinata
Valerianella dentata
Valerianella eriocarpa
Valerianella locusta
Valerianella rimosa
Vallisneria spiralis
Ventenata dubia
Verbascum blattaria
Verbascum densiflorum
Verbascum longifolium
Verbascum lychnitis
Verbascum nigrum
Verbascum phlomoides
Verbascum phoeniceum
Verbascum sp.
Verbascum speciosum
Verbascum thapsus
Verbascum virgatum
Verbena officinalis

Veronica agrestis
Veronica anagalis-aquatica
Veronica anagalloides
Veronica arvensis
Veronica austriaca
Veronica austriaca ssp. *teucrium*
Veronica beccabunga
Veronica catenata
Veronica chamaedrys
Veronica filiformis
Veronica hederifolia
Veronica hederifolia agg.
Veronica hederifolia ssp. *lucorum*
Veronica montana
Veronica officinalis
Veronica peregrina
Veronica peregrina ssp. *peregrina*
Veronica persica
Veronica polita
Veronica serpyllifolia
Veronica spp.
Veronica teucrium = *Veronica austriaca*
Veronica triphyllos
Viburnum bodnantense 'Dawn'
Viburnum davidi
Viburnum farreri
Viburnum lantana
Viburnum opulus
Viburnum rhytidophyllum
Viburnum sp.
Viburnum tinus
Vicia cracca
Vicia dumetorum
Vicia faba
Vicia grandiflora
Vicia hirsuta
Vicia lathyroides
Vicia parviflora
Vicia sativa
Vicia sativa sensu lato
Vicia sativa ssp. *sativa*
Vicia sativa ssp. *segetalis*
Vicia sepium
Vicia tenuifolia
Vicia tetrasperma
Vinca herbacea
Vinca minor
Vinca sp.
Vincetoxicum hirundinaria
Viola ambigua
Viola arvensis
Viola canina
Viola hirta
Viola mirabilis
Viola odorata
Viola palustris
Viola persicifolia
Viola pumila
Viola reichenbachiana
Viola riviniana
Viola sororia
Viola spp.
Viola x *wittrockiana*
Viscaria viscosa = *Silene suecica*
Viscum album
Viscum album ssp. *album*
Vitex agnus-castus
Vitis sp.
Vitis sylvestris
Vitis vinifera
Vitis vinifera 'Leo Millot'
Vitis vinifera ssp. *sylvestris*
Volutaria lippii
Vulpia bromoides
Vulpia myuros

Washingtonia spp.
Weigela hybrids
Weigela japonica
Wisteria sinensis
Withania frutescens
Wolffia arrhiza

Xanthium albinum
Xanthium riparium
Xanthium sp.
Xanthium spinosum
Xanthium strumarium

Yucca aloifolia
Yucca elephantipes
Yucca gloriosa

Zannichellia palustris
Zea mays
Zizania latifolia
Ziziphus lotus
Zygophyllum fabago

Appendix II

Algae Taxa Referred to in the Chapters

Nomenclature as supplied by the authors

Anabaena flos-aquae
Ankistrodesmus falcatus
Ankistrodesmus sp.
Apatococcus lobatus
Aphanizomenon flos-aquae
Astasia bulgarica
Astasia sophiensis

Ceratium hirundinella
Chara coronata var. *maxima*
Chara foetida forma *thermalis*
Chara foetida var. *subinermis* forma *normalis*
Chara foetidaa forma *variabilis*
Chara hispida
Chara spp.
Chara vulgaris
Chilomonas sp.
Chlamydomonas sp.
Chlorella sp.
Chromulina sp.
Cladophora spp.
Coccomyxa sp.
Colacium sp.
Cryptaulax sp.
Cryptomonas erosa
Cryptomonas marssonii
Cryptomonas obovata
Cryptomonas ovata
Cryptomonas reflexa
Cyclotella/Stephanodiscus

Dinobryon sp.
Diplopsalis acuta

Entermorpha intestinalis
Euglena sp.

Mallomonas pyriformis
Merismopedia sp.
Microcystis aeruginosa
Microspora sp.
Monoraphidium sp.

Navicula sp.
Nitella spp.
Nitella syncarpa
Nitzschia sp.

Oedogonium cardiacum forma *thermalis*
Oedogonium parvulum
Ooycystis sp.
Oscillatoria sp.

Pediastrum sp.
Peridinium cinctum
Peridinium sp.

Phacus sp.
Plantktothrix agardhii
Pseudanabaena limnetica

Rhizoclonium sp.
Rhodomonas lacustris
Rhodomonas lens
Rhodomonas spp.

Scenedesmus quadricauda

Scenedesmus sp.
Spirogyra sp.

Trachelomonas sp.
Trebouxia arboricola
Tribonema sp.

Ulothrix sp.

Zygonema sp.

Appendix III

Bryophyte Taxa Referred to in the Chapters

Nomenclature as supplied by the authors

Acaulon muticum
Acrocladium cuspidatum = *Calliergonella cuspidata*
Amblystegium serpens
Aneura pinguis
Anomodon viticulosus
Aphanorhegma patens = *Physcomitrella patens*
Asterella saccata
Athalamia hyalina
Atrichum undulatum
Aulacomnium androgynum

Barbula convoluta
Barbula fallax = *Didymodon fallax*
Barbula hornschuchiana = *Pseudocrossidium hornschuchianum*
Barbula unguiculata
Brachythecium albicans
Brachythecium populeum
Brachythecium reflexum
Brachythecium rivulare
Brachythecium rutabulum
Brachythecium salebrosum
Brachythecium velutinum
Bryoerythrophyllum recurvirostrum
Bryum argenteum
Bryum bicolour = *B. dichotomum*
Bryum caespiticium
Bryum capillare
Bryum ruderale
Buxbaumia aphylla

Calliergonella cuspidata
Calypogeia azurea
Calypogeia muelleriana
Camptothecium lutescens = *Homalothecium lutescens*
Camptothecium sericeum = *Homalothecium sericeum*
Campyliadelphus elodes = *Camphylium elodes*
Campylium polgamum
Campylium protensum
Campylium stellatum var. *stellatum*
Campylopus flexuosus
Campylopus pyriformis
Capylopegia muelleriana
Ceratodon purpureus
Cinclidotus fontinaloides
Cinclidotus riparius
Cirriphyllum piliferum
Climacium dendroides
Conocephalum conicum
Conocephalum salebrosum
Cratoneuron filicinum
Cryphaea heteromalla

Dicranella heteromalla
Dicranella varia
Dicranodontium denudatum
Dicranum montanum = *Orthodicranum montanum*
Dicranum polysetum
Dicranum scoparium

Dicronoweisia cirrata
Didymodon fallax
Diphyscium foliosum
Diplophyllum obtusifolium
Distichium capillaceum
Drapenocladus aduncus
Drepanocladus sendtnerii

Encalypta streptocarpa
Encalypta vulgaris
Ephemerum stellatum
Eurhynchium conferatum
Eurhynchium hians = Oxyrrhynchium hians
Eurhynchium praelogum = Kindbergia praelonga
Eurhynchium schleicheri = Oxyrrhynchium schleicheri
Eurhynchium swartzii

Fissidens adianthoides
Fissidens bryoides
Fissidens crassipes
Fissidens taxifolius
Fontinalis antipyretica
Frullania dilatata
Funaria hygrometrica
Funaria pulchella

Grimmia apocarpa = Schistidium apocarpum
Grimmia pulvinata
Gymnocolea inflate

Helodium blandowii
Hookeria lucens
Hylocomium splendens
Hypnum cuppresiforme
Hypnum cuppresiforme var. *resupinatum*
Hypnum filiforme
Hypnum pallescens

Isoptergium elegans
Isothecium alopecuroides

Jungermannia gracillima

Leptobryum pyriforme
Leska polycarpa
Leskella nervosa
Leucobryum glaucum
Leucodon sciuroides
Lophocolea bidentata
Lophocolea cuspidatum
Lophocolea heterophylla
Lophozia bicrenata
Lunularia cruciata

Marchantia polymorpha
Marchantia polymorpha ssp. *ruderalis*
Metzgeria furcata
Microbryum curvicollum = Phascum curvicollum
Mnium hornum
Mnium longirostrum = Plagimnium undulatum
Mnium undulatum

Orthotrichum affine
Orthotrichum anomalum
Orthotrichum obtusifolium
Orthotrichum pallens
Orthotrichum pulchellum
Orthotrichum pumilum
Orthotrichum speciosum
Oxyrrhynchium hians

Paludella squarrosa
Pellia endiviifolia
Pellia spp.
Phascum curvicollum
Phascum cuspidatum
Phascum floerkeanum
Philonotis fontana
Physcomitrium pyriflorme
Plagiochila porelloides
Plagiomnium cuspidatum
Plagiomnium ellipticum
Plagiomnium spp.
Plagiomnium undulatum
Plagiothecium cavifolium
Plagiothecium curvifolium
Plagiothecium denticulatum
Plagiothecium latebricola
Plagiothecium spp.
Plagiothecium succulentum
Plagiothecium undulatum
Pleuridium subulatum
Pleurozium shreberii
Pogonatum nanum
Pohlia annotina
Pohlia nutans
Pohlia wahlenbergii

Polytrichum commune
Polytrichum formosum
Polytrichum piliferum
Polytrichum spp.
Porella platyphylla
Pottia davalliana
Pseudephemerum nitidum
Pseudocrossidium hornschuchianum
Pterygoneuron ovatum
Ptilium crista-castrensis
Pylaisia polyantha

Radula complanata
Reboulia hemisphaerica
Rhodobryum roseum
Rhynchostegiella curviseta
Rhynchostegium murale
Rhynchostegium rotundifolium
Rhytidiadelphus squarrosus
Rhytidiadelphus triquetrus
Riccardia pinguis = Aneura pinguis
Riccia cavernosa
Riccia fluitans
Riccia subbifurca

Schistidium apocarpum
Schistidium spp.
Scleropodium purum
Sphagnum angustifolium
Sphagnum balticum
Sphagnum centrale = S. palustre var. *centrale*
Sphagnum fimbriatum
Sphagnum flexuosum
Sphagnum girgensohnii
Sphagnum riparium
Sphagnum squarrosum
Sphagnum subnitens
Streblotrichum convulatum = Barbula convoluta var. *convoluta*
Syntrichia ruralis

Thamnium alopercurum = Thamnobryum alopecurum
Thuidium assimile
Thuidium tamariscinum
Tortula laevipila = Syntrichia laeviplia
Tortula muralis
Tortula papillosa = Syntrichia papillosa
Tortula virescens = Syntrichia virescens var. *virescens*
Totula muralis
Trichocolea tomentella
Trichostomum sinuosum = Barbula sinuosa

Appendix IV

Fungal Taxa Referred to in the Chapters

Nomenclature as supplied by the authors

Abortiporus fractipes
Acarospora fuscata
Agaricus arvensis
Agaricus bitorquis
Agaricus campestris
Agaricus campestris var. *campestris*
Agaricus xanthodermus
Amantia strobiliformis
Armillaria mellea
Auricularia auricula-judae

Bolbitius vitellinus
Boletus impolitus
Boletus rubellus
Boletus satanas
Boletus sp.
Botryobasidium aureum

Coltricia cinnamomea
Coltricia montagnei
Coprinellus disseminatus
Coprinellus micaceus
Coprinus atramentarius
Coprinus comatus
Coprinus disseminatus
Coprinus micaceus
Coprinus niveus
Coprinus plicatilis
Coprinus strossmayeri
Crepidotus crocophyllus

Daldinia concentrica

Entoloma byssisedum
Erysiphe aquilegiae
Erysiphe sp.

Fistulina hepatica
Flammulina fennae
Fomes fomentarius

Ganoderma applanatum
Ganoderma lipsiense
Ganoderma lucidum
Geastrum melanocephalum
Geastrum striatum
Geopora sumneriana
Gyromitra fastigiata

Hebeloma crustuliniforme
Helvella phephora
Hericium ramosum
Hyphodontia latitans
Hypoxylon fragiforme
Hypoxylon ticinense

Laccaria amethystine
Lactarius glyciosmus
Lactarius torminosus
Lactarius turpis
Laetiporus sulphureus

Langermannia gigantean
Lentinus degener
Lepiota rhacodes
Lepista nebularis
Letiporus sp.
Letiporus sulphureus
Leucopaxillus lepistoides

Macrotyphula fistulosa
Marasmius rotula
Melanoleuca brevipes
Microsphaera alni
Microsphaera alphitoides
Morchella conica
Morchella elata
Morchella esculenta
Mycena inclinata

Naucaria eschariodes
Nectria cinnabarina
Nectria galligena

Omphalina rickenii
Ophiostoma novo-ulmi

Paxillus invlutus
Phallus hadriani
Phallus impudicus
Phellinus pomaceus
Phlebia ryvardenii
Pholiota ochrochlora
Phyllactinia suffulta
Pluteus aurantiorugosus
Pluteus cervinus
Polyporus squamosus
Protomyces buerenianus

Rhodotus palmatus
Russula ochroleuca

Schizophyllum commune
Scleroderma citrinum
Scleroderma verrucosum
Sparassis crispa
Sphaerotheca fuliginea
Splanchnonema platani
Stereum hirsutum
Stereum subtomentosum
Stropharia coronilla

Taaphrina rhizophora
Trametes versicolor
Trichocladia euonymi
Tulostoma brumale

Uncinula salicis
Urnula craterium

Volvariella bombycina
Volvariella surrecta

Xerocomus parasiticus
Xylaria hypoxylon

Appendix V

Lichenised Fungi (=Lichens) Taxa Referred to in the Chapters

Nomenclature as supplied by the authors

Acarospora fuscata
Amandinea punctata = *Buellia punctata*
Anaptychia ciliaris
Aspicilia calcarea

Biatorella fossarum
Bryoria sp.
Buellia punctata

Caloplaca albolutescens = *Calcoplaca aurantia*
Caloplaca aurantia
Caloplaca citrina
Caloplaca crenulatella
Caloplaca decipiens
Caloplaca heppiana
Caloplaca holocarpa
Caloplaca saxicola
Caloplaca sp.
Candelaria concolor
Candelariella aurella
Candelariella coralizza
Candelariella vitellina
Candelariella xanthostigma
Cataphyrenium squamulosum = *Placidium squamulasum*
Cetraria glauca
Chaenotheca ferruginea
Cladonia botrytis
Cladonia chlorophaea
Cladonia coniocraea
Cladonia digitata
Cladonia frimbriata
Cladonia humilis
Cladonia magyarica
Cladonia ochrochlora
Cladonia spp.

Evernia mesomorpha
Evernia prunastri

Flavocetraria nivalis
Flavoparmelia caperata
Fulgensia fulgens

Graphis scripta

Hypogymnia phycodes
Hypogymnia tubulosa

Lecania erysibe
Lecanora albescens
Lecanora argentata
Lecanora carpinea
Lecanora conizaeoides
Lecanora crenulata
Lecanora dispersa
Lecanora hagenii
Lecanora muralis
Lecanora sp.

Lecidea scraba
Lecidea sulphurea
Lecidea tumida
Lecidella anomaloides
Lecidella elaechroma
Lecidella stigmatea
Lepraria cf. *neglecta*
Lepraria incana
Leptogium schraderi
Letiporus sulphureus
Lobaria pulmonaria
Lobaria sp.

Melanelia exaspereata = Melanohalea exaspereata
Melanelia fuliginosa = Melanelixia fuliginosa
Melanelia glabra
Melanelia sp.

Neofuscelia pullao = Xanthopermelia pulla
Nephroma sp.

Parmelia saxatilis
Parmelia sulcata
Peltigera canina
Peltigera sp.
Phaeophyscia nigrans
Phaeophyscia orbicularis
Physcia adscendens
Physcia caesia
Physcia dubia
Physcia sp.
Physcia stellaris
Physcia tenella
Physconia grisea
Physconia sp.
Platismatia glauca
Pleurosticta acetabulum

Protoparmeliopsis muralis = Lecanora muralis
Pseudevernia furfuracea
Psora decipiens
Punctelia subrudecta

Ramalina farinacea
Ramalina fastigiata
Ramalina fraxinea
Ramalina sp.
Rhizocarpon geographicum
Rinodina bishoffii
Rinodina exigua = R. oleae
Rinodina subexigua

Sarcogyne regularis
Scoliciosporum chloroccum
Shaerophorus fragilis
Solorinella asteriscus
Squamarina cartilaginea
Squamarina lentigera
Stereocaulon sp.

Toninia sedifolia

Usnea cf. *subfloridana*
Usnea florida
Usnea hirta
Usnea spp.

Verrucaria muralis
Verrucaria nigrescens
Vezdaea aestivalis

Xanthoparmelia conspersa
Xanthoria aureola = X. calicicold
Xanthoria candelaria
Xanthoria fallax = X. ulophyllodes
Xanthoria parietina
Xanthoria polycarpa

Appendix VI

Animals Referred to in the Chapters

Birds

- *Athene noctua*
- *Carduelis chloris*
- *Erinaceus europaeus*
- *Hirundo rustica*
- *Turdus merula*
- *Tyto alba*
- *Upupa epops*

Mammals (E = extinct)

- *Bos taurus primigenius* E
- *Canis lupus*
- *Cervus elephus*
- *Coelodonta antiquitatis* E
- *Equus ferus*
- *Lepus europaeus*
- *Mammuthus primigenius* E
- *Palaeodoxodon antiquus* E
- *Rangifer tarandus*
- *Sus scrofa*
- *Trogontherium cuvieri* E
- *Vulpes yulpes*

Reptiles

- *Mosasaurus giganteus*
- *Podacarus muralis*

Invertebrates

Andrena lathyri
Cameraria ohridella
Chelostoma distinctum
Hoplites adunca
Rhynchophorus ferrugineus
Melitta tricincta

Appendix VII

Numbers and Titles of the European Commission Habitats Directive 92/43/EEC on the conservation of natural habitats and of wild flora and fauna.

Annex No	Title
I.	Natural habitat types of community interest whose conservation requires the designation of Special Areas of Conservation.
II.	Animal and plant species of community interest whose conservation requires the designation of Special Areas of Conservation.
III.	Criteria for selecting sites eligible for identification as sites community importance and designation as Special Areas of Conservation.
IV.	Animal and plant species of community interest in need of strict protection.
V.	Animal and plant species of community interest whose taking in the wild and exploitation may be subject to management measures.
VI.	Prohibited methods and means of capture and killing and modes of transport.

Appendix VIII

IUCN Red List Threat Categories

1. 1994

Threat category	Definition (detailed criteria not included)
Extinct (EX)	A taxon is *Extinct* when there is no reasonable doubt that the last individual has died.
Extinct in the wild (EW)	A taxon is *Extinct in the wild* when it is known to survive only in cultivation, in captivity or as a naturalised population (or populations) well outside the past range. A taxon is presumed extinct in the wild when exhaustive surveys in known and/or expected habitat, at appropriate times (diurnal, seasonal, annual) throughout its range have failed to record an individual. Surveys should be over a time frame appropriate to the taxon's life cycle and life form.
Critically endangered (CR)	A taxon is *Critically Endangered* when it is facing an extremely high risk of extinction in the wild in the immediate future, as detailed by any of the criteria A to E.
Endangered (EN)	A taxon is *Endangered* when it is not *Critically Endangered,* but is facing a very high risk of extinction in the wild in the near future, as defined by any of the criteria A to E.
Vulnerable (VU)	A taxon is *Vulnerable* when it is not *Critically Endangered* or *Endangered,* but is facing a high risk of extinction in the wild in the medium-term future, as defined by any of the detailed criteria A to D.
Lower risk (LR)	A taxon is *Lower Risk* when it has been evaluated, but does not satisfy the criteria for any of the categories *Critically Endangered, Endangered* or *Vulnerable.* Taxa included in the Lower Risk category can be separated into three sub-categories.
Conservation dependent (cd)	Taxa which are the focus of a continuing taxon-specific or habitat-specific conservation programme targeted towards the taxon in question, the cessation of which would result in the taxon qualifying for one of the threatened categories above within a period of 5 years.

Near threatened (nt)	Taxa which do not qualify for *Lower Risk* (*conservation dependent*), but which are close to qualifying for *Vulnerable.*
Least concern (lc)	Taxa which do not qualify for *Lower Risk* (*conservation dependent*) or *Lower Risk* (*near threatened*).
Data deficient (DD)	A taxon is *Data Deficient* when there is inadequate information to make a direct or indirect assessment of its risk of extinction based on its distribution and/or population status. A taxon in this category may be well studied, and its biology well known, but appropriate data on abundance and/or distribution are lacking. Data Deficient is therefore not a category of threat or Lower Risk. Listing of taxa in this category indicates that more information is required and acknowledges the possibility that future research will show that a threatened category is appropriate.
Not evaluated (NE)	A taxon is *Not Evaluated* when it has not been assessed against the criteria.

2. 2000

Threat category	Definition (detailed criteria not included)
Extinct (EX)	A taxon is *Extinct* when there is no reasonable doubt that the last individual has died.
Extinct in the wild (EW)	A taxon is *Extinct in the Wild* when it is known only to survive in cultivation, in captivity or as a naturalised population (or populations) well outside the past range. A taxon is presumed extinct in the wild when exhaustive surveys in known and/or expected habitat, at appropriate times (diurnal, seasonal, annual), throughout its historic range have failed to record an individual. Surveys should be over a time frame appropriate to the taxon's life cycle and life form.
Critically endangered (CR)	A taxon is *Critically Endangered* when it is facing an extremely high risk of extinction in the wild in the immediate future, as defined by any of the criteria (A to E).
Endangered (EN)	A taxon is *Endangered* when it is not Critically Endangered, but is facing a very high risk of extinction in the wild in the near future, as defined by any of the criteria (A to E).
Vulnerable (VU)	A taxon is *Vulnerable* when it is not *Critically Endangered* or *Endangered,* but is facing a high risk of extinction in the wild in the medium-term future, as defined by any of the criteria (A to E).
Lower risk (LR)	A taxon is *Lower Risk* when it has been evaluated, but does not satisfy the criteria for any of the categories *Critically Endangered, Endangered* or *Vulnerable*. Taxa included in the Lower Risk category can be separated into three sub-categories.
Conservation dependent (cd)	Taxa which are the focus of a continuing taxon-specific or habitat-specific conservation programme targeted towards the taxon in question, the cessation of which would result in the taxon qualifying for one of the threatened categories above within a period of 5 years.

Near threatened (nt)	Taxa which do not qualify for *Conservation Dependent*, but which are close to qualifying for *Vulnerable.*
Least concern (le)	Taxa which do not qualify for *Conservation Dependent* or *Near Threatened.*
Data deficient (DD)	A taxon is *Data Deficient* when there is inadequate information to make a direct, or indirect, assessment of its risk of extinction based on its distribution and/or population status. A taxon in this category may be well studied, and its biology well known, but appropriate data on abundance and/or distribution is lacking. Data Deficient is therefore not a category of threat or Lower Risk. Listing of taxa in this category indicates that more information is required and acknowledges the possibility that future research will show that threatened classification is appropriate. It is important to make positive use of whatever data are available. In many cases, great care should be exercised in choosing between DD and threatened status. If the range of a taxon is suspected to be relatively circumscribed, if a considerable period of time has elapsed since the last record of the taxon, threatened status may well be justified.
Not evaluated (NE)	A taxon is *Not Evaluated* when it is has not yet been assessed against the criteria.

Appendix IX

Explanation of Some Terms

The definitions given below have been simplified and may vary slightly from definitions given in other literature, particularly the time scales given for the geological and archaeological periods. The latter will vary across Europe with a difference of up to 3,000 years between the South-East and North-West. The archaeological years are given in "years before present" to convert them to "years BC" deduct 2000

Achene	Dry, one seeded fruit that does not open.
Acidophytic	Plants that prefer acidic soils.
Acrocarpous	Growth habitat of some moss species – an erect main stem and a terminal inflorescence.
Adventive	Species not belonging to the original native flora. Not native to and not fully established in a new habitat or environment; a synonym for "alien species".
Agaric	Type of fungus – typical mushroom with a "stem" and cap with gills below.
Agriophyte	Non-native species which have completely naturalised not only in anthropogenic habitats, but also in indigenous plant communities.
Albedo	The proportion of light or radiation reflected from a surface.
Alien	A species that has been directly or indirectly introduced into an area by human activity.
Alkoliphile	(See calcicole).
Allochthonous	"Brought in" – material originating from outside the area in which it occurs.
Amphibolite	Metamorphic rock comprising magnesium-iron silicates.
Anecophyte	Taxa that have evolved in secondary habitats of cultural (man-made) landscapes.
Anemochorous	Dispersal of diaspores by wind.
Angiosperm	Flowering plants.
Anglian Glaciation	Glacial period 450,000 years ago.
Anthropochory	See hemerochory.
Anthropogenic	Directly or indirectly created by human activity.
Anthropophyte	Plants growing in artificial habitats (for example, segetal and ruderal species); alien species not indigenous to a given territory.

Anthropopressure	Impacts caused by human activity.
Anthrosol	Soil type that has been formed or heavily modified by long-term human activity such as irrigation, addition of organic waste, soil in paddy fields.
Apomictic/Apomixis	Process or type of plant that produces seed without fertilisation.
Apophyte	Plant permanently occurring in habitats strongly transformed by human activity; native plants whose occurrence in an area is supported by human economic influence.
Archaean	Geological era 600+ million years ago.
Archaeophyte	Species introduced directly or indirectly by human-related activities before AD 1500.
Arenosol	Sandy soil with very weak or no soil development.
Ascomycetes	Class of Fungi in which the spores are enclosed in a sac-like structure.
Autochthonous	Material generated from within the area in which it occurs.
Balancing lakes	(Also impoundment reservoirs and detention basins) man-made or modified water bodies used to control the discharge of run-off to watercourses.
Biocoenosis/ses	All the interacting organisms living together in a specific habitat or biotope cf. plant community.
Biotope	An area where an organism lives and having all the essential resources that it needs.
Boleachorous	Diaspores dispersed by the wind using ballistic mechanisms.
Boreal	The more southerly part of the, *Tiaga,* biome, relating to the forest areas of the northern North Temperate Zone, dominated by coniferous trees such as spruce, fir and pine.
Bosquet	Formal plantation of trees.
Bronze age	Archaeological period 4,500–2,800 years before present.
Brown earths	Soil type characteristic of deciduous forests in North-West Europe.
Bryophytes	Liverworts, mosses and hornworts.
Bryozoans	Small aquatic animals that form fixed colonies with chitinous or calcareous skeletons.
Calciphilous	Preferring lime-rich substrates.
Calcicole	Plants with a preference for lime-rich substrates.
Calcifuge	Plants with a preference for acidic substrates.
Cambisol	Soil type in the early stage of soil formation with weak horizon differentiation.
Cambrian	Geological period 570–500 million years ago.
Carboniferous	Geological period 345–280 million years ago.
Chernisol	Soil type in the early stages of soil formation; the horizon differentiation is weak.
Chernozem	A black fertile soil rich in humus, nitrates, phosphates and carbonates, in cool or temperate semiarid regions, as the grasslands of Russia (from Russian, contraction of *chernaya zemlya* black earth).
Coenosis	See biocoenosis.
Colluvium	Loose sediment that has been deposited and accumulated at the bottom of a low-grade slope or against a barrier on that slope, transported by gravity.
Coloniophyte	Species whose populations are stable as the result of seed or vegetative reproduction but only occur in places where they first appeared and have not colonised other sites.

Colonophytes (Epecophytes)	Plants which have naturalised but are limited in their distribution and occurring only at the point of invasion.
Constructivism	Russian art movement in which assorted objects are combined into non-representational and mobile structural forms. Theory of knowledge (epistemology) stating that humans generate knowledge and meaning from their experiences.
Corticolous	See Lichen.
Cretaceous	Geological period 135–65 million years ago.
Crustaceous	See Lichen.
Cryptogams	Plants without what are normally considered to be flowers and seeds such as mosses and algae and which reproduce by spores and non-flowering means.
Cryptophyte	Perennial herbaceous plants wintering in the form of underground organs.
Cultivar	(Abbreviation cv) a variety of a species deliberately created by plant breeding.
Cycads	Gymnosperm, tropical/sub-tropical woody plants usually with a single stem and resembling palm trees.
Dendrology	Study of trees; dendrochronology = often used more specifically in relation to the study of dating trees.
Detention basins	See balancing lake.
Devonian	Geological period 400–345 million years ago.
Diaspores	Plant propagules of all types, for example, seeds, fruits, bulbs, stolons, suckers and rhizomes.
Diorite	Rock type of volcanic origin intermediate between granite (acidic) and gabbros (basic).
Dolomite	Sedimentary calcareous rock type, mixture of magnesium and calcium carbonates.
Ecotone	The junction or zone between two different habitat/vegetation types, for example, scrub and grassland.
Endemic	"Taxa" whose occurrence is restricted to a particular area, usually a country, island or mountain range.
Endozoochores	Seed dispersed via the digestive systems of animals.
Eocene	Geological period 59–28 million years ago.
Epecophytes (Epoecophytes)	Naturalised adventive plants; alien species introduced after the fifteenth century and permanently established in man-made habitats.
Epekophytes	Adventive species that settle in disturbed and man-made habitats.
Ephemerophytes	Plants that appear for only 1 or 2 years.
Epilithic	See Lichen.
Epiphytic	See Lichen.
Ergasiolipophytes	Relics of old cultures.
Ergasiophygophytes	Cultivated species that have temporarily escaped from present-day cultivation.
Esker	Glacial structure that forms a winding ridge of current bedded sand, gravel and boulders up to 30 m high and several kilometers long.
Eurytopic	(Of a species) able to tolerate a wide range of environments; having a wide geographical distribution: able to tolerate different conditions and growing in different habitats.
Eutric – cambisols	Soils generally moderately deep to deep dark reddish brown to reddish brown sandy to medium textured soils formed in re-worked colluvial sediments.

Eutrophic	Very rich in nutrients (nitrate and phosphates).
Eutrophication	Process of nutrient enrichment by nitrates and phosphates.
Exotic species	Visually attractive species also used as a synonym for introduced or non-native species.
Fluvio-glacial	Water discharged by melting glaciers; material carried in rivers emerging from glaciers.
Flysch	A sequence of sedimentary rocks that is deposited in a deep marine facies in the foreland basin; deposits of dark fine-grained thinly bedded sandstone, shale and clay.
Foraminifera	Free-living microscopic marine animals – part of the plankton.
Freatic (=Phreatic)	Underground water below the "static" watertable; water from the spring well.
Gallery (Woodland)	Woodlands along watercourses.
Gastromycete	A diverse group of Fungi that produce spores within the fruiting body, for example, puff-balls.
Gastropods	Snails.
Geophyte	Herbs with buds below the surface, for example, *Narcissus* bulbs.
Glacio-lacustrine	Sediments deposited into lakes that have originated from glaciers.
Gleyic Fluvisol	Calcareous gley soils originating from marine sediments.
Gleyi-Haplic Chernozem	Chernozem with a slight gley horizon.
Gneiss	A type of metamorphic rock formed by heat and pressure in the earth's crust.
Granodiorite	Acidic volcanic rock related to granite.
Gymnosperm	Cone-bearing woody plants, often "evergreen", for example, *Pinus*.
Habitat	A place or environment in which a "species" can normally be found.
Halophilous	See halophytes.
Halophyte	Plants with a preference for salty substrates.
Haplic Chernozem	Typical Chernozem without different horizons.
Heliophilous	Plants with a preference for exposure to the sun.
Helophyte	Plant found in marshes.
Hemeroby	Influence of human culture.
Hemerochore (Anthropochory)	Plant dispersal by human-related activities; plants spread as the result of human economic activity.
Hemerophyte	Plants whose distribution is favoured by human culture: plants that preferentially colonise cultivated land.
Hemicryptophyte	Herbs (rarely woody plants) with buds at soil level.
Hepatics	Liverworts.
Holocene	The current interglacial period, which started 15,000/10,000 years ago.
Hortisol	Soil type formed by cultivation.
Hoxian interglacial	An interglacial period that occurred about 400,000 years ago.
Hydrobiont	Different organisms (animals, plants, bacteria) that live in water.
Hydromorphic	Relating to or typical of a soil that has built up in the presence of excessive water.
Hydrophytes	Aquatic plants.
Hygrophilous	Adapted to growing or thriving in damp to wet conditions.
Hypertrophy	The increase in the volume of an organ or tissue due to the enlargement of its component cells.
Hypsometrical	Measurement of elevation relative to sea level.
Idiochorophyte	Native species.

Indigenous	Native (species).
Impoundment reservoir	See balancing lake.
Introduced species	A "species" that has been directly or indirectly introduced into an area by human activity.
Invasive species	Generally a species that has been directly or indirectly introduced into a new territory and has spread rapidly to out-compete native species and generally form extensive dense stands. For example, *Fallopia japonica* and *Impatiens glandulifera*.
Iron age	Archaeological period starting about 2,800 years ago until 1st century AD.
Jurassic	Geological period 190–135 million years ago.
Kame	Low hills of glacial origin comprising mixture of coarse sand and gravel.
Lacustrine	Relating to lakes; species growing in or adjacent to still water bodies.
Leptosol	Very shallow soil over hard rock or highly calcareous material or a deeper soil that is extremely gravelly and/or stony.
Lichen (growth forms)	Corticolous on tree and shrub bark.
	Crustaceous forming an inseparable crust on a substrate.
	Epilithic – on stones and rocks.
	Epiphytic – on plants, leaves.
	Foliose – leaf-like, horizontal, underside usually attached to substrate.
	Fruticose – erect, bushy, attached to substrate at base.
	Saxicolous – on rocks, walls, cement, etc.
	Terricolous – on soil (including peat).
Lithic leptosol	Very shallow (<10 cm deep) soil over continuous hard rock, especially in mountainous regions.
Leptosol	Soils with a shallow profile often contains large mounts of gravel.
Litter meadows	Grassland managed to provide bedding for cattle.
Littoral	Related to species growing on the shore of a lake or the sea.
Luvi cambisol	Brown soil with a luvi-horizon (=horizon with accumulation of translocated clay particles).
Luvisol	A characteristic soil of forested regions; identified by the presence of horizons where silicate clay is accumulated and by the presence of a leafy, humus surface horizon that is separated from the mineral horizon.
Mastodon	Extinct mammal related to early elephants and mammoths.
Mesolithic	Archaeological period 12,000–5,000 years ago.
Mesotrophic (mesic)	Moderately nutrient-rich, between oligotrophic and eutrophic.
Mesozoic	Geological era covering the period 225–65 million years ago.
Mesotraphent	Plants growing in mesic habitats.
Micro-habitat	Very small "structures" or local conditions in which a species occurs, for example, dung or dead wood.
Mollic Fluvisol	A young alluvial humus silty azonal soil with a well-structured dark coloured surface horizon.
Monolectic	A type of bee.
Moraines	Mounds or ridges of unsorted sand, gravel and boulders deposited by glaciers.
Mull-moder	Type of humus.
Musci	Mosses.
Myxomycete	Type of Fungi that includes slime moulds.

Native (species)	Species that has colonised an area by natural processes and *not* directly or indirectly introduced by human activity.
Naturalised species (see spontaneous species)	An introduced species that has become established in a new area and can self-perpetuate.
Nemoral	Pertaining to woodland or grasses. From French *némoral* or Latin *nemoralis*, from nemus "grove". Usually relating to broad-leaved forests (oak, maple, ash, elm, lime, beech, etc.) of the Northern Hemisphere.
Neogene	Geological period that includes the Miocene and Pliocene, 25–1.5 million years ago.
Neoindigenophyte	Fully naturalised species that colonise natural peat communities; syn. agriophytes.
Neolithic	Archaeological period beginning in the Middle East about 9500 BC and much later in North-West Europe. Considered to be the last part of the Stone Age.
Neophyte	Plant species introduced directly or indirectly by human-related activities since 1500 and has become established in the new territory.
Neozoans	Animal species introduced directly or indirectly by human-related activities.
Non-native species	See alien or introduced.
Oligolectic	Refers to bees that exhibit a narrow, specialised preference for pollen sources, typically a single genus of flowering plants.
Oligotrophic	Substrate with a low nutrient content/availability.
Oreophilous	Preferring, or confined to living in mountainous habitats/regions.
Ornamental species	Alien species escaped from gardens.
Palaeolithic	Form the beginning of human history to the end of the Neolithic.
Palaeozoic	Geological era covering 600–280 million years ago.
Para-rendzina	Lime-rich soil type similar to rendzina.
Parterre	Level space in a garden occupied by formal flowerbeds.
Pedology	Study of soils.
Pelagic	Free-floating, for example, aquatic plankton.
Petroleophobe	Species intolerant of soil conditions containing soil derivatives.
Petroleotolerant	Species tolerant of soil conditions containing soil derivatives.
Phaeozem	Soils with a thick dark top soil rich in organic matter and evidence of removal of carbonates; characterised by a humus-rich surface layer, in a natural state it is covered by abundant grass and deciduous forest vegetation.
Phanerophyte	Woody plants with buds >25 cm above the ground.
Phenological	Study of the time of growth events, for example, leaf and flower emergence and leaf fall.
Phycoflora	Algae.
Physiognomy	External features of a subject or object.
Phytotope	Biotope areas.
Plagiosere	Plant succession deflected from its normal course by biotic/human factors.
Pleistocene	Geological period that started 1.0 million years ago.
Pliocene	Geological period from 12 to 1.0 million years ago.
Podzol	Soil type with a surface layer of acidic humus over a severely leached mineral layer. Typically found under conifer forests.

Polemobotany	See stratiobotany.
Polemochores	Plants that have spread during wars.
Proterozoic	Geological term relating to the later part of the Pre-Cambrian (>500 million years) when the earliest known forms of life appeared; period before the appearance of abundant Complex life forms.
Protisol	Class of the soil comprising litosol, regosol and alluviosol types.
Psammophilous	Preferring, confined to loose sandy substrates (sand-loving).
Pteridophytes	Ferns, horsetails and allied species.
Quarternary	Geological period from 1.0 million years ago to present.
Red Data Lists (or Books)	Lists of species that are considered to be threatened. International Lists are published by the International Union for Nature Conservation (IUCN), others are published by national nature conservation agencies and others by local authorities or Non-Governmental Organisations.
Rendzic leptosol	Soils overlying or containing calcareous material and contain >40% calcium carbonate.
Rendzina	Shallow, lime-rich soils developed over limestone and chalk.
Rupicolous	Living or growing on or among rocks, syn. Saxicolous.
Saprophyte	Plants and Fungi that live on dying, dead or decaying organic matter.
Saxicolous	See Lichen.
Schist	A type of metamorphic rock formed by heat and pressure in the earth's crust.
Sciophilous	Plants preferring shady places; capable of thriving in shade.
Segetal	Plants associated with cereal crops in a wider sense growing in arable land.
Sensu lato (s.l.)	In the broad sense.
Sensu stricto (s.s.)	In the strict sense.
Serpentinite	Ultrabasic igneous rock in which serpentine is dominant.
Siliclastic	Clastic non-carbonate silicon-bearing sediment comprising particles composed of silicate minerals and rock fragments.
Species diversae	(Abbreviation spec.div.) various species of the genus concerned.
Species	See taxonomic classification. "Species" is often used as a non-technical, vernacular synonym that includes sub-species, varieties and forms.
Spontaneous	Self-propagation.
Stagnosol	The precise scientific term. A soil with strong mottling of the soil profile due to redox processes caused by stagnating surface water.
Stenotopic	Plants able to tolerate or adapt to a narrow range of environmental conditions.
Stone age	A broad prehistoric cultural age characterised by the creation and use of stone tools which is believed to have occurred about 2 million to 15,000 years ago.
Stratiobotany (polemobotany)	Botanical discipline dealing with the destructive effects of war on plants.
Synanthropic	Related to human activity.
Syrozem	Under developed soils, shallow podzols or brown soils at high altitude.
Taxon (plural/Taxa)	A taxonomic unit of a plant, animal, fungus etc., for example, species, sub-species, variety.
Taxonomic classification	Division of a plant into a hierarchy – class, order, family, genus, species, sub-species, variety/cultivar and form.
Terricolous	See Lichen.

Tertiary	Geological era covering the period 58–12 million years ago.
Thallus	"Body" or vegetative component of a lichen, comprising a fungus and an alga.
Thermophilous	Species that prefer warm conditions.
Therophyte	Species that pass the unfavourable season as seed.
Urbanoneutral	Species that are indifferent in their preference for urban ecosystems.
Urbanophilous	Species that have a preference for urban ecosystems.
Urbanophobous	Species that do not grow in urban ecosystems.
Urbanotechnozems	(From urbano-urban, techno-technical, zems is short from Russian *zemlya*). Anthropogenic, deeply disturbed, young humified soils.
Urbanozems	Technogenic surface soils; soils found in urban areas containing the properties of natural soils with adjacent material.
Urbi-Anthropic Regosols	Very young material with very limited soil development.
Urstromtal	Wide, valley-like depressions along the ice margin formed by the discharge of glacial meltwaters.
Vascular plants	Plants with discrete conducting tissue, for example, flowering plants and ferns.
Weichselian Glaciation	The last full glacial period in North-West Europe, 100,000–14,000 years ago.
Würm Glaciation	Name given to the Weichiselian Glaciation in the Alpine region.
Xenophyte	See Anthropophyte.
Xeric	Dry conditions.
Xerothermophilous	See Xerothermic.
Xerothermic	Species that prefer and plant communities that are formed on relatively fertile soils, in dry, highly insolated sites.

Index

M

O

MIX
Papier aus verantwortungsvollen Quellen
Paper from responsible sources
FSC® C105338

If you have any concerns about our products,
you can contact us on
ProductSafety@springernature.com

In case Publisher is established outside the EU,
the EU authorized representative is:
Springer Nature Customer Service Center GmbH
Europaplatz 3, 69115 Heidelberg, Germany

Printed by Libri Plureos GmbH
in Hamburg, Germany